STUDENT'S SOLUTIONS
MANUAL TO ACCOMPANY

ATKINS'
PHYSICAL CHEMISTRY

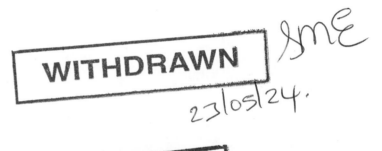

STUDENT'S SOLUTIONS MANUAL TO ACCOMPANY

ATKINS' PHYSICAL CHEMISTRY

Ninth Edition

C. A. TRAPP
Professor of Chemistry
University of Louisville, Louisville, Kentucky, USA

M. P. CADY
Professor of Chemistry
Indiana University Southeast, New Albany, Indiana, USA

C. GIUNTA
Professor of Chemistry
Le Moyne College, Syracuse, New York, USA

OXFORD
UNIVERSITY PRESS

OXFORD
UNIVERSITY PRESS

Great Clarendon Street, Oxford OX2 6DP

Oxford University Press is a department of the University of Oxford.
It furthers the University's objective of excellence in research, scholarship, and education by publishing worldwide in

Oxford New York

Auckland Cape Town Dar es Salaam Hong Kong Karachi
Kuala Lumpur Madrid Melbourne Mexico City Nairobi
New Delhi Shanghai Taipei Toronto

With offices in

Argentina Austria Brazil Chile Czech Republic France Greece
Guatemala Hungary Italy Japan Poland Portugal Singapore
South Korea Switzerland Thailand Turkey Ukraine Vietnam

Oxford is a registered trade mark of Oxford University Press
in the UK and in certain other countries

Published in the United States
by W. H. Freeman and Company, New York

First published 2010

British Library Cataloguing in Publication Data
Data available

Library of Congress Cataloging in Publication Data
Data available

Typeset by Graphicraft Limited, Hong Kong
Printed in Great Britain on acid-free paper
by CPI Antony Rowe, Chippenham, Wiltshire

ISBN 978–0–19–958397–3

1 3 5 7 9 10 8 6 4 2

Preface

This manual provides detailed solutions to all the end-of-chapter (a) Exercises, and to the odd-numbered Discussion Questions and Problems. Solutions to Exercises and Problems carried over from previous editions have been reworked, modified, or corrected when needed.

The solutions to some of the Exercises and many of the Problems in this edition relied more heavily on the mathematical, graphical, and molecular modelling software that is now generally accessible to physical chemistry students. This is particularly true for some of the new Problems that specifically request the use of such software for their solutions. We used the following software for many of the solutions in this manual: Excel™ for spreadsheet calculations and graphing, and Mathcad™ for mathematical calculations and the plotting of the results. When a quantum-chemical calculation or molecular modelling process has been called for, we have often provided the solution with PC Spartan Pro because of its common availability. However, the majority of the Exercises and many of the Problems can still be solved with a modern hand-held scientific calculator.

In general, we have adhered rigorously to the rules for significant figures in displaying the final answers. However, when intermediate answers are shown, they are often given with one more figure than would be justified by the data. These excess digits are indicated with an overline.

The solutions in this manual have been carefully cross-checked for errors not only by ourselves, but also very thoroughly by Valerie Walters, who made many helpful suggestions for improvement. We expect that most errors have been eliminated, but would be grateful to any readers who bring any remaining ones to our attention.

We warmly thank our publishers, especially Jonathan Crowe and Jessica Fiorilla, as well as Dave Quinn, for their patience in guiding this complex, detailed project to completion. We also thank Peter Atkins and Julio de Paula for the opportunity to participate in the development of their outstanding Physical Chemistry text.

Contents

Fundamentals

Exercises

F.1 Atoms

F1.1(a) The **nuclear atomic model** consists of atomic number Z protons concentrated along with all atomic neutrons within the nucleus, an extremely small central region of the atom. Z electrons occupy **atomic orbitals**, which are voluminous regions of the atom that describe where electrons are likely to be found with no more than two electrons in any orbital. The electrostatic attraction binds the negatively charged electrons to the positively charged nucleus and the so-called strong interaction binds the protons and neutrons within the nucleus.

The atomic orbitals are arranged in shells around the nucleus, each shell being characterized by a **principal quantum number**, $n = 1, 2, 3, 4. \ldots$ A shell consists of n^2 individual orbitals, which are grouped together into n subshells. The **subshells**, and the orbitals they contain, are denoted s, p, d, and f. For all neutral atoms other than hydrogen, the subshells of a given shell have slightly different energies.

F1.2(a)

	Example	Element	Ground-state electronic configuration
(a)	Group 2	Ca, calcium	$[Ar]4s^2$
(b)	Group 7	Mn, manganese	$[Ar]3d^5 4s^2$
(c)	Group 15	As, arsenic	$[Ar]3d^{10} 4s^2 4p^3$

F1.3(a) (a) chemical formula and name: $MgCl_2$, magnesium chloride
ions: Mg^{2+} and Cl^-
oxidation numbers of the elements: magnesium, +2; chlorine, −1

(b) chemical formula and name: FeO, iron(II) oxide
ions: Fe^{2+} and O^{2-}
oxidation numbers of the elements: iron, +2; oxygen, −2

(c) chemical formula and name: Hg_2Cl_2, mercury(I) chloride
ions: Cl^- and Hg_2^{2+} (a polyatomic ion)
oxidation numbers of the elements: mercury, +1; chlorine, −1

F1.4(a)

	Metals conduct electricity, have luster, and they are malleable and ductile.
	Non-metals do not conduct electricity and are neither malleable nor ductile.
	Metalloids typically have the appearance of metals but behave chemically like a non-metal.

1 IA	2 IIA	3 IIIB	4 IVB	5 VB	6 VIB	7 VIIB	8 VIIIB	9 VIIIB	10 VIIIB	11 IB	12 IIB	13 IIIA	14 IVA	15 VA	16 VIA	17 VIIA	18 VIIIA
1 H 1.008	Periodic Table of the Elements																2 He 4.003
3 Li 6.941	4 Be 9.012											5 B 10.81	6 C 12.01	7 N 14.01	8 O 16.00	9 F 19.00	10 Ne 20.18
11 Na 22.99	12 Mg 24.31											13 Al 26.98	14 Si 28.09	15 P 30.97	16 S 32.07	17 Cl 35.45	18 Ar 39.95
19 K 39.10	20 Ca 40.08	21 Sc 44.96	22 Ti 47.88	23 V 50.94	24 Cr 52.00	25 Mn 54.94	26 Fe 55.85	27 Co 58.93	28 Ni 58.69	29 Cu 63.55	30 Zn 65.38	31 Ga 69.72	32 Ge 72.59	33 As 74.92	34 Se 78.96	35 Br 79.90	36 Kr 83.80
37 Rb 85.47	38 Sr 87.62	39 Y 88.91	40 Zr 91.22	41 Nb 92.91	42 Mo 95.94	43 Tc (98)	44 Ru 101.1	45 Rh 102.9	46 Pd 106.4	47 Ag 107.9	48 Cd 112.4	49 In 114.8	50 Sn 118.7	51 Sb 121.8	52 Te 127.6	53 I 126.9	54 Xe 131.3
55 Cs 132.9	56 Ba 137.3	57 La 138.9	72 Hf 178.5	73 Ta 180.9	74 W 183.9	75 Re 186.2	76 Os 190.2	77 Ir 192.2	78 Pt 195.1	79 Au 197.0	80 Hg 200.6	81 Tl 204.4	82 Pb 207.2	83 Bi 209.0	84 Po (209)	85 At (210)	86 Rn (222)
87 Fr (223)	88 Ra 226	89 Ac (227)															
				58 Ce 140.1	59 Pr 140.9	60 Nd 144.2	61 Pm 145	62 Sm 150.4	63 Eu 152.0	64 Gd 157.3	65 Tb 158.9	66 Dy 162.5					
				90 Th 232.0	91 Pa (231)	92 U 238.0	93 Np 237	94 Pu (244)	95 Am (243)	96 Cm (247)	97 Bk (247)	98 Cf (251)					

F.2 Molecules

F2.1(a) A **single bond** is a shared pair of electrons between adjacent atoms within a molecule while a **multiple bond** involves the sharing of either two pairs of electrons (a double bond) or three pairs of electrons (a triple bond).

F2.2(a) (a) Sulfite anion, SO_3^{2-}

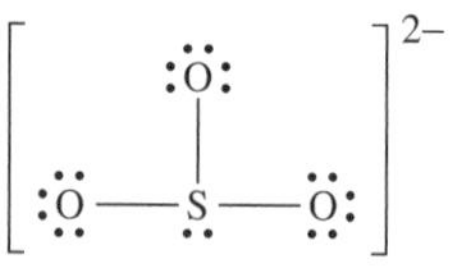

Alternatively, resonance structures may be drawn and, if desired, formal charges (shown in circles below) may be indicated.

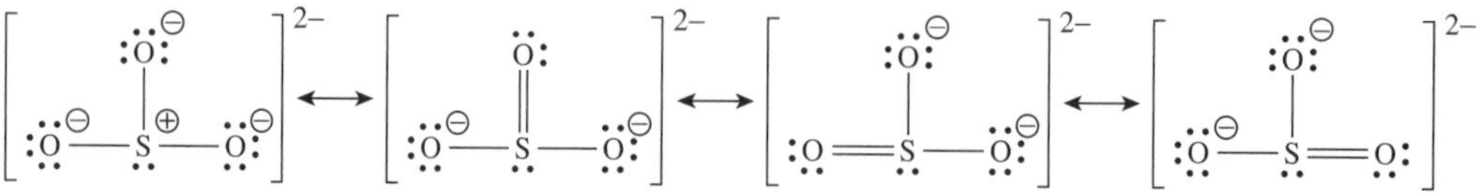

(b) Xenon tetrafluoride, XeF_4

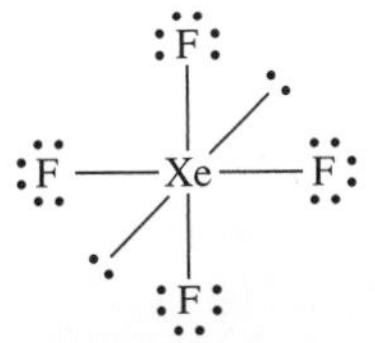

(c) White phosphorus, P_4

F2.3(a) **Valence-shell electron pair repulsion theory (VSEPR theory)** predicts molecular shape with the concept that regions of high electron density (as represented by single bonds, multiple bonds, and lone pairs) take up orientations around the central atom that maximize their separation. The resulting positions of attached atoms (not lone pairs) are used to classify the shape of the molecule. When the central atom has two or more lone pairs, the molecular geometry must minimize repulsion between the relatively diffuse orbitals of the lone pair. Furthermore, it is assumed that the repulsion between a lone pair and a bonding pair is stronger than the repulsion between two bonding pairs, thereby, making bond angles smaller than the idealized bond angles that appear in the absence of a lone pair.

F2.4(a) Molecular and polyatomic ion shape are predicted by drawing the Lewis structure and applying the concepts of VSEPR theory.

(a) PCl_3
Lewis structure:

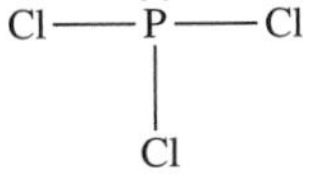

Orientations caused by repulsions between one lone pair and three bonding pairs:

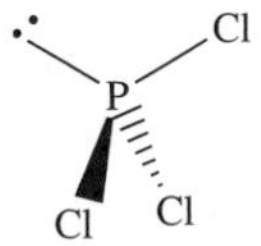

Molecular shape: trigonal pyramidal and bond angles somewhat smaller than 109.5°

(b) PCl_5
Lewis structure:

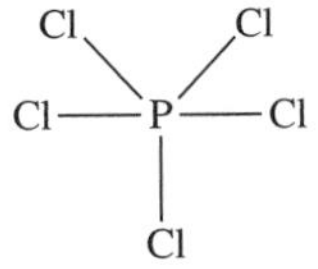

Orientations caused by repulsions between five bonding pairs (no lone pair):

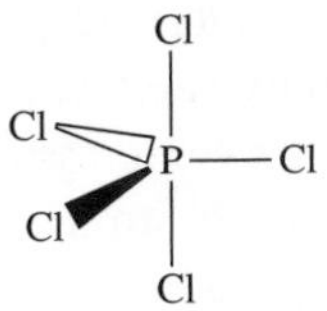

Molecular shape: trigonal bipyramidal with equatorial bond angles of 120° and axial bond angles of 90°

(c) XeF_2
Lewis structure:

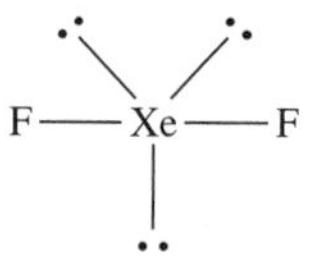

Orientations caused by repulsions between three lone pairs and two bonding pairs:

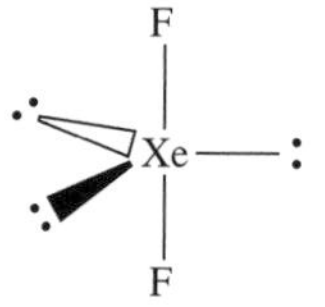

Molecular shape: linear with a 180° bond angle

(d) XeF_4
Lewis structure:

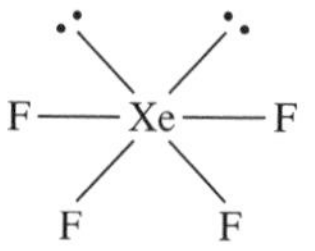

Orientations caused by repulsions between two lone pairs and four bonding pairs:

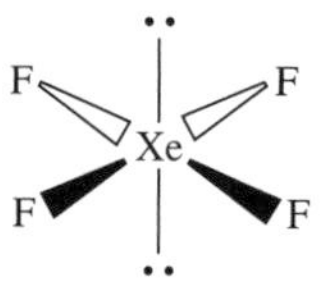

Molecular shape: square planar with 90° bond angles

F2.5(a) (a) $^{\delta+}C$——$Cl^{\delta-}$ (b) P——H Non-polar or weakly polar. (c) $^{\delta+}N$——$O^{\delta-}$

F2.6(a) (a) CO_2 is a linear, non-polar molecule.
(b) SO_2 is a bent, polar molecule.
(c) N_2O is linear, polar molecule.
(d) SF_4 has a seesaw molecule and it is a polar molecule.

F2.7(a) In the order of increasing dipole moment: CO_2, N_2O, SF_4, SO_2

F.3 Bulk matter

F3.1(a) The solid phase of matter has a shape that is independent of the container it occupies. It has a density compatible with the close proximity of either its constituent elemental atoms or its constituent molecules and, consequently, it has low compressibility. Constituent atoms, or molecules, are held firmly at specific lattice sites by relatively strong, net forces of attraction between neighbouring constituents. Solids may be characterized by terms such as brittle, ductile, tensile strength, toughness, and hardness.

A liquid adopts the shape of the part of the container that it occupies; it can flow under the influence of gravitational attraction to occupy any shape. Like a solid, it has a density caused by the close proximity of either its constituent elemental atoms or its constituent molecules and it has low compressibility. Liquids can flow because the constituent atoms or molecules have enough average kinetic energy to overcome the attractive forces between neighbouring constituents, thereby making it possible for them to slip past immediate neighbours. This causes constituents to be placed randomly in contrast to the orderly array in crystals. Liquids are characterized by terms such as surface tension, viscosity, and capillary action. Liquids within a vertical, narrow tube exhibit a meniscus that is either concave-up or concave-down, depending on the nature of the attractive or repulsive forces between the liquid and the material of the tube.

Gases have no fixed shape or volume. They expand to fill the container volume. The constituent molecules move freely and randomly. A perfect gas has a total molecular volume that is negligibly small compared to the container volume and, because of the relatively large average distance between molecules, intermolecular attractive forces are negligibly small. Gases are compressible.

F3.2(a)
(a) Mass is an extensive property.
(b) Mass density is an intensive property.
(c) Temperature is an intensive property.
(d) Number density is an intensive property.

F3.3(a)
(a) $n = \dfrac{m}{M} = 25.0\ \text{g}\left(\dfrac{1\ \text{mol}}{46.069\ \text{g}}\right) = \boxed{0.543\ \text{mol}}$ [F.1]

(b) $N = nN_A = 0.543\ \text{mol}\left(\dfrac{6.0221\times 10^{23}\ \text{molecules}}{\text{mol}}\right) = \boxed{3.27\times 10^{23}\ \text{molecules}}$

F3.4(a)
(a) $(1.45\ \text{atm})\times\left(\dfrac{101325\ \text{Pa}}{1\ \text{atm}}\right) = \boxed{1.47\times 10^{5}\ \text{Pa}}$

(b) Since 1 bar = 10^5 Pa, the above pressure is $\boxed{1.47\ \text{bar}}$.

F3.5(a)
$T/\text{K} = \theta/°\text{C} + 273.15 = 37.0 + 273.15 = 310.2$ [F.2]

$\boxed{T = 310.2\ \text{K}}$

F3.6(a)
To devise an equation relating the Fahrenheit and Celsius scales requires that consideration be given to both the degree size and a common reference point in each scale. Between the normal freezing point of water and its normal boiling point there is a degree scaling of 100°C and 180°F. Thus, the scaling ratio is 5°C per 9°F. A convenient reference point is provided by the normal freezing point of water, which is 0°C and 32°F. So, to calculate the Celsius temperature from a given Fahrenheit temperature, 32 must be subtracted from the Fahrenheit temperature (θ_F) followed by scaling to the Celsius temperature (θ) with the ratio 5°C/9°F.

$$\boxed{\theta/°\text{C} = \tfrac{5}{9}\times(\theta_F/°\text{F} - 32) \quad \text{or} \quad \theta_F/°\text{F} = \tfrac{9}{5}\times\theta/°\text{C} + 32}$$

$$\theta_F/°\text{F} = \tfrac{9}{5}\times\theta/°\text{C} + 32 = \tfrac{9}{5}\times 78.5 + 32 = 173$$

$$\boxed{\theta_F = 173\ °\text{F}}$$

F3.7(a)
$110\ \text{kPa}\times\left(\dfrac{(7.0 + 273.15)\ \text{K}}{(20.0 + 273.15)\ \text{K}}\right) = \boxed{105\ \text{kPa}}$

F.4 Energy

F4.1(a) In the classical physics of Newton the force $\boldsymbol{F}$ acting on a body so as to cause acceleration $\boldsymbol{a}$ equals the mass m of the object multiplied by its acceleration: $\boldsymbol{F}=m\boldsymbol{a}$ where the bold-faced symbols represent vector quantities. Coulomb's law describes the particularly important electrostatic force between two point charges Q_1 and Q_2 separated by the distance r:

$$F=\frac{Q_1Q_2}{4\pi\varepsilon_0 r^2} \text{ in a vacuum } (\varepsilon_0 \text{ is the vacuum permittivity})$$

and $F=\dfrac{Q_1Q_2}{4\pi\varepsilon_r\varepsilon_0 r^2}$ in a medium that has the relative permittivity ε_r (formerly, dielectric constant).

Convention assigns a negative value to the Coulomb force when it is attractive and a positive value when it is repulsive. The SI unit of force is the newton (N) and $1\text{ N}=1\text{ kg m s}^{-2}$.

An infinitesimal amount of **work**, $\mathrm{d}w$, done on a body when it experiences an infinitesimal displacement, $\mathrm{d}\boldsymbol{s}$, is defined by the scalar product (sometimes called the dot-product): $\mathrm{d}w=-\boldsymbol{F}\cdot\mathrm{d}\boldsymbol{s}$, where $\boldsymbol{F}$ is now the force that opposes the displacement. When the opposing force and displacement lie along the same direction, the z direction for example, the infinitesimal work done on the body simplifies to: $\mathrm{d}w=-F\mathrm{d}z$, where F is the magnitude of the opposing force. The SI unit of work is the joule (J) and

$$1\text{ J}=1\text{ N m}=1\text{ Pa m}^3=1\text{ kg m}^2\text{ s}^{-2}.$$

Continuing the example of displacement along the z direction alone, integration of the infinitesimal work between the initial position "i" and the final position "f" gives the total work, w, done on the object: $w=-\int_\mathrm{i}^\mathrm{f}F\mathrm{d}z$. Provided that the opposing force is constant over the displacement, the integral simplifies to give: $w=-F\times(z_\mathrm{f}-z_\mathrm{i})=-F\Delta z$ where $\Delta z=z_\mathrm{f}-z_\mathrm{i}$.

Energy is the capacity to do work. It is a property and the SI unit of energy, like the unit of work, is the joule. The **law of conservation of energy** states that the total energy E of an isolated system is conserved; that is, the total energy of an isolated system is a constant. Even in an isolated system, however, energy can be transferred from one location to another and transformed from one form to another. The transfers and transformations involve heat, work, gravitational potential energy, Coulomb potential energy, and electromagnetic radiation.

F4.2(a) We estimate the average kinetic energy of a molecule with $v=400\text{ m s}^{-1}$ and $m=(29\text{ g mol}^{-1})/N_\mathrm{A}$ in eqn F.4, $E_\mathrm{k}=\frac{1}{2}mv^2$. The number of molecules, N, is given by $N=nN_\mathrm{A}$ and, finally, the total energy stored as molecular kinetic energy is NE_k.

$$NE_\mathrm{k}=(nN_\mathrm{A})\times(\tfrac{1}{2}mv^2)=(nN_\mathrm{A})\times\left\{\tfrac{1}{2}\left(\frac{0.029\text{ kg mol}^{-1}}{N_\mathrm{A}}\right)\times v^2\right\}$$

$$=\tfrac{1}{2}(1\text{ mol})\times(0.029\text{ kg mol}^{-1})\times(400\text{ m s}^{-1})^2$$

$$=2.3\times10^3\text{ kg m}^2\text{ s}^{-2}=2.3\times10^3\text{ J}=\boxed{2.3\text{ kJ}}$$

F4.3(a) The Coulomb potential, ϕ, is

$$\phi = \frac{Q_2}{4\pi\varepsilon_0 r},$$ where r is the separation of point charge Q_1 and the nuclear charge Q_2

Q_1 interacts with two nuclei in this exercise and the interactions are additive.

$$\phi = \left(\frac{Q_2}{4\pi\varepsilon_0 r}\right)_{\text{Li nucleus}} + \left(\frac{Q_2}{4\pi\varepsilon_0 r}\right)_{\text{H nucleus}} = \left(\frac{Ze}{4\pi\varepsilon_0 r}\right)_{\text{Li nucleus}} + \left(\frac{Ze}{4\pi\varepsilon_0 r}\right)_{\text{H nucleus}}$$

$$= \frac{e}{4\pi\varepsilon_0} \times \left\{\left(\frac{Z}{r}\right)_{\text{Li nucleus}} + \left(\frac{Z}{r}\right)_{\text{H nucleus}}\right\}$$

$$= \left(\frac{1.6022 \times 10^{-19}\ \text{C}}{1.113 \times 10^{-10}\ \text{J}^{-1}\ \text{C}^2\ \text{m}^{-1}}\right) \times \left\{\frac{3}{200 \times 10^{-12}\ \text{m}} + \frac{1}{150 \times 10^{-12}\ \text{m}}\right\} = 31.2\ \text{J C}^{-1} = \boxed{31.2\ \text{V}}$$

F.5 The relationship between molecular and bulk properties

F5.1(a) Quantized energies are certain discrete values that are permitted for particles confined to a region of space.

F5.2(a) $\Delta E = E_{\text{upper}} - E_{\text{lower}} = 1\ \text{eV} = 1.6022 \times 10^{-19}\ \text{J}$

(a) $$\frac{N_{\text{upper}}}{N_{\text{lower}}} = \text{e}^{-\Delta E/kT}\ [\text{F.9}] = \text{e}^{-(1.6022\times10^{-19}\,\text{J})/\{(1.381\times10^{-23}\,\text{JK}^{-1})\times(300\,\text{K})\}} = \boxed{1.602 \times 10^{-17}}$$

(b) $$\frac{N_{\text{upper}}}{N_{\text{lower}}} = \text{e}^{-\Delta E/kT} = \text{e}^{-(1.6022\times10^{-19}\,\text{J})/\{(1.381\times10^{-23}\,\text{JK}^{-1})\times(3000\,\text{K})\}} = \boxed{2.092 \times 10^{-2}}$$

F5.3(a) Kinetic molecular theory, a model for a perfect gas, assumes that the molecules, imagined as particles of negligible size, are in ceaseless, random motion and do not interact except during their brief collisions.

F5.4(a) Molecules can survive for long periods without undergoing chemical reaction at low temperatures because few molecules have the requisite speed and corresponding kinetic energy to promote excitation and bond breakage during collisions.

F5.5(a) $v_{\text{mean}} \propto (T/M)^{1/2}$ [F.11]

$$\frac{v_{\text{mean}}(T_2)}{v_{\text{mean}}(T_1)} = \frac{(T_2/M)^{1/2}}{(T_1/M)^{1/2}} = \left(\frac{T_2}{T_1}\right)^{1/2}$$

$$\frac{v_{\text{mean}}(313\ \text{K})}{v_{\text{mean}}(273\ \text{K})} = \left(\frac{313\ \text{K}}{273\ \text{K}}\right)^{1/2} = \boxed{1.07}$$

F5.6(a) A gaseous argon atom has three translational degrees of freedom (the components of motion in the x, y, and z directions). Consequently, the equipartition theorem assigns a mean energy of $\frac{3}{2}kT$ to each atom. The molar internal energy, U_{m}, is

$$U_{\text{m}} = \tfrac{3}{2}N_{\text{A}}kT = \tfrac{3}{2}RT\ [\text{F.10}] = \tfrac{3}{2}(8.3145\ \text{J mol}^{-1}\ \text{K}^{-1})(298\ \text{K}) = 3.72\ \text{kJ mol}^{-1}$$

$$U = nU_{\text{m}} = mM^{-1}U_{\text{m}} = (5.0\ \text{g})\left(\frac{1\ \text{mol}}{39.95\ \text{g}}\right)\left(\frac{3.72\ \text{kJ}}{\text{mol}}\right) = \boxed{0.47\ \text{kJ}}$$

F5.7(a) (a) A gaseous, linear, carbon dioxide molecule has three quadratic translational degrees of freedom (the components of motion in the x, y, and z directions) but it has only two rotational quadratic degrees of freedom because there is no rotation along the internuclear line. There is a total of five quadratic degrees of freedom for the molecule. Consequently, the equipartition theorem assigns a mean energy of $\frac{5}{2}kT$ to each molecule. The molar internal energy, U_m, is

$$U_m = \tfrac{5}{2}N_AkT = \tfrac{5}{2}RT\ [\text{F.10}] = \tfrac{5}{2}(8.3145\ \text{J mol}^{-1}\ \text{K}^{-1})(293\ \text{K}) = 6.09\ \text{kJ mol}^{-1}$$

$$U = nU_m = mM^{-1}U_m = (10.0\ \text{g})\left(\frac{1\ \text{mol}}{44.0\ \text{g}}\right)\left(\frac{6.09\ \text{kJ}}{\text{mol}}\right) = \boxed{1.38\ \text{kJ}}$$

(b) A gaseous, non-linear, methane molecule has three quadratic translational degrees of freedom (the components of motion in the x, y, and z directions) and three quadratic rotational degrees of freedom. Consequently, the equipartition theorem assigns a mean energy of $\frac{6}{2}kT$ to each molecule. The molar internal energy, U_m, is

$$U_m = \tfrac{6}{2}N_AkT = 3RT\ [\text{F.10}] = 3(8.3145\ \text{J mol}^{-1}\ \text{K}^{-1})(293\ \text{K}) = 7.31\ \text{kJ mol}^{-1}$$

$$U = nU_m = mM^{-1}U_m = (10.0\ \text{g})\left(\frac{1\ \text{mol}}{16.04\ \text{g}}\right)\left(\frac{7.31\ \text{kJ}}{\text{mol}}\right) = \boxed{4.56\ \text{kJ}}$$

F.6 The electromagnetic field

F6.1(a) $$\nu = \frac{c}{\lambda}[\text{F.12}] = \frac{3.00\times10^{8}\ \text{m s}^{-1}}{230\times10^{-9}\ \text{m}} = 1.30\times10^{15}\ \text{s}^{-1} = \boxed{1.30\times10^{15}\ \text{Hz}}$$

F6.2(a) $$\tilde{\nu} = \frac{\nu}{c}[\text{F.13}] = \frac{560\times10^{12}\ \text{s}^{-1}}{3.00\times10^{8}\ \text{m s}^{-1}} = 1.87\times10^{6}\ \text{m}^{-1} = \boxed{1.87\times10^{4}\ \text{cm}^{-1}}$$

F6.3(a) (a) $$\lambda = \frac{c}{\nu}[\text{F.12}] = \frac{3.00\times10^{8}\ \text{m s}^{-1}}{91.7\times10^{6}\ \text{s}^{-1}} = \boxed{3.27\ \text{m}}$$

(b) $$\tilde{\nu} = \frac{\nu}{c}[\text{F.13}] = \frac{91.7\times10^{6}\ \text{s}^{-1}}{3.00\times10^{8}\ \text{m s}^{-1}} = \boxed{0.306\ \text{m}^{-1}}$$

F.7 Units

F7.1(a) $1.45\ \text{cm}^3 = 1.45\ (10^{-2}\ \text{m})^3 = \boxed{1.45\times10^{-6}\ \text{m}^3}$

F7.2(a) $$\left(11.2\frac{\text{g}}{\text{cm}^3}\right)\times\left(\frac{1\ \text{kg}}{10^3\ \text{g}}\right)\times\left(\frac{1\ \text{cm}}{10^{-2}\ \text{m}}\right)^3 = \boxed{11.2\times10^{3}\ \text{kg m}^{-3}}$$

F7.3(a) $$\frac{\text{Pa}}{\text{J}} = \frac{\text{N m}^{-2}}{\text{J}} = \frac{(\text{N m})\ \text{m}^{-3}}{\text{J}} = \frac{\text{J m}^{-3}}{\text{J}} = \boxed{\text{m}^{-3}}$$

F7.4(a) $$\frac{kT}{hc} = \frac{(1.381\times10^{-23}\ \mathrm{J\ K^{-1}})\times(298\ \mathrm{K})}{(6.626\times10^{-34}\ \mathrm{J\ s})\times(2.998\times10^{10}\ \mathrm{cm\ s^{-1}})} = \boxed{207.2\ \mathrm{cm^{-1}}}$$

F7.5(a) $$\left(\frac{8.3144\ \mathrm{J}}{\mathrm{K\ mol}}\right)\times\left(\frac{\mathrm{Pa\ m^3}}{\mathrm{J}}\right)\times\left(\frac{1\ \mathrm{atm}}{101325\ \mathrm{Pa}}\right)\times\left(\frac{\mathrm{dm}}{10^{-1}\ \mathrm{m}}\right)^3 = \boxed{0.08206\ \mathrm{atm\ dm^3\ K^{-1}\ mol^{-1}}}$$

F7.6(a) $$(1\ \mathrm{dm^3\ atm})\times\left(\frac{10^{-1}\ \mathrm{m}}{\mathrm{dm}}\right)^3\times\left(\frac{101325\ \mathrm{Pa}}{1\ \mathrm{atm}}\right)\times\left(\frac{\mathrm{J}}{\mathrm{Pa\ m^3}}\right) = \boxed{101.325\ \mathrm{J}}$$

F7.7(a) (a) Base unit of $$\frac{e^2}{\varepsilon_0 r^2} = \frac{\mathrm{C^2}}{(\mathrm{C^2\ J^{-1}\ m^{-1}})\times(\mathrm{m^2})} = (\mathrm{J\ m^{-1}})\times\left(\frac{\mathrm{kg\ m^2\ s^{-2}}}{\mathrm{J}}\right) = \boxed{\mathrm{kg\ m\ s^{-2}}}$$

(b) Unit of $$\frac{e^2}{\varepsilon_0 r^2} = \frac{\mathrm{C^2}}{(\mathrm{C^2\ J^{-1}\ m^{-1}})\times(\mathrm{m^2})} = (\mathrm{J\ m^{-1}})\times\left(\frac{\mathrm{kg\ m^2\ s^{-2}}}{\mathrm{J}}\right) = \mathrm{kg\ m\ s^{-2}} = \boxed{\mathrm{N}}$$

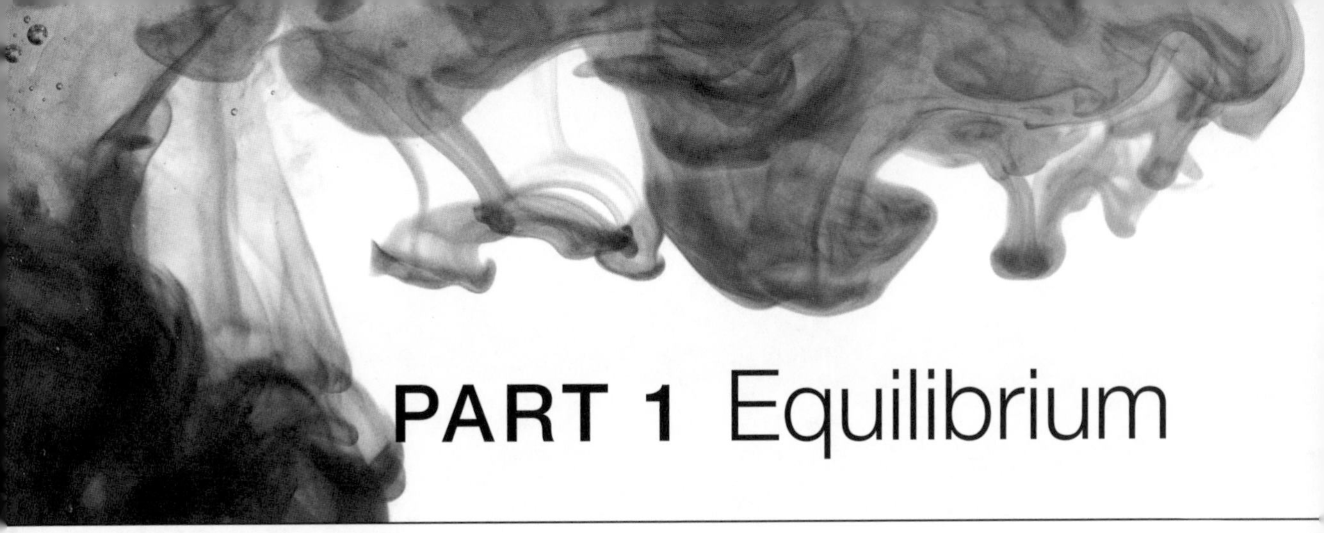

PART 1 Equilibrium

1 The properties of gases

Answers to discussion questions

D1.1 An equation of state is an equation that relates the variables that define the state of a system to each other. Boyle, Charles, and Avogadro established these relationships for gases at low pressures (perfect gases) by appropriate experiments. Boyle determined how volume varies with pressure ($V \propto 1/p$), Charles how volume varies with temperature ($V \propto T$), and Avogadro how volume varies with amount of gas ($V \propto n$). Combining all of these proportionalities into one we find

$$V \propto \frac{nT}{p}$$

Inserting the constant of proportionality, R, yields the perfect gas equation

$$V = \frac{RnT}{p} \quad \text{or} \quad pV = nRT$$

D1.3 Consider three temperature regions:

(1) $T < T_B$. At very low pressures, all gases show a compression factor, $Z \approx 1$. At high pressures, all gases have $Z > 1$, signifying that they have a molar volume greater than a perfect gas, which implies that repulsive forces are dominant. At intermediate pressures, most gases show $Z < 1$, indicating that attractive forces reducing the molar volume below the perfect value are dominant.

(2) $T \approx T_B$. $Z \approx 1$ at low pressures, slightly greater than 1 at intermediate pressures, and significantly greater than 1 only at high pressures. There is a balance between the attractive and repulsive forces at low to intermediate pressures, but the repulsive forces predominate at high pressures, where the molecules are very close to each other.

(3) $T > T_B$. $Z > 1$ at all pressures because the frequency of collisions between molecules increases with temperature.

D1.5 The van der Waals equation 'corrects' the perfect gas equation for both attractive and repulsive interactions between the molecules in a real gas. See *Justification 1.1* for a fuller explanation.

The Berthelot equation accounts for the volume of the molecules in a manner similar to the van der Waals equation but the term representing molecular attractions is modified to account for the effect of temperature. Experimentally one finds that the van der Waals a decreases with increasing temperature. Theory (see Chapter 17) also suggests that intermolecular attractions can decrease with temperature. This variation of the attractive interaction with temperature can be accounted for in the equation of state by replacing the van der Waals a with a/T.

Solutions to exercises

E1.1(a) (a) The perfect gas equation [1.8] is $pV = nRT$

Solving for the pressure gives $p = \dfrac{nRT}{V}$

The amount of xenon is $n = \dfrac{131\,\text{g}}{131\,\text{g mol}^{-1}} = 1.00\ \text{mol}$

$$p = \frac{(1.00\ \text{mol}) \times (0.0821\ \text{dm}^3\,\text{atm K}^{-1}\,\text{mol}^{-1}) \times (298.15\ \text{K})}{1.0\ \text{dm}^3} = \boxed{24\ \text{atm}}$$

That is, the sample would exert a pressure of 24 atm if it were a perfect gas, not 20 atm.

(b) The van der Waals equation [1.21*a*] for the pressure of a gas is $p = \dfrac{nRT}{V - nb} - \dfrac{an^2}{V^2}$

For xenon, Table 1.6 gives $a = 4.137\ \text{dm}^6\ \text{atm mol}^{-2}$ and $b = 5.16 \times 10^{-2}\ \text{dm}^3\ \text{mol}^{-1}$.

Inserting these constants, the terms in the equation for p become

$$\frac{nRT}{V - nb} = \frac{(1.00\ \text{mol}) \times (0.08206\ \text{dm}^3\,\text{atm K}^{-1}\,\text{mol}^{-1}) \times (298.15\ \text{K})}{1.0\ \text{dm}^3 - \{(1.00\ \text{mol}) \times (5.16 \times 10^{-2}\ \text{dm}^3\,\text{mol}^{-1})\}} = 25.\overline{8}\ \text{atm}$$

$$\frac{an^2}{V^2} = \frac{(4.137\ \text{dm}^6\,\text{atm mol}^{-2}) \times (1.00\ \text{mol})^2}{(1.0\ \text{dm}^3)^2} = 4.1\overline{37}\ \text{atm}$$

Therefore, $p = 25.\overline{8}\ \text{atm} - 4.1\overline{37}\ \text{atm} = \boxed{22\ \text{atm}}$

E1.2(a) Boyle's law [1.5] applies.

$$pV = \text{constant} \quad \text{so} \quad p_f V_f = p_i V_i$$

This equation can be solved for either initial or final pressure, hence

$$p_i = \frac{V_f}{V_i} \times p_f$$

$V_f = 4.65\ \text{dm}^3$, $V_i = (4.65 + 2.20)\ \text{dm}^3 = 6.85\ \text{dm}^3$, $p_f = 5.04\ \text{bar}$

Therefore,

(a) $p_i = \left(\dfrac{4.65\ \text{dm}^3}{6.85\ \text{dm}^3}\right) \times (5.04\ \text{bar}) = \boxed{3.42\ \text{bar}}$

(b) Since 1 atm = 1.013 bar, $p_i = 3.42\ \text{bar} \times \dfrac{1\ \text{atm}}{1.013\ \text{bar}} = \boxed{3.38\ \text{atm}}$

E1.3(a) The perfect gas law, $pV = nRT$ [1.8], can be rearranged to $\dfrac{p}{T} = \dfrac{nR}{V} = \text{constant}$, if n and V are constant. Hence, $\dfrac{p_f}{T_f} = \dfrac{p_i}{T_i}$ or, solving for p_f, $p_f = \dfrac{T_f}{T_i} \times p_i$

Note that the internal pressure is the pump pressure *plus* atmospheric pressure.

$$p_i = 24\ \text{lb in}^{-2} + 14.7\ \text{lb in}^{-2} = 38.\bar{7}\ \text{lb in}^{-2},\ T_i = 268\ \text{K}(-5°\text{C}),\ T_f = 308\ \text{K}(35°\text{C})$$

so $$p_f = \frac{308\ \text{K}}{268\ \text{K}} \times 38.\bar{7}\ \text{lb in}^{-2} = 44.\bar{5}\ \text{lb in}^{-2}$$

Therefore, $p(\text{pump}) = 44.\bar{5}\ \text{lb in}^{-2} - 14.7\ \text{lb in}^{-2} = \boxed{30\ \text{lb in}^{-2}}$

Complications are those factors that destroy the constancy of V or n, such as the change in volume of the tyre, the change in rigidity of the material from which it is made, and loss of pressure by leaks and diffusion.

E1.4(a) The perfect gas law in the form $p = \frac{nRT}{V}$ [1.8] is appropriate. T and V are given; n must be calculated.

$$n = \frac{0.255\ \text{g}}{20.18\ \text{g mol}^{-1}} = 1.26 \times 10^{-2}\ \text{mol},\quad T = 122\ \text{K},\quad V = 3.00\ \text{dm}^3$$

Therefore, on substitution,

$$p = \frac{(1.26 \times 10^{-2}\ \text{mol}) \times (0.08206\ \text{dm}^3\ \text{atm K}^{-1}\ \text{mol}^{-1}) \times (122\ \text{K})}{3.00\ \text{dm}^3} = \boxed{4.20 \times 10^{-2}\ \text{atm}}$$

E1.5(a) Boyle's law [1.5] in the form $p_f V_f = p_i V_i$ is solved for V_f:

$$V_f = \frac{p_i}{p_f} \times V_i$$

The initial pressure is 1.0 atm. The final pressure is

$$p_f = p_{ex} + \rho g h\ [1.3] = p_i + \rho g h = 1.0\ \text{atm} + \rho g h$$

where $\rho g h = 1.025\ \text{g cm}^{-3} \times \frac{1\ \text{kg}}{10^3\ \text{g}} \times \left(\frac{1\ \text{cm}}{10^{-2}\ \text{m}}\right)^3 \times 9.81\ \text{m s}^{-2} \times 50\ \text{m} = 5.0\bar{3} \times 10^5\ \text{Pa}$

Hence, $p_f = (1.0\bar{1} \times 10^5\ \text{Pa}) + (5.0\bar{3} \times 10^5\ \text{Pa}) = 6.0\bar{4} \times 10^5\ \text{Pa}$

$$V_f = \frac{1.0\bar{1} \times 10^5\ \text{Pa}}{6.0\bar{4} \times 10^5\ \text{Pa}} \times 3.0\ \text{m}^3 = \boxed{0.50\ \text{m}^3}$$

E1.6(a) The pressure in the apparatus is given by

$$p = p_{ex} + \rho g \Delta h\ [1.3]$$

where, using the fact that 760 Torr = 1 atm = 1.013×10^5 Pa,

$$p_{ex} = 770\ \text{Torr} \times \left(\frac{1.013 \times 10^5\ \text{Pa}}{760\ \text{Torr}}\right) = 1.026 \times 10^5\ \text{Pa}$$

and $$\rho g \Delta h = 0.99707\ \text{g cm}^{-3} \times \left(\frac{1\ \text{kg}}{10^3\ \text{g}}\right) \times \left(\frac{1\ \text{cm}}{10^{-2}\ \text{m}}\right)^3 \times 9.806\ \text{m s}^{-2} \times (-0.100\ \text{m}) = -978\ \text{Pa}$$

so $$p = 1.026 \times 10^5\ \text{Pa} - 978\ \text{Pa} = 1.017 \times 10^5\ \text{Pa} = \boxed{102\ \text{kPa}}$$

E1.7(a) Rearrange the perfect gas equation [1.8] to give $R = \dfrac{pV}{nT}$.

Everything on the right-hand side is related to measured quantities.

$$V = (20.000\ \text{dm}^3) \times \left(\frac{10^{-1}\ \text{m}}{1\ \text{dm}}\right)^3 = 2.0000 \times 10^{-2}\ \text{m}^3$$

$$n = \frac{m}{M} = \frac{0.25132\ \text{g}}{4.00260\ \text{g mol}^{-1}} = 0.062789\ \text{mol}$$

The gas pressure is calculated as the force per unit area that a column of water of height 206.402 cm exerts on the gas due to its weight. The manometer is assumed to have uniform cross-sectional area, A. The force is $F = mg$, where m is the mass of the column of water and g is the acceleration of free fall. As in Example 1.1, $m = \rho \times V = \rho \times h \times A$, where $h = 206.402$ cm and A is the cross-sectional area.

$$p = \frac{F}{A} = \frac{\rho h A g}{A} = \rho h g$$

$$p = (0.99707\ \text{g cm}^{-3}) \times (206.402\ \text{cm}) \times (9.8067\ \text{m s}^{-2}) \times \left(\frac{1\ \text{kg}}{10^3\ \text{g}}\right) \times \left(\frac{1\ \text{cm}}{10^{-2}\ \text{m}}\right)^2$$

$$= 2.0182 \times 10^4\ \text{Pa}$$

So $$R = \frac{(2.0182 \times 10^4\ \text{Pa}) \times (2.0000 \times 10^{-2}\ \text{m}^3)}{(0.062789\ \text{mol}) \times (773.15\ \text{K})} = \boxed{8.3147\ \text{J K}^{-1}\ \text{mol}^{-1}}$$

The accepted value is $R = 8.3145\ \text{J K}^{-1}\ \text{mol}^{-1}$. Although gas volume data should be extrapolated to zero pressure for the best value of R, helium is close to being a perfect gas under the conditions here, and thus a value of R close to the accepted value is obtained.

E1.8(a) Since $p < 1$ atm, the approximation that the vapour is a perfect gas is adequate. We introduce the molar mass into the ideal gas law [1.8] by

$$pV = nRT = \frac{m}{M}RT$$

On rearrangement

$$M = \frac{m}{V}\left(\frac{RT}{p}\right) = \rho\left(\frac{RT}{p}\right) = (3.710\ \text{kg m}^{-3}) \times \frac{(8.3145\ \text{Pa m}^3\ \text{K}^{-1}\ \text{mol}^{-1}) \times (773\ \text{K})}{9.32 \times 10^4\ \text{Pa}}$$

$$= 0.256\ \text{kg mol}^{-1} = 256\ \text{g mol}^{-1}$$

This molar mass must be an integral multiple of the molar mass of atomic sulfur; hence

$$\text{number of S atoms} = \frac{256\ \text{g mol}^{-1}}{32.1\ \text{g mol}^{-1}} = 8$$

The formula of the vapour is then $\boxed{S_8}$

E1.9(a) The partial pressure of the water vapour in the room is 60% of the equilibrium vapour pressure, which must be looked up in a handbook like the *CRC Handbook of Chemistry and Physics* or other resource such as the NIST Chemistry WebBook.

$$p_{H_2O} = (0.60) \times (26.76\ \text{torr}) = 16.1\ \text{torr}$$

Assuming that the perfect gas equation [1.8] applies, with $n = \dfrac{m}{M}$, $pV = \dfrac{m}{M}RT$ or

$$m = \frac{pVM}{RT} = \frac{(16.1\ \text{torr}) \times \left(\dfrac{1\ \text{atm}}{760\ \text{torr}}\right) \times (400\ \text{m}^3) \times \left(\dfrac{1\ \text{dm}}{10^{-1}\ \text{m}}\right)^3 \times (18.02\ \text{g mol}^{-1})}{(0.0821\ \text{dm}^3\ \text{atm K}^{-1}\ \text{mol}^{-1}) \times (300\ \text{K})}$$

$$= 6.2 \times 10^3\ \text{g} = \boxed{6.2\ \text{kg}}$$

E1.10(a) (a) For simplicity assume a container of volume 1 m³. Then the total mass is

$$m_\text{T} = n_{\text{N}_2}M_{\text{N}_2} + n_{\text{O}_2}M_{\text{O}_2} = 1146\ \text{g} \qquad (1)$$

Assuming that air is a perfect gas, $p_\text{T}V = n_\text{T}RT$, where n_T is the total amount of gas

$$n_\text{T} = \frac{p_\text{T}V}{RT} = \frac{(0.987\ \text{bar}) \times (10^5\ \text{Pa bar}^{-1}) \times (1\ \text{m}^3)}{(8.3145\ \text{Pa m}^3\ \text{K}^{-1}\ \text{mol}^{-1}) \times (300\ \text{K})} = 39.6\ \text{mol}$$

$$n_\text{T} = n_{\text{N}_2} + n_{\text{O}_2} = 39.6\ \text{mol} \qquad (2)$$

Equations (1) and (2) are simultaneous equations for the amounts of gas and may be solved for them. Inserting n_{O_2} from (2) into (1) we get

$$n_{\text{N}_2} \times (28.0136\ \text{g mol}^{-1}) + (39.6\ \text{mol} - n_{\text{N}_2}) \times (31.9988\ \text{g mol}^{-1}) = 1146\ \text{g}$$

$$(126\bar{6} - 1146)\ \text{g} = (3.9852\ \text{g mol}^{-1}) \times n_{\text{N}_2}$$

$$n_{\text{N}_2} = 30.1\bar{6}\ \text{mol}$$

$$n_{\text{O}_2} = n_\text{T} - n_{\text{N}_2} = (39.6 - 30.1\bar{6})\ \text{mol} = 9.4\bar{1}\ \text{mol}$$

The mole fractions are $x_{\text{N}_2} = \dfrac{30.1\bar{6}\ \text{mol}}{39.6\ \text{mol}} = \boxed{0.762}$ $\quad x_{\text{O}_2} = \dfrac{9.4\bar{1}\ \text{mol}}{39.6\ \text{mol}} = \boxed{0.238}$

The partial pressures are

$$p_{\text{N}_2} = (0.762) \times (0.987\ \text{bar}) = \boxed{0.752\ \text{bar}}$$

and $\quad p_{\text{O}_2} = (0.238) \times (0.987\ \text{bar}) = \boxed{0.235\ \text{bar}}$

The sum checks, $(0.752 + 0.235)\ \text{bar} = 0.987\ \text{bar}$.

(b) The simplest way to solve this part is to realize that n_T, p_T, and m_T remain the same as in part (a) as these are experimentally determined quantities. However, the simultaneous equations that need to be solved are modified as follows:

$$m_\text{T} = n_{\text{N}_2}M_{\text{N}_2} + n_{\text{O}_2}M_{\text{O}_2} + n_\text{Ar}M_\text{Ar} = 1146\ \text{g} \qquad (1')$$

$$n_\text{T} = n_{\text{N}_2} + n_{\text{O}_2} + n_\text{Ar} = 39.6\ \text{mol} \qquad (2')$$

Since $x_\text{Ar} = 0.0100$, $n_\text{Ar} = 0.396$ mol, that means

$$n_{\text{O}_2} = 39.6\ \text{mol} - n_{\text{N}_2} - n_\text{Ar} = 39.2\ \text{mol} - n_{\text{N}_2}$$

and $\quad (n_{\text{N}_2}) \times (28.0136\ \text{g mol}^{-1}) + (39.2\ \text{mol} - n_{\text{N}_2}) \times (31.9988\ \text{g mol}^{-1})$
$\quad = 1146\ \text{g} - (0.396\ \text{mol}) \times (39.948\ \text{g mol}^{-1})$

Solving the equations yields

$n_{N_2} = 30.9\bar{4}$ mol and $x_{N_2} = \boxed{0.782}$

and $n_{O_2} = 8.23$ mol and $x_{O_2} = \boxed{0.208}$

The partial pressures are:

$p_{Ar} = (0.010) \times (0.987\text{ bar}) = \boxed{0.0099\text{ bar}}$

$p_{N_2} = (0.782) \times (0.987\text{ bar}) = \boxed{0.772\text{ bar}}$

and $p_{O_2} = (0.208) \times (0.987\text{ bar}) = \boxed{0.205\text{ bar}}$

E1.11(a) This exercise uses the formula $M = \dfrac{\rho RT}{p}$, which was developed and used in Exercise 1.8(a). Substituting the data,

$$M = \frac{(1.23 \times 10^3\text{ g m}^{-3}) \times (8.3145\text{ Pa m}^3\text{ K}^{-1}\text{ mol}^{-1}) \times (330\text{ K})}{20 \times 10^3\text{ Pa}} = \boxed{169\text{ g mol}^{-1}}$$

E1.12(a) The easiest way to solve this exercise is to assume a convenient sample mass, say 1.000 g, then calculate the volume at each temperature, plot the volume against the Celsius temperature, and extrapolate to $V = 0$. Draw up the following table:

θ/°C	ρ/(g dm^{-3})	V/dm^3
–85	1.877	0.5328
0	1.294	0.7728
100	0.946	1.057

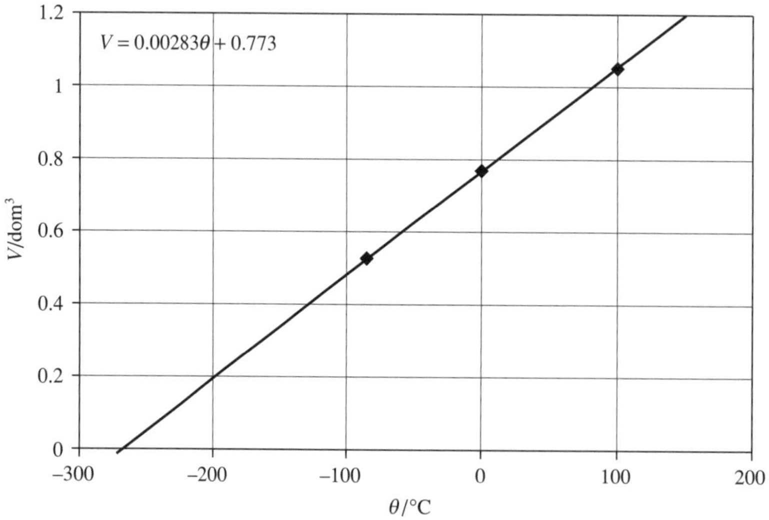

Figure 1.1

V is plotted against θ in Figure 1.1. The extrapolation gives a value for absolute zero close to –273°C. Alternatively, one could use an equation for V as a linear function of θ (the equation of the best-fit straight line through the data, also shown in Figure 1.1):

$V = V_0 + \alpha\theta$

which is Charles's law, and solve for the value of absolute zero. At absolute zero,

$$V = 0 = V_0 + \alpha\theta \quad \text{so} \quad \theta(\text{abs.zero}) = -\frac{V_0}{\alpha} = -\frac{0.773\ \text{dm}^3}{0.00283\ \text{dm}^3\ {}^\circ\text{C}^{-1}} = \boxed{-273^\circ\text{C}}$$

E1.13(a) (a) $p = \dfrac{nRT}{n}$ [1.8]

$n = 1.0\ \text{mol}, \quad T = 273.15\ \text{K (i)} \quad \text{or} \quad 1000\ \text{K (ii)}$

$V = 22.414\ \text{dm}^3\ \text{(i)} \quad \text{or} \quad 100\ \text{cm}^3\ \text{(ii)}$

$$\text{(i)} \quad p = \frac{(1.0\ \text{mol}) \times (0.08206\ \text{dm}^3\ \text{atm mol}^{-1}\ \text{K}^{-1}) \times (273.15\ \text{K})}{22.414\ \text{dm}^3} = \boxed{1.0\ \text{atm}}$$

$$\text{(ii)} \quad p = \frac{(1.0\ \text{mol}) \times (0.08206\ \text{dm}^3\ \text{atm mol}^{-1}\ \text{K}^{-1}) \times (1000\ \text{K})}{0.100\ \text{dm}^3} = \boxed{8.2 \times 10^2\ \text{atm}}$$

(b) $p = \dfrac{nRT}{V - nb} - \dfrac{an^2}{V^2}$ [1.21a]

From Table 1.6, $a = 5.507\ \text{dm}^6\ \text{atm mol}^{-2}$ and $b = 6.51 \times 10^{-2}\ \text{dm}^3\ \text{mol}^{-1}$. Therefore,

$$\text{(i)} \quad p = \frac{(1.0\ \text{mol}) \times (0.08206\ \text{dm}^3\ \text{atm mol}^{-1}\ \text{K}^{-1}) \times (273.15\ \text{K})}{[22.414 - (1.0) \times (0.0651)]\ \text{dm}^3} - \frac{(5.507\ \text{dm}^6\ \text{atm mol}^{-2}) \times (1.0\ \text{mol})^2}{(22.414\ \text{dm}^3)^2}$$

$$= 1.0\overline{03}\ \text{atm} - 0.011\ \text{atm} = 0.9\overline{92} = \boxed{1.0\ \text{atm}}$$

$$\text{(ii)} \quad p = \frac{(1.0\ \text{mol}) \times (0.08206\ \text{dm}^3\ \text{atm mol}^{-1}\ \text{K}^{-1}) \times (1000\ \text{K})}{(0.100 - 0.0651)\ \text{dm}^3} - \frac{(5.507\ \text{dm}^6\ \text{atm mol}^{-2}) \times (1.0\ \text{mol})^2}{(0.100\ \text{dm}^3)^2}$$

$$= 2.3\overline{5} \times 10^3\ \text{atm} - 5.5\overline{1} \times 10^2\ \text{atm} = \boxed{1.8 \times 10^3\ \text{atm}}$$

COMMENT. It is instructive to calculate the percentage deviation from perfect gas behaviour for (i) and (ii).

$$\text{(i)} \quad \frac{0.9\overline{92} - 1.\overline{000}}{1.000} \times 100\% = \overline{0.8}\%$$

$$\text{(ii)} \quad \frac{(18 \times 10^2) - (8.2 \times 10^2)}{8.2 \times 10^2} \times 100\% = 11\overline{9}\%$$

Deviations from perfect gas behaviour near 1 atm are observed only with very precise apparatus.

E1.14(a) The conversions needed are as follows:

$1\ \text{atm} = 1.013 \times 10^5\ \text{Pa}$, $1\ \text{Pa} = 1\ \text{kg m}^{-1}\ \text{s}^{-2}$, $1\ \text{dm}^6 = (10^{-1}\ \text{m})^6 = 10^{-6}\ \text{m}^6$, $1\ \text{dm}^3 = 10^{-3}\ \text{m}^3$.

Therefore,

$$a = 0.751\ \text{atm dm}^6\ \text{mol}^{-2} \times \frac{1.013 \times 10^5\ \text{kg m}^{-1}\ \text{s}^{-2}}{1\ \text{atm}} \times \frac{10^{-6}\ \text{m}^6}{\text{dm}^6} = \boxed{7.61 \times 10^{-2}\ \text{kg m}^5\ \text{s}^{-2}\ \text{mol}^{-2}}$$

and $b = 0.0226\ \text{dm}^3\ \text{mol}^{-1} \times \dfrac{10^{-3}\ \text{m}^3}{\text{dm}^3} = \boxed{2.26 \times 10^{-5}\ \text{m}^3\ \text{mol}^{-1}}$

E1.15(a) The definition of Z is used $Z = \dfrac{pV_m}{RT}[1.17] = \dfrac{V_m}{V_m^o}$

V_m is the actual molar volume, V_m^o is the perfect gas molar volume. $V_m^o = \dfrac{RT}{p}$. Since V_m is 12% smaller than that of a perfect gas, $V_m = 0.88V_m^o$, and

(a) $Z = \dfrac{0.88V_m^o}{V_m^o} = \boxed{0.88}$

(b) $V_m = \dfrac{ZRT}{p} = \dfrac{(0.88)\times(0.08206\ \text{dm}^3\,\text{atm mol}^{-1}\,\text{K}^{-1})\times(250\ \text{K})}{15\ \text{atm}} = \boxed{1.2\ \text{dm}^3\,\text{mol}^{-1}}$

Since $V_m < V_m^o$ attractive forces dominate.

E1.16(a) The amount of gas is first determined from its mass; then the van der Waals equation [1.21a] is used to determine its pressure at the working temperature. The initial conditions of 300 K and 100 atm are in a sense superfluous information.

$$n = \frac{92.4\times10^3\ \text{g}}{28.02\ \text{g mol}^{-1}} = 3.30\times10^3\ \text{mol}$$

$$V = 1.000\ \text{m}^3 = 1.000\times10^3\ \text{dm}^3$$

$$p = \frac{nRT}{V-nb} - \frac{an^2}{V^2} = \frac{(3.30\times10^3\ \text{mol})\times(0.08206\ \text{dm}^3\,\text{atm mol}^{-1}\,\text{K}^{-1})\times(500\ \text{K})}{(1.000\times10^3\ \text{dm}^3)-(3.30\times10^3\ \text{mol})\times(0.037\ \text{dm}^3\,\text{mol}^{-1})}$$
$$-\frac{(1.352\ \text{dm}^6\,\text{atm mol}^{-2})\times(3.30\times10^3\ \text{mol})^2}{(1.000\times10^3\ \text{dm}^3)^2} = (155-14.7)\ \text{atm} = \boxed{140\ \text{atm}}$$

E1.17(a) (a) $p = \dfrac{nRT}{V}[1.8] = \dfrac{(10.0\ \text{mol})\times(0.08206\ \text{dm}^3\,\text{atm K}^{-1}\,\text{mol}^{-1})\times(300\ \text{K})}{4.860\ \text{dm}^3} = \boxed{50.7\ \text{atm}}$

(b) $p = \dfrac{nRT}{V-nb} - a\left(\dfrac{n}{V}\right)^2$ [1.21a]

$$= \frac{(10.0\ \text{mol})\times(0.08206\ \text{dm}^3\,\text{atm mol}^{-1}\,\text{K}^{-1})\times(300\ \text{K})}{(4.860\ \text{dm}^3)-(10.0\ \text{mol})\times(0.0651\ \text{dm}^3\,\text{mol}^{-1})}$$
$$-(5.507\ \text{dm}^6\,\text{atm mol}^{-2})\times\left(\frac{10.0\ \text{mol}}{4.860\ \text{dm}^3}\right)^2 = 58.5-23.3 = \boxed{35.2\ \text{atm}}$$

The compression factor is calculated from its definition [1.17] after inserting $V_m = \dfrac{V}{n}$.

To complete the calculation of Z, a value for the pressure, p, is required. The implication in the definition [1.17] is that p is the actual pressure as determined experimentally. This pressure is neither the perfect gas pressure nor the van der Waals pressure. However, on the assumption that the van der Waals equation provides a value for the pressure close to the experimental value, we can estimate the compression factor as follows

$$Z = \frac{pV}{nRT} = \frac{(35.2\ \text{atm})\times(4.860\ \text{dm}^3)}{(10.0\ \text{mol})\times(0.08206\ \text{dm}^3\,\text{atm mol}^{-1}\,\text{K}^{-1})\times(300\ \text{K})} = \boxed{0.695}$$

E1.18(a) $n = n(\mathrm{H_2}) + n(\mathrm{N_2}) = 2.0\ \mathrm{mol} + 1.0\ \mathrm{mol} = 3.0\ \mathrm{mol}$ $\quad x_\mathrm{J} = \dfrac{n_\mathrm{J}}{n}$ [1.14]

(a) $x(\mathrm{H_2}) = \dfrac{2.0\ \mathrm{mol}}{3.0\ \mathrm{mol}} = \boxed{0.67}$ $\quad x(\mathrm{N_2}) = \dfrac{1.0\ \mathrm{mol}}{3.0\ \mathrm{mol}} = \boxed{0.33}$

(b) The perfect gas law is assumed to hold for each component individually as well as for the mixture as a whole. Hence, $p_\mathrm{J} = n_\mathrm{J}\dfrac{RT}{V}$

$$\frac{RT}{V} = \frac{(0.08206\ \mathrm{dm^3\,atm\,mol^{-1}\,K^{-1}}) \times (273.15\ \mathrm{K})}{22.4\ \mathrm{dm^3}} = 1.00\ \mathrm{atm\,mol^{-1}}$$

$$p(\mathrm{H_2}) = (2.0\ \mathrm{mol}) \times (1.00\ \mathrm{atm\,mol^{-1}}) = \boxed{2.0\ \mathrm{atm}}$$

$$p(\mathrm{N_2}) = (1.0\ \mathrm{mol}) \times (1.00\ \mathrm{atm\,mol^{-1}}) = \boxed{1.0\ \mathrm{atm}}$$

(c) $p = p(\mathrm{H_2}) + p(\mathrm{N_2})$ [1.15] $= 2.0\ \mathrm{atm} + 1.0\ \mathrm{atm} = \boxed{3.0\ \mathrm{atm}}$

Question. Does Dalton's law hold for a mixture of van der Waals gases?

E1.19(a) Equations 1.22 are solved for b and a, respectively, and yield

$b = V_\mathrm{c}/3$ and $a = 27b^2 p_\mathrm{c} = 3V_\mathrm{c}^2 p_\mathrm{c}$.

Substituting the critical constants

$$b = \frac{98.7\ \mathrm{cm^3\,mol^{-1}}}{3} = \boxed{32.9\ \mathrm{cm^3\,mol^{-1}}}$$

and $a = 3 \times (98.7 \times 10^{-3}\ \mathrm{dm^3\,mol^{-1}})^2 \times (45.6\ \mathrm{atm}) = \boxed{1.33\ \mathrm{dm^6\,atm\,mol^{-2}}}$

By interpreting b as the excluded volume of a mole of spherical molecules, we can obtain an estimate of molecular size. The centres of spherical particles are excluded from a sphere whose radius is the diameter of those spherical particles (i.e. twice their radius); that volume times the Avogadro constant is the molar excluded volume b:

$$b = N_\mathrm{A}\left(\frac{4\pi(2r)^3}{3}\right) \quad \text{so} \quad r = \frac{1}{2}\left(\frac{3b}{4\pi N_\mathrm{A}}\right)^{1/3}$$

$$r = \frac{1}{2}\left(\frac{3(32.9\ \mathrm{cm^3\,mol^{-1}})}{4\pi(6.022 \times 10^{23}\ \mathrm{mol^{-1}})}\right)^{1/3} = 1.18 \times 10^{-8}\ \mathrm{cm} = \boxed{0.118\ \mathrm{nm}}$$

But this problem is overdetermined. We have another piece of information

$$T_\mathrm{c} = \frac{8a}{27Rb}$$

According to the constants we have already determined, T_c should be

$$T_\mathrm{c} = \frac{8(1.33\ \mathrm{dm^6\,atm\,mol^{-2}})}{27(0.08206\ \mathrm{dm^3\,atm\,mol^{-1}\,K^{-1}}) \times (0.0329\ \mathrm{dm^3\,mol^{-1}})} = 146\ \mathrm{K}$$

However, the reported T_c is 190.6 K, suggesting that our computed a/b is about 25% lower than it should be.

E1.20(a) The Boyle temperature, T_B, is the temperature at which the virial coefficient $B = 0$. In order to express T_B in terms of a and b, the van der Waals equation must be recast into the form of the virial equation.

$$p = \frac{RT}{V_m - b} - \frac{a}{V_m^2} \; [1.21b]$$

Factoring out $\frac{RT}{V_m}$ yields $p = \frac{RT}{V_m}\left\{\frac{1}{1 - b/V_m} - \frac{a}{RTV_m}\right\}$

So long as $b/V_m < 1$, the first term inside the brackets can be expanded using

$$(1 - x)^{-1} = 1 + x + x^2 + \cdots,$$

which gives

$$p = \frac{RT}{V_m}\left\{1 + \left(b - \frac{a}{RT}\right) \times \left(\frac{1}{V_m}\right) + \cdots\right\}$$

We can now identify the second virial coefficient as $B = b - \frac{a}{RT}$

At the Boyle temperature

$$B = 0 = b - \frac{a}{RT_B} \quad \text{so} \quad T_B = \frac{a}{bR} = \frac{27T_c}{8}$$

(a) From Table 1.6, $a = 6.260\ \text{dm}^6\ \text{atm mol}^{-2}$ and $b = 0.0542\ \text{dm}^3\ \text{mol}^{-1}$. Therefore,

$$T_B = \frac{6.260\ \text{dm}^6\ \text{atm mol}^{-2}}{(5.42 \times 10^{-2}\ \text{dm}^3\ \text{mol}^{-1}) \times (0.08206\ \text{dm}^3\ \text{atm mol}^{-1}\ \text{K}^{-1})} = \boxed{1.41 \times 10^3\ \text{K}}$$

(b) As in Exercise 1.19(a),

$$b = N_A\left(\frac{4\pi(2r)^3}{3}\right) \quad \text{so} \quad r = \frac{1}{2}\left(\frac{3b}{4\pi N_A}\right)^{1/3}$$

$$r = \frac{1}{2}\left(\frac{3(0.0542\ \text{dm}^3\ \text{mol}^{-1})}{4\pi(6.022 \times 10^{23}\ \text{mol}^{-1})}\right)^{1/3} = 1.39 \times 10^{-9}\ \text{dm} = \boxed{0.139\ \text{nm}}$$

E1.21(a) The reduced temperature and pressure of hydrogen are calculated from the relationships

$$T_r = \frac{T}{T_c} \quad \text{and} \quad p_r = \frac{p}{p_c} \; [1.24]$$

$$T_r = \frac{298\ \text{K}}{33.23\ \text{K}} = 8.97 \; [T_c = 33.23\ \text{K, Table 1.5}]$$

$$p_r = \frac{1.0\ \text{atm}}{12.8\ \text{atm}} = 0.078 \; [p_c = 12.8\ \text{atm, Table 1.5}]$$

Hence, the gases named will be in corresponding states at $T = 8.97T_c$ and at $p = 0.078p_c$.

(a) For ammonia, $T_c = 405.5$ K and $p_c = 111.3$ atm, so

$$T = (8.97) \times (405.5\ \text{K}) = \boxed{3.64 \times 10^3\ \text{K}}$$

$$p = (0.078) \times (111.3\ \text{atm}) = \boxed{8.7\ \text{atm}}$$

(b) For xenon, $T_c = 289.75$ K and $p_c = 58.0$ atm, so

$$T = (8.97) \times (289.75\ \text{K}) = \boxed{2.60 \times 10^3\ \text{K}}$$

$$p = (0.078) \times (58.0\ \text{atm}) = \boxed{4.5\ \text{atm}}$$

(c) For helium, $T_c = 5.21$ K and $p_c = 2.26$ atm, so

$$T = (8.97) \times (5.21\ \text{K}) = \boxed{46.7\ \text{K}}$$

$$p = (0.078) \times (2.26\ \text{atm}) = \boxed{0.18\ \text{atm}}$$

E1.22(a) The van der Waals equation [1.21b] is solved for b, which yields

$$b = V_m - \frac{RT}{\left(p + \dfrac{a}{V_m^2}\right)}$$

Substituting the data

$$b = 5.00 \times 10^{-4}\ \text{m}^3\ \text{mol}^{-1} - \frac{(8.3145\ \text{J K}^{-1}\ \text{mol}^{-1}) \times (273\ \text{K})}{\left\{(3.0 \times 10^6\ \text{Pa}) + \left(\dfrac{0.50\ \text{m}^6\ \text{Pa mol}^{-2}}{(5.00 \times 10^{-4}\ \text{m}^3\ \text{mol}^{-1})^2}\right)\right\}}$$

$$= \boxed{0.46 \times 10^{-4}\ \text{m}^3\ \text{mol}^{-1}}$$

$$Z = \frac{pV_m}{RT}[1.17] = \frac{(3.0 \times 10^6\ \text{Pa}) \times (5.00 \times 10^{-4}\ \text{m}^3\ \text{mol}^{-1})}{(8.3145\ \text{J K}^{-1}\ \text{mol}^{-1}) \times (273\ \text{K})} = \boxed{0.66}$$

COMMENT. The definition of Z involves the actual pressure, volume, and temperature and does not depend upon the equation of state used to relate these variables.

Solutions to problems

Solutions to numerical problems

P1.1 Since the Neptunians know about perfect gas behaviour, we may assume that they will write $pV = nRT$ at both temperatures. We may also assume that they will establish the size of their absolute unit to be the same as the °N, just as we write 1 K = 1°C. Thus

$$pV(T_1) = 28.0\ \text{dm}^3\ \text{atm} = nRT_1 = nR \times (T_1 + 0°\text{N})$$

and $pV(T_2) = 40.0\ \text{dm}^3\ \text{atm} = nRT_2 = nR \times (T_1 + 100°\text{N})$.

So $T_1 = \dfrac{28.0\ \text{dm}^3\ \text{atm}}{nR}$ and $T_1 + 100°\text{N} = \dfrac{40.0\ \text{dm}^3\ \text{atm}}{nR}$.

Dividing yields

$$\frac{T_1 + 100°\text{N}}{T_1} = \frac{40.0\ \text{dm}^3\,\text{atm}}{28.0\ \text{dm}^3\,\text{atm}} = 1.43$$

Thus, $T_1 + 100°\text{N} = 1.43T_1$ so $T_1 = 233$ absolute units.

As in the relationship between our Kelvin scale and Celsius scale $T = \theta$ – absolute zero(°N) so absolute zero(°N) = $\boxed{-233°\text{N}}$.

COMMENT. To facilitate communication with Earth students we have converted the Neptunians' units of the pV product to units familiar to humans, namely dm^3 atm. However, we see from the solution that only the ratio of pV products is required, and that will be the same in any civilization.

Question. If the Neptunians' unit of volume is the lagoon (L), their unit of pressure is the poseidon (P), their unit of amount is the nereid (n), and their unit of absolute temperature is the titan (T), what is the value of the Neptunians' gas constant (R) in units of L, P, n, and T?

P1.3 The value of absolute zero can be expressed in terms of α by using the requirement that the volume of a perfect gas becomes zero at the absolute zero of temperature. Hence,

$$0 = V_0[1 + \alpha\theta(\text{abs. zero})]$$

Then, $\theta(\text{abs. zero}) = -\dfrac{1}{\alpha}$

All gases become perfect in the limit of zero pressure, so the best value of α and, hence, θ(abs. zero) is obtained by extrapolating α to zero pressure. This is done in Figure 1.2. Using the extrapolated value,

$$\alpha = 3.6637 \times 10^{-3}\,°\text{C}^{-1},$$

or $\theta(\text{abs. zero}) = -\dfrac{1}{3.6637 \times 10^{-3}\,°\text{C}^{-1}} = \boxed{-272.95°\text{C}}$

which is close to the accepted value of –273.15°C.

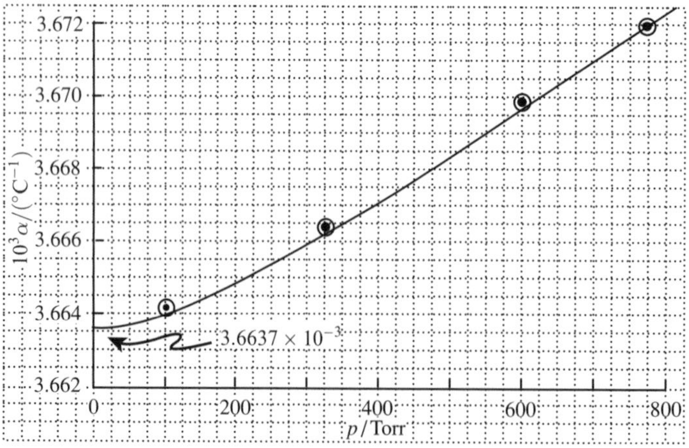

Figure 1.2

P1.5 At n and V, $\dfrac{p}{T} = \dfrac{nR}{V}$ = constant. Hence, $\dfrac{p}{T} = \dfrac{p_3}{T_3}$, where p is the measured pressure at temperature, T, and p_3 and T_3 are the triple point pressure and temperature, respectively. Rearranging, $p = \left(\dfrac{p_3}{T_3}\right)T$.

The ratio $\dfrac{p_3}{T_3}$ is a constant $= \dfrac{6.69\ \text{kPa}}{273.16\ \text{K}} = 0.0245\ \text{kPa K}^{-1}$. Thus the change in p, Δp, is proportional to the change in temperature, ΔT:

$$\Delta p = (0.0245\ \text{kPa K}^{-1}) \times (\Delta T).$$

(a) $\Delta p = (0.0245\ \text{kPa K}^{-1}) \times (1.00\ \text{K}) = \boxed{0.0245\ \text{kPa}}$

(b) Rearranging, $p = \left(\dfrac{T}{T_3}\right)p_3 = \left(\dfrac{373.16\ \text{K}}{273.16\ \text{K}}\right) \times (6.69\ \text{kPa}) = \boxed{9.14\ \text{kPa}}$

(c) Since $\dfrac{p}{T}$ is a constant at constant n and V, it always has the value $0.0245\ \text{kPa K}^{-1}$;

hence $\Delta p = p_{374.15\,\text{K}} - p_{373.15\,\text{K}} = (0.0245\ \text{kPa K}^{-1}) \times (1.00\ \text{K}) = \boxed{0.0245\ \text{kPa}}$

P1.7 (a) $V_\text{m} = \dfrac{RT}{p} = \dfrac{(0.08206\ \text{dm}^3\,\text{atm mol}^{-1}\,\text{K}^{-1}) \times (350\ \text{K})}{2.30\ \text{atm}} = \boxed{12.5\ \text{dm}^3\,\text{mol}^{-1}}$

(b) From $p = \dfrac{RT}{V_\text{m} - b} - \dfrac{a}{V_\text{m}^2}$ [1.21b], we obtain $V_\text{m} = \dfrac{RT}{p + \dfrac{a}{V_\text{m}^2}} + b$.

Then, with a and b from Table 1.6

$$V_\text{m} \approx \frac{(0.08206\ \text{dm}^3\,\text{atm mol}^{-1}\,\text{K}^{-1}) \times (350\ \text{K})}{(2.30\ \text{atm}) + \left(\dfrac{6.260\ \text{dm}^3\,\text{atm mol}^{-2}}{(12.5\ \text{dm}^3\,\text{mol}^{-1})^2}\right)} + 5.42 \times 10^{-2}\ \text{dm}^3\,\text{mol}^{-1}$$

$$\approx \frac{28.7\ \text{dm}^3\,\text{mol}^{-1}}{2.34} + 5.42 \times 10^{-2}\ \text{dm}^3\,\text{mol}^{-1} \approx \boxed{12.3\ \text{dm}^3\,\text{mol}^{-1}}$$

Substitution of $12.3\,\text{dm}^3\,\text{mol}^{-1}$ into the denominator of the first expression again results in $V_\text{m} = 12.3\,\text{dm}^3\,\text{mol}^{-1}$, so the cycle of approximation may be terminated.

P1.9 As indicated by the parts of eqn 1.19 the compression factor of a gas may be expressed as either a virial expansion in p or in $\left(\dfrac{1}{V_\text{m}}\right)$. The virial form of the van der Waals equation is derived in Exercise 1.20(a) and is $p = \dfrac{RT}{V_\text{m}}\left\{1 + \left(b - \dfrac{a}{RT}\right) \times \left(\dfrac{1}{V_\text{m}}\right) + \cdots\right\}$.

Rearranging, $Z = \dfrac{pV_\text{m}}{RT} = 1 + \left(b - \dfrac{a}{RT}\right) \times \left(\dfrac{1}{V_\text{m}}\right) + \cdots$.

On the assumption that the perfect gas expression for V_m is adequate for the second term in this expansion, we can readily obtain Z as a function of p.

$$Z = 1 + \left(\frac{1}{RT}\right) \times \left(b - \frac{a}{RT}\right) p + \cdots$$

(a) $T = T_c = 126.3\ \text{K}$

$$V_m = \left(\frac{RT}{p}\right) \times Z = \frac{RT}{p} + \left(b - \frac{a}{RT}\right) + \cdots$$

$$= \frac{(0.08206\ \text{dm}^3\,\text{atm mol}^{-1}\,\text{K}^{-1}) \times (126.3\ \text{K})}{10.0\ \text{atm}}$$

$$+ \left\{(0.0391\ \text{dm}^3\,\text{mol}^{-1}) - \left(\frac{1.390\ \text{dm}^6\,\text{atm mol}^{-2}}{(0.08206\ \text{dm}^3\,\text{atm mol}^{-1}\,\text{K}^{-1}) \times (126.3\ \text{K})}\right)\right\}$$

$$= (1.036 - 0.095)\ \text{dm}^3\,\text{mol}^{-1} = \boxed{0.941\ \text{dm}^3\,\text{mol}^{-1}}$$

$$Z = \left(\frac{p}{RT}\right) \times (V_m) = \frac{(10.0\ \text{atm}) \times (0.941\ \text{dm}^3\,\text{mol}^{-1})}{(0.08206\ \text{dm}^3\,\text{atm mol}^{-1}\,\text{K}^{-1}) \times (126.3\ \text{K})} = 0.908$$

(b) The Boyle temperature is the temperature at which the second virial coefficient is zero, hence $Z = 1$ (correct to the first power in p), and the gas is close to perfect. However, if we assume that N_2 is a van der Waals gas, when the second virial coefficient is zero

$$\left(b - \frac{a}{RT_B}\right) = 0, \quad \text{or} \quad T_B = \frac{a}{bR}$$

$$T_B = \frac{1.390\ \text{dm}^6\,\text{atm mol}^{-2}}{(0.0391\ \text{dm}^3\,\text{mol}^{-1}) \times (0.08206\ \text{dm}^3\,\text{atm mol}^{-1}\,\text{K}^{-1})} = 433\ \text{K}$$

The experimental value [Table 1.5] is 327.2 K. Two possible resolutions of the discrepancy spring to mind.

1. Terms beyond the first power in p should not be dropped in the expansion for Z.
2. Nitrogen is only approximately a van der Waals gas.

When $Z = 1$, $V_m = \frac{RT}{p}$, and using $T_B = 327.2\ \text{K}$

$$V_m = \frac{(0.08206\ \text{dm}^3\,\text{atm mol}^{-1}\,\text{K}^{-1}) \times 327.2\ \text{K}}{10.0\ \text{atm}} = \boxed{2.69\ \text{dm}^3\,\text{mol}^{-1}}$$

This is the ideal value of V_m. Using the experimental value of T_B and inserting this value into the expansion for V_m above, we have

$$V_m = \frac{0.08206\ \text{dm}^3\,\text{atm mol}^{-1}\,\text{K}^{-1} \times 327.2\ \text{K}}{10.0\ \text{atm}}$$

$$+ \left\{0.0391\ \text{dm}^3\,\text{mol}^{-1} - \left(\frac{1.390\ \text{dm}^6\,\text{atm mol}^{-2}}{0.08206\ \text{dm}^3\,\text{atm mol}^{-1}\,\text{K}^{-1} \times 327.2\ \text{K}}\right)\right\}$$

$$= (2.68\overline{5} - 0.013)\ \text{dm}^3\,\text{mol}^{-1} = \boxed{2.67\ \text{dm}^3\,\text{mol}^{-1}}$$

and $$Z = \frac{V_m}{V_m^\circ} = \frac{2.67\ \text{dm}^3\,\text{mol}^{-1}}{2.69\ \text{dm}^3\,\text{mol}^{-1}} = 0.995 \approx 1$$

(c) $T = T_I = 621$ K [Table 2.10]

$$V_m = \frac{0.08206\ \text{dm}^3\ \text{atm mol}^{-1}\ \text{K}^{-1} \times 621\ \text{K}}{10.0\ \text{atm}}$$

$$+\left\{0.0391\ \text{dm}^3\ \text{mol}^{-1} - \left(\frac{1.390\ \text{dm}^6\ \text{atm mol}^{-2}}{0.08206\ \text{dm}^3\ \text{atm K}^{-1}\ \text{mol}^{-1} \times 621\ \text{K}}\right)\right\}$$

$$= (5.09\overline{6} + 0.012)\ \text{dm}^3\ \text{mol}^{-1} = \boxed{5.11\ \text{dm}^3\ \text{mol}^{-1}}$$

and $$Z = \frac{5.11\ \text{dm}^3\ \text{mol}^{-1}}{5.10\ \text{dm}^3\ \text{mol}^{-1}} = 1.002 \approx 1$$

Based on the values of T_B and T_I given in Tables 1.5 and 2.10 and assuming that N_2 is a van der Waals gas, the calculated value of Z is closest to 1 at the $\boxed{\text{inversion temperature}}$, but the difference from the value at T_B is less than the accuracy of the method.

P1.11 (a) $$V_m = \frac{V}{n} = V \times \frac{M}{m} = \frac{M}{\rho} = \frac{18.02\ \text{g mol}^{-1}}{133.2\ \text{g dm}^{-3}} = \boxed{0.1353\ \text{dm}^3\ \text{mol}^{-1}}$$

(b) $$Z = \frac{pV_m}{RT}[1.17] = \frac{(327.6\ \text{atm}) \times (0.1353\ \text{dm}^3\ \text{mol}^{-1})}{(0.08206\ \text{dm}^3\ \text{atm K}^{-1}\ \text{mol}^{-1}) \times (776.4\ \text{K})} = \boxed{0.6957}$$

(c) Two expansions for Z based on the van der Waals equation are given in Problem 1.9. They are

$$Z = 1 + \left(b - \frac{a}{RT}\right) \times \left(\frac{1}{V_m}\right) + \cdots$$

$$= 1 + \left\{(0.03049\ \text{dm}^3\ \text{mol}^{-1}) - \left(\frac{5.464\ \text{dm}^6\ \text{atm mol}^{-2}}{(0.08206\ \text{dm}^3\ \text{atm K}^{-1}\ \text{mol}^{-1}) \times (776.4\ \text{K})}\right)\right\}$$

$$\times \frac{1}{0.1353\ \text{dm}^3\ \text{mol}^{-1}} = 1 - 0.4085 = 0.5915$$

and $$Z = 1 + \left(\frac{1}{RT}\right) \times \left(b - \frac{a}{RT}\right) \times (p) + \cdots$$

$$= 1 + \frac{1}{(0.08206\ \text{dm}^3\ \text{atm K}^{-1}\ \text{mol}^{-1}) \times (776.4\ \text{K})}$$

$$\times \left\{(0.03049\ \text{dm}^3\ \text{mol}^{-1}) - \left(\frac{5.464\ \text{dm}^6\ \text{atm mol}^{-2}}{(0.08206\ \text{dm}^3\ \text{atm K}^{-1}\ \text{mol}^{-1}) \times (776.4\ \text{K})}\right)\right\} \times 327.6\ \text{atm}$$

$$= 1 - 0.2842 \approx \boxed{0.7158}$$

In this case the expansion in p gives a value close to the experimental value; the expansion in $\frac{1}{V_m}$ is not as good. However, when terms beyond the second are included the results from the two expansions for Z converge.

P1.13 $$V_c = 2b, \quad T_c = \frac{a}{4bR}\ [\text{Table 1.7}]$$

Hence, with V_c and T_c from Table 1.5,

$$b = \tfrac{1}{2}V_c = \tfrac{1}{2} \times (118.8\ \text{cm}^3\ \text{mol}^{-1}) = \boxed{59.4\ \text{cm}^3\ \text{mol}^{-1}}$$

$$\begin{aligned} a &= 4bRT_c = 2RT_cV_c \\ &= (2) \times (0.08206\ \text{dm}^3\ \text{atm mol}^{-1}\ \text{K}^{-1}) \times (289.75\ \text{K}) \times (118.8 \times 10^{-3}\ \text{dm}^3\ \text{mol}^{-1}) \\ &= \boxed{5.649\ \text{dm}^6\ \text{atm mol}^{-2}} \end{aligned}$$

Hence,

$$\begin{aligned} p &= \frac{RT}{V_m - b} e^{-a/RTV_m} \\ &= \frac{(8.206 \times 10^{-2}\ \text{dm}^3\ \text{atm K}^{-1}\ \text{mol}^{-1}) \times (298\ \text{K})}{(1.0\ \text{dm}^3\ \text{mol}^{-1}) - (59.4 \times 10^{-3}\ \text{dm}^3\ \text{mol}^{-1})} \\ &\quad \times \exp\left(\frac{-(5.649\ \text{dm}^6\ \text{atm mol}^{-2})}{(0.08206\ \text{dm}^3\ \text{atm mol}^{-1}\ \text{K}^{-1}) \times (298\ \text{K}) \times (1.0\ \text{dm}^3\ \text{mol}^{-1})}\right) \\ &= 26.\bar{0}\ \text{atm} \times e^{-0.23\bar{1}} = \boxed{21\ \text{atm}} \end{aligned}$$

Solutions to theoretical problems

P1.15 This expansion has already been given in the solution to Exercise 1.20(a); the result is

$$p = \frac{RT}{V_m}\left(1 + \left[b - \frac{a}{RT}\right]\frac{1}{V_m} + \frac{b^2}{V_m^2} + \cdots\right)$$

Compare this expansion with the virial expansion

$$p = \frac{RT}{V_m}\left(1 + \frac{B}{V_m} + \frac{C}{V_m^2} + \cdots\right) \quad [1.19b]$$

and hence find $\boxed{B = b - \frac{a}{RT}}$ and $\boxed{C = b^2}$

Since $C = 1200\ \text{cm}^6\ \text{mol}^{-2}$, $b = C^{1/2} = \boxed{34.6\ \text{cm}^3\ \text{mol}^{-1}}$

$$\begin{aligned} a &= RT(b - B) = (8.206 \times 10^{-2}) \times (273\ \text{dm}^3\ \text{atm mol}^{-1}) \times (34.6 + 21.7)\ \text{cm}^3\ \text{mol}^{-1} \\ &= (22.4\bar{0}\ \text{dm}^3\ \text{atm mol}^{-1}) \times (56.3 \times 10^{-3}\ \text{dm}^3\ \text{mol}^{-1}) = \boxed{1.26\ \text{dm}^6\ \text{atm mol}^{-2}} \end{aligned}$$

P1.17 The critical point corresponds to a point of zero slope, which is simultaneously a point of inflection in a plot of pressure versus molar volume. A critical point exists if there are values of p, V, and T that result in a point that satisfies these conditions.

$$p = \frac{RT}{V_m} - \frac{B}{V_m^2} + \frac{C}{V_m^3}$$

$$\left.\begin{aligned} \left(\frac{\partial p}{\partial V_m}\right)_T &= -\frac{RT}{V_m^2} + \frac{2B}{V_m^3} - \frac{3C}{V_m^4} = 0 \\ \left(\frac{\partial^2 p}{\partial V_m^2}\right)_T &= \frac{2RT}{V_m^3} - \frac{6B}{V_m^4} + \frac{12C}{V_m^5} = 0 \end{aligned}\right\}\text{at the critical point}$$

That is, at the critical point

$$\left.\begin{aligned} -RT_cV_c^2 + 2BV_c - 3C = 0 \\ RT_cV_c^2 - 3BV_c + 6C = 0 \end{aligned}\right\}$$

which solve to $V_c = \boxed{\dfrac{3C}{B}}$, $\boxed{T_c = \dfrac{B^2}{3RC}}$

Now use the equation of state to find p_c

$$p_c = \frac{RT_c}{V_c} - \frac{B}{V_c^2} + \frac{C}{V_c^3} = \left(\frac{RB^2}{3RC}\right) \times \left(\frac{B}{3C}\right) - B\left(\frac{B}{3C}\right)^2 + C\left(\frac{B}{3C}\right)^3 = \boxed{\frac{B^3}{27C^2}}$$

It follows that $Z_c = \dfrac{p_cV_c}{RT_c} = \left(\dfrac{B^3}{27C^2}\right) \times \left(\dfrac{3C}{B}\right) \times \left(\dfrac{1}{R}\right) \times \left(\dfrac{3RC}{B^2}\right) = \boxed{\dfrac{1}{3}}$

P1.19 For a real gas we may use the virial expansion in terms of p [1.19a]

$$p = \frac{nRT}{V}(1 + B'p + \cdots) = \rho\frac{RT}{M}(1 + B'p + \cdots) \quad \left[\frac{n}{V} = \frac{\rho}{M}\right]$$

which rearranges to $\dfrac{p}{\rho} = \dfrac{RT}{M} + \dfrac{RTB'}{M}p + \cdots$

Therefore, the limiting slope of a plot of $\dfrac{p}{\rho}$ against p is $\dfrac{B'RT}{M}$.

Draw up the following table. Bear in mind that 1 kPa = 10^3 kg m^{-1} s^{-2}.

p/(kPa)	12.223	25.20	36.97	60.37	85.23	101.3
ρ/(kg m^{-3})	0.225	0.456	0.664	1.062	1.468	1.734
$\dfrac{p/\rho}{10^3\ \text{m}^2\ \text{s}^{-2}}$	54.3	55.3	55.7	56.8	58.1	58.4

$\dfrac{p}{\rho}$ is plotted in Figure 1.3.

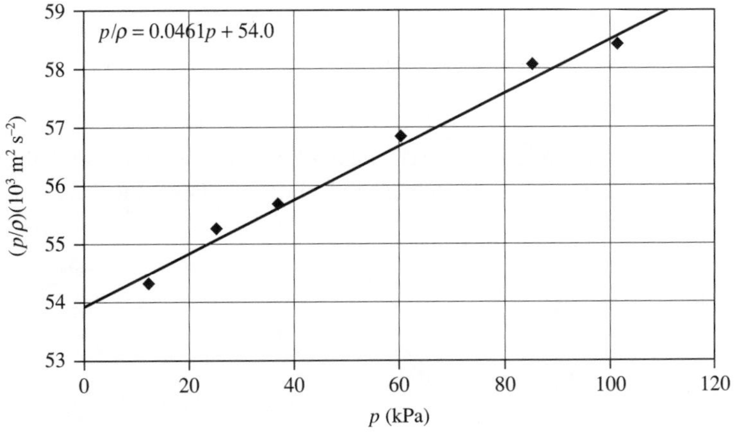

Figure 1.3

A straight line fits the data rather well. From Figure 1.3 the slope is

$$\frac{B'RT}{M} = 0.0461 \times 10^3\ \mathrm{m^2\,s^{-2}\,kPa^{-1}} = 0.0461\ \mathrm{m^3\,kg^{-1}}$$

and the y-intercept is

$$\frac{RT}{M} = 5.40 \times 10^4\ \mathrm{m^2\,s^{-2}}$$

hence $B' = \dfrac{4.61 \times 10^{-2}\ \mathrm{kg^{-1}\,m^3}}{5.40 \times 10^4\ \mathrm{m^2\,s^{-2}}} = 0.854 \times 10^{-6}\ \mathrm{Pa^{-1}}$

or $B' = (0.854 \times 10^{-6}\ \mathrm{Pa^{-1}}) \times (1.0133 \times 10^5\ \mathrm{Pa\ atm^{-1}}) = \boxed{0.0866\ \mathrm{atm^{-1}}}$.

Comparing equations 1.19a and 1.19b, we see [Problem 1.18]

$$B = RTB' = (0.08206\ \mathrm{dm^3\,atm\,mol^{-1}\,K^{-1}}) \times (298\ \mathrm{K}) \times (0.0866\ \mathrm{atm^{-1}})$$
$$= \boxed{2.12\ \mathrm{dm^3\,mol^{-1}}}$$

P1.21 The critical temperature is that temperature above which the gas cannot be liquefied by the application of pressure alone. Below the critical temperature two phases, liquid and gas, may coexist at equilibrium, and in the two-phase region there is more than one molar volume corresponding to the same conditions of temperature and pressure. Therefore, any equation of state that can even approximately describe this situation must allow for more than one real root for the molar volume at some values of T and p, but as the temperature is increased above T_c, allows only one real root. Thus, appropriate equations of state must be equations of odd degree in V_m.

The equation of state for gas A may be rewritten $V_m^2 - \dfrac{RT}{p}V_m - \dfrac{RTb}{p} = 0$, which is a quadratic and never has just one real root. Thus, this equation can never model critical behaviour. It could possibly model in a very crude manner a two-phase situation, since there are some conditions under which a quadratic has two real positive roots, but not the process of liquefaction.

The equation of state of gas B is a first-degree equation in V_m and therefore can never model critical behaviour, the process of liquefaction, or the existence of a two-phase region.

A cubic equation is the equation of lowest degree that can show a cross-over from more than one real root to just one real root as the temperature increases. The van der Waals equation is a cubic equation in V_m.

P1.23 The two masses represent the same volume of gas under identical conditions, and therefore the same number of molecules (Avogadro's principle) and moles, n. Thus, the masses can be expressed as

$$nM_N = 2.2990\ \mathrm{g}$$

for 'chemical nitrogen' and

$$n_{Ar}M_{Ar} + n_N M_N = n\{x_{Ar}M_{Ar} + (1 - x_{Ar})M_N\} = 2.3102\ \mathrm{g}$$

for 'atmospheric nitrogen'. Dividing the latter expression by the former yields

$$\frac{x_{Ar}M_{Ar}}{M_N}+(1-x_{Ar})=\frac{2.3102}{2.2990} \quad \text{so} \quad x_{Ar}\left(\frac{M_{Ar}}{M_N}-1\right)=\frac{2.3102}{2.2990}-1$$

and $$x_{Ar}=\frac{\frac{2.3102}{2.2990}-1}{\frac{M_{Ar}}{M_N}-1}=\frac{\frac{2.3102}{2.2990}-1}{\frac{39.95\text{ g mol}^{-1}}{28.013\text{ g mol}^{-1}}-1}=\boxed{0.011}$$

COMMENT. This value for the mole fraction of argon in air is close to the modern value.

Solutions to applications

P1.25 1 t = 10^3 kg Assume 250 t per day.

$$n(SO_2)=\frac{250\times10^3\text{ kg}}{64\times10^{-3}\text{ kg mol}^{-1}}=3.9\times10^6\text{ mol}$$

$$V=\frac{nRT}{p}=\frac{(3.9\times10^6\text{ mol})\times(0.08206\text{ dm}^3\text{ atm mol}^{-1}\text{ K}^{-1})\times1073\text{ K}}{1.0\text{ atm}}=\boxed{3.4\times10^8\text{ dm}^3}$$

P1.27 The pressure at the base of a column of height H is $p=\rho gH$ [1.3], but the pressure at any altitude h within the atmospheric column of height H depends only on the air above it; therefore

$$p=\rho g(H-h) \quad \text{and} \quad dp=-\rho g dh$$

Since $\rho=\frac{pM}{RT}$ [Exercise 1.8(b)], $dp=-\frac{pMgdh}{RT}$, implying that $\frac{dp}{p}=-\frac{Mgdh}{RT}$

This relationship integrates to $p=p_0e^{-Mgh/RT}$

For air, $M=29\text{ g mol}^{-1}$, and at 298 K

$$\frac{Mg}{RT}\approx\frac{(29\times10^{-3}\text{ kg mol}^{-1})\times(9.81\text{ m s}^{-2})}{(8.3145\text{ J K}^{-1}\text{mol}^{-1})\times(298\text{ K})}=1.1\bar{5}\times10^{-4}\text{m}^{-1}\ [1\text{ J}=1\text{ kg m}^2\text{ s}^{-2}]$$

(a) $h=15$ cm

$$p=p_0\times e^{(-0.15\text{ m})\times(1.1\bar{5}\times10^{-4}\text{m}^{-1})}=0.99\overline{998}\,p_0$$

That is, the pressure at the top is indistinguishable from the pressure at the bottom. To find the small pressure difference, expand the exponential function:

$$e^{-x}=1-x+\cdots \quad \text{so} \quad \frac{p}{p_0}=e^{-Mgh/RT}\approx1-\frac{Mgh}{RT}$$

Thus, the relative pressure drop is

$$\frac{p_0-p}{p_0}\approx\frac{Mgh}{RT}=1.1\bar{5}\times10^{-4}\text{m}^{-1}\times0.15\text{ m}=\boxed{1.7\times10^{-5}}$$

(b) $h=11\text{ km}=1.1\times10^4$ m

$$p=p_0\times e^{(-1.1\times10^4)\times(1.1\bar{5}\times10^{-4}\text{m}^{-1})}=0.28p_0 \quad \text{so} \quad \frac{p_0-p}{p_0}=\boxed{0.72}$$

P1.29 Refer to Figure 1.4.

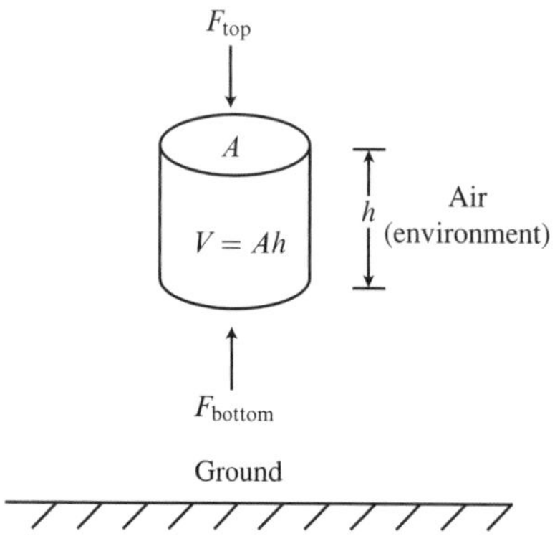

Figure 1.4

The buoyant force on the cylinder is

$$F_{buoy} = F_{bottom} - F_{top} = A(p_{bottom} - p_{top})$$

According to the barometric formula,

$$p_{top} = p_{bottom}\,e^{-Mgh/RT}$$

where M is the molar mass of the environment (air). Since h is small, the exponential can be expanded in a Taylor series around $h = 0$:

$$e^{-x} = 1 - x + \cdots$$

Keeping the first-order term only yields

$$p_{top} = p_{bottom}\left(1 - \frac{Mgh}{RT}\right)$$

The buoyant force becomes

$$F_{buoy} = Ap_{bottom}\left(1 - 1 + \frac{Mgh}{RT}\right) = Ah\left(\frac{p_{bottom}M}{RT}\right)g$$

$$= \left(\frac{p_{bottom}VM}{RT}\right)g = nMg \quad \left[n = \frac{p_{bottom}V}{RT}\right]$$

where n is the number of moles of the environment (air) displaced by the balloon and $nM = m$, the mass of the displaced environment. Thus, $F_{buoy} = mg$. The net force is the difference between the buoyant force and the weight of the balloon. Thus,

$$F_{net} = mg - m_{balloon}g = (m - m_{balloon})g$$

This is Archimedes' principle.

P1.31 We use the barometric formula [1.16] to relate the partial pressures of nitrogen and oxygen:

$$p = p_0 e^{-hMg/RT}.$$

The pressures of N_2 and O_2 are, respectively,

$$p_N = p_{0N}e^{-hgM_N/RT} \quad \text{and} \quad p_O = p_{0O}e^{-hgM_O/RT}.$$

Divide the two expressions and solve for h:

$$\frac{p_N}{p_O} = \frac{p_{0N}e^{-hgM_N/RT}}{p_{0O}e^{-hgM_O/RT}} = \frac{p_{0N}}{p_{0O}}e^{hg(M_O-M_N)/RT}$$

$$h = \frac{RT}{g(M_O - M_N)}\ln\left(\frac{p_N}{p_O}\times\frac{p_{0O}}{p_{0N}}\right)$$

$$= \frac{(8.3145\text{ J K}^{-1}\text{ mol}^{-1})\times(298\text{ K})}{(9.807\text{ m s}^{-2})\times(32.00-28.02)\times10^{-3}\text{ kg mol}^{-1}}\ln\left(\frac{90}{10}\times\frac{20}{80}\right)$$

$$= 5.1\times10^4\text{ m} = \boxed{51\text{ km}}$$

At this altitude, the partial pressure of N_2 is

$$p_N = (0.80\text{ atm})\times\exp\left(\frac{-(5.1\times10^4\text{ m})\times(28.02\times10^{-3}\text{ kg mol}^{-1})\times(9.807\text{ m s}^{-2})}{(8.3145\text{ J K}^{-1}\text{ mol}^{-1})\times(298\text{ K})}\right)$$

$$= 0.0027\text{ atm}$$

By hypothesis, this accounts for 90% of the atmosphere, so the total pressure would be

$$p_{\text{total}} = \frac{p_N}{0.90} = \boxed{0.0029\text{ atm}}$$

COMMENT. This treatment takes no account of the effect of entropy *against* the tendency of gases to fractionate under the influence of gravity. See Section 5.2(a) for a treatment of the thermodynamics of mixing. The effect of gravity alone, as suggested in this problem, is to encourage heavier gases to be more concentrated at lower altitudes and lighter ones higher; however, the effect of entropy is to encourage the mixing of gases.

2 The first law

Answers to discussion questions

D2.1 Work is a precisely defined mechanical concept. It is produced from the application of a force through a distance. The technical definition is based on the realization that both force and displacement are vector quantities and it is the component of the force acting in the direction of the displacement that is used in the calculation of the amount of work, that is, work is the scalar product of the two vectors. In vector notation $w = -\boldsymbol{F} \cdot \boldsymbol{d} = -fd\cos\theta$, where θ is the angle between the force and the displacement. The negative sign is inserted to conform to the standard thermodynamic convention.·

Heat is associated with a non-adiabatic process and is defined as the difference between the adiabatic work and the non-adiabatic work associated with the same change in state of the system. This is the formal (and best) definition of heat and is based on the definition of work. A less precise definition of heat is the statement that heat is the form of energy that is transferred between bodies in thermal contact with each other by virtue of a difference in temperature.

At the molecular level, work is a transfer of energy that results in orderly motion of the atoms and molecules in a system; heat is a transfer of energy that results in disorderly motion. See Section 2.1(b) for a more detailed discussion of the molecular interpretation of work and heat.

D2.3 The difference results from the definition $H = U + pV$; hence $\Delta H = \Delta U + \Delta(pV)$. As $\Delta(PV)$ is not usually zero, except for isothermal processes in a perfect gas, the difference between ΔH and ΔU is a non-zero quantity. As shown in Sections 2.4 and 2.5 of the text, ΔH can be interpreted as the heat associated with a process at constant pressure, and ΔU as the heat at constant volume.

D2.5 In the Joule experiment, the change in internal energy of a gas at low pressures (a perfect gas) is zero. Hence, in the calculation of energy changes for processes in a perfect gas one can ignore any effect due to a change in volume. This greatly simplifies the calculations involved because one can drop the first term of eqn 2.39 and work only with $\mathrm{d}U = C_V\,\mathrm{d}T$. In a more sensitive apparatus, Joule would have observed a small temperature change on expansion of the 'real' gas. Joule's result holds exactly only in the limit of zero pressure where all gases can be considered perfect.

The solution to Problem 2.37 shows that the Joule–Thomson coefficient can be expressed in terms of the parameters representing the attractive and repulsive interactions in a real gas. If the attractive forces predominate, then expanding the gas will reduce its energy and hence its temperature. This reduction in temperature could continue until the temperature of the gas falls below its condensation point. This is the principle underlying the liquefaction of gases with the Linde Refrigerator, which utilizes the Joule–Thomson effect. See Section 2.12 for a more complete discussion.

Solutions to exercises

E2.1(a) The physical definition of work is $\mathrm{d}w = -F\,\mathrm{d}z$ [2.5].

In a gravitational field the force is the weight of the object, which is $F = mg$.

If g is constant over the distance the mass moves, $\mathrm{d}w$ may be integrated to give the total work

$$w = -\int_{z_i}^{z_f} F\,\mathrm{d}z = -\int_{z_i}^{z_f} mg\,\mathrm{d}z = -mg(z_f - z_i) = -mgh, \quad \text{where} \quad h = (z_f - z_i)$$

On earth: $w = -(65\text{ kg}) \times (9.81\text{ m s}^{-2}) \times (4.0\text{ m}) = -2.6 \times 10^3\text{ J} = \boxed{2.6 \times 10^3\text{ J needed}}$

On the moon: $w = -(65\text{ kg}) \times (1.60\text{ m s}^{-2}) \times (4.0\text{ m}) = -4.2 \times 10^2\text{ J} = \boxed{4.2 \times 10^2\text{ J needed}}$

E2.2(a) This is an expansion against a constant external pressure; hence $w = -p_{ex}\Delta V$ [2.8]

$$p_{ex} = (1.0\text{ atm}) \times (1.013 \times 10^5\text{ Pa atm}^{-1}) = 1.0\bar{1} \times 10^5\text{ Pa}$$

The change in volume is the cross-sectional area times the linear displacement:

$$\Delta V = (100\text{ cm}^2) \times (10\text{ cm}) \times \left(\frac{1\text{ m}}{100\text{ cm}}\right)^3 = 1.0 \times 10^{-3}\text{ m}^3,$$

so $w = -(1.0\bar{1} \times 10^5\text{ Pa}) \times (1.0 \times 10^{-3}\text{ m}^3) = \boxed{-1.0 \times 10^2\text{ J}}$ as $1\text{ Pa m}^3 = 1\text{ J}$.

E2.3(a) For all cases $\Delta U = 0$, since the internal energy of a perfect gas depends only on temperature. (See Section 2.2(a) and Section 2.11(b) for a more complete discussion.) From the definition of enthalpy, $H = U + pV$, so $\Delta H = \Delta U + \Delta(pV) = \Delta U + \Delta(nRT)$ (perfect gas). Hence, $\Delta H = 0$ as well, at constant temperature for all processes in a perfect gas.

(a) $\boxed{\Delta U = \Delta H = 0}$

$$w = -nRT\ln\left(\frac{V_f}{V_i}\right) \text{ [2.10]}$$

$$= -(1.00\text{ mol}) \times (8.314\text{ J K}^{-1}\text{ mol}^{-1}) \times (273\text{ K}) \times \ln\left(\frac{44.8\text{ dm}^3}{22.4\text{ dm}^3}\right)$$

$$= -1.57 \times 10^3\text{ J} = \boxed{-1.57\text{ kJ}}$$

$$q = \Delta U - w\text{ [first law]} = 0 + 1.57\text{ kJ} = \boxed{+1.57\text{ kJ}}$$

(b) $\boxed{\Delta U = \Delta H = 0}$

$w = -p_{ex}\Delta V$ [2.8] $\quad \Delta V = (44.8 - 22.4)\text{ dm}^3 = 22.4\text{ dm}^3$

p_{ex} can be computed from the perfect gas law

$$pV = nRT$$

so $$p_{ex} = p_f = \frac{nRT}{V_f} = \frac{(1.00\text{ mol}) \times (0.08206\text{ dm}^3\text{ atm K}^{-1}\text{ mol}^{-1}) \times (273\text{ K})}{44.8\text{ dm}^3} = 0.500\text{ atm}$$

$$w = -(0.500\text{ atm}) \times \left(\frac{1.013 \times 10^5\text{ Pa}}{1\text{ atm}}\right) \times (22.4\text{ dm}^3) \times \left(\frac{1\text{ m}^3}{10^3\text{ dm}^3}\right)$$

$$= -1.13 \times 10^3\text{ Pa m}^3 = -1.13 \times 10^3\text{ J} = \boxed{-1.13\text{ kJ}}$$

$$q = \Delta U - w = 0 + 1.13\text{ kJ} = \boxed{+1.13\text{ kJ}}$$

(c) $\boxed{\Delta U = \Delta H = 0}$

Free expansion is expansion against no force, so $\boxed{w = 0}$ and $q = \Delta U - w = 0 - 0 = \boxed{0}$.

COMMENT. An isothermal free expansion of a perfect gas is also adiabatic.

E2.4(a) For a perfect gas at constant volume

$$\frac{p}{T} = \frac{nR}{V} = \text{constant}, \quad \text{hence,} \quad \frac{p_1}{T_1} = \frac{p_2}{T_2}$$

$$p_2 = \left(\frac{T_2}{T_1}\right) \times p_1 = \left(\frac{400\text{ K}}{300\text{ K}}\right) \times (1.00\text{ atm}) = \boxed{1.33\text{ atm}}$$

$$\Delta U = nC_{V,\text{m}}\Delta T\ [2.16\text{b}] = (n) \times (\tfrac{3}{2}R) \times (400\text{ K} - 300\text{ K})$$
$$= (1.00\text{ mol}) \times (\tfrac{3}{2}) \times (8.314\text{ J K}^{-1}\text{mol}^{-1}) \times (100\text{ K})$$
$$= 1.25 \times 10^3\text{ J} = \boxed{+1.25\text{ kJ}}$$

$\boxed{w = 0}$ [constant volume] $\quad q = \Delta U - \text{w[first law]} = 1.25\text{kJ} - 0 = \boxed{+1.25\text{ kJ}}$

E2.5(a) (a) $w = -p_{ex}\Delta V\,[2.8]$

$$p_{ex} = (200\text{ Torr}) \times (133.3\text{ Pa Torr}^{-1}) = 2.66\overline{6} \times 10^4\text{ Pa}$$

$$\Delta V = 3.3\text{ dm}^3 = 3.3 \times 10^{-3}\text{ m}^3$$

Therefore, $w = (-2.66\overline{6} \times 10^4\text{ Pa}) \times (3.3 \times 10^{-3}\text{ m}^3) = \boxed{-88\text{ J}}$

(b) $w = -nRT\ln\left(\frac{V_f}{V_i}\right)$ [2.10]

$$n = \frac{4.50\text{ g}}{16.04\text{ g mol}^{-1}} = 0.280\overline{5}\text{ mol},\ RT = 2.577\text{ kJ mol}^{-1},\ V_i = 12.7\text{ dm}^3,\ V_f = 16.0\text{ dm}^3$$

$$w = -(0.280\overline{5}\text{ mol}) \times (2.577\text{ kJ mol}^{-1}) \times \ln\left(\frac{16.0\text{ dm}^3}{12.7\text{ dm}^3}\right) = \boxed{-167\text{ J}}$$

E2.6(a) $$\Delta H = \Delta_{\text{cond}}H = -\Delta_{\text{vap}}H = -(1\text{ mol}) \times (40.656\text{ kJ mol}^{-1}) = \boxed{-40.656\text{ kJ}}$$

Since the condensation is done isothermally and reversibly, the external pressure is constant at 1.00 atm. Hence,

$$q = q_p = \Delta H = \boxed{-40.656\text{ kJ}}$$

$w = -p_{ex}\Delta V$ [2.8], where $\Delta V = V_{liq} - V_{vap} \approx -V_{vap}$ because $V_{liq} \ll V_{vap}$.

On the assumption that $H_2O(g)$ is a perfect gas, $V_{vap} = \dfrac{nRT}{p}$ and $p = p_{ex}$, since the condensation is done reversibly. Hence,

$$w \approx nRT = (1.00\text{ mol}) \times (8.314\text{ J K}^{-1}\text{ mol}^{-1}) \times (373\text{ K}) = +3.10 \times 10^3\text{ J} = \boxed{+3.10\text{ kJ}}$$

From eqn 2.21 $\Delta U = \Delta H - \Delta n_g RT$ $\Delta n_g = -1.00$ mol

$$\Delta U = (-40.656\text{ kJ}) + (1.00\text{ mol}) \times (8.314\text{ J K}^{-1}\text{ mol}^{-1}) \times (373.15\text{ K}) = \boxed{-37.55\text{ kJ}}$$

E2.7(a) The chemical reaction that occurs is

$$Mg(s) + 2\,HCl(aq) \rightarrow H_2(g) + MgCl_2(aq), \quad M(Mg) = 24.31\text{ g mol}^{-1}$$

Work is done against the atmosphere by the expansion of the hydrogen gas product in the reaction.

$w = -p_{ex}\Delta V$ [2.8], where $V_i \approx 0$, $V_f = \dfrac{nRT}{p_f}$, $p_f = p_{ex}$

so $w = -p_{ex}(V_f - V_i) \approx -p_{ex}V_f = -p_{ex} \times \dfrac{nRT}{p_{ex}} = -nRT.$

$$n = \frac{15\text{ g}}{24.31\text{ g mol}^{-1}} = 0.61\overline{7}\text{ mol}, \quad RT = 2.479\text{ kJ mol}^{-1}$$

Hence, $w = (-0.61\overline{7}\text{ mol}) \times (2.479\text{ kJ mol}^{-1}) = \boxed{-1.5\text{ kJ}}$.

E2.8(a) $q = \Delta H$, since pressure is constant

$$\Delta H = \int_{T_i}^{T_f} dH, \quad dH = nC_{p,m}\,dT$$

$$d(H/J) = \{20.17 + 0.3665(T/K)\}d(T/K)$$

$$\Delta(H/J) = \int_{T_i}^{T_f} (H/J) = \int_{298}^{473} \{20.17 + 0.3665(T/K)\}d(T/K)$$

$$= (20.17) \times (473 - 298) + \left(\frac{0.3665}{2}\right) \times \left(\frac{T}{K}\right)^2 \Bigg|_{298}^{473}$$

$$= (3.53\overline{0} \times 10^3) + (2.47\overline{25} \times 10^4) = 2.83 \times 10^4$$

$$q = \Delta H = \boxed{2.83 \times 10^4\text{ J}} = \boxed{+28.3\text{ kJ}}$$

$w = -p_{ex}\Delta V$ [2.8], where $p_{ex} = p$

$$w = -p\Delta V = -\Delta(pV)\,[\text{constant pressure}] = -\Delta(nRT)\,[\text{perfect gas}] = -nR\Delta T$$
$$= (-1.00\text{ mol}) \times (8.314\text{ J K}^{-1}\text{ mol}^{-1}) \times (473\text{ K} - 298\text{ K}) = \boxed{-1.45 \times 10^3\text{ J}} = \boxed{-1.45\text{ kJ}}$$

$$\Delta U = q + w = (28.3\text{ kJ}) - (1.45\text{ kJ}) = \boxed{+26.8\text{ kJ}}$$

(b) The energy and enthalpy of a perfect gas depend on temperature alone (Section 2.2(a) and Exercises 2.3), hence it does not matter whether the temperature change is brought about at constant volume or constant pressure; ΔH and ΔU are the same.

$$\Delta H = \boxed{+28.3\ \text{kJ}},\quad \Delta U = \boxed{+26.8\ \text{kJ}}$$

Under constant volume, $w = \boxed{0}$.

$$q = \Delta U - w = \boxed{+26.8\ \text{kJ}}$$

E2.9(a) For reversible adiabatic expansion

$$T_{\text{f}} = T_{\text{i}}\left(\frac{V_{\text{i}}}{V_{\text{f}}}\right)^{1/c}\ [2.28\text{a}]$$

$$\text{where } c = \frac{C_{V,\text{m}}}{R} = \frac{C_{p,\text{m}} - R}{R} = \frac{(20.786 - 8.3145)\ \text{J K}^{-1}\,\text{mol}^{-1}}{8.3145\ \text{J K}^{-1}\,\text{mol}^{-1}} = 1.500,$$

so the final temperature is

$$T_{\text{f}} = (273.15\ \text{K}) \times \left(\frac{1.0\ \text{dm}^3}{3.0\ \text{dm}^3}\right)^{1/1.500} = \boxed{13\bar{1}\ \text{K}}$$

E2.10(a) Reversible adiabatic work is

$$w = C_V \Delta T\ [2.27] = n(C_{p,\text{m}} - R) \times (T_{\text{f}} - T_{\text{i}})$$

where the temperatures are related by

$$T_{\text{f}} = T_{\text{i}}\left(\frac{V_{\text{i}}}{V_{\text{f}}}\right)^{1/c}\ [2.28\text{a}], \quad \text{where} \quad c = \frac{C_{V,\text{m}}}{R} = \frac{C_{p,\text{m}} - R}{R} = 3.463$$

$$\text{So}\quad T_{\text{f}} = [(27.0 + 273.15)\ \text{K}] \times \left(\frac{500 \times 10^{-3}\ \text{dm}^3}{3.00\ \text{dm}^3}\right)^{1/3.463} = 179\ \text{K}$$

$$\text{and}\quad w = \left(\frac{2.45\ \text{g}}{44.0\ \text{g mol}^{-1}}\right) \times [(37.11 - 8.3145)\ \text{J K}^{-1}\,\text{mol}^{-1}] \times (179 - 300)\ \text{K} = \boxed{-194\ \text{J}}$$

E2.11(a) For reversible adiabatic expansion

$$p_{\text{f}}V_{\text{f}}^{\gamma} = p_{\text{i}}V_{\text{i}}^{\gamma}\ [2.29] \quad \text{so} \quad p_{\text{f}} = p_{\text{i}}\left(\frac{V_{\text{i}}}{V_{\text{f}}}\right)^{\gamma} = (57.4\ \text{kPa}) \times \left(\frac{1.0\ \text{dm}^3}{2.0\ \text{dm}^3}\right)^{1.4} = \boxed{22\ \text{kPa}}$$

E2.12(a) $$C_p = \frac{q_p}{\Delta T}\ [2.24] = \frac{229\ \text{J}}{2.55\ \text{K}} = 89.8\ \text{J K}^{-1}$$

so $C_{p,\text{m}} = (89.8\ \text{J K}^{-1})/(3.0\ \text{mol}) = \boxed{30\ \text{J K}^{-1}\,\text{mol}^{-1}}$.

For a perfect gas $C_{p,\text{m}} - C_{V,\text{m}} = R\ [2.26]$

$$C_{V,\text{m}} = C_{p,\text{m}} - R = (30 - 8.3)\ \text{J K}^{-1}\,\text{mol}^{-1} = \boxed{22\ \text{J K}^{-1}\,\text{mol}^{-1}}$$

E2.13(a) $q_p = C_p\Delta T\ [2.24] = nC_{p,\mathrm{m}}\Delta T = (3.0\ \mathrm{mol}) \times (29.4\ \mathrm{J\ K^{-1}\ mol^{-1}}) \times (25\ \mathrm{K}) = \boxed{+2.2\ \mathrm{kJ}}$

$\Delta H = q_p\ [2.23\mathrm{b}] = \boxed{+2.2\ \mathrm{kJ}}$

$\Delta U = \Delta H - \Delta(pV)\ [\text{from } H \equiv U + pV] = \Delta H - \Delta(nRT)\ [\text{perfect gas}] = \Delta H - nR\Delta T$

$= (2.2\ \mathrm{kJ}) - (3.0\ \mathrm{mol}) \times (8.314\ \mathrm{J\ K^{-1}\ mol^{-1}}) \times (25\ \mathrm{K}) = (2.2\ \mathrm{kJ}) - (0.62\ \mathrm{kJ}) = \boxed{+1.6\ \mathrm{kJ}}$

E2.14(a) In an adiabatic process, $q = \boxed{0}$. Work against a constant external pressure is

$$w = -p_{\mathrm{ex}}\Delta V = (-600\ \mathrm{Torr}) \times \left(\frac{1.013 \times 10^5\ \mathrm{Pa}}{760\ \mathrm{Torr}}\right) \times (40 \times 10^{-3}\ \mathrm{m^3}) = \boxed{-3.2\ \mathrm{kJ}}$$

$\Delta U = q + w = \boxed{-3.2\ \mathrm{kJ}}$

One can also relate adiabatic work to ΔT (eqn 2.27):

$$w = C_V\Delta T = n(C_{p,\mathrm{m}} - R)\Delta T \quad \text{so} \quad \Delta T = \frac{w}{n(C_{p,\mathrm{m}} - R)},$$

$$\Delta T = \frac{-3.2 \times 10^3\ \mathrm{J}}{(4.0\ \mathrm{mol}) \times (29.355 - 8.3145)\ \mathrm{J\ K^{-1}\ mol^{-1}}} = \boxed{-38\ \mathrm{K}}$$

$\Delta H = \Delta U + \Delta(pV) = \Delta U + nR\Delta T$

$= (-3.2\ \mathrm{kJ}) + (4.0\ \mathrm{mol}) \times (8.3145\ \mathrm{J\ K^{-1}\ mol^{-1}}) \times (-38\ \mathrm{K}) = \boxed{-4.5\ \mathrm{kJ}}$

Question. Calculate the final pressure of the gas.

E2.15(a) In an adiabatic process, the initial and final pressures are related by (eqn 2.29)

$$p_{\mathrm{f}}V_{\mathrm{f}}^{\gamma} = p_{\mathrm{i}}V_{\mathrm{i}}^{\gamma}$$

where $\gamma = \dfrac{C_p}{C_V} = \dfrac{C_V + nR}{C_V} = \dfrac{20.8\ \mathrm{J\ K^{-1}} + (1.0\ \mathrm{mol})(8.31\ \mathrm{J\ K^{-1}\ mol^{-1}})}{20.8\ \mathrm{J\ K^{-1}}} = 1.40$

Find V_{i} from the perfect gas law:

$$V_{\mathrm{i}} = \frac{nRT_{\mathrm{i}}}{p_{\mathrm{i}}} = \frac{(1.0\ \mathrm{mol})(8.31\ \mathrm{J\ K^{-1}\ mol^{-1}})(310\ \mathrm{K})}{3.25\ \mathrm{atm}} \times \frac{1\ \mathrm{atm}}{1.013 \times 10^5\ \mathrm{Pa}}$$

$V_{\mathrm{i}} = 7.8\bar{3} \times 10^{-3}\ \mathrm{m^3}$

so $V_{\mathrm{f}} = V_{\mathrm{i}}\left(\dfrac{p_{\mathrm{i}}}{p_{\mathrm{f}}}\right)^{1/\gamma} = (7.8\bar{3} \times 10^{-3}\ \mathrm{m^3})\left(\dfrac{3.25\ \mathrm{atm}}{2.50\ \mathrm{atm}}\right)^{1/1.40} = \boxed{0.0094\bar{4}\ \mathrm{m^3}}$.

Find the final temperature from the perfect gas law:

$$T_{\mathrm{f}} = \frac{p_{\mathrm{f}}V_{\mathrm{f}}}{nR} = \frac{(2.50\ \mathrm{atm}) \times (0.0094\bar{4}\ \mathrm{m^3})}{(1.0\ \mathrm{mol})(8.31\ \mathrm{J\ K^{-1}\ mol^{-1}})} \times \frac{1.013 \times 10^5\ \mathrm{Pa}}{1\ \mathrm{atm}}$$

$T_{\mathrm{f}} = \boxed{28\bar{8}\ \mathrm{K}}$

Adiabatic work is (eqn 2.27)

$w = C_V\Delta T = (20.8\ \mathrm{J\ K^{-1}})(28\bar{8} - 310)\ \mathrm{K} = \boxed{-4.\bar{6} \times 10^2\ \mathrm{J}}$

E2.16(a) At constant pressure

$$q = \Delta H = n\Delta_{vap}H^{\ominus} = (0.50\ \text{mol}) \times (26.0\ \text{kJ mol}^{-1}) = \boxed{13.0\ \text{kJ}}$$

and $w = -p\Delta V \approx -pV_{vapor} = -nRT = -(0.50\ \text{mol}) \times (8.3145\ \text{J K}^{-1}\text{mol}^{-1}) \times (250\ \text{K})$

$$w = -1.0 \times 10^3\ \text{J} = \boxed{-1.0\ \text{kJ}}$$

$$\Delta U = w + q = 13.0 - 1.0\ \text{kJ} = \boxed{12.0\ \text{kJ}}$$

COMMENT. Because the vapour is here treated as a perfect gas, the specific value of the external pressure provided in the statement of the exercise does not affect the numerical value of the answer.

E2.17(a) The lattice enthalpy is the difference in enthalpy between an ionic solid and the corresponding isolated ions. In this exercise, it is the enthalpy corresponding to the process

$$SrI_2(s) \rightarrow Sr^{2+}(g) + 2I^-(g)$$

The standard lattice enthalpy can be computed from the standard enthalpies given in the exercise by considering the formation of $SrI_2(s)$ from its elements as occurring through the following steps: sublimation of Sr(s), removing two electrons from Sr(g), sublimation of $I_2(s)$, atomization of $I_2(g)$, electron attachment to I(g), and formation of the solid SrI_2 lattice from gaseous ions

$$\Delta_f H^{\ominus}(SrI_2,s) = \Delta_{sub}H^{\ominus}(Sr,s) + \Delta_{ion}H^{\ominus}(Sr,g) + \Delta_{sub}H^{\ominus}(I_2,s) + \Delta_{at}H^{\ominus}(I_2,g) + 2\Delta_{eg}H^{\ominus}(I,g) - \Delta_L H^{\ominus}(SrI_2,s)$$

So the lattice enthalpy is

$$\Delta_L H^{\ominus}(SrI_2,s) = \Delta_{sub}H^{\ominus}(Sr,s) + \Delta_{ion}H^{\ominus}(Sr,g) + \Delta_{sub}H^{\ominus}(I_2,s) + \Delta_{at}H^{\ominus}(I_2,g) + 2\Delta_{eg}H^{\ominus}(I,g) - \Delta_f H^{\ominus}(SrI_2,s)$$

$$\Delta_L H^{\ominus}(SrI_2,s) = [164 + 1626 + 62 + 151 - 2(304) + 558]\ \text{kJ mol}^{-1} = \boxed{1953\ \text{kJ mol}^{-1}}$$

E2.18(a) The reaction is

$$C_6H_5C_2H_5(l) + \tfrac{21}{2}O_2(g) \rightarrow 8\,CO_2(g) + 5\,H_2O(l)$$

$$\begin{aligned}\Delta_c H^{\ominus} &= 8\Delta_f H^{\ominus}(CO_2,g) + 5\Delta_f H^{\ominus}(H_2O,l) - \Delta_f H^{\ominus}(C_6H_5C_2H_5,l)\\ &= [(8)\times(-393.51) + (5)\times(-285.83) - (-12.5)]\ \text{kJ mol}^{-1}\\ &= \boxed{-4564.7\ \text{kJ mol}^{-1}}\end{aligned}$$

E2.19(a) First $\Delta_f H[(CH_2)_3,g]$ is calculated, and then that result is used to calculate $\Delta_r H$ for the isomerization

$$(CH_2)_3(g) + \tfrac{9}{2}O_2(g) \rightarrow 3\,CO_2(g) + 3\,H_2O(l) \quad \Delta_c H = -2091\ \text{kJ mol}^{-1}$$

$$\begin{aligned}\Delta_f H[(CH_2)_3,g] &= -\Delta_c H + 3\Delta_f H(CO_2,g) + 3\Delta_f H(H_2O,g)\\ &= [+2091 + (3)\times(-393.51) + (3)\times(-285.83)]\ \text{kJ mol}^{-1}\\ &= \boxed{+53\ \text{kJ mol}^{-1}}\end{aligned}$$

$$(CH_2)_3(g) \rightarrow C_3H_6(g) \quad \Delta_r H = ?$$

$$\begin{aligned}\Delta_r H &= \Delta_f H(C_3H_6,g) - \Delta_f H[(CH_2)_3,g]\\ &= (20.42 - 53)\ \text{kJ mol}^{-1} = \boxed{-33\ \text{kJ mol}^{-1}}\end{aligned}$$

E2.20(a) For naphthalene the reaction is $C_{10}H_8(s) + 12\,O_2(g) \rightarrow 10\,CO_2(g) + 4\,H_2O(l)$

A bomb calorimeter gives $q_V = n\Delta_c U^\ominus$ rather than $q_p = n\Delta_c H^\ominus$, thus we need

$$\Delta_c U^\ominus = \Delta_c H^\ominus - \Delta n_g RT\ [2.21], \quad \Delta n_g = -2\ \text{mol}$$

$$\Delta_c H^\ominus = -5157\ \text{kJ mol}^{-1}\ [\text{Table 2.6}]$$

$$\Delta_c U^\ominus = (-5157\ \text{kJ mol}^{-1}) - (-2) \times (8.3 \times 10^{-3}\ \text{kJ K}^{-1}\,\text{mol}^{-1}) \times (298\ \text{K})$$

$$= \boxed{-5152\ \text{kJ mol}^{-1}}$$

$$|q| = |q_V| = |n\Delta_c U^\ominus| = \left(\frac{120 \times 10^{-3}\ \text{g}}{128.18\ \text{g mol}^{-1}}\right) \times (5152\ \text{kJ mol}^{-1}) = 4.82\overline{3}\ \text{kJ}$$

$$C = \frac{|q|}{\Delta T} = \frac{4.82\overline{3}\ \text{kJ}}{3.05\ \text{K}} = \boxed{1.58\ \text{kJ K}^{-1}}$$

When phenol is used the reaction is

$$C_6H_5OH(s) + \tfrac{15}{2}\,O_2(g) \rightarrow 6\,CO_2(g) + 3\,H_2O(l)$$

$$\Delta_c H^\ominus = -3054\ \text{kJ mol}^{-1}\ [\text{Table 2.6}]$$

$$\Delta_c U = \Delta_c H - \Delta n_g RT,\ \Delta n_g = -\tfrac{3}{2}$$
$$= (-3054\ \text{kJ mol}^{-1}) + (\tfrac{3}{2}) \times (8.314 \times 10^{-3}\ \text{kJ K}^{-1}\,\text{mol}^{-1}) \times (298\ \text{K})$$
$$= -3050\ \text{kJ mol}^{-1}$$

$$|q| = \frac{10 \times 10^{-3}\ \text{g}}{94.12\ \text{g mol}^{-1}} \times (3050\ \text{kJ mol}^{-1}) = 0.324\overline{1}\ \text{kJ}$$

$$\Delta T = \frac{|q|}{C} = \frac{0.324\overline{1}\ \text{kJ}}{1.58\ \text{kJ K}^{-1}} = \boxed{+0.205\ \text{K}}$$

COMMENT. In this case $\Delta_c U^\ominus$ and $\Delta_c H^\ominus$ differed by about 0.1%. Thus, to within three significant figures, it would not have mattered if we had used $\Delta_c H^\ominus$ instead of $\Delta_c U^\ominus$, but for very precise work it would.

E2.21(a) The reaction is $AgCl(s) \rightarrow Ag^+(aq) + Cl^-(aq)$

$$\Delta_{sol} H^\ominus = \Delta_f H^\ominus(Ag^+,aq) + \Delta_f H^\ominus(Cl^-,aq) - \Delta_f H^\ominus(AgCl,s)$$
$$= [(105.58) - (167.16) - (-127.07)]\ \text{kJ mol}^{-1}$$
$$= \boxed{+65.49\ \text{kJ mol}^{-1}}$$

E2.22(a) $NH_3SO_2(s) \rightarrow NH_3(g) + SO_2(g) \quad \Delta_r H^\ominus = +40\ \text{kJ mol}^{-1}$

$$\Delta_r H^\ominus = \Delta_f H^\ominus(NH_3,g) + \Delta_f H^\ominus(SO_2,g) - \Delta_f H^\ominus(NH_3SO_2,s)$$

Solving for $\Delta_f H^\ominus(NH_3SO_2,s)$ yields

$$\Delta_f H^\ominus(NH_3SO_2,s) = \Delta_f H^\ominus(NH_3,g) + \Delta_f H^\ominus(SO_2,g) - \Delta_r H^\ominus$$
$$= (-46.11 - 296.83 - 40)\ \text{kJ mol}^{-1}$$
$$= \boxed{-383\ \text{kJ mol}^{-1}}$$

E2.23(a) (a) reaction(3) = (−2) × reaction(1) + reaction(2) and $\Delta n_g = -1$

The enthalpies of reactions are combined in the same manner as the equations (Hess's law).

$$\Delta_r H^\ominus(3) = (-2)\times\Delta_r H^\ominus(1) + \Delta_r H^\ominus(2)$$
$$= [(-2)\times(-184.62) + (-483.64)]\ \text{kJ mol}^{-1}$$
$$= \boxed{-114.40\ \text{kJ mol}^{-1}}$$

$$\Delta_r U^\ominus = \Delta_r H^\ominus - \Delta n_g RT\ [2.21] = (-114.40\ \text{kJ mol}^{-1}) - (-1)\times(2.48\ \text{kJ mol}^{-1})$$
$$= \boxed{-111.92\ \text{kJ mol}^{-1}}$$

(b) $\Delta_f H^\ominus$ refers to the formation of one mole of the compound, hence

$$\Delta_f H(\text{J}) = \frac{\Delta_r H(\text{J})}{\nu_\text{J}}$$

$$\Delta_f H^\ominus(\text{HCl,g}) = \frac{-184.62}{2}\ \text{kJ mol}^{-1} = \boxed{-92.31\ \text{kJ mol}^{-1}}$$

$$\Delta_f H^\ominus(\text{H}_2\text{O,g}) = \frac{-483.64}{2}\ \text{kJ mol}^{-1} = \boxed{-241.82\ \text{kJ mol}^{-1}}$$

E2.24(a)

$$\Delta_r H^\ominus = \Delta_r U^\ominus + \Delta n_g RT\ [2.21];\quad \Delta n_g = +2$$
$$= (-1373\ \text{kJ mol}^{-1}) + 2\times(2.48\ \text{kJ mol}^{-1}) = \boxed{-1368\ \text{kJ mol}^{-1}}$$

COMMENT. As a number of these exercises have shown, the use of $\Delta_r H^\ominus$ as an approximation for $\Delta_r U^\ominus$ is often valid.

E2.25(a) In each case, the strategy is to combine reactions in such a way that the combination corresponds to the formation reaction desired. The enthalpies of the reactions are then combined in the same manner as the equations to yield the enthalpies of formation.

(a)

	$\Delta_r H^\ominus/(\text{kJ mol}^{-1})$
$\text{K(s)} + \frac{1}{2}\text{Cl}_2\text{(g)} \rightarrow \text{KCl(s)}$	−436.75
$\text{KCl(s)} + \frac{3}{2}\text{O}_2\text{(g)} \rightarrow \text{KClO}_3\text{(s)}$	$\frac{1}{2}\times(89.4)$
$\text{K(s)} + \frac{1}{2}\text{Cl}_2\text{(g)} + \frac{3}{2}\text{O}_2\text{(g)} \rightarrow \text{KClO}_3\text{(s)}$	−392.1

Hence, $\Delta_f H^\ominus(\text{KClO}_3\text{,s}) = \boxed{-392.1\ \text{kJ mol}^{-1}}$

(b)

	$\Delta_r H^\ominus/(\text{kJ mol}^{-1})$
$\text{Na(s)} + \frac{1}{2}\text{O}_2\text{(g)} + \frac{1}{2}\text{H}_2\text{(g)} \rightarrow \text{NaOH(s)}$	−425.61
$\text{NaOH(s)} + \text{CO}_2\text{(g)} \rightarrow \text{NaHCO}_3\text{(s)}$	−127.5
$\text{C(s)} + \text{O}_2\text{(g)} \rightarrow \text{CO}_2\text{(g)}$	−393.51
$\text{Na(s)} + \text{C(s)} + \frac{1}{2}\text{H}_2\text{(g)} + \frac{3}{2}\text{O}_2\text{(g)} \rightarrow \text{NaHCO}_3\text{(s)}$	−946.6

Hence, $\Delta_f H^\ominus(\text{NaHCO}_3\text{,s}) = \boxed{-946.6\ \text{kJ mol}^{-1}}$

E2.26(a) When the heat capacities of all substances participating in a chemical reaction are assumed to be constant over the range of temperatures involved, Kirchoff's law [2.36a] integrates to

$$\Delta_r H^\ominus(T_2) = \Delta_r H^\ominus(T_1) + \Delta_r C_p^\ominus(T_2 - T_1) \text{ [Example 2.6]}$$

$$\Delta_r C_p^\ominus = \sum_{\text{Products}} \nu C_{p,m}^\ominus - \sum_{\text{Reactants}} \nu C_{p,m}^\ominus \quad [2.36b]$$

$$\Delta_r C_p^\ominus = C_{p,m}^\ominus(N_2O_4,g) - 2C_{p,m}^\ominus(NO_2,g) = (77.28) - (2) \times (37.20\text{ J K}^{-1}\text{ mol}^{-1})$$
$$= +2.88\text{ J K}^{-1}\text{ mol}^{-1}$$

$$\Delta_r H^\ominus(373\text{ K}) = \Delta_r H^\ominus(298\text{ K}) + \Delta_r C_p^\ominus \Delta T$$
$$= (-57.20\text{ kJ mol}^{-1}) + (2.88\text{ J K}^{-1}) \times (75\text{ K})$$
$$= [(-57.20) + (0.22)]\text{ kJ mol}^{-1}$$
$$= \boxed{-56.98\text{ kJ mol}^{-1}}$$

E2.27(a) (a) $\Delta_r H^\ominus = \sum_{\text{Products}} \nu\Delta_f H^\ominus - \sum_{\text{Reactants}} \nu\Delta_f H^\ominus$ [2.34]

$$\Delta_r H^\ominus(298\text{ K}) = [(-110.53) - (-241.82)]\text{ kJ mol}^{-1} = \boxed{+131.29\text{ kJ mol}^{-1}}$$
$$\Delta_r U^\ominus(298\text{ K}) = \Delta_r H^\ominus(298\text{ K}) - \Delta n_g RT \text{ [2.21]}$$
$$= (131.29\text{ kJ mol}^{-1}) - (1) \times (2.48\text{ kJ mol}^{-1}) = \boxed{+128.81\text{ kJ mol}^{-1}}$$

(b) $\Delta_r H^\ominus(378\text{ K}) = \Delta_r H^\ominus(298\text{ K}) + (T_2 - T_1)\Delta_r C_p^\ominus$ [Example 2.6]

$$\Delta_r C_p^\ominus = C_{p,m}^\ominus(CO,g) + C_{p,m}^\ominus(H_2,g) - C_{p,m}^\ominus(C,gr) - C_{p,m}^\ominus(H_2O,g)$$
$$= (29.14 + 28.82 - 8.53 - 33.58) \times 10^{-3}\text{ kJ K}^{-1}\text{ mol}^{-1}$$
$$= 15.85 \times 10^{-3}\text{ kJ K}^{-1}\text{ mol}^{-1}$$

$$\Delta_r H^\ominus(378\text{ K}) = (131.29\text{ kJ mol}^{-1}) + (15.85 \times 10^{-3}\text{ kJ K}^{-1}\text{ mol}^{-1}) \times (80\text{ K})$$
$$= (131.29 + 1.27)\text{ kJ mol}^{-1} = \boxed{+132.56\text{ kJ mol}^{-1}}$$

$$\Delta_r U^\ominus(378\text{ K}) = \Delta_r H^\ominus(378\text{ K}) - (1) \times (8.31 \times 10^{-3}\text{ kJ K}^{-1}\text{ mol}^{-1}) \times (378\text{ K})$$
$$= (132.56 - 3.14)\text{ kJ mol}^{-1} = \boxed{+129.42\text{ kJ mol}^{-1}}$$

COMMENT. The differences in both $\Delta_r H^\ominus$ and $\Delta_r U^\ominus$ between the two temperatures are small and justify the use of the approximation that $\Delta_r C_p^\ominus$ is a constant.

E2.28(a) $CuSO_4$ and $ZnSO_4$ are strong electrolytes; therefore the net ionic equation is

$$Zn(s) + Cu^{2+}(aq) \rightarrow Zn^{2+}(aq) + Cu(s)$$

$$\Delta_r H^\ominus = \Delta_f H^\ominus(Zn^{2+}) + \Delta_f H^\ominus(Cu) - \Delta_f H^\ominus(Zn) - \Delta_f H^\ominus(Cu^{2+})$$
$$= [(-153.89) + (0) - (0) - (64.77)]\text{ kJ mol}^{-1}$$
$$= \boxed{-218.66\text{ kJ mol}^{-1}}$$

COMMENT. SO_4^{2-} is a spectator ion and was ignored in the determination above.

E2.29(a) Since enthalpy is a state function, $\Delta_r H$ for the process (see Figure 2.1)

$$Mg^{2+}(g) + 2\,Cl(g) + 2\,e^- \rightarrow MgCl_2(aq)$$

is independent of path, therefore the change in enthalpy for the path on the left is equal to the change in enthalpy for the path on the right. All numerical values are in kJ mol^{-1}.

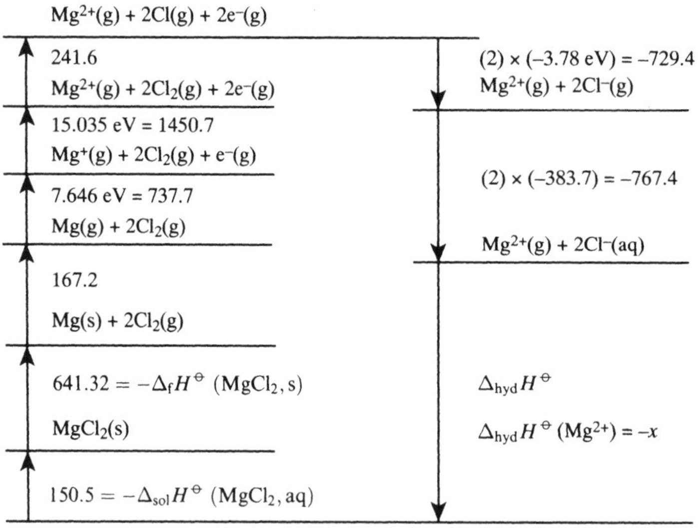

Figure 2.1

The cycle is the distance traversed upward along the left plus the distance traversed downward on the right. The sum of these distances is zero. Note that $E_{ea} = -\Delta_{eg}H$. Therefore, following the cycle up the left and down the right and using kJ units,

$$-(-150.5) - (-641.32) + (167.2) + (241.6) + (737.7 + 1450.7) + 2 \times (-364.\overline{7}) + 2 \times (-383.7) + \Delta_{hyd}H(Mg^{2+}) = 0$$

which yields $\Delta_{hyd}H(Mg^{2+}) = \boxed{-1892 \text{ kJ mol}^{-1}}$

E2.30(a) The Joule–Thomson coefficient μ is the ratio of temperature change to pressure change under conditions of isenthalpic expansion. So

$$\mu = \left(\frac{\partial T}{\partial p}\right)_H [2.50] = \lim_{\Delta p \to 0}\left(\frac{\Delta T}{\Delta p}\right)_H \approx \frac{\Delta T}{\Delta p} \; [\mu \text{ constant over temperature range}]$$

$$\mu = \frac{-22 \text{ K}}{-31 \text{ atm}} = \boxed{0.71 \text{ K atm}^{-1}}$$

E2.31(a) The internal energy is a function of temperature and volume, $U_m = U_m(T, V_m)$, so

$$dU_m = \left(\frac{\partial U_m}{\partial T}\right)_{V_m} dT + \left(\frac{\partial U_m}{\partial V_m}\right)_T dV_m \quad [\pi_T = (\partial U_m / \partial V)_T]$$

For an isothermal expansion $dT = 0$; hence

$$dU_m = \left(\frac{\partial U_m}{\partial V_m}\right)_T dV_m = \pi_T \, dV_m = \frac{a}{V_m^2} dV_m$$

$$\Delta U_m = \int_{V_{m,1}}^{V_{m,2}} dU_m = \int_{V_{m,1}}^{V_{m,2}} \frac{a}{V_m^2} dV_m = a \int_{1.00 \text{ dm}^3 \text{ mol}^{-1}}^{24.8 \text{ dm}^3 \text{ mol}^{-1}} \frac{dV_m}{V_m^2} = -\frac{a}{V_m}\Bigg|_{1.00 \text{ dm}^3 \text{ mol}^{-1}}^{24.8 \text{ dm}^3 \text{ mol}^{-1}}$$

$$= -\frac{a}{24.8 \text{ dm}^3 \text{ mol}^{-1}} + \frac{a}{1.00 \text{ dm}^3 \text{ mol}^{-1}} = \frac{23.8a}{24.8 \text{ dm}^3 \text{ mol}^{-1}} = 0.959\overline{7}a \text{ mol dm}^{-3};$$

From Table 1.6, $a = 1.352\ \text{dm}^6\ \text{atm mol}^{-1}$

$$\Delta U_{\text{m}} = (0.959\overline{7}\ \text{mol L}^{-1}) \times (1.352\ \text{dm}^6\,\text{atm mol}^{-2})$$

$$= (1.30\ \text{dm}^3\,\text{atm mol}^{-1}) \times \left(\frac{1\ \text{m}}{10\ \text{dm}}\right)^3 \times \left(\frac{1.013\times 10^5\ \text{Pa}}{\text{atm}}\right) = \boxed{+131\ \text{J mol}^{-1}}$$

$$w = -\int p\,\text{d}V_{\text{m}},\quad \text{where}\quad p = \frac{RT}{V_{\text{m}}-b} - \frac{a}{V_{\text{m}}^2}\quad \text{for a van der Waals gas. Hence,}$$

$$w = -\int\left(\frac{RT}{V_{\text{m}}-b}\right)\text{d}V_{\text{m}} + \int\frac{a}{V_{\text{m}}^2}\text{d}V_{\text{m}} = -q + \Delta U_{\text{m}}$$

Therefore,

$$q = \int_{1.00\ \text{dm}^3\,\text{mol}^{-1}}^{24.8\ \text{dm}^3\,\text{mol}^{-1}}\left(\frac{RT}{V_{\text{m}}-b}\right)\text{d}V_{\text{m}} = RT\ln(V_{\text{m}}-b)\Big|_{1.00\ \text{dm}^3\,\text{mol}^{-1}}^{24.8\ \text{dm}^3\,\text{mol}^{-1}}$$

$$= (8.314\ \text{J K}^{-1}\,\text{mol}^{-1}) \times (298\ \text{K}) \times \ln\left(\frac{24.8 - 3.9\times 10^{-2}}{1.00 - 3.9\times 10^{-2}}\right)$$

$$= \boxed{+8.05\times 10^3\ \text{J mol}^{-1}}$$

and $w = -q + \Delta U_{\text{m}} = -(8.05\times 10^3\ \text{J mol}^{-1}) + (131\ \text{J mol}^{-1}) = \boxed{-7.92\times 10^3\,\text{J mol}^{-1}}$

E2.32(a) $\alpha = \left(\frac{1}{V}\right)\left(\frac{\partial V}{\partial T}\right)_p$ [2.42]; $\alpha_{320} = \left(\frac{1}{V_{320}}\right)\left(\frac{\partial V}{\partial T}\right)_{p,320}$

$$\left(\frac{\partial V}{\partial T}\right)_p = V_{300}(3.9\times 10^{-4}/\text{K} + 2.96\times 10^{-6}T/\text{K}^2)$$

$$\left(\frac{\partial V}{\partial T}\right)_{p,320} = V_{300}(3.9\times 10^{-4}/\text{K} + 2.96\times 10^{-6}\times 320/\text{K}) = 1.34\times 10^{-3}\ \text{K}^{-1}V_{300}$$

$$V_{320} = V_{300}\{(0.75) + (3.9\times 10^{-4})\times(320) + (1.48\times 10^{-6})\times(320)^2\} = (V_{300})\times(1.02\overline{6})$$

so $\alpha_{320} = \left(\frac{1}{V_{320}}\right)\left(\frac{\partial V}{\partial T}\right)_{p,320} = \left(\frac{1}{1.02\overline{6}V_{300}}\right)\times(1.34\times 10^{-3}\ \text{K}^{-1}V_{300})$

$$\alpha_{320} = \frac{1.34\times 10^{-3}\ \text{K}^{-1}}{1.02\overline{6}} = \boxed{1.31\times 10^{-3}\ \text{K}^{-1}}$$

COMMENT. Knowledge of the density at 300 K is not required to solve this exercise, but it would be required to obtain numerical values of the volumes at the two temperatures.

E2.33(a) The isothermal compressibility is

$$\kappa_T = -\left(\frac{1}{V}\right)\left(\frac{\partial V}{\partial p}\right)_T \ [2.43],\quad \text{so}\quad \left(\frac{\partial V}{\partial p}\right)_T = -\kappa_T V$$

At constant temperature

$$\mathrm{d}V = \left(\frac{\partial V}{\partial p}\right)_T \mathrm{d}p, \quad \text{so} \quad \mathrm{d}V = -\kappa_T V\,\mathrm{d}p \quad \text{or} \quad \frac{\mathrm{d}V}{V} = -\kappa_T\,\mathrm{d}p$$

Substituting $V = \dfrac{m}{\rho}$ yields $\mathrm{d}V = -\dfrac{m}{\rho^2}\mathrm{d}\rho;\ \dfrac{\mathrm{d}V}{V} = -\dfrac{\mathrm{d}\rho}{\rho} = -\kappa_T \mathrm{d}p$

Therefore, $\dfrac{\delta\rho}{\rho} \approx \kappa_T \delta p$

For $\dfrac{\delta\rho}{\rho} = 0.08 \times 10^{-2} = 8 \times 10^{-4}$, $\quad \delta p \approx \dfrac{8 \times 10^{-4}}{\kappa_T} = \dfrac{8 \times 10^{-4}}{7.35 \times 10^{-7}\,\text{atm}^{-1}} = \boxed{1.\bar{1} \times 10^3\,\text{atm}}$

E2.34(a) The isothermal Joule–Thomson coefficient is

$$\left(\frac{\partial H_\mathrm{m}}{\partial p}\right)_T = -\mu C_{p,\mathrm{m}} = (-0.25\ \text{K atm}^{-1}) \times (29\ \text{J K}^{-1}\,\text{mol}^{-1}) = \boxed{-7.2\ \text{J atm}^{-1}\,\text{mol}^{-1}}$$

$$\mathrm{d}H = n\left(\frac{\partial H_\mathrm{m}}{\partial p}\right)_T \mathrm{d}p = -n\mu C_{p,\mathrm{m}}\,\mathrm{d}p$$

$$\Delta H = \int_{p_1}^{p_2} (-n\mu C_{p,\mathrm{m}})\mathrm{d}p = -n\mu C_{p,\mathrm{m}}(p_2 - p_1) \quad [\mu \text{ and } C_p \text{ are constant}]$$

$$\Delta H = -(15\ \text{mol}) \times (+7.2\ \text{J atm}^{-1}\,\text{mol}^{-1}) \times (-75\ \text{atm}) = +8.1\ \text{kJ}$$

so $q(\text{supplied}) = +\Delta H = \boxed{+8.1\ \text{kJ}}$

Solutions to problems

Assume all gases are perfect unless stated otherwise. Unless otherwise stated, thermochemical data are for 298 K.

Solutions to numerical problems

P2.1 The temperatures are readily obtained from the perfect gas equation, $T = \dfrac{pV}{nR}$

$$T_1 = \frac{(1.00\ \text{atm}) \times (22.4\ \text{dm}^3)}{(1.00\ \text{mol}) \times (0.0821\ \text{dm}^3\,\text{atm mol}^{-1}\,\text{K}^{-1})} = \boxed{273\ \text{K}} = T_3\ [\text{isotherm}]$$

Similarly, $T_2 = \boxed{546\ \text{K}}$

In the solutions that follow all steps in the cycle are considered to be reversible.

Step $1 \rightarrow 2$

$$w = -p_{\text{ex}}\Delta V = -p\Delta V = -nR\Delta T \quad [\Delta(pV) = \Delta(nRT)]$$

$$w = -(1.00\text{ mol}) \times (8.314\text{ J K}^{-1}\text{mol}^{-1}) \times (546 - 273)\text{ K} = \boxed{-2.27 \times 10^3\text{ J.}}$$

$$\Delta U = nC_{V,\text{m}}\Delta T = (1.00\text{ mol}) \times \tfrac{3}{2} \times (8.314\text{ J K}^{-1}\text{mol}^{-1}) \times (273\text{ K}) = \boxed{+3.40 \times 10^3\text{ J.}}$$

$$q = \Delta U - w = +3.40 \times 10^3\text{J} - (-2.27 \times 10^3\text{ J}) = \boxed{+5.67 \times 10^3\text{ J.}}$$

$$\Delta H = q_p = \boxed{+5.67 \times 10^3\text{ J}}$$

If this step were not reversible, then w, q, and ΔH would be indeterminate.

Step $2 \rightarrow 3$

$\boxed{w = 0}$ [constant volume]

$$q_V = \Delta U = nC_{V,\text{m}}\Delta T = (1.00\text{ mol}) \times (\tfrac{3}{2}) \times (8.314\text{ J K}^{-1}\text{mol}^{-1}) \times (-273\text{ K})$$

$$= \boxed{-3.40 \times 10^3\text{ J}}$$

From $H \equiv U + pV$

$$\Delta H = \Delta U + \Delta(pV) = \Delta U + \Delta(nRT) = \Delta U + nR\Delta T$$

$$= (-3.40 \times 10^3\text{ J}) + (1.00\text{ mol}) \times (8.314\text{ J K}^{-1}\text{mol}^{-1}) \times (-273\text{ K}) = \boxed{-5.67 \times 10^3\text{ J}}$$

Step $3 \rightarrow 1$

ΔU and ΔH are $\boxed{\text{zero}}$ for an isothermal process in a perfect gas; hence for the reversible compression

$$-q = w = -nRT\ln\frac{V_1}{V_3} = (-1.00\text{ mol}) \times (8.314\text{ J K}^{-1}\text{mol}^{-1}) \times (273\text{ K}) \times \ln\left(\frac{22.4\text{ dm}^3}{44.8\text{ dm}^3}\right)$$

$$= \boxed{+1.57 \times 10^3\text{ J}} \quad q = -1.57 \times 10^3\text{ J}$$

If this step were not reversible, then q and w would have different values, which would be determined by the details of the process.

Total cycle

State	p/atm	V/dm^3	T/K
1	1.00	22.44	273
2	1.00	44.8	546
3	0.50	44.8	273

Thermodynamic quantities calculated for reversible steps.

Step	Process	q/kJ	w/kJ	ΔU/kJ	ΔH/kJ
$1 \rightarrow 2$	p constant $= p_{\text{ex}}$	+5.67	−2.27	+3.40	+5.67
$2 \rightarrow 3$	V constant V constant	−3.40	0	−3.40	−5.67
$3 \rightarrow 1$	Isothermal, reversible	−1.57	+1.57	0	0
Cycle		+0.70	−0.70	0	0

COMMENT. All values can be determined unambiguously for the reversible cycle. The net result of the overall process is that 700 J of heat has been converted to work.

P2.3 Since the volume is fixed, $\boxed{w = 0.}$

Since $\Delta U = q$ at constant volume, $\boxed{\Delta U = +2.35\ \text{kJ}}$

$$\Delta H = \Delta U + \Delta(pV) = \Delta U + V\Delta p\ [\Delta V = 0]$$

From the van der Waals equation [1.21b]

$$p = \frac{RT}{V_m - b} - \frac{a}{V_m^2} \quad \text{so} \quad \Delta p = \frac{R\Delta T}{V_m - b} \quad [\Delta V_m = 0 \text{ at constant volume}]$$

Therefore, $\Delta H = \Delta U + \dfrac{RV\Delta T}{V_m - b}$

From the data,

$$V_m = \frac{15.0\ \text{dm}^3}{2.0\ \text{mol}} = 7.5\ \text{dm}^3\ \text{mol}^{-1},\ \Delta T = (341 - 300)\ \text{K} = 41\ \text{K}$$

$$V_m - b = (7.5 - 4.3 \times 10^{-2})\ \text{dm}^3\ \text{mol}^{-1} = 7.4\overline{6}\ \text{dm}^3\ \text{mol}^{-1}$$

$$\frac{RV\Delta T}{V_m - b} = \frac{(8.314\ \text{J K}^{-1}\ \text{mol}^{-1}) \times (15.0\ \text{dm}^3) \times (41\ \text{K})}{7.4\overline{6}\ \text{dm}^3\ \text{mol}^{-1}} = 0.68\ \text{kJ}$$

Therefore, $\Delta H = (2.35\ \text{kJ}) + (0.68\ \text{kJ}) = \boxed{+3.03\ \text{kJ}}$

P2.5 This cycle is represented in Figure 2.2. Assume that the initial temperature is 298 K.

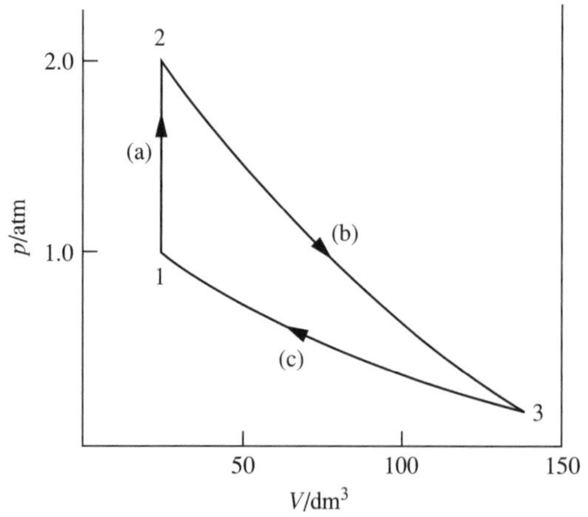

Figure 2.2

(a) First, note that $\boxed{w = 0}$ (constant volume). Then calculate ΔU since ΔT is known ($\Delta T = 298$ K) and then calculate q from the first law.

$$\Delta U = nC_{V,m}\Delta T\ [2.16b];\ C_{V,m} = C_{p,m} - R = \tfrac{7}{2}R - R = \tfrac{5}{2}R$$

$$\Delta U = (1.00\ \text{mol}) \times (\tfrac{5}{2}) \times (8.314\ \text{J K}^{-1}\ \text{mol}^{-1}) \times (298\ \text{K}) = 6.19 \times 10^3\ \text{J} = \boxed{+6.19\ \text{kJ}}$$

$$q = q_V = \Delta U - w = 6.19\ \text{kJ} - 0 = \boxed{+6.19\ \text{kJ}}$$

$$\Delta H = \Delta U + \Delta(pV) = \Delta U + \Delta(nRT) = \Delta U + nR\Delta T$$
$$= (6.19\ \text{kJ}) + (1.00\ \text{mol}) \times (8.31 \times 10^{-3}\ \text{kJ mol}^{-1}) \times (298\ \text{K}) = \boxed{+8.67\ \text{kJ}}$$

(b) $\boxed{q = 0}$ (adiabatic)

Because the energy and enthalpy of a perfect gas depend on temperature alone,

$$\Delta U(b) = -\Delta U(a) = \boxed{-6.19\text{ kJ}}, \text{ since } \Delta T(b) = -\Delta T(a).$$

Likewise $\Delta H(b) = -\Delta H\,(a) = \boxed{-8.67\text{ kJ}}$

$$w = \Delta U = \boxed{-6.19\text{ kJ}}\ [\text{first law with } q = 0]$$

(c) $\Delta U = \Delta H = 0$ [isothermal process in perfect gas]

$$q = -w\ [\text{first law with } \Delta U = 0]\text{: } w = -nRT_1 \ln \frac{V_1}{V_3}\ [2.10]$$

$$V_2 = V_1 = \frac{nRT_1}{p_1} = \frac{(1.00\text{ mol}) \times (0.08206\text{ dm}^3\text{ atm K}^{-1}\text{ mol}^{-1}) \times (298\text{ K})}{1.00\text{ atm}} = 24.4\bar{5}\text{ dm}^3$$

$$V_2 T_2^c = V_3 T_3^c\ [2.28\text{b}] \text{ hence } \quad V_3 = V_2\left(\frac{T_2}{T_3}\right)^c \quad \text{where} \quad c = \frac{C_{V,\text{m}}}{R} = \frac{5}{2}$$

so $$V_3 = (24.4\bar{5}\text{ dm}^3) \times \left(\frac{(2) \times (298\text{ K})}{298\text{ K}}\right)^{5/2} = 138.\bar{3}\text{ dm}^3$$

$$w = (-1.00\text{ mol}) \times (8.314\text{ J K}^{-1}\text{ mol}^{-1}) \times (298\text{ K}) \times \ln\left(\frac{24.4\bar{5}\text{ dm}^3}{138.3\text{ dm}^3}\right) = 4.29 \times 10^3\text{ J} = \boxed{+4.29\text{ kJ}}$$

$$q = \boxed{-4.29\text{ kJ}}$$

P2.7 The formation reaction is

$$2\text{ C(s)} + 3\text{ H}_2\text{(g)} \rightarrow 2\text{ C}_2\text{H}_6\text{(g)} \qquad \Delta_f H^{\ominus}(298\text{ K}) = -84.68\text{ kJ mol}^{-1}$$

In order to determine $\Delta_f H^{\ominus}(350\text{ K})$ we employ Kirchhoff's law [2.36a] with $T_2 = 350$ K, $T_1 = 298$ K

$$\Delta_f H^{\ominus}(T_2) = \Delta_f H^{\ominus}(T_1) + \int_{T_1}^{T_2} \Delta_r C_p^{\ominus} \text{d}T$$

where $$\Delta_r C_p^{\ominus} = \sum_J \nu_J C_{p,\text{m}}^{\ominus}(\text{J}) = C_{p,\text{m}}^{\ominus}(\text{C}_2\text{H}_6) - 2C_{p,\text{m}}^{\ominus}(\text{C}) - 3C_{p,\text{m}}^{\ominus}(\text{H}_2)$$

From Table 2.2

$$C_{p,\text{m}}^{\ominus}(\text{C}_2\text{H}_6)/(\text{J K}^{-1}\text{ mol}^{-1}) = 14.73 + \left(\frac{0.1272}{\text{K}}\right)T$$

$$C_{p,\text{m}}^{\ominus}(\text{C,s})/(\text{J K}^{-1}\text{ mol}^{-1}) = 16.86 + \left(\frac{4.77 \times 10^{-3}}{\text{K}}\right)T - \left(\frac{8.54 \times 10^5\text{ K}^2}{T^2}\right)$$

$$C_{p,\text{m}}^{\ominus}(\text{H}_2\text{,g})/(\text{J K}^{-1}\text{ mol}^{-1}) = 27.28 + \left(\frac{3.26 \times 10^{-3}}{\text{K}}\right)T - \left(\frac{0.50 \times 10^5\text{ K}^2}{T^2}\right)$$

$$\Delta_r C_p^{\ominus}/(\text{J K}^{-1}\text{ mol}^{-1}) = -100.83 + \left(\frac{0.1079T}{\text{K}}\right) + \left(\frac{1.86 \times 10^6\text{ K}^2}{T^2}\right)$$

$$\int_{T_1}^{T_2} \frac{\Delta_r C_p^{\ominus}\,dT}{\text{J K}^{-1}\,\text{mol}^{-1}} = -100.83 \times (T_2 - T_1) + (\tfrac{1}{2})(0.1079\ \text{K}^{-1})(T_2^2 - T_1^2) + (1.86 \times 10^6\ \text{K}^2)\left(\frac{1}{T_2} - \frac{1}{T_1}\right)$$

$$= -100.83 \times (52\ \text{K}) + (\tfrac{1}{2})(0.1079)(350^2 - 298^2)\ \text{K} + (1.86 \times 10^6)\left(\frac{1}{350} - \frac{1}{298}\right)\text{K}$$

$$= -4.35 \times 10^3\ \text{K}$$

Multiplying by the units J K^{-1} mol^{-1}, we obtain

$$\int_{T_1}^{T_2} \Delta_r C_p^{\ominus}\,dT = -(4.35 \times 10^3\ \text{K}) \times (\text{J K}^{-1}\,\text{mol}^{-1}) = -4.35 \times 10^3\ \text{J mol}^{-1}$$

$$= -4.35\ \text{kJ mol}^{-1}$$

Hence $\Delta_f H^{\ominus}(350\ \text{K}) = \Delta_f H^{\ominus}(298\ \text{K}) - 4.35\ \text{kJ mol}^{-1}$

$$= -84.68\ \text{kJ mol}^{-1} - 4.35\ \text{kJ mol}^{-1} = \boxed{-89.03\ \text{kJ mol}^{-1}}$$

P2.9 $Cr(C_6H_6)_2(s) \rightarrow Cr(s) + 2C_6H_6(g) \quad \Delta n_g = +2\ \text{mol}$

$\Delta_r H^{\ominus} = \Delta_r U^{\ominus} + 2RT$, from [2.21]

$$= (8.0\ \text{kJ mol}^{-1}) + (2) \times (8.314\ \text{J K}^{-1}\,\text{mol}^{-1}) \times (583\,\text{K}) = \boxed{+17.7\ \text{kJ mol}^{-1}}$$

In terms of enthalpies of formation

$$\Delta_r H^{\ominus} = (2) \times \Delta_f H^{\ominus}(\text{benzene, 583 K}) - \Delta_f H^{\ominus}(\text{metallocene, 583 K})$$

or $\Delta_r H^{\ominus}(\text{metallocene, 583 K}) = 2\Delta_f H^{\ominus}(\text{benzene, 583 K}) - 17.7\ \text{kJ mol}^{-1}$

The enthalpy of formation of benzene gas at 583 K is related to its value at 298 K by

$$\begin{aligned}\Delta_f H^{\ominus}(\text{benzene, 583 K}) = {} & \Delta_f H^{\ominus}(\text{benzene, 298 K}) \\ & + (T_b - 298\ \text{K})C_{p,m}^{\ominus}(C_6H_6, l) + \Delta_{vap}H^{\ominus} + (583\ \text{K} - T_b)C_{p,m}^{\ominus}(C_6H_6, g) \\ & - 6 \times (583\ \text{K} - 298\ \text{K})C_{p,m}^{\ominus}(\text{C, gr}) - 3 \times (583\ \text{K} - 298\ \text{K})C_{p,m}^{\ominus}(H_2, g)\end{aligned}$$

where T_b is the boiling temperature of benzene (353 K). We shall assume that the heat capacities of graphite and hydrogen are approximately constant in the range of interest and use their values from Table 2.8.

$$\begin{aligned}\Delta_f H^{\ominus}(\text{benzene, 583 K}) &= (49.0\ \text{kJ mol}^{-1}) + (353 - 298)\ \text{K} \times (136.1\ \text{J K}^{-1}\,\text{mol}^{-1}) \\ &\quad + (30.8\ \text{kJ mol}^{-1}) + (583 - 353)\ \text{K} \times (81.67\ \text{J K}^{-1}\,\text{mol}^{-1}) \\ &\quad - (6) \times (583 - 298)\ \text{K} \times (8.53\ \text{J K}^{-1}\,\text{mol}^{-1}) \\ &\quad - (3) \times (583 - 298)\ \text{K} \times (28.82\ \text{J K}^{-1}\,\text{mol}^{-1}) \\ &= \{(49.0) + (7.49) + (18.78) + (30.8) - (14.59) - (24.64)\}\ \text{kJ mol}^{-1} \\ &= +66.8\ \text{kJ mol}^{-1}\end{aligned}$$

Therefore $\Delta_f H^{\ominus}(\text{metallocene, 583 K}) = (2 \times 66.8 - 17.7)\ \text{kJ mol}^{-1} = \boxed{+116.0\ \text{kJ mol}^{-1}}$

P2.11 The needed data are the enthalpy of vapourization and heat capacity of water, available in the *Data section*.

$$C_{p,m}^{\ominus}(H_2O, l) = 75.3\ \text{J K}^{-1}\,\text{mol}^{-1} \qquad \Delta_{vap}H^{\ominus}(H_2O) = 44.0\ \text{kJ mol}^{-1}$$

$$n(H_2O) = \frac{65\ \text{kg}}{0.018\ \text{kg mol}^{-1}} = 3.6 \times 10^3\ \text{mol}$$

From $\Delta H = nC_{p,\mathrm{m}}\Delta T$ we obtain

$$\Delta T = \frac{\Delta H}{nC_{p,\mathrm{m}}} = \frac{1.0\times10^4\,\mathrm{kJ}}{(3.6\times10^3\,\mathrm{mol})\times(0.0753\,\mathrm{kJ\,K^{-1}\,mol^{-1}})} = \boxed{+37\,\mathrm{K}}$$

From $\Delta H = n\Delta_{\mathrm{vap}}H^{\ominus} = \dfrac{m}{M}\Delta_{\mathrm{vap}}H^{\ominus}$

$$m = \frac{M\times\Delta H}{\Delta_{\mathrm{vap}}H^{\ominus}} = \frac{(0.018\,\mathrm{kg\,mol^{-1}})\times(1.0\times10^4\,\mathrm{kJ})}{44.0\,\mathrm{kJ\,mol^{-1}}} = \boxed{4.09\,\mathrm{kg}}$$

COMMENT. This estimate would correspond to about 30 glasses of water per day, which is much higher than the average consumption. The discrepancy may be a result of our assumption that evaporation of water is the main mechanism of heat loss.

P2.13 (a) and (b) The table displays computed enthalpies of formation (semi-empirical, PM3 level, PC Spartan Pro™), enthalpies of combustion based on them (and on experimental enthalpies of formation of $H_2O(l)$ and $CO_2(g)$, −285.83 and −393.51 kJ mol^{-1}, respectively), experimental enthalpies of combustion (Table 2.6), and the relative error in enthalpy of combustion.

Compound	$\Delta_f H^{\ominus}$/kJ mol^{-1}	$\Delta_c H^{\ominus}$/kJ mol^{-1}(calc.)	$\Delta_c H^{\ominus}$/kJ mol^{-1} (expt.)	% error
$CH_4(g)$	−54.45	−910.72	−890	2.33
$C_2H_6(g)$	−75.88	−1568.63	−1560	0.55
$C_3H_8(g)$	−98.84	−2225.01	−2220	0.23
$C_4H_{10}(g)$	−121.60	−2881.59	−2878	0.12
$C_5H_{12}(g)$	−142.11	−3540.42	−3537	0.10

The combustion reactions can be expressed as:

$$C_nH_{2n+2}(g) + \left(\frac{3n+1}{2}\right)O_2(g) \rightarrow nCO_2(g) + (n+1)H_2O(l)$$

The enthalpy of combustion, in terms of enthalpies of reaction, is

$$\Delta_c H^{\ominus} = n\Delta_f H^{\ominus}(CO_2) + (n+1)\Delta_f H^{\ominus}(H_2O) - \Delta_f H^{\ominus}(C_nH_{2n+2}),$$

Where we have left out $\Delta_f H^{\ominus}(O_2) = 0$. The % error is defined as:

$$\%\ \text{error} = \frac{\Delta_c H^{\ominus}(\text{calc}) - \Delta_c H^{\ominus}(\text{expt.})}{\Delta_c H^{\ominus}(\text{expt.})}\times 100\%$$

The agreement is quite good.

(c) If the enthalpy of combustion is related to the molar mass by

$$\Delta_c H^{\ominus} = k[M/(\mathrm{g\,mol^{-1}})]^n$$

then one can take the natural log of both sides to obtain:

$$\ln|\Delta_c H^{\ominus}| = \ln|k| + n\ln M/(\mathrm{g\,mol^{-1}}).$$

Thus, if one plots $\ln|\Delta_c H^{\ominus}|$ vs. $\ln[M/(\mathrm{g\,mol^{-1}})]$, one ought to obtain a straight line with slope n and y-intercept $\ln|k|$. Draw up the following table:

Compound	$M/(\text{g mol}^{-1})$	$\Delta_c H^\ominus/\text{kJ mol}^{-1}$	$\ln M/(\text{g mol}^{-1})$	$\ln \lvert\Delta_c H^\ominus/\text{kJ mol}^{-1}\rvert$
$CH_4(g)$	16.04	−910.72	2.775	6.814
$C_2H_6(g)$	30.07	−1568.63	3.404	7.358
$C_3H_8(g)$	44.10	−2225.01	3.786	7.708
$C_4H_{10}(g)$	58.12	−2881.59	4.063	7.966
$C_5H_{12}(g)$	72.15	−3540.42	4.279	8.172

The plot is shown in Figure 2.3.

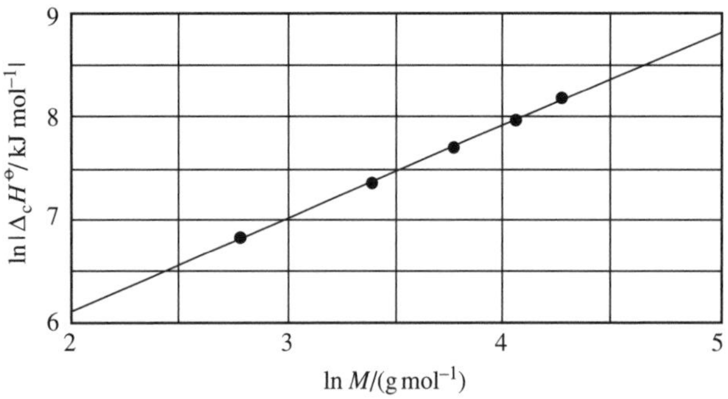

Figure 2.3

The linear least-squares fit equation is:

$$\ln \lvert\Delta_c H^\ominus/\text{kJ mol}^{-1}\rvert = 4.30 + 0.903 \ln M/(\text{g mol}^{-1}) \quad R^2 = 1.00$$

These compounds support the proposed relationships, with

$$n = \boxed{0.903} \quad \text{and} \quad k = -e^{4.30}\,\text{kJ mol}^{-1} = \boxed{-73.7\ \text{kJ mol}^{-1}}$$

The agreement of these theoretical values of k and n with the experimental values obtained in P2.10 is rather good.

P2.15 The reaction is

$$C_{60}(s) + 60O_2(g) \rightarrow 60CO_2(g)$$

Because the reaction does not change the number of moles of gas, $\Delta_r H = \Delta_r U$ [2.21]. Therefore

$$\Delta_c H^\ominus = (-36.0334\ \text{kJ g}^{-1}) \times (60 \times 12.011\ \text{g mol}^{-1}) = \boxed{-25\,968\ \text{kJ mol}^{-1}}$$

Now relate the enthalpy of combustion to enthalpies of formation and solve for that of C_{60}.

$$\Delta_c H^\ominus = 60\Delta_f H^\ominus(CO_2) - 60\Delta_f H^\ominus(O_2) - \Delta_f H^\ominus(C_{60})$$

$$\begin{aligned}\Delta_f H^\ominus(C_{60}) &= 60\Delta_f H^\ominus(CO_2) - 60\Delta_f H^\ominus(O_2) - \Delta_c H^\ominus \\ &= [60(-393.51) - 60(0) - (-25968)]\ \text{kJ mol}^{-1} \\ &= \boxed{2357\ \text{kJ mol}^{-1}}\end{aligned}$$

P2.17 (a) $$\begin{aligned}\Delta_r H^\ominus &= \Delta_f H^\ominus(SiH_2) + \Delta_f H^\ominus(H_2) - \Delta_f H^\ominus(SiH_4) \\ &= (274 + 0 - 34.3)\ \text{kJ mol}^{-1} = \boxed{240\ \text{kJ mol}^{-1}}\end{aligned}$$

(b) $\Delta_r H^\ominus = \Delta_f H^\ominus(\mathrm{SiH_2}) + \Delta_f H^\ominus(\mathrm{SiH_4}) - \Delta_f H^\ominus(\mathrm{Si_2H_6})$

$$= (274 + 34.3 - 80.3)\ \mathrm{kJ\ mol^{-1}} = \boxed{228\ \mathrm{kJ\ mol^{-1}}}$$

P2.19 The temperatures and volumes in reversible adiabatic expansion are related by eqn 2.28a:

$$T_f = T_i\left(\frac{V_i}{V_f}\right)^{1/c}, \quad \text{where} \quad c = \frac{C_{V,m}}{R}$$

From eqn 2.29, we can relate the pressures and volumes:

$$p_f = p_i\left(\frac{V_i}{V_f}\right)^{\gamma}, \quad \text{where} \quad \gamma = \frac{C_{p,m}}{C_{V,m}}$$

We are looking for $C_{p,m}$, which can be related to c and γ:

$$c\gamma = \left(\frac{C_{V,m}}{R}\right)\times\left(\frac{C_{p,m}}{C_{V,m}}\right) = \frac{C_{p,m}}{R}$$

Solving both relationships for the ratio of volumes, we have:

$$\left(\frac{p_f}{p_i}\right)^{1/\gamma} = \frac{V_i}{V_f} = \left(\frac{T_f}{T_i}\right)^{c} \quad \text{so} \quad \frac{p_f}{p_i} = \left(\frac{T_f}{T_i}\right)^{c\gamma}$$

Therefore

$$C_{p,m} = R\frac{\ln\left(\frac{p_f}{p_i}\right)}{\ln\left(\frac{T_f}{T_i}\right)} = (8.314\ \mathrm{J\ K^{-1}\,mol^{-1}})\times\left(\frac{\ln\left(\frac{81.840\ \mathrm{kPa}}{202.94\ \mathrm{kPa}}\right)}{\ln\left(\frac{248.44\ \mathrm{K}}{298.15\ \mathrm{K}}\right)}\right) = \boxed{41.40\ \mathrm{J\ K^{-1}\,mol^{-1}}}$$

P2.21 $H_m = H_m(T, p)$

$$dH_m = \left(\frac{\partial H_m}{\partial T}\right)_p dT + \left(\frac{\partial H_m}{\partial p}\right)_T dp$$

Since $dT = 0$,

$$dH_m = \left(\frac{\partial H_m}{\partial p}\right)_T dp \quad \text{where} \quad \left(\frac{\partial H_m}{\partial p}\right)_T = -\mu C_{p,m}[2.52\ \text{with}\ 2.53] = -\left(\frac{2a}{RT} - b\right)$$

$$\Delta H_m = \int_{p_i}^{p_f} dH_m = -\int_{p_i}^{p_f}\left(\frac{2a}{RT} - b\right)dp = -\left(\frac{2a}{RT} - b\right)(p_f - p_i)$$

$$= -\left(\frac{(2)\times(1.352\ \mathrm{dm^6\,atm\,mol^{-2}})}{(0.08206\ \mathrm{dm^3\,atm\,K^{-1}\,mol^{-1}})\times(300\ \mathrm{K})} - (0.0387\ \mathrm{dm^3\,mol^{-1}})\right)\times(1.00\ \mathrm{atm} - 500\ \mathrm{atm})$$

$$= (35.5\ \mathrm{atm})\times\left(\frac{1\ \mathrm{m}}{10\ \mathrm{dm}}\right)^3\times\left(\frac{1.013\times10^5\ \mathrm{Pa}}{1\ \mathrm{atm}}\right) = 3.60\times10^3\ \mathrm{J} = \boxed{+3.60\ \mathrm{kJ}}$$

COMMENT. Note that it is not necessary to know the value of $C_{p,m}$.

Solutions to theoretical problems

P2.23 (a) $\mathrm{d}z = \left(\frac{\partial z}{\partial x}\right)_y \mathrm{d}x + \left(\frac{\partial z}{\partial y}\right)_x \mathrm{d}y$ [definition of total differential]

$$\left(\frac{\partial z}{\partial x}\right)_y = (2x - 2y + 2) \quad \left(\frac{\partial z}{\partial y}\right)_x = (4y - 2x - 4)$$

$$\mathrm{d}z = \boxed{(2x - 2y + 2)\mathrm{d}x + (4y - 2x - 4)\mathrm{d}y}$$

(b) $\frac{\partial}{\partial y}\left(\frac{\partial z}{\partial x}\right) = \frac{\partial}{\partial y}(2x - 2y + 2) = -2 \quad \frac{\partial}{\partial x}\left(\frac{\partial z}{\partial y}\right) = \frac{\partial}{\partial x}(4y - 2x - 4) = -2$

(c) $\left(\frac{\partial z}{\partial x}\right)_y = \left(y - \frac{y}{x}\right) \quad \left(\frac{\partial z}{\partial y}\right)_x = (x - \ln x)$

$$\mathrm{d}z = \boxed{\left(y + \frac{1}{x}\right)\mathrm{d}x + (x - 1)\mathrm{d}y}$$

A differential is exact if it satisfies the condition

$$\frac{\partial}{\partial x}\left(\frac{\partial z}{\partial y}\right) = \frac{\partial}{\partial y}\left(\frac{\partial z}{\partial x}\right)$$

$$\frac{\partial}{\partial y}\left(\frac{\partial z}{\partial x}\right) = \frac{\partial}{\partial y}\left(y - \frac{y}{x}\right) = 1 - \frac{1}{x} \quad \frac{\partial}{\partial x}\left(\frac{\partial z}{\partial y}\right) = \frac{\partial}{\partial x}(x - \ln x) = 1 - \frac{1}{x}$$

COMMENT. The total differential of a function is necessarily exact.

P2.25 (a) $U = U(T,V)$ so $\mathrm{d}U = \left(\frac{\partial U}{\partial T}\right)_V \mathrm{d}T + \left(\frac{\partial U}{\partial V}\right)_T \mathrm{d}V = C_V\,\mathrm{d}T + \left(\frac{\partial U}{\partial V}\right)_T \mathrm{d}V$

For $U = \text{constant}$, $\mathrm{d}U = 0$, and

$$C_V\,\mathrm{d}T = -\left(\frac{\partial U}{\partial V}\right)_T \mathrm{d}V \quad \text{or} \quad C_V = -\left(\frac{\partial U}{\partial V}\right)_T \left(\frac{\mathrm{d}V}{\mathrm{d}T}\right)_U = -\left(\frac{\partial U}{\partial V}\right)_T \left(\frac{\partial V}{\partial T}\right)_U$$

This relationship is essentially Euler's chain relationship [MB2.4].

(b) $H = H(T,p)$ so $\mathrm{d}H = \left(\frac{\partial H}{\partial T}\right)_p \mathrm{d}T + \left(\frac{\partial H}{\partial p}\right)_T \mathrm{d}p = C_p\mathrm{d}T + \left(\frac{\partial H}{\partial p}\right)_T \mathrm{d}p$

According to Euler's chain relationship

$$\left(\frac{\partial H}{\partial p}\right)_T \left(\frac{\partial p}{\partial T}\right)_H \left(\frac{\mathrm{d}T}{\mathrm{d}H}\right)_p = -1$$

so, using the reciprocal identity [MB2.3c],

$$\left(\frac{\partial H}{\partial p}\right)_T = -\left(\frac{\partial T}{\partial p}\right)_H \left(\frac{\mathrm{d}H}{\mathrm{d}T}\right)_p = -\mu C_p.$$

P2.27 (a) $H = U + pV$ so $\left(\frac{\partial H}{\partial U}\right)_p = 1 + p\left(\frac{\partial V}{\partial U}\right)_p = \boxed{1 + \frac{p}{(\partial U/\partial V)_p}}$

(b) $$\left(\frac{\partial H}{\partial U}\right)_p = \frac{(\partial H/\partial V)_p}{(\partial U/\partial V)_p} = \frac{\left(\frac{\partial(U + pV)}{\partial V}\right)_p}{(\partial U/\partial V)_p} = \frac{(\partial U/\partial V)_p + p}{(\partial U/\partial V)_p}$$

so $$\left(\frac{\partial H}{\partial U}\right)_p = 1 + \frac{p}{(\partial U/\partial V)_p} = \boxed{1 + p\left(\frac{\partial V}{\partial U}\right)_p}$$

P2.29 It follows from Relationship 3 (MB2.3c) that $\left(\frac{\partial p}{\partial T}\right)_V\left(\frac{\partial T}{\partial V}\right)_p\left(\frac{\partial V}{\partial p}\right)_T = -1$ and therefore that

$$\left(\frac{\partial p}{\partial T}\right)_V = -\frac{1}{\left(\frac{\partial T}{\partial V}\right)_p\left(\frac{\partial V}{\partial p}\right)_T}.$$

Using Relationship 2 (MB2.3b) and eqns 2.42 and 2.43 for α and κ_T this can be written

$$\left(\frac{\partial p}{\partial T}\right)_V = -\frac{\left(\frac{\partial V}{\partial T}\right)_p}{\left(\frac{\partial V}{\partial p}\right)_T} = \frac{\alpha}{\kappa_T}$$

For a perfect gas, $pV = nRT$, we find $\left(\frac{\partial p}{\partial T}\right)_V = \frac{nR}{V}$, $\left(\frac{\partial V}{\partial T}\right)_p = \frac{nR}{p}$, and $\left(\frac{\partial V}{\partial p}\right)_T = -\frac{nRT}{p^2}$.

Dividing the last two partial derivatives we find

$$\left(\frac{\partial p}{\partial T}\right)_V = -\frac{\left(\frac{\partial V}{\partial T}\right)_p}{\left(\frac{\partial V}{\partial p}\right)_T} = -\frac{nR/p}{-nRT/p^2} = \frac{p}{T} = \frac{nR}{V}$$ and the relationship is confirmed.

(b) For a van der Waals gas, $p = \frac{nRT}{V - nb} - \frac{n^2a}{V^2}$, hence

$$\left(\frac{\partial p}{\partial T}\right)_V = \frac{nR}{V - nb}$$

Using Relationships 3 (MB2.3c) and 2 (MB2.3b) this derivative becomes

$$\left(\frac{\partial p}{\partial T}\right)_V = -\frac{\left(\frac{\partial p}{\partial V}\right)_T}{\left(\frac{\partial T}{\partial V}\right)_p}$$

From the van der Waals equation we obtain

$$\left(\frac{\partial p}{\partial V}\right)_T = \frac{-nRT}{(V-nb)^2} + \frac{2n^2a}{V^3}.$$

Solving for T we obtain $T = \left(\frac{p}{nR}\right) \times (V-nb) + \left(\frac{na}{RV^2}\right) \times (V-nb)$ and

$$\left(\frac{\partial T}{\partial V}\right)_p = \left(\frac{p}{nR}\right) + \left(\frac{na}{RV^2}\right) - \left(\frac{2na}{RV^3}\right) \times (V-nb) = \left(\frac{T}{V-nb}\right) - \left(\frac{2na}{RV^3}\right) \times (V-nb).$$

Dividing the last two partial derivatives and simplifying the expression we obtain

$\left(\frac{\partial p}{\partial T}\right)_V = \frac{nR}{V-nb}$, and the relationship is confirmed.

P2.31 $$w = -\int_{V_1}^{V_2} p\,\mathrm{d}V$$

Inserting $\frac{V}{n} = V_{\mathrm{m}}$ into the virial equation for p we obtain

$$p = nRT\left(\frac{1}{V} + \frac{nB}{V^2} + \frac{n^2C}{V^3} + \cdots\right)$$

Therefore, $w = -nRT\int_{V_1}^{V_2}\left(\frac{1}{V} + \frac{nB}{V^2} + \frac{n^2C}{V^3} + \cdots\right)\mathrm{d}V$

$$w = -nRT\ln\frac{V_2}{V_1} + n^2RTB\left(\frac{1}{V_2} - \frac{1}{V_1}\right) + \frac{1}{2}n^3RTC\left(\frac{1}{V_2^2} - \frac{1}{V_2^2}\right) + \cdots$$

For $n = 1$ mol: $nRT = (1.0\text{ mol}) \times (8.314\text{ J K}^{-1}\text{ mol}^{-1}) \times (273\text{ K}) = 2.2\overline{7}\text{ kJ}$

From Table 1.4, $B = -21.7\text{ cm}^3\text{ mol}^{-1}$ and $C = 1200\text{ cm}^6\text{ mol}^{-2}$, so

$$n^2BRT = (1.0\text{ mol}) \times (-21.7\text{ cm}^3\text{ mol}^{-1}) \times (2.2\overline{7}\text{ kJ}) = -49.\overline{3}\text{ kJ cm}^3$$
$$\tfrac{1}{2}n^3CRT = \tfrac{1}{2}(1.0\text{ mol})^2 \times (1200\text{ cm}^6\text{ mol}^{-2}) \times (2.2\overline{7}\text{ kJ}) = +1362\text{ kJ cm}^6$$

Therefore,

(a) $$w = -2.2\overline{7}\text{ kJ}\ln 2 - (49.\overline{3}\text{ kJ}) \times \left(\frac{1}{1000} - \frac{1}{500}\right) + (13\overline{62}\text{ kJ}) \times \left(\frac{1}{1000^2} - \frac{1}{500^2}\right)$$

$$= (-1.5\overline{7}) + (0.049) - (4.1\times10^{-3})\text{ kJ} = -1.5\overline{2}\text{ kJ} = \boxed{-1.5\text{ kJ}}$$

(b) A perfect gas corresponds to the first term of the expansion of p, so

$$w = -1.5\overline{7}\text{ kJ} = \boxed{-1.6\text{ kJ}}.$$

P2.33 $$\mu = \left(\frac{\partial T}{\partial p}\right)_H = -\frac{1}{C_p}\left(\frac{\partial H}{\partial p}\right)_T \quad [\textit{Justification } 2.2]$$

$$\mu = \frac{1}{C_p}\left\{T\left(\frac{\partial V}{\partial T}\right)_p - V\right\} \text{ [See Problem 2.38 for this result]}$$

But $V = \frac{nRT}{p} + nb$ or $\left(\frac{\partial V}{\partial T}\right)_p = \frac{nR}{p}$

Therefore,

$$\mu = \frac{1}{C_p}\left\{\frac{nRT}{p} - V\right\} = \frac{1}{C_p}\left\{\frac{nRT}{p} - \frac{nRT}{p} - nb\right\} = \frac{-nb}{C_p}$$

Since $b > 0$ and $C_p > 0$, we conclude that for this gas $\mu < 0$ or $\left(\frac{\partial T}{\partial p}\right)_H < 0$. This says that when the pressure drops during a Joule–Thomson expansion the temperature must $\boxed{\text{increase}}$.

P2.35 $$p = \frac{nRT}{V - nb} - \frac{n^2a}{V^2} \quad [1.21\text{b}]$$

Hence, $\boxed{T = \left(\frac{p}{nR}\right) \times (V - nb) + \left(\frac{na}{RV^2}\right) \times (V - nb)}$

$$\boxed{\left(\frac{\partial T}{\partial p}\right)_V = \frac{V - nb}{nR}} = \frac{V_m - b}{R} = \frac{1}{\left(\frac{\partial p}{\partial T}\right)_V}$$

For Euler's chain relationship, we need to show that $\left(\frac{\partial T}{\partial p}\right)_V \left(\frac{\partial p}{\partial V}\right)_T \left(\frac{\partial V}{\partial T}\right)_p = -1$

Hence, in addition to $\left(\frac{\partial T}{\partial p}\right)_V$ we need $\left(\frac{\partial p}{\partial V}\right)_T$ and $\left(\frac{\partial V}{\partial T}\right)_p = \frac{1}{\left(\frac{\partial T}{\partial V}\right)_p}$

$$\left(\frac{\partial p}{\partial V}\right)_T = \frac{-nRT}{(V - nb)^2} + \frac{2n^2a}{V^3}$$

which can be found from

$$\left(\frac{\partial T}{\partial V}\right)_p = \left(\frac{p}{nR}\right) + \left(\frac{na}{RV^2}\right) - \left(\frac{2na}{RV^3}\right) \times (V - nb)$$

$$\left(\frac{\partial T}{\partial V}\right)_p = \left(\frac{T}{V - nb}\right) - \left(\frac{2na}{RV^3}\right) \times (V - nb)$$

Therefore,

$$\left(\frac{\partial T}{\partial p}\right)_V\left(\frac{\partial p}{\partial V}\right)_T\left(\frac{\partial V}{\partial T}\right)_p=\frac{\left(\frac{\partial T}{\partial p}\right)_V\left(\frac{\partial p}{\partial V}\right)_T}{\left(\frac{\partial T}{\partial V}\right)_p}$$

$$=\frac{\left(\frac{V-nb}{nR}\right)\times\left(\frac{-nRT}{(V-nb)^2}+\frac{2n^2a}{V^3}\right)}{\left(\frac{T}{V-nb}\right)-\left(\frac{2na}{RV^3}\right)\times(V-nb)}=\frac{\left(\frac{-T}{V-nb}\right)+\left(\frac{2na}{RV^3}\right)\times(V-nb)}{\left(\frac{T}{V-nb}\right)-\left(\frac{2na}{RV^3}\right)\times(V-nb)}$$

$$=-1$$

P2.37 $$\mu C_p=T\left(\frac{\partial V}{\partial T}\right)_p-V=\frac{T}{\left(\frac{\partial T}{\partial V}\right)_p}-V \text{ [reciprocal identity, MB2.3b]}$$

$$\left(\frac{\partial T}{\partial V}\right)_p=\frac{T}{V-nb}-\frac{2na}{RV^3}(V-nb) \text{ [Problem 2.35]}$$

Introduction of this expression followed by rearrangement leads to

$$\mu C_p=\frac{(2na)\times(V-nb)^2-nbRTV^2}{RTV^3-2na(V-nb)^2}\times V$$

Then, introducing $\zeta=\dfrac{RTV^3}{2na(V-nb)^2}$ to simplify the appearance of the expression

$$\boxed{\mu C_p=\left(\frac{1-\frac{nb\zeta}{V}}{\zeta-1}\right)V}=\left(\frac{1-\frac{b\zeta}{V_{\mathrm{m}}}}{\zeta-1}\right)V$$

For xenon, $V_{\mathrm{m}}=24.6\ \mathrm{dm^3\ mol^{-1}}$, $T=298\ \mathrm{K}$, $a=4.137\ \mathrm{dm^6\ atm\ mol^{-2}}$, $b=5.16\times10^{-2}\ \mathrm{dm^3\ mol^{-1}}$,

$$\frac{nb}{V}=\frac{b}{V_{\mathrm{m}}}=\frac{5.16\times10^{-1}\ \mathrm{dm^3\ mol^{-1}}}{24.6\ \mathrm{dm^3\ mol^{-1}}}=2.09\times10^{-3}$$

$$\zeta=\frac{(8.206\times10^{-2}\ \mathrm{dm^3\ atm\ K^{-1}\ mol^{-1}})\times(298\ \mathrm{K})\times(24.6\ \mathrm{dm^3\ mol^{-1}})^3}{(2)\times(4.137\ \mathrm{dm^6\ atm\ mol^{-2}})\times(24.6\ \mathrm{dm^3\ mol^{-1}}-5.16\times10^{-2}\ \mathrm{dm^3\ mol^{-1}})^2}=73.0$$

Therefore, $\mu C_p=\dfrac{1-(73.0)\times(2.09\times10^{-3})}{72.0}\times(24.6\ \mathrm{dm^3\ mol^{-1}})=0.290\ \mathrm{dm^3\ mol^{-1}}$

$C_p=20.79\ \mathrm{J\ K^{-1}\ mol^{-1}}$ [Table 2.8], so

$$\mu=\frac{0.290\ \mathrm{dm^3\ mol^{-1}}}{20.79\ \mathrm{J\ K^{-1}\ mol^{-1}}}=\frac{0.290\times10^{-3}\ \mathrm{m^3\ mol^{-1}}}{20.79\ \mathrm{J\ K^{-1}\ mol^{-1}}}$$

$$=1.39\overline{3}\times10^{-5}\ \mathrm{K\ m^3\ J^{-1}}=1.39\overline{3}\times10^{-5}\ \mathrm{K\ Pa^{-1}}$$

$$=(1.39\overline{3}\times10^{-5})\times(1.013\times10^{5}\ \mathrm{K\ atm^{-1}})=\boxed{1.41\ \mathrm{K\ atm^{-1}}}$$

The value of μ changes at $T = T_I$ and when the sign of the numerator $1 - \frac{nb\zeta}{V}$ changes sign ($\zeta - 1$ is positive). Hence,

$$\frac{b\zeta}{V_m} = 1 \text{ at } T = T_I \quad \text{or} \quad \frac{RTbV^3}{2na(V-nb)^2V_m} = 1 \quad \text{implying that } T_I = \frac{2a(V_m - b)^2}{RbV_m^2}$$

that is, $$T_I = \left(\frac{2a}{Rb}\right) \times \left(1 - \frac{b}{V_m}\right)^2 = \boxed{\frac{27}{4}T_c\left(1 - \frac{b}{V_m}\right)^2}$$

For xenon, $$\frac{2a}{Rb} = \frac{(2) \times (4.137\ \text{dm}^6\ \text{atm mol}^{-2})}{(8.206 \times 10^{-2}\ \text{dm}^3\ \text{atm K}^{-1}\ \text{mol}^{-1}) \times (5.16 \times 10^{-2}\ \text{dm}^3\ \text{mol}^{-1})} = 1954\ \text{K}$$

and so $$T_I = (1954\ \text{K}) \times \left(1 - \frac{5.16 \times 10^{-2}}{24.6}\right)^2 = \boxed{1946\ \text{K}}$$

Question. An approximate relationship for μ of a van der Waals gas was obtained in Problem 2.34. Use it to obtain an expression for the inversion temperature, calculate it for xenon, and compare to the result above.

P2.39 $$C_{p,m} - C_{V,m} = \frac{\alpha^2 TV}{n\kappa_T}\ [2.48] = \frac{\alpha TV}{n}\left(\frac{\partial p}{\partial T}\right)_V \ [\text{using } 2.59]$$

$$\left(\frac{\partial p}{\partial T}\right)_V = \frac{nR}{V - nb}\ [\text{Problem } 2.35]$$

$$\alpha V = \left(\frac{\partial V}{\partial T}\right)_p = \frac{1}{\left(\frac{\partial T}{\partial V}\right)_p}$$

Substituting,

$$C_{p,m} - C_{V,m} = \frac{T\left(\frac{\partial p}{\partial T}\right)_V}{n\left(\frac{\partial T}{\partial V}\right)_p}$$

Also $$\left(\frac{\partial T}{\partial V}\right)_p = \frac{T}{V - nb} - \frac{2na}{RV^3}(V - nb)\ [\text{Problem } 2.35]$$

Substituting,

$$C_{p,m} - C_{V,m} = \frac{\frac{RT}{(V-nb)}}{\frac{T}{(V-nb)} - \frac{2na}{RV^3} \times (V - nb)} = \lambda R$$

with $$\lambda = \frac{1}{1 - \frac{2na}{(RTV^3)} \times (V - nb)^2} \quad \text{or} \quad \frac{1}{\lambda} = 1 - \frac{2a(V_m - b)^2}{RTV_m^3}$$

Now introduce the reduced variables and use $T_c = \dfrac{8a}{27Rb}$, $V_c = 3b$.

After rearrangement,

$$\boxed{\frac{1}{\lambda} = 1 - \frac{(3V_r - 1)^2}{4T_rV_r^3}}$$

For xenon, $V_c = 118.1\ \text{cm}^3\ \text{mol}^{-1}$, $T_c = 289.8$ K. The perfect gas value for V_m may be used as any error introduced by this approximation occurs only in the correction term for $\dfrac{1}{\lambda}$.

Hence, $V_m \approx 2.45\ \text{dm}^3$, $V_c = 118.8\ \text{cm}^3\ \text{mol}^{-1}$, $T_c = 289.8$ K, and $V_r = 20.6$ and $T_r = 1.03$, therefore

$$\frac{1}{\lambda} = 1 - \frac{(61.8 - 1)^2}{(4) \times (1.03) \times (20.6)^3} = 0.90, \text{ giving } \lambda \approx 1.1$$

and

$$C_{p,m} - C_{V,m} \approx 1.1R = \boxed{9.2\ \text{J K}^{-1}\ \text{mol}^{-1}}$$

P2.41 (a) $\mu = -\dfrac{1}{C_p}\left(\dfrac{\partial H}{\partial p}\right)_T = \dfrac{1}{C_p}\left\{T\left(\dfrac{\partial V_m}{\partial T}\right)_p - V_m\right\}$ [*Justification* 2.2 and Problem 2.38]

$$V_m = \frac{RT}{p} + aT^2 \quad \text{so} \quad \left(\frac{\partial V_m}{\partial T}\right)_p = \frac{R}{p} + 2aT$$

$$\mu = \frac{1}{C_p}\left\{\frac{RT}{p} + 2aT^2 - \frac{RT}{p} - aT^2\right\} = \boxed{\frac{aT^2}{C_p}}$$

(b) $C_V = C_p - \alpha T V_m \left(\dfrac{\partial p}{\partial T}\right)_V = C_p - T\left(\dfrac{\partial V_m}{\partial T}\right)_p \left(\dfrac{\partial p}{\partial T}\right)_V$

But, $p = \dfrac{RT}{V_m - aT^2}$

$$\left(\frac{\partial p}{\partial T}\right)_V = \frac{R}{V_m - aT^2} - \frac{RT(-2aT)}{(V_m - aT^2)^2} = \frac{R}{(RT/p)} + \frac{2aRT^2}{(RT/p)^2} = \frac{p}{T} + \frac{2ap^2}{R}$$

Therefore

$$C_V = C_p - T\left(\frac{R}{p} + 2aT\right) \times \left(\frac{p}{T} + \frac{2ap^2}{R}\right)$$

$$= C_p - \frac{RT}{p}\left(1 + \frac{2apT}{R}\right) \times \left(1 + \frac{2apT}{R}\right) \times \left(\frac{p}{T}\right)$$

$$\boxed{C_V = C_p - R\left(1 + \frac{2apT}{R}\right)^2}$$

Solutions to applications

P2.43 The three possible fates of the radical are

(a) *tert*-$C_4H_9 \rightarrow$ *sec*-C_4H_9,
(b) *tert*-$C_4H_9 \rightarrow C_3H_6 + CH_3$,
(c) *tert*-$C_4H_9 \rightarrow C_2H_4 + C_2H_5$.

The three corresponding enthalpy changes are

(a) $\Delta_r H^{\ominus} = \Delta_f H^{\ominus}(sec\text{-}C_4H_9) - \Delta_f H^{\ominus}(tert\text{-}C_4H_9) = (67.5 - 51.3)\ \text{kJ mol}^{-1}$

$$= \boxed{16.2\ \text{kJ mol}^{-1}}$$

(b) $\Delta_r H^{\ominus} = \Delta_f H^{\ominus}(C_3H_6) + \Delta_f H^{\ominus}(CH_3) - \Delta_f H^{\ominus}(tert\text{-}C_4H_9)$

$$= (20.42 + 145.69 - 51.3)\ \text{kJ mol}^{-1} = \boxed{114.8\ \text{kJ mol}^{-1}}$$

(c) $\Delta_r H^{\ominus} = \Delta_f H^{\ominus}(C_2H_4) + \Delta_f H^{\ominus}(C_2H_5) - \Delta_f H^{\ominus}(tert\text{-}C_4H_9)$

$$= (52.26 + 121.0 - 51.3)\ \text{kJ mol}^{-1} = \boxed{122.0\ \text{kJ mol}^{-1}}$$

P2.45 (a) The Joule–Thomson coefficient is related to the given data by

$$\mu = -(1/C_p)(\partial H/\partial p)_T = -(-3.29 \times 10^3\ \text{J mol}^{-1}\ \text{MPa}^{-1})/(110.0\ \text{J K}^{-1}\ \text{mol}^{-1})$$
$$= \boxed{29.9\ \text{K MPa}^{-1}}$$

(b) The Joule–Thomson coefficient is defined as

$$\mu = (\partial T/\partial p)_H \approx (\Delta T/\Delta p)_H$$

Assuming that the expansion is a Joule–Thomson constant-enthalpy process, we have

$$\Delta T = \mu \Delta p = (29.9\ \text{K MPa}^{-1}) \times [(0.5 - 1.5) \times 10^{-1}\ \text{MPa}] = \boxed{-2.99\ \text{K}}$$

P2.47 The thermogram shows that the protein retains its native structure up to about 45°C and then undergoes an endothermic conformational change at higher temperatures that results in the loss of its three-dimensional structure. The enthalpy change associated with this process can be determined by computing the area under the curve of $C_{p,\text{ex}}$ against T.

3 The second law

Answers to discussion questions

D3.1 We must remember that the second law of thermodynamics states only that the total entropy of both the system (here, the molecules organizing themselves into cells) and the surroundings (here, the medium) must increase in a naturally occurring process. It does not state that entropy must increase in a portion of the universe that interacts with its surroundings. In this case, the cells grow by using chemical energy from their surroundings (the medium) and in the process the increase in the entropy of the medium outweighs the decrease in entropy of the system. Hence, the second law is not violated.

D3.3 All of these expressions are obtained from a combination of the first law of thermodynamics with the Clausius inequality in the form $TdS \geq dq$ (as was done at the start of *Justification 3.2*). It may be written as

$$-dU - p_{ex}dV + dw_{add} + TdS \geq 0$$

where we have divided the work into pressure–volume work and additional work. Under conditions of constant energy and volume and no additional work, that is, an isolated system, this relationship reduces to

$$dS \geq 0$$

which is equivalent to $\Delta S_{tot} = \Delta S_{universe} \geq 0$. (The universe is an isolated system.)

Under conditions of constant entropy and volume and no additional work, the fundamental relationship reduces to

$$dU \leq 0$$

Under conditions of constant temperature and volume, with no additional work, the relationship reduces to

$$dA \leq 0,$$

where A is defined as $U - TS$.

Under conditions of constant temperature and pressure, with no additional work, the relationship reduces to

$$dG \leq 0,$$

where G is defined as $U + pV - TS = H - TS$.

In all of these relationships, choosing the inequality provides the criteria for *spontaneous change*. Choosing the equal sign gives us the criteria for *equilibrium* under the conditions specified.

D3.5 The Maxwell relationships are relationships between partial derivatives all of which are expressed in terms of functions of state (properties of the system). Partial derivatives can be thought of as a kind of shorthand for an experiment. Therefore, the partial derivative $\left(\frac{\partial S}{\partial V}\right)_T$ tells us how the entropy of the system changes when we change its volume under constant temperature conditions. But as entropy is not a property that can be measured directly (there are no entropy meters), it is important that the derivative (and hence the experiment) be transformed into a form that involves directly measurable properties. That is what the following Maxwell relationship does for us:

$$\left(\frac{\partial S}{\partial V}\right)_T = \left(\frac{\partial p}{\partial T}\right)_V$$

Pressure, temperature, and volume are easily measured properties.

D3.7 The relationship $(\partial G/\partial p)_T = V$, eqn 3.53, shows that the Gibbs function of a system increases with p at constant T in proportion to the magnitude of its volume. This makes good sense when one considers the definition of G, which is $G = U + pV - TS$. Hence, G is expected to increase with p in proportion to V when T is constant.

Solutions to exercises

Assume that all gases are perfect and that data refer to 298.15 K unless otherwise stated.

E3.1(a) Assume that the block is so large that its temperature does not change significantly as a result of the heat transfer. Then

$$\Delta S = \int_{\mathrm{i}}^{\mathrm{f}} \frac{\mathrm{d}q_{\mathrm{rev}}}{T}[3.2] = \frac{1}{T}\int_{\mathrm{i}}^{\mathrm{f}} \mathrm{d}q_{\mathrm{rev}}\ [\text{constant } T] = \frac{q_{\mathrm{rev}}}{T}$$

$$\text{(a)}\quad \Delta S = \frac{25\times10^3\ \mathrm{J}}{273.15\ \mathrm{K}} = \boxed{92\ \mathrm{J\,K^{-1}}} \qquad \text{(b)}\quad \Delta S = \frac{25\times10^3\ \mathrm{J}}{373.15\ \mathrm{K}} = \boxed{67\ \mathrm{J\,K^{-1}}}$$

$$S_{\mathrm{m}}(T_{\mathrm{f}}) = S_{\mathrm{m}}(T_{\mathrm{i}}) + \int_{T_{\mathrm{i}}}^{T_{\mathrm{f}}} \frac{C_{V,\mathrm{m}}}{T}\mathrm{d}T\ [3.23, \text{ with } C_{V,\mathrm{m}} \text{ in place of } C_p]$$

E3.2(a) If we assume that neon is a perfect gas then $C_{V,\mathrm{m}}$ may be taken to be constant and given by

$$C_{V,\mathrm{m}} = C_{p,\mathrm{m}} - R;$$

$$C_{p,\mathrm{m}} = 20.786\ \mathrm{J\,K^{-1}\,mol^{-1}}\ [\text{Table 2.8}]$$

$$C_{V,\mathrm{m}} = (20.786 - 8.314)\ \mathrm{J\,K^{-1}\,mol^{-1}} = 12.472\ \mathrm{J\,K^{-1}\,mol^{-1}}$$

Integrating, we obtain

$$S_m(500\,K) = S_m(298\,K) + C_{V,m}\ln\frac{T_f}{T_i}$$

$$= (146.22\ J\ K^{-1}\,mol^{-1}) + (12.472\ J\ K^{-1}\,mol^{-1})\ln\left(\frac{500\ K}{298\ K}\right)$$

$$= (146.22 + 6.45)\ J\ K^{-1}\,mol^{-1} = \boxed{152.67\ J\ K^{-1}\,mol^{-1}}$$

E3.3(a) Since entropy is a state function, ΔS may be calculated from the most convenient path, which in this case corresponds to constant-pressure heating followed by constant-temperature compression.

$$\Delta S = nC_{p,m}\ln\left(\frac{T_f}{T_i}\right)[3.23, \text{at } p_i] + nR\ln\left(\frac{V_f}{V_i}\right)[3.17, \text{at } T_f]$$

Since pressure and volume are inversely related (Boyle's law), $\frac{V_f}{V_i} = \frac{p_i}{p_f}$. Hence,

$$\Delta S = nC_{p,m}\ln\left(\frac{T_f}{T_i}\right) - nR\ln\left(\frac{p_f}{p_i}\right) = (3.00\ mol)\times\frac{5}{2}\times(8.314\ J\ K^{-1}\,mol^{-1})\times\ln\left(\frac{398\ K}{298\ K}\right)$$

$$-(3.00\ mol)\times(8.314\ J\ K^{-1}\,mol^{-1})\times\ln\left(\frac{5.00\ atm}{1.00\ atm}\right)$$

$$= (18.0\bar{4} - 40.1\bar{4})\ J\ K^{-1} = -22.1\ J\ K^{-1}$$

Though ΔS(system) is negative, the process can still occur spontaneously if ΔS(total) is positive.

E3.4(a) For an adiabatic reversible process, $q = q_{rev} = \boxed{0}$.

$$\Delta S = \int_i^f \frac{dq_{rev}}{T} = \boxed{0}$$

$$\Delta U = nC_{V,m}\Delta T\ [2.16b] = (3.00\ mol)\times(27.5\ J\ K^{-1}\,mol^{-1})\times(50\ K) = 4.1\times10^3\ J$$

$$= \boxed{+4.1\ kJ}$$

$w = \Delta U$ [first law with $q = 0$]

$\Delta H = nC_{p,m}\Delta T$ [2.23b]

$C_{p,m} = C_{V,m} + R$ [2.26] $= (27.5 + 8.3)\ J\ K^{-1}\,mol^{-1} = 35.8\ J\ K^{-1}\,mol^{-1}$

$$\Delta H = (3.00\ mol)\times(35.8\ J\ K^{-1}\,mol^{-1})\times(50\ K) = 5.4\times10^3\ J = \boxed{+5.4\ kJ}$$

COMMENT. Neither initial nor final pressures and volumes are needed for the solution to this exercise.

E3.5(a) Since the container is isolated, the heat flow is zero and therefore $\boxed{\Delta H = 0}$; since the masses of the blocks are equal, the final temperature must be their mean temperature, 50°C. Specific heat capacities are heat capacities per gram and are related to the molar heat capacities by

$$C_s = \frac{C_m}{M}\ [C_{p,m} \approx C_{V,m} = C_m]$$

So $nC_m = mC_s\ [nM = m]$

$$\Delta H(\text{individual}) = mC_s\Delta T = 1.00 \times 10^4\ \text{g} \times 0.385\ \text{J K}^{-1}\ \text{g}^{-1} \times (\pm 50\ \text{K}) = \pm 1.9 \times 10^2\ \text{kJ}$$

These two enthalpy changes add up to zero: $\boxed{\Delta H_{\text{tot}} = 0}$

$$\Delta S = mC_s \ln\left(\frac{T_f}{T_i}\right) [3.23]$$

$$\Delta S_1 = (10.0 \times 10^3\ \text{g}) \times (0.385\ \text{J K}^{-1}\ \text{g}^{-1}) \times \ln\left(\frac{323\ \text{K}}{373\ \text{K}}\right) = -5.541 \times 10^2\ \text{J K}^{-1}$$

$$\Delta S_2 = (10.0 \times 10^3\ \text{g}) \times (0.385\ \text{J K}^{-1}\ \text{g}^{-1}) \times \ln\left(\frac{323\ \text{K}}{273\ \text{K}}\right) = 6.475 \times 10^2\ \text{J K}^{-1}$$

$$\Delta S_{\text{tot}} = \Delta S_1 + \Delta S_2 = \boxed{+93.4\ \text{J K}^{-1}}$$

COMMENT. The positive value of ΔS_{tot} corresponds to a spontaneous process.

E3.6(a) (a) $q = \boxed{0}$ [adiabatic]

(b) $w = -p_{\text{ex}}\Delta V\ [2.8] = -(1.01 \times 10^5\ \text{Pa}) \times (20\ \text{cm}) \times (10\ \text{cm}^2) \times \left(\frac{10^{-6}\ \text{m}^3}{\text{cm}^3}\right) = -20\ \text{J}$

(c) $\Delta U = q + w = 0 - 20\ \text{J} = -20\ \text{J}$

(d) $\Delta U = nC_{V,\text{m}}\Delta T\ [2.16\text{b}]$

$$\Delta T = \frac{-20\ \text{J}}{(2.0\ \text{mol}) \times (28.8\ \text{J K}^{-1}\ \text{mol}^{-1})} = -0.34\overline{7}\ \text{K}$$

(e) Entropy is a state function, so we can compute it by any convenient path. Although the specified transformation is adiabatic, a more convenient path is constant-volume cooling followed by isothermal expansion. The entropy change is the sum of the entropy changes of these two steps:

$$\Delta S = \Delta S_1 + \Delta S_2 = nC_{V,\text{m}} \ln\left(\frac{T_f}{T_i}\right) + nR \ln\left(\frac{V_f}{V_i}\right) [3.23\ \&\ 3.17]$$

$$T_f = T_i - 0.34\overline{7}\ \text{K} = (298.15\ \text{K}) - (0.34\overline{7}\ \text{K}) = 297.80\overline{3}\ \text{K}$$

$$V_i = \frac{nRT}{p_i} = \frac{(2.0\ \text{mol}) \times (0.08206\ \text{dm}^3\ \text{atm K}^{-1}\ \text{mol}^{-1}) \times (298.15\ \text{K})}{10\ \text{atm}} = 4.8\overline{93}\ \text{dm}^3$$

$$V_f = V_i + \Delta V = (4.8\overline{93} + 0.20)\ \text{dm}^3 = 5.0\overline{93}\ \text{dm}^3$$

Substituting these values into the expression for ΔS above gives

$$\Delta S = (2.0\ \text{mol}) \times (28.8\ \text{J K}^{-1}\ \text{mol}^{-1}) \times \ln\left(\frac{297.80\overline{3}\ \text{K}}{298.15\ \text{K}}\right) + (2.0\ \text{mol}) \times (8.314\ \text{J K}^{-1}\ \text{mol}^{-1}) \ln\left(\frac{5.0\overline{93}\ \text{dm}^3}{4.8\overline{93}\ \text{dm}^3}\right)$$

$$= (-0.067\overline{1} + 0.66\overline{6})\ \text{J K}^{-1} = \boxed{+0.60\ \text{J K}^{-1}}$$

E3.7(a) (a) $\Delta_{\text{vap}}S = \dfrac{\Delta_{\text{vap}}H}{T_{\text{b}}} = \dfrac{29.4 \times 10^3\ \text{J mol}^{-1}}{334.88\ \text{K}} = \boxed{+87.8\ \text{J K}^{-1}\ \text{mol}^{-1}}$

(b) If the vaporization occurs reversibly, $\Delta S_{\text{tot}} = 0$, so $\Delta S_{\text{surr}} = \boxed{-87.8\ \text{J K}^{-1}\ \text{mol}^{-1}}$

E3.8(a) In each case $\Delta_r S^\ominus = \sum_{\text{Products}} \nu S_m^\ominus - \sum_{\text{Products}} \nu S_m^\ominus$ [3.25a]

with S_m values obtained from Tables 2.6 and 2.8.

(a) $\Delta_r S = 2S_m^\ominus(\text{CH}_3\text{COOH, l}) - 2S_m^\ominus(\text{CH}_3\text{CHO, g}) - S_m^\ominus(\text{O}_2\text{, g})$
$= [(2 \times 159.8) - (2 \times 250.3) - 205.14]\ \text{J K}^{-1}\ \text{mol}^{-1} = \boxed{-386.1\ \text{J K}^{-1}\ \text{mol}^{-1}}$

(b) $\Delta_r S^\ominus = 2S_m^\ominus(\text{AgBr, s}) + S_m^\ominus(\text{Cl}_2\text{, g}) - 2S_m^\ominus(\text{AgCl, s}) - S_m^\ominus(\text{Br}_2\text{, l})$
$= [(2 \times 107.1) + (223.07) - (2 \times 96.2) - (152.23)]\ \text{J K}^{-1}\ \text{mol}^{-1}$
$= \boxed{+92.6\ \text{J K}^{-1}\ \text{mol}^{-1}}$

(c) $\Delta_r S^\ominus = S_m^\ominus(\text{HgCl}_2\text{, s}) - S_m^\ominus(\text{Hg, l}) - S_m^\ominus(\text{Cl}_2\text{, g})$
$= [146.0 - 76.02 - 223.07]\ \text{J K}^{-1}\ \text{mol}^{-1} = \boxed{-153.1\ \text{J K}^{-1}\ \text{mol}^{-1}}$

E3.9(a) In each case we use

$$\Delta_r G^\ominus = \Delta_r H^\ominus - T\Delta_r S^\ominus \ [3.42]$$

along with

$$\Delta_r H^\ominus = \sum_{\text{Products}} \nu \Delta_f H^\ominus - \sum_{\text{Reactants}} \nu \Delta_f H^\ominus \ [2.32]$$

(a) $\Delta_r H^\ominus = 2\Delta_f H^\ominus(\text{CH}_3\text{COOH, l}) - 2\Delta_f H^\ominus(\text{CH}_3\text{CHO, g})$
$= [2 \times (-484.5) - 2 \times (-166.19)]\ \text{kJ mol}^{-1} = -636.6\bar{2}\ \text{kJ mol}^{-1}$

$\Delta_r G^\ominus = -636.6\bar{2}\ \text{kJ mol}^{-1} - (298.15\ \text{K}) \times (-386.1\ \text{J K}^{-1}\ \text{mol}^{-1}) = \boxed{-521.5\ \text{kJ mol}^{-1}}$

(b) $\Delta_r H^\ominus = 2\Delta_f H^\ominus(\text{AgBr, s}) - 2\Delta_f H^\ominus(\text{AgCl, s})$
$= [2 \times (-100.37) - 2 \times (-127.07)]\ \text{kJ mol}^{-1} = +53.40\ \text{kJ mol}^{-1}$

$\Delta_r G^\ominus = +53.40\ \text{kJ mol}^{-1} - (298.15\ \text{K}) \times (+92.6)\ \text{J K}^{-1}\ \text{mol}^{-1} = \boxed{+25.8\ \text{kJ mol}^{-1}}$

(c) $\Delta_r H^\ominus = \Delta_f H^\ominus(\text{HgCl}_2\text{, s}) = -224.3\ \text{kJ mol}^{-1}$

$\Delta_r G^\ominus = -224.3\ \text{kJ mol}^{-1} - (298.15\ \text{K}) \times (-153.1\ \text{J K}^{-1}\ \text{mol}^{-1}) = \boxed{-178.7\ \text{kJ mol}^{-1}}$

E3.10(a) In each case $\Delta_r G^\ominus = \sum_{\text{Products}} \nu \Delta_f G^\ominus - \sum_{\text{Reactants}} \nu \Delta_f G^\ominus$ [3.43a]

with $\Delta_f G^\ominus(\text{J})$ values from Tables 2.6 and 2.8.

(a) $\Delta_r G^\ominus = 2\Delta_f G^\ominus(\text{CH}_3\text{COOH, l}) - 2\Delta_f G^\ominus(\text{CH}_3\text{CHO, g})$
$= [2 \times (-389.9) - 2 \times (-128.86)]\ \text{kJ mol}^{-1} = \boxed{-522.1\ \text{kJ mol}^{-1}}$

(b) $\Delta_r G^\ominus = 2\Delta_f G^\ominus(\text{AgBr, s}) - 2\Delta_f G^\ominus(\text{AgCl, s}) = [2 \times (-96.90) - 2 \times (-109.79)]\ \text{kJ mol}^{-1}$
$= \boxed{+25.78\ \text{kJ mol}^{-1}}$

(c) $\Delta_r G^\ominus = \Delta_f G^\ominus(\text{HgCl}_2\text{, s}) = \boxed{-178.6\ \text{kJ mol}^{-1}}$

COMMENT. In each case these values of $\Delta_r G^\ominus$ agree closely with the calculated values in Exercise 3.9(a).

E3.11(a) $\Delta_r G^\ominus = \Delta_r H^\ominus - T\Delta_r S^\ominus$ [3.42] $\Delta_r H^\ominus = \sum_{\text{Products}} \nu\Delta_f H^\ominus - \sum_{\text{Reactants}} \nu\Delta_f H^\ominus$ [2.32]

$$\Delta_r S^\ominus = \sum_{\text{Products}} \nu S_m^\ominus - \sum_{\text{Reactants}} \nu S_m^\ominus \text{ [3.25a]}$$

$$\Delta_r H^\ominus = 2\Delta_f H^\ominus(\text{H}_2\text{O, l}) - 4\Delta_f H^\ominus(\text{HCl, g}) = \{2 \times (-285.83) - 4 \times (-92.31)\} \text{ kJ mol}^{-1}$$
$$= -202.42 \text{ kJ mol}^{-1}$$

$$\Delta_r S^\ominus = 2S_m^\ominus(\text{Cl}_2, \text{g}) + 2S_m^\ominus(\text{H}_2\text{O, l}) - 4S_m^\ominus(\text{HCl, g}) - S_m^\ominus(\text{O}_2, \text{g})$$
$$= [(2 \times 69.91) + (2 \times 223.07) - (4 \times 186.91) - (205.14)] \text{ J K}^{-1}\text{ mol}^{-1}$$
$$= -366.82 \text{ J K}^{-1}\text{ mol}^{-1} = -0.36682 \text{ kJ K}^{-1}\text{ mol}^{-1}$$

$$\Delta_r G^\ominus = -202.42 \text{ kJ mol}^{-1} - (298.15 \text{ K}) \times (-0.36682 \text{ kJ K}^{-1}\text{ mol}^{-1}) = \boxed{-93.05 \text{ kJ mol}^{-1}}$$

Question. Repeat the calculation based on $\Delta_f G^\ominus$ data of Table 2.8. What difference, if any, is there from the value above?

E3.12(a) The formation reaction for phenol is

$$6\text{ C(s)} + 3\text{ H}_2\text{(g)} + \tfrac{1}{2}\text{ O}_2\text{(g)} \rightarrow \text{C}_6\text{H}_5\text{OH(s)}$$

$$\Delta_f G^\ominus = \Delta_f H^\ominus - T\Delta_f S^\ominus \text{ [3.42]}$$

$\Delta_f H^\ominus$ is to be obtained from $\Delta_c H^\ominus$ for phenol and data from Tables 2.6 and 2.8. Thus

$$\text{C}_6\text{H}_5\text{OH(s)} + 7\text{ O}_2\text{(g)} \rightarrow 6\text{ CO}_2\text{(g)} + 3\text{ H}_2\text{O(l)}$$

$$\Delta_c H = 6\Delta_f H^\ominus(\text{CO}_2, \text{g}) + 3\Delta_f H^\ominus(\text{H}_2\text{O, l}) - \Delta_f H^\ominus(\text{C}_6\text{H}_5\text{OH, s})$$

$$\Delta_f H^\ominus(\text{C}_6\text{H}_5\text{OH, s}) = 6\Delta_f H^\ominus(\text{CO}_2, \text{g}) + 3\Delta_f H^\ominus(\text{H}_2\text{O, l}) - \Delta_c H^\ominus$$
$$= [6 \times (-393.51) + 3 \times (-285.83) - (-3054)] \text{ kJ mol}^{-1}$$
$$= -164.5\overline{5} \text{ kJ mol}^{-1}$$

$$\Delta_r S^\ominus = \sum_{\text{Products}} \nu S_m^\ominus - \sum_{\text{Reactants}} \nu S_m^\ominus \text{ [3.25a]}$$

$$\Delta_f S^\ominus = S_m^\ominus(\text{C}_6\text{H}_5\text{OH, s}) - 6S_m^\ominus(\text{C, s}) - 3S_m^\ominus(\text{H}_2, \text{g}) - \tfrac{1}{2}S_m^\ominus(\text{O}_2, \text{g})$$
$$= [144.0 - (6 \times 5.740) - (3 \times 130.68) - (\tfrac{1}{2} \times 205.14)] \text{ J K}^{-1}\text{ mol}^{-1}$$
$$= -385.0\overline{5} \text{ J K}^{-1}\text{ mol}^{-1}$$

Hence, $\Delta_r G^\ominus = -164.5\overline{5} \text{ kJ mol}^{-1} - (298.15 \text{ K}) \times (-385.0\overline{5} \text{ J K}^{-1}\text{ mol}^{-1}) = \boxed{-50 \text{ kJ mol}^{-1}}$

E3.13(a) (a) $\Delta S(\text{gas}) = nR\ln\dfrac{V_f}{V_i}$ [3.17] $= \left(\dfrac{14 \text{ g}}{28.02 \text{ g mol}^{-1}}\right) \times (8.314 \text{ J K}^{-1}\text{ mol}^{-1}) \times (\ln 2)$

$$= \boxed{+2.9 \text{ J K}^{-1}}$$

$\Delta S(\text{surroundings}) = \boxed{-2.9 \text{ J K}^{-1}}$ [overall zero entropy production]

$\Delta S(\text{total}) = \boxed{0}$ [reversible process]

(b) $\Delta S(\text{gas}) = \boxed{+2.9 \text{ J K}^{-1}}$ [S is a state function]

$\Delta S(\text{surroundings}) = \boxed{0}$ [surroundings do not change]

$\Delta S(\text{total}) = \boxed{+2.9 \text{ J K}^{-1}}$

(c) $\Delta S(\text{gas}) = \boxed{0}\ [q_{\text{rev}} = 0]$

$\Delta S(\text{surroundings}) = \boxed{0}$ [no heat transferred to surroundings]

$\Delta S(\text{total}) = \boxed{0}$

E3.14(a) $CH_4(g) + 2\,O_2(g) \rightarrow CO_2(g) + 2\,H_2O(l)$

$$\Delta_r G^\ominus = \sum_{\text{Products}} \nu\Delta_f G^\ominus - \sum_{\text{Rectants}} \nu\Delta_f G^\ominus \text{ [3.43a]}$$

$$\Delta_r G^\ominus = \Delta_f G^\ominus(CO_2, g) + 2\Delta_f G^\ominus(H_2O, l) - \Delta_f G^\ominus(CH_4, g)$$
$$= \{-394.36 + (2 \times -237.13) - (-50.72)\}\ \text{kJ mol}^{-1} = -817.90\ \text{kJ mol}^{-1}$$

Therefore, the maximum non-expansion work is $\boxed{817.90\ \text{kJ mol}^{-1}}$ since $|w_{\text{add}}| = |\Delta G|$.

E3.15(a) $$\eta_{\text{rev}} = 1 - \frac{T_c}{T_h} \text{ [3.10]}$$

$$\eta = 1 - \frac{333\ \text{K}}{373\ \text{K}} = \boxed{0.11} \quad \text{(11\% efficiency for the old steam engine)}$$

$$\eta = 1 - \frac{353\ \text{K}}{573\ \text{K}} = \boxed{0.38} \quad \text{(38\% efficiency for the modern turbine)}$$

E3.16(a) $$\Delta G = nRT \ln\left(\frac{p_f}{p_i}\right) \text{ [3.59]} = nRT \ln\left(\frac{V_i}{V_f}\right) \text{ [Boyle's law]}$$

$$\Delta G = (3.0 \times 10^{-3}\ \text{mol}) \times (8.314\ \text{J K}^{-1}\ \text{mol}^{-1}) \times (300\ \text{K}) \times \ln\left(\frac{36}{60}\right) = \boxed{-3.8\ \text{J}}$$

E3.17(a) $$\left(\frac{\partial G}{\partial T}\right)_p = -S \text{ [3.53]; hence } \left(\frac{\partial G_f}{\partial T}\right)_p = -S_f, \text{ and } \left(\frac{\partial G_i}{\partial T}\right)_p = -S_i$$

$$\Delta S = S_f - S_i = -\left(\frac{\partial G_f}{\partial T}\right)_p + \left(\frac{\partial G_i}{\partial T}\right)_p = -\left(\frac{\partial (G_f - G_i)}{\partial T}\right)_p$$
$$= -\left(\frac{\partial \Delta G}{\partial T}\right)_p = -\frac{\partial}{\partial T}\left(-85.40\ \text{J} + 36.5\ \text{J} \times \frac{T}{\text{K}}\right)$$
$$= \boxed{-36.5\ \text{J K}^{-1}}$$

E3.18(a) $dG = -SdT + Vdp$ [3.52]; at constant T, $dG = Vdp$, therefore

$$\Delta G = \int_{p_i}^{p_f} V dp$$

The change in volume of a condensed phase under isothermal compression is given by the isothermal compressibility (eqn 2.43).

$$\kappa_T = -\frac{1}{V}\left(\frac{\partial V}{\partial p}\right)_T = 76.8 \times 10^{-6}\ \text{atm}^{-1} \text{ [Table 2.9]}$$

This small isothermal compressibility (typical of condensed phases) tells us that we can expect a small change in volume from even a large increase in pressure. So we can make the following approximations to obtain a simple expression for the volume as a function of the pressure

$$\kappa_T \approx -\frac{1}{V}\left(\frac{V-V_\mathrm{i}}{p-p_\mathrm{i}}\right) \approx -\frac{1}{V_\mathrm{i}}\left(\frac{V-V_\mathrm{i}}{p}\right) \quad \text{so} \quad V = V_\mathrm{i}(1-\kappa_T p),$$

where V_i is the volume at 1 atm, namely the sample mass over the density, m/ρ.

$$\begin{aligned}\Delta G &= \int_{1\,\mathrm{atm}}^{3000\,\mathrm{atm}} \frac{m}{\rho}(1-\kappa_T p)\mathrm{d}p \\ &= \frac{m}{\rho}\left(\int_{1\,\mathrm{atm}}^{3000\,\mathrm{atm}} \mathrm{d}p - \kappa_T \int_{1\,\mathrm{atm}}^{3000\,\mathrm{atm}} p\mathrm{d}p\right) \\ &= \frac{m}{\rho}\left(p\Big|_{1\,\mathrm{atm}}^{3000\,\mathrm{atm}} - \frac{1}{2}\kappa_T p^2\Big|_{1\,\mathrm{atm}}^{3000\,\mathrm{atm}}\right) \\ &= \frac{35\,\mathrm{g}}{0.789\,\mathrm{g\,cm^{-3}}}\left(2999\,\mathrm{atm} - \frac{1}{2}(76.8\times10^{-6}\,\mathrm{atm^{-1}})\times(9.00\times10^{6}\,\mathrm{atm^2})\right) \\ &= 44.\bar{4}\,\mathrm{cm^3}\times\left(\frac{1\,\mathrm{m}}{100\,\mathrm{cm}}\right)^3 \times 2653\,\mathrm{atm}\times(1.013\times10^5\,\mathrm{Pa\,atm^{-1}}) \\ &= 1.1\bar{9}\times10^4\,\mathrm{J} = \boxed{12\,\mathrm{kJ}}\end{aligned}$$

E3.19(a) $\Delta G_\mathrm{m} = G_\mathrm{m,f} - G_\mathrm{m,i} = RT\ln\left(\frac{p_\mathrm{f}}{p_\mathrm{i}}\right)[3.59] = (8.314\,\mathrm{J\,K^{-1}\,mol^{-1}})\times(313\,\mathrm{K})\times\ln\left(\frac{29.5}{1.8}\right)$

$= \boxed{+7.3\,\mathrm{kJ\,mol^{-1}}}$

E3.20(a) For an ideal gas, $G_\mathrm{m}^\circ = G_\mathrm{m}^\ominus + RT\ln\left(\frac{p}{p^\ominus}\right)$ [3.59 with $G_\mathrm{m} = G_\mathrm{m}^\circ$]

But for a real gas, $G_\mathrm{m} = G_\mathrm{m}^\ominus + RT\ln\left(\frac{f}{p^\ominus}\right)$ [3.61]

So $G_\mathrm{m} - G_\mathrm{m}^\circ = RT\ln\frac{f}{p}$ [3.61 minus 3.59]; $\frac{f}{p} = \phi$

$= RT\ln\phi = (8.314\,\mathrm{J\,K^{-1}\,mol^{-1}})\times(200\,\mathrm{K})\times(\ln 0.72) = \boxed{-0.55\,\mathrm{kJ\,mol^{-1}}}$

E3.21(a) $\Delta G = nV_\mathrm{m}\Delta p$ [3.58] $= V\Delta p$

$\Delta G = (1.0\,\mathrm{dm^3})\times\left(\frac{1\,\mathrm{m^3}}{10^3\,\mathrm{dm^3}}\right)\times(99\,\mathrm{atm})\times(1.013\times10^5\,\mathrm{Pa}) = 10\,\mathrm{kPa\,m^3} = \boxed{+10\,\mathrm{kJ}}$

E3.22(a) $\Delta G_\mathrm{m} = RT\ln\frac{p_\mathrm{f}}{p_\mathrm{i}}$ [3.59] $= (8.314\,\mathrm{J\,K^{-1}\,mol^{-1}})\times(298\,\mathrm{K})\times\ln\left(\frac{100.0}{1.0}\right) = \boxed{+11\,\mathrm{kJ\,mol^{-1}}}$

Solutions to problems

Solutions to numerical problems

P3.1 (a) Because entropy is a state function $\Delta_{trs}S(l \rightarrow s, -5°C)$ may be determined indirectly from the following cycle

$$\begin{array}{ccc} H_2O(l, 0°C) & \xrightarrow{\Delta_{trs}S(l\rightarrow s, 0°C)} & H_2O(s, 0°C) \\ \Delta S_l \uparrow & & \downarrow \Delta S_s \\ H_2O(l, -5°C) & \xrightarrow{\Delta_{trs}S(l\rightarrow s, -5°C)} & H_2O(s, -5°C) \end{array}$$

Thus $\Delta_{trs}S(l \rightarrow s, -5°C) = \Delta S_l + \Delta_{trs}S(l \rightarrow s, 0°C) + \Delta S_s$,

where $\Delta S_l = C_{p,m}(l) \ln \dfrac{T_f}{T}$ $[3.23;\ \theta_f = 0°C, \theta = -5°C]$

and $\Delta S_s = C_{p,m}(s) \ln \dfrac{T}{T_f}$.

$$\Delta S_l + \Delta S_s = -\Delta C_p \ln \frac{T}{T_f} \quad \text{with } \Delta C_p = C_{p,m}(l) - C_{p,m}(s) = +37.3\ \mathrm{J\ K^{-1}\ mol^{-1}}$$

$$\Delta_{trs}S(l \rightarrow s, T_f) = \frac{-\Delta_{fus}H}{T_f} \ [3.20]$$

Thus, $\Delta_{trs}S(l \rightarrow s, T) = \dfrac{-\Delta_{fus}H}{T_f} - \Delta C_p \ln \dfrac{T}{T_f}$

$$\Delta_{trs}S(l \rightarrow s, -5°C) = \frac{-6.01 \times 10^3\ \mathrm{J\ mol^{-1}}}{273\ \mathrm{K}} - (37.3\ \mathrm{J\ K^{-1}\ mol^{-1}}) \times \ln \frac{268}{273} = \boxed{-21.3\ \mathrm{J\ K^{-1}\ mol^{-1}}}$$

$$\Delta S_{sur} = \frac{\Delta_{fus}H(T)}{T}$$

$$\Delta_{fus}H(T) = -\Delta H_l + \Delta_{fus}H(T_f) - \Delta H_s$$

$$\Delta H_l + \Delta H_s = C_{p,m}(l)(T_f - T) + C_{p,m}(s)(T - T_f) = \Delta C_p(T_f - T)$$

$$\Delta_{fus}H(T) = \Delta_{fus}H(T_f) - \Delta C_p(T_f - T)$$

Thus, $\Delta S_{sur} = \dfrac{\Delta_{fus}H(T)}{T} = \dfrac{\Delta_{fus}H(T_f)}{T} + \Delta C_p \dfrac{(T - T_f)}{T}$

$$\Delta S_{sur} = \frac{6.01\ \mathrm{kJ\ mol^{-1}}}{268\ \mathrm{K}} + (37.3\ \mathrm{J\ K^{-1}\ mol^{-1}}) \times \left(\frac{268 - 273}{268}\right) = \boxed{+21.7\ \mathrm{J\ K^{-1}\ mol^{-1}}}$$

$$\Delta S_{total} = \Delta S_{sur} + \Delta S = (21.7 - 21.3)\ \mathrm{J\ K^{-1}\ mol^{-1}} = \boxed{+0.4\ \mathrm{J\ K^{-1}\ mol^{-1}}}$$

Since $\Delta S_{total} > 0$, the transition $l \rightarrow s$ is spontaneous at $-5°C$.

(b) A similar cycle and analysis can be set up for the transition liquid $\rightarrow$ vapour at 95°C. However, since the transformation here is to the high temperature state (vapour) from the low temperature state (liquid), which is the opposite of part (a), we can expect that the analogous equations will occur with a change of sign.

$$\Delta_{\rm trs}S(\mathrm{l}\to\mathrm{g},T)=\Delta_{\rm trs}S(\mathrm{l}\to\mathrm{g},T_{\rm b})+\Delta C_p\ln\frac{T}{T_{\rm b}}=\frac{\Delta_{\rm vap}H}{T_{\rm b}}+\Delta C_p\ln\frac{T}{T_{\rm b}},\quad \Delta C_p=-41.9\ \mathrm{J\ K^{-1}\ mol^{-1}}$$

$$\Delta_{\rm trs}S(\mathrm{l}\to\mathrm{g},T)=\frac{40.7\ \mathrm{kJ\ mol^{-1}}}{373\ \mathrm{K}}-(41.9\ \mathrm{J\ K^{-1}\ mol^{-1}})\times\ln\left(\frac{368}{373}\right)=\boxed{+109.7\ \mathrm{J\ K^{-1}\ mol^{-1}}}$$

$$\Delta S_{\rm sur}=\frac{-\Delta_{\rm vap}H(T)}{T}=-\frac{\Delta_{\rm vap}H(T_{\rm b})}{T}-\frac{\Delta C_p(T-T_{\rm b})}{T}$$

$$=\left(\frac{-40.7\ \mathrm{kJ\ mol^{-1}}}{368\ \mathrm{K}}\right)-(-41.9\ \mathrm{J\ K^{-1}\ mol^{-1}})\times\left(\frac{368-373}{368}\right)=\boxed{-111.2\ \mathrm{J\ K^{-1}\ mol^{-1}}}$$

$$\Delta S_{\rm total}=(109.7-111.2)\ \mathrm{J\ K^{-1}\ mol^{-1}}=\boxed{-1.5\ \mathrm{J\ K^{-1}\ mol^{-1}}}$$

Since $\Delta S_{\rm total}<0$, the reverse transition, $\mathrm{g}\to\mathrm{l}$, is spontaneous at 95°C.

P3.3 (a) $q(\text{total})=q(\mathrm{H_2O})+q(\mathrm{Cu})=0$, hence $-q(\mathrm{H_2O})=q(\mathrm{Cu})$

$$q(\mathrm{H_2O})=n(-\Delta_{\rm vap}H)+nC_{p,\rm m}(\mathrm{H_2O,l})\times(\theta-100°\mathrm{C})$$

where θ is the final temperature of the water and copper.

$$q(\mathrm{Cu})=mC_{\rm s}(\theta-0)=mC_{\rm s}\theta,\quad C_{\rm s}=0.385\ \mathrm{J\ K^{-1}\ g^{-1}}$$

Setting $-q(\mathrm{H_2O})=q(\mathrm{Cu})$ allows us to solve for θ.

$$n(\Delta_{\rm vap}H)-nC_{p,\rm m}(\mathrm{H_2O,l})\times(\theta-100°\mathrm{C})=mC_{\rm s}\theta$$

Solving for θ yields:

$$\theta=\frac{n\{\Delta_{\rm vap}H+C_{p,\rm m}(\mathrm{H_2O,l})\times100°\mathrm{C}\}}{mC_{\rm s}+nC_{p,\rm m}(\mathrm{H_2O,l})}$$

$$=\frac{(1.00\ \mathrm{mol})\times(40.656\times10^3\ \mathrm{J\ mol^{-1}}+75.3\ \mathrm{J\ °C^{-1}\ mol^{-1}}\times100°\mathrm{C})}{2.00\times10^3\ \mathrm{g}\times0.385\ \mathrm{J\ °C^{-1}\ g^{-1}}+1.00\ \mathrm{mol}\times75.3\ \mathrm{J\ °C^{-1}\ mol^{-1}}}$$

$$=57.0°\mathrm{C}=330.2\ \mathrm{K}$$

Note: °C = K

$$q(\mathrm{Cu})=(2.00\times10^3\ \mathrm{g})\times(0.385\ \mathrm{J\ K^{-1}\ g^{-1}})\times(57.0\ \mathrm{K})=4.39\times10^4\ \mathrm{J}=\boxed{43.9\ \mathrm{kJ}}$$

$$q(\mathrm{H_2O})=\boxed{-43.9\ \mathrm{kJ}}$$

$$\Delta S(\text{total})=\Delta S(\mathrm{H_2O})+\Delta S(\mathrm{Cu})$$

$$\Delta S(\mathrm{H_2O})=\frac{-n\Delta_{\rm vap}H}{T_{\rm b}}\,[3.20]+nC_{p,\rm m}\ln\left(\frac{T_{\rm f}}{T_{\rm i}}\right)[3.23]$$

$$=-\frac{(1.00\ \mathrm{mol})\times(40.656\times10^3\ \mathrm{J\ mol^{-1}})}{373.2\ \mathrm{K}}$$

$$+(1.00\ \mathrm{mol})\times(75.3\ \mathrm{J\ K^{-1}\ mol^{-1}})\times\ln\left(\frac{330.2\ \mathrm{K}}{373.2\ \mathrm{K}}\right)$$

$$=-108.\bar{9}\ \mathrm{J\ K^{-1}}-9.22\ \mathrm{J\ K^{-1}}=\boxed{-118.\bar{1}\ \mathrm{J\ K^{-1}}}$$

$$\Delta S(\text{Cu}) = mC_s \ln\frac{T_f}{T_i} = (2.00\times10^3\ \text{g})\times(0.385\ \text{J K}^{-1}\text{g}^{-1})\times\ln\left(\frac{330.2\ \text{K}}{273.2\ \text{K}}\right) = \boxed{145.\overline{9}\ \text{J K}^{-1}}$$

$$\Delta S(\text{total}) = -118.\overline{1}\ \text{J K}^{-1} + 145.\overline{9}\ \text{J K}^{-1} = \boxed{28\ \text{J K}^{-1}}$$

This process is spontaneous since ΔS(surroundings) is zero and, hence,

$$\Delta S(\text{universe}) = \Delta S(\text{total}) > 0$$

(b) The volume of the container may be calculated from the perfect gas law.

$$V = \frac{nRT}{p} = \frac{(1.00\ \text{mol})\times(0.08206\ \text{dm}^3\,\text{atm}\,\text{K}^{-1}\,\text{mol}^{-1})\times(373.2\ \text{K})}{1.00\ \text{atm}} = 30.6\ \text{dm}^3$$

At 57°C the vapour pressure of water is 130 Torr (*CRC Handbook of Chemistry and Physics*). The amount of water vapour present at equilibrium is then

$$n = \frac{pV}{RT} = \frac{(130\ \text{Torr})\times\left(\dfrac{1\ \text{atm}}{760\ \text{Torr}}\right)\times(30.6\ \text{dm}^3)}{(0.08206\ \text{L atm K}^{-1}\,\text{mol}^{-1})\times(330.2\ \text{K})} = 0.193\ \text{mol}$$

This is a substantial fraction of the original amount of water and cannot be ignored. Consequently, the calculation needs to be redone taking into account the fact that only a part, n_1, of the vapour condenses into a liquid, while the remainder ($1.00\ \text{mol} - n_1$) remains gaseous. The heat flow involving water, then, becomes

$$q(\text{H}_2\text{O}) = -n_1\Delta_{\text{vap}}H + n_1C_{p,\text{m}}(\text{H}_2\text{O, l})\Delta T(\text{H}_2\text{O}) + (1.00\,\text{mol} - n_1)C_{p,\text{m}}(\text{H}_2\text{O, g})\Delta T(\text{H}_2\text{O})$$

Because n_1 depends on the equilibrium temperature through $n_1 = 1.00\ \text{mol} - \dfrac{pV}{RT}$, where p is the vapour pressure of water, we will have two unknowns (p and T) in the equation $-q(\text{H}_2\text{O}) = q(\text{Cu})$. There are two ways out of this dilemma: (1) p may be expressed as a function of T by use of the Clapeyron equation (Chapter 4) or (2) by use of successive approximations. Redoing the calculation yields:

$$\theta = \frac{n_1\Delta_{\text{vap}}H + n_1C_{p,\text{m}}(\text{H}_2\text{O, l})\times100°\text{C} + (1.00 - n_1)C_{p,\text{m}}(\text{H}_2\text{O, g})\times100°\text{C}}{mC_s + nC_{p,\text{m}}(\text{H}_2\text{O, l}) + (1.00 - n_1)C_{p,\text{m}}(\text{H}_2\text{O, g})}$$

With

$$n_1 = (1.00\ \text{mol}) - (0.193\ \text{mol}) = 0.80\overline{7}\ \text{mol}$$

(noting that $C_{p,\text{m}}(\text{H}_2\text{O, g}) = 33.6\ \text{J mol}^{-1}\ \text{K}^{-1}$ [Table 2.8]) $\theta = 47.2°\text{C}$. At this temperature, the vapour pressure of water is 80.41 Torr, corresponding to

$$n_1 = (1.00\ \text{mol}) - (0.123\ \text{mol}) = 0.87\overline{7}\ \text{mol}$$

This leads to $\theta = 50.8°\text{C}$. The successive approximations eventually converge to yield a value of $\theta = \boxed{49.9°\text{C} = 323.1\ \text{K}}$ for the final temperature. (At this temperature, the vapour pressure is 0.123 bar.) Using this value of the final temperature, the heat transferred and the various entropies are calculated as in part (a).

$$q(\mathrm{Cu}) = (2.00 \times 10^3\ \mathrm{g}) \times (0.385\ \mathrm{J\ K^{-1}\ g^{-1}}) \times (49.9\ \mathrm{K}) = \boxed{38.4\ \mathrm{kJ}} = -q(\mathrm{H_2O})$$

$$\Delta S(\mathrm{H_2O}) = \frac{-n\Delta_{\mathrm{vap}}H}{T_\mathrm{b}} + nC_{p,\mathrm{m}} \ln\left(\frac{T_\mathrm{f}}{T_\mathrm{i}}\right) = \boxed{-119.\overline{8}\ \mathrm{J\ K^{-1}}}$$

$$\Delta S(\mathrm{Cu}) = mC_\mathrm{s} \ln \frac{T_\mathrm{f}}{T_\mathrm{i}} = \boxed{129.\overline{2}\ \mathrm{J\ K^{-1}}}$$

$$\Delta S(\mathrm{total}) = -119.\overline{8}\ \mathrm{J\ K^{-1}} + 129.\overline{2}\ \mathrm{J\ K^{-1}} = \boxed{9\ \mathrm{J\ K^{-1}}}$$

P3.5

	Step 1	Step 2	Step 3	Step 4	Cycle
q	+11.5 kJ	0	−5.74 kJ	0	−5.8 kJ
w	−11.5 kJ	−3.74 kJ	+5.74 kJ	+3.74 kJ	−5.8 kJ
ΔU	0	−3.74 kJ	0	+3.74 kJ	0
ΔH	0	−6.23 kJ	0	+6.23 kJ	0
ΔS	+19.1 J K^{-1}	0	−19.1 J K^{-1}	0	0
ΔS_{tot}	0	0	0	0	0
ΔG	−11.5 kJ	?	+5.73 kJ	?	0

Step 1

$$\Delta U = \Delta H = \boxed{0}\ [\text{isothermal}]$$

$$w = -nRT \ln\left(\frac{V_\mathrm{f}}{V_\mathrm{i}}\right) = nRT \ln\left(\frac{p_\mathrm{f}}{p_\mathrm{i}}\right) [2.10, \text{and Boyle's law}]$$

$$= (1.00\ \mathrm{mol}) \times (8.314\ \mathrm{J\ K^{-1}\ mol^{-1}}) \times (600\ \mathrm{K}) \times \ln\left(\frac{1.00\ \mathrm{atm}}{10.0\ \mathrm{atm}}\right) = \boxed{-11.5\ \mathrm{kJ}}$$

$$q = -w = \boxed{11.5\ \mathrm{kJ}}$$

$$\Delta S = nR \ln\left(\frac{V_\mathrm{f}}{V_\mathrm{i}}\right) [3.17] = -nR \ln\left(\frac{p_\mathrm{f}}{p_\mathrm{i}}\right) [\text{Boyle's law}]$$

$$= -(1.00\ \mathrm{mol}) \times (8.314\ \mathrm{J\ K^{-1}\ mol^{-1}}) \times \ln\left(\frac{1.00\ \mathrm{atm}}{10.0\ \mathrm{atm}}\right) = \boxed{+19.1\ \mathrm{J\ K^{-1}}}$$

$$\Delta S(\mathrm{sur}) = -\Delta S(\mathrm{system})\ [\text{reversible process}] = -19.1\ \mathrm{J\ K^{-1}}$$

$$\Delta S_{\mathrm{tot}} = \Delta S(\mathrm{system}) + \Delta S(\mathrm{sur}) = \boxed{0}$$

$$\Delta G = \Delta H - T\Delta S = 0 - (600\ \mathrm{K}) \times (19.1\ \mathrm{J\ K^{-1}}) = \boxed{-11.5\ \mathrm{kJ\ mol^{-1}}}$$

Step 2

$$q = \boxed{0}\ [\text{adiabatic}]$$

$$\Delta U = nC_{V,\mathrm{m}}\Delta T\ [2.16\mathrm{b}]$$
$$= (1.00\ \mathrm{mol}) \times (\tfrac{3}{2}) \times (8.314\ \mathrm{J\ K^{-1}\ mol^{-1}}) \times (300\ \mathrm{K} - 600\ \mathrm{K}) = \boxed{-3.74\ \mathrm{kJ}}$$

$$w = \Delta U = \boxed{-3.74\ \mathrm{kJ}}$$

$$\Delta H = \Delta U + \Delta(pV) = \Delta U + nR\Delta T$$
$$= (-3.74\text{ kJ}) + (1.00\text{ mol}) \times (8.314\text{ J K}^{-1}\text{mol}^{-1}) \times (-300\text{ K}) = \boxed{-6.23\text{ kJ}}$$

$\Delta S = \Delta S(\text{sur}) = \boxed{0}$ [reversible adiabatic process]

$\Delta S_{\text{tot}} = \boxed{0}$

$\Delta G = \Delta(H - TS) = \Delta H - S\Delta T$ [no change in entropy]

Although the change in entropy is known to be zero, the entropy itself is not known, so ΔG is $\boxed{\text{indeterminate}}$.

Step 3

These quantities may be calculated in the same manner as for *Step 1* or more easily as follows

$\Delta U = \Delta H = \boxed{0}$ [isothermal]

$$\eta_{\text{rev}} = 1 - \frac{T_{\text{c}}}{T_{\text{h}}}\,[3.10] = 1 - \frac{300\text{ K}}{600\text{ K}} = 0.500 = 1 + \frac{q_{\text{c}}}{q_{\text{h}}}\,[3.9]$$

$$q_{\text{c}} = -0.500q_{\text{h}} = -(0.500) \times (11.5\text{ kJ}) = -5.74\text{ kJ}$$

$q_{\text{c}} = \boxed{-5.74\text{ kJ}}$ $\quad w = -q_{\text{c}} = \boxed{5.74\text{ kJ}}$

$$\Delta S = \frac{q_{\text{rev}}}{T}\,[\text{isothermal}] = \frac{-5.74 \times 10^3\text{ J}}{300\text{ K}} = \boxed{-19.1\text{ J K}^{-1}}$$

$\Delta S(\text{sur}) = -\Delta S(\text{system}) = +19.1\text{ J K}^{-1}$

$\Delta S_{\text{tot}} = \boxed{0}$

$\Delta G = \Delta H - T\Delta S = 0 - (300\text{ K}) \times (-19.1\text{ J K}^{-1}) = \boxed{+5.73\text{ kJ}}$

Step 4

ΔU and ΔH are the negative of their values in *Step 2*. (Initial and final temperatures reversed.)

$\Delta U = \boxed{+3.74\text{ kJ}}$, $\quad \Delta H = \boxed{+6.23\text{ kJ}}$, $\quad q = \boxed{0}$ [adiabatic]

$w = \Delta U = \boxed{+3.74\text{ kJ}}$

$\Delta S = \Delta S(\text{sur}) = \boxed{0}$ [reversible adiabatic process]

$\Delta S_{\text{tot}} = \boxed{0}$

Again $\Delta G = \Delta(H - TS) = \Delta H - S\Delta T$ [no change in entropy]

but S is not known, so ΔG is $\boxed{\text{indeterminate}}$.

Cycle

$\Delta U = \Delta H = \Delta S = \Delta G = \boxed{0}$ $\quad$ [Δ(state function) = 0 for any cycle]

$\Delta S(\text{sur}) = 0$ [all reversible processes]

$\Delta S_{\text{tot}} = \boxed{0}$

$q(\text{cycle}) = (11.5 - 5.74)\text{ kJ} = \boxed{5.8\text{ kJ}}$ $\quad w(\text{cycle}) = -q(\text{cycle}) = \boxed{-5.8\text{ kJ}}$

P3.7 $S_m^\ominus(T) = S_m^\ominus(298\text{ K}) + \Delta S$

$$\Delta S = \int_{T_1}^{T_2} C_{p,m}\frac{\mathrm{d}T}{T} = \int_{T_1}^{T_2}\left(\frac{a}{T} + b + \frac{c}{T^3}\right)\mathrm{d}T = a\ln\frac{T_2}{T_1} + b(T_2 - T_1) - \frac{1}{2}c\left(\frac{1}{T_2^2} - \frac{1}{T_1^2}\right)$$

(a) $$S_m^\ominus(373\text{ K}) = (192.45\text{ J K}^{-1}\text{ mol}^{-1}) + (29.75\text{ J K}^{-1}\text{ mol}^{-1}) \times \ln\left(\frac{373}{298}\right)$$

$$+ (25.10\times10^{-3}\text{ J K}^{-2}\text{ mol}^{-1}) \times (75.0\text{ K})$$

$$+ (\tfrac{1}{2}) \times (1.55\times10^{5}\text{ J K}^{-1}\text{ mol}^{-1}) \times \left(\frac{1}{(373.15)^2} - \frac{1}{(298.15)^2}\right)$$

$$= \boxed{200.7\text{ J K}^{-1}\text{ mol}^{-1}}$$

(b) $$S_m^\ominus(773\text{ K}) = (192.45\text{ J K}^{-1}\text{ mol}^{-1}) + (29.75\text{ J K}^{-1}\text{ mol}^{-1}) \times \ln\left(\frac{773}{298}\right)$$

$$+ (25.10\times10^{-3}\text{ J K}^{-2}\text{ mol}^{-1}) \times (475\text{ K})$$

$$+ (\tfrac{1}{2}) \times (1.55\times10^{5}\text{ J K}^{-1}\text{ mol}^{-1}) \times \left(\frac{1}{773^2} - \frac{1}{298^2}\right)$$

$$= \boxed{232.0\text{ J K}^{-1}\text{ mol}^{-1}}$$

P3.9 $\Delta S = \boxed{nC_{p,m}\ln\frac{T_f}{T_h} + nC_{p,m}\ln\frac{T_f}{T_c}}$ [3.23] [T_f is the final temperature, $T_f = \tfrac{1}{2}(T_h + T_c)$]

In the present case, $T_f = \tfrac{1}{2}(500\text{ K} + 250\text{ K}) = 375\text{ K}$

$$\Delta S = nC_{p,m}\ln\frac{T_f^2}{T_hT_c} = nC_{p,m}\ln\frac{(T_h + T_c)^2}{4T_hT_c} = \left(\frac{500\text{ g}}{63.54\text{ g mol}^{-1}}\right) \times (24.4\text{ J K}^{-1}\text{ mol}^{-1})$$

$$\times \ln\left(\frac{375^2}{500\times250}\right) = \boxed{+22.6\text{ J K}^{-1}}$$

P3.11 $$S_m(T) = S_m(0) + \int_0^T \frac{C_{p,m}\,\mathrm{d}T}{T}\ [3.22]$$

From the data, draw up the following table

T/K	10	15	20	25	30	50
$\frac{C_{p,m}}{T}$/(J K^{-2} mol^{-1})	0.28	0.47	0.540	0.564	0.550	0.428
T/K	70	100	150	200	250	298
$\frac{C_{p,m}}{T}$/(J K^{-2} mol^{-1})	0.333	0.245	0.169	0.129	0.105	0.089

Plot $C_{p,m}/T$ against T (Figure 3.1). This has been done on two scales. The region 0–10 K has been constructed using $C_{p,m} = aT^3$, fitted to the point at $T = 10$ K, at which $C_{p,m} = 2.8$ J K^{-1} mol^{-1}, so $a = 2.8\times10^{-3}$ J K^{-4} mol^{-1}. The area can be determined (primitively) by counting squares. Area A = 38.28 J K^{-1} mol^{-1}. Area B up to 0°C = 25.60 J K^{-1} mol^{-1}; area B up to 25°C = 27.80 J K^{-1} mol^{-1}. Hence,

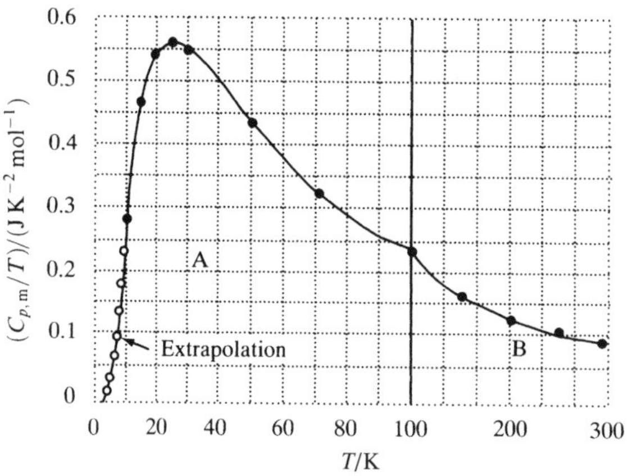

Figure 3.1

(a) $S_m(273\text{ K}) = S_m(0) + \boxed{63.88\text{ J K}^{-1}\text{ mol}^{-1}}$ (b) $S_m(298\text{ K}) = S_m(0) + \boxed{66.08\text{ J K}^{-1}\text{ mol}^{-1}}$

P3.13 $S_m(T) = S_m(0) + \int_0^T \frac{C_{p,m}\,dT}{T}$ [3.22]

Perform a graphical integration by plotting $C_{p,m}/T$ against T and determining the area under the curve. Draw up the following table. (The last two columns come from determining areas under the curves described below.)

T/K	$C_{p,m}$ / J K^{-1} mol^{-1}	$C_{p,m}/T$ / J K^{-2} mol^{-1}	$(S^\ominus_m - S^\ominus_m(0))$ / J K^{-1} mol^{-1}	$(H^\ominus_m - H^\ominus_m(0))$ / kJ mol^{-1}
0.00	0.00	0.00	0.00	0.00
10.00	2.09	0.21	0.80	0.01
20.00	14.43	0.72	5.61	0.09
30.00	36.44	1.21	15.60	0.34
40.00	62.55	1.56	29.83	0.85
50.00	87.03	1.74	46.56	1.61
60.00	111.00	1.85	64.62	2.62
70.00	131.40	1.88	83.29	3.84
80.00	149.40	1.87	102.07	5.26
90.00	165.30	1.84	120.60	6.84
100.00	179.60	1.80	138.72	8.57
110.00	192.80	1.75	156.42	10.44
150.00	237.60	1.58	222.91	19.09
160.00	247.30	1.55	238.54	21.52
170.00	256.50	1.51	253.79	24.05
180.00	265.10	1.47	268.68	26.66
190.00	273.00	1.44	283.21	29.35
200.00	280.30	1.40	297.38	32.13

Plot $C_{p,\mathrm{m}}/T$ against T (Figure 3.2(a)). Extrapolate to $T = 0$ using $C_{p,\mathrm{m}} = aT^3$ fitted to the point at $T = 10$ K, which gives $a = 2.09$ mJ K^{-2} mol^{-1}. Determine the area under the graph up to each T and plot S_m against T (Figure 3.2(b)).

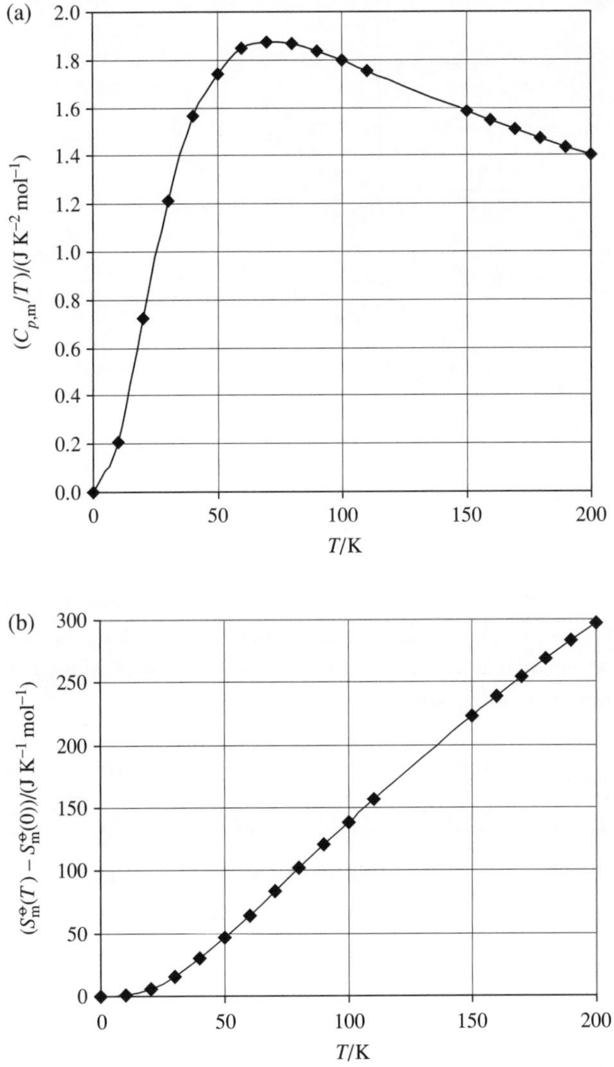

Figure 3.2

The molar enthalpy is determined in a similar manner from a plot of $C_{p,\mathrm{m}}$ against T by determining the area under the curve (Figure 3.3).

$$H^{\ominus}_{\mathrm{m}}(200\text{ K}) - H^{\ominus}_{\mathrm{m}}(0) = \int_0^{200\text{ K}} C_{p,\mathrm{m}}\,\mathrm{d}T = \boxed{32.1\text{ kJ mol}^{-1}}$$

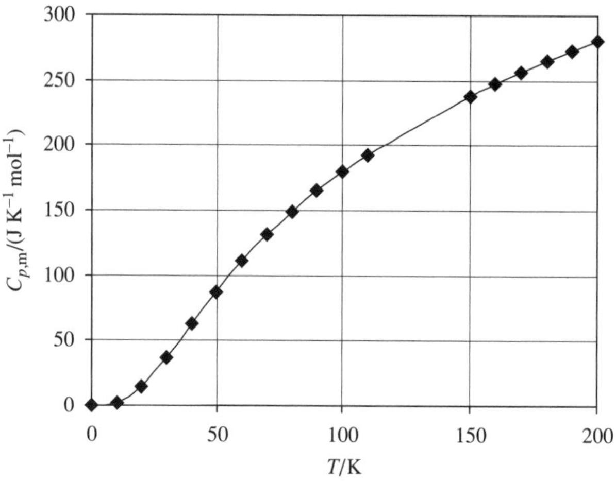

Figure 3.3

P3.15 The entropy at 200 K is calculated from

$$S^{\ominus}_{m}(200\ K) = S^{\ominus}_{m}(100\ K) + \int_{100\ K}^{200\ K} \frac{C_{p,m}\,dT}{T}$$

The integrand may be evaluated at each of the data points; the transformed data appear below. The numerical integration can be carried out by a standard procedure such as the trapezoid rule (taking the integral within any interval as the mean value of the integrand times the length of the interval). Programs for performing this integration are readily available for personal computers. Many graphing calculators will also perform this numerical integration.

T/K	100	120	140	150	160	180	200
$C_{p,m}/(J\ K^{-1}\ mol^{-1})$	23.00	23.74	24.25	24.44	24.61	24.89	25.11
$\frac{C_{p,m}}{T}\Big/(J\ K^{-2}\ mol^{-1})$	0.230	0.1978	0.1732	0.1629	0.1538	0.1383	0.1256

Integration by the trapezoid rule yields

$$S^{\ominus}_{m}(200\ K) = (29.79 + 16.81)\ J\ K^{-1}\ mol^{-1} = \boxed{46.60\ J\ K^{-1}\ mol^{-1}}$$

Taking $C_{p,m}$ constant yields

$$S^{\ominus}_{m}(200\ K) = S^{\ominus}_{m}(100\ K) + C_{p,m}\ln(200\ K/100\ K)$$
$$= [29.79 + 24.44\ln(200/100\ K)]\ J\ K^{-1}\ mol^{-1} = \boxed{46.73\ J\ K^{-1}\ mol^{-1}}$$

The difference is slight.

P3.17 The Gibbs–Helmholtz equation [3.55] may be recast into an analogous equation involving ΔG and ΔH, since

$$\left(\frac{\partial \Delta G}{\partial T}\right)_p = \left(\frac{\partial G_f}{\partial T}\right)_p - \left(\frac{\partial G_i}{\partial T}\right)_p$$

and $\Delta H = H_f - H_i$

Thus, $\left(\dfrac{\partial}{\partial T}\dfrac{\Delta_r G^\ominus}{T}\right)_p = -\dfrac{\Delta_r H^\ominus}{T^2}$

$$\mathrm{d}\left(\frac{\Delta_r G^\ominus}{T}\right) = \left(\frac{\partial}{\partial T}\frac{\Delta_r G^\ominus}{T}\right)_p \mathrm{d}T \text{ [constant pressure]} = -\frac{\Delta_r H^\ominus}{T^2}\mathrm{d}T$$

$$\Delta\left(\frac{\Delta_r G^\ominus}{T}\right) = -\int_{T_c}^{T}\frac{\Delta_r H^\ominus \mathrm{d}T}{T^2} \approx -\Delta_r H^\ominus \int_{T_c}^{T}\frac{\mathrm{d}T}{T^2} = \Delta_r H^\ominus\left(\frac{1}{T}-\frac{1}{T_c}\right) \quad [\Delta_r H^\ominus \text{ assumed constant}]$$

Therefore, $\dfrac{\Delta_r G^\ominus(T)}{T} - \dfrac{\Delta_r G^\ominus(T_c)}{T_c} \approx \Delta_r H^\ominus\left(\dfrac{1}{T}-\dfrac{1}{T_c}\right)$

and so $\Delta_r G^\ominus(T) = \dfrac{T}{T_c}\Delta_r G^\ominus(T_c) + \left(1-\dfrac{T}{T_c}\right)\Delta_r H^\ominus(T_c)$

$$= \tau\Delta_r G^\ominus(T_c) + (1-\tau)\Delta_r H^\ominus(T_c), \quad \text{where} \quad \tau = \frac{T}{T_c}$$

For the reaction

$N_2(g) + 3\,H_2(g) \rightarrow 2\,NH_3(g) \quad \Delta_r G^\ominus = 2\,\Delta_f G^\ominus(NH_3, g)$

(a) At 500 K, $\tau = \dfrac{500}{298} = 1.67\overline{8}$,

so $\Delta_r G^\ominus(500\text{ K}) = \{(1.67\overline{8}) \times 2 \times (-16.45) + (1-1.67\overline{8}) \times 2 \times (-46.11)\}\text{ kJ mol}^{-1}$

$\boxed{= 7\text{ kJ mol}^{-1}}$

(b) At 1000 K, $\tau = \dfrac{1000}{298} = 3.35\overline{6}$,

so $\Delta_r G^\ominus(1000\text{ K}) = \{(3.35\overline{6}) \times 2 \times (-16.45) + (1-3.35\overline{6}) \times 2 \times (-46.11)\}\text{ kJ mol}^{-1}$

$\boxed{= +107\text{ kJ mol}^{-1}}$

Solutions to theoretical problems

P3.19 The isotherms correspond to T = constant, and the reversibly traversed adiabats correspond to S = constant. Thus, we can represent the cycle as in Figure 3.4.

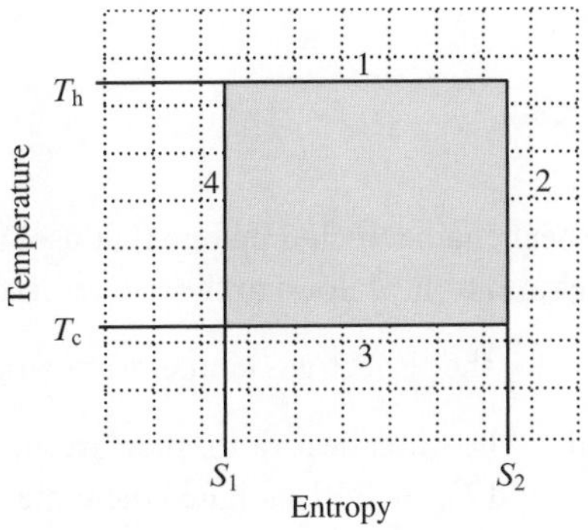

Figure 3.4

In this figure, paths 1, 2, 3, and 4 correspond to the four stages of the Carnot cycle listed in the text following eqn 3.6. The area within the rectangle is

$$\text{area} = \int_{\text{cyclic}} T\,\mathrm{d}S = (T_\mathrm{h} - T_\mathrm{c}) \times (S_2 - S_1) = (T_\mathrm{h} - T_\mathrm{c})\Delta S = (T_\mathrm{h} - T_\mathrm{c})nR\ln\frac{V_\mathrm{B}}{V_\mathrm{A}}$$

(isothermal expansion from V_A to V_B, stage 1).

But, $w(\text{cycle}) = \eta q_\mathrm{h} = \left(\dfrac{T_\mathrm{h} - T_\mathrm{c}}{T_\mathrm{h}}\right) nRT_\mathrm{h}\ln\dfrac{V_\mathrm{B}}{V_\mathrm{A}}$ [text Figure 3.6] $= nR(T_\mathrm{h} - T_\mathrm{c})\ln\dfrac{V_\mathrm{B}}{V_\mathrm{A}}$

Therefore, the area is equal to the net work done in the cycle.

P3.21 The thermodynamic temperature scale defines a temperature T^a (where the superscript a is used to distinguish this absolute thermodynamic temperature from the perfect gas temperature) in terms of the reversible heat flows of a heat engine operating between it and an arbitrary fixed temperature T_h^a (eqn 3.11)

$$T^\mathrm{a} = (1 - \eta_\mathrm{rev})T_\mathrm{h}^\mathrm{a}$$

where the efficiency of a heat engine is given in terms of work and heat flows:

$$\eta = \frac{|w|}{|q_\mathrm{h}|}\,[3.8] = 1 - \frac{|q_\mathrm{c}|}{|q_\mathrm{h}|}\,[3.9]$$

The problem asks to show that the thermodynamic and perfect gas temperatures differ by at most a constant numerical factor. That amounts to showing that

$$\frac{T_\mathrm{c}^\mathrm{a}}{T_\mathrm{h}^\mathrm{a}} = \frac{T_\mathrm{c}^\mathrm{g}}{T_\mathrm{h}^\mathrm{g}}$$

where the superscipt g indicates the perfect gas temperature defined by the perfect gas law. The subscripts c and h represent two reservoirs between which one might run a heat engine and whose temperatures one might characterize using either temperature scale. Justification 3.1 relates the ratio of two perfect-gas temperatures to reversible isothermal heat flows in a Carnot cycle run between the two temperatures.

$$\frac{q_\mathrm{h}}{q_\mathrm{c}} = -\frac{T_\mathrm{h}^\mathrm{g}}{T_\mathrm{c}^\mathrm{g}}\ [3.7,\ \textit{Justification}\ 3.1] \quad \text{so} \quad \frac{T_\mathrm{c}^\mathrm{g}}{T_\mathrm{h}^\mathrm{g}} = -\frac{q_\mathrm{c}}{q_\mathrm{h}}$$

The corresponding ratio of thermodynamic temperatures is:

$$\frac{T_\mathrm{c}^\mathrm{a}}{T_\mathrm{h}^\mathrm{a}} = 1 - \eta_\mathrm{rev} = \left(1 - \frac{|w|}{|q_\mathrm{h}|}\right)_\mathrm{rev} = -\left(\frac{q_\mathrm{c}}{q_\mathrm{h}}\right)_\mathrm{rev}$$

As Section 3.2(c) shows, the efficiency of any reversible heat engine (including one that uses a perfect gas as a working fluid) is the same, and therefore the ratio of heat flows to the two reservoirs is the same. That is, the ratio $\dfrac{q_\mathrm{c}}{q_\mathrm{h}}$ is the same in the expression for the perfect-gas temperature ratio and the thermodynamic ratio; since the two ratios are equal to the same heat ratio, they are equal to each other. The constant numerical factor becomes 1 if T_h and T_h^a are both assigned the same value, say 273.16 at the triple point of water.

P3.23 $$\left(\frac{\partial S}{\partial V}\right)_T = \left(\frac{\partial p}{\partial T}\right)_V \text{ [Table 3.5]}$$

(a) For a van der Waals gas

$$p = \frac{nRT}{V - nb} - \frac{n^2 a}{V^2} = \frac{RT}{V_m - b} - \frac{a}{V_m^2}$$

Hence, $$\left(\frac{\partial S}{\partial V}\right)_T = \left(\frac{\partial p}{\partial T}\right)_V = \boxed{\frac{R}{V_m - b}}$$

(b) For a Dieterici gas

$$p = \frac{RT e^{-a/RTV_m}}{V_m - b}$$

$$\left(\frac{\partial S}{\partial V}\right)_T = \left(\frac{\partial p}{\partial T}\right)_V = \boxed{\frac{R\left(1 + \frac{a}{RV_m T}\right)e^{-a/RV_m T}}{V_m - b}}$$

For an isothermal expansion, $\Delta S = \int_{V_i}^{V_f} dS = \int_{V_i}^{V_f} \left(\frac{\partial S}{\partial V}\right)_T dV,$

so we can simply compare $\left(\frac{\partial S}{\partial V}\right)_T$ expressions for the three gases. For a perfect gas,

$$p = \frac{nRT}{V} = \frac{RT}{V_m} \quad \text{so} \quad \left(\frac{\partial S}{\partial V}\right)_T = \left(\frac{\partial p}{\partial T}\right)_V = \frac{R}{V_m}$$

$\left(\frac{\partial S}{\partial V}\right)_T$ is certainly greater for a van der Waals gas than for a perfect gas, for the denominator is smaller for the van der Waals gas. To compare the van der Waals gas to the Dieterici gas, we assume that both have the same parameter b. (That is reasonable, for b is an excluded volume in both equations of state.) In that case,

$$\left(\frac{\partial S}{\partial V}\right)_{T,\text{Die}} = \frac{R\left(1 + \frac{a}{RV_m T}\right)e^{-a/RV_m T}}{V_m - b} = \left(\frac{\partial S}{\partial V}\right)_{T,\text{vdW}} \left(1 + \frac{a}{RV_m T}\right)e^{-a/RV_m T}$$

Now notice that the additional factor in $\left(\frac{\partial S}{\partial V}\right)_{T,\text{Die}}$ has the form $(1 + x)e^{-x}$, where $x > 0$. This factor is always less than 1. Clearly $(1 + x)e^{-x} < 1$ for large x, for then the exponential dominates. But $(1 + x)e^{-x} < 1$ even for small x, as can be seen by using the power series expansion for the exponential: $(1 + x)(1 - x + x^2/2 + \cdots) = 1 - x^2/2 + \cdots$ So $\left(\frac{\partial S}{\partial V}\right)_{T,\text{Die}} < \left(\frac{\partial S}{\partial V}\right)_{T,\text{vdW}}$

To summarize, for isothermal expansions:

$$\boxed{\Delta S_{\text{vdW}} > \Delta S_{\text{Die}}} \quad \text{and} \quad \boxed{\Delta S_{\text{vdW}} > \Delta S_{\text{perfect}}}$$

The comparison between a perfect gas and a Dieterici gas depends on particular values of the constants a and b and on the physical conditions.

P3.25 $H \equiv U + pV$

$$\mathrm{d}H = \mathrm{d}U + p\mathrm{d}V + V\mathrm{d}p = T\mathrm{d}S - p\mathrm{d}V\,[3.46] + p\mathrm{d}V + V\mathrm{d}p = T\mathrm{d}S + V\mathrm{d}p$$

Since H is a state function, $\mathrm{d}H$ is exact, and it follows that

$$\left(\frac{\partial H}{\partial S}\right)_p = T \quad \text{and} \quad \boxed{\left(\frac{\partial V}{\partial S}\right)_p = \left(\frac{\partial T}{\partial p}\right)_S} \text{ applying eqn 3.49.}$$

Similarly, $A \equiv U - TS$

$$\mathrm{d}A = \mathrm{d}U - T\mathrm{d}S - S\mathrm{d}T = T\mathrm{d}S - p\mathrm{d}V\,[3.46] - T\mathrm{d}S - S\mathrm{d}T = -p\mathrm{d}V - S\mathrm{d}T$$

Since $\mathrm{d}A$ is exact,

$$\boxed{\left(\frac{\partial S}{\partial V}\right)_T = \left(\frac{\partial p}{\partial T}\right)_V} \text{ applying eqn 3.49.}$$

P3.27 $\left(\frac{\partial S}{\partial V}\right)_T = \left(\frac{\partial p}{\partial T}\right)_V$ [Maxwell relation]; $\left(\frac{\partial p}{\partial T}\right)_V = \left\{\frac{\partial}{\partial T}\left(\frac{nRT}{V}\right)\right\}_V = \frac{nR}{V}$

$$\mathrm{d}S = \left(\frac{\partial S}{\partial V}\right)_T \mathrm{d}V \text{ [constant temperature]} = nR\frac{\mathrm{d}V}{V} = nR\,\mathrm{d}\ln V$$

$$S = \int \mathrm{d}S = \int nR\,\mathrm{d}\ln V$$

$$S = nR\ln V + \text{constant} \quad \text{or} \quad S \propto R\ln V$$

P3.29 $\pi_T = T\left(\frac{\partial p}{\partial T}\right)_V - p$ [3.51]

$$p = \frac{RT}{V_\mathrm{m}} + \frac{BRT}{V_\mathrm{m}^2} \text{ [first two terms of the virial expansion, 1.19b]}$$

$$\left(\frac{\partial p}{\partial T}\right)_V = \frac{R}{V_\mathrm{m}} + \frac{BR}{V_\mathrm{m}^2} + \frac{RT}{V_\mathrm{m}^2}\left(\frac{\partial B}{\partial T}\right)_V = \frac{p}{T} + \frac{RT}{V_\mathrm{m}^2}\left(\frac{\partial B}{\partial T}\right)_V$$

Hence, $\pi_T = \frac{RT^2}{V_\mathrm{m}^2}\left(\frac{\partial B}{\partial T}\right)_V \approx \frac{RT^2\Delta B}{V_\mathrm{m}^2\Delta T}$

Since π_T represents a (usually) small deviation from perfect gas behaviour, we may approximate V_m.

$$V_\mathrm{m} \approx \frac{RT}{p} \quad \boxed{\pi_T \approx \frac{p^2}{R} \times \frac{\Delta B}{\Delta T}}$$

From the data $\Delta B = \{(-15.6) - (-28.0)\}\ \mathrm{cm^3\,mol^{-1}} = +12.4\ \mathrm{cm^3\,mol^{-1}}$

Hence,

(a) $\pi_T = \frac{(1.0\ \mathrm{atm})^2 \times (12.4 \times 10^{-3}\ \mathrm{dm^3\,mol^{-1}})}{(8.206 \times 10^{-2}\ \mathrm{dm^3\,atm\,K^{-1}\,mol^{-1}}) \times (50\ \mathrm{K})} = \boxed{3.0 \times 10^{-3}\ \mathrm{atm}}$

(b) $\pi_T \propto p^2$; so at $p = 10.0$ atm, $\pi_T = \boxed{0.30\ \mathrm{atm}}$

COMMENT. In (a) π_T is 0.3% of p; in (b) it is 3%. Hence, at these pressures the approximation for V_m is justified. At 100 atm it would not be.

Question. How would you obtain a reliable estimate of π_T for argon at 100 atm?

P3.31 $$\pi_T = T\left(\frac{\partial p}{\partial T}\right)_V - p \ [3.51]$$

$$p = \frac{nRT}{V-nb} \times e^{-an/RTV} \ [\text{Table 1.7}]$$

$$T\left(\frac{\partial p}{\partial T}\right)_V = \frac{nRT}{V-nb} \times e^{-an/RTV} + \frac{na}{RTV} \times \frac{nRT}{V-nb} \times e^{-an/RTV} = p + \frac{nap}{RTV}$$

Hence, $\boxed{\pi_T = \dfrac{nap}{RTV}}$

$\pi_T \to 0$ as $p \to 0$, $V \to \infty$, $a \to 0$, and $T \to \infty$. The fact that $\pi_T > 0$ (because $a > 0$) is consistent with a representing attractive contributions, since it implies that $\left(\dfrac{\partial U}{\partial V}\right)_T > 0$ and the internal energy rises as the gas expands (so decreasing the average attractive interactions).

P3.33 If $S = S(T,p)$

then $$\mathrm{d}S = \left(\frac{\partial S}{\partial T}\right)_p \mathrm{d}T + \left(\frac{\partial S}{\partial p}\right)_T \mathrm{d}p$$

$$T\,\mathrm{d}S = T\left(\frac{\partial S}{\partial T}\right)_p \mathrm{d}T + T\left(\frac{\partial S}{\partial p}\right)_T \mathrm{d}p$$

Use $$\left(\frac{\partial S}{\partial T}\right)_p = \left(\frac{\partial S}{\partial H}\right)_p \left(\frac{\partial H}{\partial T}\right)_p = \frac{1}{T} \times C_p \quad \left[\left(\frac{\partial H}{\partial S}\right)_p = T, \text{ Problem 3.25}\right]$$

$$\left(\frac{\partial S}{\partial p}\right)_T = -\left(\frac{\partial V}{\partial T}\right)_p \ [\text{Maxwell relationship}]$$

Hence, $$T\,\mathrm{d}S = C_p\,\mathrm{d}T - T\left(\frac{\partial V}{\partial T}\right)_p \mathrm{d}p = \boxed{C_p\,\mathrm{d}T - \alpha TV\,\mathrm{d}p}$$

For reversible, isothermal compression, $T\mathrm{d}S = \mathrm{d}q_{\text{rev}}$ and $\mathrm{d}T = 0$; hence

$$\mathrm{d}q_{\text{rev}} = -\alpha TV\,\mathrm{d}p$$

$$q_{\text{rev}} = \int_{p_i}^{p_f} -\alpha TV\,\mathrm{d}p = \boxed{-\alpha TV\Delta p} \ [\alpha \text{ and } V \text{ assumed constant}]$$

For mercury

$$q_{\text{rev}} = (-1.82 \times 10^{-4}\,\text{K}^{-1}) \times (273\,\text{K}) \times (1.00 \times 10^{-4}\,\text{m}^3) \times (1.0 \times 10^8\,\text{Pa})$$
$$= \boxed{-0.50\,\text{kJ}}$$

P3.35 $\ln \phi = \int_0^p \left(\frac{Z-1}{p}\right) \mathrm{d}p$ [3.63]

$$Z = 1 + \frac{B}{V_\mathrm{m}} + \frac{C}{V_\mathrm{m}^2} = 1 + B'p + C'p^2 + \cdots$$

with $B' = \frac{B}{RT}$, $C' = \frac{C - B^2}{R^2T^2}$ [Problem 1.18]

$$\frac{Z-1}{p} = B' + C'p + \cdots$$

Therefore, $\ln \phi = \int_0^p B'\mathrm{d}p + \int_0^p C'p\mathrm{d}p + \cdots = B'p + \tfrac{1}{2}C'p^2 + \cdots$

$$= \boxed{\frac{Bp}{RT} + \frac{(C-B^2)p^2}{2R^2T^2} + \cdots}$$

For argon, $\frac{Bp}{RT} = \frac{(-21.13 \times 10^{-3}\,\mathrm{dm^3\,mol^{-1}}) \times (1.00\,\mathrm{atm})}{(8.206 \times 10^{-2}\,\mathrm{dm^3\,atm\,K^{-1}\,mol^{-1}}) \times (100\,\mathrm{K})} = -2.58 \times 10^{-3}$

$$\frac{(C-B^2)p^2}{2R^2T^2} = \frac{\{(1.054 \times 10^{-3}\,\mathrm{dm^6\,mol^{-2}}) - (-21.13 \times 10^{-3}\,\mathrm{dm^3\,mol^{-1}})^2\} \times (1.00\,\mathrm{atm})^2}{(2) \times \{(8.206 \times 10^{-2}\,\mathrm{dm^3\,atm\,K^{-1}\,mol^{-1}}) \times (100\,\mathrm{K})\}^2}$$

$$= 4.51 \times 10^{-6}$$

Therefore, $\ln \phi = (-2.58 \times 10^{-3}) + (4.51 \times 10^{-6}) = -2.57 \times 10^{-3}$; $\phi = 0.9974$

Hence, $f = (1.00\,\mathrm{atm}) \times (0.9974) = \boxed{0.99\overline{74}\,\mathrm{atm}}$

Solutions to applications

P3.37 $w_{\mathrm{add,max}} = \Delta_\mathrm{r} G$ [3.41b]

$$\Delta_\mathrm{r} G^\ominus(37°\mathrm{C}) = \tau\,\Delta_\mathrm{r} G^\ominus(T_\mathrm{c}) + (1-\tau)\Delta_\mathrm{r} H^\ominus(T_\mathrm{c}) \quad \left[\text{Problem 3.17, } \tau = \frac{T}{T_\mathrm{c}}\right]$$

$$= \left(\frac{310\,\mathrm{K}}{298.15\,\mathrm{K}}\right) \times (-6333\,\mathrm{kJ\,mol^{-1}}) + \left(1 - \frac{310\,\mathrm{K}}{298.15\,\mathrm{K}}\right) \times (-5797\,\mathrm{kJ\,mol^{-1}})$$

$$= -6354\,\mathrm{kJ\,mol^{-1}}$$

The difference is

$\Delta_\mathrm{r} G^\ominus(37°\mathrm{C}) - \Delta_\mathrm{r} G^\ominus(T_\mathrm{c}) = \{-6354 - (-6333)\}\,\mathrm{kJ\,mol^{-1}} = \boxed{-21\,\mathrm{kJ\,mol^{-1}}}$ Therefore, additional 21 kJ mol^{-1} of non-expansion work may be done at the higher temperature.

COMMENT. As shown by Problem 3.16, increasing the temperature does not necessarily increase the maximum non-expansion work. The relative magnitude of $\Delta_\mathrm{r} G^\ominus$ and $\Delta_\mathrm{r} H^\ominus$ is the determining factor.

P3.39 The relative increase in water vapour in the atmosphere at constant relative humidity is the same as the relative increase in the equilibrium vapour pressure of water. Examination of the molar Gibbs function will help us estimate this increase. At equilibrium, the vapour and liquid have the same molar Gibbs function. So, at the current temperature

$$G_{\mathrm{m,liq}}(T_0) = G_{\mathrm{m,vap}}(T_0) \quad \text{so} \quad G^{\ominus}_{\mathrm{m,liq}}(T_0) = G^{\ominus}_{\mathrm{m,vap}}(T_0) + RT_0 \ln p_0,$$

where the subscript 0 refers to the current equilibrium and p is the pressure divided by the standard pressure. The Gibbs function changes with temperature as follows

$$(\partial G/\partial T) = -S \quad \text{so} \quad G^{\ominus}_{\mathrm{m,liq}}(T_1) = G^{\ominus}_{\mathrm{m,liq}}(T_0) - (\Delta T)S^{\ominus}_{\mathrm{liq}}$$

and similarly for the vapour. Thus, at the higher temperature

$$G^{\ominus}_{\mathrm{m,liq}}(T_0) - (\Delta T)S^{\ominus}_{\mathrm{liq}} = G^{\ominus}_{\mathrm{m,vap}}(T_0) - (\Delta T)S^{\ominus}_{\mathrm{vap}} + R(T_0 + \Delta T)\ln p$$

Solving both of these expressions for $G^{\ominus}_{\mathrm{m,liq}}(T_0) - G^{\ominus}_{\mathrm{m,vap}}(T_0)$ and equating them leads to

$$(\Delta T)(S^{\ominus}_{\mathrm{liq}} - S^{\ominus}_{\mathrm{vap}}) + R(T_0 + \Delta T)\ln p = RT_0 \ln p_0$$

Isolating p leads to

$$\ln p = \frac{(\Delta T)(S^{\ominus}_{\mathrm{vap}} - S^{\ominus}_{\mathrm{liq}})}{R(T_0 + \Delta T)} + \frac{T_0 \ln p_0}{T_0 + \Delta T}$$

$$p = \exp\left(\frac{(\Delta T)(S^{\ominus}_{\mathrm{vap}} - S^{\ominus}_{\mathrm{liq}})}{R(T_0 + \Delta T)}\right) p_0^{(T_0/(T_0+\Delta T))}$$

So $$p = \exp\left(\frac{(2.0\ \mathrm{K}) \times (188.83 - 69.91)\ \mathrm{J\ mol^{-1}\ K^{-1}}}{(8.3145\ \mathrm{J\ mol^{-1}\ K^{-1}}) \times (290 + 2.0)\ \mathrm{K}}\right) \times (0.0189)^{(290\ \mathrm{K}/(290+2.0)\ \mathrm{K})},$$

$p = 0.0214$ bar, which represents a 13% increase.

P3.41 The change in the Helmholtz energy equals the maximum work associated with stretching the polymer. Then

$$\mathrm{d}w_{\mathrm{max}} = \mathrm{d}A = -f\,\mathrm{d}l$$

For stretching at constant T

$$f = -\left(\frac{\partial A}{\partial l}\right)_T = -\left(\frac{\partial U}{\partial l}\right)_T + T\left(\frac{\partial S}{\partial l}\right)_T$$

assuming that $(\partial U/\partial l)_T = 0$ (valid for rubbers)

$$f = T\left(\frac{\partial S}{\partial l}\right)_T = T\left(\frac{\partial}{\partial l}\right)_T \left\{-\frac{3kl^2}{2Na^2} + C\right\} = T\left\{-\frac{3kl}{Na^2}\right\} = -\left(\frac{3kT}{Na^2}\right)l$$

This tensile force has the Hooke's law form $f = -k_{\mathrm{f}}l$ with $k_{\mathrm{f}} = 3kT/Na^2$.

P3.43 The Otto cycle is represented in Figure 3.5. Assume one mole of air.

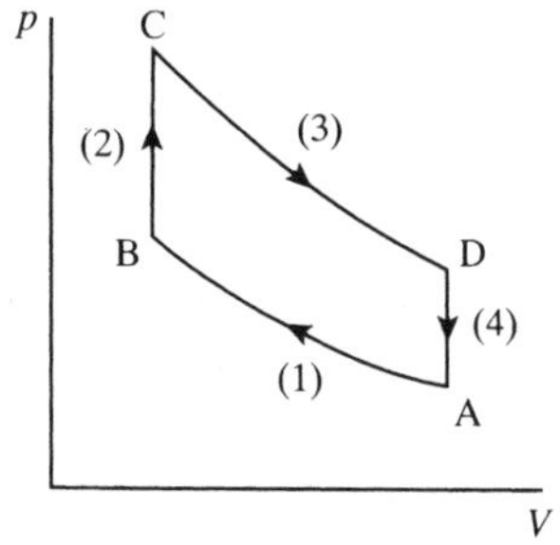

Figure 3.5

$$\eta = \frac{|w|_{\text{cycle}}}{|q_2|} \ [3.8]$$

$$w_{\text{cycle}} = w_1 + w_3 = \Delta U_1 + \Delta U_3 \ [q_1 = q_3 = 0] = C_V(T_B - T_A) + C_V(T_D - T_C) \ [2.27]$$

$$q_2 = \Delta U_2 = C_V(T_C - T_B)$$

$$\eta = \frac{|T_B - T_A + T_D - T_C|}{|T_C - T_B|} = 1 - \left(\frac{T_D - T_A}{T_C - T_B}\right)$$

We know that

$$\frac{T_A}{T_B} = \left(\frac{V_B}{V_A}\right)^{1/c} \quad \text{and} \quad \frac{T_D}{T_C} = \left(\frac{V_C}{V_D}\right)^{1/c} \ [2.28a]$$

Since $V_B = V_C$ and $V_A = V_D$, $\frac{T_A}{T_B} = \frac{T_D}{T_C}$, or $T_D = \frac{T_A T_C}{T_B}$.

Then, $\eta = 1 - \dfrac{\frac{T_A T_C}{T_B} - T_A}{T_C - T_B} = 1 - \dfrac{T_A}{T_B}$ or $\boxed{\eta = 1 - \left(\frac{V_B}{V_A}\right)^{1/c}}$.

Given that $C_{p,m} = \frac{7}{2}R$, we have $C_{V,m} = \frac{5}{2}R$ [2.26] and $c = \frac{2}{5}$.

For $\frac{V_A}{V_B} = 10$, $\eta = 1 - \left(\frac{1}{10}\right)^{2/5} = \boxed{0.47}$

$$\Delta S_1 = \Delta S_3 = \Delta S_{\text{sur},1} = \Delta S_{\text{sur},3} = \boxed{0} \quad \text{[adiabatic reversible steps]}$$

$$\Delta S_2 = C_{V,m} \ln\left(\frac{T_C}{T_B}\right)$$

At constant volume $\left(\frac{T_C}{T_B}\right) = \left(\frac{p_C}{p_B}\right) = 5.0$

$$\Delta S_2 = (\tfrac{5}{2}) \times (8.314\ \mathrm{J\ K^{-1}\ mol^{-1}}) \times (\ln 5.0) = \boxed{+33\ \mathrm{J\ K^{-1}}}$$

$$\Delta S_{\mathrm{sur},2} = -\Delta S_2 = \boxed{-33\ \mathrm{J\ K^{-1}}}$$

$$\Delta S_4 = -\Delta S_2 = \boxed{-33\ \mathrm{J\ K^{-1}}} \quad \left[\frac{T_\mathrm{C}}{T_\mathrm{D}} = \frac{T_\mathrm{B}}{T_\mathrm{A}}\right]$$

$$\Delta S_{\mathrm{sur},4} = -\Delta S_4 = \boxed{+33\ \mathrm{J\ K^{-1}}}$$

P3.45 In case (a), the electric heater converts 1.00 kJ of electrical energy into heat, providing $\boxed{1.00\ \mathrm{kJ}}$ of energy as heat to the room. (The second law places no restriction on the complete conversion of work to heat—only on the reverse process.) In case (b), we want to find the heat deposited in the room $|q_\mathrm{h}|$:

$$|q_\mathrm{h}| = |q_\mathrm{c}| + |w|, \quad \text{where} \quad \frac{|q_\mathrm{c}|}{|w|} = c = \frac{T_\mathrm{c}}{T_\mathrm{h} - T_\mathrm{c}} \quad [\textit{Impact}\ \mathrm{I3.1}]$$

so $$|q_\mathrm{c}| = \frac{|w|\,T_\mathrm{c}}{T_\mathrm{h} - T_\mathrm{c}} = \frac{1.00\ \mathrm{kJ} \times 260\ \mathrm{K}}{(295 - 260)\ \mathrm{K}} = 7.4\ \mathrm{kJ}$$

The heat transferred to the room is $|q_\mathrm{h}| = |q_\mathrm{c}| + |w| = 7.4\ \mathrm{kJ} + 1.00\ \mathrm{kJ} = \boxed{8.4\ \mathrm{kJ}}$. Most of the thermal energy the heat pump deposits into the room comes from outdoors. Difficult as it is to believe on a cold winter day, the intensity of thermal energy (that is, the absolute temperature) outdoors is a substantial fraction of that indoors. The work put into the heat pump is not simply converted to heat, but is 'leveraged' to transfer additional heat from outdoors.

4 Physical transformations of pure substances

Answers to discussion questions

D4.1 Chemical potential is the single function that governs phase stability. The phase whose chemical potential is least under a set of given conditions is the most stable. Conditions under which two or more phases have equal chemical potentials are conditions under which those phases are in equilibrium. Understanding how chemical potential varies with physical conditions such as temperature, pressure, and (in later chapters) composition allows one to compute chemical potentials for various phases and to map out the conditions for stability of those phases and for equilibrium between them.

D4.3 A differential scanning calorimeter (DSC) can be used to identify phase transitions, whether first-order, second-order, or lambda transitions. (See Section 4.6 for classification of phase transitions.) DSC measures constant-pressure heat capacity (C_p) as a function of temperature by electrically heating a sample and comparing it to a reference sample. (See *Impact I2.1* for more information on DSC.) Phase transitions are characterized by discontinuities in C_p as a function of temperature. Familiar first-order phase transitions such as fusion and evaporation exhibit latent enthalpies—non-zero changes in enthalpy at constant temperature. Thus, these phase changes have infinite spikes in C_p. Second-order transitions have discontinuities in C_p, albeit not infinite ones. Either way, such changes in C_p will stand out against an otherwise constant or slowly changing plot of C_p vs. temperature.

D4.5 A supercritical fluid extractor consists of a pump to pressurize the solvent (e.g. CO_2), an oven with extraction vessel, and a trapping vessel. Extractions are performed dynamically or statically. Supercritical fluid flows continuously through the sample within the extraction vessel when operating in dynamic mode. Analytes extracted into the fluid are released through a pressure-maintaining restrictor into a trapping vessel. In static mode the supercritical fluid circulates repetitively through the extraction vessel until being released into the trapping vessel after a period of time. Supercritical carbon dioxide volatilizes when decompression occurs on release into the trapping vessel.

Advantages	Disadvantages	Current uses
Dissolving power of SCF can be adjusted with selection of T and p	Elevated pressures are required and the necessary apparatus expensive	Extraction of caffeine, fatty acids, spices, aromas, flavours, and biological materials from natural sources
Select SCFs are inexpensive and non-toxic. They reduce pollution	Cost may prohibit large-scale applications	Extraction of toxic salts (with a suitable chelation agent) and organics from contaminated water
Thermally unstable analytes may be extracted at low temperature	Modifiers like methanol (1–10%) may be required to increase solvent polarity	Extraction of herbicides from soil
The volatility of $scCO_2$ makes it easy to isolate analyte	$scCO_2$ is toxic to whole cells in biological applications (CO_2 is not toxic to the environment)	scH_2O oxidation of toxic, intractable organic waste during water treatment
SCFs have high diffusion rates, low viscosity, and low surface tension		Synthetic chemistry, polymer synthesis and crystallization, textile processing
O_2 and H_2 are completely miscible with $scCO_2$. This reduces multi-phase reaction problems		Heterogeneous catalysis for green chemistry processes

SCF, supercritical fluid; sc, supercritical.

Solutions to exercises

E4.1(a) The phase rule [eqn 4.1] relates the number of phases (P), components (C), and degrees of freedom (F) of a thermodynamic system:

$$F = C - P + 2$$

Restricting to pure substances ($C = 1$) and rearranging for phases gives

$$P = 3 - F$$

Areas in the phase diagram have two degrees of freedom; one can vary pressure and temperature independently (within limits) and stay within the area. Thus, $F = 2$ and $P = 1$ in areas. Lines have one degree of freedom; one can vary pressure or temperature, but to stay on the line the value of the other is determined by the line. Thus, $F = 1$ and $P = 2$ on lines. Points on the phase diagram have zero degrees of freedom; one can vary neither pressure nor temperature and on a given point. Thus, $F = 0$ and $P = 3$ on points.

(a) is in an area, so there is a single phase. (b) and (d) are on lines, so there are two phases present.

(c) is a point, so there are three phases present.

E4.2(a) For pure substances (one-component systems), the chemical potential is the molar Gibbs energy:

$$dG = (\mu_2 - \mu_1)dn$$

so $\Delta G = (\mu_2 - \mu_1)n = (+7.1\text{ kJ mol}^{-1})(0.10 \times 10^{-3}\text{ mol}) = +0.71 \times 10^{-3}\text{ kJ} = \boxed{0.71\text{ J}}$.

E4.3(a) The difference between the definition of normal and standard transition temperatures is the pressure at which the transition takes place: normal refers to exactly 1 atm (101325 Pa), while standard refers to exactly 1 bar (10^5 Pa). At the standard melting temperature and pressure, the liquid and solid phases are in equilibrium, so their chemical potentials are equal:

$$\mu_{\text{liquid}}(T_{\text{std}}, p_{\text{std}}) = \mu_{\text{solid}}(T_{\text{std}}, p_{\text{std}})$$

The same can be said at the normal melting temperature and pressure:

$$\mu_{\text{liquid}}(T_{\text{norm}}, p_{\text{norm}}) = \mu_{\text{solid}}(T_{\text{norm}}, p_{\text{norm}})$$

Equations 4.2 and 4.3 show how the chemical potential changes with temperature and pressure, so for small changes we can write

$$d\mu = \left(\frac{\partial \mu}{\partial T}\right)_p dT + \left(\frac{\partial \mu}{\partial p}\right)_T dp = -S_m dT + V_m dp$$

Assuming that the differences between standard and normal melting point are small enough, we can equate the differences in the chemical potentials of the two phases:

$$\Delta\mu_{\text{solid}} = -S_{\text{m,solid}}\Delta T + V_{\text{m,solid}}\Delta p = -S_{\text{m,liquid}}\Delta T + V_{\text{m,liquid}}\Delta p = \Delta\mu_{\text{liquid}},$$

where Δp is defined as $p_{\text{norm}} - p_{\text{std}}$. Rearrange to isolate ΔT:

$$(S_{\text{m,liquid}} - S_{\text{m,solid}})\Delta T = (V_{\text{m,liquid}} - V_{\text{m,solid}})\Delta p,$$

$$(\Delta_{\text{fus}}S)\Delta T = (V_{\text{m,liquid}} - V_{\text{m,solid}})\Delta p \quad \text{so} \quad \Delta T = \frac{(V_{\text{m,liquid}} - V_{\text{m,solid}})\Delta p}{\Delta_{\text{fus}}S}$$

The molar volume of each phase is the molar mass divided by the density. We need to look up the density of ice and of liquid water near the melting temperature. We also need to find $\Delta_{\text{fus}}S$ or to use eqn 3.20.

$$\Delta T = \left(\frac{1}{\rho_{\text{liquid}}} - \frac{1}{\rho_{\text{solid}}}\right)\frac{M\Delta p}{\Delta_{\text{fus}}S} = \left(\frac{1}{\rho_{\text{liquid}}} - \frac{1}{\rho_{\text{solid}}}\right)\frac{MT_f\Delta p}{\Delta_{\text{fus}}H}$$

$$= \left(\frac{1\text{ cm}^3}{1.000\text{ g}} - \frac{1\text{ cm}^3}{0.915\text{ g}}\right)\frac{(18.02\text{ g mol}^{-1})(273.15\text{ K})(1325\text{ Pa})}{6008\text{ J mol}^{-1}} \times \frac{1\text{ m}^3}{10^6\text{ cm}^3}$$

$$= \boxed{-1.0 \times 10^{-4}\text{ K}}$$

That is, the normal melting temperature is 0.1 mK lower than the standard melting temperature.

E4.4(a) Use the phase rule [eqn 4.1]

$$F = C - P + 2$$

to solve for the number of phases:

$$P = C - F + 2 = 2 - F + 2 = 4 - F \le \boxed{4}$$

The maximum number of phases in equilibrium occurs when the number of degrees of freedom is at a minimum, namely zero; that number is four.

E4.5(a) Eqn 4.2 shows how the chemical potential changes with temperature

$$\mathrm{d}\mu = \left(\frac{\partial \mu}{\partial T}\right)_p \mathrm{d}T = -S_\mathrm{m}\mathrm{d}T$$

so $\Delta\mu = -\int S_\mathrm{m}\mathrm{d}T = -S_\mathrm{m}\Delta T = -69.91\ \mathrm{J\ K^{-1}\ mol^{-1}} \times (100 - 25)\ \mathrm{K}$

$$= \boxed{5.2\times10^3\ \mathrm{J\ mol^{-1}}} = \boxed{5.2\ \mathrm{kJ\ mol^{-1}}}$$

Note: We assumed that the entropy is constant over the temperature range.

E4.6(a) Eqn 4.3 shows how the chemical potential changes with pressure

$$\mathrm{d}\mu = \left(\frac{\partial \mu}{\partial p}\right)_T \mathrm{d}p = V_\mathrm{m}\mathrm{d}p = \frac{M}{\rho}\mathrm{d}p$$

so $$\Delta\mu = \int \frac{M}{\rho}\mathrm{d}p = \frac{M}{\rho}\Delta p = \frac{63.55\ \mathrm{g\ mol^{-1}}}{8.960\ \mathrm{g\ cm^{-3}}} \times (10\times10^6 - 100\times10^3)\,\mathrm{Pa} \times \frac{1\ \mathrm{m^3}}{10^6\ \mathrm{cm^3}} = \boxed{70\ \mathrm{J\ mol^{-1}}}$$

Note: we assumed that the sample is incompressible.

E4.7(a) The effect on vapour pressure of a change in applied external pressure on a liquid is given by eqn 4.4:

$$p = p^*\mathrm{e}^{V_\mathrm{m}(\mathrm{l})\Delta P/RT}.$$

For water, the molar volume at 20°C (based on density from the *CRC Handbook of Chemistry and Physics*) is

$$V_\mathrm{m} = \frac{M}{\rho} = \frac{18.02\ \mathrm{g\ mol^{-1}}}{0.9982\ \mathrm{g\ cm^{-3}}} = 18.05\ \mathrm{cm^3\ mol^{-1}}$$

so $$\frac{V_\mathrm{m}(\mathrm{l})\Delta P}{RT} = \frac{18.05\ \mathrm{cm^3\ mol^{-1}} \times (20\times10^6 - 1.0\times10^5)\ \mathrm{Pa}}{8.3145\ \mathrm{J\ K^{-1}\ mol^{-1}} \times 293\ \mathrm{K}} \times \frac{1\ \mathrm{m^3}}{10^6\ \mathrm{cm^3}} = 0.147$$

and $p = p^*\mathrm{e}^{V_\mathrm{m}(\mathrm{l})\Delta P/RT} = (2.34\ \mathrm{kPa})\mathrm{e}^{0.147} = \boxed{2.71\ \mathrm{kPa}}$.

E4.8(a) Use the Clapeyron equation [4.6]

$$\frac{\mathrm{d}p}{\mathrm{d}T} = \frac{\Delta_\mathrm{trs}S}{\Delta_\mathrm{trs}V}$$

Assume that $\Delta_\mathrm{fus}S$ and $\Delta_\mathrm{fus}T$ are independent of temperature:

$$\Delta_\mathrm{fus}S = \Delta_\mathrm{fus}V \times \left(\frac{\mathrm{d}p}{\mathrm{d}T}\right) \approx \Delta_\mathrm{fus}V \times \frac{\Delta p}{\Delta T}$$

$$\Delta_\mathrm{fus}S = [(163.3 - 161.0)\times10^{-6}\ \mathrm{m^3\ mol^{-1}}] \times \left(\frac{(100-1)\times(1.013\times10^5\ \mathrm{Pa})}{(351.26 - 350.75)\ \mathrm{K}}\right) = \boxed{+45.2\bar{3}\ \mathrm{J\ K^{-1}\ mol^{-1}}}$$

At the melting temperature

$$\Delta_\mathrm{fus}H = T_\mathrm{f}\Delta_\mathrm{fus}S = (350.75\ \mathrm{K}) \times (45.23\ \mathrm{J\ K^{-1}\ mol^{-1}}) = \boxed{+16\ \mathrm{kJ\ mol^{-1}}}$$

E4.9(a) On the assumption that the vapour is a perfect gas and that $\Delta_{\text{vap}}H$ is independent of temperature, we may write

$$p = p^*e^{-\chi}, \quad \chi = \left(\frac{\Delta_{\text{vap}}H}{R}\right) \times \left(\frac{1}{T} - \frac{1}{T^*}\right) [4.12], \quad \ln\frac{p^*}{p} = \chi$$

$$\frac{1}{T} = \frac{1}{T^*} + \frac{R}{\Delta_{\text{vap}}H}\ln\frac{p^*}{p}$$

$$= \frac{1}{297.25\text{ K}} + \left(\frac{8.3145\text{ J K}^{-1}\text{ mol}^{-1}}{28.7\times10^3\text{ J mol}^{-1}}\right)\ln\frac{53.3\text{ kPa}}{70.0\text{ kPa}} = 3.28\overline{5}\times10^{-3}\text{ K}^{-1}$$

Hence, $T = \boxed{304\text{ K}} = \boxed{31°\text{C}}$.

E4.10(a) Integrating the Clausius–Clapeyron equation [4.11] yields an expression for $\ln p$:

$$\int \text{d}\ln p = \int \frac{\Delta_{\text{vap}}H}{RT^2}\text{d}T$$

so $\quad \ln p = \text{constant} - \dfrac{\Delta_{\text{vap}}H}{RT}$

Therefore, $\Delta_{\text{vap}}H = (2501.8\text{ K}) \times R = (2501.8\text{ K}) \times (8.3145\text{ J K}^{-1}\text{ mol}^{-1}) = \boxed{+20.80\text{ kJ mol}^{-1}}$

E4.11(a) (a) The indefinitely integrated form of eqn 4.11 is used as in Exercise 4.10(a).

$$\ln p = \text{constant} - \frac{\Delta_{\text{vap}}H}{RT}, \quad \text{or} \quad \log p = \text{constant} - \frac{\Delta_{\text{vap}}H}{2.303\,RT}$$

Therefore, $\Delta_{\text{vap}}H = (2.303)\times(1780\text{ K})\times R = (2.303)\times(1780\text{ K})\times(8.3145\text{ J K}^{-1}\text{ mol}^{-1})$

$$= \boxed{+34.08\text{ kJ mol}^{-1}}$$

(b) The normal boiling point corresponds to $p = 1.000\text{ atm} = 760\text{ Torr}$,

so $\quad \log 760 = 7.960 - \dfrac{1780\text{ K}}{T_\text{b}}$

and $\quad T_\text{b} = \boxed{350.5\text{ K}}$.

E4.12(a) $\quad \Delta T \approx \dfrac{\Delta_{\text{fus}}V}{\Delta_{\text{fus}}S}\times\Delta p$ [4.6 and Exercise 4.8(a)]

$$\approx \frac{T_\text{f}\Delta_{\text{fus}}V}{\Delta_{\text{fus}}H}\times\Delta p = \frac{T_\text{f}M\Delta p}{\Delta_{\text{fus}}H}\times\Delta\left(\frac{1}{\rho}\right) \quad [V_\text{m} = M/\rho]$$

$$\approx \frac{278.6\text{ K}\times 999\text{ atm}\times 78.11\text{ g mol}^{-1}}{10.59\times10^3\text{ J mol}^{-1}}\times\left(\frac{1}{0.879\text{ g cm}^{-3}} - \frac{1}{0.891\text{ g cm}^{-3}}\right)$$

$$\times\left(\frac{1.013\times10^5\text{ Pa}}{1\text{ atm}}\right)\times\left(\frac{1\text{ m}^3}{10^6\text{ cm}^3}\right)$$

$$\approx 3.2\text{ K}$$

Therefore, at 1000 atm, $T_\text{f} = (278.6 + 3.2)\text{ K} = \boxed{281.8\text{ K}}$ or $\boxed{8.7°\text{C}}$.

E4.13(a) The rate of loss of mass of water may be expressed as

$$\frac{dm}{dt} = \frac{d}{dt}(nM), \quad \text{where} \quad n = \frac{q}{\Delta_{vap}H}$$

Thus, $$\frac{dn}{dt} = \frac{dq/dt}{\Delta_{vap}H} = \frac{(1.2 \times 10^3\ \text{W m}^{-2}) \times (50\ \text{m}^2)}{44.0 \times 10^3\ \text{J mol}^{-1}} = 1.4\ \text{mol s}^{-1}$$

and $$\frac{dm}{dt} = (1.4\ \text{mol s}^{-1}) \times (18.02\ \text{g mol}^{-1}) = \boxed{25\ \text{g s}^{-1}}$$

E4.14(a) Assume perfect-gas behaviour. The volume is 75 m³.

$$n = \frac{pV}{RT} = \frac{m}{M} \quad \text{so} \quad m = \frac{pVM}{RT}$$

(a) $$m = \frac{(3.2 \times 10^3\ \text{Pa}) \times (75\ \text{m}^3) \times (18.02\ \text{g mol}^{-1})}{(8.3145\ \text{J K}^{-1}\text{mol}^{-1}) \times (298\ \text{K})} = \boxed{1.7 \times 10^3\ \text{g}}\ \text{water}$$

(b) $$m = \frac{(13.1 \times 10^3\ \text{Pa}) \times (75\ \text{m}^3) \times (78.11\ \text{g mol}^{-1})}{(8.3145\ \text{J K}^{-1}\text{mol}^{-1}) \times (298\ \text{K})} = \boxed{31 \times 10^3\ \text{g}}\ \text{benzene}$$

(c) $$m = \frac{(0.23\ \text{Pa}) \times (75\ \text{m}^3) \times (200.59\ \text{g mol}^{-1})}{(8.3145\ \text{J K}^{-1}\text{mol}^{-1}) \times (298\ \text{K})} = \boxed{1.4\ \text{g}}\ \text{mercury}$$

COMMENT. Note that an ordinary reagent bottle of benzene would evaporate completely before saturating the air of the laboratory with benzene vapour.

Question. Assuming all the mercury vapour breathed remains in the body, how long would it take to accumulate 1.4 g? Make reasonable assumptions about the volume and frequency of a breath.

E4.15(a) The Clausius–Clapeyron equation [4.11] integrates to the form [4.12] that may be rewritten as

$$\ln\left(\frac{p_2}{p_1}\right) = \frac{\Delta_{vap}H}{R} \times \left(\frac{1}{T_1} - \frac{1}{T_2}\right)$$

(a) $$\Delta_{vap}H = R \ln\left(\frac{p_2}{p_1}\right) \times \left(\frac{1}{T_1} - \frac{1}{T_2}\right)^{-1}$$

$$= (8.3145\ \text{J K}^{-1}\text{mol}^{-1}) \times \ln\left(\frac{5.3\ \text{kPa}}{1.3\ \text{kPa}}\right) \times \left(\frac{1}{359.0\ \text{K}} - \frac{1}{392.5\ \text{K}}\right)^{-1}$$

$$= \boxed{+4.9 \times 10^4\ \text{J mol}^{-1}} = \boxed{+49\ \text{kJ mol}^{-1}}$$

(b) Rearrange the equation above to isolate a temperature

$$\frac{R}{\Delta_{vap}H} \ln\left(\frac{p_2}{p_1}\right) + \frac{1}{T_2} = \frac{1}{T_1}$$

The normal boiling point (T_1 in this equation) corresponds to a vapour pressure of 101.3 kPa. Using the data at 119.3°C (T_2), we have

$$\left(\frac{8.3145\ \text{J K}^{-1}\text{mol}^{-1}}{+4.9 \times 10^4\ \text{J mol}^{-1}}\right) \ln\left(\frac{5.3\ \text{kPa}}{101.3\ \text{kPa}}\right) + \frac{1}{392.5\ \text{K}} = 2.0\overline{5} \times 10^{-3}\ \text{K}^{-1} = \frac{1}{48\overline{8}\ \text{K}}$$

Thus, the estimated boiling point is $48\overline{8}$ K = $\boxed{21\overline{5}°C}$

The accepted value is 218°C.

(c) At the boiling point

$$\Delta_{vap}S = \frac{\Delta_{vap}H}{T_b} \approx \frac{+49\times10^3\ \text{J mol}^{-1}}{48\overline{8}\ \text{K}} = \boxed{+101\ \text{J K}^{-1}\ \text{mol}^{-1}}$$

E4.16(a)

$$\Delta T \approx \frac{T_f M\Delta p}{\Delta_{fus}H}\times\Delta\left(\frac{1}{\rho}\right)\ \text{[Exercise 4.12(a)]}$$

$$\Delta T \approx \left(\frac{(273.15\ \text{K})\times(49\times10^5\ \text{Pa})\times(18.0\ \text{g mol}^{-1})}{6.01\times10^3\ \text{J mol}^{-1}}\right)\times\left(\frac{1}{1.00\ \text{g cm}^{-3}}-\frac{1}{0.92\ \text{g cm}^{-3}}\right)\times\frac{1\ \text{m}^3}{10^6\ \text{cm}^3}$$

$$= -0.35\ \text{K}$$

$T_f(50\ \text{bar}) = (273.15-0.35)\ \text{K} = \boxed{272.80\ \text{K}}$.

E4.17(a)

$$\Delta_{vap}H = \Delta_{vap}U + \Delta_{vap}(pV) = 40.656\ \text{kJ mol}^{-1}$$

$$\Delta_{vap}(pV) = p\Delta_{vap}V = p(V_{gas}-V_{liq}) \approx pV_{gas} = RT\ \text{[perfect gas]}$$

$$\Delta_{vap}(pV) \approx (8.3145\ \text{J K}^{-1}\ \text{mol}^{-1})\times(373.2\ \text{K}) = 3102\ \text{J mol}^{-1}$$

$$\text{Fraction} = \frac{\Delta_{vap}(pV)}{\Delta_{vap}H} = \frac{3.102\ \text{kJ mol}^{-1}}{40.656\ \text{kJ mol}^{-1}} = \boxed{0.0763} = 7.63\%$$

Solutions to problems

Solutions to numerical problems

P4.1 At the triple point, T_3, the vapour pressures of liquid and solid are equal, hence

$$10.5916 - \frac{1871.2\ \text{K}}{T_3} = 8.3186 - \frac{1425.7\ \text{K}}{T_3};\quad T_3 = \boxed{196.0\ \text{K}}$$

$$\log(p_3/\text{Torr}) = \frac{-1871.2\ \text{K}}{196.0\ \text{K}} + 10.5916 = 1.044\overline{7};\quad p_3 = \boxed{11.1\ \text{Torr}}$$

P4.3 (a)

$$\frac{dp}{dT} = \frac{\Delta_{vap}S}{\Delta_{vap}V}\ [4.6] = \frac{\Delta_{vap}H}{T_b\Delta_{vap}V}\ [4.10]$$

$$= \frac{14.4\times10^3\ \text{J mol}^{-1}}{(180\ \text{K})\times(14.5\times10^{-3}-115\times10^{-6})\ \text{m}^3\ \text{mol}^{-1}} = \boxed{+5.56\times10^3\ \text{Pa K}^{-1}}$$

(b)

$$\frac{dp}{dT} = \frac{\Delta_{vap}H}{RT^2}\times p\left[4.11,\ \text{with d ln}\ p = \frac{dp}{p}\right] = \frac{(14.4\times10^3\ \text{J mol}^{-1})\times(1.013\times10^5\ \text{Pa})}{(8.3145\ \text{J K}^{-1}\ \text{mol}^{-1})\times(180\ \text{K})^2}$$

$$= +5.41\times10^3\ \text{Pa K}^{-1}$$

The percentage error is $\boxed{2.6\%}$.

P4.5 (a) $\left(\frac{\partial\mu(\mathrm{l})}{\partial p}\right)_T - \left(\frac{\partial\mu(\mathrm{s})}{\partial p}\right)_T = V_\mathrm{m}(\mathrm{l}) - V_\mathrm{m}(\mathrm{s})\,[4.13] = M\Delta\left(\frac{1}{\rho}\right)$

$$= (18.02\ \mathrm{g\ mol^{-1}}) \times \left(\frac{1}{1.000\ \mathrm{g\ cm^{-3}}} - \frac{1}{0.917\ \mathrm{g\ cm^{-3}}}\right)$$

$$= \boxed{-1.63\ \mathrm{cm^3\ mol^{-1}}}$$

(b) $\left(\frac{\partial\mu(\mathrm{g})}{\partial p}\right)_T - \left(\frac{\partial\mu(\mathrm{l})}{\partial p}\right)_T = V_\mathrm{m}(\mathrm{g}) - V_\mathrm{m}(\mathrm{l})$

$$= (18.02\ \mathrm{g\ mol^{-1}}) \times \left(\frac{1}{0.598\ \mathrm{g\ dm^{-3}}} - \frac{1}{0.958\times10^3\ \mathrm{g\ dm^{-3}}}\right)$$

$$= \boxed{+30.1\ \mathrm{dm^3\ mol^{-1}}}$$

(c) $\mu(\mathrm{g},1.2\ \mathrm{atm}) - \mu(\mathrm{l},1.2\ \mathrm{atm}) = \mu(\mathrm{g},1.2\ \mathrm{atm}) - \mu(\mathrm{g},1\ \mathrm{atm}) - \{\mu(\mathrm{l},1.2\ \mathrm{atm}) - \mu(\mathrm{l},1\ \mathrm{atm})\}$ because $\mu(\mathrm{g},1\ \mathrm{atm}) = \mu(\mathrm{l},1\ \mathrm{atm})$

Thus, $\mu(\mathrm{g},1.2\ \mathrm{atm}) - \mu(\mathrm{l},1.2\ \mathrm{atm}) = \Delta\mu(\mathrm{g}) - \Delta\mu(\mathrm{l})$

where $\Delta\mu$ is the difference in chemical potential of a given phase at 1.2 atm compared to that at 1 atm.

$$\Delta\mu \approx \left(\frac{\partial\mu}{\partial p}\right)_T \Delta p = V_\mathrm{m}\Delta p\ \ [4.3]$$

so $\mu(\mathrm{g},1.2\ \mathrm{atm}) - \mu(\mathrm{l},1.2\ \mathrm{atm}) = \{V_\mathrm{m}(\mathrm{g}) - V_\mathrm{m}(\mathrm{l})\}\Delta p \approx V_\mathrm{m}(\mathrm{g})\Delta p$

$$\Delta\mu(\mathrm{g},1.2\ \mathrm{atm}) - \Delta\mu(\mathrm{g},1.2\ \mathrm{atm}) \approx (30.1\times10^{-3}\ \mathrm{m^3\ mol^{-1}}) \times (0.2\ \mathrm{atm}) \times \frac{1.013\times10^5\ \mathrm{Pa}}{1\ \mathrm{atm}}$$

$$\approx \boxed{+6\times10^2\ \mathrm{J\ mol^{-1}}}$$

Since $\mu(\mathrm{g},1.2\ \mathrm{atm}) > \mu(\mathrm{l},1.2\ \mathrm{atm})$, the gas tends to condense into a liquid.

P4.7 The amount (moles) of water evaporated is $n_\mathrm{g} = \frac{p_{\mathrm{H_2O}}V}{RT}$

The heat leaving the water is $q = n_\mathrm{g}\Delta_\mathrm{vap}H$.

The temperature change of the water is $\Delta T = \frac{-q}{n_\mathrm{liq}C_{p,\mathrm{m}}} = \frac{-p_{\mathrm{H_2O}}V\Delta_\mathrm{vap}H}{RTn_\mathrm{liq}C_{p,\mathrm{m}}}$.

Therefore, $$\Delta T = \frac{-(3.17\ \mathrm{kPa})\times(50.0\ \mathrm{dm^3})\times(44.0\times10^3\ \mathrm{J\ mol^{-1}})}{(8.3145\ \mathrm{kPa\ dm^3\ K^{-1}\ mol^{-1}})\times(298\ \mathrm{K})\times(75.5\ \mathrm{J\ K^{-1}\ mol^{-1}})\times\left(\frac{250\ \mathrm{g}}{18.02\ \mathrm{g\ mol^{-1}}}\right)}$$

$$= -2.7\ \mathrm{K}$$

The final temperature will be about $\boxed{22°\mathrm{C}}$.

P4.9 (a) T_b is the temperature at which the vapour pressure is 1 atm (760 Torr); $T_b = \boxed{227.5°C}$.

(b) Integrating the Clausius–Clapeyron eqation (4.11) yields an expression for ln p:

$$\ln p = \text{constant} - \frac{\Delta_{vap}H}{RT}$$

Therefore, plot ln p against $1/T$ and identify $-\Delta_{vap}H/R$ as the slope of the plot. Construct the following table:

θ/°C	57.4	100.4	133.0	157.3	203.5	227.5
T/K	330.6	373.6	406.2	430.5	476.7	500.7
1000 K/T	3.02	2.68	2.46	2.32	2.10	2.00
ln (p/Torr)	0.00	2.30	3.69	4.61	5.99	6.63

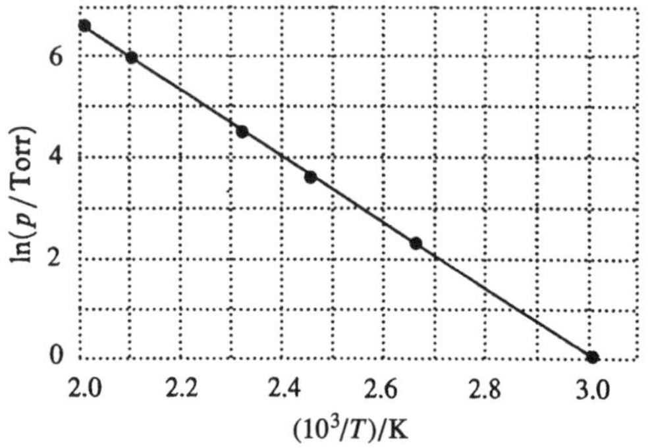

Figure 4.1

The points are plotted in Figure 4.1. The slope is -6.4×10^3 K, so $\Delta_{vap}H = \boxed{+53 \text{ kJ mol}^{-1}}$.

P4.11 (a) The phase diagram is shown in Figure 4.2.

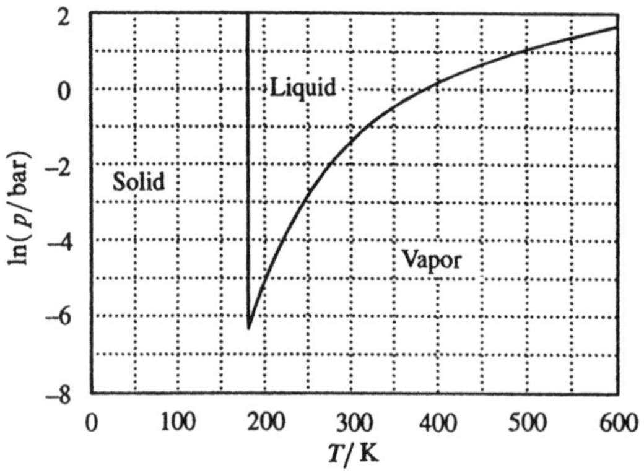

Figure 4.2

(b) The standard melting point is the temperature at which solid and liquid are in equilibrium at 1 bar. That temperature can be found by solving the equation of the solid–liquid coexistence curve for the temperature:

$$1 = p_3/\text{bar} + 1000(5.60 + 11.727x)x.$$

Put the equation into standard form:

$$11\,727x^2 + 5600x + (4.362 \times 10^{-7} - 1) = 0$$

The quadratic formula yields

$$x = \frac{-5600 \pm \{(5600)^2 - 4 \times 11727 \times (-1)\}^{1/2}}{2 \times 11727} = \frac{-1 \pm \left(1 + \frac{4\times11727}{5600^2}\right)^{1/2}}{2 \times \left(\frac{11727}{5600}\right)}$$

The square root is rewritten to make it clear that the square root is of the form $\{1 + a\}^{1/2}$, with $a \ll 1$; thus the numerator is approximately $-1 + (1 + \frac{1}{2}a) = \frac{1}{2}a$, and the whole expression reduces to

$$x \approx 1/5600 = 1.79 \times 10^{-4}.$$

Thus, the melting point is

$$T = (1 + x)T_3 = (1.000179) \times (178.15\ \text{K}) = \boxed{178.18\ \text{K}}.$$

(c) The standard boiling point is the temperature at which the liquid and vapour are in equilibrium at 1 bar. That temperature can be found by solving the equation of the liquid–vapour coexistence curve for the temperature. This equation is too complicated to solve analytically, but not difficult to solve numerically with a spreadsheet. The calculated answer is $y = 0.6459$,

so $T = 0.6459 \times 593.95\ \text{K} = \boxed{383.6\ \text{K}}$.

(d) The slope of the liquid–vapour coexistence curve is given by

$$\frac{dp}{dT} = \frac{\Delta_{\text{vap}}H}{T\Delta_{\text{vap}}V}\ [4.10] \quad \text{so} \quad \Delta_{\text{vap}}H = T\Delta_{\text{vap}}V\frac{dp}{dT}$$

The slope can be obtained by differentiating the equation for the coexistence curve.

$$\frac{dp}{dT} = p\frac{d\ln p}{dT} = p\frac{d\ln p}{dy}\frac{dy}{dT}$$

$$\frac{dp}{dT} = \left(\frac{10.413}{y^2} - 15.996 + 2(14.015)y - 3(5.0120)y^2 - (1.70) \times (4.7224) \times (1-y)^{0.70}\right) \times \left(\frac{p}{T_c}\right)$$

Substituting the value of y at the boiling point yields

$$\frac{dp}{dT} = 2.848 \times 10^{-2}\ \text{bar K}^{-1} = 2.848\ \text{kPa K}^{-1}$$

and $\Delta_{\text{vap}}H = (383.6\ \text{K}) \times \left(\frac{(30.3 - 0.12)\ \text{dm}^3\ \text{mol}^{-1}}{1000\ \text{dm}^3\ \text{m}^{-3}}\right) \times (2.848\ \text{kPa K}^{-1}) = \boxed{33.0\ \text{kJ mol}^{-1}}$

Solutions to theoretical problems

P4.13 $$\left(\frac{\partial \Delta G}{\partial p}\right)_T = \left(\frac{\partial G_\beta}{\partial p}\right)_T - \left(\frac{\partial G_\alpha}{\partial p}\right)_T = V_\beta - V_\alpha$$

Therefore, if $V_\alpha = V_\beta$, ΔG is independent of pressure. In general, $V_\alpha \neq V_\beta$, so that ΔG is non-zero, though small, since $V_\beta - V_\alpha$ is small.

P4.15 The amount (moles) of gas bubbled through liquid is $\dfrac{PV}{RT}$

(P is the initial pressure of the gas and the pressure of the emerging gaseous mixture).

The amount (moles) of vapour carried away is $\dfrac{m}{M}$

Mole fraction of vapour in the gaseous mixture is

$$x = \frac{\dfrac{m}{M}}{\dfrac{m}{M} + \dfrac{PV}{RT}} = \frac{MRT}{mRT + PVM} \times \frac{m}{M} = \frac{mRT}{mRT + PVM}$$

The partial pressure of the vapour (i.e. the vapour pressure) is

$$p = xP = \frac{mRT}{mRT + PVM} \times P = \frac{P\left(\dfrac{mRT}{PVM}\right)}{\left(\dfrac{mRT}{PVM}\right) + 1} = \frac{mPA}{mA + 1}, \quad \text{where } A = \frac{RT}{PVM}$$

Inserting the data yields

For geraniol, $M = 154.2\ \text{g mol}^{-1}$, $T = 383\ \text{K}$, $V = 5.00\ \text{dm}^3$, $P = 1.00\ \text{atm}$, and $m = 0.32\ \text{g}$.
$M = 154.2\ \text{g mol}^{-1}$, $T = 383\ \text{K}$, $V = 5.00\ \text{dm}^3$, $p = 1.00\ \text{atm}$, and $m = 0.32\ \text{g}$, so

$$A = \frac{(0.08206\ \text{dm}^3\ \text{atm mol}^{-1}\ \text{K}^{-1}) \times (383\ \text{K})}{(1.00\ \text{atm}) \times (5.00\ \text{dm}^3) \times (154.2\ \text{g mol}^{-1})} = 0.0408\ \text{g}^{-1}$$

Therefore

$$p = \frac{(0.32\ \text{g}) \times (760\ \text{Torr}) \times (0.0408\ \text{g}^{-1})}{(0.32\ \text{g}) \times (0.0408\ \text{g}^{-1}) + 1} = \boxed{9.8\ \text{Torr}}$$

P4.17 The barometric formula for ambient pressure as a function of altitude h is

$$p = p_0 e^{-Mgh/RT}\ [1.16]$$

Equation 4.12 for the dependence of vapour pressure on temperature comes from integrating the Clausius–Clapeyron equation:

$$p = p^* e^{-\chi} \qquad \chi = \frac{\Delta_{\text{vap}}H}{R} \times \left(\frac{1}{T} - \frac{1}{T^*}\right)$$

Boiling occurs when the vapour pressure is equal to the ambient pressure at altitude h and ambient temperature T, so we can set these two equations equal to each other. Let us choose the normal boiling point as the reference conditions in eqn 4.12; that means $T^* = T_b$ and p^* is exactly 1 atm. Note that the other temperature T in eqn 4.12 is the boiling temperature at altitude h (call it T_h), not the ambient temperature. Let us choose exactly 1 atm as the reference pressure p_0 in the barometric formula. Thus

$$\exp\left\{-\frac{Mgh}{RT}\right\} = \exp\left\{-\frac{\Delta_{\mathrm{vap}}H}{R}\times\left(\frac{1}{T_h}-\frac{1}{T_{\mathrm{b}}}\right)\right\}$$

It follows that $\dfrac{1}{T_h} = \dfrac{1}{T_{\mathrm{b}}} + \dfrac{Mgh}{T\Delta_{\mathrm{vap}}H}$

For water at 3000 m, taking the average molar mass of air to be $M = 29\ \mathrm{g\ mol^{-1}}$,

$$\frac{1}{T_h} = \frac{1}{373\ \mathrm{K}} + \frac{(29\times10^{-3}\ \mathrm{kg\ mol^{-1}})\times 9.81\ \mathrm{m\ s^{-2}}\times(3000\ \mathrm{m})}{(293\ \mathrm{K})\times(40.7\times10^{3}\ \mathrm{J\ mol^{-1}})}$$

$$= \frac{1}{373\ \mathrm{K}} + \frac{1}{1.397\times10^{4}\ \mathrm{K}} = \frac{2.73\times10^{-3}}{\mathrm{K}}$$

Hence, $T_h = \boxed{363\ \mathrm{K}}$ (90°C).

P4.19 The Clapeyron equation is inadequate because both V and S are continuous through a second-order transition. Thus, ΔV and ΔS vanish, resulting in an indeterminate form 0/0. Just as with l'Hospital's rule in calculus, examining derivatives helps resolve the indeterminate form.

(1) $V = V(T, p)$

$$\text{so}\quad \mathrm{d}V = \left(\frac{\partial V}{\partial T}\right)_p \mathrm{d}T + \left(\frac{\partial V}{\partial p}\right)_T \mathrm{d}p$$

$$\left(\frac{\partial V}{\partial T}\right)_p = \alpha V\ [2.42],\quad \left(\frac{\partial V}{\partial p}\right)_T = -\kappa_T V\ [2.43]$$

hence, $\mathrm{d}V = \alpha V\,\mathrm{d}T - \kappa_T V\,\mathrm{d}p$.

This equation applies to both phases 1 and 2, and since V is continuous through a second-order transition

$$\alpha_1\,\mathrm{d}T - \kappa_{T,1}\,\mathrm{d}p = \alpha_2\,\mathrm{d}T - \kappa_{T,2}\,\mathrm{d}p$$

Solving for $\dfrac{\mathrm{d}p}{\mathrm{d}T}$ yields $\boxed{\dfrac{\mathrm{d}p}{\mathrm{d}T} = \dfrac{\alpha_2-\alpha_1}{\kappa_{T,2}-\kappa_{T,1}}}$.

(2) $S_{\mathrm{m}} = S_{\mathrm{m}}(T,p)$

$$\text{so}\quad \mathrm{d}S_{\mathrm{m}} = \left(\frac{\partial S_{\mathrm{m}}}{\partial T}\right)_p \mathrm{d}T + \left(\frac{\partial S_{\mathrm{m}}}{\partial p}\right)_T \mathrm{d}p.$$

To evaluate the derivative $\partial S_{\mathrm{m}}/\partial T$, consider $\mathrm{d}S$ at constant p:

$$\mathrm{d}S = \frac{\mathrm{d}q_{\mathrm{rev}}}{T} = \frac{\mathrm{d}H}{T} = \frac{C_p\mathrm{d}T}{T}\quad\text{so}\quad \left(\frac{\partial S_{\mathrm{m}}}{\partial T}\right)_p = \frac{C_{p,\mathrm{m}}}{T}$$

Also, $\left(\dfrac{\partial S_{\mathrm{m}}}{\partial p}\right)_T = -\left(\dfrac{\partial V_{\mathrm{m}}}{\partial T}\right)_p$ [Maxwell relation; Table 3.5] $= -\alpha V_{\mathrm{m}}$

Thus, $\mathrm{d}S_{\mathrm{m}} = \dfrac{C_{p,\mathrm{m}}}{T}\mathrm{d}T - \alpha V_{\mathrm{m}}\,\mathrm{d}p$

This relationship applies to both phases. For second-order transitions both S_m and V_m are continuous through the transition, so that

$$\frac{C_{p,\text{m1}}}{T}\,dT - \alpha_1 V_m\,dp = \frac{C_{p,\text{m2}}}{T}\,dT - \alpha_2 V_m\,dp$$

Solving for $\dfrac{dp}{dT}$ yields $\boxed{\dfrac{dp}{dT} = \dfrac{C_{p,\text{m2}} - C_{p,\text{m1}}}{T V_m (\alpha_2 - \alpha_1)}}$

Solutions to applications

P4.21 (a) $\Delta G_m = (n-4)\Delta_{\text{hb}}H_m - (n-2)T_m\Delta_{\text{hb}}S_m$ (1)

Enthalpy and entropy terms give a Gibbs energy change of $\Delta G = \Delta H - T\Delta S$ for a constant temperature process. The enthalpy term is justified by $n-4$ independent hydrogen bonds for which each requires $\Delta_{\text{hb}}H_m$ of heat to break during melting dissociation. The entropy term is justified by $n-2$ highly ordered, but independent, structures for which each experiences an entropy increase of $\Delta_{\text{hb}}H_m$ during the melting process.

(b) $\Delta_{\text{trs}}S = \dfrac{\Delta_{\text{trs}}H}{T_{\text{trs}}}$ [3.20] yields $T_{\text{trs}} = \dfrac{\Delta_{\text{trs}}H}{\Delta_{\text{trs}}S}$, which here becomes

$$T_m = \frac{(n-4)\Delta_{\text{hb}}H_m}{(n-2)\Delta_{\text{hb}}S_m}$$

$$\frac{T_m\Delta_{\text{hb}}S_m}{\Delta_{\text{hb}}H_m} = \frac{n-4}{n-2}$$

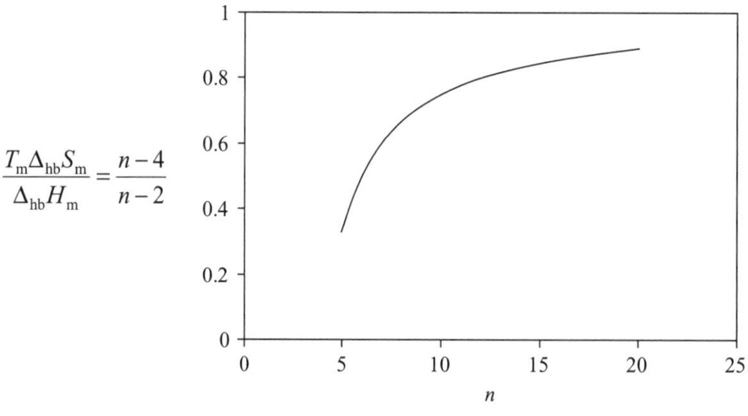

Figure 4.3

(c) See Figure 4.3.

Consider $\dfrac{1}{T_m}\dfrac{dT_m}{dn} = \dfrac{\Delta_{\text{hb}}H_m}{T_m\Delta_{\text{hb}}S_m}\dfrac{d(T_m\Delta_{\text{hb}}S_m/\Delta_{\text{hb}}H_m)}{dn} = \left(\dfrac{n-2}{n-4}\right)\dfrac{d}{dn}\left(\dfrac{n-4}{n-2}\right)$

$$= \left(\frac{n-2}{n-4}\right)\left(\frac{2}{(n-2)^2}\right) = \frac{2}{(n-4)(n-2)}$$

This expression will be less than 1% when $\dfrac{2}{(n-4)(n-2)} < 0.01$ or when n equals, or is larger than the value given by $n^2 - 6n + 8 = 200$. The positive root of this quadratic is $n \equiv \boxed{17}$. T_m changes by about 1% or less upon addition of another amino acid residue when the polypeptide consists of 17 or more residues.

P4.23 (a) The phase boundary is plotted in Figure 4.4.

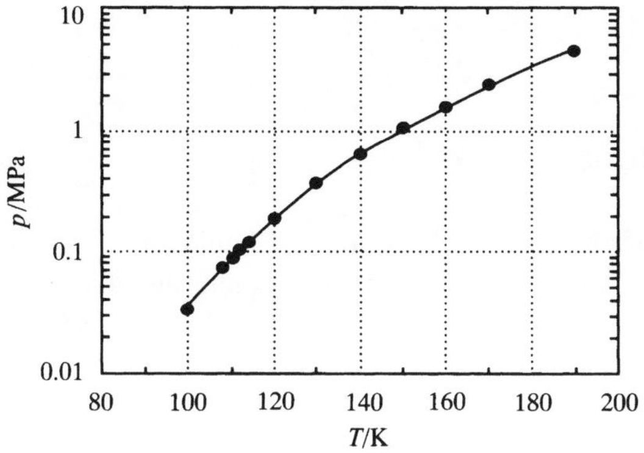

Figure 4.4

(b) The standard boiling point is the temperature at which the liquid is in equilibrium with the standard pressure of 1 bar (0.1 MPa). Interpolation of the plotted points gives $T_b = \boxed{112\text{ K}}$.

(c) The slope of the liquid–vapour coexistence curve is given by

$$\frac{dp}{dT} = \frac{\Delta_{vap}H}{T\Delta_{vap}V} \quad \text{so} \quad \Delta_{vap}H = (T\Delta_{vap}V)\frac{dp}{dT}$$

The slope can be obtained graphically or by fitting the points nearest the boiling point. Then

$$\frac{dp}{dT} = 8.14 \times 10^{-3}\text{ MPa K}^{-1}$$

$$\text{so} \quad \Delta_{vap}H = (112\text{ K}) \times \left(\frac{(8.89 - 0.0380)\text{ dm}^3\text{ mol}^{-1}}{1000\text{ dm}^3\text{ m}^{-3}}\right) \times (8.14\text{ kPa K}^{-1}) = \boxed{8.07\text{ kJ mol}^{-1}}$$

5 Simple mixtures

Answers to discussion questions

D5.1 At equilibrium, the chemical potentials of any component in both the liquid and vapour phases must be equal. This is justified by the requirement that for systems at equilibrium under constant temperature and pressure conditions, with no additional work, $\Delta G = 0$ (see eqn 3.36(b) and the answer to Discussion question 3.3). Here $\Delta G = \mu_i(\text{v}) - \mu_i(\text{l})$, for all components, i, of the solution; hence their chemical potentials must be equal in the liquid and vapour phases.

D5.3 All the colligative properties are a result of the lowering of the chemical potential of the solvent due to the presence of the solute. This reduction takes the form $\mu_A = \mu_A^* + RT\ln x_A$ or $\mu_A = \mu_A^* + RT\ln a_A$, depending on whether or not the solution can be considered ideal. The lowering of the chemical potential results in a freezing point depression and a boiling point elevation, as illustrated in Figure 5.20 of the text. Both of these effects can be explained by the lowering of the vapour pressure of the solvent in solution due to the presence of the solute. The solute molecules get in the way of the solvent molecules, reducing their escaping tendency.

D5.5 The Debye–Hückel theory is a theory of the activity coefficients of ions in solution. It is the coulombic (electrostatic) interaction of the ions in solution with each other and also the interaction of the ions with the solvent that is responsible for the deviation of their activity coefficients from the ideal value of 1. The electrostatic ion–ion interaction is the stronger of the two and is fundamentally responsible for the deviation. Because of this interaction there is a build up of charge of opposite sign around any given ion in the overall electrically neutral solution. The energy, and hence the chemical potential, of any given ion is lowered as a result of the existence of this ionic atmosphere. The lowering of the chemical potential below its ideal value is identified with a non-zero value of $RT\ln\gamma_\pm$. This non-zero value implies that $\gamma_\pm$ will have a value different from unity, which is its ideal value. The role of the solvent is more indirect. The solvent determines the dielectric constant, ε, of the solution. Looking at the details of the theory as outlined in *Further Information 5.1* we see that ε enters into a number of the basic equations, in particular Coulomb's law, Poisson's equation, and the equation for the Debye length. The larger the dielectric constant, the smaller (in magnitude) is $\ln\gamma_\pm$.

D5.7(a) See Figure 5.1(a). Congruent melting is evident from the presence of liquid AB wherever liquids are present (along with whichever of A or B happens to be in excess).

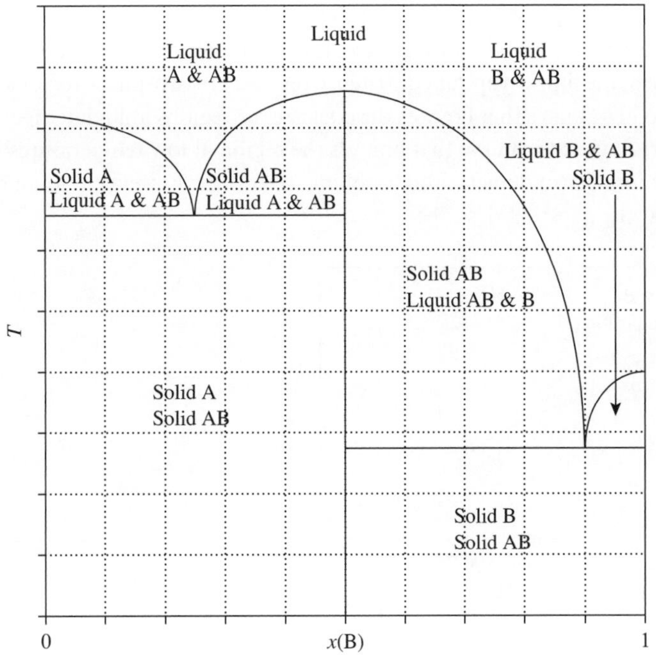

Figure 5.1(a)

(b) See Figure 5.1(b). Incongruent melting is reflected in the absence of liquid AB_2. When solid AB_2 melts, it decomposes.

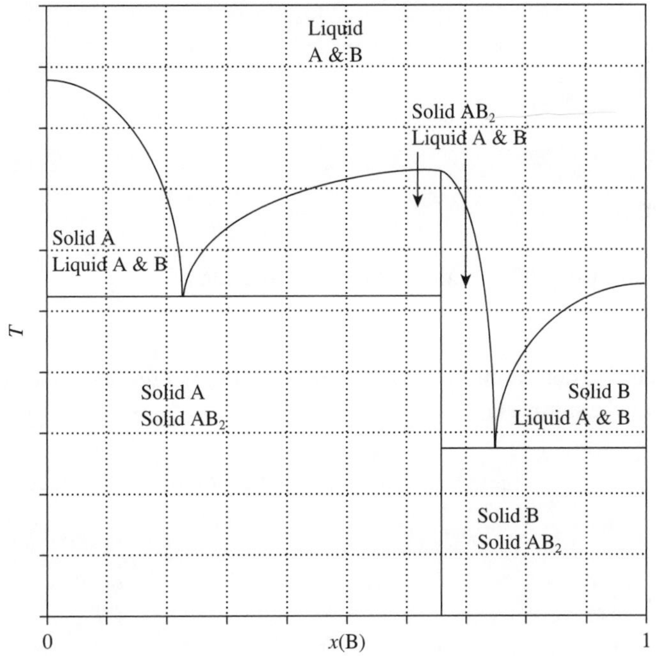

Figure 5.1(b)

(c) See Figure 5.1(c). Vapours are always miscible and these liquids are miscible, so both the vapour region at high temperature and the liquid region at low temperature are two-component single-phase regions. In between, the liquid-vapour coexistence regions are two-phase regions, each of which has both components. The curve that bounds the coexistence regions at high temperature is the vapour composition curve, while the curve that bounds the region at low temperatures is the liquid boiling temperature. Figure 5.1(c) depicts a high-boiling azeotrope; however, a low-boiling azeotrope could illustrate what the question describes.

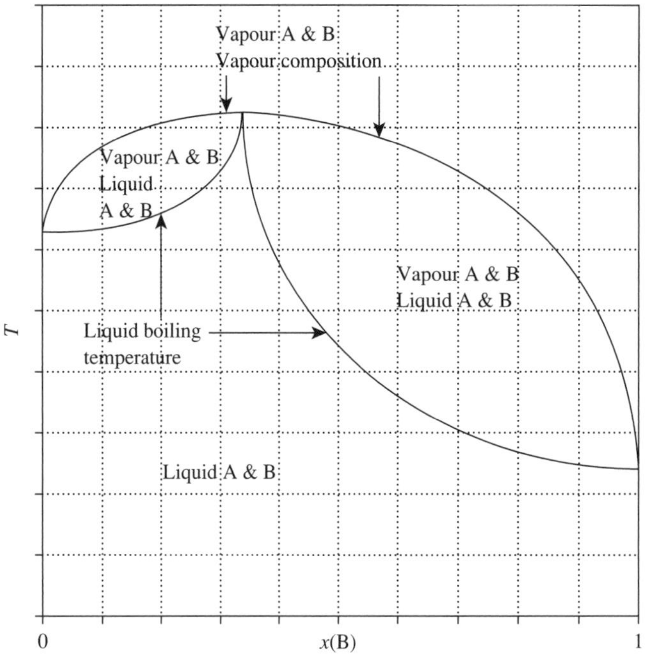

Figure 5.1(c)

Solutions to exercises

E5.1(a) Let A denote acetone and C chloroform. The total volume of the solution is

$$V = n_A V_A + n_C V_C$$

V_A and V_C are given; hence we need to determine n_A and n_C in 1.000 kg of the solution with the stated mole fraction. The total mass of the sample is

(a) $m = n_A M_A + n_C M_C$

We also know that

$$x_A = \frac{n_A}{n_A + n_C}$$

implying that $(x_A - 1)n_A + x_A n_C = 0$,

and hence that

(b) $-x_C n_A + x_A n_C = 0$

On solving (a) and (b), we find

$$n_A = \left(\frac{x_A}{x_C}\right) \times n_C, \quad n_C = \frac{m x_C}{x_A M_A + x_C M_C}$$

Since $x_C = 0.4693$, $x_A = 1 - x_C = 0.5307$,

$$n_C = \frac{(0.4693) \times (1000\ \text{g})}{[(0.5307) \times (58.08) + (0.4693) \times (119.37)]\ \text{g mol}^{-1}} = 5.404\ \text{mol},$$

and $n_A = \left(\dfrac{0.5307}{0.4693}\right) \times (5.404)\ \text{mol} = 6.111\ \text{mol}$

The total volume, $V = n_A V_A + n_C V_C$, is therefore

$$V = (6.111\ \text{mol}) \times (74.166\ \text{cm}^3\ \text{mol}^{-1}) + (5.404\ \text{mol}) \times (80.235\ \text{cm}^3\ \text{mol}^{-1})$$
$$= \boxed{886.8\ \text{cm}^3}$$

E5.2(a) Let W denote water and E ethanol. The total volume of the solution is

$$V = n_W V_W + n_E V_E$$

We are given V_W, we need to determine n_W and n_E in order to solve for V_E, for

$$V_E = \frac{V - n_W V_W}{n_E}$$

Take 100 cm^3 of solution as a convenient sample. The mass of this sample is

$$m = \rho V = (0.914\ \text{g cm}^{-3}) \times (100\ \text{cm}^3) = 91.4\ \text{g}$$

Half of this mass is water and half ethanol, so the moles of each component are

$$n_W = \frac{(0.50) \times (91.4\ \text{g})}{18.02\ \text{g mol}^{-1}} = 2.5\ \text{mol} \quad \text{and} \quad n_E = \frac{(0.50) \times (91.4\ \text{g})}{46.07\ \text{g mol}^{-1}} = 0.99\ \text{mol}$$

$$V_E = \frac{V - n_W V_W}{n_E} = \frac{100\ \text{cm}^3 - (2.5\ \text{mol}) \times (17.4\ \text{cm}^3\ \text{mol}^{-1})}{0.99\ \text{mol}} = \boxed{56\ \text{cm}^3\ \text{mol}^{-1}}$$

E5.3(a) Check whether p_B/x_B is equal to a constant (K_B)

x	0.005	0.012	0.019
$(p/\text{kPa})/x$	6.4×10^3	6.4×10^3	6.4×10^3

Hence, $K_B = \boxed{6.4 \times 10^3\ \text{kPa}}$.

E5.4(a) In Exercise 5.3(a), the Henry's law constant was determined for concentrations expressed in mole fractions, $K_B = 6.4 \times 10^3$ kPa. Thus, the concentration must be converted from molality to mole fraction.

$$m(\text{GeCl}_4) = 1000\ \text{g, corresponding to } n(\text{GeCl}_4) = \frac{1000\ \text{g}}{214.39\ \text{g mol}^{-1}} = 4.664\ \text{mol}$$

Therefore, $x_{HCl} = \dfrac{0.10\text{ mol}}{(0.10\text{ mol}) + (4.664\text{ mol})} = 0.021$

The pressure is

$$p_{HCl} = K_B x_{HCl} = (0.021) \times (6.4 \times 10^3\text{ kPa}) = \boxed{1.3 \times 10^2\text{ kPa}}$$

E5.5(a) We assume that the solvent, benzene, is ideal and obeys Raoult's law.

Let B denote benzene and A the solute; then

$$p_B = x_B p_B^* \quad \text{and} \quad x_B = \frac{n_B}{n_A + n_B}$$

Hence $p_B = \dfrac{n_B p_B^*}{n_A + n_B}$, which solves to

$$n_A = \frac{n_B(p_B^* - p_B)}{p_B}$$

Then, since $n_A = \dfrac{m_A}{M_A}$, where m_A is the mass of A present,

$$M_A = \frac{m_A p_B}{n_B(p_B^* - p_B)} = \frac{m_A M_B p_B}{m_B(p_B^* - p_B)}$$

From the data

$$M_A = \frac{(19.0\text{ g}) \times (78.11\text{ g mol}^{-1}) \times (51.5\text{ kPa})}{(500\text{ g}) \times (53.3 - 51.5)\text{ kPa}} = \boxed{85\text{ g mol}^{-1}}$$

E5.6(a) Let B denote the compound and A the solvent, CCl_4. $K_f = 30\text{ K kg mol}^{-1}$ [Table 5.2]

$$M_B = \frac{m_B}{n_B}$$

$$n_B = m_A b_B, \quad \text{where} \quad b_B = \frac{\Delta T}{K_f}\ [5.35]$$

Thus, $M_B = \dfrac{m_B K_f}{m_A \Delta T} = \dfrac{(100\text{ g}) \times (30\text{ K kg mol}^{-1})}{(0.750\text{ kg}) \times (10.5\text{ K})} = \boxed{3.8 \times 10^2\text{ g mol}^{-1}}$

E5.7(a) From the osmotic pressure, compute the concentration, and from the concentration the freezing point. According to the van't Hoff equation (5.38), the osmotic pressure is

$$\Pi = [B]RT \quad \text{so} \quad [B] = \frac{\Pi}{RT} = \frac{n_B}{V_{soln}}$$

The expression for freezing point depression (eqn 5.35) includes the molality b_B rather than the molarity [B]. In dilute solutions, the two concentration measures are readily related:

$$b_B = \frac{n_B}{m_A} \approx \frac{n_B}{V_{soln}\rho_{soln}} = \frac{[B]}{\rho_{soln}} = \frac{\Pi}{RT\rho_{soln}}$$

The freezing point depression is

$$\Delta T = K_f b_B \approx \frac{K_f \Pi}{RT\rho_{soln}}, \quad \text{where} \quad K_f = 1.86 \text{ K mol}^{-1} \text{ kg [Table 5.2]}$$

The density of a dilute aqueous solution is approximately that of water:

$$\rho \approx 1.0 \text{ g cm}^{-3} = 1.0 \times 10^3 \text{ kg m}^{-3}$$

So, $$\Delta T \approx \frac{(1.86 \text{ K kg mol}^{-1}) \times (120 \times 10^3 \text{ Pa})}{(8.3145 \text{ J K}^{-1} \text{mol}^{-1}) \times (300 \text{ K}) \times (10^3 \text{ kg m}^{-3})} = 0.089 \text{ K}$$

Therefore, the solution will freeze at about $\boxed{-0.09°\text{C}}$.

COMMENT. Osmotic pressures are inherently large. Even dilute solutions with small freezing point depressions have large osmotic pressures.

E5.8(a) The Gibbs energy of mixing perfect gases is

$$\Delta_{mix}G = nRT(x_A \ln x_A + x_B \ln x_B) \text{ [5.16]} = pV(x_A \ln x_A + x_B \ln x_B) \text{ [ideal gas]}$$

Because the compartments are of equal size, each contains half of the gas; therefore,

$$\begin{aligned}\Delta_{mix}G &= (pV) \times (\tfrac{1}{2} \ln \tfrac{1}{2} + \tfrac{1}{2} \ln \tfrac{1}{2}) = -pV \ln 2 \\ &= (-1.0) \times (1.013 \times 10^5 \text{ Pa}) \times (5.0 \times 10^{-3} \text{ m}^3) \times (\ln 2) \\ &= -3.5 \times 10^2 \text{ J} = \boxed{-0.35 \text{ kJ}}\end{aligned}$$

The entropy of mixing perfect gases is

$$\Delta_{mix}S = -nR(x_A \ln x_A + x_B \ln x_B) \text{ [5.17]} = \frac{-\Delta_{mix}G}{T} = \frac{+350 \text{ kJ}}{298 \text{ K}} = \boxed{+1.2 \text{ J K}^{-1}}$$

E5.9(a) $$\Delta_{mix}S = -nR \sum_J x_J \ln x_J \text{ [5.17]}$$

Therefore, for a total amount of one mole,

$$\begin{aligned}\Delta_{mix}S &= -R\sum_J x_J \ln x_J \\ &= -R\{(0.780 \ln 0.780) + (0.210 \ln 0.210) + (0.0096 \ln 0.0096)\} \\ &= 0.566R = \boxed{+4.71 \text{ J K}^{-1} \text{mol}^{-1}}\end{aligned}$$

E5.10(a) (a) Hexane and heptane form nearly ideal solutions, therefore eqn 5.17 applies.

$$\Delta_{mix}S = -nR(x_A \ln x_A + x_B \ln x_B) \text{ [5.17]}$$

We need to differentiate eqn 5.17 with respect to x_A and look for the value of x_A at which the derivative is zero. Since $x_B = 1 - x_A$, we need to differentiate

$$\Delta_{mix}S = -nR\{x_A \ln x_A + (1 - x_A)\ln(1 - x_A)\}$$

This gives $\left(\text{using} \dfrac{\text{d} \ln x}{\text{d}x} = \dfrac{1}{x}\right)$

$$\frac{\text{d}\Delta_{mix}S}{\text{d}x_A} = -nR\{\ln x_A + 1 - \ln(1 - x_A) - 1\} = -nR \ln \frac{x_A}{1 - x_A}$$

which is zero when $x_A = \boxed{\frac{1}{2}}$. Hence, the maximum entropy of mixing occurs for the preparation of a mixture that contains equal mole fractions of the two components.

(b) Because entropy of mixing is maximized when $n_{hexane} = n_{heptane}$,

$$\frac{m_{hexane}}{M_{hexane}} = \frac{m_{heptane}}{M_{heptane}}$$

This makes the mass ratio

$$\frac{m_{hexane}}{m_{heptane}} = \frac{M_{hexane}}{M_{heptane}} = \frac{86.17\ \text{g mol}^{-1}}{100.20\ \text{g mol}^{-1}} = \boxed{0.8600}$$

E5.11(a) With concentrations expressed in molalities, Henry's law [5.23] becomes $p_B = b_B K_B$. Solving for b_B, the solubility, we have $b_B = \frac{p_B}{K}$.

(a) $p_B = 0.10\ \text{atm} = 10.1\ \text{kPa}$

$$b_B = \frac{10.1\ \text{kPa}}{3.01 \times 10^3\ \text{kPa kg mol}^{-1}} = \boxed{3.4 \times 10^{-3}\ \text{mol kg}^{-1}}$$

(b) $p_B = 1.00\ \text{atm} = 101.3\ \text{kPa}$

$$b_B = \frac{101.3\ \text{kPa}}{3.01 \times 10^3\ \text{kPa kg mol}^{-1}} = \boxed{3.37 \times 10^{-2}\ \text{mol kg}^{-1}}$$

E5.12(a) As in Exercise 5.11(a), we have $b_B = \frac{p_B}{K_B} = \frac{5.0 \times 101.3\ \text{kPa}}{3.01 \times 10^3\ \text{kPa kg mol}^{-1}} = 0.17\ \text{mol kg}^{-1}$.

Hence, the molality of the solution is about 0.17 mol kg^{-1}. Since molalities and molar concentrations (molarities) for dilute aqueous solutions are numerically approximately equal, the molar concentration is about $\boxed{0.17\ \text{mol dm}^{-3}}$.

E5.13(a) The ideal solubility in terms of mole fraction is given by eqn 5.37, where B stands for the solute (anthracene):

$$\ln x_B = \frac{\Delta_{fus}H}{R} \times \left(\frac{1}{T_f} - \frac{1}{T}\right)$$

$$= \left(\frac{28.8 \times 10^3\ \text{J mol}^{-1}}{8.3145\ \text{J K}^{-1}\text{mol}^{-1}}\right) \times \left(\frac{1}{490\ \text{K}} - \frac{1}{298\ \text{K}}\right) = -4.55$$

Therefore, $x_B = e^{-4.35} = 0.0105$.

$$x_B = \frac{n_B}{n_A + n_B} \approx \frac{n_B}{n_A}\left[\textit{i.e.,}\ \frac{n(\text{anthracene})}{n(\text{benzene})}\right]$$

(The approximation is justified because x_B is small.) Therefore, in 1 kg of benzene,

$$n_{anth} = x_{anth} \times n_{benz} \approx x_{anth} \times \frac{m_{benz}}{M_{benz}} = (0.0105) \times \left(\frac{1000\ \text{g}}{78.11\ \text{g mol}^{-1}}\right) = 0.135\ \text{mol}$$

The molality of the solution is therefore $\boxed{0.135 \text{ mol kg}^{-1}}$. Since $M = 178 \text{ g mol}^{-1}$, 0.135 mol corresponds to $\boxed{24.0 \text{ g anthracene}}$ per kilogram of benzene.

E5.14(a) The best value of the molar mass is obtained from values of the data extrapolated to zero concentration, since it is under this condition that the van't Hoff equation [5.38] applies.

$$\Pi V = n_B RT \text{ [5.38]}, \quad \text{so} \quad \Pi = \frac{mRT}{MV} = \frac{cRT}{M}, \quad \text{where } c = \frac{m}{V}$$

But the osmotic pressure is also equal to the hydrostatic pressure

$$\Pi = \rho g h \text{ [1.3]}, \quad \text{so} \quad h = \left(\frac{RT}{\rho g M}\right)c$$

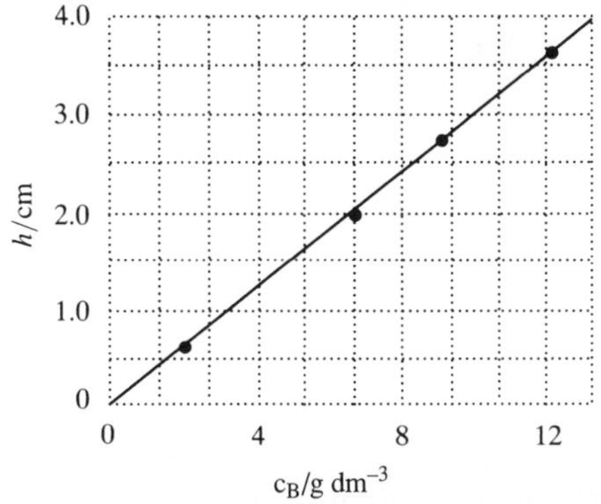

Figure 5.2

Hence, plot h against c and identify the slope as $\frac{RT}{\rho g M}$. Figure 5.2 shows the plot of the data. The slope of the line is 0.29 cm/(g dm^{-3}), so

$$\frac{RT}{\rho g M} = \frac{0.29 \text{ cm}}{\text{g dm}^{-3}} = 0.29 \text{ cm dm}^3 \text{ g}^{-1} = 2.9 \times 10^{-3} \text{ m}^4 \text{ kg}^{-1}$$

Therefore,

$$M = \frac{RT}{(\rho g) \times (2.9 \times 10^{-3} \text{ m}^4 \text{ kg}^{-1})}$$

$$= \frac{(8.3145 \text{ J K}^{-1} \text{mol}^{-1}) \times (298 \text{ K})}{(1.004 \times 10^3 \text{ kg m}^{-3}) \times (9.81 \text{ m s}^{-2}) \times (2.9 \times 10^{-3} \text{ m}^4 \text{ kg}^{-1})} = \boxed{87 \text{ kg mol}^{-1}}$$

E5.15(a) For A (Raoult's law basis; concentration in mole fraction):

$$a_A = \frac{p_A}{p_A^*} \text{ [5.49]} = \frac{250 \text{ Torr}}{300 \text{ Torr}} = \boxed{0.833}; \quad \gamma_A = \frac{a_A}{x_A} = \frac{0.833}{0.90} = \boxed{0.93}$$

For B (Henry's law basis; concentration in mole fraction):

$$a_B = \frac{p_B}{K_B} \text{ [5.57]} = \frac{25 \text{ Torr}}{200 \text{ Torr}} = \boxed{0.125}; \quad \gamma_B = \frac{a_B}{x_B} = \frac{0.125}{0.10} = \boxed{1.2\overline{5}}$$

For B (Henry's law basis; concentration in molality), the activity coefficient remains the same, but the activity is

$$a_B = \gamma_B \times \frac{b_B}{b^\ominus} [5.61] = 1.2\bar{5} \times 2.22 = \boxed{2.8}$$

COMMENT. The two methods for the 'solute' B give different values for the activities. This is reasonable since the reference states are different. However, the activity coefficient is the same: it is the ratio between activity and a concentration unit, and it incorporates all deviations from an ideality defined by Henry's law.

Question. What are the activity and activity coefficient of B in the Raoult's law basis?

E5.16(a) In an ideal dilute solution the solvent (CCl_4, A) obeys Raoult's law and the solute (Br_2, B) obeys Henry's law; hence

CCl_4: $p_A = x_A p_A^* [5.21] = (0.950) \times (33.85\ \text{Torr}) = \boxed{32.2\ \text{Torr}}$

Br_2: $p_B = x_B K_B [5.23] = (0.050) \times (122.36\ \text{Torr}) = \boxed{6.1\ \text{Torr}}$

$$p_{\text{total}} = (32.2 + 6.1)\ \text{Torr} = \boxed{38.3\ \text{Torr}}$$

The composition of the vapour in equilibrium with the liquid is

CCl_4: $y_A = \dfrac{p_A}{p_{\text{total}}} = \dfrac{32.2\ \text{Torr}}{38.3\ \text{Torr}} = \boxed{0.840}$

Br_2: $y_B = \dfrac{p_B}{p_{\text{total}}} = \dfrac{6.1\ \text{Torr}}{38.3\ \text{Torr}} = \boxed{0.160}$

E5.17(a) From Dalton's law of partial pressures and y_A we can compute the partial pressures:

$$y_A = \frac{p_A}{p_A + p_M} = \frac{p_A}{101.3\ \text{kPa}} = 0.516$$

So, $p_A = 101.3\ \text{kPa} \times 0.516 = 52.3\ \text{kPa}$

and $p_M = 101.3\ \text{kPa} - 52.3\ \text{kPa} = 49.0\ \text{kPa}$

$$a_A = \frac{p_A}{p_A^*} [5.49] = \frac{52.3\ \text{kPa}}{105\ \text{kPa}} = \boxed{0.498} \quad \text{and} \quad a_M = \frac{p_M}{p_M^*} = \frac{49.0\ \text{kPa}}{73.5\ \text{kPa}} = \boxed{0.667}$$

$$\gamma_A = \frac{a_A}{x_A} [5.51] = \frac{0.498}{0.400} = \boxed{1.24} \quad \text{and} \quad \gamma_M = \frac{a_M}{x_M} = \frac{0.667}{0.600} = \boxed{1.11}$$

E5.18(a) The definition of ionic strength is

$$I = \frac{1}{2}\sum_i \left(\frac{b_i}{b^\ominus}\right) z_i^2 \ [5.76]$$

and if b is the molal concentration of an M_pX_q salt, the molal concentrations of the ions are

$$b_M = p \times b \quad \text{and} \quad b_X = q \times b$$

Hence, $I = \dfrac{1}{2}(pz_+^2 + qz_-^2)\left(\dfrac{b}{b^\ominus}\right)$

For KCl (or any other compound of monovalent ions)

$$I = \frac{1}{2}(1\times 1 + 1\times 1)\left(\frac{b}{b^{\ominus}}\right) = \left(\frac{b}{b^{\ominus}}\right)$$

For $CuSO_4$ (or any other compound of divalent ions)

$$I = \frac{1}{2}(1\times 2^2 + 1\times 2^2)\left(\frac{b}{b^{\ominus}}\right) = 4\left(\frac{b}{b^{\ominus}}\right)$$

Thus, for this mixture

$$I = I(\mathrm{KCl}) + I(\mathrm{CuSO_4}) = \left(\frac{b(\mathrm{KCl})}{b^{\ominus}}\right) + 4\left(\frac{b(\mathrm{CuSO_4})}{b^{\ominus}}\right) = 0.10 + 4\times 0.20 = \boxed{0.90}$$

COMMENT. Note that the strength of a solution of more than one electrolyte may be calculated by summing the ionic strengths of each electrolyte considered as a separate solution, as in the solution to this exercise, or by summing the product $\frac{1}{2}\left(\frac{b_i}{b^{\ominus}}\right)z_i^2$ for each individual ion as in the definition of I [5.76].

E5.19(a) The original KNO_3 solution has an ionic strength of 0.150. (For compounds of monovalent ions, the ionic strength is numerically equal to the molal concentration, as shown in Exercise 5.18(a).) Therefore, the ionic strengths of the added salts must be 0.100.

(a) For $Ca(NO_3)_2$ $I = \frac{1}{2}(1\times 2^2 + 2\times 1^2)\left(\frac{b}{b^{\ominus}}\right)$ [5.76] $= 3\left(\frac{b}{b^{\ominus}}\right)$

Therefore, the solution should be made $0.100\ \mathrm{mol\ kg^{-1}}/3 = 0.0333\ \mathrm{mol\ kg^{-1}}$ in $Ca(NO_3)_2$. The mass that should be added to 500 g of the solution is therefore

$$(0.500\ \mathrm{kg})\times(0.0333\ \mathrm{mol\ kg^{-1}})\times(164\ \mathrm{g\ mol^{-1}}) = \boxed{2.73\ \mathrm{g}}$$

(b) For NaCl, the ionic strength is numerically equal to the molality; therefore, the solution should be made $0.100\ \mathrm{mol\ kg^{-1}}$ in NaCl. The mass that should be added to 500 g of the solution is therefore

$$(0.500\ \mathrm{kg})\times(0.100\ \mathrm{mol\ kg^{-1}})\times(58.44\ \mathrm{g\ mol^{-1}}) = \boxed{2.92\ \mathrm{g}}$$

E5.20(a) The solution is dilute, so use the Debye–Hückel limiting law:

$$\log \gamma_{\pm} = -|z_+z_-|AI^{1/2}\ [5.75]$$

$$I = \frac{1}{2}\sum_i\left(\frac{b_i}{b^{\ominus}}\right)z_i^2 = \frac{1}{2}\{(0.010\times 2^2) + (2\times 0.010\times 1^2) + (0.030\times 1^2) + (0.030\times 1^2)\} = 0.060$$

For $CaCl_2$:

$$\log \gamma_{\pm} = -2\times 1\times 0.509\times(0.060)^{1/2} = -0.25 \quad \text{so} \quad \gamma_{\pm} = \boxed{0.56}$$

The activities of the ions are

$$a(\mathrm{Ca^{2+}}) = \gamma_{\pm} b(\mathrm{Ca^{2+}})/b^{\ominus} = 0.56\times 0.010 = \boxed{0.0056}$$

and $a(\mathrm{Cl^-}) = \gamma_{\pm} b(\mathrm{Cl^-})/b^{\ominus} = 0.56\times(2\times 0.010) = \boxed{0.011}$

Question. What are the activity coefficients and activities of NaF in the same solution?

E5.21(a) The extended Debye–Hückel law (eqn 5.78), with the parameter C set equal to zero, is

$$\log\gamma_\pm = -\frac{A|z_+z_-|I^{1/2}}{1+BI^{1/2}}$$

Solve for B.

$$B = -\left(\frac{1}{I^{1/2}} + \frac{A|z_+z_-|}{\log\gamma_\pm}\right) = -\left(\frac{1}{(b/b^\ominus)^{1/2}} + \frac{0.509}{\log\gamma_\pm}\right)$$

Draw up the following table

$b/(\text{mol kg}^{-1})$	5.0×10^{-3}	10.0×10^{-3}	20.0×10^{-3}
$\gamma_\pm$	0.930	0.907	0.879
B	2.01	2.01	2.02

The values of B are constant, illustrating that the extended law fits these activity coefficients with $B = \boxed{2.01}$.

E5.22(a) Let subscript 1 denote the methylbenzene and 2 the dimethylbenzene. The partial pressures of the two liquids sum to 0.50 atm = 0.50 atm × 101.3 kPa atm^{-1} = 50.7 kPa.

$$p_1 + p_2 = p = x_1p_1^* + x_2p_2^* \text{ [Raoult's law]} = x_1p_1^* + (1-x_1)p_2^*$$

Solve for x_1:

$$x_1 = \frac{p - p_2^*}{p_1^* - p_2^*} = \frac{(50.7-20.0)\text{ kPa}}{(53.3-20.0)\text{ kPa}} = \boxed{0.92}$$

and $x_2 = 1 - 0.92 = \boxed{0.08}$

The vapour phase mole fractions are given by Dalton's law:

$$y_1 = \frac{p_1}{p} = \frac{x_1p_1^*}{p} = \frac{(0.92)\times 53.3\text{ kPa}}{50.7\text{ kPa}} = \boxed{0.97}$$

and $y_2 = \dfrac{x_2p_2^*}{p} = \dfrac{(0.08)\times 20.0\text{ kPa}}{50.7\text{ kPa}} = \boxed{0.03}$

E5.23(a) The partial vapour pressures are given by Raoult's law:

$$p_A = x_Ap_A^* \quad\text{and}\quad p_B = x_Bp_B^* = (1-x_B)p_B^*$$

Dalton's law relates these vapour pressures to the vapour-phase mole fractions:

$$y_A = \frac{p_A}{p_{\text{total}}} = \frac{x_Ap_A^*}{x_Ap_A^* + (1-x_A)p_B^*}$$

Solve for x_A:

$$x_Ap_A^* + (1-x_A)p_B^* = \frac{x_Ap_A^*}{y_A}x_A;\text{ hence}\left(p_A^* - p_B^* - \frac{p_A^*}{y_A}\right) = -p_B^*$$

$$x_A = \frac{p_B^*}{p_B^* + \dfrac{p_A^*}{y_A} - p_A^*} = \frac{52.0\text{ kPa}}{\left(52.0 + \dfrac{76.7}{0.350} - 76.7\right)\text{kPa}} = \boxed{0.267}$$

and $x_B = 1 - x_A = 1 - 0.267 = \boxed{0.733}$

The total vapour pressure is

$$p_{\text{total}} = x_A p_A^* + x_B p_B^* = 0.267 \times 76.7\text{ kPa} + 0.733 \times 52.0\text{ kPa} = \boxed{58.6\text{ kPa}}$$

E5.24(a) (a) If the solution is ideal, then the partial vapour pressures are given by Raoult's law:

$$p_A^\circ = x_A p_A^* = 0.6589 \times 127.6\text{ kPa} = 84.1\text{ kPa}$$

and $p_B^\circ = x_B p_B^* = (1 - x_B)p_B^* = (1 - 0.6589) \times 50.60\text{ kPa} = 17.26\text{ kPa}$

(Note the use of the symbol ° to emphasize that these are idealized quantities; we do not yet know if they are the actual partial vapour pressures.) At the normal boiling temperature, the partial vapour pressures must add up to 1 atm (101.3 kPa). These ideal partial vapour pressures **do** add up to 101.3 kPa, so the $\boxed{\text{solution is ideal}}$.

(b) Because the solution is ideal, the ideal partial vapour pressures are the actual partial vapour pressures. We can use them to find the vapour phase composition:

$$y_A = \frac{p_A}{p_{\text{total}}} = \frac{84.1\text{ kPa}}{101.3\text{ kPa}} = \boxed{0.830}$$

and $y_B = \dfrac{p_B}{p_{\text{total}}} = \dfrac{17.26\text{ kPa}}{101.3\text{ kPa}} = \boxed{0.1703}$

E5.25(a) (a) If the solution is ideal, then the partial vapour pressures are given by Raoult's law:

$$p_{DE} = x_{DE} p_{DE}^* = 0.60 \times 22.9\text{ kPa} = 13.7\text{ kPa } [x_A = z_A \text{ when mixture is all liquid}]$$

and $p_{DP} = x_{DP} p_{DP}^* = 0.40 \times 17.1\text{ kPa} = 6.8\text{ kPa}$

The total pressure is

$$p_{\text{total}} = p_{DE} + p_{DP} = (13.7 + 6.8)\text{ kPa} = \boxed{20.6\text{ kPa}}.$$

(b) The composition of the vapour is given by

$$y_{DE} = \frac{p_{DE}}{p_{\text{total}}} = \frac{13.7\text{ kPa}}{20.6\text{ kPa}} = \boxed{0.668}$$

and $y_{DP} = \dfrac{p_{DP}}{p_{\text{total}}} = \dfrac{6.8\text{ kPa}}{20.6\text{ kPa}} = \boxed{0.332}$

E5.26(a) Add the boiling point of methylbenzene to the table at $x_M = y_M = 1$ and the boiling point of octane at $x_M = y_M = 0$. Plot the boiling temperatures against liquid mole fractions and the same boiling temperatures against vapour mole fractions on the same plot.

The phase diagram is shown in Figure 5.3. The phase boundary curves are polynomial fits to the data points.

(a) Find $x_M = 0.25$ on the lower curve and draw a horizontal tie line to the upper curve. The mole fraction at that point is $\boxed{y_M = 0.36}$.

(b) Find $x_M = 0.75$ (i.e. $x_O = 0.25$) on the lower curve and draw a horizontal tie line to the upper curve. The mole fraction at that point is $\boxed{y_M = 0.80 \text{ (i.e. } y_O = 0.20)}$.

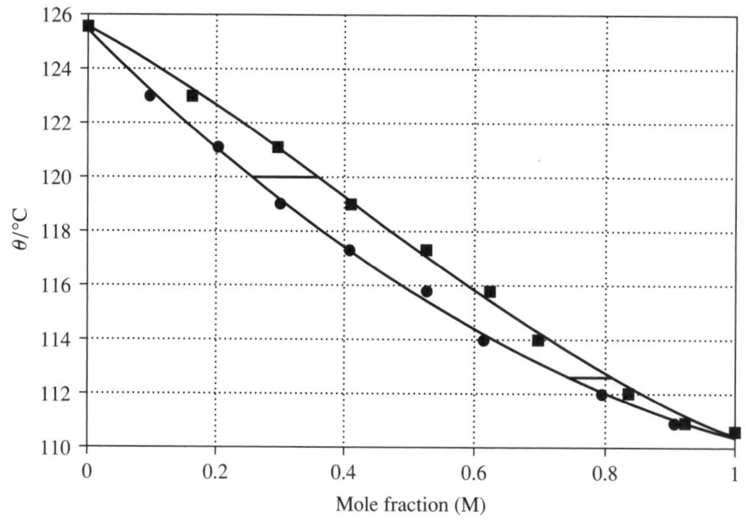

Figure 5.3

E5.27(a) The phase diagram is shown in Figure 5.4.

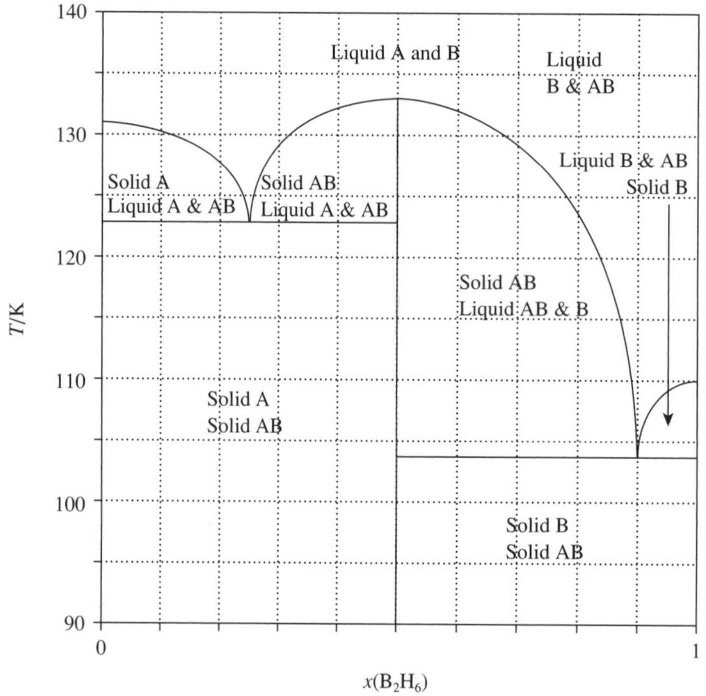

Figure 5.4

E5.28(a) Refer to Figure 5.64 of the main text. At the lowest temperature shown on the phase diagram, there are two liquid phases, a water-rich phase ($x_B = 0.07$) and a methylpropanol-rich phase ($x_B = 0.88$); the latter phase is about 10 times as abundant as the former (lever rule). On heating, the compositions of the two phases change, the water-rich phase increasing significantly in methylpropanol and the methylpropanol-rich phase more gradually increasing in water. (Note how the composition of the left side of the diagram changes more with temperature than the right.) The relative proportions of the phases continue to be given by the lever rule. Just before the isopleth intersects the phase boundary, the methylpropanol-rich phase ($x_B = 0.8$) is in equilibrium with a vanishingly small water-rich phase ($x_B = 0.36$). Then the phases merge, and the single-phase region is encountered with $x_B = 0.8$.

E5.29(a) The feature that indicates incongruent melting (Section 5.9(c)) is circled in Figure 5.5. The incongruent melting point is marked as $T_1 \approx 350°C$. The composition of the eutectic is $x_B \approx \boxed{0.26}$ and its melting point is labeled $T_2 \approx \boxed{200°C}$.

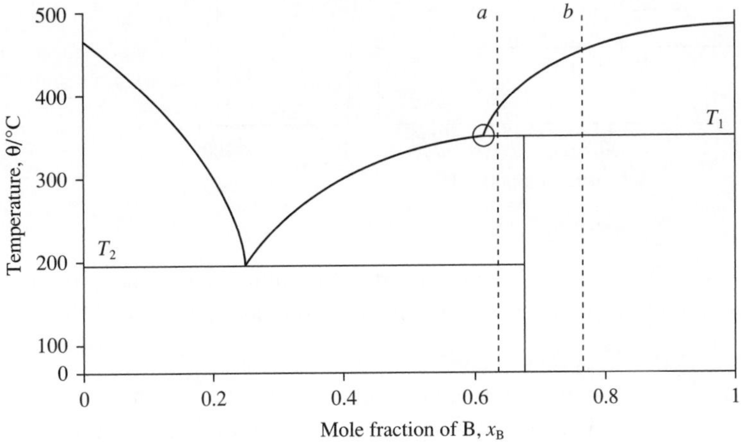

Figure 5.5

E5.30(a) The cooling curves are shown in Figure 5.6(b). Note the breaks (abrupt change in slope) at temperatures corresponding to points a_1, a_2, b_1, and b_2 of the phase diagram (Figure 5.6(a)). Also note the eutectic halt at a_3.

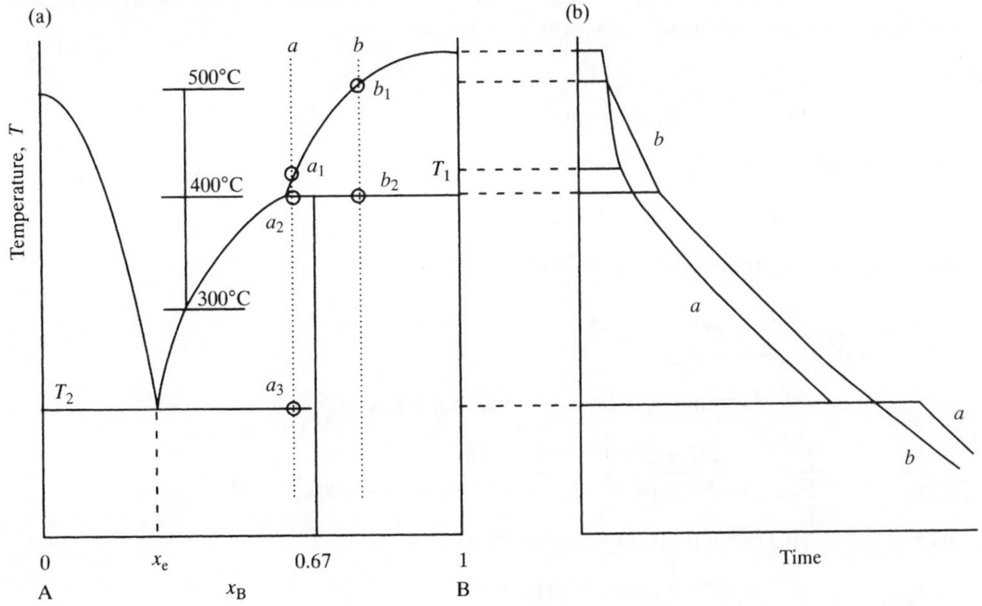

Figure 5.6

E5.31(a) Refer to Figure 5.7. Dotted horizontal lines have been drawn at the relevant temperatures.

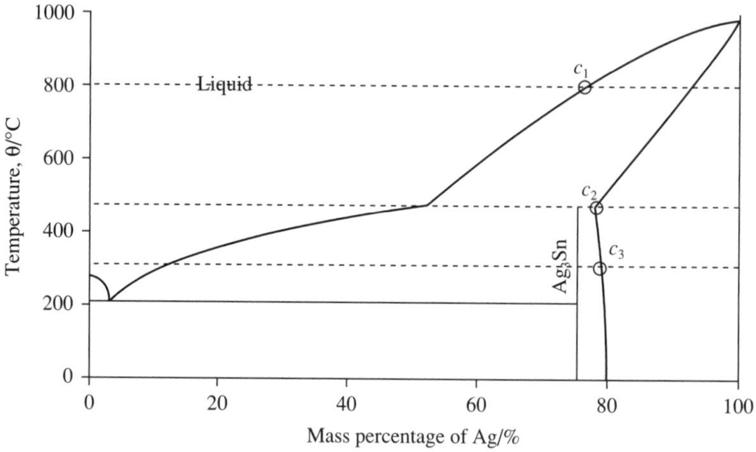

Figure 5.7

(a) The solubility of silver in tin at 800°C is determined by the point c_1. (At higher proportions of silver, the system separates into two phases, a liquid and a solid phase rich in silver.) The point c_1 corresponds to $\boxed{76\%}$ silver by mass.

(b) The compound Ag_3Sn decomposes at this temperature. Three phases are in equilibrium here: a liquid containing atomic Ag and Sn about 52% Ag by mass; a solid solution of Ag_3Sn in Ag; and solid Ag_3Sn. See point c_2.

(c) At point c_3, two phases coexist: solid Ag_3Sn and a solid solution of the compound and atomic silver. Because this point is close to the Ag_3Sn composition, the solid solution is mainly Ag_3Sn—at least when measured in mass terms. Let us express the composition of the solid solution as a ratio of moles of compound (n_c) to moles of atomic silver (n_a). We can relate these quantities to the silver mass fraction c_{Ag} by employing the definition of mass fraction, namely the mass of silver (from the compound and from atomic silver) over the total sample mass:

$$c_{Ag} = \frac{m_{Ag}}{m_{Ag} + m_{Sn}} = \frac{(3n_c + n_a)M_{Ag}}{(3n_c + n_a)M_{Ag} + n_c M_{Sn}}.$$

Rearrange this relationship, collecting terms in n_c on one side and n_a on the other:

$$n_c\{3M_{Ag}(c_{Ag} - 1) + M_{Sn}c_{Ag}\} = n_a M_{Ag}(1 - c_{Ag})$$

The mole ratio of compound to atomic silver is given by

$$\frac{n_c}{n_a} = \frac{M_{Ag}(1 - c_{Ag})}{3M_{Ag}(c_{Ag} - 1) + M_{Sn}c_{Ag}}$$

At 300°C, $c_{Ag} = 0.78$ (point c_3 on the coexistence curve), so

$$\frac{n_c}{n_a} = \frac{(107.9\ \text{g mol}^{-1}) \times (1 - 0.78)}{3 \times (107.9\ \text{g mol}^{-1}) \times (0.78 - 1) + (118.7\ \text{g mol}^{-1}) \times 0.78} = \boxed{1.11}$$

At 460°C, $c_{Ag} = 0.77$ (point c_2 on the coexistence curve), so

$$\frac{n_c}{n_a} = \frac{(107.9\ \text{g mol}^{-1}) \times (1 - 0.77)}{3 \times (107.9\ \text{g mol}^{-1}) \times (0.77 - 1) + (118.7\ \text{g mol}^{-1}) \times 0.77} = \boxed{1.46}$$

E5.32(a) (a) See Figures 5.8(a) and (b).

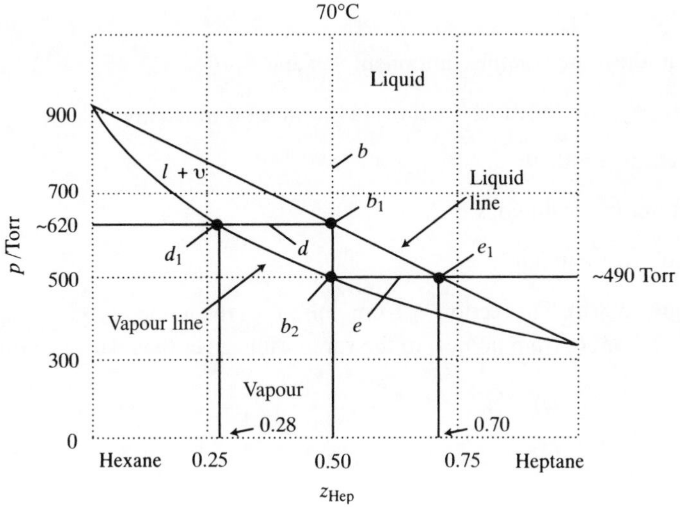

Figure 5.8(a)

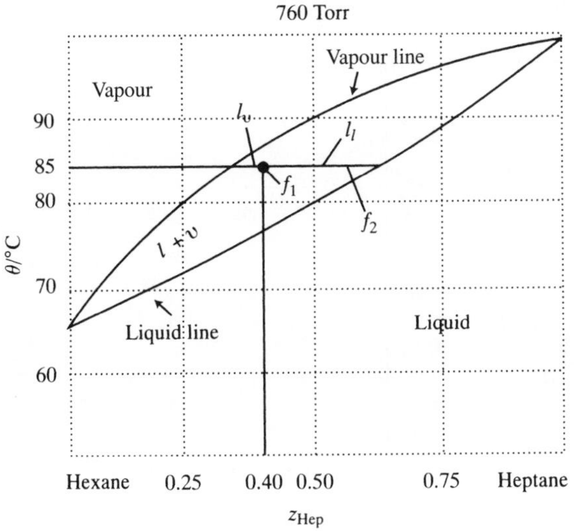

Figure 5.8(b)

(b) Follow line b in Figure 5.8(a) down to the liquid line, which intersects at point b_1. The vapour pressure at b_1 is $\boxed{620\ \text{Torr}}$.

(c) Follow line b in Figure 5.8(a) down to the vapour line, which intersects at point b_2. The vapour pressure at b_2 is $\boxed{490\ \text{Torr}}$. From points b_1 to b_2, the system changes from essentially all liquid to essentially all vapour.

(d) Consider tie line d; point b_1 gives the mole fractions of the liquid, which are

$$x_{\text{Hep}} = 0.50 = 1 - x_{\text{Hex}}, \quad \text{so} \quad x_{\text{Hex}} = \boxed{0.50}$$

Point d_1 gives the mole fractions in the vapour, which are

$$y_{\text{Hep}} = 0.28 = 1 - y_{\text{Hex}}, \quad \text{so} \quad y_{\text{Hex}} = \boxed{0.72}$$

The initial vapour is richer in the more volatile component, hexane.

(e) Consider tie line e; point b_2 gives the mole fractions in the vapour, which are

$$y_{\text{Hep}} = 0.50 = 1 - y_{\text{Hex}}, \quad \text{so} \quad y_{\text{Hex}} = \boxed{0.50}$$

Point e_1 gives the mole fractions in the liquid, which are

$$x_{\text{Hep}} = 0.70 = 1 - x_{\text{Hex}}, \quad \text{so} \quad x_{\text{Hex}} = \boxed{0.30}$$

(f) Consider tie line in Figure 5.8(b). The section, l_l, from point f_1 to the liquid line gives the relative amount of vapour; the section, l_v, from point f_1 to the vapour line gives the relative amount of liquid. That is

$$n_v l_v = n_l l_l \ [5.46] \quad \text{or} \quad \frac{n_v}{n_l} = \frac{l_l}{l_v} \approx \frac{7}{2}$$

Since the total amount is 2 mol, $n_v = 1.6$ mol and $n_l = 0.4$ mol.

E5.33(a) The phase diagram is drawn in Figure 5.9.

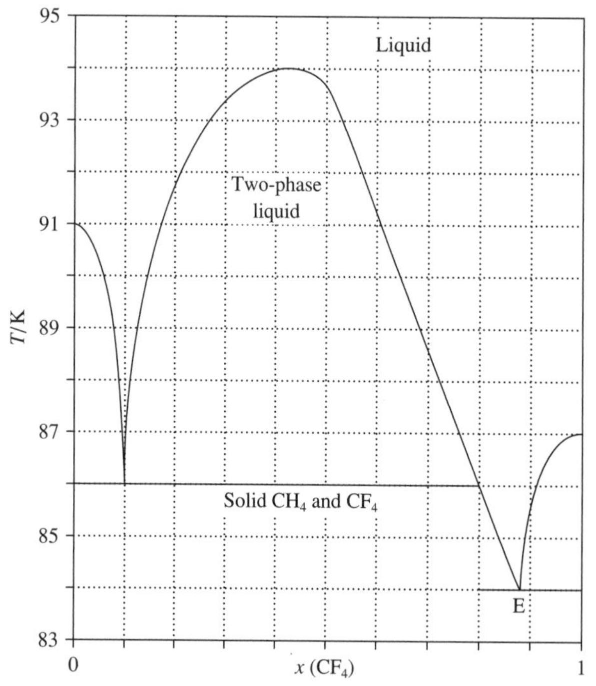

Figure 5.9

E5.34(a) The cooling curves are sketched in Figure 5.10. Note the breaks and halts. The breaks correspond to changes in the rate of cooling due to the freezing out of a solid that releases its heat of fusion and thus slows down the cooling process. The halts correspond to the existence of three phases and hence no variance until one of the phases disappears.

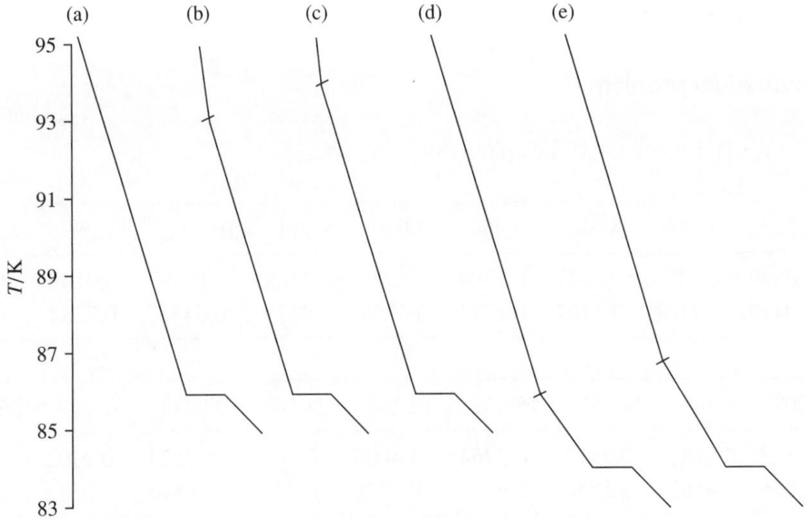

Figure 5.10

E5.35(a) The phase diagram is sketched in Figure 5.11.

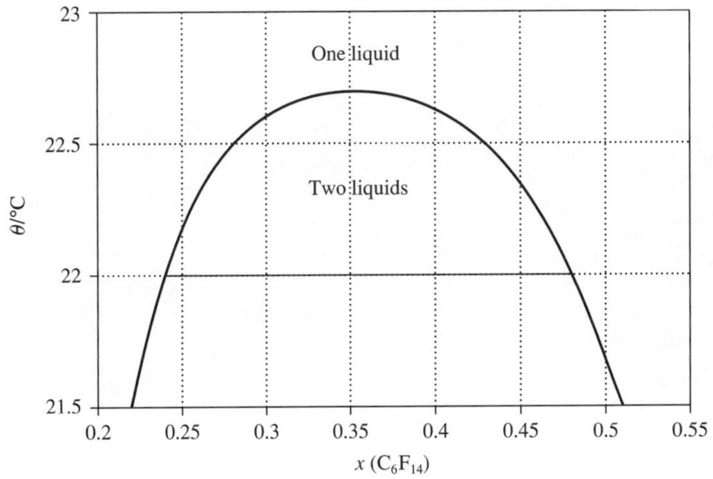

Figure 5.11

(a) The mixture has a single liquid phase at all compositions.

(b) Adding C_6F_{14} amounts to moving from left to right on the phase diagram on the 22.0°C line. When the overall composition reaches $x(C_6F_{14}) = 0.24$ the mixture separates into two liquid phases, the more abundant phase having $x = 0.24$ and an initially minuscule phase of $x = 0.48$. The compositions of the two phases remain the same but their relative amounts change as more C_6F_{14} is added until the overall composition reaches $x = 0.48$. At that point, the phase having $x = 0.24$ vanishes, and from there on the mixture forms a single liquid phase.

Solutions to problems

Solutions to numerical problems

P5.1 $p_A = y_A p$ and $p_B = y_B p$ [1.13]. Hence, draw up the following table:

p_A/kPa	0	1.399	3.566	5.044	6.996	7.940	9.211	10.105	11.287	12.295
x_A	0	0.0898	0.2476	0.3577	0.5194	0.6036	0.7188	0.8019	0.9105	1
y_A	0	0.0410	0.1154	0.1762	0.2772	0.3393	0.4450	0.5435	0.7284	1

p_B/kPa	0	4.209	8.487	11.487	15.462	18.243	23.582	27.334	32.722	36.066
x_B	0	0.0895	0.1981	0.2812	0.3964	0.4806	0.6423	0.7524	0.9102	1
y_B	0	0.2716	0.4565	0.5550	0.6607	0.7228	0.8238	0.8846	0.9590	1

The data are plotted in Figure 5.12. We can assume, at the lowest concentrations of both A and B, that Henry's law (eqn 5.23) will hold. The Henry's law constants are then given by:

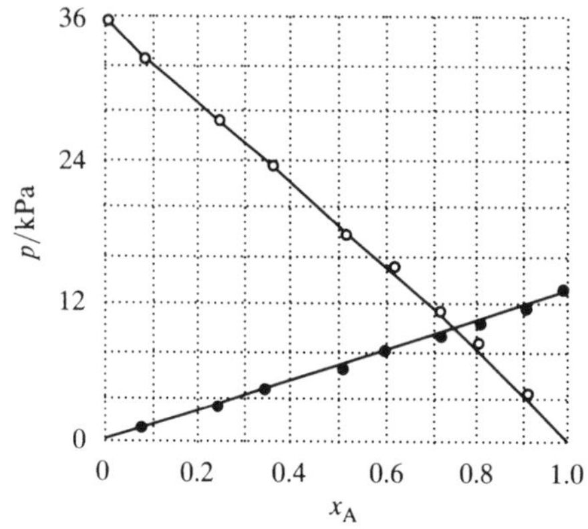

Figure 5.12

$$K_A = \frac{p_A}{x_A} = \boxed{15.58\ \text{kPa}} \text{ from the point at } x_A = 0.0898$$

$$K_B = \frac{p_B}{x_B} = \boxed{47.03\ \text{kPa}} \text{ from the point at } x_B = 0.0895$$

P5.3 Let A stand for water and B for $MgSO_4$(aq):

$$V_B = \left(\frac{\partial V}{\partial n_B}\right)_{p,T,n_A} [5.1] = \left(\frac{\partial V}{\partial x}\right)_{p,T,n_A} \text{mol}^{-1}\ [x \equiv b/b^{\ominus}]$$

$$= 2 \times 34.69 \times (x - 0.070)\ \text{cm}^3\,\text{mol}^{-1} = \boxed{-1.4\ \text{cm}^3\,\text{mol}^{-1}}$$

for this solution consisting of 0.050 mol of $MgSO_4$ and 1.000 kg (55.49 mol) water. The total volume at this composition is

$$V = 1001.21 + 34.69 \times (0.050 - 0.070)^2 = 1001.23\ \text{cm}^3$$

The total volume is also equal to

$V = V_A n_A + V_B n_B$ [5.3]

Therefore, $V_A = \dfrac{V - V_B n_B}{n_A} = \dfrac{1001.23\ \text{cm}^3 - (-1.4\ \text{cm}^3) \times (0.050\ \text{mol})}{55.49\ \text{mol}} = \boxed{18.0\overline{4}\ \text{cm}^3\ \text{mol}^{-1}}$

Question. What meaning can be ascribed to a negative partial molar volume?

P5.5 Let E denote ethanol and W water; then

$V = V_E n_E + V_W n_W$ [5.3]

For a 50% mixture by mass, $m_E = m_W$, implying that

$$M_E n_E = M_W n_W \quad \text{or} \quad n_W = \frac{n_E M_E}{M_W}$$

Hence, $V = n_E V_E + \dfrac{n_E M_E V_W}{M_W}$

Solving for n_E and then for n_W yields

$$n_E = \frac{V}{V_E + \dfrac{M_E V_W}{M_W}} \quad \text{and} \quad n_W = \frac{M_E V}{V_E M_W + M_E V_W}$$

Furthermore, $x_E = \dfrac{n_E}{n_E + n_W} = \dfrac{1}{1 + \dfrac{M_E}{M_W}}$

Now put in the numbers:

$M_E = 46.07\ \text{g mol}^{-1}$ and $M_W = 18.02\ \text{g mol}^{-1}$, so $\dfrac{M_E}{M_W} = 2.557$

Therefore, $x_E = 0.2812$ and $x_W = 1 - x_E = 0.7188$. At this composition

$V_E = 56.3\ \text{cm}^3\ \text{mol}^{-1}$ and $V_W = 17.6\ \text{cm}^3\ \text{mol}^{-1}$ [Figure 5.1 of main text]

Therefore, $n_E = \dfrac{100\ \text{cm}^3}{(56.3\ \text{cm}^3\ \text{mol}^{-1}) + (2.557) \times (17.6\ \text{cm}^3\ \text{mol}^{-1})} = 0.987\ \text{mol}$

and $n_W = 2.557 \times 0.987\ \text{mol} = 2.52\ \text{mol}$

The fact that these amounts correspond to a mixture containing 50% by mass of both components is easily checked as follows.

$m_E = n_E M_E = 0.987\ \text{mol} \times 46.07\ \text{g mol}^{-1} = 45.5\ \text{g}$ ethanol

and $m_W = n_W M_W = 2.52\ \text{mol} \times 18.02\ \text{g mol}^{-1} = 45.5\ \text{g}$ water

At 20°C the densities of ethanol and water are, $\rho_E = 0.789\ \text{g cm}^{-3}$ and $\rho_W = 0.997\ \text{g cm}^{-3}$. Hence, the volumes of pure ethanol and water that must be mixed are

$$V_E^* = \frac{m_E}{\rho_E} = \frac{45.5\text{ g}}{0.789\text{ g cm}^{-3}} = \boxed{57.6\text{ cm}^3}\text{ of ethanol}$$

$$V_W^* = \frac{m_W}{\rho_W} = \frac{45.5\text{ g}}{0.997\text{ g cm}^{-3}} = \boxed{45.6\text{ cm}^3}\text{ of water}$$

The change in volume on adding a small amount of ethanol can be approximated by

$$\Delta V = \int dV = \int V_E dn_E \approx V_E \Delta n_E$$

where we have assumed that V_E is constant over this small range of n_E. Hence

$$\Delta V \approx (56.3\text{ cm}^3\text{ mol}^{-1}) \times \left(\frac{(0.987\text{ cm}^3)\times(0.789\text{ g cm}^{-3})}{(46.07\text{ g mol}^{-1})}\right) = \boxed{+0.95\text{ cm}^3}$$

COMMENT. Note that the volumes of pure ethanol and water before mixing do not add up to the volume of the mixture.

P5.7 The apparent molality is

$$b_{app} = \frac{\Delta T}{K_f}\,[5.35] = \frac{0.0703\text{ K}}{1.86\text{ K/(mol kg}^{-1})} = 0.0378\text{ mol kg}^{-1}$$

Since the solution molality is nominally 0.0096 mol kg^{-1} in $Th(NO_3)_4$, each formula unit supplies $\frac{0.0378}{0.0096} \approx \boxed{4\text{ ions}}$. (More careful data, as described in the original reference gives 5 to 6 ions.)

P5.9 The data are plotted in Figure 5.13. The dotted lines correspond to Henry's law vapour pressures and the dashed lines to Raoult's law; the solid curves represent the experimental data.

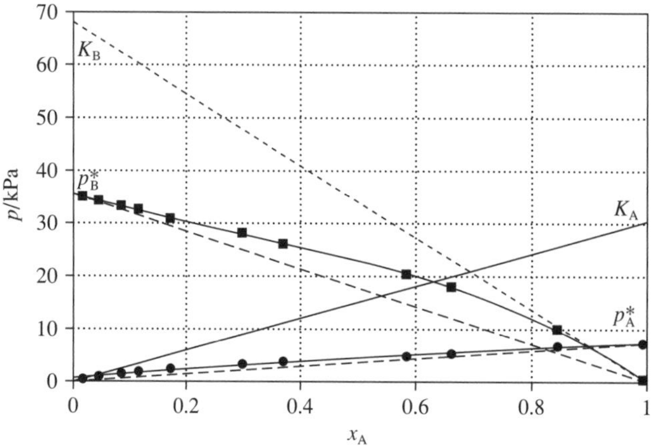

Figure 5.13

On a Raoult's law basis, $a = \frac{p}{p^*}$ [5.49] and $a = \gamma x$ [5.51], so $\gamma = \frac{p}{xp^*}$. On a Henry's law basis, $a = \frac{p}{K}$ [5.57], so $\gamma = \frac{p}{xK}$. The vapour pressures of the pure components are not given in the table of data, so we extrapolate the experimental data to obtain $p_A^* = 7.3\text{ kPa}$ and $p_B^* = 35.6\text{ kPa}$. The Henry's law constant for benzene is determined by extrapolating the low-B data to $x_B = 1$, i.e. to $x_A = 0$.

(The Henry's law constant for acetic acid can also be determined by extrapolating the low-A data to $x_A = 1$) The values obtained are $K_B = 68.1$ kPa and $K_A = 30.3$ kPa. Then draw up the following table based on the partial pressures given in the data.

x_A	0.016	0.0439	0.0835	0.1138	0.1714
p_A/kPa	0.484	0.967	1.535	1.89	2.45
p_B/kPa	35.05	34.29	33.28	32.64	30.9
a_A(R)	0.066	0.132	0.210	0.259	0.336 $[p_A/p_A^*]$
a_B(R)	0.985	0.963	0.935	0.917	0.868 $[p_B/p_B^*]$
γ_A(R)	4.144	3.017	2.518	2.275	1.958 $[p_A/x_A p_A^*]$
γ_B(R)	1.001	1.007	1.020	1.035	1.048 $[p_B/x_B p_B^*]$
a_B(H)	0.515	0.504	0.489	0.479	0.454 $[p_B/K_B]$
γ_B(H)	0.523	0.527	0.533	0.541	0.548 $[p_B/x_B K_B]$

x_A	0.2973	0.3696	0.5834	0.6604	0.8437	0.9931
p_A/kPa	3.31	3.83	4.84	5.36	6.76	7.29
p_B/kPa	28.16	26.08	20.42	18.01	10	0.47
a_A(R)	0.453	0.525	0.663	0.734	0.926	0.999
a_B(R)	0.791	0.733	0.574	0.506	0.281	0.013
γ_A(R)	1.525	1.420	1.136	1.112	1.098	1.006
γ_B(R)	1.126	1.162	1.377	1.490	1.797	1.913
a_B(H)	0.414	0.383	0.300	0.264	0.147	0.007
γ_B(H)	0.588	0.607	0.720	0.779	0.939	1.000

G^E is defined [Section 5.4(b)] as

$$G^E = \Delta_{mix}G - \Delta_{mix}G^{ideal} = nRT(x_A \ln a_A + x_B \ln a_B) - nRT(x_A \ln x_A + x_B \ln x_B)$$

and with $a = \gamma x$

$$G^E = nRT(x_A \ln \gamma_A + x_B \ln \gamma_B)$$

For $n = 1$, we can draw up the following table from the information above and

$$RT = 8.3145 \text{ J mol}^{-1} \text{ K}^{-1} \times 323 \text{ K} = 2.69 \times 10^3 \text{ J mol}^{-1} = 2.69 \text{ kJ mol}^{-1}$$

x_A	0.016	0.0439	0.0835	0.1138	0.1714
$x_A \ln \gamma_A$	0.023	0.0485	0.077	0.094	0.115
$x_B \ln \gamma_B$(R)	0.001	0.0071	0.018	0.030	0.038
G^E/kJ mol^{-1}	0.0626	0.1492	0.256	0.332	0.413

x_A	0.2973	0.3696	0.5834	0.6604	0.8437	0.9931
$x_A \ln \gamma_A$	0.125	0.129	0.075	0.070	0.079	0.006
$x_B \ln \gamma_B$(R)	0.083	0.095	0.133	0.135	0.092	0.004
G^E/kJ mol^{-1}	0.560	0.602	0.558	0.551	0.457	0.027

Question. In this problem both A and B were treated as solvents, but only B as a solute. Extend the table by including a row for γ_A(H).

P5.11 (a) Let V_1^* be the molar volume of *pure* propionic acid and V_1 be its *partial* molar volume in the mixture (and V_2^* and V_2 the analogous quantities for oxane). The volume of an ideal mixture is additive

$$V^{\text{ideal}} = n_1V_1^* + n_2V_2^*,$$

so the volume of a real mixture is

$$V = V^{\text{ideal}} + V^{\text{E}}.$$

We have an expression for excess molar volume in terms of mole fractions. To compute partial molar volumes, we need an expression for the excess volume as a function of moles

$$V^{\text{E}} = (n_1 + n_2)V_{\text{m}}^{\text{E}} = \frac{n_1n_2}{n_1+n_2}\left(a_0 + \frac{a_1(n_1-n_2)}{n_1+n_2}\right)$$

so $$V = n_1V_1^* + n_2V_2^* + \frac{n_1n_2}{n_1+n_2}\left(a_0 + \frac{a_1(n_1-n_2)}{n_1+n_2}\right)$$

The partial molar volume of propionic acid is

$$V_1 = \left(\frac{\partial V}{\partial n_1}\right)_{n_2} = V_1^* + \frac{a_0n_2^2}{(n_1+n_2)^2} + \frac{a_1(3n_1-n_2)n_2^2}{(n_1+n_2)^3} = \boxed{V_1^* + a_0x_2^2 + a_1(3x_1-x_2)x_2^2}$$

That of oxane is

$$V_2 = \left(\frac{\partial V}{\partial n_2}\right)_{n_1} = V_2^* + \frac{a_0n_1^2}{(n_1+n_2)^2} + \frac{a_1(n_1^3-3n_1^2n_2^2)}{(n_1+n_2)^3} = \boxed{V_2^* + a_0x_1^2 + a_1(x_1-3x_2)x_1^2}$$

(b) We need the molar volumes of the pure liquids

$$V_1^* = \frac{M_1}{\rho_1} = \frac{74.08\text{ g mol}^{-1}}{0.97174\text{ g cm}^{-3}} = 76.23\text{ cm}^3\text{ mol}^{-1}$$

and $$V_2^* = \frac{M_2}{\rho_2} = \frac{86.13\text{ g mol}^{-1}}{0.86398\text{ g cm}^{-3}} = 99.69\text{ cm}^3\text{ mol}^{-1}$$

In an equimolar mixture, the partial molar volume of propionic acid is

$$V_1 = 76.23 + (-2.4697)\times(0.5)^2 + (0.0608)\times\{3(0.5)-0.5\}\times(0.5)^2\text{ cm}^3\text{ mol}^{-1}$$
$$= \boxed{75.63\text{ cm}^3\text{ mol}^{-1}}$$

and that of oxane is

$$V_2 = 99.69 + (-2.4697)\times(0.5)^2 + (0.0608)\times\{0.5-3(0.5)\}\times(0.5)^2\text{ cm}^3\text{ mol}^{-1}$$
$$= \boxed{99.06\text{ cm}^3\text{ mol}^{-1}}$$

P5.13 The Henry's law constant of the solute (here CO_2) is the slope of a plot of p_B versus x_B in the limit of zero x_B. The partial pressures of CO_2 are almost but not quite equal to the total pressures reported:

$$p_{CO_2} = py_{CO_2} = p(1-y_{\text{cyc}})$$

CO_2 partial pressures are shown in the table below and plotted in Figure 5.14.

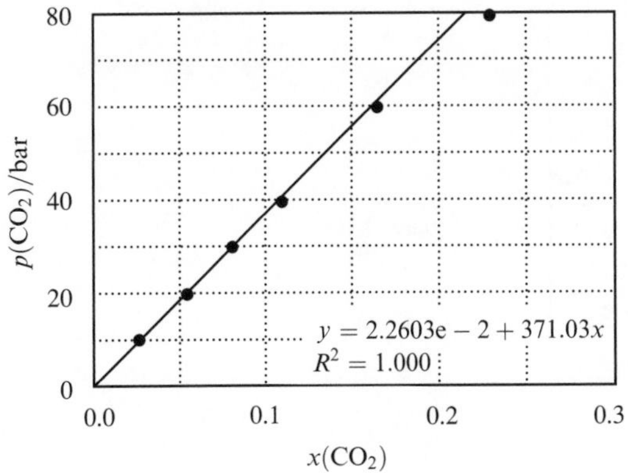

Figure 5.14

Linear regression of the low-pressure points gives $K_B = \boxed{371\text{ bar}}$.

The activity of a solute is

$$a_B = \frac{p_B}{K_B}\ [5.57] = x_B\gamma_B\ [5.58]$$

so the activity coefficient is

$$\gamma_B = \frac{p_B}{x_B K_B} = \frac{y_B p}{x_B K_B} = \frac{(1-y_A)p}{(1-x_A)K_B}$$

where the last equality applies Dalton's law of partial pressures to the vapour phase. A spreadsheet applied this equation to the above data to yield

p/bar	p_{CO_2}/bar	y_{cyc}	x_{cyc}	γ_{CO_2}
10.0	9.7	0.0267	0.9741	1.01
20.0	19.7	0.0149	0.9464	0.99
30.0	29.7	0.0112	0.9204	1.00
40.0	39.6	0.009 47	0.892	0.99
60.0	59.5	0.008 35	0.836	0.98
80.0	79.3	0.009 21	0.773	0.94

P5.15 In this mixture, $x = 0.250$, so $G^E = RTx(1-x)\{0.4857 - 0.1077(2x-1) + 0.0191(2x-1)^2\}$

$$G^E = RT(0.250)(1-0.250)\{0.4857 - 0.1077(0.500-1) + 0.0191(0.500-1)^2\} = 0.1021RT$$

Therefore, since

$$\Delta_{mix}G = \Delta_{mix}G^{ideal} + nG^E = nRT(x_A \ln x_A + x_B \ln x_B) + nG^E$$

$$\Delta_{mix}G = nRT(0.250 \ln 0.250 + 0.750 \ln 0.750) + 0.1021nRT = -0.460nRT$$

$$\Delta_{mix}G = -0.460 \times 4.00\text{ mol} \times 8.3145\text{ J mol}^{-1}\text{ K}^{-1} \times 303.15\text{ K} = \boxed{-4.64\text{ kJ mol}^{-1}}$$

P5.17 (a) The data, including that for pure chlorobenzene, are plotted in Figure 5.15.

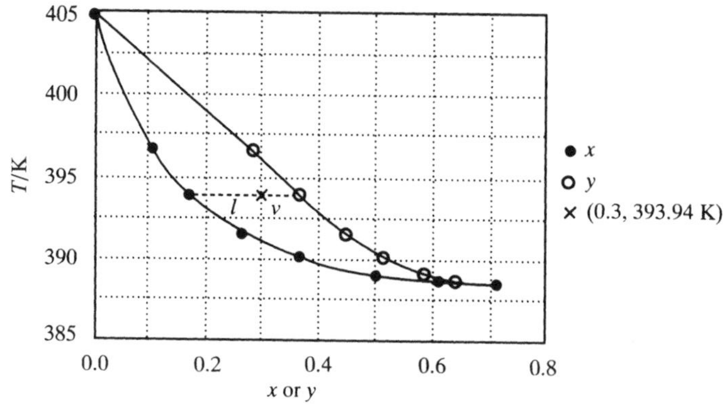

Figure 5.15

(b) The smooth curve through the (x, T) data crosses $x = 0.300$ at $\boxed{391.0\ \text{K}}$, the boiling point of the mixture.

(c) We need not interpolate data, for 393.94 K is a temperature for which we have experimental data. The mole fraction of 1-butanol in the liquid phase is 0.1700 and in the vapour phase 0.3691. According to the lever rule, the proportions of the two phases are in an inverse ratio of the distances their mole fractions are from the composition point in question. That is

$$\frac{n_{\text{liq}}}{n_{\text{vap}}} = \frac{v}{l} = \frac{0.3691 - 0.300}{0.300 - 0.1700} = \boxed{0.532}$$

P5.19 $p_{\text{A}} = a_{\text{A}} p_{\text{A}}^{*}$ [5.49] $= \gamma_{\text{A}} x_{\text{A}} p_{\text{A}}^{*}$ [5.51]

so $\gamma_{\text{A}} = \dfrac{p_{\text{A}}}{x_{\text{A}} p_{\text{A}}^{*}} = \dfrac{y_{\text{A}} p}{x_{\text{A}} p_{\text{A}}^{*}}$

Sample calculation at 80 K:

$$\gamma(\text{O}_2) = \frac{0.11 \times 100\ \text{kPa}}{0.34 \times 225\ \text{Torr}} \times \left(\frac{760\ \text{Torr}}{101.325\ \text{kPa}}\right) = 1.079$$

Summary

T/K	77.3	78	80	82	84	86	88	90.2
$\gamma(\text{O}_2)$	—	0.877	1.079	1.039	0.995	0.993	0.990	0.987

To within the experimental uncertainties the solution appears to be ideal ($\gamma = 1$). The low value at 78 K may be caused by non-ideality; however, the larger relative uncertainty in $y(\text{O}_2)$ is probably the origin of the low value.

A temperature–composition diagram is shown in Figure 5.16(a). The near ideality of this solution is, however, best shown in the pressure–composition diagram of Figure 5.16(b). The liquid line is essentially a straight line as predicted for an ideal solution.

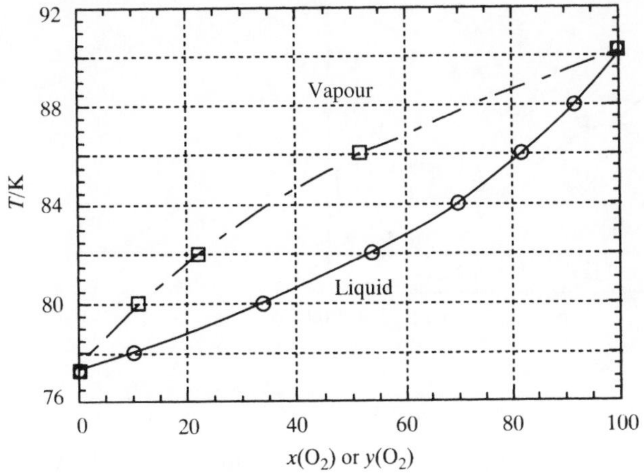

Figure 5.16(a)

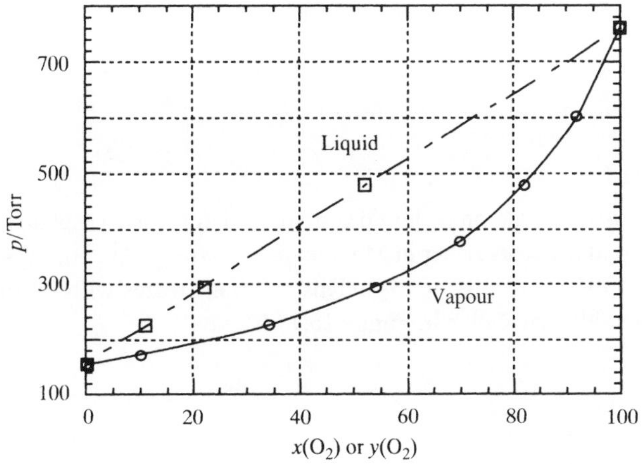

Figure 5.16(b)

P5.21 Figure 5.17 displays the phase diagram. A compound with probable formula A_3B exists. It melts incongruently at 700°C, undergoing the peritectic reaction

$$A_3B(s) \rightarrow A(s) + (A + B, l)$$

The proportions of A and B in the product are dependent upon the overall composition and the temperature. A eutectic exists at 400°C and $x_B = 0.83$.

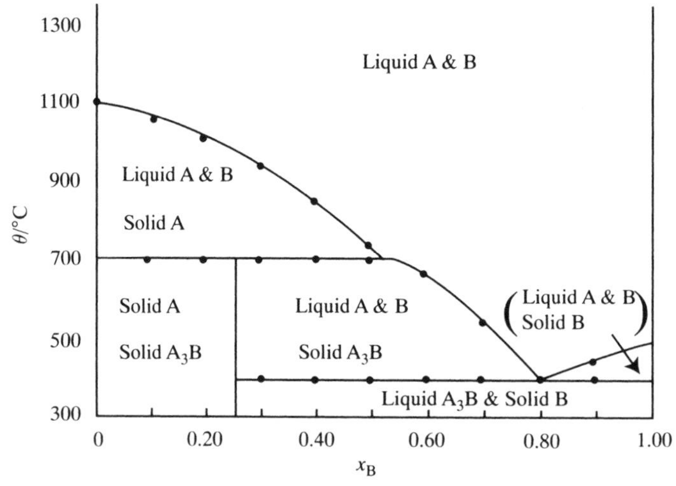

Figure 5.17

P5.23 The information has been used to construct the phase diagram in Figure 5.18(a). In $MgCu_2$ the mass percentage of Mg is

$$(100) \times \frac{24.3}{24.3 + 127} = \boxed{16}$$

and in Mg_2Cu it is

$$(100) \times \frac{48.6}{48.6 + 63.5} = \boxed{43}$$

The initial point is a_1, corresponding to a single-phase liquid system. At a_2 (at 720°C) $MgCu_2$ begins to come out of solution and the liquid becomes richer in Mg, moving toward e_2. At a_3 there is solid $MgCu_2$ + liquid of composition e_2 (33% by mass of Mg). This solution freezes without further change. The cooling curve will resemble that shown in Figure 15.18(b).

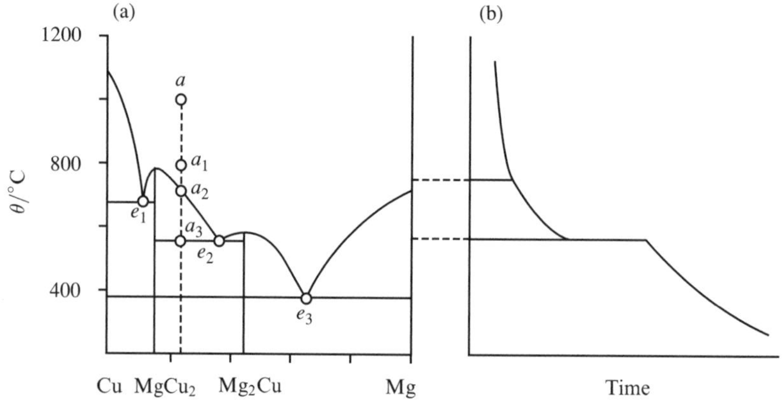

Figure 5.18

P5.25 (a) **eutectics:** $x_{Si} = 0.402$ at 1268°C $x_{Si} = 0.694$ at 1030°C

congruent melting compounds: Ca_2Si $T_f = 1314°C$

$CaSi$ $T_f = 1324°C$

incongruent melting compound: $CaSi_2$ $T_f = 1040°C$ (melts into CaSi(s) and Si-rich liquid with x_{Si} around 0.69)

(b) At 1000°C the phases in equilibrium are $\boxed{Ca_2Si \text{ and a Ca-rich liquid } (x_{Si} = 0.13)}$. The lever rule (eqn 5.46) gives the relative amounts:

$$\frac{n_{Ca_2Si}}{n_{liq}} = \frac{l_{liq}}{l_{Ca_2Si}} = \frac{0.2 - 0.13}{0.333 - 0.2} = \boxed{0.5}$$

(c) Si(s) begins to appear at about 1300°C. Further cooling causes more Si(s) to freeze out of the melt so that the melt becomes more concentrated in Ca. There is a eutectic at $x_{Si} = 0.694$ and 1030°C. Just before the eutectic is reached (i.e. at a temperature just higher than 1030°C), the lever rule says that the relative amounts of the Si(s) and liquid ($x_{Si} = 0.694$) phases are:

$$\frac{n_{Si}}{n_{liq}} = \frac{l_{liq}}{l_{Si}} = \frac{0.80 - 0.694}{1.0 - 0.80} = \boxed{0.53}$$

At the eutectic temperature a third phase appears, $CaSi_2$(s). As the melt cools at this temperature, both Si(s) and $CaSi_2$(s) freeze out of the melt while the composition of the melt remains constant. At a temperature slightly below 1030°C, all the melt will have frozen to Si(s) and $CaSi_2$(s) with the relative amounts:

$$\frac{n_{Si}}{n_{CaSi_2}} = \frac{l_{CaSi_2}}{l_{Si}} = \frac{0.80 - 0.667}{1.0 - 0.80} = \boxed{0.67}$$

Plotting the mole fraction of Si(s) and of $CaSi_2$(s) vs. that of the eutectic liquid is a convenient way to show relative amounts of the three phases as the eutectic liquid freezes. See Figure 5.19. Equations for the graph are derived from the law of conservation of atoms. For silicon atoms,

$$nz_{Si} = n_{liq}w_{Si} + n_{Si}x_{Si} + n_{CaSi_2}y_{Si}$$

where n = total number of moles of all phases
w_{Si} = Si fraction in eutectic liquid = 0.694
x_{Si} = Si fraction in Si(s) = 1.000
y_{Si} = Si fraction in $CaSi_2$(s) = 0.667
z_{Si} = Si fraction in melt = 0.800

This equation may be rewritten in mole fractions of each phase by dividing by n:

$$z_{Si} = (\text{mol fraction liq})w_{Si} + (\text{mol fraction Si})x_{Si} + (\text{mol fraction } CaSi_2)y_{Si}$$

Since (mol fraction liq) + (mol fraction Si) + (mol fraction $CaSi_2$) = 1

or (mol fraction $CaSi_2$) = 1 − (mol fraction liq) − (mol fraction Si)

we may write:

$$z_{Si} = (\text{mol fraction liq})w_{Si} + (\text{mol fraction Si})x_{Si} + [1 - (\text{mol fraction liq} + \text{mol fraction Si})]y_{Si}$$

Solving for mol fraction Si:

$$\text{mol fraction Si} = \frac{(z_{\text{Si}} - y_{\text{Si}}) - (w_{\text{Si}} - y_{\text{Si}})(\text{mol fraction liq})}{x_{\text{Si}} - y_{\text{Si}}}$$

and mol fraction $CaSi_2$ = 1 − (mol fraction liq + mol fraction Si)

These two eqns are used to prepare plots of the mol fraction of Si and mol fraction of $CaSi_2$ against the mol fraction of the melt in the range 0–0.65.

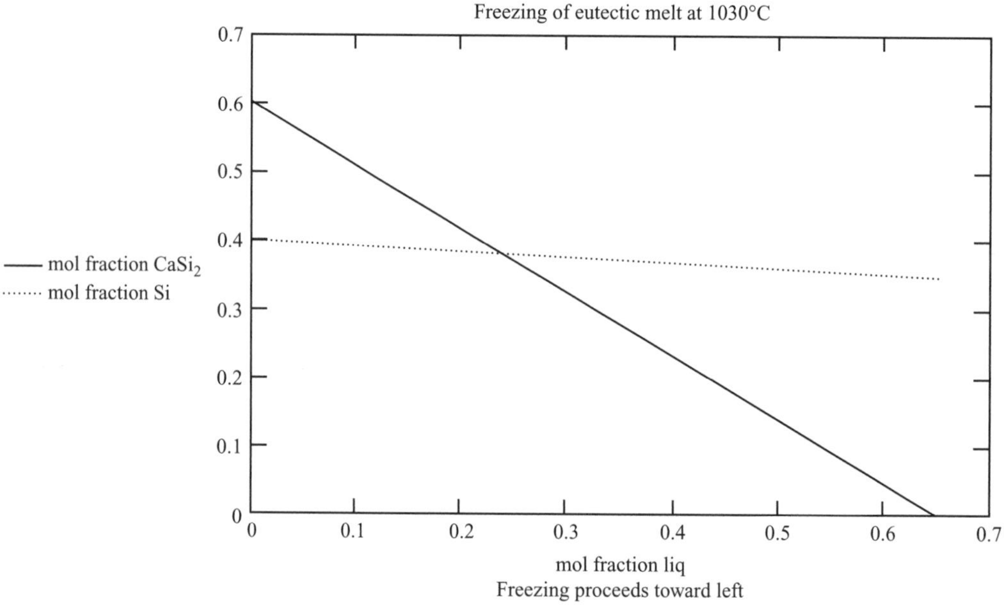

Figure 5.19

Solutions to theoretical problems

P5.27 $$\mu_{\text{A}} = \left(\frac{\partial G}{\partial n_{\text{A}}}\right)_{n_{\text{B}}} [5.4] = \mu_{\text{A}}^{\text{ideal}} + \left(\frac{\partial}{\partial n_{\text{A}}}(nG^{\text{E}})\right)_{n_{\text{B}}}, \quad \text{where} \quad \mu_{\text{A}}^{\text{ideal}} = \mu_{\text{A}}^* + RT \ln x_{\text{A}}$$

$$\left(\frac{\partial nG^{\text{E}}}{\partial n_{\text{A}}}\right)_{n_{\text{B}}} = G^{\text{E}} + n\left(\frac{\partial G^{\text{E}}}{\partial n_{\text{A}}}\right)_{n_{\text{B}}} = G^{E} + n\left(\frac{\partial x_{\text{A}}}{\partial n_{\text{A}}}\right)_{n_{\text{B}}}\left(\frac{\partial G^{\text{E}}}{\partial x_{\text{A}}}\right)_{n_{\text{B}}}$$

where $$\left(\frac{\partial x_{\text{A}}}{\partial n_{\text{A}}}\right)_{n_{\text{B}}} = \frac{\partial}{\partial n_{\text{A}}}\left(\frac{n_{\text{A}}}{n_{\text{A}} + n_{\text{B}}}\right) = \frac{1}{n_{\text{A}} + n_{\text{B}}} - \frac{n_{\text{A}}}{(n_{\text{A}} + n_{\text{B}})^2} = \frac{n_{\text{B}}}{(n_{\text{A}} + n_{\text{B}})^2} = \frac{x_{\text{B}}}{n_{\text{A}} + n_{\text{B}}} = \frac{x_{\text{B}}}{n}.$$

Hence, $$\left(\frac{\partial nG^{\text{E}}}{\partial n_{\text{A}}}\right)_{n_{\text{B}}} = G^{\text{E}} + x_{\text{B}}\left(\frac{\partial G^{\text{E}}}{\partial x_{\text{A}}}\right)_{n_{\text{B}}} = gRTx_{\text{A}}(1 - x_{\text{A}}) + (1 - x_{\text{A}})gRT(1 - 2x_{\text{A}})$$

After expanding and collecting terms, we arrive at

$$\left(\frac{\partial nG^{\text{E}}}{\partial n_{\text{A}}}\right)_{n_{\text{B}}} = gRT(1 - x_{\text{A}})^2 = gRTx_{\text{B}}^2$$

Therefore, $\mu_A = \boxed{\mu_A^* + RT \ln x_A + gRTx_B^2}$

This function is plotted for several values of the parameter g in Figure 5.20.

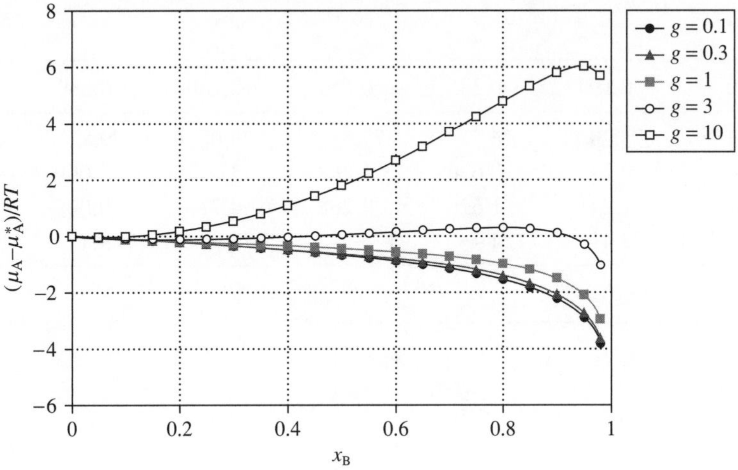

Figure 5.20

P5.29 The Gibbs–Duhem equation applies to any partial molar quantity, so we start, as in Example 5.1, with

$$n_A \, dV_A + n_B \, dV_B = 0$$

Hence, $dV_B = -\frac{n_A}{n_B} dV_A$

Therefore, by integration,

$$V_B(x_A, x_B) - V_B(0, 1) = \int_{V_B(0,1)}^{V_B(x_A,x_B)} dV_B = -\int_{V_A(0,1)}^{V_A(x_A,x_B)} \frac{n_A}{n_B} dV_A = -\int_{V_A(0,1)}^{V_A(x_A,x_B)} \frac{nx_A dV_A}{n(1-x_A)}$$

The notation V_B^* means the molar volume of pure B, which is the same as the partial molar volume of B when $x_B = 1$. Therefore,

$$V_B(x_A,x_B) = V_B^* - \int_{V_A(0,1)}^{V_A(x_A,x_B)} \frac{x_A dV_A}{1-x_A}.$$

We must now plot $x_A/(1-x_A)$ against V_A and estimate the integral. That means we must first find the partial molar volumes of chloroform (V_A) that corresponds to various chloroform mole fractions (x_A). At constant temperature and pressure,

$$V_A = \left(\frac{\partial V}{\partial n_A}\right)_{n_B} [5.1] = \left(\frac{\partial(nV_m)}{\partial(nx_A)}\right)_{n_B}, \quad \text{where } n = n_A + n_B \text{ is the total number of moles.}$$

Thus, V_A is the tangent line to the curve of a plot of V vs. n_A for a constant value of n_B. For convenience, let $n_B = 1$ mol. Then we can draw up a table of V, n, and n_A values using the V_m, x_A data given and the relationship

$$x_A = \frac{n_A}{n} = \frac{n_A}{n_A + 1\text{ mol}}.$$

Solving for n_A yields $n_A = \frac{x_A}{1 - x_A} \times 1\text{ mol}$

x_A	0	0.194	0.385	0.559	0.788	0.889	1
$V_m/(\text{cm}^3\text{ mol}^{-1})$	73.99	75.29	76.5	77.55	79.08	79.82	80.67
n_A/mol	0	0.241	0.626	1.268	3.717	8.009	
n/mol	1	1.241	1.626	2.268	4.717	9.009	
V/cm^3	73.99	93.41	124.4	175.9	373.0	719.1	

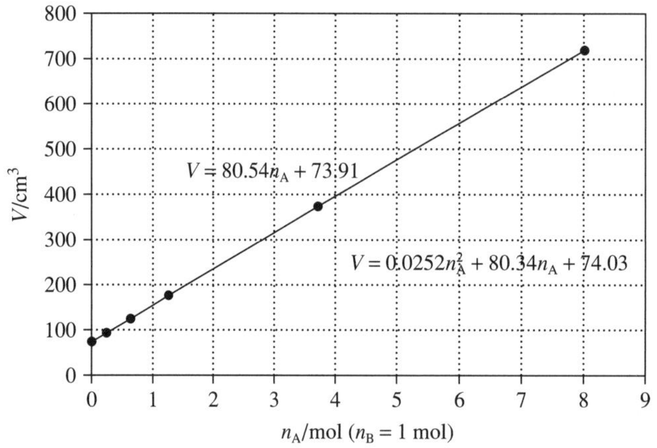

Figure 5.21(a)

In Figure 5.21(a), we plot V against n_A. Both linear and quadratic fits to the data are shown. The data fit a straight line quite well; however, the slope of a straight line is constant, which would imply V_A is constant (at 80.54 $\text{cm}^3\text{ mol}^{-1}$) over this range of compositions. We require some variation in V_A, so we use the quadratic fit,

$$V/\text{cm}^3 = 0.0252(n_A/\text{mol})^2 + 80.34(n_A/\text{mol}) + 74.03,$$

which leads to

$$V_A = \left(\frac{\partial V}{\partial n_A}\right)_{n_B} = \{2 \times 0.0252(n_A/\text{mol}) + 80.34\}\text{ cm}^3\text{ mol}^{-1}$$

Finally, we can draw up the table, including $x_A = 0.500$

x_A	0	0.194	0.385	0.500	0.559	0.788	0.889
$V_A/(\text{cm}^3\text{ mol}^{-1})$	80.34	80.35	80.37	80.39	80.40	80.53	80.74

For the present purpose we integrate up to $V_A(0.5, 0.5) = 84.39\text{ cm}^3\text{ mol}^{-1}$.

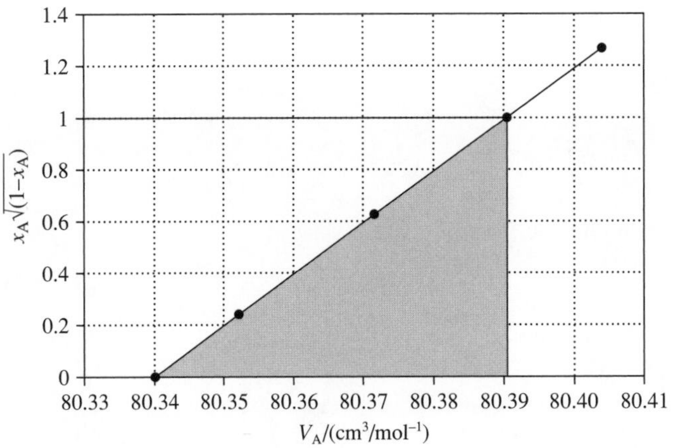

Figure 5.21(b)

The points are plotted in Figure 5.21(b), and the area required is 0.025 cm³ mol⁻¹. Hence,

$$V_B(0.5, 0.5) = 73.99\ \text{cm}^3\ \text{mol}^{-1} - 0.025\ \text{cm}^3\ \text{mol}^{-1} = \boxed{73.96\ \text{cm}^3\ \text{mol}^{-1}}.$$

COMMENT. The integral derived at the start of this problem is most useful for computing the partial molar quantity of one component given that of the other. In this case, the data given were overall molar volumes, from which we had to compute V_A before we could apply the integral to compute V_B. In such a case, it would have been easier to compute V_B directly in the same way we computed V_A.

P5.31 $$\phi = -\frac{x_A}{x_B}\ln a_A = -\frac{\ln a_A}{r} \qquad \text{(a)}$$

Therefore, $d\phi = -\frac{1}{r}d\ln a_A + \frac{1}{r^2}\ln a_A dr$

and $$d\ln a_A = \frac{1}{r}\ln a_A dr - r d\phi. \qquad \text{(b)}$$

Now, the Gibbs–Duhem equation (eqn 5.12a), implies

$$x_A d\mu_A + x_B d\mu_B = 0$$

Since $\mu = \mu^* + RT\ln a$,

$$x_A\, d\ln a_A + x_B\, d\ln a_B = 0$$

Therefore, $d\ln a_B = -\frac{x_A}{x_B}d\ln a_A = -\frac{d\ln a_A}{r}$

$$= -\frac{1}{r^2}\ln a_A dr + d\phi \text{ [from (b)]}$$

$$= \frac{\phi}{r}dr + d\phi \text{ [from (a)]} = \phi\, d\ln r + d\phi$$

Subtract $d\ln r$ from both sides, to obtain

$$d\ln\frac{a_B}{r} = (\phi - 1)d\ln r + d\phi = \frac{(\phi-1)}{r}dr + d\phi$$

Integrate both sides of the equality from pure A (where $r = 0$) to an arbitrary composition:

$$\int \mathrm{d}\ln\frac{a_\mathrm{B}}{r} = \int \frac{(\phi-1)}{r}\mathrm{d}r + \int \mathrm{d}\phi$$

The lower limit of the left-hand integral is:

$$\lim_{r\to 0}\ln\left(\frac{a_\mathrm{B}}{r}\right) = \lim_{r\to 0}\ln\left(\frac{\gamma_\mathrm{B}x_\mathrm{B}}{r}\right) = \lim_{r\to 0}\ln(\gamma_\mathrm{B}x_\mathrm{A}) = \ln 1 = 0,$$

leaving the desired expression

$$\ln\frac{a_\mathrm{B}}{r} = \phi - \phi(0) + \int_0^r \left(\frac{\phi-1}{r}\right)\mathrm{d}r$$

P5.33 $\mathrm{A(s)} \rightleftharpoons \mathrm{A(l)}$

$\mu_\mathrm{A}^*(\mathrm{s}) = \mu_\mathrm{A}^*(\mathrm{l}) + RT\ln a_\mathrm{A}$

so $\Delta_\mathrm{fus}G = \mu_\mathrm{A}^*(\mathrm{l}) - \mu_\mathrm{A}^*(\mathrm{s}) = -RT\ln a_\mathrm{A}$

Hence, $\ln a_\mathrm{A} = \dfrac{-\Delta_\mathrm{fus}G}{RT}$

Differentiate with respect to T and use the Gibbs-Helmholtz equation (3.55):

$$\frac{\mathrm{d}\ln a_\mathrm{A}}{\mathrm{d}T} = -\frac{1}{R}\frac{\mathrm{d}}{\mathrm{d}T}\left(\frac{\Delta_\mathrm{fus}G}{T}\right) = \frac{\Delta_\mathrm{fus}H}{RT^2}$$

The freezing point depression is $\Delta T = T_\mathrm{f} - T$, so $\mathrm{d}\Delta T = -\mathrm{d}T$

and $\dfrac{\mathrm{d}\ln a_\mathrm{A}}{\mathrm{d}\Delta T} = \dfrac{-\Delta_\mathrm{fus}H}{RT^2} \approx \dfrac{-\Delta_\mathrm{fus}H}{RT_\mathrm{f}^2}$

But $K_\mathrm{f} = \dfrac{RT_\mathrm{f}^2M_\mathrm{A}}{\Delta_\mathrm{fus}H}$ [from eqns 5.34 and 5.35]

Therefore,

$$\frac{\mathrm{d}\ln a_\mathrm{A}}{\mathrm{d}\Delta T} = \frac{-M_\mathrm{A}}{K_\mathrm{f}} \quad \text{and} \quad \mathrm{d}\ln a_\mathrm{A} = \frac{-M_\mathrm{A}\mathrm{d}\Delta T}{K_\mathrm{f}}$$

According to the Gibbs–Duhem equation (5.12a)

$$n_\mathrm{A}\,\mathrm{d}\mu_\mathrm{A} + n_\mathrm{B}\,\mathrm{d}\mu_\mathrm{B} = 0.$$

which implies that

$$n_\mathrm{A}\,\mathrm{d}\ln a_\mathrm{A} + n_\mathrm{B}\,\mathrm{d}\ln a_\mathrm{B} = 0\ [\mu = \mu^* + RT\ln a],$$

and hence that $\mathrm{d}\ln a_\mathrm{A} = -\dfrac{n_\mathrm{B}}{n_\mathrm{A}}\mathrm{d}\ln a_\mathrm{B}$.

Hence, $\dfrac{\mathrm{d}\ln a_\mathrm{B}}{\mathrm{d}\Delta T} = \dfrac{n_\mathrm{A}M_\mathrm{A}}{n_\mathrm{B}K_\mathrm{f}} = \dfrac{1}{b_\mathrm{B}K_\mathrm{f}}$ [for $n_\mathrm{A}M_\mathrm{A} = 1\,\mathrm{kg}$]

We know from the Gibbs–Duhem equation that

$$x_\mathrm{A}\,\mathrm{d}\mu_\mathrm{A} + x_\mathrm{B}\,\mathrm{d}\mu_\mathrm{B} = 0$$

and hence that $\int \mathrm{d}\ln a_\mathrm{A} = -\int \frac{x_\mathrm{B}}{x_\mathrm{A}}\mathrm{d}\ln a_\mathrm{B} = \ln a_\mathrm{A}$.

The osmotic coefficient was defined in Problem 5.31 as

$$\phi = -\frac{1}{r}\ln a_\mathrm{A} = -\frac{x_\mathrm{A}}{x_\mathrm{B}}\ln a_\mathrm{A}$$

Therefore,

$$\phi = \frac{x_\mathrm{A}}{x_\mathrm{B}}\int \frac{x_\mathrm{B}}{x_\mathrm{A}}\ln a_\mathrm{B} = \frac{1}{b}\int_0^b b\mathrm{d}\ln a_\mathrm{B} = \frac{1}{b}\int_0^b b\mathrm{d}\ln \gamma b = \frac{1}{b}\int_0^b b\mathrm{d}\ln b + \frac{1}{b}\int_0^b b\mathrm{d}\ln \gamma = 1 + \frac{1}{b}\int_0^b b\mathrm{d}\ln \gamma$$

From the Debye-Hückel limiting law (eqn 5.75)

$$\ln \gamma = -A'b^{1/2}\ [A' = A\ln 10 = 2.303A]$$

Hence, $\mathrm{d}\ln\gamma = -\frac{1}{2}A'b^{-1/2}\,\mathrm{d}b$

$$\phi = 1 + \frac{1}{b}\left(-\frac{1}{2}A'\right)\int_0^b b^{1/2}\,\mathrm{d}b = 1 - \frac{1}{2}\left(\frac{A'}{b}\right)\times\frac{2}{3}b^{3/2} = 1 - \frac{1}{3}A'b^{1/2}$$

COMMENT. For the depression of the freezing point in the solution of a 1,1-electrolyte

$$\ln a_\mathrm{A} = \frac{-\Delta_\mathrm{fus}G}{RT} + \frac{\Delta_\mathrm{fus}G}{RT^*}$$

and hence $-r\phi = \frac{-\Delta_\mathrm{fus}H}{R}\left(\frac{1}{T} - \frac{1}{T^*}\right)$

Therefore, $\phi = \frac{\Delta_\mathrm{fus}Hx_\mathrm{A}}{Rx_\mathrm{B}}\left(\frac{1}{T} - \frac{1}{T^*}\right) = \frac{\Delta_\mathrm{fus}Hx_\mathrm{A}}{Rx_\mathrm{B}}\left(\frac{T^* - T}{TT^*}\right) \approx \frac{\Delta_\mathrm{fus}Hx_\mathrm{A}\Delta T}{Rx_\mathrm{B}T^{*2}} \approx \frac{\Delta_\mathrm{fus}H\Delta T}{vRb_\mathrm{B}T^{*2}M_\mathrm{A}}$

where v is the number of ions per formula unit (here 2). Therefore, since $K_\mathrm{f} = \frac{MRT^{*2}}{\Delta_\mathrm{fus}H}$

$$\phi = \frac{\Delta T}{2b_\mathrm{B}K_\mathrm{f}}$$

Solutions to applications

P5.35 In this case it is convenient to rewrite the Henry's law expression as

mass of $N_2 = p_{N_2} \times$ mass of $H_2O \times K_{N_2}$

(1) At $p_{N_2} = 0.78 \times 4.0\ \mathrm{atm} = 3.1\ \mathrm{atm}$

mass of $N_2 = 3.1\ \mathrm{atm} \times 100\ \mathrm{g\ H_2O} \times 0.18\ \mu\mathrm{g\ N_2/(g\ H_2O\ atm)} = \boxed{56\ \mu\mathrm{g\ N_2}}$

(2) At $p_{N_2} = 0.78$ atm, mass of $N_2 = \boxed{14\ \mu\mathrm{g\ N_2}}$

(3) In fatty tissue the increase in N_2 concentration from 1 atm to 4 atm is

$4 \times (56 - 14)\mu\mathrm{g\ N_2} = \boxed{1.7 \times 10^2\ \mu\mathrm{g\ N_2}}$

P5.37 (a) The sum has just one term, so

$$\frac{v}{[A]_{out}} = \frac{NK}{1+K[A]_{out}} = \frac{4.0\times10^{7}\,dm^{3}\,mol^{-1}}{1+(1.0\times10^{7}\,dm^{3}\,mol^{-1})[A]_{out}} = \frac{40\,dm^{3}\,\mu mol^{-1}}{1+(10\,dm^{3}\,\mu mol^{-1})[A]_{out}}$$

The plot is shown in Figure 5.22(a).

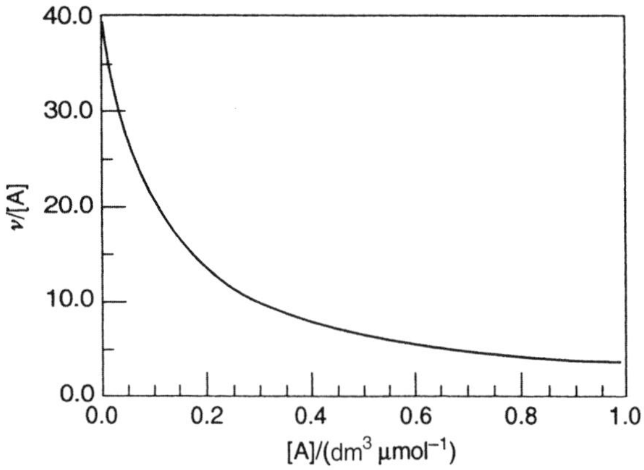

Figure 5.22(a)

(b) There are two terms in the sum here

$$\frac{v}{[A]_{out}} = \frac{4\times(1\times10^{5}\,dm^{3}\,mol^{-1})}{1+(1\times10^{5}\,dm^{3}\,mol^{-1})\times[A]_{out}} + \frac{2\times(2\times10^{6}\,dm^{3}\,mol^{-1})}{1+(2\times10^{6}\,dm^{3}\,mol^{-1})\times[A]_{out}}$$

$$= \frac{0.4\,dm^{3}\,\mu mol^{-1}}{1+(0.1\,dm^{3}\,\mu mol^{-1})\times[A]_{out}} + \frac{4\,dm^{3}\,\mu mol^{-1}}{1+(2\,dm^{3}\,\mu mol^{-1})\times[A]_{out}}$$

The plot is shown in Figure 5.22(b).

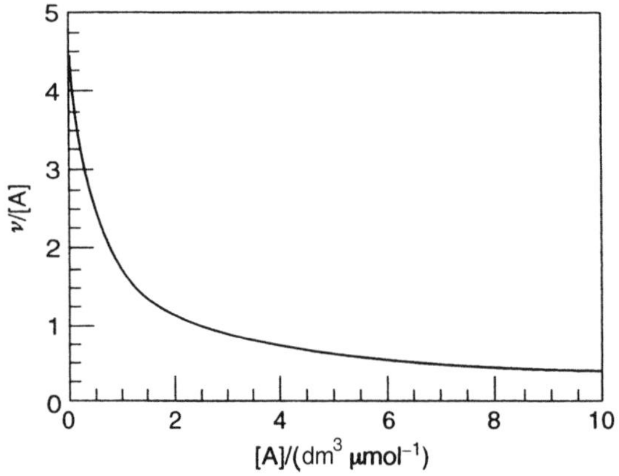

Figure 5.22(b)

P5.39 By the van't Hoff equation (5.38)

$$\Pi = [B]RT = \frac{cRT}{M}$$

Division by the standard acceleration of free fall, g, gives

$$\frac{\Pi}{g} = \frac{c(R/g)T}{M}$$

(a) This expression may be written in the form

$$\Pi' = \frac{cR'T}{M}$$

which has the same form as the van't Hoff equation, but the unit of osmotic pressure (Π') is now

$$\frac{\text{force/area}}{\text{length/time}^2} = \frac{\text{(mass length)/(area time}^2\text{)}}{\text{length/time}^2} = \frac{\text{mass}}{\text{area}}$$

This ratio can be specified in g cm^{-2}. Likewise, the constant of proportionality (R') would have the units of R/g.

$$\frac{\text{energy K}^{-1}\text{ mol}^{-1}}{\text{length/time}^2} = \frac{\text{(mass length}^2\text{/time}^2\text{) K}^{-1}\text{ mol}^{-1}}{\text{length/time}^2} = \text{mass length K}^{-1}\text{ mol}^{-1}$$

This result may be specified in $\boxed{\text{g cm K}^{-1}\text{ mol}^{-1}}$.

$$R' = \frac{R}{g} = \frac{8.314\,47\text{ J K}^{-1}\text{ mol}^{-1}}{9.806\,65\text{ m s}^{-2}} = 0.847\,840\text{ kg m K}^{-1}\text{ mol}^{-1}\left(\frac{10^3\text{ g}}{\text{kg}}\right)\times\left(\frac{10^2\text{ cm}}{\text{m}}\right)$$

$$= \boxed{84\,784.0\text{ g cm K}^{-1}\text{ mol}^{-1}}$$

In the following we will drop the primes, giving

$$\Pi = \frac{cRT}{M}$$

and use the Π units of g cm^{-2} and the R units g cm K^{-1} mol^{-1}.

(b) By extrapolating the low concentration plot of Π/c versus c (Figure 5.23(a)) to $c = 0$ we find the intercept 230 g cm^{-2}/(g cm^{-3}). In this limit the van't Hoff equation is valid so

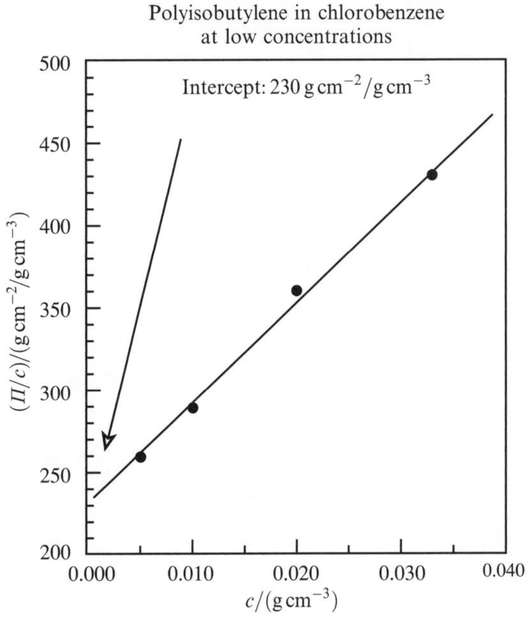

Figure 5.23(a)

$$\frac{RT}{M} = \text{intercept} \quad \text{or} \quad M = \frac{RT}{\text{intercept}}$$

$$M = \frac{RT}{\text{intercept}} = \frac{(84\,784.0 \text{ g cm K}^{-1}\text{ mol}^{-1}) \times (298.15 \text{ K})}{(230 \text{ g cm}^{-2})/(\text{g cm}^{-3})} = \boxed{1.1 \times 10^5 \text{ g mol}^{-1}}$$

(c) The plot of Π/c versus c for the full concentration range (Figure 5.23(b)) is very non-linear. We may conclude that the solvent is good. This may be due to the non-polar nature of both solvent and solute.

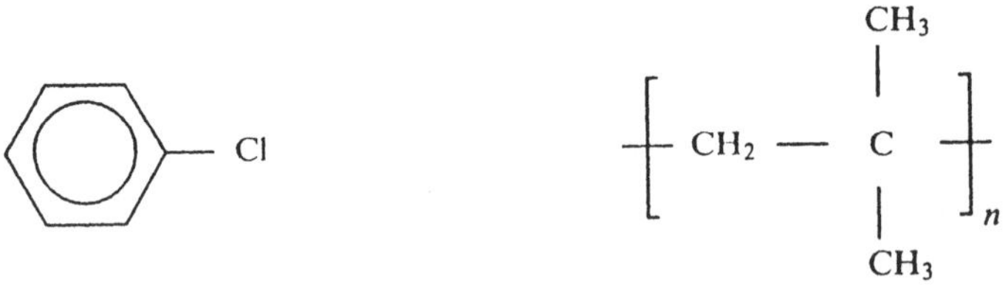

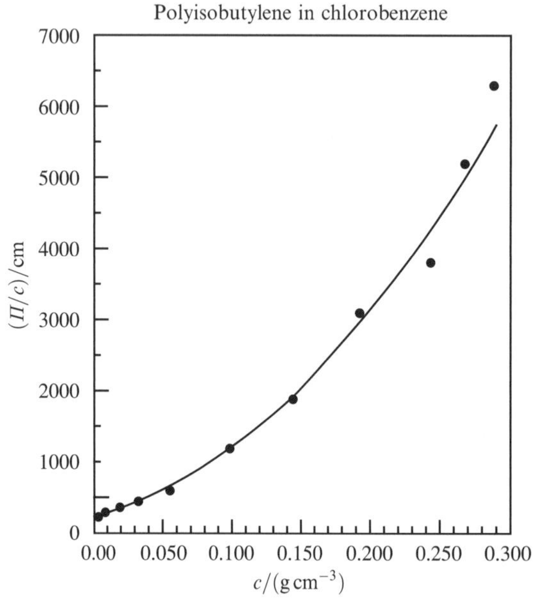

Figure 5.23(b)

(d) $\Pi/c = (RT/M)(1 + B'c + C'c^2)$

Since RT/M has been determined in part (b) by extrapolation to $c = 0$, it is best to determine the second and third virial coefficients with the linear regression fit

$$\frac{(\Pi/c)/(RT/M)^{-1}}{c} = B' + C'c$$

$R = 0.9791$

$B' = \boxed{21.4 \text{ cm}^3 \text{ g}^{-1}}$; standard deviation $= 2.4 \text{ cm}^3 \text{ g}^{-1}$

$C' = \boxed{211 \text{ cm}^6 \text{ g}^{-2}}$; standard deviation $= 15 \text{ cm}^6 \text{ g}^{-2}$

(e) Using $\frac{1}{4}$ for g and neglecting terms beyond the second power, we may write

$$\left(\frac{\Pi}{c}\right)^{1/2} = \left(\frac{RT}{M}\right)^{1/2}(1+\tfrac{1}{2}B'c)$$

We can solve for B', then $g(B')^2 = C'$

$$\frac{\left(\frac{\Pi}{c}\right)^{1/2}}{\left(\frac{RT}{M}\right)^{1/2}} - 1 = \tfrac{1}{2}B'c$$

RT/M has been determined above as 230 g cm^{-2}/(g cm^{-3}). We may analytically solve for B' from one of the data points, say, $\Pi/c = 430$ g cm^{-2}/g cm^{-3} at $c = 0.033$ g cm^{-3}.

$$\left(\frac{430\text{ g cm}^{-2}/\text{g cm}^{-3}}{230\text{ g cm}^{-2}/\text{g cm}^{-3}}\right)^{1/2} - 1 = \tfrac{1}{2}B' \times (0.033\text{ g cm}^{-3})$$

$$B' = \frac{2\times(1.367-1)}{0.033\text{ g cm}^{-3}} = 22\text{ cm}^3\text{ g}^{-1}$$

$$C' = g(B')^2 = 0.25 \times (22\text{ cm}^3\text{ g}^{-1})^2 = 12\overline{3}\text{ cm}^6\text{ g}^{-2}$$

Better values of B' and C' can be obtained by plotting $\left(\frac{\Pi}{c}\right)^{1/2} \Big/ \left(\frac{RT}{M}\right)^{1/2}$ against c. This plot is shown in Figure 5.23(c). The slope is 14 cm^3 g^{-1}. $B' = 2\times\text{slope} = \boxed{28\text{ cm}^3\text{ g}^{-1}}$. C' is then $\boxed{19\overline{6}\text{ cm}^6\text{ g}^{-2}}$. The intercept of this plot should theoretically be 1.00, but it is in fact 0.916 with a standard deviation of 0.066. The overall consistency of the values of the parameters confirms that g is roughly $\frac{1}{4}$, as assumed.

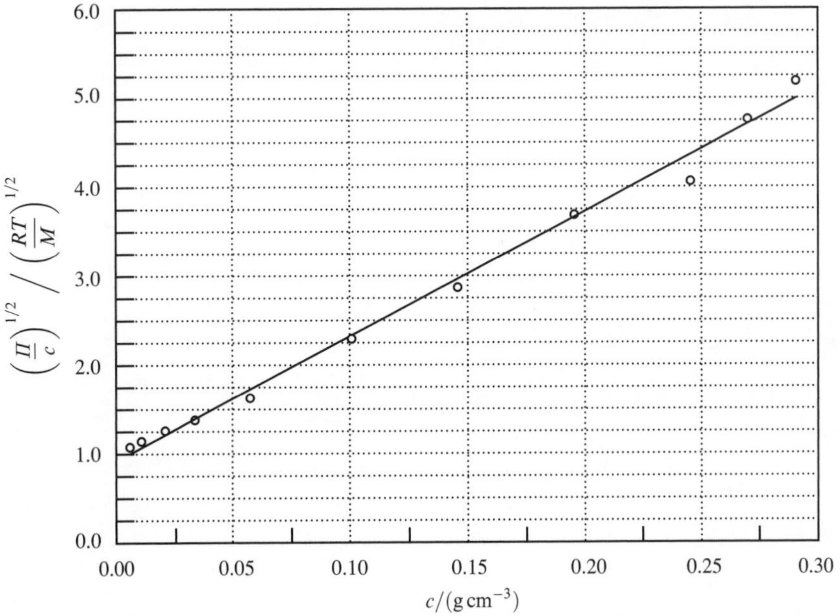

Figure 5.23(c)

P5.41 $C = 1$; hence, according to the phase rule (eqn 4.1) $F = C - P + 2 = 3 - P$.

Since the tube is sealed there will always be some gaseous compound in equilibrium with the condensed phases. Thus, when liquid begins to form upon melting, $P = 3$ (s, l, and g) and $F = 0$, corresponding to a definite melting temperature. At the transition to a normal liquid, $P = 3$ (l, l′, and g) as well, so again $F = 0$.

6 Chemical equilibrium

Answers to discussion questions

D6.1 The position of equilibrium is always determined by the condition that the **reaction quotient**, defined by the expression

$$Q = \frac{\text{activities of products}}{\text{activities of reactants}} = \prod_J a_J^{\nu_J} \text{ [6.12] with the } \pm \text{ convention for stoichometric numbers,}$$

must equal the **thermodynamic equilibrium constant, *K***, a ratio that depends on temperature but is independent of both composition and pressure. Mixing in of an additional amount of reactant or product to an equilibrium system destroys that equality and the reaction system spontaneously responds by shifting in such a way as to restore the equality. That implies that some of the added reactant or product must be removed by the reacting system and the amounts of other components will also be affected. These adjustments restore the concentrations to new equilibrium values but the value of the equilibrium constant is unchanged.

Chemical additions to an equilibrium system increases Gibbs energy, a condition that provides the impetus to spontaneously return to the equilibrium thermodynamic state, the state of minimum Gibbs energy, for which

$\Delta_r G = 0$ [6.3, the **condition for equilibrium**] and, therefore, $\Delta_r G^{\ominus} = -RT \ln Q_{eq} = -RT \ln K$ [6.14].

If the mixing of reactants and products gives $\Delta_r G < 0$, reactant activities will spontaneously diminish to increase product activities until $\Delta_r G = 0$. If the mixing gives $\Delta_r G > 0$, product activities will spontaneously diminish to increase reactant activities until $\Delta_r G = 0$. For the general reaction

$$a\,\mathrm{A} + b\,\mathrm{B} \rightleftharpoons c\,\mathrm{C} + d\,\mathrm{D} \quad Q = a_C^c a_D^d / a_A^a a_B^b \quad [6.12]$$

we say that, when the mixing gives $\Delta_r G < 0$, the reaction proceeds spontaneously to the right (*forward* reaction) until equilibrium is achieved but, when $\Delta_r G > 0$, the reaction proceeds spontaneously to the left (*reverse* direction). If reactant or product is added to an equilibrium mixture, the reaction spontaneously shifts in the direction that lowers the Gibbs energy of the reaction mixture (see text Figure 6.1, remember that $\Delta_r G$ is the *slope* of G plotted against the extent of reaction [6.1]). The reaction spontaneously shifts to the right upon addition of reactant to an equilibrium mixture; left upon addition of product. We must also remember that thermodynamics says nothing about the rate at which the reaction occurs or shifts. A spontaneous reaction may occur very rapidly, infinitely slowly, or at any intermediate speed.

D6.3 The definition of the thermodynamic equilibrium constant involves activities rather than pressures:

$$K = \left(\prod_J a_J^{\nu_J}\right)_{\text{equilibrium}} \quad [6.13] \text{ with the } \pm \text{ convention for stoichiometric numbers}$$

At low pressures, the activities of gases may be replaced with their partial pressures with little error, but at high pressures that is not a good approximation. The difference between the equilibrium constant expressed in activities and the constant expressed in pressures is dependent on two factors: the stoichiometry of the reaction and the magnitude of the partial pressures. Thus, there is no one answer to this question. For the example of the ammonia synthesis reaction, in a range of pressures where the activity coefficients are greater than one, an increase in pressure results in a greater shift to the product side than would be predicted by the constant expressed in partial pressures.

The activity coefficients of real gases give an indication of dominant intermolecular forces. Coefficients greater than 1 are observed when repulsions dominate; coefficients less than one dominate when attractions dominate. In the limit of zero pressure all gases behave as perfect gases with activity coefficients equal to 1 and, therefore $a_{J(\text{gas})} = \gamma_J p_J / p^{\ominus} = p_J / p^{\ominus}$ in the perfect gas case only. Thus, we conclude that intermolecular forces cause the thermodynamic equilibrium constant to respond differently to changes in pressure from the equilibrium constant expressed in terms of partial pressures.

For an exothermic reaction, such as the ammonia synthesis, an increase in temperature will shift the reaction to the reactant side, but the relative shift is independent of the activity coefficients. The ratio $\ln(K_2/K_1)$ depends on $\Delta_r H^{\ominus}$ [see eqn 6.23].

D6.5 The physical dependence of the equilibrium constant on temperature as predicted by the van't Hoff equation, $\dfrac{d\ln K}{dT} = \dfrac{\Delta_r H^{\ominus}}{RT^2}$ [6.21(a)], can be seen when the expression $\Delta_r G^{\ominus} = \Delta_r H^{\ominus} - T\Delta_r S^{\ominus}$ is written in the form $R\ln K = -\Delta_r H^{\ominus}/T + \Delta_r S^{\ominus}$ when the reaction is exothermic and the temperature is raised, $\ln K$ and hence K decrease, since T occurs in the denominator, and the reaction shifts to favour the reactants. When the reaction is endothermic, increasing T makes $\ln K$ less negative, or K more positive and products are favoured. Another factor of importance when the reaction is endothermic is the increasing entropy of their reacting system resulting in a more positive $\ln K$, favouring products.

The molecular basis of the van't Hoff equation stems from the Boltzmann distribution of molecules over the available energy levels. When the temperature of an equilibrium system is changed, the Boltzmann distribution adjusts. For an endothermic reaction a temperature increase causes an increased population of the higher energy states of products at the expense of the population of the lower energy states of reactants and the reaction shifts to the right. The converse applies to a temperature increase for an exothermic reaction and the reaction shifts to the left.

D6.7 Construct a cell using a standard hydrogen electrode and an electrode designed around the redox couple of interest. The cell potential E is measured with a high impedance voltmeter under zero current conditions. When using SHE as a reference electrode, E_{cell} is the desired half-reaction potential [6.29 and 6.35]. Should the redox couple have one or more electroactive species (i) that are solvated with concentration b_i, E_{cell} must be measured over a range of b_i values.

The Nernst equation [6.27], with Q being the cell reaction quotient, is the starting point for analysis of the $E_{\text{cell}}(b_i)$ data.

$$E_{cell} = E^{\ominus}_{cell} - \frac{RT}{\nu F}\ln Q \, [6.27]$$

It would seem that substitution of E_{cell} and Q values would allow the computation of the standard redox potential $E^{\ominus}_{cell}$ for the couple. However, a problem arises because the calculation of Q requires not only knowledge of the concentrations of the species involved in the cell reaction but also of their activity coefficients. These coefficients are not usually available, so the calculation cannot be directly completed. However, at very low concentrations, the Debye–Hückel limiting law for the coefficients holds. The procedure then is to substitute the Debye–Hückel law for the activity coefficients into the specific form of the Nernst equation for the cell under investigation and carefully examine the equation to determine what kind of plot to make of the $E_{cell}(b_i)$ data so that extrapolation of the plot to zero concentration, where the Debye−Hückel law is valid, gives a plot intercept that equals $E^{\ominus}_{cell}$. See Section 6.8 for the details of this procedure and an example for which the relevant graph involves a plot of $E_{cell} + (2RT/F)\ln b$ against $b^{1/2}$ [see eqn 6.31].

Solutions to exercises

E6.1(a) $A \rightarrow 2\,B$

$n_J = n_J(0) + \nu_J\Delta\xi$, where ξ is the extent of reaction; ν_J is negative for reactants and positive for products.

$$n_A = 1.5\text{ mol} - 1 \times (0.60\text{ mol}) = \boxed{0.9\text{ mol}}$$

$$n_B = 0 + 2 \times (0.60\text{ mol}) = \boxed{1.2\text{ mol}}$$

E6.2(a) $A \rightarrow 2\,B \quad \Delta_r G = -6.4\text{ kJ mol}^{-1}$

$$\Delta_r G = \left(\frac{\partial G}{\partial \xi}\right)_{p,T} \quad [6.1]$$

With the approximation that $\left(\frac{\partial G}{\partial \xi}\right)_{p,T} \simeq \frac{\Delta G}{\Delta \xi}$, which is valid when $\Delta\xi$ is very small, we find that

$$\Delta G \simeq \Delta_r G \times \Delta\xi$$
$$\simeq (-6.4\text{ kJ mol}^{-1}) \times (+0.10\text{ mol}) = \boxed{-0.64\text{ kJ}}$$

E6.3(a) $N_2(g) + 3\,H_2(g) \rightarrow 2\,NH_3(g) \quad \Delta_r G^{\ominus} = -32.9\text{ kJ mol}^{-1}$

$$\Delta_r G = \Delta_r G^{\ominus} + RT\ln Q \, [6.10]$$
$$= (-32.9\text{ kJ mol}^{-1}) + (2.4790\text{ kJ mol}^{-1}) \times \ln Q \quad \text{at } 298.15\text{ K}$$

Part	Q	$\Delta_r G$/kJ mol^{-1}
(a)	0.010	−44.3
(b)	1.00	−32.9
(c)	10.0	−27.2
(d)	100 000	−4.4
(e)	1 000 000	+1.3

Figure 6.1 shows a plot of $\Delta_r G$ against Q. $K = Q_{\text{equilibrium}}$ when $\Delta_r G = 0$ so we find the equilibrium constant by interpolating the plot to the point at which $\Delta_r G = 0$ and read the value of Q at that point to find that $K \sim \boxed{6 \times 10^5}$. Although the plot is very interesting, the estimate for the equilibrium constant can also be made with a linear interpolation between the points of parts (d) and (e). The interpolation approximation agrees with the value calculated with eqn 6.14:

$$K = e^{-\Delta_r G^{\ominus}/RT} \; [6.14]$$
$$= e^{-(-32.9\,\text{kJ mol}^{-1})/(2.4790\,\text{kJ mol}^{-1})} = 5.8 \times 10^5$$

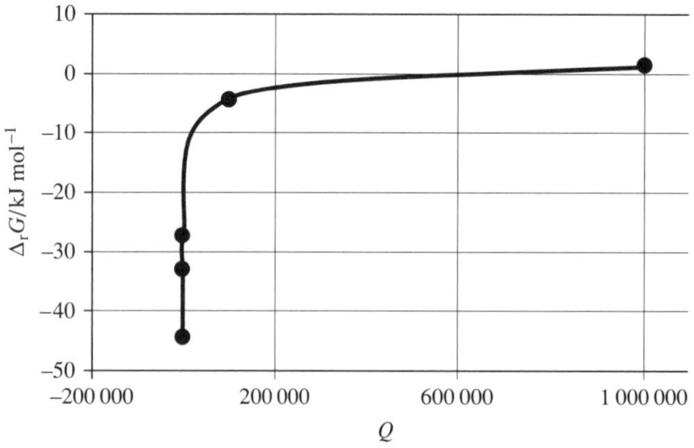

Figure 6.1

E6.4(a) $2\,H_2O(g) \rightleftharpoons 2\,H_2(g) + O_2(g)$ $\quad T = 2257$ K, $p = 1$ bar $= p^{\ominus}$, $\alpha = 0.0177$ at equilibrium

We draw up the following equilibrium table (Example 6.2).

	$H_2O(g)$	$H_2(g)$	$O_2(g)$
Amount at equilibrium	$(1-\alpha)n$	αn	$\frac{1}{2}\alpha n$
Mole fraction	$\dfrac{1-\alpha}{1+\frac{1}{2}\alpha}$	$\dfrac{\alpha}{1+\frac{1}{2}\alpha}$	$\dfrac{\frac{1}{2}\alpha}{1+\frac{1}{2}\alpha}$
Partial pressure	$\dfrac{(1-\alpha)p}{1+\frac{1}{2}\alpha}$	$\dfrac{\alpha p}{1+\frac{1}{2}\alpha}$	$\dfrac{\frac{1}{2}\alpha p}{1+\frac{1}{2}\alpha}$

$$K = \left(\prod_J a_J^{\nu_J}\right)_{\text{equilibrium}} [6.13] = \left(\prod_J (p_J/p^{\ominus})^{\nu_J}\right)_{\text{equilibrium}} \quad \text{(perfect gas assumption)}$$

$$= \frac{(p_{H_2}/p^{\ominus})^2(p_{O_2}/p^{\ominus})}{(p_{H_2O}/p^{\ominus})^2} = \frac{p_{H_2}^2 p_{O_2}}{p_{H_2O}^2 p^{\ominus}} = \frac{\left(\dfrac{\alpha p}{1+\frac{1}{2}\alpha}\right)^2 \left(\dfrac{\frac{1}{2}\alpha p}{1+\frac{1}{2}\alpha}\right)}{\left(\dfrac{(1-\alpha)p}{1+\frac{1}{2}\alpha}\right)^2 p^{\ominus}}$$

$$= \frac{p}{2p^{\ominus}}\left\{\frac{\alpha^3}{(1-\alpha)^2 \times (1+\frac{1}{2}\alpha)}\right\} = \frac{1}{2}\left\{\frac{(0.0177)^3}{(1-0.0177)^2 \times (1+\frac{1}{2}\times 0.0177)}\right\} = \boxed{2.85 \times 10^{-6}}$$

E6.5(a) (a) $N_2O_4(g) \rightleftharpoons 2\,NO_2(g) \quad T = 298.15\ \text{K}, p = 1\ \text{bar} = p^\ominus, \alpha = 0.1846$ at equilibrium

We draw up the following equilibrium table (Example 6.2).

	$N_2O_4(g)$	$NO_2(g)$
Amount at equilibrium	$(1-\alpha)n$	$2\alpha n$
Mole fraction	$\dfrac{1-\alpha}{1+\alpha}$	$\dfrac{2\alpha}{1+\alpha}$
Partial pressure	$\dfrac{(1-\alpha)p}{1+\alpha}$	$\dfrac{2\alpha p}{1+\alpha}$

$$K = \left(\prod_J a_J^{\nu_J}\right)_{\text{equilibrium}} [6.13] = \left(\prod_J (p_J/p^\ominus)^{\nu_J}\right)_{\text{equilibrium}} \quad \text{(perfect gas assumption)}$$

$$= \frac{(p_{NO_2}/p^\ominus)^2}{p_{N_2O_4}/p^\ominus} = \frac{p_{NO_2}^2}{p_{N_2O_4}p^\ominus} = \frac{\left(\dfrac{2\alpha p}{1+\alpha}\right)^2}{\left(\dfrac{(1-\alpha)p}{1+\alpha}\right)p^\ominus}$$

$$= \frac{4p}{p^\ominus}\left\{\frac{\alpha^2}{(1-\alpha)\times(1+\alpha)}\right\} = 4\left\{\frac{(0.1846)^2}{(1-0.1846)\times(1+0.1846)}\right\}$$

$$= \boxed{0.141} \text{ at } 25°\text{C}$$

(b) $$\ln K_2 = \ln K_1 - \frac{\Delta_r H^\ominus}{R}\left(\frac{1}{T_2} - \frac{1}{T_1}\right) [6.23]$$

$$\ln K_{100°\text{C}} = \ln K_{25°\text{C}} - \frac{\Delta_r H^\ominus}{R}\left(\frac{1}{373.15\ \text{K}} - \frac{1}{298.15\ \text{K}}\right)$$

$$= \ln(0.141) - \left(\frac{+56.2\times 10^3\ \text{J mol}^{-1}}{8.3145\ \text{J K}^{-1}\ \text{mol}^{-1}}\right)\times\left(\frac{1}{373.15\ \text{K}} - \frac{1}{298.15\ \text{K}}\right) = 2.60$$

$$K_{100°\text{C}} = e^{2.60} = \boxed{13.5}$$

As expected, the temperature increase causes a shift to the right when the reaction is endothermic.

E6.6(a) $PbO(s) + CO(g) \rightleftharpoons Pb(s) + CO_2(g)$

(a) Using data tables at 25°C, we find

$$\begin{aligned}\Delta_r G^\ominus &= \Delta_f G^\ominus(\text{Pb, s}) + \Delta_f G^\ominus(\text{CO}_2\text{, g}) - \Delta_f G^\ominus(\text{PbO, s, red}) - \Delta_f G^\ominus(\text{CO, g})\\ &= (-394.36\ \text{kJ mol}^{-1}) - (-188.93\ \text{kJ mol}^{-1}) - (-137.17\ \text{kJ mol}^{-1})\\ &= \boxed{-68.26\ \text{kJ mol}^{-1}}\end{aligned}$$

$$\begin{aligned}\Delta_r H^\ominus &= \Delta_f H^\ominus(\text{Pb, s}) + \Delta_f H^\ominus(\text{CO}_2\text{, g}) - \Delta_f H^\ominus(\text{PbO, s, red}) - \Delta_f H^\ominus(\text{CO, g})\\ &= (-393.51\ \text{kJ mol}^{-1}) - (-218.99\ \text{kJ mol}^{-1}) - (-110.53\ \text{kJ mol}^{-1})\\ &= -63.99\ \text{kJ mol}^{-1}\end{aligned}$$

$$\ln K = \frac{-\Delta_r G^\ominus}{RT} [6.14] = \frac{+68.26\times 10^3\ \text{J mol}^{-1}}{(8.3145\ \text{J K}^{-1}\text{mol}^{-1})\times(298.15\ \text{K})} = 27.54$$

$$K = e^{27.54} = \boxed{9.13\times 10^{11}}$$

(b) $\ln K_2 = \ln K_1 - \frac{\Delta_r H^\ominus}{R}\left(\frac{1}{T_2} - \frac{1}{T_1}\right)$ [6.23]

$$\ln K_{400\,\mathrm{K}} = \ln K_{298.15\,\mathrm{K}} - \frac{\Delta_r H^\ominus}{R}\left(\frac{1}{400\ \mathrm{K}} - \frac{1}{298.15\ \mathrm{K}}\right)$$

$$= \ln(9.13\times10^{11}) - \left(\frac{-63.99\times10^3\ \mathrm{J\ mol^{-1}}}{8.3145\ \mathrm{J\ K^{-1}\ mol^{-1}}}\right)\times\left(\frac{1}{400\ \mathrm{K}} - \frac{1}{298.15\ \mathrm{K}}\right) = 21.0$$

$$K_{400\,\mathrm{K}} = \mathrm{e}^{21.0} = \boxed{1.32\times10^9}$$

As expected, the temperature increase causes a shift to the left when the reaction is exothermic.

$\Delta_r G^\ominus = -RT\ln K$ [6.14]

$$\Delta_r G^\ominus_{400\,\mathrm{K}} = -(8.3145\ \mathrm{J\ K^{-1}\ mol^{-1}})\times(400\ \mathrm{K})\ln(1.32\times10^9) = \boxed{-69.8\ \mathrm{kJ\ mol^{-1}}}$$

E6.7(a) $H_2CO(g) \rightarrow CO(g) + H_2(g)$ For this gas-phase reaction $\Delta\nu = \sum_J \nu_J = 1$

$$K = \left(\prod_J a_J^{\nu_J}\right)_{\text{equilibrium}} [6.13] = \left(\prod_J (\gamma_J p_J/p^\ominus)^{\nu_J}\right)_{\text{equilibrium}} = \left(\prod_J \gamma_J^{\nu_J}\right)_{\text{equilibrium}}\left(\prod_J (p_J/p^\ominus)^{\nu_J}\right)_{\text{equilibrium}}$$

$$= K_\gamma K_p \quad \text{where} \quad K_\gamma = \left(\prod_J \gamma_J^{\nu_J}\right)_{\text{equilibrium}} \quad \text{and} \quad K_p = \left(\prod_J (p_J/p^\ominus)^{\nu_J}\right)_{\text{equilibrium}}$$

Let us assume that the gases are perfect gases. Then, the activity coefficients equal 1 and $K_\gamma = 1$. Additionally, $p_J = n_J RT/V = [\mathrm{J}]RT$. Substitution gives

$$K = K_p = \left(\prod_J ([\mathrm{J}]RT/p^\ominus)^{\nu_J}\right)_{\text{equilibrium}} = \left(\prod_J ([\mathrm{J}]/c^\ominus)^{\nu_J}\right)_{\text{equilibrium}}\left(\prod_J (c^\ominus RT/p^\ominus)^{\nu_J}\right)_{\text{equilibrium}}$$

$$= K_c(c^\ominus RT/p^\ominus)^{\Delta\nu}, \quad \text{where} \quad K_c = \left(\prod_J ([\mathrm{J}]/c^\ominus)^{\nu_J}\right)_{\text{equilibrium}} \quad [6.17(a)]$$

$K = K_c \times (c^\ominus RT/p^\ominus)$ because $\Delta\nu = 1$ for this reaction

Since $c^\ominus R/p^\ominus = 0.0831451\ \mathrm{K^{-1}}$, this expression may be written in the form

$K = (0.0831451\ \mathrm{K^{-1}})\times K_c T$

E6.8(a) Draw up the following equilibrium table for the reaction equation: $2\,A + B \rightleftharpoons 3\,C + 2\,D$.

	A	B	C	D	Total
Initial amounts/mol	1.00	2.00	0	1.00	4.00
Stated change/mol			+0.90		
Implied change/mol	−0.60	−0.30	+0.90	+0.60	
Equilibrium amounts/mol	0.40	1.70	0.90	1.60	4.60
Mole fractions	0.087	0.370	0.196	0.348	1.001

(a) The mole fractions are given in the table.

(b) $K_x = \prod_J x_J^{\nu_J}$

$$K_x = \frac{(0.196)^3 \times (0.348)^2}{(0.087)^2 \times (0.370)} = 0.32\overline{6} = \boxed{0.33}$$

(c) $p_J = x_J p, \quad p = 1 \text{ bar}, \quad p^{\ominus} = 1 \text{ bar}$

Assuming that the gases are perfect, $a_J = p_J/p^{\ominus}$, hence

$$K = \frac{(p_C/p^{\ominus})^3 \times (p_D/p^{\ominus})^2}{(p_A/p^{\ominus})^2 \times (p_B/p^{\ominus})}$$

$$= \frac{x_C^3 x_D^2}{x_A^2 x_B} \times \left(\frac{p}{p^{\ominus}}\right)^2 = K_x \quad \text{when } p = 1.00 \text{ bar} = \boxed{0.33}$$

(d) $\Delta_r G^{\ominus} = -RT \ln K = -(8.3145 \text{ J K}^{-1} \text{ mol}^{-1}) \times (298 \text{ K}) \times (\ln 0.32\overline{6}) = \boxed{+2.8 \text{ kJ mol}^{-1}}$

E6.9(a) At 1280 K, $\Delta_r G^{\ominus} = +33 \times 10^3 \text{ J mol}^{-1}$; thus

$$\ln K_1(1280 \text{ K}) = -\frac{\Delta_r G^{\ominus}}{RT} = -\frac{33 \times 10^3 \text{ J mol}^{-1}}{(8.3145 \text{ J K}^{-1} \text{ mol}^{-1}) \times (1280 \text{ K})} = -3.1\overline{0}$$

$$K_1 = \boxed{0.045}$$

$$\ln K_2 = \ln K_1 - \frac{\Delta_r H^{\ominus}}{R}\left(\frac{1}{T_2} - \frac{1}{T_1}\right) [6.23]$$

We look for the temperature T_2 that corresponds to $\ln K_2 = \ln(1) = 0$. This is the crossover temperature. Solving for T_2 from eqn 6.23 with $\ln K_2 = 0$, we obtain

$$\frac{1}{T_2} = \frac{R \ln K_1}{\Delta_r H^{\ominus}} + \frac{1}{T_1}\left(\frac{(8.3145 \text{ J K}^{-1} \text{ mol}^{-1}) \times (-3.1\overline{0})}{224 \times 10^3 \text{ J mol}^{-1}}\right) + \left(\frac{1}{1280 \text{ K}}\right) = 6.6\overline{6} \times 10^{-4} \text{ K}^{-1}$$

$$T_2 = \boxed{15\overline{00} \text{ K}}$$

E6.10(a) $2\,C_3H_6(g) \rightarrow C_2H_4(g) + C_4H_8(g) \quad T = 400 \text{ K}$

$$\ln K = A + \frac{B}{T} + \frac{C}{T^2} \quad \text{where} \quad A = -1.04,\ B = -1088 \text{ K},\ C = 1.51 \times 10^5 \text{ K}^2$$

$$\Delta_r G^{\ominus} = -RT \ln K\ [6.14] = RT \times \left(A + \frac{B}{T} + \frac{C}{T^2}\right)$$

$$= -(8.3145 \text{ J K}^{-1} \text{ mol}^{-1}) \times (400 \text{ K}) \times \left(-1.04 - \frac{1088 \text{ K}}{400 \text{ K}} + \frac{1.51 \times 10^5 \text{ K}^2}{(400 \text{ K})^2}\right)$$

$$= +9.37 \text{ kJ mol}^{-1}$$

$$\Delta_r H^{\ominus} = -R\frac{\mathrm{d}\ln K}{\mathrm{d}(1/T)}[6.21\text{(b)}] = -R\frac{\mathrm{d}}{\mathrm{d}(1/T)}\left(A + \frac{B}{T} + \frac{C}{T^2}\right) = -R \times \left(B + \frac{2C}{T}\right)$$

$$= -(8.3145 \text{ J K}^{-1} \text{ mol}^{-1}) \times \left\{(-1088 \text{ K}) + 2 \times \frac{1.51 \times 10^5 \text{ K}^2}{400 \text{ K}}\right\} = \boxed{+2.77 \text{ kJ mol}^{-1}}$$

$$\Delta_r G^\ominus = \Delta_r H^\ominus - T\Delta_r S^\ominus$$

$$\Delta_r S^\ominus = \frac{\Delta_r H^\ominus - \Delta_r G^\ominus}{T} = \frac{2.77\ \text{kJ mol}^{-1} - 9.37\ \text{kJ mol}^{-1}}{400\ \text{K}} = \boxed{-16.5\ \text{J K}^{-1}\,\text{mol}^{-1}}$$

E6.11(a) $H_2CO(g) \rightarrow CO(g) + H_2(g)$ For this gas-phase reaction $\Delta\nu = \sum_J \nu_J = 1$

$$K = \left(\prod_J a_J^{\nu_J}\right)_{\text{equilibrium}} [6.13] = \left(\prod_J (\gamma_J p_J/p^\ominus)^{\nu_J}\right)_{\text{equilibrium}} = \left(\prod_J \gamma_J^{\nu_J}\right)_{\text{equilibrium}} \left(\prod_J (p_J/p^\ominus)^{\nu_J}\right)_{\text{equilibrium}}$$

$$= K_\gamma K_p, \quad \text{where} \quad K_\gamma = \left(\prod_J \gamma_J^{\nu_J}\right)_{\text{equilibrium}} \quad \text{and } K_p = \left(\prod_J (p_J/p^\ominus)^{\nu_J}\right)_{\text{equilibrium}}$$

Let us assume that the gases are perfect gases. Then, the activity coefficients equal 1 and $K_\gamma = 1$. Additionally, $p_J = n_J RT/V = [J]RT$. Substitution gives

$$K = K_p = \left(\prod_J ([J]RT/p^\ominus)^{\nu_J}\right)_{\text{equilibrium}} = \left(\prod_J ([J]/c^\ominus)^{\nu_J}\right)_{\text{equilibrium}} \left(\prod_J (c^\ominus RT/p^\ominus)^{\nu_J}\right)_{\text{equilibrium}}$$

$$= K_c (c^\ominus RT/p^\ominus)^{\Delta\nu}, \quad \text{where} \quad K_c = \left(\prod_J ([J]/c^\ominus)^{\nu_J}\right)_{\text{equilibrium}} \quad [6.17(b)]$$

$$K = K_c \times (c^\ominus RT/p^\ominus) \quad \text{because } \Delta\nu = 1 \text{ for this reaction}$$

Since $c^\ominus R/p^\ominus = 0.0831451\ \text{K}^{-1}$, this expression may be written in the form

$$\boxed{K = (0.0831451\ \text{K}^{-1}) \times K_c T}$$

E6.12(a) $H_2CO(g) \rightarrow CO(g) + H_2(g)$ For this gas-phase reaction $\Delta\nu = \sum_J \nu_J = 1$

From the *CRC Handbook of Chemistry and Physics*,

$\Delta_f H^\ominus(H_2CO,g) = -108.57\ \text{kJ mol}^{-1}$ and $\Delta_f G^\ominus(H_2CO,g) = -102.53\ \text{kJ mol}^{-1}$. Thus,

$$\Delta_r H^\ominus = \Delta_f H^\ominus(CO,g) + \Delta_f H^\ominus(H_2,g) - \Delta_f H^\ominus(H_2CO,g)$$
$$= \{(-110.53) + 0 - (-108.57)\}\ \text{kJ mol}^{-1} = -1.96\ \text{kJ mol}^{-1}$$

$$\Delta_r G^\ominus = \Delta_f G^\ominus(CO,g) + \Delta_f G^\ominus(H_2,g) - \Delta_f G^\ominus(H_2CO,g)$$
$$= \{(-137.17) + 0 - (-102.53)\}\ \text{kJ mol}^{-1} = -34.64\ \text{kJ mol}^{-1}$$

Also, from Exercise 6.11(a) we have the relationship

$$K_c = K/\{(0.0831451\ \text{K}^{-1}) \times T\}$$

(a) At 25°C

$$K_{25°C} = e^{-\Delta_r G^\ominus/RT}\ [6.14]$$
$$= e^{-(-34.64\ \text{kJ mol}^{-1})/(2.4790\ \text{kJ mol}^{-1})} = \boxed{1.17 \times 10^6}$$

$$K_c(25°C) = K_{25°C}/\{(0.0831451\ \text{K}^{-1}) \times (298.15\ \text{K})\}$$
$$= (1.17 \times 10^6)/\{(0.0831451\ \text{K}^{-1}) \times (298.15\ \text{K})\} = \boxed{4.72 \times 10^4}$$

(b) At 100°C

$$\ln K_2 = \ln K_1 - \frac{\Delta_r H^\ominus}{R}\left(\frac{1}{T_2} - \frac{1}{T_1}\right) [6.23]$$

$$\ln K_{100°C} = \ln K_{25°C} - \frac{\Delta_r H^\ominus}{R}\left(\frac{1}{373.15\ \text{K}} - \frac{1}{298.15\ \text{K}}\right)$$

$$= \ln(1.17\times 10^6) - \left(\frac{-1.96\times 10^3\ \text{J mol}^{-1}}{8.3145\ \text{J K}^{-1}\,\text{mol}^{-1}}\right)\times\left(\frac{1}{373.15\ \text{K}} - \frac{1}{298.15\ \text{K}}\right)$$

$$= 13.8\overline{1}$$

$$K_{100°C} = e^{13.8\overline{1}} = \boxed{9.95\times 10^5}$$

$$K_c(100°C) = K_{100°C}/\{(0.0831451\ \text{K}^{-1})\times(373.15\ \text{K})\}$$

$$= (9.95\times 10^5)/\{(0.0831451\ \text{K}^{-1})\times(373.15\ \text{K})\} = \boxed{3.21\times 10^4}$$

E6.13(a) borneol $\rightleftharpoons$ isoborneol

$$\Delta_r G = \Delta G^\ominus + RT\ln Q\ [6.10] \quad \text{where} \quad Q = \frac{p_{\text{isoborneol}}}{p_{\text{borneol}}}\ [\text{perfect gas assumption}]$$

$$p_{\text{borneol}} = x_{\text{borneol}}\,p = \frac{0.15\ \text{mol}}{0.15\ \text{mol} + 0.30\ \text{mol}} \times 600\ \text{Torr} = 200\ \text{Torr}$$

$$p_{\text{isoborneol}} = p - p_{\text{borneol}} = 600\ \text{Torr} - 200\ \text{Torr} = 400\ \text{Torr}$$

$$Q = \frac{400\ \text{Torr}}{200\ \text{Torr}} = 2.00$$

$$\Delta_r G = (+9.4\ \text{kJ mol}^{-1}) + (8.314\ \text{J K}^{-1}\,\text{mol}^{-1})\times(503\ \text{K})\times(\ln 2.00) = \boxed{+12.3\ \text{kJ mol}^{-1}}$$

This mixture reacts spontaneously to the left under these conditions.

E6.14(a) $H_2CO(g) \rightarrow CO(g) + H_2(g)$ For this gas-phase reaction $\Delta\nu = \sum_J \nu_J = 1$

$$K = \left(\prod_J a_J^{\nu_J}\right)_{\text{equilibrium}} [6.13] = \left(\prod_J (\gamma_J p_J/p^\ominus)^{\nu_J}\right)_{\text{equilibrium}} = \left(\prod_J \gamma_J^{\nu_J}\right)_{\text{equilibrium}}\left(\prod_J (p_J/p^\ominus)^{\nu_J}\right)_{\text{equilibrium}}$$

$$= K_\gamma K_p, \quad \text{where} \quad K_\gamma = \left(\prod_J \gamma_J^{\nu_J}\right)_{\text{equilibrium}} \quad \text{and } K_p = \left(\prod_J (p_J/p^\ominus)^{\nu_J}\right)_{\text{equilibrium}}$$

Let us assume that the gases are perfect gases. Then, the activity coefficients equal 1 and $K_\gamma = 1$. Additionally, $p_J = x_J p$. Substitution gives

$$K = K_p = \left(\prod_J (x_J p/p^\ominus)^{\nu_J}\right)_{\text{equilibrium}} = \left(\prod_J x_J^{\nu_J}\right)_{\text{equilibrium}}\left(\prod_J (p/p^\ominus)^{\nu_J}\right)_{\text{equilibrium}}$$

$$= K_x (p/p^\ominus)^{\Delta\nu}, \quad \text{where} \quad K_x = \left(\prod_J x_J^{\nu_J}\right)_{\text{equilibrium}}$$

For this reaction: $K = K_x \times (p/p^{\ominus})$ or $K_x = K \times (p/p^{\ominus})^{-1}$ because $\Delta\nu = 1$

$$\text{Percentage change in } K_x = \frac{K_x(1\text{ bar}) - K_x(2\text{ bar})}{K_x(1\text{ bar})} \times 100\%$$

$$= \frac{K \times (1\text{ bar}/p^{\ominus})^{-1} - K \times (2\text{ bar}/p^{\ominus})^{-1}}{K \times (1\text{ bar}/p^{\ominus})^{-1}} \times 100\%$$

(Remember that K is independent of pressure.)

$$= \frac{1-\frac{1}{2}}{1} \times 100\% = \boxed{50\%}$$

E6.15(a) borneol(g) $\rightleftharpoons$ isoborneol(g) $K = 0.106$ at 503 K; $M = 154.24$ g mol^{-1}

Let z be the number of moles of isoborneol that convert to borneol during the establishment of equilibrium.

	Borneol	Isoborneol	Total
Initial amounts/mol	$n_{\text{borneol}} = 7.50/154.24 = 0.0486$	$n_{\text{iso}} = 14.0/154.24 = 0.0908$	$n = 0.1394$
Equilibrium amounts/mol	$n_{\text{borneol}} + z$	$n_{\text{iso}} - z$	$n = 0.1394$
Equilibrium mole fractions	$(n_{\text{borneol}} + z)/n$	$(n_{\text{iso}} - z)/n$	

$$K = \left(\frac{p_{\text{isoborneol}}}{p_{\text{borneol}}}\right)_{\text{equilibrium}} \quad \text{(perfect gas assumption)}$$

$$= \left(\frac{x_{\text{isoborneol}}}{x_{\text{borneol}}}\right)_{\text{equilibrium}} \quad (\text{because } p_{\text{A}} = x_{\text{A}}\,p)$$

$$= \frac{n_{\text{iso}} - z}{n_{\text{borneol}} + z}$$

$$z = \frac{n_{\text{iso}} - n_{\text{borneol}}K}{1+K} = \frac{0.0908 - 0.0486 \times 0.106}{1 + 0.106} = 0.0861$$

Thus, the equilibrium mole fractions are

$$x_{\text{borneol}} = \frac{n_{\text{borneol}} + z}{n} = \frac{0.0486 + 0.0861}{0.1394} = \boxed{0.9663}$$

$$x_{\text{iso}} = \frac{n_{\text{borneol}} - z}{n} = \frac{0.0908 - 0.0861}{0.1394} = \boxed{0.0337}$$

E6.16(a) $\ln\dfrac{K_2}{K_1} = -\dfrac{\Delta_r H^\ominus}{R}\left(\dfrac{1}{T_2} - \dfrac{1}{T_1}\right)$ [6.23]

$$\Delta_r H^\ominus = R \times \left(\frac{1}{T_1} - \frac{1}{T_2}\right)^{-1} \times \ln\left(\frac{K_2}{K_1}\right)$$

$$= (8.3145\ \mathrm{J\ K^{-1}\ mol^{-1}}) \times \left(\frac{1}{298\ \mathrm{K}} - \frac{1}{308\ \mathrm{K}}\right)^{-1} \times \ln\left(\frac{K_2}{K_1}\right)$$

$$= (76.31\ \mathrm{kJ\ mol^{-1}}) \times \ln\left(\frac{K_2}{K_1}\right)$$

(a) $K_2/K_1 = 2.00$

$\Delta_r H^\ominus = (76.31\ \mathrm{kJ\ mol^{-1}}) \times \ln(2.00) = \boxed{52.89\ \mathrm{kJ\ mol^{-1}}}$

(b) $K_2/K_1 = 0.500$

$\Delta_r H^\ominus = (76.31\ \mathrm{kJ\ mol^{-1}}) \times \ln(0.500) = \boxed{-52.89\ \mathrm{kJ\ mol^{-1}}}$

E6.17(a) $\frac{1}{2}N_2(g) + \frac{3}{2}H_2(g) \rightarrow NH_3(g) \quad \Delta_f G^\ominus = -16.5\ \mathrm{kJ\ mol^{-1}}$

$\Delta_f G = \Delta_f G^\ominus + RT\ln Q$ [6.10], where $Q = \prod_J a_J^{\nu_J}$ [6.12b]

$$= \Delta_f G^\ominus + RT\ln\frac{p_{NH_3}/p^\ominus}{(p_{N_2}/p^\ominus)^{1/2}(p_{H_2}/p^\ominus)^{3/2}} \text{ [perfect gas assumption]}$$

$$= (-16.5\ \mathrm{kJ\ mol^{-1}}) + (8.3145\times10^{-3}\ \mathrm{kJ\ K^{-1}\ mol^{-1}}) \times (298.15\ \mathrm{K}) \times \ln\left(\frac{4}{3^{1/2}\times1^{3/2}}\right)$$

$$= \boxed{-14.4\ \mathrm{kJ\ mol^{-1}}}$$

Since $\Delta_f G < 0$, the spontaneous direction of reaction is $\boxed{\text{toward the ammonia product}}$.

E6.18(a) The decomposition reaction is $CaCO_3(s) \rightleftharpoons CaO(s) + CO_2(g)$.

For the purposes of this exercise we may assume that the required temperature is that temperature at which $K = 1$, which corresponds to a pressure of 1 bar for the gaseous product. For $K = 1$, $\ln K = 0$ and $\Delta_r G^\ominus = 0$.

$\Delta_r G^\ominus = \Delta_r H^\ominus - T\Delta_r S^\ominus = 0$ when $\Delta_r H^\ominus = T\Delta_r S^\ominus$

Therefore, the decomposition temperature (when $K = 1$) is

$$T = \frac{\Delta_r H^\ominus}{\Delta_r S^\ominus}$$

$$\Delta_r H^\ominus = \sum_J \nu_J \Delta_{f,J} H^\ominus = \{(-635.09) - (393.51) - (-1206.9)\}\ \mathrm{kJ\ mol^{-1}} = +178.3\ \mathrm{kJ\ mol^{-1}}$$

$$\Delta_r S^\ominus = \sum_J \nu_J S_J^\ominus = \{(39.75) + (213.74) - (92.9)\}\ \mathrm{J\ K^{-1}\ mol^{-1}} = +160.6\ \mathrm{J\ K^{-1}\ mol^{-1}}$$

$$T = \frac{178.3\times10^3\ \mathrm{J\ mol^{-1}}}{160.6\ \mathrm{J\ K^{-1}\ mol^{-1}}} = \boxed{1110\ \mathrm{K}}\ (837°\mathrm{C})$$

E6.19(a) $CaF_2(s) \rightleftharpoons Ca^{2+}(aq) + 2F^-(aq) \quad K_s = 3.9 \times 10^{-11}$

$$\Delta_r G^\ominus = -RT \ln K_s$$
$$= -(8.3145 \text{ J K}^{-1} \text{ mol}^{-1}) \times (298.15 \text{ K}) \times \ln(3.9 \times 10^{-11}) = +59.4 \text{ kJ mol}^{-1}$$
$$= \Delta_f G^\ominus(CaF_2, aq) - \Delta_f G^\ominus(CaF_2, s)$$

$$\Delta_f G^\ominus(CaF_2, aq) = \Delta_r G^\ominus + \Delta_f G^\ominus(CaF_2, s) = (59.4 - 1167) \text{ kJ mol}^{-1} = \boxed{-1108 \text{ kJ mol}^{-1}}$$

E6.20(a) The cell notation specifies the right and left electrodes. Note that for proper cancellation we must equalize the number of electrons in half-reactions being combined. To calculate the standard cell potential we have used $E^\ominus_{cell} = E^\ominus_R - E^\ominus_L$, with standard electrode potentials from the data table.

		$E^\ominus$
(a)	R: $2\,Ag^+(aq) + 2\,e^- \rightarrow 2\,Ag(s)$	+0.80 V
	L: $Zn^{2+}(aq) + 2\,e^- \rightarrow Zn(s)$	−0.76 V
	Overall (R – L): $2\,Ag^+(aq) + Zn(s) \rightarrow 2\,Ag(s) + Zn^{2+}(aq)$	+1.56 V
(b)	R: $2\,H^+(aq) + 2\,e^- \rightarrow H_2(g)$	0
	L: $Cd^{2+}(aq) + 2\,e^- \rightarrow Cd(s)$	−0.40 V
	Overall (R – L): $Cd(s) + 2\,H^+(aq) \rightarrow Cd^{2+}(aq) + H_2(g)$	+0.40 V
(c)	R: $Cr^{3+}(aq) + 3\,e^- \rightarrow Cr(s)$	−0.74 V
	L: $3\,[Fe(CN_6)]^{3-}(aq) + 3\,e^- \rightarrow 3\,[Fe(CN)_6]^{4-}(aq)$	+0.36 V
	Overall (R – L): $Cr^{3+}(aq) + 3\,[Fe(CN)_6]^{4-}(aq) \rightarrow Cr(s) + 3\,[Fe(CN)_6]^{3-}(aq)$	−1.10 V

COMMENT. Those cells for which $E^\ominus_{cell} > 0$ may operate as spontaneous galvanic cells under standard conditions. Those for which $E^\ominus_{cell} > 0$ may operate as non-spontaneous electrolytic cells. Recall that $E^\ominus_{cell}$ informs us of the spontaneity of a cell under standard conditions only. For other conditions we require E_{cell}.

E6.21(a) The conditions (concentrations, etc.) under which these reactions occur are not given. For the purposes of this exercise we assume standard conditions. The specification of the right and left electrodes is determined by the direction of the reaction as written. As always, in combining half-reactions to form an overall cell reaction we must write the half-reactions with equal numbers of electrons to ensure proper cancellation. We first identify the half-reactions, and then set up the corresponding cell.

		$E^\ominus$
(a)	R: $2\,Cu^{2+}(aq) + 2\,e^- \rightarrow Cu(s)$	+0.34 V
	L: $Zn^{2+}(aq) + 2\,e^- \rightarrow Zn(s)$	−0.76 V
	Hence, the cell is	
	$Zn(s)\vert ZnSO_4(aq)\Vert CuSO_4(aq)\vert Cu(s)$	$\boxed{+1.10 \text{ V}}$
(b)	R: $AgCl(s) + e^- \rightarrow Ag(s) + Cl^-(aq)$	+0.22 V
	L: $H^+(aq) + e^- \rightarrow \frac{1}{2}\,H_2(g)$	0
	and the cell is	
	$Pt\vert H_2(g)\vert H^+(aq)\vert AgCl(s)\vert Ag(s)$	
	or $Pt\vert H_2(g)\vert HCl(aq)\vert AgCl(s)\vert Ag(s)$	$\boxed{+0.22 \text{ V}}$
(c)	R: $O_2(g) + 4\,H^+(aq) + 4\,e^- \rightarrow 2\,H_2O(l)$	+1.23 V
	L: $4\,H^+(aq) + 4\,e^- \rightarrow 2\,H_2(g)$	0
	and the cell is	
	$Pt\vert H_2(g)\vert H^+(aq)\vert O_2(g)\vert Pt$	$\boxed{+1.23 \text{ V}}$

COMMENT. All of these cells have $E^{\ominus}_{\text{cell}} > 0$, corresponding to a spontaneous cell reaction under standard conditions. If $E^{\ominus}_{\text{cell}}$ had turned out to be negative, the spontaneous reaction would have been the reverse of the one given, with the right and left electrodes of the cell also reversed.

E6.22(a) $Ag|AgBr(s)|KBr(aq, 0.050\ mol\ kg^{-1})||Cd(NO_3)_2(aq, 0.010\ mol\ kg^{-1})|Cd(s)$

(a) R: $Cd^{2+}(aq) + 2\,e^- \rightarrow Cd(s)$ $\quad E^{\ominus} = -0.40\ V$

L: $2\,AgBr(s) + 2\,e^- \rightarrow 2\,Ag(s) + 2\,Br^-(aq)$ $\quad E^{\ominus} = +0.713\ V$

R–L: $\boxed{Cd^{2+}(aq) + 2\,Br^-(aq) + 2\,Ag(s) \rightarrow Cd(s) + 2\,AgBr(s)}$ $\quad E^{\ominus}_{\text{cell}} = -0.47\ V$

The cell reaction is not spontaneous toward the right under standard conditions because $E^{\ominus}_{\text{cell}} < 0$.

(b) The Nernst equation for the above cell reaction is:

$$E_{\text{cell}} = E^{\ominus}_{\text{cell}} - \frac{RT}{\nu F}\ln Q\ [6.27]$$

where $\nu = 2$ and $Q = \dfrac{1}{a_{Cd^{2+}}a^2_{Br^-}} = \dfrac{1}{\gamma_{Cd^{2+}}\gamma^2_{Br^-}} \times \dfrac{(b^{\ominus})^3}{b_{Cd^{2+}}b^2_{Br^-}} = \dfrac{1}{\gamma_{\pm,R}\gamma^2_{\pm,L}} \times \dfrac{(b^{\ominus})^3}{b_{Cd^{2+}}b^2_{Br^-}}$

$b_{Cd^{2+}} = 0.010\ mol\ kg^{-1}$ for the right-hand electrode and $b_{Br^-} = 0.050\ mol\ kg^{-1}$ for the left-hand electrode.

(c) The ionic strength and mean activity coefficient at the right-hand electrode are:

$$I_R = \tfrac{1}{2}\sum_i z_i^2(b_i/b^{\ominus})\ [5.76] = \tfrac{1}{2}\{4(0.010) + 1(.020)\} = 0.030$$

$$\log \gamma_{\pm,R} = -|z_+z_-|AI^{1/2}\ [5.75] = -2 \times (0.509) \times (0.030)^{1/2} = -0.176$$

$$\gamma_{\pm,R} = 0.667$$

The ionic strength and mean acitivity coefficient at the left-hand electrode are:

$$I_L = \tfrac{1}{2}\sum_i z_i^2(b_i/b^{\ominus})\ [5.76] = \tfrac{1}{2}\{1(0.050) + 1(.050)\} = 0.050$$

$$\log \gamma_{\pm,L} = -|z_+z_-|AI^{1/2}\ [5.75] = -1 \times (0.509) \times (0.050)^{1/2} = -0.114$$

$$\gamma_{\pm,L} = 0.769$$

Therefore,

$$Q = \left(\frac{1}{(0.667) \times (0.769)^2}\right) \times \left(\frac{1}{(0.010) \times (0.050)^2}\right)$$

$$= 1.01 \times 10^5$$

and

$$E_{\text{cell}} = -0.47\ V - \left(\frac{25.693 \times 10^{-3}\ V}{2}\right)\ln(1.01 \times 10^5)$$

$$= \boxed{-0.62\ V}$$

E6.23(a) In each case the equilibrium constant is calculated with the expression $\ln K = \dfrac{\nu F E^{\ominus}_{\text{cell}}}{RT}$ [6.28].

(a) $\text{Sn(s)} + \text{Sn}^{4+}\text{(aq)} \rightleftharpoons 2\,\text{Sn}^{2+}\text{(aq)}$

$$\left.\begin{array}{ll}\text{R: Sn}^{4+} + 2\,\text{e}^- \to \text{Sn}^{2+}\text{(aq)} & +0.15\text{ V} \\ \text{L: Sn}^{2+}\text{(aq)} + 2\,\text{e}^- \to \text{Sn(s)} & -0.14\text{ V}\end{array}\right\} E^{\ominus}_{\text{cell}} = +0.29\text{ V}$$

$$\ln K = \frac{\nu F E^{\ominus}_{\text{cell}}}{RT}\,[6.28] = \frac{(2)\times(0.29\text{ V})}{25.693\text{ mV}} = 22.\overline{6}$$

$$K = \text{e}^{22.\overline{6}} = \boxed{6.5\times10^{9}}$$

(b) $\text{Sn(s)} + 2\,\text{AgCl(s)} \rightleftharpoons \text{SnCl}_2\text{(aq)} + 2\,\text{Ag(s)}$

$$\left.\begin{array}{ll}\text{R: 2 AgCl(s)} + 2\,\text{e}^- \to 2\,\text{Ag(s)} + 2\,\text{Cl}^-\text{(aq)} & +0.22\text{ V} \\ \text{L: Sn}^{2+}\text{(aq)} + 2\,\text{e}^- \to \text{Sn(s)} & -0.14\text{ V}\end{array}\right\} E^{\ominus}_{\text{cell}} = +0.36\text{ V}$$

$$\ln K = \frac{\nu F E^{\ominus}_{\text{cell}}}{RT}\,[6.28] = \frac{(2)\times(0.36\text{ V})}{25.693\text{ mV}} = +2\overline{8}.0$$

$$K = \text{e}^{2\overline{8}.0} = \boxed{1.4\times10^{12}}$$

E6.24(a)

$$\left.\begin{array}{ll}\text{R: Ag}^+\text{(aq)} + \text{e}^- \to \text{Ag(s)} & +0.80\text{ V} \\ \text{L: AgI(s)} + \text{e}^- \to \text{Ag(s)} + \text{I}^-\text{(aq)} & -0.15\text{ V}\end{array}\right\} E^{\ominus}_{\text{cell}} = E^{\ominus}_{\text{R}} - E^{\ominus}_{\text{L}} = 0.95\text{ V}$$

Overall: $\text{Ag}^+\text{(aq)} + \text{I}^-\text{(aq)} \to \text{AgI(s)} \quad \upsilon = 1$

(a) $$\ln K = \frac{\nu F E^{\ominus}_{\text{cell}}}{RT}\,[6.28] = \frac{1\times(+0.95\text{ V})}{25.693\text{ mV}} = 37.\overline{0}$$

$$K = \text{e}^{37.\overline{0}} = 1.1\overline{7}\times10^{16}$$

The solubility equilibrium is written as the reverse of the cell reaction. Therefore, the solubility product of AgI(s) is $K_{\text{sp}} = K^{-1} = 1/1.17\times10^{16} = \boxed{8.5\times10^{-17}}$.

(b) The solubility product of AgI(s) is very small. Consequently, the molar solubility, s, of silver iodide must also be very low and we can reasonably take the activity coefficients of the aqueous ions to equal 1.

$$K_{\text{sp}} = [\text{Ag}^+][\text{I}^-]/(c^{\ominus})^2 = s^2/(c^{\ominus})^2$$

$$s = K_{\text{sp}}^{1/2}c^{\ominus} = (8.5\times10^{-17})^{1/2}\text{ mol dm}^{-3} = \boxed{9.2\times10^{-9}\text{ mol dm}^{-3}\text{ or }2.2\ \mu\text{g dm}^{-3}}$$

Solutions to problems

Solutions to numerical problems

P6.1 $\text{I}_2\text{(s)} + \text{Br}_2\text{(g)} \to 2\,\text{IBr(g)} \quad K = 0.164\text{ at }25°\text{C}$

(a) $$\begin{aligned}\Delta_r G^{\ominus} &= -RT\ln K\,[6.14] \\ &= -(8.3145\text{ J K}^{-1}\text{ mol}^{-1})\times(298\text{ K})\times(\ln 0.164) = 4.48\times10^{3}\text{ J mol}^{-1} \\ &= \boxed{+4.48\text{ kJ mol}^{-1}}\end{aligned}$$

(b) Draw up the following equilibrium table:

	I_2	Br_2	IBr
Amounts	—	$(1-\alpha)n$	$2\alpha n$
Mole fractions	—	$\dfrac{(1-\alpha)}{(1+\alpha)}$	$\dfrac{2\alpha}{(1+\alpha)}$
Partial pressure $p_J = x_J p$	—	$\dfrac{(1-\alpha)p}{(1+\alpha)}$	$\dfrac{2\alpha p}{(1+\alpha)}$

$$K = \prod_J a_J^{\nu_J} [6.13] = \frac{(p_{IBr}/p^{\ominus})^2}{p_{Br_2}/p^{\ominus}} \text{ [perfect gases]}$$

$$= \frac{\{(2\alpha)^2 p/p^{\ominus}\}}{(1-\alpha)\times(1+\alpha)} = \frac{(4\alpha^2 p/p^{\ominus})}{1-\alpha^2} = 0.164$$

With $p = 0.164$ atm, $4\alpha^2 = 1-\alpha^2 \quad \alpha^2 = 1/5 \quad \alpha = 0.447$

$$p_{IBr} = \frac{2\alpha}{1+\alpha}\times p = \frac{(2)\times(0.447)}{1+0.447}\times(0.164\text{ atm}) = \boxed{0.101\text{ atm}}$$

(c) The equilibrium table needs to be modified as follows

$$p = p_{I_2} + p_{Br_2} + p_{IBr}, \quad p_{Br_2} = x_{Br_2}p, \quad p_{IBr} = x_{IBr}p, \quad p_{I_2} = x_{I_2}p$$

with $x_{Br_2} = \dfrac{(1-\alpha)n}{(1+\alpha)n + n_{I_2}}$ [n = amount of Br_2 introduced into the container]

and $x_{IBr} = \dfrac{2\alpha n}{(1+\alpha)n + n_{I_2}}$.

K is constructed as above [6.13], but with these modified partial pressures. In order to complete the calculation additional data are required, namely, the amount of Br_2 introduced, n, and the equilibrium vapour pressure of $I_2(s)$. n_{I_2} can be calculated from a knowledge of the volume of the container at equilibrium, which is most easily determined by successive approximations since p_{I_2} is small.

What is the partial pressure of IBr(g) if 0.0100 mol of $Br_2(g)$ is introduced into the container? The partial pressure of $I_2(s)$ at 25°C is 0.305 Torr.

P6.3 $U(s) + \frac{3}{2}H_2(g) \rightleftharpoons UH_3(s)$

$$K = a_{H_2}^{-3/2} = (p_{H_2}/p^{\ominus})^{-3/2} \text{ [perfect gas]}$$
$$= (p/p^{\ominus})^{-3/2} \qquad [p_{H_2} = p]$$

$$\Delta_f H^{\ominus} = RT^2\frac{d\ln K}{dT}[6.21(a)] = RT^2\frac{d}{dT}\ln(p/p^{\ominus})^{-3/2} = -\tfrac{3}{2}RT^2\frac{d}{dT}(\ln p - \ln p^{\ominus})$$

$$= -\tfrac{3}{2}RT^2\frac{d}{dT}(\ln p)$$

$$= -\tfrac{3}{2}RT^2\frac{d}{dT}(A + B/T + C\ln(T/K)) = -\tfrac{3}{2}RT^2\times\left(\frac{-B}{T^2}+\frac{C}{T}\right)$$

$$= \boxed{\tfrac{3}{2}R\times(B-CT)}, \quad \text{where} \quad B = -1.464\times10^4\text{ K} \quad \text{and} \quad C = -5.65$$

$$d(\Delta_f H^\ominus) = \Delta_r C_p^\ominus dT \text{ [from 2.36a]}$$

$$\text{or } \Delta_r C_p^\ominus = \left(\frac{\partial \Delta_f H^\ominus}{\partial T}\right)_p = -\tfrac{3}{2}CR = \boxed{70.5\text{ J K}^{-1}\text{ mol}^{-1}}$$

P6.5 $CaCl_2 \cdot NH_3(s) \rightleftharpoons CaCl_2(s) + NH_3(g) \quad K = \frac{p}{p^\ominus}$ and $\Delta_r H^\ominus = +78\text{ kJ mol}^{-1}$

$$\Delta_r G^\ominus = -RT\ln K = -RT\ln\frac{p}{p^\ominus}$$

$$= -(8.3145\text{ J K}^{-1}\text{ mol}^{-1}) \times (400\text{ K}) \times \ln\left(\frac{1.71\text{ kPa}}{100.0\text{ kPa}}\right) \quad [p^\ominus = 1\text{ bar} = 100.0\text{ kPa}]$$

$$= +13.5\text{ kJ mol}^{-1}\text{ at 400 K}$$

$$\frac{\Delta_r G^\ominus(T_2)}{T_2} - \frac{\Delta_r G^\ominus(T_1)}{T_1} = \Delta_r H^\ominus\left(\frac{1}{T_2} - \frac{1}{T_1}\right) \text{[6.23 and 6.14]}$$

Therefore, taking $T_1 = 400$ K and letting $T = T_2$ be any temperature in the range 350 K to 470 K,

$$\Delta_r G^\ominus(T) = \left(\frac{T}{400\text{ K}}\right) \times (13.5\text{ kJ mol}^{-1}) + (78\text{ kJ mol}^{-1}) \times \left(1 - \frac{T}{400\text{ K}}\right)$$

$$= (78\text{ kJ mol}^{-1}) + \left(\frac{(13.5 - 7.8)\text{ kJ mol}^{-1}}{400}\right) \times \left(\frac{T}{\text{K}}\right)$$

That is, $\Delta_r G^\ominus(T)/(\text{kJ mol}^{-1}) = \boxed{78 - 0.161 \times (T/\text{K})}$.

P6.7 The equilibrium we need to consider is $A_2(g) \rightleftharpoons 2\,A(g)$, where A is acetic acid. It is convenient to express the equilibrium constant in terms of α, the degree of dissociation of the dimer, which is the predominant species at low temperatures. Consequently, we let n represent the amount of dimer present should all the contained mass be dimerized. Thus, $n = m_{\text{toal}}/M_{A_2}$ which $M_{A_2} = 120.1\text{ g mol}^{-1}$. We take the experimental masses to be: $m_{\text{total}} = 0.463$ g at 437 K and $m_{\text{total}} = 0.0379$ g at 471 K.

	A	A_2	Total
At equilibrium	$2\alpha n$	$(1-\alpha)n$	$(1+\alpha)n$
Mole fraction	$\frac{2\alpha}{1+\alpha}$	$\frac{1-\alpha}{1+\alpha}$	1
Partial pressure	$\frac{2\alpha p}{1+\alpha}$	$\left(\frac{1-\alpha}{1+\alpha}\right)p$	p

The equilibrium constant for the dissociation is

$$K = \frac{(p_A/p^\ominus)^2}{p_{A_2}/p^\ominus} = \frac{p_A^2}{p_{A_2}p^\ominus} = \frac{4\alpha^2(p/p^\ominus)}{1-\alpha^2}.$$

We also know that

$$pV = n_{\text{total}}RT = (1+\alpha)nRT, \text{ implying that } \alpha = \frac{pV}{nRT} - 1 = \frac{pVM_{A_2}}{m_{\text{total}}RT} - 1.$$

In the first experiment,

$$\alpha = \frac{pVM_{A_2}}{m_{\text{total}}RT} - 1 = \frac{(101.9\text{ kPa})\times(21.45\times10^{-3}\text{ dm}^3)\times(120.1\text{ g mol}^{-1})}{(0.0463\text{ g})\times(8.3145\text{ kPa dm}^3\text{ K}^{-1}\text{ mol}^{-1})\times(437\text{ K})} - 1 = 0.560$$

$$\text{Hence, } K = \frac{4\times(0.560)^2\times(101.9\text{ kPa}/100.0\text{ kPa})}{1-(0.560)^2} = \boxed{1.86}\ [p^{\ominus} = 1\text{ bar} = 100.0\text{ kPa}]$$

In the second experiment,

$$\alpha = \frac{pVM_{A_2}}{m_{\text{total}}RT} - 1 = \frac{(101.9\text{ kPa})\times(21.45\times10^{-3}\text{ dm}^3)\times(120.1\text{ g mol}^{-1})}{(0.0379\text{ g})\times(8.3145\text{ kPa dm}^3\text{ K}^{-1}\text{ mol}^{-1})\times(471\text{ K})} - 1 = 0.769$$

$$\text{Hence, } K = \frac{4\times(0.769)^2\times(101.9\text{ kPa}/100.0\text{ kPa})}{1-(0.769)^2} = \boxed{5.90}\ [p^{\ominus} = 1\text{ bar} = 100.0\text{ kPa}].$$

The enthalpy of dissociation is

$$\Delta_r H^{\ominus} = \frac{R\ln(K_2/K_1)}{\left(\frac{1}{T_1} - \frac{1}{T_2}\right)}[6.23] = \frac{(8.3145\text{ J mol}^{-1}\text{ K}^{-1})\ln(5.90/1.86)}{\left(\frac{1}{437\text{ K}} - \frac{1}{471\text{ K}}\right)} = +58.1\text{ kJ mol}^{-1}.$$

The enthalpy of dimerization is the negative of this value, or $\boxed{-58.1\text{ kJ mol}^{-1}}$ (i.e. per mole of dimer).

P6.9 The equilibrium $I_2(g) \rightleftharpoons 2\,I(g)$ is described by the equilibrium constant

$$K = \frac{x_I^2}{x_{I_2}} \times \frac{p}{p^{\ominus}}\ [\text{perfect gas, } p_J = x_J p] = \frac{4\alpha^2(p/p^{\ominus})}{1-\alpha^2}\ [\text{Problem 6.7; } p^{\ominus} = 1\text{ bar} = 0.98692\text{ atm}]$$

Let $p^{\circ} = \dfrac{n_I RT}{V}$, then $p = (1+\alpha)p^{\circ}$, implying that $\alpha = \dfrac{p - p^{\circ}}{p^{\circ}}$.

We therefore draw up the following table:

	973 K	1073 K	1173 K
p/atm	0.06244	0.07500	0.09181
$10^4 n_I$	2.4709	2.4555	2.4366
p°/atm	0.05757	0.06309	0.06844
α	0.08459	0.1888	0.3415
K	$\boxed{1.824\times10^{-3}}$	$\boxed{1.123\times10^{-2}}$	$\boxed{4.911\times10^{-2}}$

The standard enthalpy of dissociation is measured by determining the slope of a plot of ln K against $1/T$ (see Example 6.3). It is found that the plot is highly linear, with slope −18799 K. Thus, we conclude that the reaction enthalpy is constant over the temperature range of the data and that the reaction enthalpy can be calculated with the van't Hoff equation [6.21(b)].

$$\Delta_r H^\ominus = -R\frac{\mathrm{d}\ln K}{\mathrm{d}(1/T)}$$
$$= -(8.3145\ \mathrm{J\ mol^{-1}\ K^{-1}}) \times (-187\overline{99}\ \mathrm{K})$$
$$= \boxed{156\ \mathrm{kJ\ mol^{-1}}}$$

P6.11 The reaction is $\mathrm{Si(s) + H_2(g) \rightleftharpoons SiH_2(g)}$.

The equilibrium constant is $K = \exp(-\Delta_f G^\ominus/RT) = \exp(-\Delta_f H^\ominus/RT)\exp(\Delta_f S^\ominus/RT)$.

Let h be the uncertainty in $\Delta_f H^\ominus$ so that the high value is the low value plus h. The K based on the low value is

$$K_{\mathrm{low}\ H} = \exp\left(\frac{-\Delta_f H^\ominus_{\mathrm{low}}}{RT}\right)\exp\left(\frac{\Delta_f S^\ominus}{R}\right) = \exp\left(\frac{-\Delta_f H^\ominus_{\mathrm{high}}}{RT}\right)\exp\left(\frac{h}{RT}\right)\exp\left(\frac{\Delta_f S^\ominus}{R}\right)$$
$$= \exp\left(\frac{h}{RT}\right)K_{\mathrm{high}\ H}$$

So, $\dfrac{K_{\mathrm{low}\ H}}{K_{\mathrm{high}\ H}} = \exp\left(\dfrac{h}{RT}\right)$

(a) At 298 K, $\dfrac{K_{\mathrm{low}\ H}}{K_{\mathrm{high}\ H}} = \exp\left(\dfrac{(289-243)\ \mathrm{kJ\ mol^{-1}}}{(8.3145\times10^{-3}\ \mathrm{kJ\ K^{-1}\ mol^{-1}})\times(298\ \mathrm{K})}\right) = \boxed{1.2\times10^{8}}$

(b) At 700 K, $\dfrac{K_{\mathrm{low}\ H}}{K_{\mathrm{high}\ H}} = \exp\left(\dfrac{(289-243)\ \mathrm{kJ\ mol^{-1}}}{(8.3145\times10^{-3}\ \mathrm{kJ\ K^{-1}\ mol^{-1}})\times(700\ \mathrm{K})}\right) = \boxed{2.7\times10^{3}}$

P6.13 (a) $I = \dfrac{1}{2}\left\{\left(\dfrac{b}{b^\ominus}\right)_+ z_+^2 + \left(\dfrac{b}{b^\ominus}\right)_- z_-^2\right\}$ [5.76] $= 4\left(\dfrac{b}{b^\ominus}\right)$

For $\mathrm{CuSO_4}$, $I = (4)\times(1.0\times10^{-3}) = \boxed{4.0\times10^{-3}}$

For $\mathrm{ZnSO_4}$, $I = (4)\times(3.0\times10^{-3}) = \boxed{1.2\times10^{-2}}$

(b) $\log\gamma_\pm = -|z_+z_-|AI^{1/2}$ [5.75]

$\log\gamma_\pm(\mathrm{CuSO_4}) = -(4)\times(0.509)\times(4.0\times10^{-3})^{1/2} = -0.12\overline{88}$

$\gamma_\pm(\mathrm{CuSO_4}) = \boxed{0.74}$

$\log\gamma_\pm(\mathrm{ZnSO_4}) = -(4)\times(0.509)\times(1.2\times10^{-2})^{1/2} = -0.22\overline{30}$

$\gamma_\pm(\mathrm{ZnSO_4}) = \boxed{0.60}$

(c) The reaction in the Daniell cell is

$\mathrm{Cu^{2+}(aq) + SO_4^{2-}(aq) + Zn(s) \rightarrow Cu(s) + Zn^{2+}(aq) + SO_4^{2-}(aq)}$

Hence, $Q = \dfrac{a(\mathrm{Zn}^{2+})a(\mathrm{SO}_4^{2-},\mathrm{R})}{a(\mathrm{Cu}^{2+})a(\mathrm{SO}_4^{2-},\mathrm{L})}$

$$= \frac{\gamma_+ b_+(\mathrm{Zn}^{2+})\gamma_- b_-(\mathrm{SO}_4^{2-},\mathrm{R})}{\gamma_+ b_+(\mathrm{Cu}^{2+})\gamma_- b_-(\mathrm{SO}_4^{2-},\mathrm{L})}\left[b \equiv \frac{b}{b^{\ominus}} \text{ here and below}\right]$$

where the designations R and L refer to the right and left sides of the equation for the cell reaction and all b are assumed to be unitless, that is, $\dfrac{b}{b^{\ominus}}$.

$$b_+(\mathrm{Zn}^{2+}) = b_-(\mathrm{SO}_4^{2-},\mathrm{R}) = b(\mathrm{ZnSO}_4)$$

$$b_+(\mathrm{Cu}^{2+}) = b_-(\mathrm{SO}_4^{2-},\mathrm{L}) = b(\mathrm{CuSO}_4)$$

Therefore,

$$Q = \frac{\gamma_\pm^2(\mathrm{ZnSO}_4)b^2(\mathrm{ZnSO}_4)}{\gamma_\pm^2(\mathrm{CuSO}_4)b^2(\mathrm{CuSO}_4)} = \frac{(0.60)^2 \times (3.0\times10^{-3})^2}{(0.74)^2\times(1.0\times10^{-3})^2} = 5.9\overline{2} = \boxed{5.9}$$

(d) $$E^{\ominus}_{\mathrm{cell}} = -\frac{\Delta_r G^{\ominus}}{\nu F}\,[6.26] = \frac{-(-212.7\times10^3\ \mathrm{J\ mol^{-1}})}{(2)\times(9.6485\times10^4\ \mathrm{C\ mol^{-1}})} = \boxed{+1.102\ \mathrm{V}}$$

(e) $$E_{\mathrm{cell}} = E^{\ominus}_{\mathrm{cell}} - \frac{25.693\times10^{-3}\ \mathrm{V}}{\nu}\ln Q = (1.102\ \mathrm{V}) - \left(\frac{25.693\times10^{-3}\ \mathrm{V}}{2}\right)\ln(5.9\overline{2})$$

$$= (1.102\ \mathrm{V}) - (0.023\ \mathrm{V}) = \boxed{+1.079\ \mathrm{V}}$$

P6.15 The electrode half-reactions and their potentials are

		$E^{\ominus}$
R:	$\mathrm{Q(aq)} + 2\,\mathrm{H^+(aq)} + 2\,\mathrm{e^-} \rightarrow \mathrm{QH_2(aq)}$	0.6994 V
L:	$\mathrm{Hg_2Cl_2(s)} + 2\,\mathrm{e^-} \rightarrow 2\,\mathrm{Hg(l)} + 2\,\mathrm{Cl^-(aq)}$	0.2676 V
Overall (R – L):	$\mathrm{Q(aq)} + 2\,\mathrm{H^+(aq)} + 2\,\mathrm{Hg(l)} + 2\,\mathrm{Cl^-(aq)} \rightarrow \mathrm{QH_2(aq)} + \mathrm{Hg_2Cl_2(s)}$	0.4318 V

$$Q = \frac{a(\mathrm{QH_2})}{a(\mathrm{Q})a^2(\mathrm{H^+})a^2(\mathrm{Cl^-})}$$

Since quinhydrone is an equimolecular complex of Q and $\mathrm{QH_2}$, $b(\mathrm{Q}) = b(\mathrm{QH_2})$, and since their activity coefficients are assumed to be 1 or to be equal, we have $a(\mathrm{QH_2}) \approx a(\mathrm{Q})$. Thus,

$$Q = \frac{1}{a^2(\mathrm{H^+})a^2(\mathrm{Cl^-})} \quad \text{and} \quad E_{\mathrm{cell}} = E^{\ominus}_{\mathrm{cell}} - \frac{25.7\ \mathrm{mV}}{\nu}\ln Q\ [6.27,\ 25°\mathrm{C}]$$

$$\ln Q = \frac{\nu(E^{\ominus}_{\mathrm{cell}} - E_{\mathrm{cell}})}{25.7\ \mathrm{mV}} = \frac{(2)\times(0.4318-0.190)\ \mathrm{V}}{25.7\times10^{-3}\ \mathrm{V}} = 18.8\overline{2} \quad \text{and} \quad Q = 1.\overline{49}\times10^8$$

$$a^2(\mathrm{H^+}) = (\gamma_+ b_+)^2;\quad a^2(\mathrm{Cl^-}) = (\gamma_- b_-)^2 \quad [b \equiv b/b^{\ominus}]$$

For HCl(aq), $b_+ = b_- = b$, and if the activity coefficients are assumed equal, $a^2(\mathrm{H^+}) = a^2(\mathrm{Cl^-})$, hence

$$Q = \frac{1}{a^2(\mathrm{H^+})a^2(\mathrm{Cl^-})} = \frac{1}{a^4(\mathrm{H^+})}$$

Thus, $a(\mathrm{H^+}) = \left(\dfrac{1}{Q}\right)^{1/4} = \left(\dfrac{1}{1.49\times10^8}\right)^{1/4} = 9\times10^{-3}$

$\text{pH} = -\log a(\text{H}^+) = \boxed{2.0}$

P6.17 $\text{Pt} \mid \text{H}_2(\text{g}) \mid \text{HCl}(\text{aq},b) \mid \text{Hg}_2\text{Cl}_2(\text{s}) \mid \text{Hg}(1)$

$\frac{1}{2}\,\text{Hg}_2\text{Cl}_2(\text{s}) + \frac{1}{2}\,\text{H}_2(\text{g}) \rightarrow \text{Hg(l)} + \text{HCl(aq)}$ and $\nu = 1$

$$E_{\text{cell}} = E^{\ominus}_{\text{cell}} - \frac{RT}{F}\ln a(\text{H}^+)a(\text{Cl}^-)\ [6.27]$$

$a(\text{H}^+) = \gamma_+ b_+ = \gamma_+ b$; $a(\text{Cl}^-) = \gamma_- b_- = \gamma_- b$ $[b = b/b^{\ominus}$ here and below$]$

$a(\text{H}^+)a(\text{Cl}^-) = \gamma_+\gamma_- b^2 = \gamma_\pm^2 b^2$

$$E_{\text{cell}} = E^{\ominus}_{\text{cell}} - \frac{2RT}{F}\ln b - \frac{2RT}{F}\ln\gamma_\pm \qquad \text{(a)}$$

Converting from natural logarithms to common logarithms (base 10) in order to introduce the Debye–Hückel expression, we obtain

$$\begin{aligned} E_{\text{cell}} &= E^{\ominus}_{\text{cell}} - \frac{(2.303)\times 2RT}{F}\log b - \frac{(2.303)\times 2RT}{F}\log\gamma_\pm \\ &= E^{\ominus}_{\text{cell}} - (0.1183\ \text{V})\log b - (0.1183\ \text{V})\log\gamma_\pm \\ &= E^{\ominus}_{\text{cell}} - (0.1183\ \text{V})\log b - (0.1183\ \text{V})[-|z_+z_-|AI^{1/2}]\,[5.75] \\ &= E^{\ominus}_{\text{cell}} - (0.1183\ \text{V})\log b + (0.1183\ \text{V})\times A\times b^{1/2}\ [5.76,\ I = b] \end{aligned}$$

Rearranging,

$$E_{\text{cell}} + (0.1183\ \text{V})\,|\log|\,b = E^{\ominus}_{\text{cell}} + \text{constant}\times b^{1/2}$$

Therefore, plot $E_{\text{cell}} + (0.1183\ \text{V})\log b$ against $b^{1/2}$, and the intercept at $b = 0$ is $E^{\ominus}_{\text{cell}}/\text{V}$. Draw up the following table:

$b/(\text{mmol kg}^{-1})$	1.6077	3.0769	5.0403	7.6938	10.9474
$(b/b^{\ominus})^{1/2}$	0.04010	0.05547	0.07100	0.08771	0.1046
$E_{\text{cell}}/\text{V} + (0.1183)\,|\log|\,b$	0.27029	0.27109	0.27186	0.27260	0.27337

The points are plotted in Figure 6.2. The intercept is at 0.26840, so $E^{\ominus}_{\text{cell}} = +0.26840$ V. A least-squares best fit gives $E^{\ominus}_{\text{cell}} = \boxed{+0.26843\ \text{V}}$ and a coefficient of determination equal to 0.99895.

For the activity coefficients we obtain from equation (a)

$$\ln\gamma_\pm = \frac{E^{\ominus}_{\text{cell}} - E_{\text{cell}}}{2RT/F} - \ln\frac{b}{b^{\ominus}} = \frac{0.26843 - E_{\text{cell}}/V}{0.05139} - \ln\frac{b}{b^{\ominus}}$$

and we draw up the following table:

$b/(\text{mmol kg}^{-1})$	1.6077	3.0769	5.0403	7.6938	10.9474
E_{cell}/V	0.60080	0.56825	0.54366	0.52267	0.50532
$\ln\gamma_\pm$	−0.03465	−0.05038	−0.06542	−0.07993	−0.09500
$\gamma_\pm$	0.9659	0.9509	0.9367	0.9232	0.9094

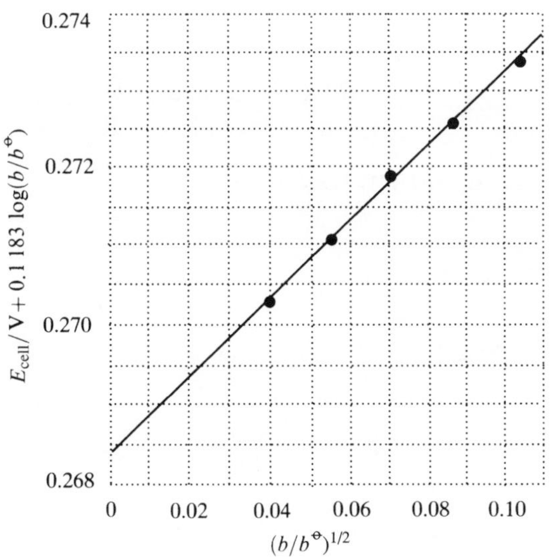

Figure 6.2

P6.19 The cells described in the problem are back-to-back pairs of cells each of the type $M_xHg(s) \mid MX(b_2) \mid AgX(s) \mid Ag(s)$.

R: $AgX(s) + e^- \rightarrow Ag(s) + X^-(b_2)$
L: $M^+(b_2) + e^- \xrightarrow{Hg} M_xHg(s)$ (reduction of M^+ and formation of amalgam)
R − L: $M_xHg(s) + AgX(s) \xrightarrow{Hg} Ag(s) + M^+(b_2) + X^-(b_2)$ and $\nu = 1$

$$E_{cell} = E^{\ominus}_{cell} - \frac{RT}{F}\ln Q \,[6.27] \text{ where } Q = \frac{a(M^+)a(X^-)}{a(M_xHg)} \,[6.12(b)]$$

For a pair of such cells back to back,

$$Ag(s) \mid AgX(s) \mid MX(b_1) \mid M_xHg(s) \mid MX(b_2) \mid AgX(s) \mid Ag(s)$$

$$E_R = E^{\ominus}_{cell} - \frac{RT}{F}\ln Q_R \quad \text{and} \quad E_L = E^{\ominus}_{cell} - \frac{RT}{F}\ln Q_L$$

$$E_{cell} = E_R - E_L = \frac{-RT}{F}\ln\frac{Q_R}{Q_L} = \frac{RT}{F}\ln\frac{(a(M^+)a(X^-))_L}{(a(M^+)a(X^-))_R}$$

(Note that the unknown quantity $a(M_xHg)$ drops out of the expression for E.)

$$(a(M^+)a(X^-))_{L\text{ or }R} = \left(\left(\frac{\gamma_+ b_+}{b^{\ominus}}\right)\left(\frac{\gamma_- b_-}{b^{\ominus}}\right)\right)_{L\text{ or }R} = \gamma_{\pm}^2\left(\frac{b_{L\text{ or }R}}{b^{\ominus}}\right)^2 \quad (b_+ = b_- \equiv b_{L\text{ or }R})$$

With L = (1) and R = (2) we have

$$E_{cell} = \frac{2RT}{F}\ln\frac{b_1}{b_2} + \frac{2RT}{F}\ln\frac{\gamma_{\pm}(1)}{\gamma_{\pm}(2)}$$

Take $b_2 = 0.09141$ mol kg^{-1} (the reference value), and write $b = b_1/b^{\ominus}$.

$$E_{cell} = \frac{2RT}{F}\left\{\ln\frac{b/\text{mol kg}^{-1}}{0.09141} + \ln\frac{\gamma_{\pm}}{\gamma_{\pm}(\text{ref})}\right\}$$

For $b = 0.09141$ mol kg^{-1}, the extended Debye–Hückel law gives

$$\log \gamma_\pm = -\frac{AI^{1/2}}{1+BI^{1/2}} + CI \text{ [5.78]} \quad \text{with} \quad A = 1.461,\ B = 1.70,\ C = 0.20, \text{ and } I = b/b^\ominus$$

$$\log \gamma_\pm(\text{ref}) = \frac{(-1.461)\times(0.09141)^{1/2}}{(1)+(1.70)\times(0.09141)^{1/2}} + (0.20)\times(0.09141) = -0.273\overline{5}$$

$$\gamma_\pm(\text{ref}) = 0.532\overline{8}$$

$$\text{Then, } E_{\text{cell}} = (0.05139\text{ V})\left\{\ln\frac{b/\text{mol kg}^{-1}}{0.09141} + \ln\frac{\gamma_\pm}{0.5328}\right\}$$

$$\ln\gamma_\pm = \frac{E_{\text{cell}}}{0.05139\text{ V}} - \ln\frac{b/\text{mol kg}^{-1}}{(0.09141)\times(0.05328)}$$

We then draw up the following table:

$b/(\text{mol kg}^{-1})$	0.0555	0.09141	0.1652	0.2171	1.040	1.350
E/V	−0.0220	0.0000	0.0263	0.0379	0.1156	0.1336
$\gamma_\pm$	0.572	0.533	0.492	0.469	0.444	0.486

P6.21 Electrochemical cell equation: $\frac{1}{2}\,H_2(g,1\text{ bar}) + AgCl(s) \rightleftharpoons H^+(aq) + Cl^-(aq) + Ag(s)$

with $a(H_2) = 1\text{ bar} = p^\ominus$ and $a_{Cl^-} = \gamma_{Cl^-}b$.

Weak acid equilibrium: $BH^+ \rightleftharpoons B + H^+$ with $b_{BH^+} = b_B = b$

$$K_a = a_B a_{H^+}/a_{BH^+} = \gamma_B b a_{H^+}/(\gamma_{BH^+}b) = \gamma_B a_{H^+}/\gamma_{BH^+}$$

Thus, $a_{H^+} = \gamma_{BH^+}K_a/\gamma_B$.

Ionic strength (neglect b_{H^+} because $b_{H^+} \ll b$): $I = \frac{1}{2}\{z_{BH^+}^2 b_{BH^+} + z_{Cl^-}^2 b_{Cl^-}\} = b$.

According to the Nernst equation [6.27]

$$E_{\text{cell}} = E_{\text{cell}}^\ominus - \frac{RT}{F}\ln\left(\frac{a_{H^+}a_{Cl^-}}{p(H_2)/p^\ominus}\right) = E_{\text{cell}}^\ominus - \frac{RT\ln(10)}{F}\log(a_{H^+}a_{Cl^-})$$

$$\frac{F}{RT\ln(10)}(E_{\text{cell}} - E_{\text{cell}}^\ominus) = -\log(a_{H^+}\gamma_{Cl^-}b) = -\log\left(\frac{K_a\gamma_{BH^+}\gamma_{Cl^-}b}{\gamma_B}\right)$$

$$= \text{p}K_a - \log(b) - 2\log(\gamma_\pm), \quad \text{where} \quad \gamma_\pm^2 \equiv \gamma_{BH^+}\gamma_{Cl^-}/\gamma_B$$

Substitution of the extended Debye−Hückel law,

$$\log\gamma_\pm = -\frac{A|z_+z_-|I^{1/2}}{1+BI^{1/2}} + CI \text{ [5.78]},$$

gives

$$\frac{F}{RT\ln(10)}(E_{\text{cell}} - E_{\text{cell}}^\ominus) = \text{p}K_a - \log(b) + \frac{2A\sqrt{b}}{1+B\sqrt{b}} - 2Cb, \quad \text{where } A = 0.5091.$$

The expression to the left of the above equality is experimental data that is a function of b. The parameters pK_a, B, and C on the right side are systematically varied with a mathematical regression software package like Mathcad until the right side fits the left side in a least squares sense. The results are:

$\boxed{pK_a = 6.736}$, $\boxed{B = 1.997}$, and $\boxed{C = -0.121}$

The mean activity coefficient is calculated with the equation $\gamma_\pm = 10^{\left(\frac{-AI^{1/2}}{1+BI^{1/2}} + Cb\right)}$ for desired values of b and I. Figure 6.3 shows a $\gamma_\pm$ against I plot for $b = 0.04$ mol kg^{-1} and $0 \le I \le 0.1$.

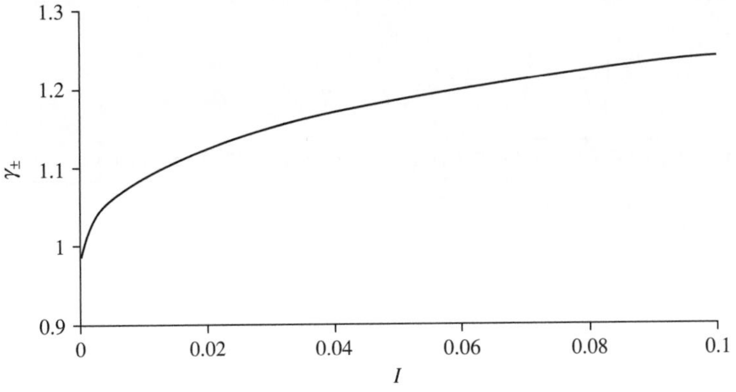

Figure 6.3

Solutions to theoretical problems

P6.23 $\Delta_r H(T') = \Delta_r H(T) + \int_T^{T'} \Delta_r C_p(T)\mathrm{d}T$ [2.36a; $\Delta_r C_p(T) = \sum_J \nu_J C_{p,J}$]

$$\Delta_r S(T') = \Delta_r S(T) + \int_T^{T'} \frac{\Delta_r C_p(T)}{T}\mathrm{d}T \text{ [3.22 applied to reaction equations]}$$

$$\Delta_r G(T') = \Delta_r H(T') - T'\Delta_r S(T')$$

$$= \left\{\Delta_r H(T) + \int_T^{T'} \Delta_r C_p(T)\mathrm{d}T\right\} - T'\left\{\Delta_r S(T) + \int_T^{T'} \frac{\Delta_r C_p(T)}{T}\mathrm{d}T\right\}$$

$$= \Delta_r H(T) - T\Delta_r S(T) - (T'-T)\Delta_r S(T) + \int_T^{T'} \Delta_r C_p(T)\mathrm{d}T - T'\int_T^{T'} \frac{\Delta_r C_p(T)}{T}\mathrm{d}T$$

$$= \Delta_r G(T) - (T'-T)\Delta_r S(T) + \int_T^{T'} \Delta_r C_p(T)\mathrm{d}T - T'\int_T^{T'} \frac{\Delta_r C_p(T)}{T}\mathrm{d}T$$

$$= \Delta_r G(T) - (T'-T)\Delta_r S(T) + \int_T^{T'} \left(1 - \frac{T'}{T}\right)\Delta_r C_p(T)\mathrm{d}T$$

$$\Delta_r C_p(T) = \Delta a + T\Delta b + \frac{\Delta c}{T^2}$$

$$\left(1 - \frac{T'}{T}\right)\Delta_r C_p(T) = \Delta a + T\Delta b + \frac{\Delta c}{T^2} - \frac{T'\Delta a}{T} - T'\Delta b - \frac{T'\Delta c}{T^3}$$

$$= \Delta a - T'\Delta b + T\Delta b - \frac{T'\Delta a}{T} + \frac{\Delta c}{T^2} - \frac{T'\Delta c}{T^3}$$

$$\int_T^{T'}\left(1 - \frac{T'}{T}\right)\Delta_r C_p(T)\mathrm{d}T = (\Delta a - T'\Delta b)(T' - T) + \tfrac{1}{2}(T'^2 - T^2)\Delta b - T'\Delta a \ln\frac{T'}{T}$$

$$+ \Delta c\left(\frac{1}{T} - \frac{1}{T'}\right) - \tfrac{1}{2}T'\Delta c\left(\frac{1}{T^2} - \frac{1}{T'^2}\right)$$

Therefore, $\boxed{\Delta_r G(T') = \Delta_r G(T) + (T - T')\Delta_r S(T) + \alpha(T', T) \times \Delta a + \beta(T', T) \times \Delta b + \gamma(T', T) \times \Delta c}$

where $\alpha(T', T) = T' - T - T'\ln\frac{T'}{T}$, $\beta(T', T) = \tfrac{1}{2}(T'^2 - T^2) - T'(T' - T)$,

and $\gamma(T', T) = \frac{1}{T} - \frac{1}{T'} + \tfrac{1}{2}T'\left(\frac{1}{T'^2} - \frac{1}{T^2}\right)$.

For water,

$$H_2(g) + \tfrac{1}{2}O_2(g) \rightarrow H_2O(l)$$

$\Delta_f G^\ominus(298\ \mathrm{K}) = -237.13\ \mathrm{kJ\ mol^{-1}}$ and $\Delta_r S^\ominus(298\ \mathrm{K}) = -163.34\ \mathrm{J\ K^{-1}\ mol^{-1}}$

$\Delta a = a(H_2O) - a(H_2) - \tfrac{1}{2}a(O_2) = (75.29 - 27.88 - 14.98)\ \mathrm{J\ K^{-1}\ mol^{-1}}$
$= +33.03\ \mathrm{J\ K^{-1}\ mol^{-1}}$

$\Delta b = [(0) - (3.26 \times 10^{-3}) - (2.09 \times 10^{-3})]\ \mathrm{J\ K^{-2}\ mol^{-1}} = -5.35 \times 10^{-3}\ \mathrm{J\ K^{-2}\ mol^{-1}}$

$\Delta c = [(0) - (0.50 \times 10^5) + (0.83 \times 10^5)]\ \mathrm{J\ K\ mol^{-1}} = +0.33 \times 10^5\ \mathrm{J\ K\ mol^{-1}}$

For $T = 298$ K and $T' = 372$ K,

$\alpha = -8.5\ \mathrm{K}$, $\beta = -2738\ \mathrm{K^2}$, $\gamma = -8.288 \times 10^{-5}\ \mathrm{K^{-1}}$

and so

$$\begin{aligned}\Delta_f G^\ominus(372\ \mathrm{K}) &= (-237.13\ \mathrm{kJ\ mol^{-1}}) + (-74\ \mathrm{K}) \times (-163.34\ \mathrm{J\ K^{-1}\ mol^{-1}}) \\ &\quad + (-8.5\ \mathrm{K}) \times (33.03 \times 10^{-3}\ \mathrm{kJ\ K^{-1}\ mol^{-1}}) \\ &\quad + (-2738\ \mathrm{K^2}) \times (-5.35 \times 10^{-6}\ \mathrm{kJ\ K^{-2}\ mol^{-1}}) \\ &\quad + (-8.288 \times 10^{-5}\ \mathrm{K^{-1}}) \times (0.33 \times 10^2\ \mathrm{kJ\ K\ mol^{-1}}) \\ &= [(-237.13) + (12.09) - (0.28) + (0.015) - (0.003)]\ \mathrm{kJ\ mol^{-1}} \\ &= \boxed{-225.31\ \mathrm{kJ\ mol^{-1}}}\end{aligned}$$

Note that the β and γ terms are not significant (for this reaction and temperature range).

Solutions to applications: biology, environmental science, and chemical engineering

P6.25 (a) ATP hydrolysis at physiological pH, $ATP(aq) + H_2O(l) \rightarrow ADP(aq) + P_i^-(aq) + H_3O^+(aq)$, converts two reactant moles into three product moles. The increased number of chemical species present in solution increases the disorder of the system by increasing the number of molecular rotational, vibrational, and translational degrees of freedom. This is an effective increase in the number of available molecular states and an increase in entropy.

(b) At physiological pH the oxygen atoms of ATP are deprotonated, negatively charged, and the molecule is best represented as ATP^{4-}. The electrostatic repulsions between the highly charged oxygen atoms of ATP^{4-} is expected to give it an exergonic hydrolysis free energy by making the hydrolysis enthalpy negative. Also, the deprotonated phosphate species, $P_i(aq)$, produced in the hydrolysis ATP has more resonance structures than ATP^{4-}. Resonance lowers the energy of the dissociated phosphate, making the hydrolysis enthalpy more negative and contributing to the exergonicity of the hydrolysis.

The electrostatic repulsion between the highly charged oxygen atoms of ATP^{4-} is a hypothesis that is consistent with the observation that protonated ATP, H_4ATP, has an exergonic hydrolysis free energy of smaller magnitude because the negative repulsions of oxygen atoms are not present. Likewise for $MgATP^{2-}$ because the Mg^{2+} ion lies between negatively charged oxygen atoms, thereby reducing repulsions and stabilizing the ATP molecule.

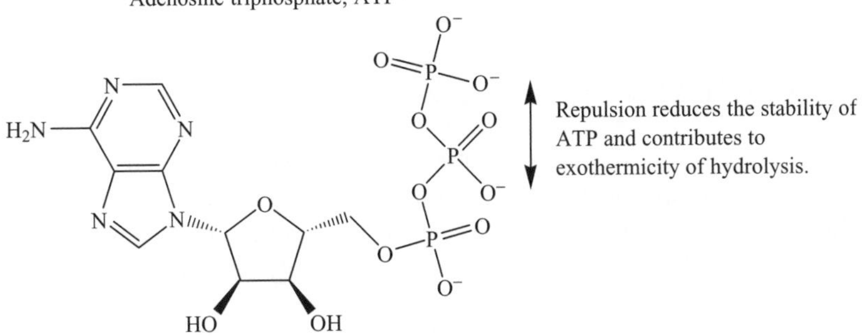
Adenosine triphosphate, ATP4−
Repulsion reduces the stability of ATP and contributes to exothermicity of hydrolysis.

P6.27 $C_6H_{12}O_6(aq) + 6\,O_2(g) \rightarrow 6\,CO_2(g) + 6\,H_2O(l)$ $\quad \Delta_r G^{\oplus} = \Delta_r G^{\ominus} = -2880\ \mathrm{kJ\ mol^{-1}}$ (i.e. pH independent)

$ATP(aq) + H_2O(l) \rightarrow ADP(aq) + P_i^-(aq) + H_3O^+(aq)$ $\quad \Delta_r G^{\oplus} = -31\ \mathrm{kJ\ mol^{-1}}$

Refer to *Impact On Biochemistry I6.1* for information necessary to the solution of this problem.

(a) If we assume that each mole of ATP formed during the aerobic breakdown of glucose produces about $-31\ \mathrm{kJ\ mol^{-1}}$ and 38 moles of ATP are produced per mole of glucose consumed, then

$$\text{percentage efficiency} = \frac{38 \times (-31\ \mathrm{kJ\ mol^{-1}})}{-2880\ \mathrm{kJ\ mol^{-1}}} \times 100\% = \boxed{41\%}.$$

(b) For the oxidation of glucose under the biological conditions of $T = 310\ \mathrm{K}$, $p_{CO_2} = 5.3 \times 10^{-2}$ atm, $p_{O_2} = 0.132$ atm, and $[\text{glucose}] = 5.6 \times 10^{-2}\ \mathrm{mol\ dm^{-3}}$ we have

$$\Delta_r G = \Delta_r G^{\ominus} + RT\ln Q = \Delta_r G^{\ominus} + RT\ln\left(\frac{(p_{CO_2}/p^{\ominus})^6 \times 1\,M}{[\text{glucose}]\times(p_{O_2}/p^{\ominus})^6}\right)$$

$$= -2880\ \text{kJ mol}^{-1} + (8.3145\ \text{J K}^{-1}\ \text{mol}^{-1})\times(310\ \text{K})\ln\left(\frac{0.053^6}{0.056\times 0.132^6}\right)$$

$$= -2887\ \text{kJ mol}^{-1}$$

This is not much different from the standard value of -2880 kJ mol^{-1}.

For the ATP hydrolysis under the given conditions of $T = 310$ K, $p_{CO_2} = 5.3\times10^{-2}$ atm, $p_{O_2} = 0.132$ atm, [glucose] $= 5.6\times10^{-2}$ mol dm^{-3}, [ATP] = [ADP] = [P$_i$] $= 1.0\times10^{-4}$ mol dm^{-3}, and pH = 7.4 we have

$$\Delta_r G = \Delta_r G^{\oplus} + RT\ln Q^{\oplus} = \Delta_r G^{\oplus} + RT\ln\left(\frac{[\text{ADP}]\times[\text{P}_i^-]}{[\text{ATP}]\times(1\,\text{M})}\times\frac{[\text{H}^+]}{10^{-7}\,\text{M}}\right)$$

[i.e. $[\text{H}^+]^{\oplus} = 10^{-7}$ M, $[\text{J}]^{\oplus} = 1$ M]

$$= -31\ \text{kJ mol}^{-1} + (8.3145\ \text{J K}^{-1}\ \text{mol}^{-1})\times(310\ \text{K})\ln(1.0\times10^{-4}\times10^{-7.4}/10^{-7})$$

$$= -57\ \text{kJ mol}^{-1}$$

With this value for $\Delta_r G$ the efficiency becomes

$$\text{efficiency} = \frac{38\times(-57\ \text{kJ mol}^{-1})}{-2887\ \text{kJ mol}^{-1}} = \boxed{75\%}$$

(c) The theoretical limit of the diesel engine is

$$\varepsilon = 1 - \frac{T_c}{T_h} = 1 - \frac{873\ \text{K}}{1923\ \text{K}} = \boxed{55\%}$$

With an efficiency of 75% of the theoretical limit the energy conversion is reduced to 41% in a heat engine. We see that the biological efficiency (part (b), 75%) under the conditions given is greater than that of the diesel engine. What limits the efficiency of the diesel engine, or any heat engine, is that heat engines must convert heat ($q \approx \Delta_c H$) into useful work ($w_{\text{add,max}} = \Delta_r G$). Because of the second law, a substantial fraction of that heat is wasted. The biological process involves $\Delta_r G$ directly and does not go through a heat step.

P6.29 The half-reactions involved are:

$$\text{R: cyt}_{ox} + e^- \rightarrow \text{cyt}_{red} \qquad E^{\ominus}_{cyt}$$
$$\text{L: D}_{ox} + e^- \rightarrow \text{D}_{red} \qquad E^{\ominus}_{D}$$

The overall cell reaction is:

$$\text{R} - \text{L: cyt}_{ox} + \text{D}_{red} \rightleftharpoons \text{cyt}_{red} + \text{D}_{ox} \qquad E^{\ominus}_{cell} = E^{\ominus}_{cyt} - E^{\ominus}_{D}$$

(a) The Nernst equation for the cell reaction is

$$E_{cell} = E^{\ominus}_{cell} - \frac{RT}{F}\ln\frac{[\text{cyt}_{red}][\text{D}_{ox}]}{[\text{cyt}_{ox}][\text{D}_{red}]}.$$

At equilibrium, $E_{\text{cell}} = 0$; therefore

$$\ln\left(\frac{[\text{cyt}_{\text{red}}]_{\text{eq}}[\text{D}_{\text{ox}}]_{\text{eq}}}{[\text{cyt}_{\text{ox}}]_{\text{eq}}[\text{D}_{\text{red}}]_{\text{eq}}}\right) = \frac{F}{RT}(E^{\ominus}_{\text{cyt}} - E^{\ominus}_{\text{D}})$$

$$\ln\left(\frac{[\text{D}_{\text{ox}}]_{\text{eq}}}{[\text{D}_{\text{red}}]_{\text{eq}}}\right) = \ln\left(\frac{[\text{cyt}_{\text{ox}}]_{\text{eq}}}{[\text{cyt}_{\text{red}}]_{\text{eq}}}\right) + \frac{F}{RT}(E^{\ominus}_{\text{cyt}} - E^{\ominus}_{\text{D}})$$

Therefore a plot of $\ln\left(\frac{[\text{D}_{\text{ox}}]_{\text{eq}}}{[\text{D}_{\text{red}}]_{\text{eq}}}\right)$ against $\ln\left(\frac{[\text{cyt}_{\text{ox}}]_{\text{eq}}}{[\text{cyt}_{\text{red}}]_{\text{eq}}}\right)$ is linear with a slope of one and an intercept of $\frac{F}{RT}(E^{\ominus}_{\text{cyt}} - E^{\ominus}_{\text{D}})$.

(b) Draw up the following table:

$\ln\left(\frac{[\text{D}_{\text{ox}}]_{\text{eq}}}{[\text{D}_{\text{red}}]_{\text{eq}}}\right)$	−5.882	−4.776	−3.661	−3.002	−2.593	−1.436	−0.6274
$\ln\left(\frac{[\text{cyt}_{\text{ox}}]_{\text{eq}}}{[\text{cyt}_{\text{red}}]_{\text{eq}}}\right)$	−4.547	−3.772	−2.415	−1.625	−1.094	−0.2120	−0.3293

The plot of $\ln\left(\frac{[\text{D}_{\text{ox}}]_{\text{eq}}}{[\text{D}_{\text{red}}]_{\text{eq}}}\right)$ against $\ln\left(\frac{[\text{cyt}_{\text{ox}}]_{\text{eq}}}{[\text{cyt}_{\text{red}}]_{\text{eq}}}\right)$ is shown in Figure 6.4. The intercept is −1.2124.

Hence

$$\begin{aligned} E^{\ominus}_{\text{cyt}} &= \frac{RT}{F} \times (-1.2124) + 0.237\ \text{V} \\ &= 0.0257\ \text{V} \times (-1.2124) + 0.237\ \text{V} \\ &= \boxed{+0.206\ \text{V}}. \end{aligned}$$

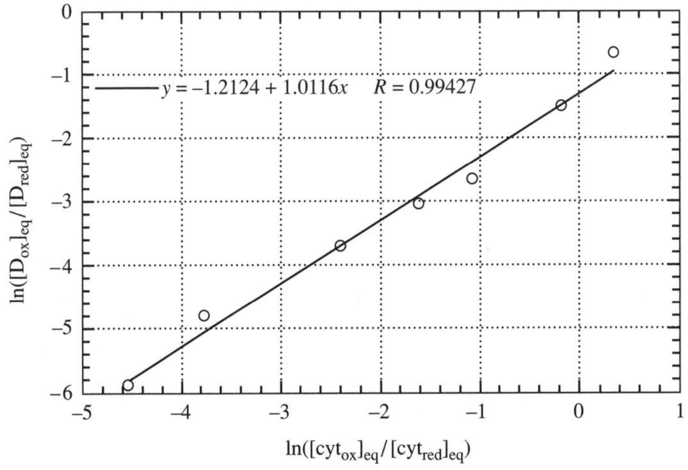

Figure 6.4

P6.31 A reaction proceeds spontaneously if its reaction Gibbs function is negative.

$$\Delta_r G = \Delta_r G^\ominus + RT \ln Q \,[6.10]$$

Note that under the given conditions, $RT = 1.58\ \text{kJ mol}^{-1}$

(i) $\Delta_r G/(\text{kJ mol}^{-1}) = \{\Delta_r G^\ominus(\text{i}) - RT \ln p_{H_2O}\}/\text{kJ mol}^{-1} = -23.6 - 1.58 \ln (1.3 \times 10^{-7})$
$= +1.5$

(ii) $\Delta_r G/(\text{kJ mol}^{-1}) = \{\Delta_r G^\ominus(\text{ii}) - RT \ln p_{H_2O} p_{HNO_3}\}/(\text{kJ mol}^{-1})$
$= -57.2 - 1.58 \ln [(1.3 \times 10^{-7}) \times (4.1 \times 10^{-10})]$
$= +2.0$

(iii) $\Delta_r G/(\text{kJ mol}^{-1}) = \{\Delta_r G^\ominus(\text{iii}) - RT \ln p_{H_2O}^2 p_{HNO_3}\}/(\text{kJ mol}^{-1})$
$= -85.6 - 1.58 \ln [(1.3 \times 10^{-7})^2 \times (4.1 \times 10^{-10})]$
$= -1.3$

(iv) $\Delta_r G/(\text{kJ mol}^{-1}) = \{\Delta_r G^\ominus(\text{iv}) - RT \ln p_{H_2O}^2 p_{HNO_3}\}/(\text{kJ mol}^{-1})$
$= -112.8 - 1.58 \ln [(1.3 \times 10^{-7})^3 \times (4.1 \times 10^{-10})]$
$= -3.5$

So both the dihydrate and trihydrate form spontaneously from the vapour. Does one convert spontaneously into the other? Consider the reaction

$$HNO_3 \cdot 2\,H_2O(s) + H_2O(g) \rightleftharpoons HNO_3 \cdot 3\,H_2O(s)$$

which may be considered as reaction(iv) – reaction(iii). $\Delta_r G$ for this reaction is

$$\Delta_r G = \Delta_r G(\text{iv}) - \Delta_r G(\text{iii}) = -2.2\ \text{kJ mol}^{-1}.$$

We conclude that the dihydrate converts spontaneously to the $\boxed{\text{trihydrate}}$, the most stable solid (at least of the four we considered).

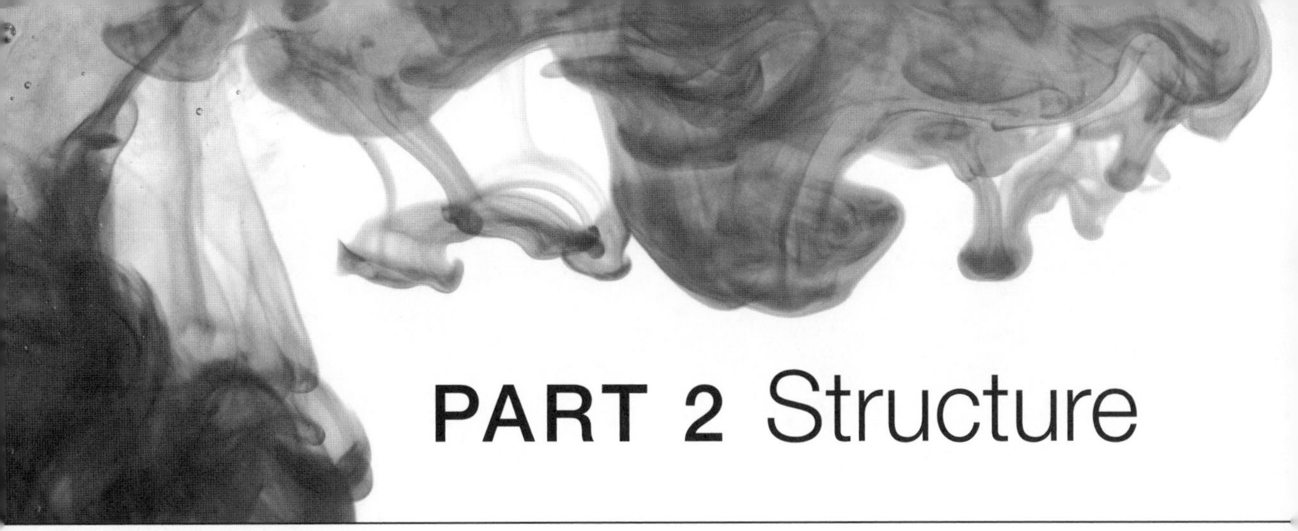

PART 2 Structure

7 Quantum theory: introduction and principles

Answers to discussion questions

D7.1 At the end of the nineteenth century and the beginning of the twentieth, there were many experimental results on the properties of matter and radiation that could not be explained on the basis of established physical principles and theories. Here we list only some of the most significant.

(1) The photoelectric effect revealed that electromagnetic radiation, classically considered to be a wave, also exhibits the particle-like behaviour of photons. Each photon is a discrete unit, or quantum, of energy that is absorbed during collisions with electrons. Photons are never partially absorbed. They either completely give up their energy or they are not absorbed. The energy of a photon can be calculated if either the radiation frequency or wavelength is known: $\boldsymbol{E_{\text{photon}} = h\nu = hc/\lambda}$.

(2) Absorption and emission spectra indicated that atoms and molecules can only absorb or emit discrete packets of energy (i.e. photons). This means that an atom or molecule has specific allowed energy levels and we say that their energies are quantized. During a spectroscopic transition the atom or molecule gains or loses the energy ΔE by either absorption of a photon or emission of a photon, respectively. Thus, spectral lines must satisfy the **Bohr frequency condition:** $\boldsymbol{\Delta E = h\nu}$.

(3) Neutron and electron diffraction studies indicated that these particles also possess wave-like properties of constructive and destructive interference. The joint particle and wave character of matter and radiation is called **wave-particle duality**. The **de Broglie relationship**, $\lambda_{\text{de Broglie}} = h/p$, connects the wave character of a particle ($\lambda_{\text{de Broglie}}$) with its particulate momentum (p).

(4) The concept of classical physics that all energies are permitted incorrectly predicts that the ideal emitter of electromagnetic radiation, an object called a **black body**, which emits and absorbs all wavelengths of radiation uniformly, should emit an infinite amount of high-frequency radiation at all temperatures. This **ultraviolet catastrophe** sharply contrasts with the experimental evidence that at a given temperature the emissions from a pinhole in an empty container have a maximum intensity at intermediate frequencies and little-to-no intensity at both low and high frequencies. Max Planck (in 1900) correctly accounted for the experimental observations by postulating that the permitted energies of an electromagnetic oscillator of frequency ν are **quantized** as integer multiples of $h\nu$.

(5) Classical physics also fails to explain why the heat capacities of monatomic solids such as copper metal vary from zero in the limit of the absolute zero of temperature to a value of about $3R$ at high temperature. In order to explain these observations Einstein (in 1905) found it necessary to postulate that each atom of the solid oscillates about its equilibrium position

with a frequency v and that the oscillation energy is restricted to quantized values that must equal an integer multiple of hv.

D7.3 The heat capacities of monatomic solids are primarily a result of the energy acquired by vibrations of the atoms about their equilibrium positions. If this energy can be acquired continuously, we expect that the **equipartition of energy principle** should apply. This principle states that for each direction of motion and for each kind of energy (potential and kinetic) the associated energy should be $\frac{1}{2}kT$. Hence, for three directions and both kinds of motion, a total of $3kT$, which gives a heat capacity of $3k$ per atom, or $3R$ per mole, independent of temperature. But the experiments show a temperature dependence. The heat capacity falls steeply below $3R$ at low temperatures. Einstein showed that by allowing the energy of the atomic oscillators to be quantized according to Planck's formula $E = nhv = nhc/\lambda$, where the **quantum number *n*** can equal zero or any positive integer, rather than the continuous variability of energy prescribed by classical physics, this temperature dependence could be explained. The physical reason is that at low temperatures only a few atomic oscillators have enough energy to populate the higher quantized levels; at higher temperatures more of them can acquire the energy to become active.

D7.5 In quantum mechanics all dynamical properties of a physical system have associated with them a corresponding operator. The system itself is described by a wavefunction. The observable properties of the system can be obtained in one of two ways from the wavefunction, depending on whether or not the wavefunction is an eigenfunction of the operator for the observable.

When the function representing the state of the system is an eigenfunction of the operator $\hat{\Omega}$, we solve the eigenvalue equation [eqn 7.28b] $\hat{\Omega}\psi = \omega\psi$ in order to obtain the observable values, ω, of the dynamical property. The time-independent Schrödinger equation, $\hat{H}\psi = E\psi$ [7.27a], is an especially important example: the wavefunction ψ is an eigenfunction of the system hamiltonian operator, while the system energy E is an eigenvalue. When ψ is not an eigenfunction of $\hat{\Omega}$, we can only find the average or expectation value of the dynamical property by performing the integration shown in eqn 7.37: $\langle\Omega\rangle = \int \psi^*\hat{\Omega}\psi\,\mathrm{d}\tau$.

D7.7 According to eqns 7.31 and 7.37 the expectation value for the kinetic energy of a particle equals the expectation value of the 'curvature' operator, $\mathrm{d}^2/\mathrm{d}x^2$:

$$\langle E_\mathrm{k}\rangle = -\frac{\hbar^2}{2m}\left\langle \frac{\mathrm{d}^2}{\mathrm{d}x^2}\right\rangle$$

Thus, a wavefunction that is sharply curved will have a higher expectation value for kinetic energy than a wavefunction of low curvature. See the text discussion about eqn 7.31 and examine text Figures 7.16 and 7.25–7.27.

Solutions to exercises

E7.1(a) The de Broglie relationship is $\lambda = \frac{h}{p}[7.16] = \frac{h}{mv}$.

Hence, $v = \frac{h}{m_\mathrm{e}\lambda} = \frac{6.626\times10^{-34}\,\mathrm{J\,s}}{(9.109\times10^{-31}\,\mathrm{kg})\times(0.030\,\mathrm{m})} = \boxed{0.024\,\mathrm{m\,s^{-1}}}$, which is very slow!

E7.2(a) $$\lambda = \frac{h}{p}[7.16] = \frac{h}{mv} = \frac{h}{m_e\alpha c} = \frac{\alpha^{-1}h}{m_e c} = \frac{137 \times (6.626 \times 10^{-34}\,\text{J s})}{(9.109 \times 10^{-31}\,\text{kg}) \times (2.998 \times 10^{8}\,\text{m s}^{-1})} = \boxed{332\,\text{pm}}$$

COMMENT. One wavelength of the matter wave of an electron with this velocity just fits in the first Bohr orbit. The velocity of the electron in the first Bohr classical orbit is thus $\frac{1}{137}c$.

Question. What is the wavelength of an electron with velocity approaching the speed of light? Such velocities can be achieved with particle accelerators.

E7.3(a) The desired uncertainty in the proton momentum is

$$\Delta p = 1.00 \times 10^{-4}\, p = 1.00 \times 10^{-4}\, m_p v$$
$$= (1.00 \times 10^{-4}) \times (1.673 \times 10^{-27}\,\text{kg}) \times (0.45 \times 10^{6}\,\text{m s}^{-1}) = 7.5\overline{3} \times 10^{-26}\,\text{kg m s}^{-1}$$

Thus, the minimum uncertainty in position must be

$$\Delta x = \frac{\hbar}{2\Delta p}[7.39a] = \frac{1.055 \times 10^{-34}\,\text{J s}}{2 \times (7.5\overline{3} \times 10^{-26}\,\text{kg m s}^{-1})} = \boxed{700\,\text{pm}}$$

E7.4(a) $$E = h\nu = \frac{hc}{\lambda}[7.14 \text{ and } 7.1] = \frac{(6.626 \times 10^{-34}\,\text{J s}) \times (2.998 \times 10^{8}\,\text{m s}^{-1})}{\lambda} = \frac{1.986 \times 10^{-25}\,\text{J m}}{\lambda}$$

$$E_m = N_A E = \frac{N_A hc}{\lambda} = \frac{0.1196\,\text{J m}}{\lambda}$$

We can therefore draw up the following table:

λ/nm	E/J	E_m/(kJ mol^{-1})
(a) 600	3.31×10^{-19}	199
(b) 550	3.61×10^{-19}	218
(c) 400	4.97×10^{-19}	299

E7.5(a) On absorption of a photon by a free hydrogen atom, the law of conservation of energy requires that the acquired kinetic energy, E_k, of the atom equal the energy of the absorbed photon: $E_k = E_{\text{photon}} = \frac{1}{2}m_H v^2$. The values of E_{photon} are calculated in E7.4a so the atom is accelerated to the speed

$$v = \sqrt{\frac{2E_{\text{photon}}}{m_H}} = \sqrt{\frac{2N_A E_{\text{photon}}}{M_H}}$$
$$= \sqrt{\frac{2 \times (6.022 \times 10^{23}\,\text{mol}^{-1}) \times E_{\text{photon}}}{1.0079 \times 10^{-3}\,\text{kg mol}^{-1}}} = (3.457 \times 10^{13}\,\text{m s}^{-1}) \times \sqrt{E_{\text{photon}}/\text{J}}$$

We can therefore draw up the following table:

λ/nm	E_{photon}/J	v/(km s^{-1})
(a) 600	3.31×10^{-19}	19.9
(b) 550	3.61×10^{-19}	20.8
(c) 400	4.97×10^{-19}	24.4

E7.6(a) The total energy emitted in time Δt is $P\Delta t$ where P is the power of the emission. The energy of each emitted photon $E = hc/\lambda$ [7.1]. The total number of photons emitted in an interval Δt is then the total energy divided by the energy per photon.

$$N = \frac{P\Delta t}{E} = \frac{P\Delta t\lambda}{hc}$$

Assuming that de Broglie's relationship applies to each photon, and recognizing that the law of conservation of linear momentum specifies that the loss of a photon imparts an equivalent momentum (in the opposite direction) to the glow-worm, the total momentum imparted to the glow-worm in time Δt is

$$p = Np_{\text{photon}} = \frac{Nh}{\lambda}\,[7.16] = \frac{P\Delta t\lambda}{hc} \times \frac{h}{\lambda} = \frac{P\Delta t}{c}$$

Since $p = (mv)_{\text{glow-worm}}$, the final speed of the glow-worm is

$$v = \frac{P\Delta t}{cm_{\text{glow-worm}}} = \frac{(0.10\ \text{W}) \times (10\ \text{y}) \times (3.1557 \times 10^{7}\ \text{s y}^{-1})}{(2.9979 \times 10^{8}\ \text{m s}^{-1}) \times (0.0050\ \text{kg})} = \boxed{21\ \text{m s}^{-1}}$$

COMMENT. Note that the answer is independent of the wavelength of the radiation emitted: the greater the wavelength the smaller the photon momentum, but the greater the number of photons emitted.

Question. If this glow-worm eventually turns into a firefly that glows for 1 s intervals while flying with a speed of 0.1 m s^{-1} what additional speed does the 1 s glowing impart to the firefly? Ignore any frictional effects of air.

E7.7(a) The total energy emitted in time Δt is $P\Delta t$ where P is the power of the emission. The energy of each emitted photon is given by $E = hc/\lambda$ [7.1]. The total number of photons emitted in an interval Δt is then the total energy divided by the energy per photon.

$$N = \frac{P\Delta t}{E} = \frac{P\Delta t\lambda}{hc}$$

$$= \frac{(1.00\ \text{s}) \times (550 \times 10^{-9}\ \text{m})}{(6.626 \times 10^{-34}\ \text{J s}) \times (2.998 \times 10^{8}\ \text{m s}^{-1})} P = (2.77 \times 10^{18}) \times P/\text{W}$$

(a) When $P = 1.0$ W, $N = \boxed{2.77 \times 10^{18}}$.

(b) When $P = 100$ W, $N = \boxed{2.77 \times 10^{20}}$.

E7.8(a) $E_{\text{k}} = h\nu - \Phi = \dfrac{hc}{\lambda} - \Phi\,[7.15]$ and, since $E_{\text{k}} = \frac{1}{2}m_{\text{e}}v^2$, $v = \sqrt{2E_{\text{k}}/m_{\text{e}}}$

$$\Phi = 2.14\ \text{eV} = (2.14) \times (1.602 \times 10^{-19}\ \text{J}) = 3.43 \times 10^{-19}\ \text{J}$$

(a) For $\lambda = 700$ nm

$$E_{\text{k}} = \frac{(6.626 \times 10^{-34}\ \text{J s}) \times (2.998 \times 10^{8}\ \text{m s}^{-1})}{700 \times 10^{-9}\ \text{m}} - 3.43 \times 10^{-19}\ \text{J} = 2.84 \times 10^{-19}\ \text{J} - 3.43 \times 10^{-19}\ \text{J}$$

Inspection of the above equation reveals that the photon energy of the radiation is less than the work function and we conclude that $\boxed{\text{no electron ejection}}$ occurs.

(b) For $\lambda = 300$ nm

$$E_k = \frac{(6.626\times10^{-34}\,\text{J s})\times(2.998\times10^{8}\,\text{m s}^{-1})}{300\times10^{-9}\,\text{m}} - 3.43\times10^{-19}\,\text{J} = 6.62\times10^{-19}\,\text{J} - 3.43\times10^{-19}\,\text{J}$$

$$= \boxed{3.19\times10^{-19}\,\text{J}}$$

$$v = \sqrt{2\times(3.19\times10^{-19}\,\text{J})/(9.109\times10^{-31}\,\text{kg})} = \boxed{837\,\text{km s}^{-1}}$$

E7.9(a) $\Delta E = h\nu = h/T\ [T = \text{period} = 1/\nu]$ and $\Delta E_m = N_A\Delta E$

(a) For $T = 1.0$ fs

$$\Delta E = (6.626\times10^{-34}\,\text{J s})/(1.0\times10^{-15}\,\text{s}) = \boxed{6.6\times10^{-19}\,\text{J}}$$

$$\Delta E_m = (6.022\times10^{23}\,\text{mol}^{-1})\times(6.6\times10^{-19}\,\text{J}) = \boxed{4.0\times10^{2}\,\text{kJ mol}^{-1}}$$

(b) For $T = 10$ fs

$$\Delta E = (6.626\times10^{-34}\,\text{J s})/(10\times10^{-15}\,\text{s}) = \boxed{6.6\times10^{-20}\,\text{J}}$$

$$\Delta E_m = (6.022\times10^{23}\,\text{mol}^{-1})\times(6.6\times10^{-20}\,\text{J}) = \boxed{40\,\text{kJ mol}^{-1}}$$

(c) For $T = 1.0$ s

$$\Delta E = (6.626\times10^{-34}\,\text{J s})/(1.0\,\text{s}) = \boxed{6.6\times10^{-34}\,\text{J}}$$

$$\Delta E_m = (6.022\times10^{23}\,\text{mol}^{-1})\times(6.6\times10^{-34}\,\text{J}) = \boxed{4.0\times10^{-13}\,\text{kJ mol}^{-1}}$$

E7.10(a) The de Broglie wavelength is $\lambda = \frac{h}{p} = \frac{h}{mv}$ [7.16]

(a) $$\lambda = \frac{6.626\times10^{-34}\,\text{J s}}{(1.0\times10^{-3}\,\text{kg})\times(1.0\times10^{-2}\,\text{m s}^{-1})} = \boxed{6.6\times10^{-29}\,\text{m}}$$

(b) $$\lambda = \frac{6.626\times10^{-34}\,\text{J s}}{(1.0\times10^{-3}\,\text{kg})\times(1.00\times10^{5}\,\text{m s}^{-1})} = \boxed{6.6\times10^{-36}\,\text{m}}$$

(c) $$\lambda = \frac{6.626\times10^{-34}\,\text{J s}}{(4.003)\times(1.6605\times10^{-27}\,\text{kg})\times(1000\,\text{m s}^{-1})} = \boxed{99.7\,\text{pm}}$$

COMMENT. The wavelengths in (a) and (b) are smaller than the dimensions of any known particle, whereas that in (c) is comparable to atomic dimensions.

Question. For stationary particles, $v = 0$, corresponding to an infinite de Broglie wavelength. What meaning can be ascribed to this result?

E7.11(a) The normalized wavefunction has the form $\psi(\phi) = Ne^{i\phi}$ where N is the normalization constant.

$$\int_0^{2\pi} \psi^*\psi\,d\phi = 1\ [7.20a]$$

$$N^2\int_0^{2\pi} e^{-i\phi}e^{i\phi}\,d\phi = N^2\int_0^{2\pi} d\phi = 2\pi N^2 = 1 \quad \boxed{N = \left(\frac{1}{2\pi}\right)^{1/2}}$$

E7.12(a) $\psi(\phi) = (1/2\pi)^{1/2}\,e^{i\phi}$ so $|\psi(\phi)|^2 = (1/2\pi)e^{i\phi}e^{-i\phi} = 1/2\pi$. Thus, the probability of finding the atom in an infinitesimal volume element at any angle is $\boxed{(1/2\pi)\,d\phi}$.

E7.13(a) The normalized wavefunction is $\psi = \left(\dfrac{1}{2\pi}\right)^{1/2} e^{i\phi}$.

$$\text{Probability that } \pi/2 \le \phi \le 3\pi/2 = \int_{\pi/2}^{3\pi/2} \psi^*\psi\,d\phi$$

$$= \left(\frac{1}{2\pi}\right)\int_{\pi/2}^{3\pi/2} e^{-i\phi}e^{i\phi}\,d\phi = \left(\frac{1}{2\pi}\right)\int_{\pi/2}^{3\pi/2} d\phi = \left(\frac{1}{2\pi}\right)\phi\Big|_{\phi=\pi/2}^{\phi=3\pi/2} = \left(\frac{1}{2\pi}\right)\times\left(\frac{3\pi}{2}-\frac{\pi}{2}\right)$$

$$= \boxed{\frac{1}{2}}$$

E7.14(a) $$\int_0^{2\pi} \psi_i^*\hat{l}_z\psi_j\,d\phi = \frac{\hbar}{i}\int_0^{2\pi}\psi_i^*\frac{d\psi_j}{d\phi}\,d\phi$$

Application of 'integration by parts' yields

$$\int_0^{2\pi} \psi_i^*\hat{l}_z\psi_j\,d\phi = \frac{\hbar}{i}\psi_i^*\psi_j\Big|_0^{2\pi} - \frac{\hbar}{i}\int_0^{2\pi}\psi_j\frac{d\psi_i^*}{d\phi}\,d\phi$$

The first term to the right vanishes because the wavefunction must repeat itself every 2π radian (i.e. $\psi_i(0) = \psi_i(2\pi)$)

$$\int_0^{2\pi} \psi_i^*\hat{l}_z\psi_j\,d\phi = -\frac{\hbar}{i}\int_0^{2\pi}\psi_j\frac{d\psi_i^*}{d\phi}\,d\phi = \left\{\frac{\hbar}{i}\int_0^{2\pi}\psi_j^*\frac{d\psi_i}{d\phi}\,d\phi\right\}^* = \left\{\int_0^{2\pi}\psi_j^*\hat{l}_z\psi_i\,d\phi\right\}^*$$

This is the condition of eqn 7.33 so $\hat{l}_z$ is an hermitian operator.

E7.15(a) The minimum uncertainty in position and momentum is given by the uncertainty principle in the form $\Delta p\Delta q \ge \tfrac{1}{2}\hbar$ [7.39a], with the choice of the equality. The uncertainty in momentum is $\Delta p = m\Delta v$ so:

$$\Delta v_{\min} = \frac{\hbar}{2m\Delta q} = \frac{1.055\times10^{-34}\text{ J s}}{(2)\times(0.500\text{ kg})\times(1.0\times10^{-6}\text{ m})} = \boxed{1.1\times10^{-28}\text{ m s}^{-1}}$$

The minimum uncertainty in position of a 5.0 g mass is

$$\Delta q_{\min} = \frac{\hbar}{2m\Delta v} = \frac{1.055\times10^{-34}\text{ J s}}{(2)\times(5.0\times10^{-3}\text{ kg})\times(1\times10^{-5}\text{ m s}^{-1})} = \boxed{1\times10^{-27}\text{ m}}$$

COMMENT. These uncertainties are extremely small; thus, the ball and bullet are effectively classical particles.

Question. If the ball were stationary (no uncertainty in position) the uncertainty in speed would be infinite. Thus, the ball could have a very high speed, contradicting the fact that it is stationary. What is the resolution of this apparent paradox?

E7.16(a) $$E_{\text{binding}} = E_{\text{photon}} - E_{\text{k}} = h\nu - \tfrac{1}{2}m_{\text{e}}v^2 = \frac{hc}{\lambda} - \tfrac{1}{2}m_{\text{e}}v^2$$

$$E_{\text{binding}} = \frac{hc}{\lambda} - \tfrac{1}{2}m_{\text{e}}v^2$$

$$= \frac{(6.626\times10^{-34}\,\text{J s})\times(2.998\times10^{8}\,\text{m s}^{-1})}{150\times10^{-12}\,\text{m}} - \tfrac{1}{2}(9.109\times10^{-31}\,\text{kg})\times(2.14\times10^{7}\,\text{m s}^{-1})^2$$

$$= (1.12\times10^{-15}\,\text{J})\times\left(\frac{1\,\text{eV}}{1.602\times10^{-19}\,\text{J}}\right) = \boxed{6.96\,\text{keV}}$$ without a relativist mass correction

Note: The photoelectron is moving at 7.1% of the speed of light. So, in order to calculate a more accurate value of the binding energy, it would be necessary to use the relativistic mass in place of the rest mass.

$$m = \frac{m_{\text{e}}}{(1-(v/c)^2)^{1/2}} = \frac{9.109\times10^{-31}\,\text{kg}}{(1-(2.14\times10^{7}\,\text{m s}^{-1}/2.998\times10^{8}\,\text{m s}^{-1})^2)^{1/2}} = 9.13\times10^{-31}\,\text{kg}$$

$$E_{\text{binding}} = \frac{hc}{\lambda} - \tfrac{1}{2}mv^2$$

$$= \frac{(6.626\times10^{-34}\,\text{J s})\times(2.998\times10^{8}\,\text{m s}^{-1})}{150\times10^{-12}\,\text{m}} - \tfrac{1}{2}(9.13\times10^{-31}\,\text{kg})\times(2.14\times10^{7}\,\text{m s}^{-1})^2$$

$$= (1.12\times10^{-15}\,\text{J})\times\left(\frac{1\,\text{eV}}{1.602\times10^{-19}\,\text{J}}\right) = \boxed{6.96\,\text{keV}}$$ with the relativistic mass correction.

The relativistic mass correction did not make a difference in this exercise.

E7.17(a) The quantity $[\hat{\Omega}_1,\hat{\Omega}_2] = \hat{\Omega}_1\hat{\Omega}_2 - \hat{\Omega}_2\hat{\Omega}_1$ [7.41] is referred to as the commutator of the operators $\hat{\Omega}_1$ and $\hat{\Omega}_2$. In obtaining the commutator it is necessary to realize that the operators operate on functions; thus we find expressions for $[\hat{\Omega}_1,\hat{\Omega}_2]\psi(x) = \hat{\Omega}_1\hat{\Omega}_2\psi(x) - \hat{\Omega}_2\hat{\Omega}_1\psi(x)$.

(a) $$\left[\frac{\text{d}}{\text{d}x},\frac{1}{x}\right]\psi = \frac{\text{d}}{\text{d}x}\times\left(\frac{1}{x}\psi\right) - \frac{1}{x}\frac{\text{d}}{\text{d}x}\psi = \left(-\frac{1}{x^2}\right)\psi + \frac{1}{x}\frac{\text{d}}{\text{d}x}\psi - \frac{1}{x}\frac{\text{d}}{\text{d}x}\psi = \left(-\frac{1}{x^2}\right)\psi$$

Thus, $$\left[\frac{\text{d}}{\text{d}x},\frac{1}{x}\right] = \boxed{-\frac{1}{x^2}}$$

(b) $$\left[\frac{\text{d}}{\text{d}x},x^2\right]\psi = \frac{\text{d}}{\text{d}x}\times(x^2\psi) - x^2\frac{\text{d}}{\text{d}x}\psi = (2x)\psi + x^2\frac{\text{d}}{\text{d}x}\psi - x^2\frac{\text{d}}{\text{d}x}\psi = (2x)\psi$$

Thus, $$\left[\frac{\text{d}}{\text{d}x},x^2\right] = \boxed{2x}$$

Solutions to problems

Solutions to numerical problems

P7.1 A cavity approximates an ideal black body; hence the Planck distribution applies

$$\rho = \frac{8\pi hc}{\lambda^5}\left(\frac{1}{e^{hc/\lambda kT}-1}\right) \; [7.8]$$

Since the wavelength range is small (5 nm) we may write as a good approximation

$$\Delta E = \rho\Delta\lambda, \quad \lambda \approx 652.5\ \text{nm}$$

$$\frac{hc}{\lambda k} = \frac{(6.626\times10^{-34}\ \text{J s})\times(2.998\times10^{8}\ \text{m s}^{-1})}{(652.5\times10^{-9}\ \text{m})\times(1.381\times10^{-23}\ \text{J K}^{-1})} = 2.205\times10^{4}\ \text{K}$$

$$\frac{8\pi hc}{\lambda^5} = \frac{8\pi\times(6.626\times10^{-34}\ \text{J s})\times(2.998\times10^{8}\ \text{m s}^{-1})}{(652.5\times10^{-9}\ \text{m})^5} = 4.221\times10^{7}\ \text{J m}^{-4}$$

$$\Delta E = (4.221\times10^{7}\ \text{J m}^{-4})\times\left(\frac{1}{e^{(2.205\times10^4\ \text{K})/T}-1}\right)\times(5\times10^{-9}\ \text{m}) = \frac{0.21\bar{1}\ \text{J m}^{-3}}{e^{(2.205\times10^4\ \text{K})/T}-1}$$

(a) $T = 298\ \text{K},\ \Delta E = \dfrac{0.21\bar{1}\ \text{J m}^{-3}}{e^{(2.205\times10^4\ \text{K})/(298\ \text{K})}-1} = \boxed{1.6\times10^{-33}\ \text{J m}^{-3}}$

(b) $T = 3273\ \text{K},\ \Delta E = \dfrac{0.21\bar{1}\ \text{J m}^{-3}}{e^{(2.205\times10^4\ \text{K})/(3273\ \text{K})}-1} = \boxed{2.5\times10^{-4}\ \text{J m}^{-3}}$

COMMENT. The energy density in the cavity does not depend on the volume of the cavity, but the total energy in any given wavelength range does, as well as the total energy over all wavelength ranges.

Question. What is the total energy in this cavity within the range 650–655 nm at the stated temperatures?

P7.3 $$\theta_E = \frac{h\nu}{k}, \quad [\theta_E] = \frac{(\text{J s})\times(\text{s}^{-1})}{\text{J K}^{-1}} = \text{K}$$

In terms of θ_E the Einstein equation [7.11] for the heat capacity of solids is

$$C_{V,m} = 3R\left(\frac{\theta_E}{T}\right)^2\times\left(\frac{e^{\theta_E/2T}}{e^{\theta_E/T}-1}\right)^2 \quad [7.11,\ \text{classical value} = \lim_{T\to\infty} C_{V,m} = 3R]$$

The Einstein expression for $C_{V,m}$ approaches the classical value when $T \gg \theta_E$ or when $\dfrac{h\nu}{kT} \ll 1$ as demonstrated in the text (Section 7.1(b)). The criterion for classical behaviours is therefore that $\boxed{T \gg \theta_E}$.

$$\theta_E = \frac{h\nu}{k} = \frac{(6.626\times10^{-34}\ \text{J Hz}^{-1})\times\nu}{1.381\times10^{-23}\ \text{J K}^{-1}} = (4.798\times10^{-11}\ \text{K})\times(\nu/\text{Hz})$$

(a) For diamond $\nu = 46.5 \times 10^{12}$ Hz, $\theta_E = (4.798 \times 10^{-11}\ \text{K}) \times (46.5 \times 10^{12}) = \boxed{223\bar{1}\ \text{K}}$

$$\text{At 298 K,}\ \frac{C_{V,\text{m}}}{3R} = \left(\frac{223\bar{1}\ \text{K}}{298\ \text{K}}\right)^2 \times \left(\frac{e^{223\bar{1}/(2\times298)}}{e^{223\bar{1}/298}-1}\right)^2 = \boxed{0.0315}$$

(b) For copper $\nu = 7.15 \times 10^{12}$ Hz, $\theta_E = (4.798 \times 10^{-11}\ \text{K}) \times (7.15 \times 10^{12}) = \boxed{343\ \text{K}}$

$$\text{At 298 K,}\ \frac{C_{V,\text{m}}}{3R} = \left(\frac{343\ \text{K}}{298\ \text{K}}\right)^2 \times \left(\frac{e^{343/(2\times298)}}{e^{343/298}-1}\right)^2 = \boxed{0.897}$$

COMMENT. For many metals the classical value is approached at room temperature; consequently, the failure of classical theory became apparent only after methods for achieving temperatures well below 25°C were developed in the latter part of the nineteenth century.

P7.5 $$\psi(r) = \left(\frac{1}{\pi a_0^3}\right)^{1/2} e^{-r/a_0} \quad \text{and} \quad \psi(r)^2 = \frac{1}{\pi a_0^3} e^{-2r/a_0} \quad \text{where } a_0 = 53\ \text{pm}$$

The probability P that the particle will be found in the region is the integral summation of all the infinitesimal probabilities of finding the particle within the region: $P = \int_{\text{region}} \psi(r)^2 d\tau$. In this problem we are to integrate over the region of a sphere for which the radius ($b = 1.0$ pm $= a_0/53$) is a very small fraction of the characteristic wavefunction distance a_0. Consequently, we may assume that ψ does not vary much within the sphere. Then, the probability is given by

$$P = \int_{\text{region}} \psi(r)^2 d\tau = \psi(a)^2 \int_{\text{region}} d\tau, \quad \text{where } a \text{ is the position of the centre of the sphere}$$

$$= \left(\frac{4\pi b^3}{3}\right)\left(\frac{1}{\pi a_0^3}\right) e^{-2a/a_0} = \left(\frac{4}{3}\right)\left(\frac{b}{a_0}\right)^3 e^{-2a/a_0}$$

(a) $b = 1.0$ pm $= a_0/53$ is the radius of a sphere that is centered on the origin ($a = 0$).

$$P(0,b) = \frac{4}{3 \times (53)^3} e^{-0} = \boxed{9.0 \times 10^{-6}}$$

(b) $b = 1.0$ pm $= a_0/53$ is the radius of a sphere that is centered on the origin ($a = a_0$).

$$P = \left(\frac{4}{3 \times (53)^3}\right) e^{-2} = \boxed{1.2 \times 10^{-6}}$$

Question. If there is a nonzero probability that the electron can be found at $r = 0$ how does it avoid destruction at the nucleus?

P7.7 The most probable location occurs when the probability density, $|\psi|^2$, is a maximum. Thus, we wish to find the value $x = x_{\text{max}}$ such that $d|\psi|^2/dx = 0$.

$$\psi(x) = Nxe^{-x^2/2a^2}$$
$$|\psi^2| = N^2x^2e^{-x^2/a^2}$$
$$d|\psi^2|/dx = N^2\{2xe^{-x^2/a^2} - (2x^3/a^2)e^{-x^2/a^2}\} = 2N^2x\{1 - x^2/a^2\}e^{-x^2/a^2}$$

The above derivative equals zero when the factor $1 - x^2/a^2$ equals zero, so we conclude that $\boxed{x_{\text{max}} = a}$.

P7.9 $$\psi = \left(\frac{2a}{\pi}\right)^{1/4} e^{-ax^2}, \quad \text{where} \quad -\infty \le x \le \infty$$

To verify the uncertainty principle, $\Delta x \Delta p \ge \frac{1}{2}\hbar$ [7.39a], we must use the above wavefunction to evaluate the root mean squares deviations for the uncertainties in position and momentum. If multiplication of the deviations yields a result that is greater than, or equal to, $\frac{1}{2}\hbar$, the uncertainty principle is verified.

We begin by finding the quantum-mechanical averages $\langle x\rangle$, $\langle x^2\rangle$, $\langle p\rangle$, and $\langle p^2\rangle$.

$$\langle x\rangle = \int \psi^* x \psi \, d\tau = \int_{-\infty}^{\infty} \left(\frac{2a}{\pi}\right)^{1/4} e^{-ax^2} x \left(\frac{2a}{\pi}\right)^{1/4} e^{-ax^2} dx$$

$$= \left(\frac{2a}{\pi}\right)^{1/2} \int_{-\infty}^{\infty} x e^{-2ax^2} dx$$

$= 0$ because the integration of an odd function around a centre of symmetry equals zero.

$$\langle x^2\rangle = \int_{-\infty}^{\infty} \left(\frac{2a}{\pi}\right)^{1/4} e^{-ax^2} x^2 \left(\frac{2a}{\pi}\right)^{1/4} e^{-ax^2} dx = \left(\frac{2a}{\pi}\right)^{1/2} \int_{-\infty}^{\infty} x^2 e^{-2ax^2} dx = \left(\frac{2a}{\pi}\right)^{1/2} \left(\frac{\pi^{1/2}}{2(2a)^{3/2}}\right) = \frac{1}{4a}$$

$$\langle p\rangle = \int_{-\infty}^{\infty} \psi^* \hat{p} \psi \, dx = \int_{-\infty}^{\infty} \psi^* \left(\frac{\hbar d}{i dx}\right) \psi \, dx$$

$$= \int_{-\infty}^{\infty} \left(\frac{2a}{\pi}\right)^{1/4} e^{-ax^2} \left(\frac{\hbar d}{i dx}\right) \left(\frac{2a}{\pi}\right)^{1/4} e^{-ax^2} dx = \left(\frac{\hbar}{i}\right) \times \left(\frac{2a}{\pi}\right)^{1/2} \int_{-\infty}^{\infty} e^{-ax^2} \left(\frac{d}{dx} e^{-ax^2}\right) dx$$

$$= -2a\left(\frac{\hbar}{i}\right) \times \left(\frac{2a}{\pi}\right)^{1/2} \int_{-\infty}^{\infty} x e^{-2ax^2} dx$$

$= 0$ because the integration of an odd function around a centre of symmetry equals zero.

$$\langle p^2\rangle = \int_{-\infty}^{\infty} \psi^* \hat{p}^2 \psi \, dx = \int_{-\infty}^{\infty} \psi^* \left(\frac{\hbar d}{i dx}\right)^2 \psi \, dx = \int_{-\infty}^{\infty} \left(\frac{2a}{\pi}\right)^{1/4} e^{-ax^2} \left(\frac{\hbar d}{i dx}\right)^2 \left(\frac{2a}{\pi}\right)^{1/4} e^{-ax^2} dx$$

$$= \left(\frac{\hbar}{i}\right)^2 \times \left(\frac{2a}{\pi}\right)^{1/2} \int_{-\infty}^{\infty} e^{-ax^2} \left(\frac{d^2}{dx^2} e^{-ax^2}\right) dx = -2a\left(\frac{\hbar}{i}\right)^2 \times \left(\frac{2a}{\pi}\right)^{1/2} \int_{-\infty}^{\infty} e^{-ax^2} \left(\frac{d}{dx} x e^{-ax^2}\right) dx$$

$$= -2a\left(\frac{\hbar}{i}\right)^2 \times \left(\frac{2a}{\pi}\right)^{1/2} \int_{-\infty}^{\infty} e^{-ax^2} (e^{-ax^2} - 2ax^2 e^{-ax^2}) dx$$

$$= 2a\hbar^2 \times \left(\frac{2a}{\pi}\right)^{1/2} \left\{ \int_{-\infty}^{\infty} e^{-2ax^2} dx - 2a \int_{-\infty}^{\infty} x^2 e^{-2ax^2} dx \right\}$$

$$= 2a\hbar^2 \times \left(\frac{2a}{\pi}\right)^{1/2} \left\{ \left(\frac{\pi}{2a}\right)^{1/2} - 2a\left(\frac{\pi}{32a^3}\right)^{1/2} \right\}$$

$$= a\hbar^2$$

We now substitute the above expressions for $\langle x\rangle$, $\langle x^2\rangle$, $\langle p\rangle$, and $\langle p^2\rangle$ into the expressions for the root mean squares deviations [7.39b].

$$\Delta x\Delta p = \{\langle x^2\rangle - \langle x\rangle^2\}^{1/2} \times \{\langle p^2\rangle - \langle p\rangle^2\}^{1/2} \text{ [7.39b]}$$
$$= \left\{\frac{1}{4a} - 0^2\right\}^{1/2} \times \{a\hbar^2 - 0^2\}^{1/2} = \tfrac{1}{2}\hbar$$

The above expression for $\Delta x\Delta p$ verifies the uncertainty principle in the sense that the quantum-mechanical expressions yield the minimum value allowed by the uncertainty principle ($\tfrac{1}{2}\hbar$).

Solutions to theoretical problems

P7.11 The Planck distribution is $\rho(\lambda,T) = \dfrac{8\pi hc}{\lambda^5\left(e^{\frac{hc}{\lambda kT}} - 1\right)}$ [7.8].

As λ increases, $\dfrac{hc}{\lambda kT}$ decreases, and at very long wavelength $hc/\lambda kT \ll 1$. Hence, we can expand the exponential in a power series. Let $x = hc/\lambda kT$, then $e^x = 1 + x + \frac{1}{2!}x^2 + \frac{1}{3!}x^3 + \cdots$ and the Planck distribution becomes

$$\rho = \frac{8\pi hc}{\lambda^5\left(1 + x + \frac{1}{2!}x^2 + \frac{1}{3!}x^3 + \cdots - 1\right)} = \frac{8\pi hc}{\lambda^5\left(x + \frac{1}{2!}x^2 + \frac{1}{3!}x^3 + \cdots\right)}$$

When x is much, much smaller than 1, second- and higher-order terms in x become negligibly small compared to x. Consequently,

$$\lim_{\lambda\to\infty} \rho = \frac{8\pi hc}{\lambda^5 x} = \frac{8\pi hc}{\lambda^5}\left(\frac{1}{hc/\lambda kT}\right) = \boxed{\frac{8\pi kT}{\lambda^4}}$$

This is the Rayleigh–Jeans law [7.6].

P7.13 The total energy density of black-body radiation is

$$\mathcal{E} = \int_0^\infty \rho(\lambda)\,d\lambda = 8\pi hc\int_0^\infty \frac{d\lambda}{\lambda^5(e^{hc/\lambda kT} - 1)} \text{ [7.8]}$$

Let $x = \dfrac{hc}{\lambda kT}$. Then, $dx = -\dfrac{hc}{\lambda^2 kT}d\lambda$ or $d\lambda = -\dfrac{\lambda^2 kT}{hc}dx$

$$\mathcal{E} = 8\pi kT\int_0^\infty \frac{\lambda^2\,dx}{\lambda^5(e^x - 1)} = 8\pi kT\int_0^\infty \frac{dx}{\lambda^3(e^x - 1)} = 8\pi kT\left(\frac{kT}{hc}\right)^3\int_0^\infty \frac{x^3\,dx}{(e^x - 1)} = 8\pi kT\left(\frac{kT}{hc}\right)^3\left(\frac{\pi^4}{15}\right)$$
$$= \left(\frac{8\pi^5 k^4}{15h^3c^3}\right)T^4 = \boxed{\left(\frac{4}{c}\right)\sigma T^4}, \text{ where } \sigma = \frac{2\pi^5 k^4}{15h^3c^2} \text{ is the Stefan–Boltzmann constant.}$$

This is the Stefan–Boltzmann law that specifies that the total energy density of black-body radiation is proportional to T^4.

P7.15 We require $\int \psi^*\psi \, d\tau = 1$, and so write $\psi = Nf$ and find N for the given f.

(a) $$N^2\int_0^L \sin^2\frac{n\pi x}{L}dx = \frac{1}{2}N^2\int_0^L\left(1-\cos\frac{2n\pi x}{L}\right)dx \text{ [trigonometric identity]}$$

$$= \frac{1}{2}N^2\left(x - \frac{L}{2n\pi}\sin\frac{2n\pi x}{L}\right)\Bigg|_0^L$$

$$= \frac{L}{2}N^2 = 1 \quad \text{if } \boxed{N = \left(\frac{2}{L}\right)^{1/2}}$$

(b) $$N^2\int_{-L}^{L} c^2\,dx = 2N^2c^2L = 1 \quad \text{if } \boxed{N = \frac{1}{c(2L)^{1/2}}}$$

(c) $$N^2\int_0^\infty e^{-2r/a}r^2\,dr\int_0^\pi \sin\theta\, d\theta\int_0^{2\pi} d\phi \quad [d\tau = r^2\sin\theta\, dr\, d\theta\, d\phi]$$

$$= N^2\left(\frac{a^3}{4}\right)\times(2)\times(2\pi) = 1 \quad \text{if } \boxed{N = \frac{1}{(\pi a^3)^{1/2}}}$$

(d) $$N^2\int_0^\infty r^2\times r^2 e^{-r/a}\,dr\int_0^\pi \sin^3\theta\, d\theta\int_0^{2\pi}\cos^2\phi\, d\phi \quad [x = r\cos\phi\sin\theta]$$

$$= N^2 4!a^5\times\tfrac{4}{3}\times\pi = 32\pi a^5N^2 = 1 \quad \text{if } \boxed{N = \frac{1}{(32\pi a^5)^{1/2}}}$$

We have used $\int \sin^3\theta\, d\theta = -\frac{1}{3}(\cos\theta)(\sin^2\theta + 2)$, as found in tables of integrals and

$$\int_0^{2\pi}\cos^2\phi\, d\phi = \int_0^{2\pi}\sin^2\phi\, d\phi$$

by symmetry with $\int_0^{2\pi}(\cos^2\phi + \sin^2\phi)\,d\phi = \int_0^{2\pi} d\phi = 2\pi$

P7.17 In each case form $\hat{\Omega}f$. If the result is ωf where ω is a constant, then f is an eigenfunction of the operator $\hat{\Omega}$ and ω is the eigenvalue [7.28a, 7.28b, and 7.28c]. We check whether f is an eigenfunction of the $\hat{\Omega} = d/dx$ operator.

(a) $$\frac{d}{dx}f = \frac{d}{dx}e^{ikx} = ike^{ikx} = ikf$$

$\boxed{\text{Yes}}$, the function is an eigenfunction with the eigenvalue ik.

(b) $$\frac{d}{dx}f = \frac{d}{dx}\cos kx = -k\sin kx$$

$\boxed{\text{No}}$, the function is not an eigenfunction because $-k\sin kx \neq \text{constant}\times f$.

(c) $\dfrac{d}{dx}f = \dfrac{d}{dx}k = 0 = 0 \times f$

$\boxed{\text{Yes}}$; the function is an eigenfunction with the eigenvalue of zero.

(d) $\dfrac{d}{dx}f = \dfrac{d}{dx}kx = k$

$\boxed{\text{No}}$, the function is not an eigenfunction because $k \neq \text{constant} \times f$.

(e) $\dfrac{d}{dx}f = \dfrac{d}{dx}e^{-ax^2} = -2axe^{-ax^2} = -2axf$

$\boxed{\text{No}}$, the function is not an eigenfunction because $-2axf \neq \text{constant} \times f$.

P7.19 In each case form $\hat{\Omega}f$. If the result is ωf, where ω is a constant, then f is an eigenfunction of the operator $\hat{\Omega}$ and ω is the eigenvalue [7.28a, 7.28b, and 7.28c]. We first check whether f is an eigenfunction of the $\hat{\Omega} = d/dx$ operator, after which we check whether f is an eigenfunction of the $\hat{\Omega} = d^2/dx^2$ operator.

(a) $f = e^{ikx}$

$$\frac{d}{dx}f = \frac{d}{dx}e^{ikx} = ike^{ikx} = \text{constant} \times f$$

$$\frac{d^2}{dx^2}f = \frac{d}{dx}(ike^{ikx}) = (ik)^2e^{ikx} = \text{constant} \times f \ \boxed{\text{Yes}},$$

f is an eigenfunction of d^2/dx^2 with eigenvalue is $\boxed{-k^2}$.

(b) $f = \cos kx$

$$\frac{d}{dx}f = \frac{d}{dx}(\cos kx) = -k\sin kx \neq \text{constant} \times f$$

$$\frac{d^2}{dx^2}f = \frac{d}{dx}(-k\sin kx) = -k^2\cos kx = \text{constant} \times f$$

$\boxed{\text{Yes}}$, f is an eigenfunction of d^2/dx^2 with eigenvalue is $\boxed{-k^2}$.

(c) $f = k$

$$\frac{d}{dx}f = \frac{d}{dx}k = 0 = \text{constant} \times f$$

$$\frac{d^2}{dx^2}f = \frac{d}{dx}(0) = 0 = \text{constant} \times f \ \boxed{\text{Yes}}, f \text{ is a trivial eigenfunction of } d^2/dx^2.$$

(d) $f = kx$

$$\frac{d}{dx}f = \frac{d}{dx}(kx) = k \neq \text{constant} \times f$$

$$\frac{d^2}{dx^2}f = \frac{d}{dx}(k) = 0 = \text{constant} \times f \ \boxed{\text{Yes}}, f \text{ is a trivial eigenfunction of } d^2/dx^2.$$

(e) $f = e^{-ax^2}$

$$\frac{d}{dx}f = \frac{d}{dx}e^{-ax^2} = -2axe^{-ax^2} \neq \text{constant} \times f$$

$$\frac{d^2}{dx^2}f = \frac{d}{dx}(-2axe^{-ax^2}) = -2a(1-2ax^2)e^{-ax^2} \neq \text{constant} \times f \ \boxed{\text{No}},$$

f is not an eigenfunction of d^2/dx^2.

(i) In addition to being an eigenfunction of d/dx the functions of $\boxed{\text{(a) and (c)}}$ are also eigenfunctions of d/dx^2.

(ii) The functions of $\boxed{\text{(b) and (d)}}$, not being eigenfunctions of d/dx, are only eigenfunctions of d/dx^2.

The problem may be repeated with the functions of P7.18 to get the following results.

(a) $f = x^3 - kx$

$$\frac{d}{dx}f = \frac{d}{dx}(x^3 - kx) = 3x^2 - k \neq \text{constant} \times f$$ No, f is not an eigenfunction of d/dx.

$$\frac{d^2}{dx^2}f = \frac{d}{dx}(3x^2 - k) = 6x \neq \text{constant} \times f$$ No, f is not an eigenfunction of d^2/dx^2.

(b) $f = \cos kx$

$$\frac{d}{dx}f = \frac{d}{dx}(\cos kx) = -k\sin kx \neq \text{constant} \times f$$ No, f is not an eigenfunction of d/dx.

$$\frac{d^2}{dx^2}f = \frac{d}{dx}(-k\sin kx) = -k^2\cos kx = \text{constant} \times f$$

Yes, f is an eigenfunction of d^2/dx^2. Its eigenvalue is $-k^2$.

(c) $f = x^2 + 3x - 1$

$$\frac{d}{dx}f = \frac{d}{dx}(x^2 + 3x - 1) = 2x + 3 \neq \text{constant} \times f$$ No, f is not an eigenfunction of d/dx.

$$\frac{d^2}{dx^2}f = \frac{d}{dx}(2x + 3) = 2 \neq \text{constant} \times f$$ No, f is not an eigenfunction of d^2/dx^2.

Of these P7.18 functions only $f = \cos kx$ is an eigenfunction of d^2/dx^2. It is not, however, an eigenfunction of d/dx.

P7.21 Time-independent Schrödinger equation of a single particle: $\left(-\frac{\hbar^2}{2m}\frac{d^2}{dx^2} + \hat{V}\right)\psi = E\psi$ [7.32]

(a) $$\left(-\frac{\hbar^2}{2m_e}\frac{d^2}{dx^2} - \frac{e^2}{4\pi\varepsilon_0 x}\right)\psi = E\psi$$

(b) $$\left(-\frac{\hbar^2}{2m}\frac{d^2}{dx^2}\right)\psi = E\psi$$

(c) $F = c$ (a constant) implies that $V = -cx$ because $F = -\mathrm{d}V/\mathrm{d}x$.

$$\left(-\frac{\hbar^2}{2m}\frac{\mathrm{d}^2}{\mathrm{d}x^2} - cx\right)\psi = E\psi$$

P7.23 $\psi = (\cos\chi)\mathrm{e}^{+ikx} + (\sin\chi)\mathrm{e}^{-ikx} = c_1\mathrm{e}^{+ikx} + c_2\mathrm{e}^{-ikx}$

The expectation value for kinetic energy depends on the curvature of the wavefunction so we begin by finding the expression for $\mathrm{d}^2\psi/\mathrm{d}x^2$.

$$\frac{\mathrm{d}\psi}{\mathrm{d}x} = \mathrm{i}kc_1\mathrm{e}^{+ikx} - \mathrm{i}kc_2\mathrm{e}^{-ikx}$$

$$\frac{\mathrm{d}^2\psi}{\mathrm{d}x^2} = -k^2c_1\mathrm{e}^{+ikx} - k^2c_2\mathrm{e}^{-ikx} = -k^2\psi$$

Thus, ψ is an eigenfunction of the $\mathrm{d}^2\psi/\mathrm{d}x^2$ operator and it must also be an eigenfunction of the kinetic energy operator.

$$\hat{E}_{\mathrm{k}}\psi = -\frac{\hbar^2}{2m}\frac{\mathrm{d}^2}{\mathrm{d}x^2}\psi\ [7.31] = \frac{(\hbar k)^2}{2m}\psi$$

The kinetic energy of the particle is the eigenvalue $\boxed{\dfrac{(\hbar k)^2}{2m}}$.

P7.25 The normalized wavefunctions for the hydrogen atom [see P7.16] are

(a) $\psi = \left(\dfrac{1}{\pi a_0^3}\right)^{1/2}\left(2 - \dfrac{r}{a_0}\right)\mathrm{e}^{-r/a_0}$ and (b) $\psi = \left(\dfrac{1}{32\pi a_0^5}\right)^{1/2} r\sin\theta\cos\phi\,\mathrm{e}^{-r/2a_0}$

The expectation values $\langle r\rangle = \int r\,|\psi|^2\,\mathrm{d}\tau$ and $\langle r^2\rangle = \int r^2|\psi|^2\,\mathrm{d}\tau$ are best evaluated using the spherical coordinates (r, θ, ϕ) for which $0 \le r \le \infty$, $0 \le \theta \le \pi$, and $0 \le \phi \le 2\pi$. It is helpful to recognize that, when a wavefunction has the separation of variables form $\psi(r, \theta, \phi) = R(r) \times \Theta(\theta) \times \Phi(\phi)$, the expectation integral of the function $f(r)$ over the space of all variables is

$$\langle f(r)\rangle = \int f(r)\,|\psi|^2\,\mathrm{d}\tau = \int_{r=0}^{\infty}\int_{\theta=0}^{\pi}\int_{\phi=0}^{2\pi} f(r)\,|\psi|^2 r^2\sin\theta\,\mathrm{d}r\,\mathrm{d}\theta\,\mathrm{d}\phi\ [\text{Note that } \mathrm{d}\tau = r^2\sin\theta\,\mathrm{d}r\,\mathrm{d}\theta\,\mathrm{d}\phi.]$$

$$= \int_{r=0}^{\infty}\int_{\theta=0}^{\pi}\int_{\phi=0}^{2\pi} r^2 f(r) \times (R(r) \times \Theta(\theta) \times \Phi(\phi))^2 \sin\theta\,\mathrm{d}r\,\mathrm{d}\theta\,\mathrm{d}\phi$$

$$= \int_{r=0}^{\infty} r^2 f(r) \times R(r)^2\mathrm{d}r \times \int_{\theta=0}^{\pi} \sin(\theta) \times \Theta(\theta)^2\mathrm{d}\theta \times \int_{\phi=0}^{2\pi} \Phi(\phi)^2\mathrm{d}\phi$$

In the special case (a) for which $\Theta(\theta) = 1$ and $\Phi(\phi) = 1$:

$$\int_{\theta=0}^{\pi} \sin(\theta) \times \Theta(\theta)^2\mathrm{d}\theta \times \int_{\phi=0}^{2\pi} \Phi(\phi)^2\mathrm{d}\phi = \int_{\theta=0}^{\pi} \sin(\theta)\mathrm{d}\theta \times \int_{\phi=0}^{2\pi} \mathrm{d}\phi = [-\cos\theta]_{\theta=0}^{\pi} \times [\phi]_{\phi=0}^{2\pi} = 4\pi$$

and the expectation integral is

$$\langle f(r)\rangle = 4\pi\int_{r=0}^{\infty} r^2 f(r)\times R(r)^2\mathrm{d}r$$

In the special case (b) for which $\Theta(\theta) = \sin\theta$ and $\Phi(\phi) = \cos\phi$:

$$\int_{\theta=0}^{\pi}\sin(\theta)\times\Theta(\theta)^2\mathrm{d}\theta\times\int_{\phi=0}^{2\pi}\Phi(\phi)^2\mathrm{d}\phi = \int_{\theta=0}^{\pi}\sin^3(\theta)\mathrm{d}\theta\times\int_{\phi=0}^{2\pi}\cos^2(\phi)\mathrm{d}\phi$$
$$= [-\cos\theta + \tfrac{1}{3}\cos^3\theta]_{\theta=0}^{\pi}\times[\tfrac{\phi}{2} + \tfrac{1}{4}\sin 2\phi]_{\phi=0}^{2\pi}$$
$$= \tfrac{4}{3}\pi$$

and the expectation integral is

$$\langle f(r)\rangle = \tfrac{4}{3}\pi\int_{r=0}^{\infty} r^2 f(r)\times R(r)^2\mathrm{d}r$$

(a) $$\langle r\rangle = 4\pi\int_{r=0}^{\infty} r^3\times R(r)^2\mathrm{d}r$$

$$= 4\pi\int_{r=0}^{\infty} r^3\times\left\{\left(\frac{1}{\pi a_0^3}\right)^{1/2}\left(2-\frac{r}{a_0}\right)\mathrm{e}^{-r/a_0}\right\}^2\mathrm{d}r$$

$$= 4a_0\int_{\chi=0}^{\infty}\chi^3\times\{(2-\chi)\mathrm{e}^{-\chi}\}^2\mathrm{d}\chi \quad \text{where} \quad \chi = r/a_0$$

$$= 4a_0\int_{\chi=0}^{\infty}\{4\chi^3 - 4\chi^4 + \chi^5\}\mathrm{e}^{-2\chi}\mathrm{d}\chi\left[\text{use the standard integral} \int_0^{\infty}\chi^n\mathrm{e}^{-a\chi}\mathrm{d}\chi = n!/a^{n+1}\right]$$

$$= 4a_0\{4\times 3!/2^{3+1} - 4\times 4!/2^{4+1} + 5!/2^{5+1}\}$$
$$= \boxed{1.5a_0}$$

$$\langle r^2\rangle = 4\pi\int_{r=0}^{\infty} r^4\times R(r)^2\mathrm{d}r$$

$$= 4\pi\int_{r=0}^{\infty} r^4\times\left\{\left(\frac{1}{\pi a_0^3}\right)^{1/2}\left(2-\frac{r}{a_0}\right)\mathrm{e}^{-r/a_0}\right\}^2\mathrm{d}r$$

$$= 4a_0^2\int_{\chi=0}^{\infty}\chi^4\times\{(2-\chi)\mathrm{e}^{-\chi}\}^2\mathrm{d}\chi, \quad \text{where} \quad \chi = r/a_0$$

$$= 4a_0^2\int_{\chi=0}^{\infty}\{4\chi^4 - 4\chi^5 + \chi^6\}\mathrm{e}^{-2\chi}\mathrm{d}\chi\left[\text{use the standard integral} \int_0^{\infty}\chi^n\mathrm{e}^{-a\chi}\mathrm{d}\chi = n!/a^{n+1}\right]$$

$$= 4a_0^2\{4\times 4!/2^{4+1} - 4\times 5!/2^{5+1} + 6!/2^{6+1}\}$$
$$= \boxed{4.5a_0^2}$$

(b) $\langle r\rangle = \frac{4}{3}\pi\int_{r=0}^{\infty} r^3 \times R(r)^2 \mathrm{d}r$

$$= \frac{4}{3}\pi\int_{r=0}^{\infty} r^3 \times \left\{\left(\frac{1}{32\pi a_0^5}\right)^{1/2} r\mathrm{e}^{-r/2a_0}\right\}^2 \mathrm{d}r$$

$$= \frac{a_0}{24}\int_{\chi=0}^{\infty} \chi^5 \mathrm{e}^{-\chi}\mathrm{d}\chi, \quad \text{where} \quad \chi = r/a_0$$

$$= \left(\frac{a_0}{24}\right) \times (5!) \left[\text{use the standard integral } \int_0^{\infty} \chi^n \mathrm{e}^{-a\chi}\mathrm{d}\chi = n!/a^{n+1}\right]$$

$$= \boxed{5a_0}$$

$$\langle r^2\rangle = \frac{4}{3}\pi\int_{r=0}^{\infty} r^4 \times R(r)^2 \mathrm{d}r$$

$$= \frac{4}{3}\pi\int_{r=0}^{\infty} r^4 \times \left\{\left(\frac{1}{32\pi a_0^5}\right)^{1/2} r\mathrm{e}^{-r/2a_0}\right\}^2 \mathrm{d}r$$

$$= \frac{a_0^2}{24}\int_{\chi=0}^{\infty} \chi^6 \mathrm{e}^{-\chi}\mathrm{d}\chi, \quad \text{where} \quad \chi = r/a_0$$

$$= \left(\frac{a_0^2}{24}\right) \times (6!) \left[\text{use the standard integral } \int_0^{\infty} \chi^n \mathrm{e}^{-a\chi}\mathrm{d}\chi = n!/a^{n+1}\right]$$

$$= \boxed{30a_0^2}$$

P7.27 The superpositions of cosine functions of the form $\cos(nx)$ can be chosen with n equal to any integer between 1 and m. For convenience, x can be examined in the range between $-\pi/2$ and $\pi/2$. The normalization constant for each function is determined by integrating the function squared over the range of x [7.19]. Using Mathcad to perform the integration, we find:

$$\int_{-\pi/2}^{\pi/2} (\cos(n\cdot x))^2 \mathrm{d}x \rightarrow \frac{1}{2}\cdot\frac{2\cdot\cos\left(\frac{1}{2}\cdot\pi\cdot n\right)\cdot\sin\left(\frac{1}{2}\cdot\pi\cdot n\right)+\pi\cdot n}{n}$$

When n is an even integer, $\sin(\pi n/2) = 0$, and when n is an odd integer, $\cos(\pi n/2) = 0$. Consequently, when n is an integer, the above integral equals $\pi/2$ and we select $(2/\pi)^{1/2}$ as the normalization constant for the function $\cos(nx)$. The normalized function is $\phi(n, x)$. The superposition, $\psi(m, x)$, is the sum of these cosine functions from $n = 1$ to $n = m$. Since the cosine functions are orthogonal, $\psi(m, x)$ has a normalization constant equal to $(1/m)^{1/2}$. The Mathcad probability density plots of Figure 7.1 show $\psi^2(m, x)$ against x for $m = 1$, 3, and 10.

$$\phi(n,x) := \left(\frac{2}{\pi}\right)^{1/2}\cdot\cos(n\cdot x) \qquad \psi(m,x) := \left(\frac{1}{m}\right)^{1/2}\cdot\sum_{n=1}^{m}\phi(n,x) \qquad x := \frac{-\pi}{2}, \frac{-\pi}{2}+0.001 \cdots \frac{\pi}{2}$$

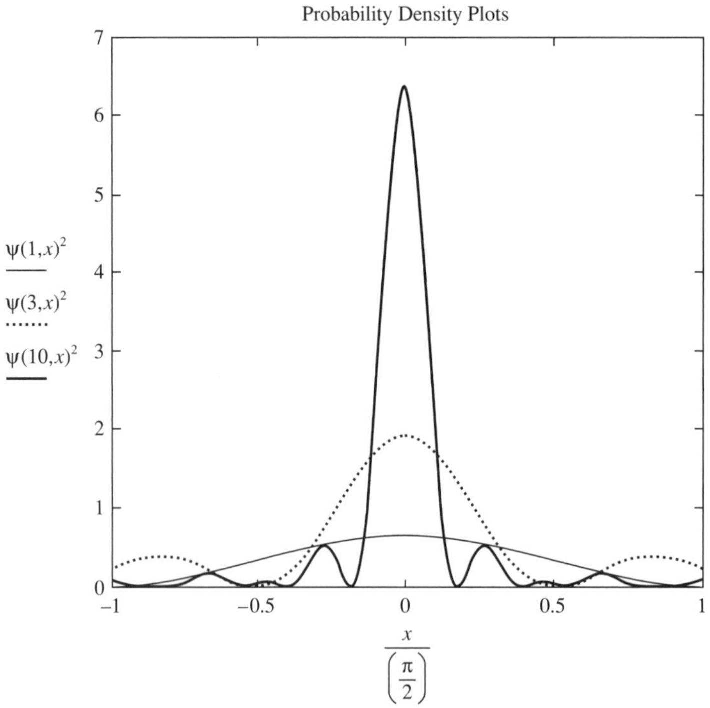

Figure 7.1

Examination of the probability density plots reveals that, when the superposition has few terms, the particle position is ill-defined. There is a great uncertainty in knowledge of position. When many terms are added to the superposition, the uncertainty narrows to a small region around $x = 0$. A plot with m greater than 10 will further confirm this conclusion.

Each function in the superposition has been assigned a weight equal to the normalization constant $(1/m)^{1/2}$. This means that each cosine function in the superposition has an identical probability contribution to the expectation value for momentum (see *Justification 7.5*). Each cosine function contributes with a probability equal to $1/m$. Furthermore, each cosine function represents a particle momentum that is proportional to the argument n (see Example 7.6). The Figure 7.2 Mathcad plot of momentum probability against momentum, as represented by n, is an interesting contrast to the plot of probability density against position.

Variables needed for the Mathcad plot: $n := 1 \ldots 12 \quad \text{Prob}(n, m) := \text{if}\left(n \le m, \dfrac{1}{m}, 0\right)$

The momentum probability plot shows momentum probabilities for superpositions of $m = 1$, 5, and 10 terms. When there are many terms in the superposition, the range of possible momentum is very broad even though the range of observed positions becomes narrow. Position and momentum are complementary variables. As location becomes more precise with the superposition of many functions, precise knowledge of momentum decreases. This illustrates the Heisenberg uncertainty principle [7.39a].

The plot of probability density against position clearly indicates that the superposition is symmetrical around the point $x = 0$. Consequently, the expectation position for all superpositions is $x = 0$. The expectation value for position is independent of the number of terms in the superposition.

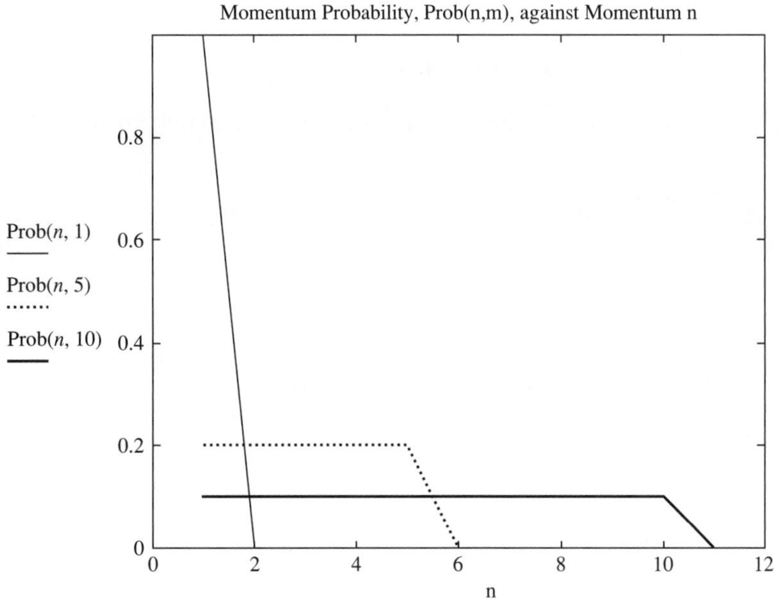

Figure 7.2

The square root of the expectation value of x^2 is called the root-mean-square value of x, x_{rms}. The Figure 7.3 Mathcad plot of x_{rms} against m indicates that this expectation value depends on the number of terms in the superposition. However, it does appear to very slowly converge to a very small value (zero?) when the superposition contains many functions.

$$x_{rms}(m) := \left(\int_{-\pi/2}^{\pi/2} x^2 \cdot \psi(m, x)^2 \mathrm{d}x \right)^{1/2} \qquad m := 1 \ldots 50$$

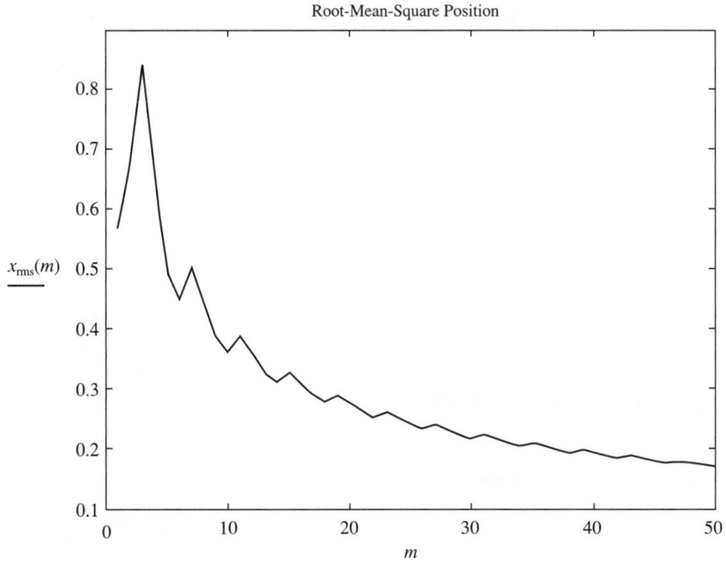

Figure 7.3

P7.29 (a) In the momentum representation $\hat{p}_x = p_x \times$, consequently

$$[\hat{x}, \hat{p}_x]\phi = [\hat{x}, p_x \times]\phi = \hat{x}p_x \times \phi - p_x \times \hat{x}\phi = \mathrm{i}\hbar\phi \text{ [7.42]}$$

Suppose that the position operator has the form $\hat{x} = a\dfrac{\mathrm{d}}{\mathrm{d}p_x}$, where a is a complex number. Then,

$$a\frac{\mathrm{d}}{\mathrm{d}p_x}(p_x \times \phi) - p_x \times \left(a\frac{\mathrm{d}}{\mathrm{d}p_x}\phi\right) = \mathrm{i}\hbar\phi$$

$$\frac{\mathrm{d}}{\mathrm{d}p_x}(p_x \times \phi) - p_x \times \left(\frac{\mathrm{d}}{\mathrm{d}p_x}\phi\right) = \frac{\mathrm{i}\hbar}{a}\phi$$

$$\frac{\mathrm{d}p_x}{\mathrm{d}p_x}\phi + p_x\frac{\mathrm{d}\phi}{\mathrm{d}p_x} - p_x\frac{\mathrm{d}\phi}{\mathrm{d}p_x} = \frac{\mathrm{i}\hbar}{a}\phi \text{ [rule for differentiation of } f(x)g(x).]$$

$$\phi = \frac{\mathrm{i}\hbar}{a}\phi$$

This is true when $a = \mathrm{i}\hbar$. We conclude that $\hat{x} = \mathrm{i}\hbar\dfrac{\mathrm{d}}{\mathrm{d}p_x}$ in the momentum representation.

(b) The fact that integration is the inverse of differentiation suggests the guess that in the momentum representation

$$\hat{x}^{-1}\phi = \left(\mathrm{i}\hbar\frac{\mathrm{d}}{\mathrm{d}p_x}\right)^{-1}\phi = \left(\frac{1}{\mathrm{i}\hbar}\int_{-\infty}^{p_x}\mathrm{d}p_x\right)\phi = \frac{1}{\mathrm{i}\hbar}\int_{-\infty}^{p_x}\phi\,\mathrm{d}p_x$$

where the symbol $\int_{-\infty}^{p_x}\mathrm{d}p_x$ is understood to be an integration operator that uses any function on its right side as an integrand. To validate the guess that $\hat{x}^{-1} = \dfrac{1}{\mathrm{i}\hbar}\int_{-\infty}^{p_x}\mathrm{d}p_x$ we need to confirm the operator relationship $\hat{x}^{-1}\hat{x} = \hat{x}\hat{x}^{-1} = \hat{1}$. Using the Leibnitz's rule for differentiation of integrals:

$$\hat{x}\hat{x}^{-1}\phi = \left(\mathrm{i}\hbar\frac{\mathrm{d}}{\mathrm{d}p_x}\right)\left(\frac{1}{\mathrm{i}\hbar}\int_{-\infty}^{p_x}\mathrm{d}p_x\right)\phi = \left(\frac{\mathrm{d}}{\mathrm{d}p_x}\right)\left(\int_{-\infty}^{p_x}\phi\,\mathrm{d}p_x\right)$$

$$= \left(\int_{-\infty}^{p_x}\frac{\mathrm{d}\phi}{\mathrm{d}p_x}\mathrm{d}p_x\right) + \phi(p_x)\lim_{c\to-\infty}\frac{\mathrm{d}c}{\mathrm{d}p_x} - \phi(-\infty)\frac{\mathrm{d}p_x}{\mathrm{d}p_x} = \left(\int_{-\infty}^{p_x}\frac{\mathrm{d}\phi}{\mathrm{d}p_x}\mathrm{d}p_x\right) - \phi(-\infty)$$

Since $\phi(-\infty)$ must equal zero, we find that

$$\hat{x}\hat{x}^{-1}\phi = \int_{-\infty}^{p_x}\frac{\mathrm{d}\phi}{\mathrm{d}p_x}\mathrm{d}p_x = \int_{-\infty}^{p_x}\mathrm{d}\phi = \phi(p_x) - \phi(-\infty) = \phi(p_x) = \phi$$

from which we conclude that $\hat{x}\hat{x}^{-1} = \hat{1}$. Likewise,

$$\hat{x}^{-1}\hat{x}\phi = \left(\frac{1}{\mathrm{i}\hbar}\int_{-\infty}^{p_x}\mathrm{d}p_x\right)\left(\mathrm{i}\hbar\frac{\mathrm{d}}{\mathrm{d}p_x}\right)\phi = \int_{-\infty}^{p_x}\mathrm{d}\phi = \phi(p_x) - \phi(-\infty) = \phi(p_x) = \phi$$

Solutions to applications: nanoscience, environmental science, and astrophysics

P7.31 (a) Let $\Delta\phi$ be the potential difference through which the electrons are accelerated. The non-relativistic and relativistic wavelengths are:

$$\lambda_{\text{non-relativistic}} = \frac{h}{(2m_e e\Delta\phi)^{1/2}} = \frac{6.626\times10^{-34}\text{ J s}}{\{2(9.109\times10^{-31}\text{ kg})\times(1.602\times10^{-19}\text{ C})\times(50.0\times10^3\text{ V})\}^{1/2}}$$

$$= 5.48\text{ pm [Example 7.2]}$$

$$\lambda_{\text{relativistic}} = \frac{h}{\left\{2m_e e\Delta\phi\left(1+\frac{e\Delta\phi}{2m_ec^2}\right)\right\}^{1/2}} = \frac{\lambda_{\text{non-relativistic}}}{\left(1+\frac{e\Delta\phi}{2m_ec^2}\right)^{1/2}}$$

$$= \frac{5.48\text{ pm}}{\left\{1+\frac{(1.602\times10^{-19}\text{ C})(50.0\times10^3\text{ V})}{2(9.109\times10^{-31}\text{ kg})(3.00\times10^8\text{ m s}^{-1})^2}\right\}^{1/2}} = \boxed{5.35\text{ pm}}$$

(b) For an electron accelerated through 50 kV the non-relativistic de Broglie wavelength is calculated to be high by 2.4%. This error may be insignificant for many applications. However, should an accuracy of 1% or better be required, use the relativistic equation at accelerations through a potential above 20.4 kV as demonstrated in the following calculation.

$$\frac{\lambda_{\text{non-relativistic}}-\lambda_{\text{relativistic}}}{\lambda_{\text{relativistic}}} = \frac{\lambda_{\text{non-relativistic}}}{\lambda_{\text{relativistic}}} - 1 = \left(1+\frac{e\Delta\phi}{2m_ec^2}\right)^{1/2} - 1$$

$$= \cancel{1} + \frac{1}{2}\left(\frac{e\Delta\phi}{2m_ec^2}\right) - \frac{1}{2\cdot4}\left(\frac{e\Delta\phi}{2m_ec^2}\right)^2 + \frac{1\cdot3}{2\cdot4\cdot6}\left(\frac{e\Delta\phi}{2m_ec^2}\right)^3 - \cdots \cancel{-1}$$

$$\simeq \frac{e\Delta\phi}{4m_ec^2} \quad \text{because 2nd- and 3rd-order terms are very small.}$$

The largest value of $\Delta\phi$ for which the non-relativistic equation yields a value that has less than 1% error:

$$\Delta\phi \simeq \left(\frac{4m_ec^2}{e}\right)\times\left(\frac{\lambda_{\text{non-relativistic}}-\lambda_{\text{relativistic}}}{\lambda_{\text{relativistic}}}\right) = \left(\frac{4m_ec^2}{e}\right)\times(0.01) = 20.4\text{ kV}$$

P7.33 (a) $CH_4(g) \rightarrow C(\text{graphite}) + 2\,H_2(g)$

$$\Delta_rG^\ominus = -\Delta_fG^\ominus(CH_4) = -(-50.72\text{ kJ mol}^{-1}) = 50.72\text{ kJ mol}^{-1} \quad \text{at } T_{\text{ref}} = 298.15\text{ K}$$
$$\Delta_rH^\ominus = -\Delta_fH^\ominus(CH_4) = -(-74.81\text{ kJ mol}^{-1}) = 74.81\text{ kJ mol}^{-1} \quad \text{at } T_{\text{ref}} = 298.15\text{ K}$$

We want to find the temperature at which $\Delta_rG^\ominus(T) = 0$. Below this temperature methane is stable with respect to decomposition into the elements. Above this temperature it is unstable. Assuming that the heat capacities are basically independent of temperature,

$$\Delta_rC_p^\ominus(T) \simeq \Delta_rC_p^\ominus(T_{\text{ref}}) = \{8.527 + 2(28.824) - 35.31\}\text{J K}^{-1}\text{ mol}^{-1} = 30.865\ \text{ J K}^{-1}\text{ mol}^{-1}$$

$$\Delta_rH^\ominus(T) = \Delta_rH^\ominus(T_{\text{ref}}) + \int_{T_{\text{ref}}}^{T}\Delta_rC_p^\ominus(T)\text{d}T\ [2.36a] = \Delta_rH^\ominus(T_{\text{ref}}) + \Delta_rC_p^\ominus\times(T - T_{\text{ref}})$$

A modification of the Gibbs–Helmholtz equation [3.55] for chemical reactions can be integrated at constant pressure ($p^\ominus$) with use of the properties listed above. The result is a computational equation for $\Delta_r G^\ominus(T)$.

$$\left(\frac{\partial}{\partial T}\left(\frac{\Delta_r G^\ominus}{T}\right)\right)_p = -\frac{\Delta_r H^\ominus}{T^2} \; [3.55]$$

$$\int_{T_{\text{ref}}}^{T} \mathrm{d}\left(\frac{\Delta_r G^\ominus}{T}\right) = -\int_{T_{\text{ref}}}^{T} \frac{\Delta_r H^\ominus}{T^2}\mathrm{d}T$$

$$\int_{T_{\text{ref}}}^{T} \frac{\Delta_r G^\ominus(T)}{T} = \frac{\Delta_r G^\ominus(T_{\text{ref}})}{T_{\text{ref}}} - \int_{T_{\text{ref}}}^{T} \left\{\frac{\Delta_r H^\ominus(T_{\text{ref}}) + \Delta_r C_p^\ominus \times (T - T_{\text{ref}})}{T^2}\right\}\mathrm{d}T$$

$$= \frac{\Delta_r G^\ominus(T_{\text{ref}})}{T_{\text{ref}}} - \{\Delta_r H^\ominus(T_{\text{ref}}) - \Delta_r C_p^\ominus \times T_{\text{ref}}\}\int_{T_{\text{ref}}}^{T} \frac{1}{T^2}\mathrm{d}T - \Delta_r C_p^\ominus \int_{T_{\text{ref}}}^{T} \frac{1}{T}\mathrm{d}T$$

$$= \frac{\Delta_r G^\ominus(T_{\text{ref}})}{T_{\text{ref}}} + \{\Delta_r H^\ominus(T_{\text{ref}}) - \Delta_r C_p^\ominus \times T_{\text{ref}}\} \times \left\{\frac{1}{T} - \frac{1}{T_{\text{ref}}}\right\} - \Delta_r C_p^\ominus \ln\left(\frac{T}{T_{\text{ref}}}\right)$$

$$= \{170.1\ \text{J K}^{-1}\,\text{mol}^{-1}\} + \{6.561 \times 10^4\ \text{J K}^{-1}\,\text{mol}^{-1}\} \times \left\{\frac{1}{T} - \frac{1}{298.15\ \text{K}}\right\}$$

$$- \{30.865\ \text{J K}^{-1}\,\text{mol}^{-1}\} \ln\left(\frac{T}{298.15\ \text{K}}\right)$$

The temperature at which $\Delta_r G^\ominus(T) = 0$ can be found by finding the root of the right-hand side of the above equation on a scientific calculator (the numeric solver can also be used). We find the temperature to be 811 K and conclude that above $\boxed{811\ \text{K}}$ methane is unstable. Alternatively, the above relationship can be used to prepare a plot of $\Delta_r G^\ominus(T)/T$ against T. Inspection of the plot for the point at which $\Delta_r G^\ominus(T)/T = 0$ yields the temperature at which methane becomes unstable.

(b) Wien's law, which is derived in P7.12, relates the temperature to the wavelength of the most intense radiation.

$$\lambda_{\max} = \frac{hc}{5kT}$$

$$= \frac{(6.626 \times 10^{-34}\ \text{J s}) \times (2.998 \times 10^8\ \text{m s}^{-1})}{5 \times (1.381 \times 10^{-23}\ \text{J K}^{-1}) \times (1000\ \text{K})}$$

$$= \boxed{2.88\ \mu\text{m}}$$

(c) We begin by reviewing the Stefan–Boltzmann law, derived in P7.13, which relates the total energy density $\mathcal{E}$ of black-body radiation to the temperature T of the body:

$$\mathcal{E} = \int_0^\infty \rho(\lambda)\mathrm{d}\lambda = 8\pi hc \int_0^\infty \frac{\mathrm{d}\lambda}{\lambda^5(\mathrm{e}^{hc/\lambda kT} - 1)} [7.8] = \left(\frac{4}{c}\right)\sigma T^4$$

where $\sigma = \dfrac{2\pi^5 k^4}{15h^3c^2} = 5.671 \times 10^{-8}\ \text{W m}^{-2}\ \text{K}^{-4}$ is the Stefan–Boltzmann constant.

Consequently, the ratio of the energy density of a star at a 1000 K surface to the energy density of the Sun's 6000 K surface is

$$\left(\frac{T_{\text{star}}}{T_{\text{Sun}}}\right)^4 = \left(\frac{1000\ \text{K}}{6000\ \text{K}}\right)^4 = \boxed{7.72 \times 10^{-4}}$$

(d) We find the fraction of the total energy density by integration of the Planck distribution over the visible range of radiation (between about 700 nm (red) and 420 nm (violet)) followed by division by the total energy density.

$$\mathcal{E}_{\text{visible}} = \int_{420\ \text{nm}}^{700\ \text{nm}} \rho(\lambda)\,\mathrm{d}\lambda = 8\pi hc \int_{420\ \text{nm}}^{700\ \text{nm}} \frac{\mathrm{d}\lambda}{\lambda^5(\mathrm{e}^{hc/\lambda kT}-1)}\ [7.8]$$

At a high temperature the integration range is small compared to the total range of emitted wavelengths so we estimate the above integrand to be a constant equal to the value of the Planck distribution at the mid-point of the visible range (560 nm).

$$\text{Integrand} = \frac{1}{(560\times10^{-9}\ \text{m})^5 \times (\mathrm{e}^{(6.626\times10^{-34}\ \text{J s})\times(2.998\times10^{8}\ \text{m s}^{-1})/\{(560\times10^{-9}\ \text{m})\times(1.381\times10^{-23}\ \text{J K}^{-1})\times(1000\ \text{K})\}}-1)}$$
$$= 1.270 \times 10^{20}\ \text{m}^{-5}$$

$$\mathcal{E}_{\text{visible}} = 8\pi hc \times (\text{integrand}) \times \Delta\lambda$$
$$= 8\pi \times (6.626\times10^{-34}\ \text{J s}) \times (2.998\times10^{8}\ \text{m s}^{-1}) \times (1.270\times10^{20}\ \text{m}^{-5}) \times (700-420)\times10^{-9}\ \text{m}$$
$$= 1.78\times10^{-10}\ \text{J m}^{-3}$$

$$\text{Fraction of total energy density} = \frac{\mathcal{E}_{\text{visible}}}{\mathcal{E}_{\text{total}}}$$
$$= \frac{\mathcal{E}_{\text{visible}}}{\frac{4}{c}\sigma T^4}$$
$$= \frac{1.78\times10^{-10}\ \text{J m}^{-3}}{\left(\frac{4}{2.998\times10^{8}\ \text{m s}^{-1}}\right)\times(5.671\times10^{-8}\ \text{W m}^{-2}\ \text{K}^{-4})\times(1000\ \text{K})^4}$$
$$= \boxed{2.35\times10^{-7}}$$

Very little of the brown dwarf's radiation is in the visible. It doesn't shine brightly.

8 Quantum theory: techniques and applications

Answers to discussion questions

D8.1 In quantum mechanics, particles are said to have wave characteristics. The fact of the existence of the particle then requires that the wavelengths of the waves representing it be such that the wave does not experience destructive interference on reflection by a barrier or in its motion around a closed loop. This requirement restricts the wavelength to values $\lambda = 2/n \times L$, where L is the length of the path and n is a positive integer. Then, using the relationships $\lambda = h/p$ and $E = p^2/2m$, the energy is quantized at $E = n^2h^2/8mL^2$. This derivation applies specifically to the particle in a box, the derivation is similar for the particle on a ring; the same principles apply (see Section 8.6).

D8.3 The lowest energy level possible for a confined quantum-mechanical system is the zero-point energy, and zero-point energy is not zero energy. The system must have at least that minimum amount of energy even at absolute zero. The physical reason is that if the particle is confined, its position is not completely uncertain, and therefore its momentum, and hence its kinetic energy, cannot be exactly zero. The particle in a box, the harmonic oscillator, the particle on a ring or on a sphere, the hydrogen atom, and many other systems we will encounter, all have zero-point energy.

D8.5 Fermions are particles with half-integral spin, $\frac{1}{2}, \frac{3}{2}, \frac{5}{2}, \ldots$, whereas bosons have integral spin, 0, 1, 2, … . All fundamental particles that make up matter have spin $\frac{1}{2}$ and are fermions, but composite particles can be either fermions or bosons.

Fermions: electrons, protons, neutrons, ^{3}He, … .

Bosons: photons, deuterons

The most familiar consequence of the difference between fermions and bosons, the Pauli exclusion principle, is well known from introductory chemistry and explored in detail in Section 9.4(b). The principle applies to fermions and not to bosons. Because electrons are fermions, electrons in a many-electron atom cannot have identical quantum numbers. The consequences of the principle include the shell structure of atoms. Bosons, by contrast, can aggregate in groups of identical particles in identical quantum states. Such aggregations, called Bose–Einstein condensates, are more in the domain of physics than of chemistry.

Solutions to exercises

E8.1(a) $$E = \frac{n^2h^2}{8m_eL^2} \text{ [8.4a]}$$

$$\frac{h^2}{8m_eL^2} = \frac{(6.626 \times 10^{-34} \text{ J s})^2}{(8) \times (9.11 \times 10^{-31} \text{ kg}) \times (1.0 \times 10^{-9} \text{ m})^2} = 6.02 \times 10^{-20} \text{ J}$$

The conversion factors required are

$$1 \text{ eV} = 1.602 \times 10^{-19} \text{ J}; \quad 1 \text{ cm}^{-1} = 1.986 \times 10^{-23} \text{ J}; \quad 1 \text{ eV} = 96.485 \text{ kJ mol}^{-1}$$

(a) $$E_2 - E_1 = (4-1)\frac{h^2}{8m_eL^2} = \frac{3h^2}{8m_eL^2} = (3) \times (6.02 \times 10^{-20} \text{ J})$$

$$= \boxed{1.81 \times 10^{-19} \text{ J}}, \boxed{1.13 \text{ eV}}, \boxed{9100 \text{ cm}^{-1}}, \boxed{109 \text{ kJ mol}^{-1}}$$

(b) $$E_6 - E_5 = (36-25)\frac{h^2}{8m_eL^2} = \frac{11h^2}{8m_eL^2} = (11) \times (6.02 \times 10^{-20} \text{ J})$$

$$= \boxed{6.6 \times 10^{-19} \text{ J}}, \boxed{4.1 \text{ eV}}, \boxed{33\,000 \text{ cm}^{-1}}, \boxed{400 \text{ kJ mol}^{-1}}$$

COMMENT. The energy-level separations increase as n increases.

Question. For what value of n is $E_{n+1} - E_n$ for the system of this exercise equal to the ionization energy of the hydrogen atom, which is 13.6 eV?

E8.2(a) The wavefunctions are

$$\psi_n = \left(\frac{2}{L}\right)^{1/2} \sin\left(\frac{n\pi x}{L}\right) \text{ [8.4b]}$$

The required probability is

$$P = \int \psi^*\psi \, dx = \frac{2}{L}\int \sin^2\left(\frac{n\pi x}{L}\right) dx \approx \frac{2\Delta x}{L} \sin^2\left(\frac{n\pi x}{L}\right)$$

where $\Delta x = 0.02L$ and the function is evaluated at $x = 0.50L$.

(a) For $n = 1$ $$P = \left(\frac{2}{L}\right) \times 0.02L \times \sin^2\left(\frac{\pi}{2}\right) = \boxed{0.04}$$

(b) For $n = 2$ $$P = \left(\frac{2}{L}\right) \times 0.02L \times \sin^2 \pi = 0 \quad \text{so} \quad P \approx \boxed{0}$$

E8.3(a) The wavefunction for a particle in the state $n = 1$ in a square-well potential is

$$\psi_1 = \left(\frac{2}{L}\right)^{1/2} \sin\left(\frac{\pi x}{L}\right) \text{ [8.4b]}$$

and the momentum operator is

$$\hat{p} = \frac{\hbar}{\mathrm{i}}\frac{\mathrm{d}}{\mathrm{d}x}$$

so $$\langle p \rangle = \int_0^L \psi_1^* \hat{p} \psi_1 \,\mathrm{d}x = \frac{2\hbar}{\mathrm{i}L}\int_0^L \sin\left(\frac{\pi x}{L}\right)\frac{\mathrm{d}}{\mathrm{d}x}\sin\left(\frac{\pi x}{L}\right)\mathrm{d}x = \frac{2\pi\hbar}{\mathrm{i}L^2}\int_0^L \sin\left(\frac{\pi x}{L}\right)\cos\left(\frac{\pi x}{L}\right)\mathrm{d}x = \boxed{0}$$

$$\hat{p}^2 = -\hbar^2\frac{\mathrm{d}^2}{\mathrm{d}x^2}$$

$$\langle p^2 \rangle = -\frac{2\hbar^2}{L}\int_0^L \sin\left(\frac{\pi x}{L}\right)\frac{\mathrm{d}^2}{\mathrm{d}x^2}\sin\left(\frac{\pi x}{L}\right)\mathrm{d}x = \left(\frac{2\hbar^2}{L}\right)\times\left(\frac{\pi}{L}\right)^2\int_0^L \sin^2 ax\,\mathrm{d}x \quad \left[a = \frac{\pi}{L}\right]$$

$$= \left(\frac{2\hbar^2}{L}\right)\times\left(\frac{\pi}{L}\right)^2\left(\frac{1}{2}x - \frac{1}{4a}\sin 2ax\right)\Bigg|_0^L = \left(\frac{2\hbar^2}{L}\right)\times\left(\frac{\pi}{L}\right)^2\times\left(\frac{L}{2}\right) = \boxed{\frac{h^2}{4L^2}}$$

COMMENT. The expectation value of $\hat{p}$ is zero because on average the particle moves to the left as often as the right.

E8.4(a) The wavefunction is

$$\psi_n = \left(\frac{2}{L}\right)^{1/2}\sin\left(\frac{n\pi x}{L}\right) \text{ [8.4b]}$$

Hence, $$\langle x \rangle = \int \psi^* x \psi \,\mathrm{d}x = \frac{2}{L}\int_0^L x\sin^2\left(\frac{n\pi x}{L}\right)\mathrm{d}x$$

An integral table says

$$\int x\sin^2 ax\,\mathrm{d}x = \frac{x^2}{4} - \frac{x\sin 2ax}{4a} - \frac{\cos 2ax}{8a^2}$$

so $$\langle x \rangle = \frac{2}{L}\left[\frac{x^2}{4} - \frac{Lx}{4n\pi}\sin\left(\frac{2n\pi x}{L}\right) - \frac{L^2}{8(n\pi)^2}\cos\left(\frac{2n\pi x}{L}\right)\right]\Bigg|_0^L = \boxed{\frac{L}{2}} \text{ for all } n.$$

$$\langle x^2 \rangle = \frac{2}{L}\int_0^L x^2\sin^2\left(\frac{n\pi x}{L}\right)\mathrm{d}x$$

An integral table says

$$\int x^2\sin^2 ax\,\mathrm{d}x = \frac{x^3}{6} - \left(\frac{x^2}{4a} - \frac{1}{8a^3}\right)\sin 2ax - \frac{x\cos 2ax}{4a^2}$$

so $$\langle x^2 \rangle = \frac{2}{L}\left[\frac{x^3}{6} - \left(\frac{Lx^2}{4n\pi} - \frac{L^3}{(2n\pi)^3}\right)\sin\left(\frac{2n\pi x}{L}\right) - \frac{L^2 x}{(2n\pi)^2}\cos\left(\frac{2n\pi x}{L}\right)\right]\Bigg|_0^L$$

$$= \frac{2}{L}\left(\frac{L^3}{6} - \frac{L^3}{(2n\pi)^2}\right) = L^2\left(\frac{1}{3} - \frac{1}{2n^2\pi^2}\right)$$

For $n = 1$, $\langle x^2 \rangle = \boxed{L^2\left(\frac{1}{3} - \frac{1}{2\pi^2}\right)}$

E8.5(a) The zero-point energy is the ground-state energy, that is, with $n = 1$:

$$E = \frac{n^2h^2}{8m_eL^2}\,[8.4a] = \frac{h^2}{8m_eL^2}$$

Set this equal to the rest energy m_ec^2 and solve for L:

$$m_ec^2 = \frac{h^2}{8m_eL^2} \quad \text{so} \quad L = \boxed{\frac{h}{8^{1/2}m_ec} = \frac{\lambda_C}{8^{1/2}}}$$

In absolute units, the length is

$$L = \frac{6.63\times10^{-34}\,\text{J s}}{8^{1/2}\times(9.11\times10^{-31}\,\text{kg})\times(3.00\times10^{8}\,\text{m s}^{-1})} = 8.58\times10^{-13}\,\text{m} = 0.858\,\text{pm}$$

In terms of the Compton wavelength of an electron, $\lambda_C = \dfrac{h}{m_ec}$, $L = \dfrac{\lambda_C}{2\sqrt{2}}$.

E8.6(a) $$\psi_3 = \left(\frac{2}{L}\right)^{1/2}\sin\left(\frac{3\pi x}{L}\right)\,[8.4b]$$

$$P(x) \propto \psi_3^2 \propto \sin^2\left(\frac{3\pi x}{L}\right)$$

The maxima and minima in $P(x)$ correspond to $\dfrac{dP(x)}{dx} = 0$.

$$\frac{dP(x)}{dx} \propto \sin\left(\frac{3\pi x}{L}\right)\cos\left(\frac{3\pi x}{L}\right) \propto \sin\left(\frac{6\pi x}{L}\right) \quad [2\sin\alpha\cos\alpha = \sin 2\alpha]$$

$\sin\theta = 0$ when $\theta = \left(\dfrac{6\pi x}{L}\right) = n'\pi$, $n' = 0, 1, 2, \ldots$, which corresponds to $x = \dfrac{n'L}{6}$, $n' \le 6$.

$n' = 0, 2, 4$, and 6 correspond to minima in ψ_3, leaving $n' = 1, 3$ and 5 for the maxima, that is

$$x = \boxed{\frac{L}{6}, \frac{L}{2} \text{ and } \frac{5L}{6}}$$

COMMENT. Maxima in ψ^2 correspond to maxima *and* minima in ψ itself, so one can also solve this exercise by finding all points where $\dfrac{d\psi}{dx} = 0$.

E8.7(a) In the original box

$$E_1 = \frac{n^2h^2}{8mL^2}$$

In the longer box

$$E_2 = \frac{n^2h^2}{8m(1.1L)^2}$$

So $\Delta E = \frac{n^2h^2}{8mL^2}\left(\frac{1}{1.1^2} - 1\right),$

and the relative change is

$$\frac{\Delta E}{E_1} = \frac{1}{1.1^2} - 1 = -0.174 = \boxed{-17.4\%}$$

E8.8(a) The energy is

$$E_n = \frac{n^2h^2}{8mL^2} \text{ [8.4a]}$$

so the difference between neighbouring levels is

$$\Delta E_n = E_{n+1} - E_1 = \frac{\{(n+1)^2 - n^2\}h^2}{8mL^2} = \frac{(2n+1)h^2}{8mL^2}.$$

Set this difference equal to the thermal energy $kT/2$ and solve for n:

$$\frac{(2n+1)h^2}{8mL^2} = \frac{kT}{2},$$

so $n = \boxed{\frac{2kTmL^2}{h^2} - \frac{1}{2}}$

E8.9(a) $$E = \left(\upsilon + \frac{1}{2}\right)\hbar\omega, \quad \omega = \left(\frac{k_f}{m}\right)^{1/2} \text{ [8.24]}$$

The zero-point energy corresponds to $\upsilon = 0$; hence,

$$E_0 = \frac{1}{2}\hbar\omega = \frac{1}{2}\hbar\left(\frac{k_f}{m}\right)^{1/2} = \left(\frac{1}{2}\right) \times (1.055 \times 10^{-34}\text{ J s}) \times \left(\frac{155\text{ N m}^{-1}}{2.33 \times 10^{-26}\text{ kg}}\right)^{1/2}$$

$$= \boxed{4.30 \times 10^{-21}\text{ J}}$$

E8.10(a) The difference in adjacent energy levels is

$$\Delta E = E_{\upsilon+1} - E_\upsilon = \hbar\omega \text{ [8.25]} = \hbar\left(\frac{k_f}{m}\right)^{1/2} \text{ [8.24]}$$

Hence, $k_f = m\left(\frac{\Delta E}{\hbar}\right)^2 = (1.33 \times 10^{-25}\text{ kg}) \times \left(\frac{4.82 \times 10^{-21}\text{ J}}{1.055 \times 10^{-34}\text{ J s}}\right)^2 = 278\text{ kg s}^{-2} = \boxed{278\text{ N m}^{-1}}$

E8.11(a) The requirement for a transition to occur is that $\Delta E(\text{system}) = E(\text{photon})$,

so $\Delta E(\text{system}) = \hbar\omega\,[8.25] = E(\text{photon}) = h\nu = \dfrac{hc}{\lambda}$

Therefore, $\dfrac{hc}{\lambda} = \dfrac{h\omega}{2\pi} = \left(\dfrac{h}{2\pi}\right) \times \left(\dfrac{k_\mathrm{f}}{m}\right)^{1/2}$ [8.24]

$$\lambda = 2\pi c\left(\frac{m}{k_\mathrm{f}}\right)^{1/2} = (2\pi) \times (2.998 \times 10^{8}\ \mathrm{m\ s^{-1}}) \times \left(\frac{1.0078 \times 1.6605 \times 10^{-27}\ \mathrm{kg}}{855\ \mathrm{N\ m^{-1}}}\right)^{1/2}$$

$$= 2.64 \times 10^{-6}\,\mathrm{m} = \boxed{2.64\ \mu\mathrm{m}}$$

E8.12(a) The frequency of a harmonic oscillator is

$$\omega = \left(\frac{k_\mathrm{f}}{m}\right)^{1/2}.$$

^{1}H (H) and ^{2}H (D) are isotopes, so we expect that the force constant is the same in H_2 and D_2. They differ in mass. So the frequencies are inversely proportional to the square root of the mass:

$$\omega_\mathrm{D} = \omega_\mathrm{H}\left(\frac{m_\mathrm{H}}{m_\mathrm{D}}\right)^{1/2}$$

But the mass is not the mass of the molecule. We need a result from Chapter 12 to incorporate the mass (eqn 12.32). The appropriate mass for the oscillation of a diatomic molecule is

$$m_\mathrm{eff} = \frac{m_1 m_2}{m_1 + m_2} = \frac{m}{2}\ [m_1 = m_2 = m]$$

For H_2: $m_\mathrm{eff} = \dfrac{m}{2} = \dfrac{1.0078 \times (1.6605 \times 10^{-27}\ \mathrm{kg})}{2} = \boxed{8.3673 \times 10^{-28}\ \mathrm{kg}}$

For D_2: $m_\mathrm{eff} = \dfrac{m}{2} = \dfrac{2.0141 \times (1.6605 \times 10^{-27}\ \mathrm{kg})}{2} = \boxed{1.6722 \times 10^{-27}\ \mathrm{kg}}$

$$\omega_\mathrm{D} = 131.9\ \mathrm{THz} \times \left(\frac{8.3673 \times 10^{-28}\ \mathrm{kg}}{1.6722 \times 10^{-27}\ \mathrm{kg}}\right)^{1/2} = \boxed{93.3\ \mathrm{THz}}$$

E8.13(a) (a) From introductory physics, we have

$$\omega = \left(\frac{g}{l}\right)^{1/2}$$

$\Delta E = \hbar\omega$ [harmonic oscillator level separations, 8.25]

$$= (1.055 \times 10^{-34}\ \mathrm{J\ s}) \times \left(\frac{9.81\ \mathrm{m\ s^{-2}}}{1.0\ \mathrm{m}}\right)^{1/2} = \boxed{3.3 \times 10^{-34}\ \mathrm{J}}$$

(b) $\Delta E = h\nu = (6.626 \times 10^{-34}\ \mathrm{J\ Hz^{-1}}) \times (5\ \mathrm{Hz}) = \boxed{3.3 \times 10^{-33}\ \mathrm{J}}$

E8.14(a) The Schrödinger equation for the linear harmonic oscillator is

$$-\frac{\hbar^2}{2m}\frac{d^2\psi}{dx^2}+\frac{1}{2}k_f x^2\psi = E\psi \text{ [8.23]}$$

The ground-state wavefunction is

$$\psi_0 = N_0 e^{-x^2/2\alpha^2} \text{ [8.28]}$$

$$\text{with} \quad \alpha = \left(\frac{\hbar^2}{mk_f}\right)^{1/4} = \left(\frac{\hbar^2}{m^2\omega^2}\right)^{1/4}; \quad k_f = \frac{\hbar^2}{m\alpha^4} \qquad \text{(a)}$$

Take derivatives of the wavefunction

$$\frac{d\psi_0}{dx} = \left(-\frac{1}{\alpha^2}x\right)\psi_0$$

$$\frac{d^2\psi_0}{dx^2} = \left(-\frac{1}{\alpha^2}x\right)\times\left(-\frac{1}{\alpha^2}x\right)\times\psi_0 + \left(-\frac{1}{\alpha^2}\psi_0\right) = \frac{x^2}{\alpha^4}\psi_0 - \frac{1}{\alpha^2}\psi_0 = \left(\frac{x^2}{\alpha^4}-\frac{1}{\alpha^2}\right)\psi_0$$

Substitute into the Schrödinger equation,

$$-\frac{\hbar^2}{2m}\left(\frac{x^2}{\alpha^4}-\frac{1}{\alpha^2}\right)\psi_0 + \frac{1}{2}k_f x^2\psi_0 = E_0\psi_0$$

which implies

$$E_0 = \frac{-\hbar^2}{2m}\left(\frac{x^2}{\alpha^4}-\frac{1}{\alpha^2}\right)+\frac{1}{2}k_f x^2 \qquad \text{(b)}$$

But E_0 is a constant, independent of x, therefore the terms that contain x must drop out, which is possible only if

$$-\frac{\hbar^2}{2m\alpha^4}+\frac{1}{2}k_f = 0$$

which is consistent with $k_f = \dfrac{\hbar^2}{m\alpha^4}$ as in (a). What is left in (b) is

$$E_0 = \frac{\hbar^2}{2m\alpha^2} = \frac{1}{2}\hbar\omega \left[\text{using } \omega = \left(\frac{k_f}{m}\right)^{1/2} \text{ and } k_f = \frac{\hbar^2}{m\alpha^4}\right]$$

Therefore, ψ_0 is a solution of the Schrödinger equation with energy $\frac{1}{2}\hbar\omega$.

E8.15(a) The harmonic oscillator wavefunctions have the form

$$\psi_v(x) = N_v H_v(y)\exp(-\tfrac{1}{2}y^2) \text{ with } y = \frac{x}{\alpha} \text{ and } \alpha = \left(\frac{\hbar^2}{mk_f}\right)^{1/4} \text{ [8.27]}$$

The exponential function approaches zero only as x approaches $\pm\infty$, so the nodes of the wavefunction are the nodes of the Hermite polynomials.

$$H_4(y) = 16y^4 - 48y^2 + 12 = 0 \text{ [Table 8.1]}$$

Dividing through by 4 and letting $z = y^2$, we have a quadratic equation

$$4z^2 - 12z + 3 = 0$$

$$\text{so} \quad z = \frac{-b \pm \sqrt{b^2 - 4ac}}{2a} = \frac{12 \pm \sqrt{12^2 - 4 \times 4 \times 3}}{2 \times 4} = \frac{3 \pm \sqrt{6}}{2}$$

Evaluating the result numerically yields $z = 0.275$ or 2.72, so $y = \pm 0.525$ or ± 1.65, therefore $x = \boxed{\pm 0.525\alpha \text{ or } \pm 1.65\alpha}$.

COMMENT. Numerical values could also be obtained graphically by plotting $H_4(y)$.

E8.16(a) The most probable displacements are the values of x that maximize ψ^2. As noted in Exercise 8.6(a), maxima in ψ^2 correspond to maxima *and* minima in ψ itself, so one can solve this exercise by finding all points where $\frac{d\psi}{dx} = 0$. The wavefunction is

$$\psi_1 = 2N_1 y \exp(-\tfrac{1}{2}y^2) \quad \text{with} \quad y = \frac{x}{\alpha} \quad \text{and} \quad \alpha = \left(\frac{\hbar^2}{mk_f}\right)^{1/4} \text{ [8.30]}$$

$$\frac{d\psi_1}{dx} = \frac{dy}{dx}\frac{d\psi_1}{dy} = \frac{2N_1}{\alpha}\{\exp(-\tfrac{1}{2}y^2) - y^2\exp(-\tfrac{1}{2}y^2)\} = 0$$

Dividing through by constants and the exponential functions yields

$$1 - y^2 = 0 \quad \text{so} \quad y = \pm 1 \quad \text{and} \quad x = \boxed{\pm\alpha}$$

E8.17(a) The zero-point energy is

$$E_0 = \frac{1}{2}\hbar\omega \text{ [8.26]} = \frac{\hbar}{2}\left(\frac{k}{m_{\text{eff}}}\right)^{1/2} \text{ [8.24]}$$

$$m_{\text{eff}} = 34.9688 \times (1.6605 \times 10^{-27}\text{ kg})/2 = 2.9033 \times 10^{-26}\text{ kg}$$

$$\text{so} \quad E_0 = \left(\frac{1.0546 \times 10^{-34}\text{ J s}}{2}\right) \times \left(\frac{329\text{ N m}^{-1}}{2.9033 \times 10^{-26}\text{ kg}}\right)^{1/2} = \boxed{5.61 \times 10^{-21}\text{ J}}$$

E8.18(a) We require

$$\int \psi^* \psi d\tau = 1 \text{ [7.20c]}$$

That is

$$\int_0^{2\pi} N^2 e^{-im\phi} e^{im\phi} d\phi = N^2 \int_0^{2\pi} d\phi = 2\pi N^2 = 1$$

$$\text{Thus,} \quad N^2 = \frac{1}{2\pi} \quad N = \boxed{\left(\frac{1}{2\pi}\right)^{1/2}}$$

E8.19(a) The energy levels of a particle on a ring are given by eqn 8.38a:

$$E = \frac{m_l^2\hbar^2}{2I} = \frac{m_l^2\hbar^2}{2mr^2}, \quad m_l = 0, \pm1, \pm2, \ldots$$

The minimum excitation energy is the energy difference between the ground state ($m_l = 0$) and the first excited level ($m_l = \pm1$). So the minimum excitation energy is

$$\Delta E = \frac{(1^2 - 0^2)\hbar^2}{2mr^2} = \frac{(1.0546 \times 10^{-34}\ \text{J s})^2}{2 \times (1.6726 \times 10^{-27}\ \text{kg}) \times (100 \times 10^{-12}\ \text{m})^2} = \boxed{3.32 \times 10^{-22}\ \text{J}}.$$

E8.20(a) In Chapter 12, we will see that the results on rotational motion developed in this chapter apply to rigid rotors of various shapes, not just to point masses. So, we can use the quantum expression for angular momentum, $J_z = m_l\hbar$, and set it equal to the classical expression, $J_z = pr = mvr$. Solving for m_l yields

$$m_l = \frac{mvr}{\hbar} = \frac{(1.0\ \text{kg}) \times (20 \times 10^3\ \text{m h}^{-1}) \times (60 \times 10^{-2}\ \text{m})}{1.0546 \times 10^{-34}\ \text{J s}} \times \frac{1\ \text{h}}{3600\ \text{s}} = \boxed{3.2 \times 10^{34}}$$

E8.21(a) In Chapter 12, we will see that the results on rotational motion developed in this chapter apply to rigid rotors of various shapes, not just to point masses. The rotor in this case is not a ring-like object (as in Exercise 8.20(a)), so we use results derived for rotation in three dimensions. The energy levels are

$$E = \frac{l(l+1)\hbar^2}{2I}, \quad l = 0, 1, 2, \ldots\ [8.53]$$

The minimum energy to start it rotating is the minimum excitation energy, the energy to take it from the motionless $l = 0$ to the rotating $l = 1$ state:

$$\Delta E = E_1 = \frac{1 \times 2 \times (1.0546 \times 10^{-34}\ \text{J s})^2}{2 \times (5.27 \times 10^{-47}\ \text{kg m}^2)} = \boxed{2.11 \times 10^{-22}\ \text{J}}$$

E8.22(a) The energy levels are

$$E = \frac{l(l+1)\hbar^2}{2I}, \quad l = 0, 1, 2, \ldots\ [8.53]$$

So the excitation energy is

$$\Delta E = E_2 - E_1 = \frac{(2 \times 3 - 1 \times 2) \times (1.0546 \times 10^{-34}\ \text{J s})^2}{2 \times (5.27 \times 10^{-47}\ \text{kg m}^2)} = \boxed{4.22 \times 10^{-22}\ \text{J}}$$

E8.23(a) The energy levels are

$$E = \frac{l(l+1)\hbar^2}{2I}, \quad l = 0, 1, 2, \ldots\ [8.53]$$

So the minimum energy *allowed* for this system is zero—but that corresponds to rest, not rotation. So the minimum energy of rotation occurs for the state that has $l = 1$. The angular momentum in that state is

$$J = \{l(l+1)\}^{1/2}\hbar\ [8.54a] = \boxed{2^{1/2}\hbar} = 2^{1/2} \times (1.0546 \times 10^{-34}\ \mathrm{J\ s}) = \boxed{1.49 \times 10^{-34}\ \mathrm{J\ s}}.$$

COMMENT. Note that the moment of inertia does not enter into the result. Thus, the minimum angular momentum is the same for a molecule of CH_4 as for a molecule of C_{60} as for a football.

E8.24(a) The diagrams are drawn by forming a vector of length $\{j(j+1)\}^{1/2}$, with $j = s$ or l as appropriate, and with a projection m_j on the z-axis (see Figure 8.1). Each vector represents the edge of a cone around the z-axis (that for $m_j = 0$ represents the side view of a disk perpendicular to z).

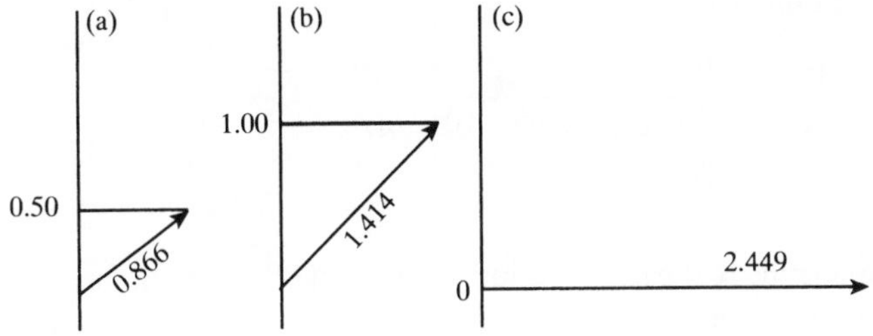

Figure 8.1

Solutions to problems

Solutions to numerical problems

P8.1 $$E = \frac{n^2h^2}{8mL^2}\ [8.4a], \quad E_2 - E_1 = \frac{3h^2}{8mL^2}$$

For O_2, $m = 32.00 \times (1.6605 \times 10^{-27}\ \mathrm{kg}) = 5.314 \times 10^{-26}\ \mathrm{kg}$,

so $$E_2 - E_1 = \frac{(3) \times (6.626 \times 10^{-34}\ \mathrm{J\ s})^2}{(8) \times (5.314 \times 10^{-26}\ \mathrm{kg}) \times (5.0 \times 10^{-2}\ \mathrm{m})^2} = \boxed{1.24 \times 10^{-39}\ \mathrm{J}}$$

We set $E = \frac{n^2h^2}{8mL^2} = \frac{kT}{2}$ and solve for n:

$$n = \frac{2L(kTm)^{1/2}}{h}$$

$$= \frac{2 \times (5.0 \times 10^{-2}\ \mathrm{m}) \times \{(1.381 \times 10^{-23}\ \mathrm{J\ K^{-1}}) \times (300\ \mathrm{K}) \times (5.314 \times 10^{-26}\ \mathrm{kg})\}^{1/2}}{6.626 \times 10^{-34}\ \mathrm{J\ s}}$$

$$= \boxed{2.2 \times 10^9}$$

At this level,

$$E_n - E_{n-1} = \{n^2 - (n-1)^2\} \times \frac{h^2}{8mL^2}$$

$$= (2n-1) \times \frac{h^2}{8mL^2} \approx (2n) \times \frac{h^2}{8mL^2}$$

From above, $\dfrac{h^2}{8mL^2} = \dfrac{E_2 - E_1}{3} = 4.13 \times 10^{-40}$ J,

so $E_n - E_{n-1} = 2 \times (2.2 \times 10^9) \times (4.13 \times 10^{-40}\,\text{J}) \approx \boxed{1.8 \times 10^{-30}\,\text{J}}$ [or 1.1 μJ mol⁻¹]

P8.3 A particle on a ring is our model for rotational motion in a plane.

$$J_l = m_l\hbar\ [8.37] \quad \text{and} \quad E = \frac{m_l^2\hbar^2}{2I}\ [8.38\text{a}] = \frac{m_l^2\hbar^2}{2mr^2} \quad [I = mr^2]$$

The lowest non-zero energy occurs when $m_l = \pm 1$:

$$E_1 = \frac{\hbar^2}{2mr^2} = \frac{(1.055 \times 10^{-34}\,\text{J s})^2}{(2) \times (1.008) \times (1.6605 \times 10^{-27}\,\text{kg}) \times (160 \times 10^{-12}\,\text{m})^2} = \boxed{1.30 \times 10^{-22}\,\text{J}}$$

The minimum angular momentum is $J_l = \boxed{\hbar}$.

P8.5 Mathematical software can animate the real part or the imaginary part of $\Psi(\phi,t)$, or you may wish to have it display $|\Psi(\phi,t)|^2$. Try a 'pure' state, that is, let $c = 1$ for one value of m_l and 0 for all others. This 'packet' does not spread, but only circulates. Also, try making all the coefficients in the sum equal (all 1, for example). Whatever your choice of coefficients, the pattern will repeat with a period T that makes all the time-dependent factors equal to the exponential of (2πi × an integer):

$$T = \frac{4\pi I}{\hbar}$$

making the exponent $\dfrac{\mathrm{i}E_{m_l}t}{\hbar}$ equal to $2\pi \mathrm{i} m_l^2$ when $t = T$ and at intervals of T thereafter. (See Problem 8.19.) An example of this approach using Mathcad is illustrated below:

Wavepacket on a Ring as a MathCad Document. Let $\tau = \dfrac{h \cdot t}{4 \cdot \pi \cdot I}$

and let each function in the superposition of $m + 1$ functions contribute with equal probability (i.e. $1/m + 1$).

The normalized angular momentum functions are:

$$\psi(m,\phi) := \left(\frac{1}{2 \cdot \pi}\right)^{1/2} \cdot e^{i \cdot m \cdot \phi} \text{ [8.38b] where } m \text{ is an integer.}$$

The normalized superposition is

$$\Psi(m_{max},\phi,\tau) := \left(\frac{1}{m+1}\right)^{1/2} \cdot \sum_{m=0}^{m_{max}} \psi(m,\phi) \cdot e^{-i \cdot m^2 \cdot \tau}$$

$N := 500 \quad j := 0..N \quad \phi_j := \dfrac{2 \cdot \pi \cdot j}{N} \quad m_{max} := 8 \quad \Delta\tau := 0.03$

The probability density of the superposition is $P(\phi,\tau) := \Psi(m_{max},\phi,\tau) \cdot \bar{\Psi}(m_{max},\phi,\tau)$

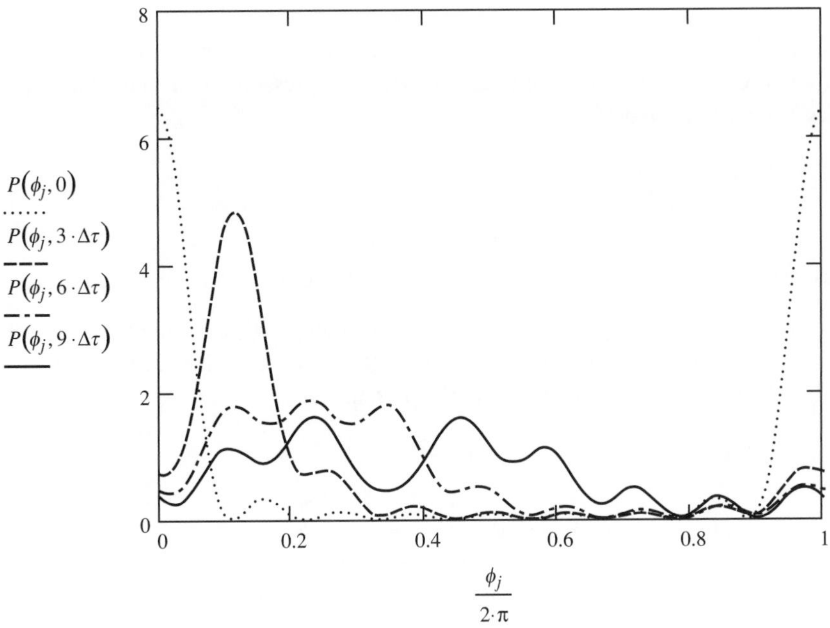

Figure 8.2

The above plots (Figure 8.2) show that as the initially localized wave propagates around the ring it spreads with time and the uncertainty in knowing particle position increases. The effect of increasing or decreasing the energies accessible to the particle may be explored by increasing or decreasing the value of m_{max} in the MathCad document.

Solutions to theoretical problems

P8.7 The energy of any given molecule is

$$E = \frac{n^2h^2}{8mL^2} \text{ [8.11b with } n^2 = n_1^2 + n_2^2 + n_3^2 \text{ and equal lengths]}$$

(The lowest energy level is $n_1 = n_2 = n_3 = 1$, so $n^2 = 3$; however, what follows applies to any allowed energy level.) So, the internal energy of a sample of N molecules is

$$U = NE = \frac{Nn^2h^2}{8mL^2} = \frac{Nn^2h^2}{8mV^{2/3}}$$

In the last step we used $V = L^3$ because we are interested in how the energy changes with volume. Consider an adiabatic change of volume, that is, a change in which no heat enters or leaves the sample. In that case, the change in energy is entirely work (first law with $q = 0$). Differentiate the expression for U:

$$\mathrm{d}w = \left(\frac{\partial U}{\partial V}\right)_{\text{adiabatic}} \mathrm{d}V = -\frac{Nn^2h^2}{12mV^{5/3}}\mathrm{d}V \qquad \text{(a)}$$

Expansion work has the form $\mathrm{d}w = -p_{ex}\mathrm{d}V$ [2.6a]. Can these expressions be reconciled and, if so, under what conditions? First, note that the expression that multiplies $\mathrm{d}V$ in equation (a) refers to the sample, so if it is some sort of pressure, it must be the sample pressure, and not an arbitrary external pressure. Hence, if the expressions can be reconciled, it must be for *reversible* adiabatic expansion or compression (in which the internal and external pressures are equal). The expression that multiplies $\mathrm{d}V$ can be expressed as

$$\frac{Nn^2h^2}{12mV^{5/3}} = \frac{2}{3}\frac{N}{V}E.$$

In fact, the kinetic model of gases (Chapter 20) says that the pressure of a gas is equal to $\frac{2}{3}\frac{N}{V}E$ where E is the average kinetic energy of the gas molecules—completely consistent with interpreting it as the average particle-in-a-box energy. To summarize, reversible adiabatic work for a gas of particle-in-a-box molecules is $\mathrm{d}w = -p\mathrm{d}V$, where the pressure is

$$p = \frac{Nn^2h^2}{12mL^5} = \frac{2}{3}\frac{N}{V}E$$

In expansion, the volume increases, meaning that the box gets bigger. Equation 8.11b tells us that the kinetic energy decreases, even as the quantum numbers remain constant. This is also consistent with what we know of adiabatic expansion and the kinetic model of gases: the temperature of the sample drops on expansion, and temperature is related to the kinetic energy ($T^2 \propto E$).

In isothermal expansion, energy must enter the system as heat to maintain the temperature. We can interpret this influx of heat as an increase in quantum numbers (an excitation of the molecules) that offsets the falling energy levels.

P8.9 (a) The wavefunctions in each region (see Figure 8.3(a)) are (eqns 8.13, 8.15, and 8.16):

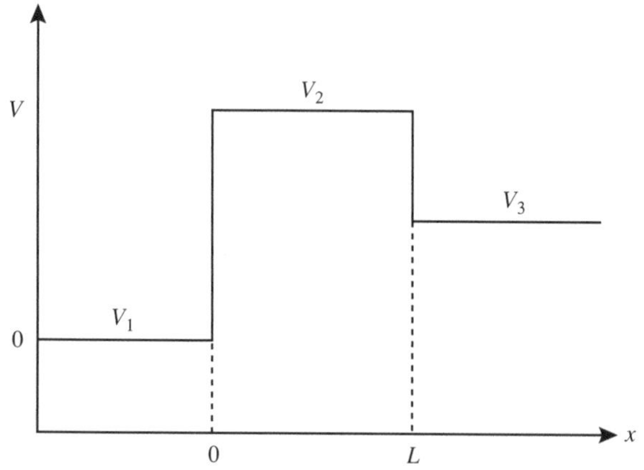

Figure 8.3(a)

$$\psi_1(x) = e^{ik_1x} + B_1e^{-ik_1x}$$

$$\psi_2(x) = A_2e^{k_2x} + B_2e^{-k_2x}$$

$$\psi_3(x) = A_3e^{ik_3x}$$

With the above choice of $A_1 = 1$ the transmission probability is simply $T = |A_3|^2$. The wavefunction coefficients are determined by the criteria that both the wavefunctions and their first derivatives with respect to x be continuous at potential boundaries

$$\psi_1(0)=\psi_2(0) \quad \text{and} \quad \psi_2(L)=\psi_3(L)$$

$$\frac{d\psi_1(0)}{dx}=\frac{d\psi_2(0)}{dx} \quad \text{and} \quad \frac{d\psi_2(L)}{dx}=\frac{d\psi_3(L)}{dx}$$

These criteria establish the algebraic relationships:

$$1+B_1-A_2-B_2=0$$
$$(-ik_1-k_2)A_2+(-ik_1+k_2)B_2+2ik_1=0$$
$$A_2e^{k_2L}+B_2e^{-k_2L}-A_3e^{ik_3L}=0$$
$$A_2k_2e^{k_2L}-B_2k_2e^{-k_2L}-iA_3k_3e^{ik_3L}=0$$

Solving the simultaneous equations for A_3 gives

$$A_3=\frac{4k_1k_2\,e^{ik_3L}}{(ia+b)e^{k_2L}-(ia-b)e^{-k_2L}}$$

where $a=k_2^2-k_1k_3$ and $b=k_1k_2+k_2k_3$

Since $\sinh(z)=(e^z-e^{-z})/2$ or $e^z=2\sinh(z)+e^{-z}$, substitute $e^{k_2L}=2\sinh(k_2L)+e^{-k_2L}$ giving

$$A_3=\frac{2k_1k_2e^{ik_3L}}{(ia+b)\sinh(k_2L)+b\,e^{-k_2L}}$$

the transmission coefficient is

$$T=|A_3|^2=A_3\times A_3^*=\frac{4k_1^2k_2^2}{(a^2+b^2)\sinh^2(k_2L)+b^2}$$

where $a^2+b^2=(k_1^2+k_2^2)(k_2^2+k_3^2)$ and $b^2=k_2^2(k_1+k_3)^2$

(b) In the special case for which $V_1=V_3=0$, eqns 8.13 and 8.16 require that $k_1=k_3$. Additionally,

$$\left(\frac{k_1}{k_2}\right)^2=\frac{E}{V_2-E}=\frac{\varepsilon}{1-\varepsilon}, \quad \text{where} \quad \varepsilon=E/V_2.$$

$$a^2+b^2=(k_1^2+k_2^2)^2=k_2^4\left\{1+\left(\frac{k_1}{k_2}\right)^2\right\}^2$$

$$b^2=4k_1^2k_2^2$$

$$\frac{a^2+b^2}{b^2}=\frac{k_2^2\left\{1+\left(\frac{k_1}{k_2}\right)^2\right\}^2}{4k_1^2}=\frac{1}{4\varepsilon(1-\varepsilon)}$$

$$T=\frac{b^2}{b^2+(a^2+b^2)\sinh^2(k_2L)}=\frac{1}{1+\left(\frac{a^2+b^2}{b^2}\right)\sinh^2(k_2L)}$$

$$=\left\{1+\frac{\sinh^2(k_2L)}{4\varepsilon(1-\varepsilon)}\right\}^{-1}=\left\{1+\frac{(e^{k_2L}-e^{-k_2L})^2}{16\varepsilon(1-\varepsilon)}\right\}^{-1}$$

This proves eqn 8.19a, where $V_1 = V_3 = 0$.

In the limit of a high and wide barrier, $k_2L \gg 1$. This implies both that e^{-k_2L} is negligibly small compared to e^{k_2L} and that 1 is negligibly small compared to $e^{2k_2L}/\{16\varepsilon(1-\varepsilon)\}$. The previous equation simplifies to

$$T = 16\varepsilon(1-\varepsilon)e^{-2k_2L} \text{ [8.19b]}$$

(c) The specified graph is shown in Figure 8.3(b).

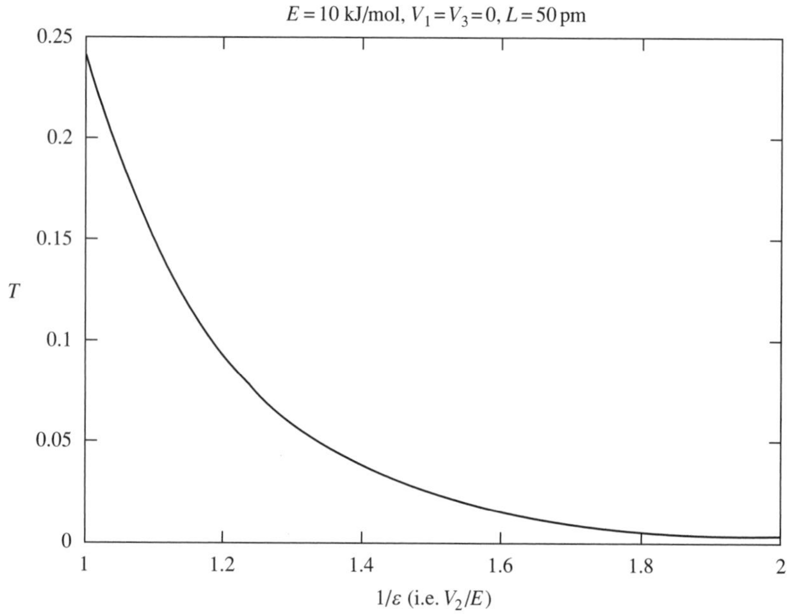

Figure 8.3(b)

P8.11 The Schrödinger equation is $-\dfrac{\hbar^2}{2m}\dfrac{d^2\psi}{dx^2} + \tfrac{1}{2}kx^2\psi = E\psi$.

We try $\psi = e^{-gx^2}$, so $\dfrac{d\psi}{dx} = -2gxe^{-gx^2}$

and $\dfrac{d^2\psi}{dx^2} = -2ge^{-gx^2} + 4g^2x^2e^{-gx^2} = -2g\psi + 4g^2x^2\psi$.

Inserting into the Schrödinger equation yields

$$\left(\frac{\hbar^2 g}{m}\right)\psi - \left(\frac{2\hbar^2 g^2}{m}\right)x^2\psi + \tfrac{1}{2}kx^2\psi = E\psi$$

$$\left(\frac{\hbar^2 g}{m} - E\right)\psi + \left(\tfrac{1}{2}k - \frac{2\hbar^2 g^2}{m}\right)x^2\psi = 0$$

This equation is satisfied if

$$E = \frac{\hbar^2 g}{m} \quad \text{and} \quad 2\hbar^2 g^2 = \tfrac{1}{2}mk, \quad \text{or} \quad \boxed{g = \tfrac{1}{2}\left(\frac{mk}{\hbar^2}\right)^{1/2}}$$

Therefore,

$$E = \tfrac{1}{2}\hbar\left(\frac{k}{m}\right)^{1/2} = \tfrac{1}{2}\hbar\omega \quad \text{if } \omega = \left(\frac{k}{m}\right)^{1/2}$$

P8.13

$$\langle x^n\rangle = \alpha^n\langle y^n\rangle = \alpha^n\int_{-\infty}^{+\infty}\psi y^n\psi\,\mathrm{d}x = \alpha^{n+1}\int_{-\infty}^{+\infty}\psi^2 y^n\,\mathrm{d}y \quad [x = \alpha y]$$

$$\langle x^3\rangle \propto \int_{-\infty}^{+\infty}\psi^2 y^3\,\mathrm{d}y = \boxed{0} \text{ by symmetry} \quad [y^3 \text{ is an odd function of } y]$$

$$\langle x^4\rangle = \alpha^5\int_{-\infty}^{+\infty}\psi y^4\psi\,\mathrm{d}y$$

$$y^4\psi = y^4 N H_v \mathrm{e}^{-y^2/2}$$

$$\begin{aligned}
y^4 H_v &= y^3(\tfrac{1}{2}H_{v+1} + vH_{v-1}) = y^2[\tfrac{1}{2}(\tfrac{1}{2}H_{v+2} + (v+1)H_v) + v(\tfrac{1}{2}H_v + (v-1)H_{v-2})] \\
&= y^2[\tfrac{1}{4}H_{v+2} + (v+\tfrac{1}{2})H_v + v(v-1)H_{v-2}] \\
&= y[\tfrac{1}{4}(\tfrac{1}{2}H_{v+3} + (v+2)H_{v+1}) + (v+\tfrac{1}{2})\times(\tfrac{1}{2}H_{v+1} + vH_{v-1}) + v(v-1)\times(\tfrac{1}{2}H_{v-1} + (v-2)H_{v-3})] \\
&= y(\tfrac{1}{8}H_{v+3} + \tfrac{3}{4}(v+1)H_{v+1} + \tfrac{3}{2}v^2H_{v-1} + v(v-1)\times(v-2)H_{v-3})
\end{aligned}$$

Only yH_{v+1} and yH_{v-1} lead to H_v and contribute to the expectation value (since H_v is orthogonal to all except H_v) [Table 8.1]; hence

$$\begin{aligned}
y^4 H_v &= \tfrac{3}{4}y\{(v+1)H_{v+1} + 2v^2H_{v-1}\} + \cdots \\
&= \tfrac{3}{4}[(v+1)(\tfrac{1}{2}H_{v+2} + (v+1)H_v) + 2v^2(\tfrac{1}{2}H_v + (v-1)H_{v-2})] + \cdots \\
&= \tfrac{3}{4}\{(v+1)^2H_v + v^2H_v\} + \cdots \\
&= \tfrac{3}{4}(2v^2 + 2v + 1)H_v + \cdots
\end{aligned}$$

Therefore,

$$\int_{-\infty}^{+\infty}\psi y^4\psi\,\mathrm{d}y = \tfrac{3}{4}(2v^2 + 2v + 1)N^2\int_{-\infty}^{+\infty}H_v^2\mathrm{e}^{-y^2}\,\mathrm{d}y = \frac{3}{4\alpha}(2v^2 + 2v + 1)$$

and so

$$\langle x^4\rangle = (\alpha^5)\times\left(\frac{3}{4\alpha}\right)\times(2v^2 + 2v + 1) = \boxed{\tfrac{3}{4}(2v^2 + 2v + 1)\alpha^4}$$

P8.15 (a) The turning points in terms of the displacement x are

$$x_{\mathrm{tp}} = \pm\left(\frac{2E}{k}\right)^{1/2} = \pm\left(\frac{2(v+\tfrac{1}{2})\hbar\omega}{k}\right)^{1/2} = \pm\left(\frac{(2v+1)\hbar}{(km)^{1/2}}\right)^{1/2} [8.24] = \pm(2v+1)^{1/2}\alpha\ [8.27]$$

In terms of the dimensionless variable $y = x/\alpha$, $y_{\mathrm{tp}} = \pm(2v+1)^{1/2}$.

(b) The probability of extension beyond the classical turning point is

$$P = \int_{|x_{tp}|}^{\infty} \psi_v^2 \mathrm{d}x = \alpha N_v^2 \int_{|y_{tp}|}^{\infty} \{H_v(y)\}^2 \mathrm{e}^{-y^2} \mathrm{d}y$$

For $v = 0$, $H_0(y) = 1$, so $\psi_0 = N_0 \mathrm{e}^{-y^2/2}$, where $N_0 = \left(\dfrac{1}{2^v \alpha \pi^{1/2} v!}\right)^{1/2} = \left(\dfrac{1}{\alpha \pi^{1/2}}\right)^{1/2}$;

the lower limit of the integral is $|y_{tp}| = \pm(2v+1)^{1/2} = 1$.

Hence, $P = \alpha N_0^2 \displaystyle\int_1^{\infty} \mathrm{e}^{-y^2} \mathrm{d}y = \frac{1}{\pi^{1/2}} \int_1^{\infty} \mathrm{e}^{-y^2} \mathrm{d}y$

Comparison of this expression to the definition of the error function, erf 1, yields

$$\mathrm{erf}\ 1 = 1 - \frac{2}{\pi^{1/2}} \int_1^{\infty} \mathrm{e}^{-y^2} \mathrm{d}y = 1 - 2P,$$

so $P = (1 - \mathrm{erf}\ 1)/2 = (1 - 0.842701)/2 = \boxed{0.0786}$.

COMMENT. Note that molecular parameters such as m and k do not enter into the calculation.

P8.17 Call the integral I:

$$I = \int \psi_{v'} x \psi_v \mathrm{d}x = \alpha^2 \int \psi_{v'} y \psi_v \mathrm{d}y \quad [x = \alpha y]$$

$$y\psi_v = N_v \left(\frac{H_{v+1}}{2} + vH_{v-1}\right) \mathrm{e}^{-y^2/2} \text{ [Table 8.1]}$$

Hence, $I = \alpha^2 N_v N_{v'} \displaystyle\int \left(\frac{1}{2} H_{v'} H_{v+1} + vH_{v'} H_{v-1}\right) \mathrm{e}^{-y^2} \mathrm{d}y = 0$ unless $v' = v \pm 1$ [Table 8.1]

For $v' = v + 1$

$$I = \frac{\alpha^2 N_v N_{v+1}}{2} \int H_{v+1}^2 \mathrm{e}^{-y^2} \mathrm{d}y = \frac{\alpha^2 N_v N_{v+1}}{2} \pi^{1/2} 2^{v+1} (v+1)! = \boxed{\alpha \left(\frac{v+1}{2}\right)^{1/2}}$$

For $v' = v - 1$

$$I = v\alpha^2 N_v N_{v-1} \int H_{v-1}^2 \mathrm{e}^{-y^2} \mathrm{d}y = v\alpha^2 N_v N_{v-1} \pi^{1/2} 2^{v-1} (v-1)! = \boxed{\alpha \left(\frac{v}{2}\right)^{1/2}}$$

No other values of v' result in a non-zero value for I, hence no other transitions are allowed.

P8.19 To address this time-dependent problem, we need a time-dependent wavefunction, made up from solutions of the time-dependent Schrödinger equation

$$\hat{H}\Psi(x,t) = \mathrm{i}\hbar \frac{\partial \Psi(x,t)}{\partial t} \text{ [Table 7.1]}$$

If $\psi(x)$ is an eigenfunction of the energy operator with energy eigenvalue E, then

$$\Psi(x,t) = \psi(x)e^{-iEt/\hbar}$$

is a solution of the time-dependent Schrödinger equation (provided the energy operator is not itself time dependent). To verify this, evaluate both sides of the time-dependent Schrödinger equation. On the left we have

$$\hat{H}\Psi(x,t) = \hat{H}\psi(x)e^{-iEt/\hbar} = E\psi(x)e^{-iEt/\hbar} = E\Psi(x,t)$$

On the right we have

$$i\hbar\frac{\partial\Psi(x,t)}{\partial t} = i\hbar\psi(x)\frac{\partial}{\partial t}e^{-iEt/\hbar} = -i^2E\psi(x)e^{-iEt/\hbar} = E\Psi(x,t)$$

the same as on the left. Our wavepacket is an arbitrary superposition of time-evolving harmonic oscillator states,

$$\Psi(x,t) = \sum_{v=0} c_v\psi_v(x)e^{-iE_vt/\hbar}$$

where $\psi_v(x)$ are time-independent harmonic-oscillator wavefunctions and

$$E_v = (v + \tfrac{1}{2})\hbar\omega \; [8.24]$$

Hence, the wavepacket is

$$\Psi(x,t) = e^{-i\omega t/2}\sum_{v=0} c_v\psi_v(x)e^{-iv\omega t}$$

The angular frequency ω is related to the period T by $T = 2\pi/\omega$, so we can evaluate the wavepacket at any whole number of periods after t, that is at a time $t + nT$, where n is any integer. (n is not a quantum number.) Note that

$$t + nT = t + 2\pi n/\omega,$$

so $$\Psi(x,t+nT) = e^{-i\omega t/2}e^{-i\omega nT/2}\sum_{v=0} c_v\psi_v(x)e^{-iv\omega t}e^{-iv\omega nT} = e^{-i\omega t/2}e^{-i\pi n}\sum_{v=0} c_v\psi_v(x)e^{-iv\omega t}e^{-2\pi ivn}$$

Since the exponential of ($2\pi i \times$ any integer) = 1, we note that the last factor inside the sum is 1 for every state. Also, since $e^{-in\pi} = (-1)^n$, we have

$$\Psi(x,t+nT) = (-1)^n\Psi(x,t)$$

At any whole number of periods after time t, the wavefunction is either the same as at time t or -1 times its value at time t. Either way, $|\Psi|^2$ returns to its original value each period, so the wavepacket returns to the same spatial distribution each period.

P8.21 In each case, if the function is an eigenfunction of the operator, the eigenvalue is also the expectation value; if it is not an eigenfunction we form

$$\langle\Omega\rangle = \int \psi^*\hat{\Omega}\psi\,d\tau \; [7.37]$$

(a) $$\hat{l}_z e^{i\phi} = \frac{\hbar}{i}\frac{d}{d\phi}e^{i\phi} \; [8.46] = \hbar e^{i\phi}; \quad \text{hence} \quad l_z = \boxed{+\hbar}$$

(b) $$\hat{l}_z e^{-2i\phi} = \frac{\hbar}{i}\frac{d}{d\phi}e^{-2i\phi} = -2\hbar e^{-2i\phi}; \quad \text{hence} \quad l_z = \boxed{-2\hbar}$$

(c) $\langle l_z \rangle \propto \int_0^{2\pi} \cos\phi \left(\frac{\hbar}{\mathrm{i}} \frac{\mathrm{d}}{\mathrm{d}\phi} \cos\phi \right) \mathrm{d}\phi \propto -\frac{\hbar}{\mathrm{i}} \int_0^{2\pi} \cos\phi \sin\phi \, \mathrm{d}\phi = \boxed{0}$

(d) $\langle l_z \rangle = N^2 \int_0^{2\pi} (\cos\chi \mathrm{e}^{\mathrm{i}\phi} + \sin\chi \mathrm{e}^{-\mathrm{i}\phi})^* \left(\frac{\hbar}{\mathrm{i}} \frac{\mathrm{d}}{\mathrm{d}\phi} \right) \times (\cos\chi \mathrm{e}^{\mathrm{i}\phi} + \sin\chi \mathrm{e}^{-\mathrm{i}\phi}) \mathrm{d}\phi$

$$= \frac{\hbar}{\mathrm{i}} N^2 \int_0^{2\pi} (\cos\chi \mathrm{e}^{-\mathrm{i}\phi} + \sin\chi \mathrm{e}^{\mathrm{i}\phi}) \times (\mathrm{i}\cos\chi \mathrm{e}^{\mathrm{i}\phi} - \mathrm{i}\sin\chi \mathrm{e}^{-\mathrm{i}\phi}) \mathrm{d}\phi$$

$$= \hbar N^2 \int_0^{2\pi} (\cos^2\chi - \sin^2\chi + \cos\chi \sin\chi [\mathrm{e}^{2\mathrm{i}\phi} - \mathrm{e}^{-2\mathrm{i}\phi}]) \mathrm{d}\phi$$

$$= \hbar N^2 (\cos^2\chi - \sin^2\chi) \times (2\pi) = 2\pi\hbar N^2 \cos 2\chi$$

We must evaluate the normalization constant:

$$N^2 \int_0^{2\pi} (\cos\chi \mathrm{e}^{\mathrm{i}\phi} + \sin\chi \mathrm{e}^{-\mathrm{i}\phi})^* (\cos\chi \mathrm{e}^{\mathrm{i}\phi} + \sin\chi \mathrm{e}^{-\mathrm{i}\phi}) \mathrm{d}\phi = 1$$

$$1 = N^2 \int_0^{2\pi} (\cos^2\chi + \sin^2\chi + \cos\chi \sin\chi [\mathrm{e}^{2\mathrm{i}\phi} + \mathrm{e}^{-2\mathrm{i}\phi}]) \mathrm{d}\phi$$

$$= 2\pi N^2 (\cos^2\chi + \sin^2\chi) = 2\pi N^2 \quad \text{so } N^2 = \frac{1}{2\pi}$$

Therefore

$$\langle l_z \rangle = \boxed{\hbar \cos 2\chi} \; [\chi \text{ is a parameter}]$$

For the kinetic energy we use $\hat{E}_{\mathrm{K}} = \frac{\hat{J}_z^2}{2I}$ [8.36] $= -\frac{\hbar^2}{2I} \frac{\mathrm{d}^2}{\mathrm{d}\phi^2}$ [8.40]

(a) $\hat{E}_{\mathrm{K}} \mathrm{e}^{\mathrm{i}\phi} = -\frac{\hbar^2}{2I} (\mathrm{i}^2 \mathrm{e}^{\mathrm{i}\phi}) = \frac{\hbar^2}{2I} \mathrm{e}^{\mathrm{i}\phi}$; hence $\langle E_{\mathrm{K}} \rangle = \boxed{\frac{\hbar^2}{2I}}$

(b) $\hat{E}_{\mathrm{K}} \mathrm{e}^{-2\mathrm{i}\phi} = -\frac{\hbar^2}{2I} (2\mathrm{i})^2 \mathrm{e}^{-2\mathrm{i}\phi} = \frac{4\hbar^2}{2I} \mathrm{e}^{-2\mathrm{i}\phi}$; hence $\langle E_{\mathrm{K}} \rangle = \boxed{\frac{2\hbar^2}{I}}$

(c) $\hat{E}_{\mathrm{K}} \cos\phi = -\frac{\hbar^2}{2I} (-\cos\phi) = \frac{\hbar^2}{2I} \cos\phi$; hence $\langle E_{\mathrm{K}} \rangle = \boxed{\frac{\hbar^2}{2I}}$

(d) $\hat{E}_{\mathrm{K}} (\cos\chi \mathrm{e}^{\mathrm{i}\phi} + \sin\chi \mathrm{e}^{-\mathrm{i}\phi}) = -\frac{\hbar^2}{2I} (-\cos\chi \mathrm{e}^{\mathrm{i}\phi} - \sin\chi \mathrm{e}^{-\mathrm{i}\phi}) = \frac{\hbar^2}{2I} (\cos\chi \mathrm{e}^{\mathrm{i}\phi} + \sin\chi \mathrm{e}^{-\mathrm{i}\phi})$

and hence $\langle E_{\mathrm{K}} \rangle = \boxed{\frac{\hbar^2}{2I}}$

COMMENT. All of these functions are eigenfunctions of the kinetic energy operator, which is also the total energy or Hamiltonian operator, since the potential energy is zero for this system.

P8.23 The Schrödinger equation is

$$-\frac{\hbar^2}{2m} \nabla^2 \psi = E\psi \; [8.49]$$

$$\nabla^2\psi = \frac{1}{r}\frac{\partial^2(r\psi)}{\partial r^2} + \frac{1}{r^2}\Lambda^2\psi \text{ [Table 7.1]}$$

since r = constant, the first term is eliminated and the Schrödinger equation may be rewritten

$$-\frac{\hbar^2}{2mr^2}\Lambda^2\psi = E\psi \quad \text{or} \quad -\frac{\hbar^2}{2I}\Lambda^2\psi = E\psi\ [I = mr^2] \quad \text{or} \quad \Lambda^2\psi = -\frac{2IE\psi}{\hbar^2}$$

where $\Lambda^2 = \dfrac{1}{\sin^2\theta}\dfrac{\partial^2}{\partial\phi^2} + \dfrac{1}{\sin\theta}\dfrac{\partial}{\partial\theta}\sin\theta\dfrac{\partial}{\partial\theta}$

Now use the specified $\psi = Y_{l,m_l}$ from Table 8.2, and see if they satisfy this equation.

(a) Because $Y_{0,0}$ is a constant, all derivatives with respect to angles are zero, so $\Lambda^2 Y_{0,0} = \boxed{0}$, implying that $E = \boxed{0}$ and angular momentum $= \boxed{0}$ [from $\{l(l+1)\}^{1/2}\hbar$].

(b) $\Lambda^2 Y_{2,-1} = \dfrac{1}{\sin^2\theta}\dfrac{\partial^2 Y_{2,-1}}{\partial\phi^2} + \dfrac{1}{\sin\theta}\dfrac{\partial}{\partial\theta}\sin\theta\dfrac{\partial Y_{2,-1}}{\partial\theta}$, where $Y_{2,-1} = N\cos\theta\sin\theta\, e^{-i\phi}$

$$\frac{\partial Y_{2,-1}}{\partial\theta} = Ne^{-i\phi}(\cos^2\theta - \sin^2\theta)$$

$$\frac{1}{\sin\theta}\frac{\partial}{\partial\theta}\sin\theta\frac{\partial Y_{2,-1}}{\partial\theta} = \frac{1}{\sin\theta}\frac{\partial}{\partial\theta}\sin\theta\, Ne^{-i\phi}(\cos^2\theta - \sin^2\theta)$$

$$= \frac{Ne^{-i\phi}}{\sin\theta}(\sin\theta(-4\cos\theta\sin\theta) + \cos\theta(\cos^2\theta - \sin^2\theta))$$

$$= Ne^{-i\phi}\left(-6\cos\theta\sin\theta + \frac{\cos\theta}{\sin\theta}\right)[\cos^3\theta = \cos\theta(1-\sin^2\theta)]$$

$$\frac{1}{\sin^2\theta}\frac{\partial^2 Y_{2,-1}}{\partial\phi^2} = \frac{-N\cos\theta\sin\theta\, e^{-i\phi}}{\sin^2\theta} = \frac{-N\cos\theta\, e^{-i\phi}}{\sin\theta}$$

so $\Lambda^2 Y_{2,-1} = Ne^{-i\phi}(-6\cos\theta\sin\theta) = -6Y_{2,-1} = -2(2+1)Y_{2,-1}$ [i.e. $l = 2$]

and hence

$$-6Y_{2,-1} = -\frac{2IE}{\hbar^2}Y_{2,-1}, \text{ implying that } \boxed{E = \frac{3\hbar^2}{I}}$$

and the angular momentum is $\{2(2+1)\}^{1/2}\hbar = \boxed{6^{1/2}\hbar}$.

(c) $\Lambda^2 Y_{3,3} = \dfrac{1}{\sin^2\theta}\dfrac{\partial^2 Y_{3,3}}{\partial\phi^2} + \dfrac{1}{\sin\theta}\dfrac{\partial}{\partial\theta}\sin\theta\dfrac{\partial Y_{3,3}}{\partial\theta}$, where $Y_{3,3} = N\sin^3\theta\, e^{3i\phi}$

$$\frac{\partial Y_{3,3}}{\partial\theta} = 3N\sin^2\theta\cos\theta\, e^{3i\phi}$$

$$\frac{1}{\sin\theta}\frac{\partial}{\partial\theta}\sin\theta\frac{\partial Y_{3,3}}{\partial\theta} = \frac{1}{\sin\theta}\frac{\partial}{\partial\theta}3N\sin^3\theta\cos\theta\, e^{3i\phi}$$

$$= \frac{3Ne^{3i\phi}}{\sin\theta}(3\sin^2\theta\cos^2\theta - \sin^4\theta) = 3Ne^{3i\phi}\sin\theta(3\cos^2\theta - \sin^2\theta)$$

$$= 3Ne^{3i\phi}\sin\theta(3 - 4\sin^2\theta) \quad [\cos^3\theta = \cos\theta(1-\sin^2\theta)]$$

$$\frac{1}{\sin^2\theta}\frac{\partial^2 Y_{3,3}}{\partial\phi^2} = \frac{-9N\sin^3\theta\, e^{3i\phi}}{\sin^2\theta} = -9N\sin\theta\, e^{3i\phi}$$

so $\Lambda^2 Y_{3,3} = -12N\sin^3\theta\, e^{3i\phi} = -12Y_{3,3} = -3(3+1)Y_{3,3}$ [i.e. $l = 3$]

and hence

$$-12Y_{3,3} = -\frac{2IE}{\hbar^2}Y_{3,3}, \text{ implying that } \boxed{E = \frac{6\hbar^2}{I}}$$

and the angular momentum is $\{3(3+1)\}^{1/2}\hbar = \boxed{12^{1/2}\hbar}$.

P8.25 From the diagram in Figure 8.4, $\cos\theta = \frac{m_l}{\{l(l+1)\}^{1/2}}$ and hence $\boxed{\theta = \arccos\frac{m_l}{\{l(l+1)\}^{1/2}}}$

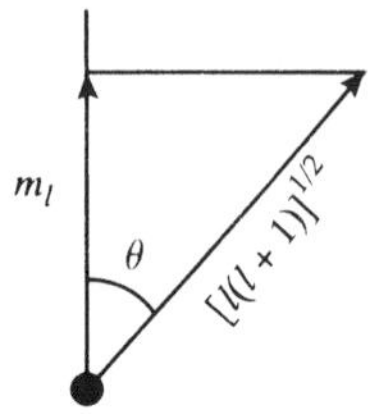

Figure 8.4

For an α electron, $m_s = +\frac{1}{2}$, $s = \frac{1}{2}$ and (with $m_l \to m_s$, $l \to s$)

$$\theta = \arccos\frac{\frac{1}{2}}{(\frac{3}{4})^{1/2}} = \arccos\frac{1}{\sqrt{3}} = \boxed{54°44'}$$

The minimum angle occurs for $m_l = l$:

$$\lim_{l\to\infty}\theta_{\min} = \lim_{l\to\infty}\arccos\left(\frac{l}{\{l(l+1)\}^{1/2}}\right) = \lim_{l\to\infty}\arccos\frac{l}{l} = \arccos 1 = \boxed{0}$$

P8.27 $\hat{\mathbf{l}} = \hat{\mathbf{r}} \times \hat{\mathbf{p}} = \begin{vmatrix} \mathbf{i} & \mathbf{j} & \mathbf{k} \\ \hat{x} & \hat{y} & \hat{z} \\ \hat{p}_x & \hat{p}_y & \hat{p}_z \end{vmatrix}$ [see any book treating the vector product of vectors]

$$= \mathbf{i}(\hat{y}\hat{p}_z - \hat{z}\hat{p}_y) + \mathbf{j}(\hat{z}\hat{p}_x - \hat{x}\hat{p}_z) + \mathbf{k}(\hat{x}\hat{p}_y - \hat{y}\hat{p}_x)$$

Therefore,

$$\hat{l}_x = (\hat{y}\hat{p}_z - \hat{z}\hat{p}_y) = \boxed{\frac{\hbar}{i}\left(y\frac{\partial}{\partial z} - z\frac{\partial}{\partial y}\right)}$$

$$\hat{l}_y = (\hat{z}\hat{p}_x - \hat{x}\hat{p}_z) = \boxed{\frac{\hbar}{i}\left(z\frac{\partial}{\partial x} - x\frac{\partial}{\partial z}\right)}$$

$$\hat{l}_z = (\hat{x}\hat{p}_y - \hat{y}\hat{p}_x) = \boxed{\frac{\hbar}{i}\left(x\frac{\partial}{\partial y} - y\frac{\partial}{\partial x}\right)}$$

We have used $\hat{p}_x = \dfrac{\hbar}{\mathrm{i}}\dfrac{\partial}{\partial x}$, etc. The commutator of $\hat{l}_x$ and $\hat{l}_y$ is $(\hat{l}_x\hat{l}_y - \hat{l}_y\hat{l}_x)$. We note that the operations always imply operation on a function. We form

$$\hat{l}_x\hat{l}_y f = -\hbar^2\left(y\frac{\partial}{\partial z} - z\frac{\partial}{\partial y}\right)\left(z\frac{\partial}{\partial x} - x\frac{\partial}{\partial z}\right)f$$

$$= -\hbar^2\left(yz\frac{\partial^2 f}{\partial z\partial x} + y\frac{\partial f}{\partial x} - yx\frac{\partial^2 f}{\partial z^2} - z^2\frac{\partial^2 f}{\partial y\partial x} + zx\frac{\partial^2 f}{\partial z\partial y}\right)$$

and $\hat{l}_y\hat{l}_x f = -\hbar^2\left(z\dfrac{\partial}{\partial x} - x\dfrac{\partial}{\partial z}\right)\left(y\dfrac{\partial}{\partial z} - z\dfrac{\partial}{\partial y}\right)f$

$$= -\hbar^2\left(zy\frac{\partial^2 f}{\partial x\partial z} - z^2\frac{\partial^2 f}{\partial x\partial y} - xy\frac{\partial^2 f}{\partial z^2} + xz\frac{\partial^2 f}{\partial z\partial y} + x\frac{\partial f}{\partial y}\right)$$

Since multiplication and differentiation are each commutative, the results of the operation $\hat{l}_x\hat{l}_y$ and $\hat{l}_y\hat{l}_x$ differ only in one term. For $\hat{l}_y\hat{l}_x f$, $x\dfrac{\partial f}{\partial y}$ replaces $y\dfrac{\partial f}{\partial x}$. Hence, the commutator of the operations, $(\hat{l}_x\hat{l}_y - \hat{l}_y\hat{l}_x)$ is $-\hbar^2\left(y\dfrac{\partial}{\partial x} - x\dfrac{\partial}{\partial y}\right)$ or $\boxed{-\dfrac{\hbar}{\mathrm{i}}\hat{l}_z}$.

COMMENT. We also would find

$$(\hat{l}_y\hat{l}_z - \hat{l}_z\hat{l}_y) = -\frac{\hbar}{\mathrm{i}}\hat{l}_x \text{ and } (\hat{l}_z\hat{l}_x - \hat{l}_x\hat{l}_z) = -\frac{\hbar}{\mathrm{i}}\hat{l}_y$$

P8.29 We are to show that $[\hat{l}^2,\hat{l}_z] = [\hat{l}_x^2 + \hat{l}_y^2 + \hat{l}_z^2,\hat{l}_z] = [\hat{l}_x^2,\hat{l}_z] + [\hat{l}_y^2,\hat{l}_z] + [\hat{l}_z^2,\hat{l}_z] = 0$

The three commutators are:

$$[\hat{l}_z^2,\hat{l}_z] = \hat{l}_z^2\hat{l}_z - \hat{l}_z\hat{l}_z^2 = \hat{l}_z^3 - \hat{l}_z^3 = 0$$

$$\begin{aligned}[\hat{l}_x^2,\hat{l}_z] &= \hat{l}_x^2\hat{l}_z - \hat{l}_z\hat{l}_x^2 = \hat{l}_x^2\hat{l}_z - \hat{l}_x\hat{l}_z\hat{l}_x + \hat{l}_x\hat{l}_z\hat{l}_x - \hat{l}_z\hat{l}_x^2\\ &= \hat{l}_x(\hat{l}_x\hat{l}_z - \hat{l}_z\hat{l}_x) + (\hat{l}_x\hat{l}_z - \hat{l}_z\hat{l}_x)\hat{l}_x = \hat{l}_x[\hat{l}_x,\hat{l}_z] + [\hat{l}_x,\hat{l}_z]\hat{l}_x\\ &= \hat{l}_x(-\mathrm{i}\hbar\hat{l}_y) + (-\mathrm{i}\hbar\hat{l}_y)\hat{l}_x = -\mathrm{i}\hbar(\hat{l}_x\hat{l}_y + \hat{l}_y\hat{l}_x)\ [8.56a]\end{aligned}$$

$$\begin{aligned}[\hat{l}_y^2,\hat{l}_z] &= \hat{l}_y^2\hat{l}_z - \hat{l}_z\hat{l}_y^2 = \hat{l}_y^2\hat{l}_z - \hat{l}_y\hat{l}_z\hat{l}_y + \hat{l}_y\hat{l}_z\hat{l}_y - \hat{l}_z\hat{l}_y^2\\ &= \hat{l}_y(\hat{l}_y\hat{l}_z - \hat{l}_z\hat{l}_y) + (\hat{l}_y\hat{l}_z - \hat{l}_z\hat{l}_y)\hat{l}_y = \hat{l}_y[\hat{l}_y,\hat{l}_z] + [\hat{l}_y,\hat{l}_z]\hat{l}_y\\ &= \hat{l}_y(\mathrm{i}\hbar\hat{l}_x) + (\mathrm{i}\hbar\hat{l}_x)\hat{l}_y = \mathrm{i}\hbar(\hat{l}_y\hat{l}_x + \hat{l}_x\hat{l}_y)\ [8.56a]\end{aligned}$$

Therefore, $[\hat{l}^2,\hat{l}_z] = -\mathrm{i}\hbar(\hat{l}_x\hat{l}_y + \hat{l}_y\hat{l}_x) + \mathrm{i}\hbar(\hat{l}_x\hat{l}_y + \hat{l}_y\hat{l}_x) + 0 = 0$

We may also conclude that $[\hat{l}^2,\hat{l}_x] = 0$ and $[\hat{l}^2,\hat{l}_y] = 0$ because $\hat{l}_x$, $\hat{l}_y$, and $\hat{l}_z$ occur symmetrically in $\hat{l}^2$.

Solutions to applications

P8.31 (a) The energy levels are given by:

$$E_n = \frac{h^2n^2}{8mL^2} \text{ [8.4a]},$$

and we are looking for the energy difference between $n = 6$ and $n = 7$:

$$\Delta E = \frac{h^2(7^2 - 6^2)}{8mL^2}.$$

Since there are 12 atoms on the conjugated backbone, the length of the box is 11 times the bond length:

$$L = 11(140 \times 10^{-12}\text{ m}) = 1.54 \times 10^{-9}\text{ m},$$

$$\text{so} \quad \Delta E = \frac{(6.626 \times 10^{-34}\text{ J s})^2(49 - 36)}{8(9.11 \times 10^{-31}\text{ kg})(1.54 \times 10^{-9}\text{ m})^2} = \boxed{3.30 \times 10^{-19}\text{ J}}$$

(b) The relationship between energy and frequency is:

$$\Delta E = h\nu \quad \text{so} \quad \nu = \frac{\Delta E}{h} = \frac{3.30 \times 10^{-19}\text{ J}}{6.626 \times 10^{-34}\text{ J s}} = \boxed{4.95 \times 10^{-14}\text{ s}^{-1}}$$

(c) Look at the terms in the energy expression that change with the number of conjugated atoms, N. The energy (and frequency) are inversely proportional to L^2 and directly proportional to $(n + 1)^2 - n^2 = 2n + 1$, where n is the quantum number of the highest occupied state. Since n is proportional to N (equal to $N/2$) and L is approximately proportional to N (strictly to $N - 1$), the energy and frequency are approximately proportional to N^{-1}. So *the absorption spectrum of a linear polyene shifts to* $\boxed{\textit{lower}}$ *frequency as the number of conjugated atoms* $\boxed{\textit{increases}}$.

P8.33 In effect, we are looking for the vibrational frequency of an O atom bound, with a force constant equal to that of free CO, to an infinitely massive and immobile protein complex. The angular frequency is

$$\omega = \left(\frac{k_f}{m}\right)^{1/2},$$

where m is the mass of the O atom

$$m = 16.0 \times (1.66 \times 10^{-27}\text{ kg}) = 2.66 \times 10^{-26}\text{ kg},$$

and k_f is the same force constant as in Problem 8.2. In that problem (unlike this one), the mass appropriate for the vibration of the free CO molecule was the effective mass.

$$k_f = \omega^2 m_{\text{eff}} = 4\pi^2c^2\tilde{\nu}^2 m_{\text{eff}} = \frac{4\pi^2c^2\tilde{\nu}^2 m_1m_2}{m_1 + m_2} = 1902\text{ N m}^{-1}.$$

$$\text{Hence,} \quad \omega = \left(\frac{1902\text{ N m}^{-1}}{2.66 \times 10^{-26}\text{ kg}}\right)^{1/2} = \boxed{2.68 \times 10^{14}\text{ s}^{-1}}.$$

P8.35 The angular momentum states are defined by the quantum number $m_l = 0, \pm 1, \pm 2$, etc. The energy of state m_l is

$$E_{m_l} = \frac{m_l^2 \hbar^2}{2I} \quad [8.38a]$$

and the angular momentum is

$$J_z = m_l \hbar$$

(a) If there are 22 electrons, two in each of the lowest 11 states, then the highest occupied states are $m_l = \pm 5$, so,

$$J_z = \pm 5\hbar = \pm 5 \times (1.0546 \times 10^{-34}\text{ J s}) = \boxed{5.275 \times 10^{-34}\text{ J s}}$$

and $E_{\pm 5} = \dfrac{25\hbar^2}{2I}$

The moment of inertia of an electron on a ring of radius 440 pm is

$$I = mr^2 = (9.11 \times 10^{-31}\text{ kg}) \times (440 \times 10^{-12}\text{ m})^2 = 1.76 \times 10^{-49}\text{ kg m}^2$$

Hence, $E_{\pm 5} = \dfrac{25 \times (1.055 \times 10^{-34}\text{ J s})^2}{2 \times (1.76 \times 10^{-49}\text{ kg m}^2)} = \boxed{7.89 \times 10^{-19}\text{ J}}$.

(b) The lowest unoccupied energy level is $m_l = \pm 6$, which has energy

$$E_{\pm 6} = \frac{36 \times (1.055 \times 10^{-34}\text{ J s})^2}{2 \times (1.76 \times 10^{-49}\text{ kg m}^2)} = 1.14 \times 10^{-18}\text{ J}.$$

Radiation that would induce a transition between these levels must have a frequency such that

$$h\nu = \Delta E \quad \text{so} \quad \nu = \frac{\Delta E}{h} = \frac{(11.4 - 7.89) \times 10^{-19}\text{ J}}{6.626 \times 10^{-34}\text{ J s}} = \boxed{5.2 \times 10^{14}\text{ Hz}}.$$

This corresponds to a wavelength of about 570 nm, a wave of visible light.

P8.37 (a) In the box, the Schrödinger equation is

$$-\frac{\hbar^2}{2m}\left(\frac{\partial^2}{\partial x^2} + \frac{\partial^2}{\partial y^2} + \frac{\partial^2}{\partial z^2}\right)\psi = E\psi$$

Assume that the solution is a product of three functions of a single variable; that is, let

$$\psi(x,y,z) = X(x)Y(y)Z(z)$$

Substituting into the Schrödinger equation gives

$$-\frac{\hbar^2}{2m}\left(YZ\frac{\partial^2 X}{\partial x^2} + XZ\frac{\partial^2 Y}{\partial y^2} + XY\frac{\partial^2 Z}{\partial z^2}\right) = EXYZ$$

Divide both sides by XYZ:

$$-\frac{\hbar^2}{2m}\left(\frac{1}{X}\frac{\partial^2 X}{\partial x^2} + \frac{1}{Y}\frac{\partial^2 Y}{\partial y^2} + \frac{1}{Z}\frac{\partial^2 Z}{\partial z^2}\right) = E$$

For the purposes of illustration, isolate the terms that depend on x on the left side of the equation:

$$-\frac{\hbar^2}{2m}\left(\frac{1}{X}\frac{\partial^2 X}{\partial x^2}\right) = E + \frac{\hbar^2}{2m}\left(\frac{1}{Z}\frac{\partial^2 Z}{\partial z^2} + \frac{1}{Y}\frac{\partial^2 Y}{\partial y^2}\right)$$

Note that the left side depends only on one variable, x, while the right side depends on two different and independent variables, y and z. The only way that the two sides can be equal to each other for all x, y, and z is if they are both equal to a constant. Call that constant E_x, and we have, from the left side of the equation:

$$-\frac{\hbar^2}{2m}\left(\frac{1}{X}\frac{\partial^2 X}{\partial x^2}\right) = E_x \quad \text{so} \quad -\frac{\hbar^2}{2m}\frac{\partial^2 X}{\partial x^2} = E_x X$$

Note that this is just the Schrödinger equation for a particle in a one-dimensional box. Note also that we could just as easily have isolated y terms or z terms, leading to similar equations:

$$-\frac{\hbar^2}{2m}\frac{\partial^2 Y}{\partial y^2} = E_y Y \quad \text{and} \quad -\frac{\hbar^2}{2m}\frac{\partial^2 Z}{\partial z^2} = E_z Z$$

The assumption that the wavefunction can be written as a product of single-variable functions is a valid one, for we can find ordinary differential equations for the assumed factors. That is what it means for a partial differential equation to be separable.

(b) Since X, Y, and Z are particle-in-a-box wavefunctions of independent variables x, y, and z, respectively, each of them has its own quantum number. The three-dimensional wavefunction is a product of the three, and therefore depends on all three quantum numbers:

$$\psi(x,y,z) = X(x)Y(y)Z(z) = \left(\frac{2}{L_1}\right)^{1/2} \sin\frac{n_x \pi x}{L_1} \times \left(\frac{2}{L_2}\right)^{1/2} \sin\frac{n_y \pi y}{L_2} \times \left(\frac{2}{L_3}\right)^{1/2} \sin\frac{n_z \pi z}{L_3}$$

Each constant of separation (E_x, E_y, and E_z) depends on its own quantum number. The three constants of separation add up to the total energy, which therefore depends on all three quantum numbers:

$$E = E_x + E_y + E_z = \frac{h^2}{8m}\left(\frac{n_x^2}{L_1^2} + \frac{n_y^2}{L_2^2} + \frac{n_z^2}{L_3^2}\right)$$

(c) For a cubic box, $L_1 = L_2 = L_3 = L$, so

$$E = \frac{h^2(n_x^2 + n_y^2 + n_z^2)}{8mL^2}$$

The energy levels are shown in Figure 8.5.

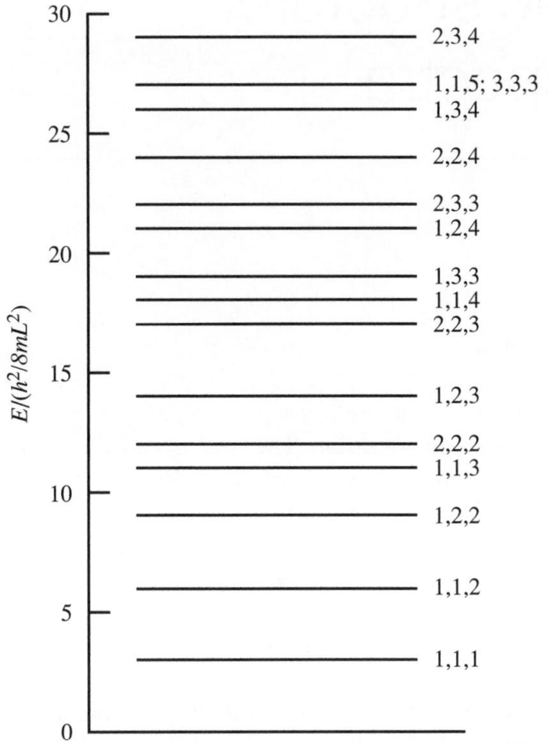

Figure 8.5

(d) Compare this energy-level diagram to Figure 8.2 of the textbook. The energy levels here are much more closely spaced. In a one-dimensional box, the 15th energy level is not reached until $\dfrac{E}{h^2/8mL^2} = 225$, and the previous level is 29 units below that. In the three-dimensional box, the first 15 energy levels fit within the range of 29 units. The energy levels in a one-dimensional box are sparse compared to those in a three-dimensional box.

P8.39 The Coulombic force is

$$F = -\frac{\mathrm{d}}{\mathrm{d}r}\frac{Q_1Q_2}{4\pi\varepsilon_0 r} = \frac{Q_1Q_2}{4\pi\varepsilon_0 r^2}$$

For two electrons 2.0 nm apart, the force is

$$F = \frac{(1.60\times10^{-19}\,\mathrm{C})^2}{4\pi\times(8.854\times10^{-12}\,\mathrm{C^2\,J^{-1}\,m^{-1}})\times(2.0\times10^{-9}\,\mathrm{m})^2} = \boxed{5.8\times10^{-11}\,\mathrm{N}}$$

9 Atomic structure and spectra

Answers to discussion questions

D9.1 When hydrogen atoms are excited into higher energy levels they emit energy at a series of discrete wavenumbers that show up as a series of lines in the spectrum. The lines in the Lyman series, for which the emission of energy corresponds to the final state $n_1 = 1$, are in the ultraviolet region of the electromagnetic spectrum. The Balmer series, for which $n_1 = 2$, is in the visible region. The Paschen series, $n_1 = 3$, is in the infrared. The Brackett series, $n_1 = 4$, is also in the infrared. The Pfund series, $n_1 = 5$, is in the far infrared. The Humphrey series, $n_1 = 6$, is in the very far infrared.

D9.3 (1) The principal quantum number, n, determines the energy of a hydrogenic atomic orbital through eqn 9.9.
(2) The azimuthal quantum number, l, determines the magnitude of the angular momentum of a hydrogenic atomic orbital through the formula $\{l(l+1)\}^{1/2}\hbar$.
(3) The magnetic quantum number, m_l, determines the z-component of the angular momentum of a hydrogenic orbital through the formula $m_l\hbar$.
(4) The spin quantum number, s, determines the magnitude of the spin angular momentum through the formula $\{s(s+1)\}^{1/2}\hbar$. For hydrogenic atomic orbitals, s can only be $\frac{1}{2}$.
(5) The spin quantum number, m_s, determines the z-component of the spin angular momentum through the formula $m_s\hbar$. For hydrogenic atomic orbitals, m_s can only be $\pm\frac{1}{2}$.

D9.5 (a) A boundary surface for a hydrogenic orbital is drawn so as to contain most (say 90%) of the probability density of an electron in that orbital. Its shape varies from orbital to orbital because the electron density distribution is different for different orbitals.
(b) The radial distribution function gives the probability that the electron will be found anywhere within a shell of radius r around the nucleus. It gives a better picture of where the electron is likely to be found with respect to the nucleus than the probability density, which is the square of the wavefunction.

D9.7 In period 2, the first ionization energies increase markedly from Li to Be, decrease slightly from Be to B, again increase markedly from B to N, again decrease slightly from N to O, and finally increase markedly from O to Ne. The general trend is an overall increase of I_1 with atomic number across the period. That is to be expected since the principal quantum number (electron shell) of the outer electron remains the same, while its attraction to the nucleus increases. The slight decrease from Be to B is a reflection of the outer electron being in a higher energy subshell (larger l value) in B than in Be. The slight decrease from N to O is due to the half-filled subshell effect; half-filled sub-shells have increased stability. O has one electron outside of the half-filled p subshell and that electron must pair with another, resulting in strong electron–electron repulsions between them.

The same kind of variation is expected for the elements of period 3 because in both periods the outer shell electrons are only s and p.

D9.9 An electron has a magnetic moment and magnetic field due to its orbital angular momentum. It also has a magnetic moment and magnetic field due to its spin angular momentum. There is an interaction energy between magnetic moments and magnetic fields. That between the spin magnetic moment and the magnetic field generated by the orbital motion is called spin–orbit coupling. The energy of interaction is proportional to the scalar product of the two vectors representing the spin and orbital angular momenta and hence depends upon the orientation of the two vectors. See Figure 9.28. The total angular momentum of an electron in an atom is the vector sum of the orbital and spin angular momenta as illustrated in Figure 9.29 and expressed in eqn 9.45. The spin–orbit coupling results in a splitting of the energy levels associated with atomic terms as shown in Figures 9.30 and 9.31. This splitting shows up in atomic spectra as a fine structure as illustrated in Figure 9.31.

Solutions to exercises

E9.1(a) Eqn. 9.1 implies that the shortest wavelength corresponds to $n_2 = \infty$, and the longest to $n_2 = 2$. Solve eqn. 9.1 for λ.

$$\lambda = \frac{(1/n_1^2 - 1/n_2^2)^{-1}}{R_H}$$

$$\text{Shortest: } \lambda = \frac{(1/1^2 - 1/\infty^2)^{-1}}{109\,677\ \text{cm}^{-1}} = \boxed{9.118 \times 10^{-6}\ \text{cm}}$$

$$\text{Longest: } \lambda = \frac{(1/1^2 - 1/2^2)^{-1}}{109\,677\ \text{cm}^{-1}} = \boxed{1.216 \times 10^{-5}\ \text{cm}}$$

E9.2(a) For atoms A, eqn. 9.9 may be rewritten in terms of the Rydberg constant R_A as

$$E_n = -\frac{Z^2\mu_A hcR_A}{m_e n^2} \approx -\frac{Z^2 hcR_\infty}{n^2}\ [9.15]$$

where to within 0.01% the ratio μ_A/m_e is unity. Eqn. 9.1 can then be rewritten as

$$\tilde{\nu} = Z^2 R_\infty\left(\frac{1}{n_1^2} - \frac{1}{n_2^2}\right) \quad \lambda = \frac{1}{\tilde{\nu}} \quad \nu = \frac{c}{\lambda}$$

$$\tilde{\nu} = 4 \times 109\,737\ \text{cm}^{-1}\left(\frac{1}{1^2} - \frac{1}{2^2}\right) = \boxed{3.292 \times 10^5\ \text{cm}^{-1}} \quad \lambda = \boxed{3.038 \times 10^{-6}\ \text{cm}}$$

$$\nu = \frac{2.9978 \times 10^{10}\ \text{cm s}^{-1}}{3.0376 \times 10^{-6}\ \text{cm}} = \boxed{9.869 \times 10^{15}\ \text{s}^{-1}}$$

E9.3(a) This is essentially the photoelectric effect [eqn. 7.15 of Section 7.2] with the ionization energy of the ejected electron being the work function Φ.

$$h\nu = \tfrac{1}{2}m_e v^2 + I$$

$$I = h\nu - \tfrac{1}{2}m_e v^2 = (6.626 \times 10^{-34}\ \mathrm{J\ Hz^{-1}}) \times \left(\frac{2.998 \times 10^8\ \mathrm{ms^{-1}}}{58.4 \times 10^{-9}\ \mathrm{m}}\right)$$

$$-\left(\tfrac{1}{2}\right) \times (9.109 \times 10^{-31}\ \mathrm{kg}) \times (1.59 \times 10^6\ \mathrm{ms^{-1}})^2$$

$$= 2.25 \times 10^{-18}\ \mathrm{J,\ corresponding\ to\ } \boxed{14.0\ \mathrm{eV}}$$

E9.4(a) The energies are $E = -\dfrac{hcR_H}{n^2}$ [9.9 with 9.14], and the orbital degeneracy g of an energy level of principal quantum number n is

$$g = \sum_{l=0}^{n-1}(2l+1) = 1 + 3 + 5 + \cdots + 2n - 1 = \frac{(1 + 2n - 1)n}{2} = n^2$$

(a) $E = -hcR_H$ implies that $n = 1$, so $\boxed{g = 1}$ [the 1s orbital].

(b) $E = -\dfrac{hcR_H}{9}$ implies that $n = 3$, so $\boxed{g = 9}$ (3s orbital, the three 3p orbitals, and the five 3d orbitals).

(c) $E = -\dfrac{hcR_H}{25}$ implies that $n = 5$, so $\boxed{g = 25}$ (the 5s orbital, the three 5p orbitals, the five 5d orbitals, the seven 5f orbitals, the nine 5g orbitals).

E9.5(a) $R_{1,0} = N\mathrm{e}^{-r/a_0}$

$$\int_0^\infty R^2 r^2 \mathrm{d}r = 1 = \int_0^\infty N^2 r^2 \mathrm{e}^{-2r/a_0}\mathrm{d}r = N^2 \times \frac{2!}{\left(\frac{2}{a_0}\right)^3} = 1 \quad \left[\int_0^\infty x^n \mathrm{e}^{-ax}\mathrm{d}x = \frac{n!}{a^{n+1}}\right]$$

$$N^2 = \frac{4}{a_0^3},\ \boxed{N = \frac{2}{a_0^{3/2}}}$$

Thus,

$$R_{1,0} = 2\left(\frac{1}{a_0}\right)^{3/2} \mathrm{e}^{-r/a_0},$$

which agrees with Table 9.1.

E9.6(a) $$R_{2,0} \propto \left(2 - \frac{\rho}{2}\right)\mathrm{e}^{-\rho/4} \quad \text{with } \rho = \frac{2r}{a_0}$$

Note: ρ defined here is twice ρ defined in Table 9.1.

$$\frac{dR}{dr}=\frac{2}{a_0}\frac{dR}{d\rho}=\frac{2}{a_0}\left(-\frac{1}{2}-\frac{1}{2}+\frac{1}{8}\rho\right)e^{-\rho/4}=0, \quad \text{when } \rho=8$$

Hence, the wavefunction has an extremum at $r=\boxed{4a_0}$. Since $2-\frac{\rho}{2}<0, \psi<0$ and the extremum is a minimum (more formally: $\frac{d^2\psi}{dr^2}>0$ at $\rho=8$).

The second extremum is at $\boxed{r=0.}$ It is not a minimum and in fact is a physical maximum, although not one that can be obtained by differentiation. To see that it is maximum substitute $\rho=0$ into $R_{2,0}$.

E9.7(a) Here we use the probability density function ψ^2, rather than the radial distribution function P, since we are seeking the probability at a point, namely $\psi^2 d\tau$

The probability density varies as

$$\psi^2=\frac{1}{\pi a_0^3}e^{-2r/a_0} \text{ [From 9.17, the ground state is assumed.]}$$

Therefore, the maximum value is at $r=0$ and ψ^2 is 50% of the maximum when

$$e^{-2r/a_0}=0.50$$

implying that $r=-\frac{1}{2}a_0\ln 0.50$, which is at $\boxed{r=0.35a_0}$ [18 pm]

E9.8(a) The radial nodes correspond to $R_{3,0}=0$. $R_{3,0}\propto 6-2\rho+\frac{1}{9}\rho^2$ (Table 9.1); the radial nodes occur at

$$6-6\rho+\rho^2=0, \quad \text{or} \quad \rho=1.27 \text{ and } 4.73.$$

Since $r=\frac{n\rho a_0}{2}$, the radial nodes occur at $\boxed{101 \text{ pm and } 376 \text{ pm}}$

E9.9(a) This exercise has already been solved in Problem 8.20 by use of the virial theorem. Here we will solve it by straightforward integration.

$$\psi_{1,0,0}=R_{1,0}Y_{0,0}=\left(\frac{1}{\pi a_0^3}\right)^{1/2}e^{-r/a_0} \text{ [Table 8.2 and 9.1]}$$

The potential energy operator is

$$V=-\frac{Ze^2}{4\pi\varepsilon_0}\times\left(\frac{1}{r}\right)=-k\left(\frac{1}{r}\right)$$

$$\langle V\rangle=-k\left\langle\frac{1}{r}\right\rangle\left[k=\frac{e^2}{4\pi\varepsilon_0}\right]=-k\int_0^\infty\int_0^\pi\int_0^{2\pi}\left(\frac{1}{\pi a_0^3}\right)e^{-r/a_0}\left(\frac{1}{r}\right)e^{-r/a_0}r^2 dr\sin\theta\, d\theta\, d\phi$$

$$=-k\times(4\pi)\times\left(\frac{1}{\pi a_0^3}\right)\int_0^\infty re^{-2r/a_0}dr$$

$$=-k\times\left(\frac{4}{a_0^3}\right)\times\left(\frac{a_0^2}{4}\right)=-k\left(\frac{1}{a_0}\right)$$

$$\left[\text{We have used } \int_0^{\pi} \sin\theta\, d\theta = 2,\ \int_0^{2\pi} d\theta = 2\pi, \text{ and } \int_0^{\infty} x^n e^{-ax} dx = \frac{n!}{a^{n+1}}\right]$$

Hence,

$$\langle V\rangle = -\frac{e^2}{4\pi\varepsilon_0 a_0} = \boxed{2E_{1s}}$$

The kinetic energy operator is $-\frac{\hbar^2}{2\mu}\nabla^2$ [*Further Information* 9.1]; hence

$$\langle E_K\rangle \equiv \langle T\rangle = \int \psi_{1s}^*\left(-\frac{\hbar^2}{2\mu}\right)\nabla^2\psi_{1s}\, d\tau$$

$$\nabla^2\psi_{1s} = \frac{1}{r}\frac{\partial^2(r\psi_{1s})}{\partial r^2} + \frac{1}{r^2}\Lambda^2\psi_{1s} \text{ [Problem 8.23]}$$

$$= \left(\frac{1}{\pi a_0^3}\right)^{1/2} \times \left(\frac{1}{r}\right) \times \left(\frac{d^2}{dr^2}\right) r e^{-r/a_0}$$

$[\Lambda^2\psi_{1s} = 0,\ \psi_{1s}$ contains no angular variables$]$

$$= \left(\frac{1}{\pi a_0^3}\right)^{1/2}\left[-\left(\frac{2}{a_0 r}\right) + \left(\frac{1}{a_0^2}\right)\right] e^{-r/a_0}$$

$$\langle T\rangle = -\left(\frac{\hbar^2}{2\mu}\right) \times \left(\frac{1}{\pi a_0^3}\right)\int_0^{\infty}\left[-\left(\frac{2}{a_0 r}\right) + \left(\frac{1}{a_0^2}\right)\right] e^{-2r/a_0} r^2\, dr \times \int_0^{\pi}\sin\theta\, d\theta \int_0^{2\pi} d\phi$$

$$= -\left(\frac{2\hbar^2}{\mu a_0^3}\right)\int_0^{\infty}\left[-\left(\frac{2r}{a_0}\right) + \left(\frac{r^2}{a_0^2}\right)\right] e^{-2r/a_0}\, dr = -\left(\frac{2\hbar^2}{\mu a_0^3}\right) \times \left(-\frac{a_0}{4}\right) = \frac{\hbar^2}{2\mu a_0^2}$$

$$= \boxed{-E_{1s}}$$

Hence, $\langle T\rangle + \langle V\rangle = 2E_{1s} - E_{1s} = E_{1s}$

COMMENT. E_{1s} may also be written as

$$E_{1s} = -\frac{\mu e^4}{32\pi^2\varepsilon_0^2\hbar^2}$$

Question. Are the three different expressions for E_{1s} given in this exercise all equivalent?

E9.10(a) $P_{2s} = 4\pi r^2\psi_{2s}^2$

$$\psi_{2s} = \left(\frac{1}{4\pi}\right)\frac{1}{2\sqrt{2}}\left(\frac{Z}{a_0}\right)^{3/2} \times \left(2 - \frac{\rho}{2}\right) e^{-\rho/4} \quad \left[\rho = \frac{2Zr}{a_0}\right]$$

[Note: ρ defined here is $n \times \rho$ as defined in Table 9.1]

$$P_{2s} = \left(\frac{1}{4\pi}\right)\left(\frac{a_0\rho}{2Z}\right)^2 \times \left(\frac{1}{8}\right) \times \left(\frac{Z}{a_0}\right)^3 \left(2 - \frac{\rho}{2}\right)^2 e^{-\rho/2}$$

$$P_{2s} = k\rho^2\left(2 - \frac{\rho}{2}\right)^2 e^{-\rho/2} \quad \left[k = \frac{Z}{128\pi a_0} = \text{constant}\right]$$

The most probable value of r, or equivalently ρ, is where

$$\frac{d}{d\rho}\left\{\rho^2\left(2 - \frac{\rho}{2}\right)^2 e^{-\rho/2}\right\} = 0$$

$$\propto \left\{2\rho\left(2 - \frac{\rho}{2}\right)^2 - 2\rho^2\left(2 - \frac{\rho}{2}\right) - \rho^2\left(2 - \frac{\rho}{2}\right)^2\right\} e^{-\rho/2} = 0$$

$$\propto \rho(\rho - 4)(\rho^2 - 12\rho + 16) = 0 \quad [e^{-\rho/2} \text{ is never zero, except as } \rho \to \infty]$$

Thus, $\rho^* = 0$, $\rho^* = 4$, $\rho^* = 6 \pm 2\sqrt{5}$

The principal (outermost) maximum is at $\rho^* = 6 \pm 2\sqrt{5}$.

Hence, $r^* = (6 + 2\sqrt{5})\dfrac{a_0}{2Z} = \boxed{5.24\dfrac{a_0}{Z}}$.

E9.11(a) The most probable radius occurs when the radial distribution function is a maximum. At this point the derivative of the function with respect to either r or ρ equals zero.

$$\left(\frac{dR_{21}}{d\rho}\right)_{\max} = 0 = \left(\frac{d(\rho e^{-\rho/2})}{d\rho}\right)_{\max} \text{[Table 9.1]} = \left(1 - \frac{\rho}{2}\right)e^{-\rho/2}$$

The function is a maximum when $\rho = 2$. Since $\rho = (2Z/na_0)r$ and $n = 2$, this corresponds to $\boxed{r = 2a_0/Z}$.

E.9.12(a) Identify l and use angular momentum $= \{l(l+1)\}^{1/2}\hbar$.

(a) $l = 0$, so angular momentum $= 0$ (b) $l = 0$, so angular momentum $= 0$

(c) $l = 2$, so angular momentum $= \sqrt{6}\hbar$

The total number of nodes is equal to $n - 1$ and the number of angular nodes is equal to l, hence the number of radial nodes is equal to $n - l - 1$. We can draw up the following table:

	1s	3s	3d
n, l	1,0	3,0	3,2
Angular nodes	0	0	2
Radial nodes	0	2	0

E9.13(a) See Figure 9.15 of the text. The number of angular nodes is the value of the quantum number l, which for p orbitals is 1. Hence, each of the three p orbitals has one angular node. To locate the angular nodes look for the value of θ that makes the wavefunction zero.

p_0 orbital: see eqn 9.20a and Figure 9.15. The nodal plane is the $\boxed{xy \text{ plane}}$ and $\boxed{\theta = \pi/2}$ is an angular node.

p_x orbital: see eqns 9.21 and 9.22 and Figure 9.15. The nodal plane is the $\boxed{yz}$ plane and $\boxed{\theta = 0}$ is an angular node.

p_y orbital: see eqns 9.21 and 9.22 and Figure 9.15. The nodal plane is the $\boxed{xz}$ plane and $\boxed{\theta = 0}$ is an angular node.

E9.14(a) The selection rules for a many-electron atom are given by the set [9.46]. For a single-electron transition these amount to Δn = any integer; $\Delta l = \pm 1$. Hence,

(a) $2s \rightarrow 1s$; $\Delta l = 0$, $\boxed{\text{forbidden}}$ (b) $2p \rightarrow 1s$; $\Delta l = -1$, $\boxed{\text{allowed}}$

(c) $3d \rightarrow 2p$; $\Delta l = -1$, $\boxed{\text{allowed}}$

E9.15(a) A source approaching an observer appears to be emitting light of frequency

$$\nu_{\text{approaching}} = \frac{\nu}{1 - \frac{s}{c}} \quad [9.37\text{b}]$$

Since $\nu \propto \dfrac{1}{\lambda}$, $\lambda_{\text{obs}} = \left(1 - \dfrac{s}{c}\right)\lambda$

$s = 60 \text{ km h}^{-1} = 16.\overline{7} \text{ m s}^{-1}$. Hence,

$$\lambda_{\text{obs}} = \left(1 - \frac{16.\overline{7} \text{ m s}^{-1}}{2.998 \times 10^8 \text{ m s}^{-1}}\right) \times (680 \text{ nm}) = \boxed{0.999999944 \times 680 \text{ nm}}$$

For all practical purposes no shift at all at this speed.

E9.16(a) $\delta\tilde{\nu} \approx \dfrac{5.31 \text{ cm}^{-1}}{\tau/\text{ps}}$ [equation in *Brief Illustration*], implying that $\tau \approx \dfrac{5.31 \text{ ps}}{\delta\tilde{\nu}/\text{cm}^{-1}}$

(a) $\tau \approx \dfrac{5.31 \text{ ps}}{0.20} = \boxed{27 \text{ ps}}$

(b) $\tau \approx \dfrac{5.31 \text{ ps}}{2.0} = \boxed{2.7 \text{ ps}}$

E9.17(a) $\delta\tilde{\nu} \approx \dfrac{5.31 \text{ cm}^{-1}}{\tau/\text{ps}}$ [equation in *Brief illustration*, Section 9.6(b)]

(a) $\tau \approx 1.0 \times 10^{-13} \text{ s} = 0.10 \text{ ps}$, implying that $\delta\tilde{\nu} \approx \boxed{53 \text{ cm}^{-1}}$

(b) $\tau \approx (100) \times (1.0 \times 10^{-13} \text{ s}) = 10 \text{ ps}$, implying that $\delta\tilde{\nu} \approx \boxed{0.53 \text{ cm}^{-1}}$

E9.18(a) Sc: $[Ar]4s^23d^1$
Ti: $[Ar]4s^23d^2$
V: $[Ar]4s^23d^3$
Cr: $[Ar]4s^23d^4$ or $[Ar]4s^13d^5$ (most probable)
Mn: $[Ar]4s^23d^5$
Fe: $[Ar]4s^23d^6$
Co: $[Ar]4s^23d^7$
Ni: $[Ar]4s^23d^8$
Cu: $[Ar]4s^23d^9$ or $[Ar]4s^13d^{10}$ (most probable)
Zn: $[Ar]4s^23d^{10}$

E9.19(a) (a) $1s^22s^22p^63s^23p^63d^8 = \boxed{[Ar]3d^8}$

(b) All subshells except 3d are filled and hence have no net spin. Applying Hund's rule to $3d^8$ shows that there are two unpaired spins. The paired spins do not contribute to the net spin, hence we consider only $s_1 = \frac{1}{2}$ and $s_2 = \frac{1}{2}$. The Clebsch–Gordan series [9.44] produces

$$S = s_1 + s_2, \ldots, |s_1 - s_2|, \quad \text{hence } \boxed{S = 1,0}$$

$$M_S = -S, -S+1, \ldots, S$$

For $S = 1$, $\boxed{M_S = -1,0,+1}$

$S = 0$, $\boxed{M_S = 0}$

E9.20(a) We use the Clebsch–Gordan series [9.45] in the form

$$j = l+s, l+s-1, \ldots, |l-s| \quad \text{[lower case for a single electron]}$$

(a) $l = 2$, $s = \frac{1}{2}$; so $j = \boxed{\frac{5}{2}, \frac{3}{2}}$ (b) $l = 3$, $s = \frac{1}{2}$; so $j = \boxed{\frac{7}{2}, \frac{5}{2}}$

E9.21(a) The Clebsch–Gordan series for $\boxed{l = 1}$ and $s = \frac{1}{2}$ leads to $j = \frac{3}{2}$ and $\frac{1}{2}$.

E9.22(a) The letter D indicates that $L = 2$, the superscript 1 is the value of $2S+1$, so $S = 0$ and the subscript 2 is the value of J. Hence, $\boxed{L = 2,\ S = 0,\ J = 2}$

E9.23(a) Use the Clebsch–Gordan series in the form

$$S' = s_1 + s_2, s_1 + s_2 - 1, \ldots, |s_1 - s_2|$$

and

$$S = S' + s_1, S' + s_1 - 1, \ldots, |S' - s_1|$$

in succession. The multiplicity is $2S+1$.

(a) $S = \frac{1}{2} + \frac{1}{2}, \frac{1}{2} - \frac{1}{2} = \boxed{1,0}$ with multiplicities $\boxed{3,1}$, respectively

(b) $S' = 1,0$; then $S = \boxed{\frac{3}{2}, \frac{1}{2}}$ [from 1], $\boxed{\text{and } \frac{1}{2}}$ [from 0], with multiplicities $\boxed{4,2,2}$

E9.24(a) These electrons are not equivalent (different subshells), hence all the terms that arise from the vector model and the Clebsch–Gordan series are allowed (Example 9.6).

$$L = l_1 + l_2, \ldots, |l - l_2| \, [9.43] = 2 \text{ only}$$
$$S = s_1 + s_2, \ldots, |s_1 - s_2| = 1,0$$

The allowed terms are then ^{3}D and ^{1}D. The possible values of J are given by

$$J = L + S, \ldots, |L - S| \, [9.45] = 3, 2, 1 \text{ for } ^3\text{D and 2 for } ^1\text{D}$$

The allowed complete term symbols are then

$$\boxed{^3D_3, \ ^3D_2, \ ^3D_1, \ ^1D_2}$$

The $\boxed{^3\text{D set of terms are the lower in energy}}$ [Hund's rule]

COMMENT. Hund's rule in the form given in the text does not allow the energies of the triplet terms to be distinguished. Experimental evidence indicates that $^3D_1$ is lowest.

E9.25(a) Use the Clebsch–Gordan series in the form

$$J = L + S, L + S - 1, \ldots, |L - S|$$

The number of states (M_J values) is $2J + 1$ in each case.

(a) $L = 0$, $S = 0$; hence $\boxed{J = 0}$ and there is only $\boxed{\text{1 state}}$ $(M_J = 0)$.

(b) $L = 1$, $S = \frac{1}{2}$; hence $\boxed{J = \frac{3}{2}, \frac{1}{2}}$ $(^2P_{3/2}, \ ^2P_{1/2})$ with $\boxed{\text{4, 2 states, respectively}}$.

(c) $L = 1$, $S = 1$; hence $\boxed{J = 2, 1, 0}$ $(^3P_2, \ ^3P_1, \ ^3P_0)$ with $\boxed{\text{5, 3, 1 states, respectively}}$.

E9.26(a) Closed shells and subshells do not contribute to either L or S and thus are ignored in what follows.

(a) Li[He]2s^1: $S = \frac{1}{2}$, $L = 0$; $J = \frac{1}{2}$, so the only term is $\boxed{^2S_{1/2}}$

(b) Na[Ne]3p^1: $S = \frac{1}{2}$, $L = 1$; $J = \frac{3}{2}, \frac{1}{2}$, so the terms are $\boxed{^2P_{3/2} \text{ and } ^2P_{1/2}}$

E9.27(a) See eqn. 9.46 for the selection rules. (a) $\boxed{\text{allowed}}$, (b) $\boxed{\text{forbidden}}$, (c) $\boxed{\text{allowed}}$.

Solutions to problems

Solutions to numerical problems

P9.1 All lines in the hydrogen spectrum fit the Rydberg formula:

$$\frac{1}{\lambda} = R_H\left(\frac{1}{n_1^2} - \frac{1}{n_2^2}\right) \quad \left[9.1, \text{ with } \tilde{\nu} = \frac{1}{\lambda}\right] \quad R_H = 109677 \text{ cm}^{-1}$$

Find n_1 from the value of λ_{max}, which arises from the transition $n_1 + 1 \rightarrow n_1$

$$\frac{1}{\lambda_{max} R_H} = \frac{1}{n_1^2} - \frac{1}{(n_1+1)^2} = \frac{2n_1+1}{n_1^2(n_1+1)^2}$$

$$\lambda_{max} R_H = \frac{n_1^2(n_1+1)^2}{2n_1+1} = (12\,368 \times 10^{-9}\,\text{m}) \times (109\,677 \times 10^2\,\text{m}^{-1}) = 135.65$$

Since $n_1 = 1, 2, 3$ and 4 have already been accounted for, try $n_1 = 5, 6, \ldots$. With $n_1 = 6$ we get $\frac{n_1^2(n_1+1)^2}{2n_1+1} = 136$. Hence, the Humphreys series is $\boxed{n_2 \rightarrow 6}$ and the transitions are given by

$$\frac{1}{\lambda} = (109\,677\,\text{cm}^{-1}) \times \left(\frac{1}{36} - \frac{1}{n_2^2}\right), \quad n_2 = 7, 8, \ldots$$

and occur at 12372 nm, 7503 nm, 5908 nm, 5129 nm, …, 3908 nm (at $n_2 = 15$), converging to 3282 nm as $n_2 \rightarrow \infty$, in agreement with the quoted experimental result.

P9.3 A Lyman series corresponds to $n_1 = 1$; hence

$$\tilde{\nu} = R_{Li^{2+}}\left(1 - \frac{1}{n^2}\right), \quad n = 2, 3, \ldots \quad \left[\tilde{\nu} = \frac{1}{\lambda}\right]$$

Therefore, if the formula is appropriate, we expect to find that $\tilde{\nu}\left(1 - \frac{1}{n^2}\right)^{-1}$ is a constant ($R_{Li^{2+}}$).

We therefore draw up the following table:

n	2	3	4
ν/cm^{-1}	740747	877924	925933
$\tilde{\nu}\left(1 - \frac{1}{n^2}\right)^{-1} / \text{cm}^{-1}$	987663	987665	987662

Hence, the formula does describe the transitions, and $\boxed{R_{Li^{2+}} = 987\,663\,\text{cm}^{-1}}$. The Balmer transitions lie at

$$\tilde{\nu} = R_{Li^{2+}}\left(\frac{1}{4} - \frac{1}{n^2}\right) \quad n = 3, 4, \ldots$$

$$= (987\,663\,\text{cm}^{-1}) \times \left(\frac{1}{4} - \frac{1}{n^2}\right) = \boxed{137\,175\,\text{cm}^{-1}}, \boxed{185\,187\,\text{cm}^{-1}}, \ldots$$

The ionization energy of the ground-state ion is given by

$$\tilde{\nu} = R_{Li^{2+}}\left(1 - \frac{1}{n^2}\right), \quad n \rightarrow \infty$$

and hence corresponds to

$$\tilde{\nu} = 987\,663\,\text{cm}^{-1} \quad \text{or} \quad \boxed{122.5\,\text{eV}}$$

P9.5 The 7p configuration has just one electron outside a closed subshell. That electron has $l = 1$, $s = 1/2$, and $j = 1/2$ or $3/2$, so the atom has $L = 1$, $S = 1/2$, and $J = 1/2$ or $3/2$. The term symbols are $\boxed{^2P_{1/2} \text{ and } ^2P_{3/2}}$, of which the former has the lower energy. The 6d configuration also has just one electron outside a closed subshell; that electron has $l = 2$, $s = 1/2$, and $j = 3/2$ or $5/2$, so the atom has $L = 2$, $S = 1/2$, and $J = 3/2$ or $5/2$. The term symbols are $\boxed{^2D_{3/2} \text{ and } ^2D_{5/2}}$, of which the former has the lower energy. According to the simple treatment of spin–orbit coupling, the energy is given by

$$E_{l,s,j} = \tfrac{1}{2}hc\tilde{A}[j(j+1) - l(l+1) - s(s+1)]$$

where $\tilde{A}$ is the spin–orbit coupling constant. So

$$E(^2P_{1/2}) = \tfrac{1}{2}hc\tilde{A}[\tfrac{1}{2}(\tfrac{1}{2}+1) - 1(1+1) - \tfrac{1}{2}(\tfrac{1}{2}+1)] = -hc\tilde{A}$$

and $E(^2D_{3/2}) = \tfrac{1}{2}hc\tilde{A}[\tfrac{3}{2}(\tfrac{3}{2}+1) - 2(2+1) - \tfrac{1}{2}(\tfrac{1}{2}+1)] = -\tfrac{3}{2}hc\tilde{A}$

This approach would predict the ground state to be $\boxed{^2D_{3/2}}$.

COMMENT. The computational study cited finds the $^2P_{1/2}$ level to be lowest, but the authors caution that the error of similar calculations on Y and Lu is comparable to the computed difference between levels.

P9.7 $R_H = k\mu_H$, $R_D = k\mu_D$, $R_\infty = k\mu$ [9.15]

where R_∞ corresponds to an infinitely heavy nucleus, with $\mu = m_e$.

Since $\mu = \dfrac{m_e m_N}{m_e + m_N}$ [N = p or d]

$$R_H = k\mu_H = \frac{km_e}{1 + \frac{m_e}{m_p}} = \frac{R_\infty}{1 + \frac{m_e}{m_p}}$$

Likewise, $R_D = \dfrac{R_\infty}{1 + \frac{m_e}{m_d}}$, where m_p is the mass of the proton and m_d the mass of the deuteron. The two lines in question lie at

$$\frac{1}{\lambda_H} = R_H(1 - \tfrac{1}{4}) = \tfrac{3}{4}R_H \qquad \frac{1}{\lambda_D} = R_D(1 - \tfrac{1}{4}) = \tfrac{3}{4}R_D$$

and hence

$$\frac{R_H}{R_D} = \frac{\lambda_D}{\lambda_H} = \frac{\tilde{\nu}_H}{\tilde{\nu}_D}$$

Then, since

$$\frac{R_H}{R_D} = \frac{1 + \frac{m_e}{m_d}}{1 + \frac{m_e}{m_p}}, \quad m_d = \frac{m_e}{\left(1 + \frac{m_e}{m_p}\right)\frac{R_H}{R_D} - 1}$$

we can calculate m_d from

$$m_d = \frac{m_e}{\left(1+\frac{m_e}{m_p}\right)\frac{\lambda_D}{\lambda_H}-1} = \frac{m_e}{\left(1+\frac{m_e}{m_p}\right)\frac{\tilde{\nu}_H}{\tilde{\nu}_D}-1}$$

$$= \frac{9.10939\times10^{-31}\text{ kg}}{\left(1+\frac{9.1039\times10^{-31}\text{ kg}}{1.67262\times10^{-27}\text{ kg}}\right)\times\left(\frac{82259.098\text{ cm}^{-1}}{82281.476\text{ cm}^{-1}}\right)-1}$$

$$= \boxed{3.3429\times10^{-27}\text{ kg}}$$

Since $I = R_\infty hc$,

$$\frac{I_D}{I_H} = \frac{R_D}{R_H} = \frac{\tilde{\nu}_D}{\tilde{\nu}_H} = \frac{82281.476\text{ cm}^{-1}}{82259.098\text{ cm}^{-1}} = \boxed{1.000272}$$

P9.9 (a) The splitting of adjacent energy levels is related to the difference in wavenumber of the spectral lines as follows:

$$hc\Delta\tilde{\nu} = \Delta E = \mu_B\mathcal{B}, \quad \text{so} \quad \Delta\tilde{\nu} = \frac{\mu_B\mathcal{B}}{hc} = \frac{(9.274\times10^{-24}\text{ J T}^{-1})(2\text{ T})}{(6.626\times10^{-34}\text{ J s})(2.998\times10^{10}\text{ cm s}^{-1})}$$

$$\Delta\tilde{\nu} = \boxed{0.9\text{ cm}^{-1}}.$$

(b) Transitions induced by absorbing visible light have wavenumbers in the tens of thousands of reciprocal centimeters, so normal Zeeman splitting is $\boxed{\text{small}}$ compared to the difference in energy of the states involved in the transition. Take a wavenumber from the middle of the visible spectrum as typical:

$$\tilde{\nu} = \frac{1}{\lambda} = \frac{1}{600\text{ nm}}\left(\frac{10^9\text{ nm m}^{-1}}{10^2\text{ cm m}^{-1}}\right) = 1.7\times10^4\text{ cm}^{-1}$$

Or take the Balmer series as an example, as suggested in the problem; the Balmer wavenumbers are (eqn 9.1):

$$\tilde{\nu} = R_H\left(\frac{1}{2^2} - \frac{1}{n^3}\right)$$

The smallest Balmer wavenumber is

$$\tilde{\nu} = (109\,677\text{ cm}^{-1})\times(1/4 - 1/9) = 15\,233\text{ cm}^{-1}$$

and the upper limit is

$$\tilde{\nu} = (109\,677\text{ cm}^{-1})\times(1/4 - 0) = 27\,419\text{ cm}^{-1}$$

P9.11 (a) ${}^2S_{1/2} \rightarrow {}^2P_{3/2}$ (or ${}^2P_{1/2}$)

(b) See the solution to Exercise 9.2(a).

For atoms A, eqn. 9.9 may be rewritten in terms of the Rydberg constant R_A as

$$E_n = -\frac{Z^2\mu_A hcR_A}{m_e n^2} \approx -\frac{Z^2 hcR_\infty}{n^2}$$

where to within 0.01% the ratio μ_A/m_e is unity. Eqn 9.1 can then be rewritten as

$$\tilde{\nu} = Z^2 R_\infty \left(\frac{1}{n_1^2} - \frac{1}{n_2^2}\right) \quad \lambda = \frac{1}{\tilde{\nu}} \quad \nu = \frac{c}{\lambda}$$

$$\tilde{\nu} = 4 \times 109737\ \text{cm}^{-1}\left(\frac{1}{1^2} - \frac{1}{4^2}\right) = \boxed{4.115 \times 10^5\ \text{cm}^{-1}} \quad \lambda = \boxed{2.430 \times 10^{-6}\ \text{cm}}$$

$$\nu = \frac{2.9978 \times 10^{10}\ \text{cm s}^{-1}}{2.430 \times 10^{-6}\ \text{cm}} = \boxed{1.234 \times 10^{16}\ \text{s}^{-1}}$$

(c) See Example 9.2. The mean radius is given by $\langle r \rangle = \int_0^\infty r^3 R_{n,l}^2 \mathrm{d}r$. After substituting eqn 9.12 for $R_{n,l}$ and integrating, the general expression for the mean radius of a hydrogenic orbital with quantum numbers l and n is obtained (see eqn. 10.19 of Atkins and de Paula, *Physical Chemistry*, 8th edn).

$$\langle r_{n,l} \rangle = n^2 \left\{1 + \frac{1}{2}\left(1 - \frac{l(l+1)}{n^2}\right)\right\} \frac{a_0}{Z}$$

$$\langle r_{4,1} \rangle = \boxed{\frac{23}{2} a_0}$$

Hence, since $\langle r_{1,0} \rangle = \boxed{\frac{3}{4} a_0}$, the mean radius has increased by $\boxed{\frac{43}{4} a_0}$.

Solutions to theoretical problems

P9.13 Consider $\psi_{2p_z} = \psi_{2,1,0}$, which extends along the z-axis. The most probable point along the z-axis is where the radial function has its maximum value (for ψ^2 is also a maximum at that point). From Table 9.1 we know that

$$R_{21} \propto \rho e^{-\rho/2}$$

and so $\dfrac{\mathrm{d}R}{\mathrm{d}\rho} = (1 - \tfrac{1}{2}\rho)e^{-\rho/4} = 0$ when $\rho = 2$.

Therefore $r^* = \dfrac{2a_0}{Z}$, and the point of maximum probability lies at $z = \pm\dfrac{2a_0}{Z} = \boxed{\pm 106\ \text{pm}}$.

COMMENT. Since the radial portion of a 2p function is the same, the same result would have been obtained for all of them. The direction of the most probable point would, however, be different.

P9.15 (a) We must show that $\int |\psi_{3p_x}|^2\,d\tau = 1$. The integrations are most easily performed in spherical coordinates (Figure 7.21).

$$\int |\psi_{3p_x}|^2\,d\tau = \int_0^{2\pi}\int_0^{\pi}\int_0^{\infty} |\psi_{3p_x}|^2\, r^2 \sin(\theta)\, dr\, d\theta\, d\phi$$

$$= \int_0^{2\pi}\int_0^{\pi}\int_0^{\infty} \left| R_{31}(\rho)\left\{\frac{Y_{1-1} - Y_{11}}{\sqrt{2}}\right\}\right|^2 r^2 \sin(\theta)\, dr\, d\theta\, d\phi \text{ [Table 9.1, eqn 9.22]}$$

where $\rho = 2r/a_0$, $r = \rho a_0/2$, $dr = (a_0/2)d\rho$. [Note that ρ defined here is $n \times \rho$ as defined in Table 9.1, with $n = 3$ here.]

$$= \frac{1}{2}\int_0^{2\pi}\int_0^{\pi}\int_0^{\infty}\left(\frac{a_0}{2}\right)^3 \left|\left[\left(\frac{1}{27(6)^{1/2}}\right)\left(\frac{1}{a_0}\right)^{3/2}\left(4 - \frac{1}{3}\rho\right)\rho e^{-\rho/6}\right]\right.$$

$$\left.\times\left[\left(\frac{3}{8\pi}\right)^{1/2} 2\sin(\theta)\cos(\phi)\right]\right|^2 \rho^2 \sin(\theta)\, d\rho\, d\theta\, d\phi$$

$$= \frac{1}{46656\pi}\int_0^{2\pi}\int_0^{\pi}\int_0^{\infty}\left|\left(4 - \frac{1}{3}\rho\right)\rho e^{-\rho/6}\sin(\theta)\cos(\phi)\right|^2 \rho^2 \sin(\theta)\, d\rho\, d\theta\, d\phi$$

$$= \frac{1}{46656\pi}\underbrace{\int_0^{2\pi}\cos^2(\phi)\,d\phi}_{\pi}\;\underbrace{\int_0^{\pi}\sin^3(\theta)\,d\theta}_{4/3}\;\underbrace{\int_0^{\infty}\left(4 - \frac{1}{3}\rho\right)^2 \rho^4 e^{-\rho/3}\,d\rho}_{34992}$$

$= 1$ Thus, ψ_{3p_x} is normalized to 1.

We must also show that $\int \psi_{3p_x}\psi_{3d_{xy}}\,d\tau = 0$.

Using Tables 8.2 and 9.1, we find that

$$\psi_{3p_x} = \frac{1}{54(2\pi)^{1/2}}\left(\frac{1}{a_0}\right)^{3/2}\left(4 - \frac{1}{3}\rho\right)\rho e^{-\rho/6}\sin(\theta)\cos(\phi)$$

$$\psi_{3d_{xy}} = R_{32}\left\{\frac{Y_{22} - Y_{2-2}}{\sqrt{2}i}\right\}$$

$$= \frac{1}{32(2\pi)^{1/2}}\left(\frac{1}{a_0}\right)^{3/2}\rho^2 e^{-\rho/6}\sin^2(\theta)\sin(2\phi)$$

where $\rho = 2r/a_0$, $r = \rho a_0/2$, $dr = (a_0/2)d\rho$.

$$\int \psi_{3p_x}\psi_{3d_{xy}}\,d\tau = \text{constant} \times \int_0^{\infty}\rho^5 e^{-\rho/3}\,d\rho\;\underbrace{\int_0^{2\pi}\cos(\phi)\sin(2\phi)\,d\phi}_{0}\;\int_0^{\pi}\sin^3(\theta)\,d\theta$$

Since the integral equals zero, ψ_{3p_x} and $\psi_{3d_{xy}}$ are orthogonal.

(b) Radial nodes are determined by finding the ρ values, ($\rho = 2r/a_0$), [ρ here is defined as in Table 9.1], for which the radial wavefunction equals zero. These values are the roots of the polynomial portion of the wavefunction. For the 3s orbital, $6 - 6\rho + \rho^2 = 0$ [Table 9.1], when

$$\boxed{\rho_{\text{node}} = 3 + \sqrt{3} \quad \text{and} \quad \rho_{\text{node}} = 3 - \sqrt{3}}$$

The 3s orbital has these two spherically symmetrical nodes. There is no node at $\rho = 0$ so we conclude that there is a finite probability of finding a 3s electron at the nucleus.

For the $3p_x$ orbital, $(4 - \rho)(\rho) = 0$ when $\boxed{\rho_{\text{node}} = 0 \quad \text{and} \quad \rho_{\text{node}} = 4}$. There is a zero probability of finding a $3p_x$ electron at the nucleus.

For the $3d_{xy}$ orbital $\boxed{\rho_{\text{node}} = 0}$ is the only radial node.

(c) $$\langle r\rangle_{3s} = \int |R_{10}Y_{00}|^2 r\, d\tau = \int |R_{10}Y_{00}|^2 r^3 \sin(\theta)\, dr\, d\theta\, d\phi$$

$$= \int_0^\infty R_{10}^2 r^3\, dr \underbrace{\int_0^{2\pi}\int_0^{\pi} |Y_{00}|^2 \sin(\theta)\, d\theta\, d\phi}_{1} = \frac{a_0}{3888} \underbrace{\int_0^\infty (6 - 2\rho + \rho^2/9)^2 \rho^3 e^{-\rho/3}\, d\rho}_{52488}$$

[Note that ρ defined here is $n \times \rho$ as defined in Table 9.1, with $n = 3$ here.]

$$\boxed{\langle r\rangle_{3s} = \frac{27a_0}{2}}$$

(d)

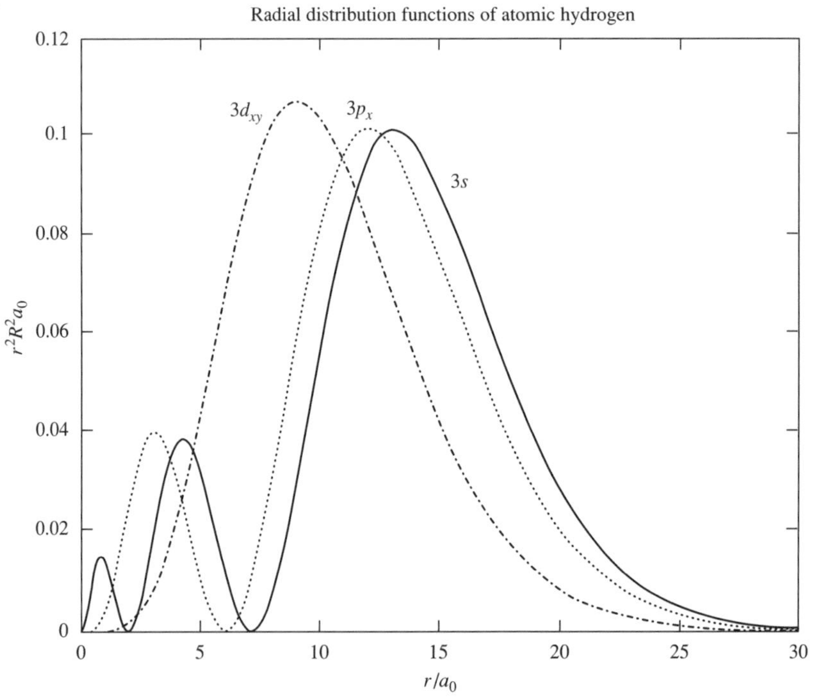

Figure 9.1(a)

The plot in Figure 9.1(a) shows that the 3s orbital has larger values of the radial distribution function for $r < a_0$. This penetration of inner core electrons of multi-electron atoms means that a 3s electron experiences a larger effective nuclear charge and, consequently, has a lower energy than either a 3p or $3d_{xy}$ electron. This reasoning also leads us to conclude that a $3p_x$ electron has less energy than a $3d_{xy}$ electron.

$$E_{3s} < E_{3p_x} < E_{3d_{xy}}.$$

(e) Polar plots with $\theta = 90°$, Figure 9.1(b).

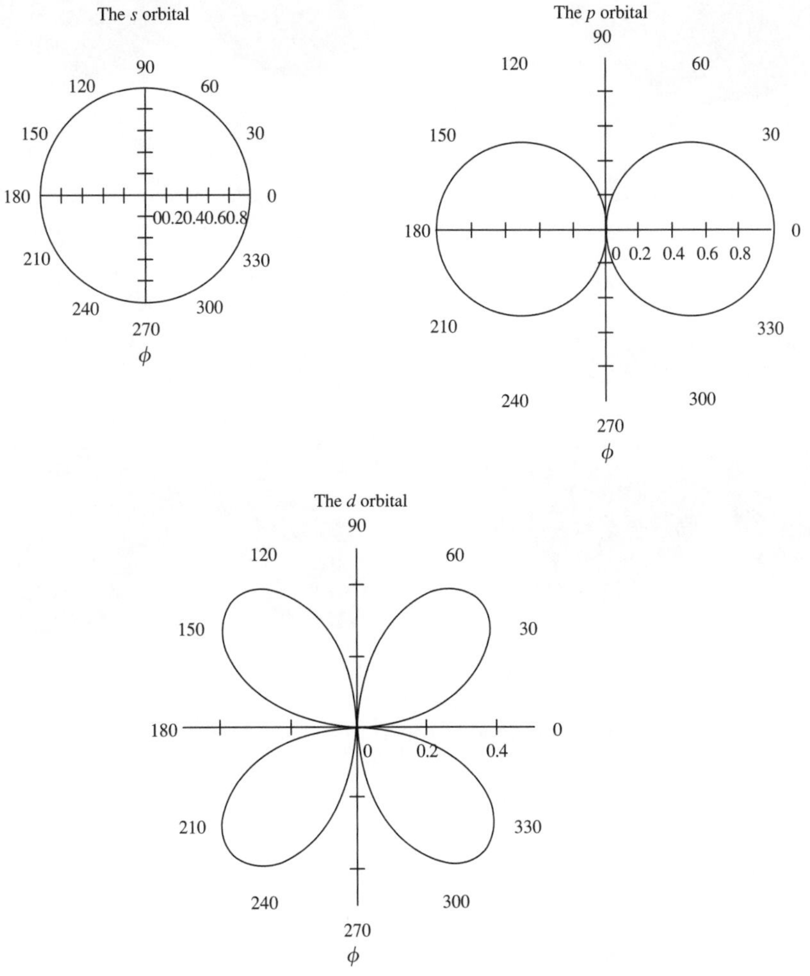

Figure 9.1(b)

Boundary surface plots, Figure 9.1(c).

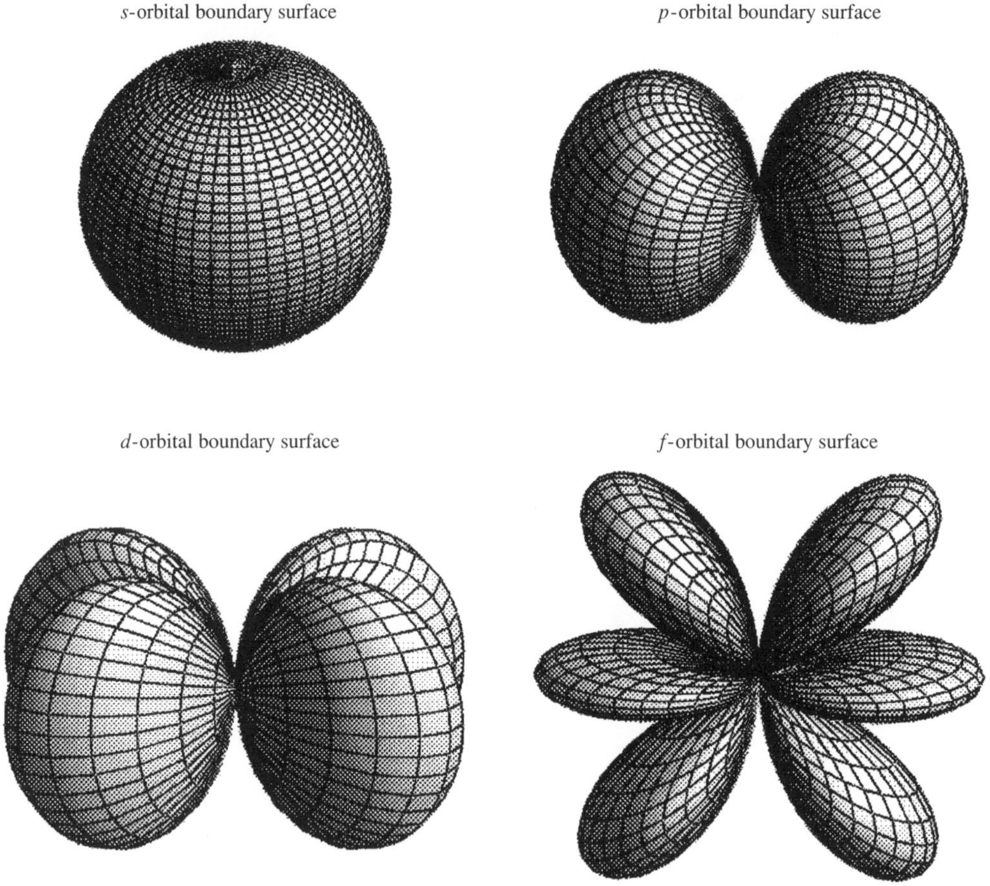

Figure 9.1(c)

P9.17 The general rule to use in deciding commutation properties is that operators having no variable in common will commute with each other. We first consider the commutation of $\hat{l}_z$ with the Hamiltonian. This is most easily solved in spherical polar coordinates.

$$\hat{l}_z = \frac{\hbar}{\mathrm{i}}\frac{\partial}{\partial\phi} \text{ [Problem 8.28, Section 8.6, and eqn 8.46]}$$

$$H = -\frac{\hbar^2}{2\mu}\nabla^2 + V \text{[\textit{Further Information} 9.1]} \quad V = -\frac{Ze^2}{4\pi\varepsilon_0 r}$$

Since V has no variable in common with $\hat{l}_z$, this part of the Hamiltonian and $\hat{l}_z$ commute.

$$\nabla^2 = \text{terms in } r \text{ only} + \text{terms in } \theta \text{ only} + \frac{1}{r^2\sin^2\theta}\frac{\partial^2}{\partial\phi^2} \text{ [\textit{Justification 8.5}]}$$

The terms in r only and θ only necessarily commute with $\hat{l}_z$ (ϕ only). The final term in ∇^2 contains $\frac{\partial^2}{\partial\phi^2}$, which commutes with $\frac{\partial}{\partial\phi}$ since an operator necessarily commutes with itself. By symmetry we can deduce that if H commutes with $\hat{l}_z$ it must also commute with $\hat{l}_x$ and $\hat{l}_y$ since they are related to each other by a simple transformation of coordinates. This proves useful in establishing the commutation of l^2 and H. We form

$$\hat{l}^2 = \hat{l}\cdot\hat{l} = (i\hat{l}_x + j\hat{l}_y + k\hat{l}_z)\cdot(i\hat{l}_x + j\hat{l}_y + k\hat{l}_z) = \hat{l}_x^2 + \hat{l}_y^2 + \hat{l}_z^2$$

If H commutes with each of $\hat{l}_x$, $\hat{l}_y$, and $\hat{l}_z$ it must commute with $\hat{l}_x^2$, $\hat{l}_y^2$, and $\hat{l}_z^2$. Therefore, it also commutes with $\hat{l}^2$. Thus, H commutes with both $\hat{l}^2$ and $\hat{l}_z$.

COMMENT. As described at the end of Section 7.6, the physical properties associated with non-commuting operators cannot be simultaneously known with precision. However, since H, $\hat{l}^2$, and $\hat{l}_z$ commute we may simultaneously have exact knowledge of the energy, the total orbital angular momentum, and the projection of the orbital angular momentum along an arbitrary axis.

P9.19

$$\langle r^m\rangle_{nl} = \int r^m|\psi_{nl}|^2\,\mathrm{d}\tau = \int_0^\infty\int_0^{2\pi}\int_0^\pi r^{m+2}|R_{nl}Y_{l0}|^2\sin(\theta)\,\mathrm{d}\theta\,\mathrm{d}\phi\,\mathrm{d}r$$

$$= \int_0^\infty r^{m+2}|R_{nl}|^2\,\mathrm{d}r\int_0^{2\pi}\int_0^\pi |Y_{l0}|^2\sin(\theta)\,\mathrm{d}\theta\,\mathrm{d}\phi = \int_0^\infty r^{m+2}|R_{nl}|^2\,\mathrm{d}r$$

With $r = (na_0/2Z)\rho$ and $m = -1$ the expectation value is

$$\langle r^{-1}\rangle_{nl} = \left(\frac{na_0}{2Z}\right)^2\int_0^\infty \rho|R_{nl}|^2\,\mathrm{d}\rho$$

(a) $$\langle r^{-1}\rangle_{1s} = \left(\frac{a_0}{2Z}\right)^2\left\{2\left(\frac{Z}{a_0}\right)^{3/2}\right\}^2\int_0^\infty \rho\mathrm{e}^{-\rho}\,\mathrm{d}\rho \quad [\text{Table 9.1}]$$

$$= \boxed{\frac{Z}{a_0}} \quad \text{because} \quad \int_0^\infty \rho\mathrm{e}^{-\rho}\,\mathrm{d}\rho = 1$$

(b) $$\langle r^{-1}\rangle_{2s} = \left(\frac{a_0}{Z}\right)^2\left\{\frac{1}{8^{1/2}}\left(\frac{Z}{a_0}\right)^{3/2}\right\}^2\int_0^\infty \rho(2-\rho)^2\mathrm{e}^{-\rho}\,\mathrm{d}\rho \quad [\text{Table 9.1}]$$

$$= \frac{Z}{8a_0}(2) \quad \text{because} \quad \int_0^\infty \rho(2-\rho)^2\mathrm{e}^{-\rho}\,\mathrm{d}\rho = 2$$

$$\langle r^{-1}\rangle_{2s} = \boxed{\frac{Z}{4a_0}}$$

(c) $$\langle r^{-1}\rangle_{2p} = \left(\frac{a_0}{Z}\right)^2\left\{\frac{1}{24^{1/2}}\left(\frac{Z}{a_0}\right)^{3/2}\right\}^2\int_0^\infty \rho^3\mathrm{e}^{-\rho}\,\mathrm{d}\rho \quad [\text{Table 9.1}]$$

$$= \frac{Z}{24a_0}(6) \quad \text{because} \quad \int_0^\infty \rho^3\mathrm{e}^{-\rho}\,\mathrm{d}\rho = 6$$

$$\langle r^{-1}\rangle_{2p} = \boxed{\frac{Z}{4a_0}}$$

The general formula for a hydrogenic orbital is $\langle r^{-1}\rangle_{nl} = \dfrac{Z}{n^2 a_0}$.

P9.21 (a) The trajectory is defined, which is not allowed according to quantum mechanics.

(b) The angular momentum of a three-dimensional system is given by $\{l(l+1)\}^{1/2}\hbar$, not by $n\hbar$. In the Bohr model, the ground state possesses orbital angular momentum ($n\hbar$, with $n = 1$), but the actual ground state has no angular momentum ($l = 0$). Moreover, the distribution of the electron is quite different in the two cases. The two models can be distinguished experimentally by (a) showing that there is zero orbital angular momentum in the ground state (by examining its magnetic properties) and (b) examining the electron distribution (such as by showing that the electron and the nucleus do come into contact, Chapter 14).

P9.23 *Justification 9.4* noted that the transition dipole moment, $\mu_{\rm fi}$, had to be non-zero for a transition to be allowed. The *Justification* examined conditions that allowed the z component of this quantity to be non-zero; now examine the x and y components.

$$\mu_{x,\rm fi} = -e\int \psi_{\rm f}^* x \psi_{\rm i} d\tau \quad \text{and} \quad \mu_{y,\rm fi} = -e\int \psi_{\rm f}^* y \psi_{\rm i} d\tau$$

As in the *Justification*, express the relevant Cartesian variables in terms of the spherical harmonics, $Y_{l,m}$. Start by expressing them in spherical polar coordinates:

$$x = r\sin\theta\cos\phi \quad \text{and} \quad y = r\sin\theta\sin\phi.$$

Note that $Y_{1,1}$ and $Y_{1,-1}$ have factors of $\sin\theta$. They also contain complex exponentials that can be related to the sine and cosine of ϕ through the identities

$$\cos\phi = 1/2(e^{i\phi} + e^{-i\phi}) \quad \text{and} \quad \sin\phi = 1/2i(e^{i\phi} - e^{-i\phi}).$$

These relationships motivate us to try linear combinations $Y_{1,1} + Y_{1,-1}$ and $Y_{1,1} - Y_{1,-1}$ (from Table 8.2; note c here corresponds to the normalization constant in the table):

$$Y_{1,1} + Y_{1,-1} = -c\sin\theta(e^{i\phi} + e^{-i\phi}) = -2c\sin\theta\cos\phi = -2cx/r,$$

so $x = -(Y_{1,1} + Y_{1,-1})r/2c$;

$$Y_{1,1} - Y_{1,-1} = c\sin\theta(e^{i\phi} - e^{-i\phi}) = 2ic\sin\theta\cos\phi = 2icy/r,$$

so $y = (Y_{1,1} - Y_{1,-1})r/2ic$.

Now we can express the integrals in terms of radial wavefunctions $R_{n,l}$ and spherical harmonics Y_{l,m_l}

$$\mu_{x,\rm fi} = \frac{e}{2c}\int_0^\infty R_{n_{\rm f},l_{\rm f}} r R_{n_{\rm i},l_{\rm i}} r^2 dr \int_0^\pi\int_0^{2\pi} Y_{l_{\rm f},m_{l\rm f}}^*(Y_{1,1} + Y_{1,-1})Y_{l_{\rm i},m_{l\rm i}} \sin\theta\, d\theta\, d\phi.$$

The angular integral can be broken into two, one of which contains $Y_{1,1}$ and the other $Y_{1,-1}$. According to the 'triple integral' relationship below Table 8.2, the integral

$$\int_0^\pi\int_0^{2\pi} Y_{l_{\rm f},m_{l\rm f}}^* Y_{1,1} Y_{l_{\rm i},m_{l\rm i}} \sin\theta\, d\theta\, d\phi$$

vanishes unless $l_f = l_i \pm 1$ and $m_{l_f} = m_{l_i} \pm 1$. The integral that contains $Y_{1,-1}$ introduces no further constraints; it vanishes unless $l_f = l_i \pm 1$ and $m_{l_f} = m_{l_i} \pm 1$. Similarly, the y component introduces no further constraints, for it involves the same spherical harmonics as the x component. The whole set of selection rules, then, is that transitions are allowed only if

$$\boxed{\Delta l = \pm 1 \text{ and } \Delta m_l = 0 \text{ or } \pm 1}.$$

P9.25 (a) The Slater wavefunction [9.30] is

$$\psi(1,2,3,\ldots,N) = \frac{1}{(N!)^{1/2}}\begin{vmatrix} \psi_a(1)\alpha(1) & \psi_a(2)\alpha(2) & \psi_a(3)\alpha(3) & \cdots & \psi_a(N)\alpha(N) \\ \psi_a(1)\beta(1) & \psi_a(2)\beta(2) & \psi_a(3)\beta(3) & \cdots & \psi_a(N)\beta(N) \\ \psi_b(1)\alpha(1) & \psi_b(2)\alpha(2) & \psi_b(3)\alpha(3) & \cdots & \psi_b(N)\alpha(N) \\ \vdots & \vdots & \vdots & \vdots & \vdots \\ \psi_z(1)\beta(1) & \psi_z(2)\beta(2) & \psi_z(3)\beta(3) & \cdots & \psi_z(N)\beta(N) \end{vmatrix}$$

Interchanging any two columns or rows leaves the function unchanged except for a change in sign. For example, interchanging the first and second columns of the above determinant gives:

$$\psi(1,2,3,\ldots,N) = \frac{-1}{(N!)^{1/2}}\begin{vmatrix} \psi_a(2)\alpha(2) & \psi_a(1)\alpha(1) & \psi_a(3)\alpha(3) & \cdots & \psi_a(N)\alpha(N) \\ \psi_a(2)\beta(2) & \psi_a(1)\beta(1) & \psi_a(3)\beta(3) & \cdots & \psi_a(N)\beta(N) \\ \psi_b(2)\alpha(2) & \psi_b(1)\alpha(1) & \psi_b(3)\alpha(3) & \cdots & \psi_b(N)\alpha(N) \\ \vdots & \vdots & \vdots & \vdots & \vdots \\ \psi_z(2)\beta(2) & \psi_z(1)\beta(1) & \psi_z(3)\beta(3) & \cdots & \psi_z(N)\beta(N) \end{vmatrix}$$

$$= -\psi(2,1,3,\ldots,N)$$

This demonstrates that a Slater determinant is antisymmetric under particle exchange.

(b) The possibility that two electrons occupy the same orbital with the same spin can be explored by making any two rows of the Slater determinant identical, thereby providing identical orbital and spin functions to two rows. Rows 1 and 2 are identical in the Slater wavefunction below. Interchanging these two rows causes the sign to change without in any way changing the determinant.

$$\psi(1,2,3,\ldots,N) = \frac{1}{(N!)^{1/2}}\begin{vmatrix} \psi_a(1)\alpha(1) & \psi_a(2)\alpha(2) & \psi_a(3)\alpha(3) & \cdots & \psi_a(N)\alpha(N) \\ \psi_a(1)\alpha(1) & \psi_a(2)\alpha(2) & \psi_a(3)\alpha(3) & \cdots & \psi_a(N)\alpha(N) \\ \psi_b(1)\alpha(1) & \psi_b(2)\alpha(2) & \psi_b(3)\alpha(3) & \cdots & \psi_b(N)\alpha(N) \\ \vdots & \vdots & \vdots & \vdots & \vdots \\ \psi_z(1)\beta(1) & \psi_z(2)\beta(2) & \psi_z(3)\beta(3) & \cdots & \psi_z(N)\beta(N) \end{vmatrix}$$

$$= -\psi(2,1,3,\ldots,N) = -\psi(1,2,3,\ldots,N)$$

Only the null function satisfies a relationship in which it is the negative of itself so we conclude that, since the null function is inconsistent with existence, the Slater determinant satisfies the Pauli exclusion principle [Section 9.4 b]: no two electrons can occupy the same orbital with the same spin.

Solutions to applications

P9.27 The wavenumber of a spectroscopic transition is related to the difference in the relevant energy levels. For a one-electron atom or ion the relationship is

$$hc\tilde{\nu} = \Delta E = \frac{Z^2\mu_{\mathrm{He}}e^4}{32\pi^2\varepsilon_0^2\hbar^2 n_1^2} - \frac{Z^2\mu_{\mathrm{He}}e^4}{32\pi^2\varepsilon_0^2\hbar^2 n_2^2} = \frac{Z^2\mu_{\mathrm{He}}e^4}{32\pi^2\varepsilon_0^2\hbar^2}\left(\frac{1}{n_2^2} - \frac{1}{n_1^2}\right)$$

Solving for $\tilde{\nu}$, using the definition $\hbar = h/2\pi$ and the fact that $Z = 2$ for He, yields

$$\tilde{\nu} = \frac{\mu_{\mathrm{He}}e^4}{2\varepsilon_0^2 h^3 c}\left(\frac{1}{n_2^2} - \frac{1}{n_1^2}\right)$$

Note that the wavenumbers are proportional to the reduced mass, which is very close to the mass of the electron for both isotopes. In order to distinguish between them, we need to carry lots of significant figures in the calculation.

$$\tilde{\nu} = \frac{\mu_{\mathrm{He}}(1.60218 \times 10^{-19}\,\mathrm{C})^4}{2(8.85419 \times 10^{-12}\,\mathrm{J^{-1}C^2\,m^{-1}})^2 \times (6.62607 \times 10^{-34}\,\mathrm{J\,s})^3 \times (2.99792 \times 10^{10}\,\mathrm{cm\,s^{-1}})} \times \left(\frac{1}{n_2^2} - \frac{1}{n^2}\right)$$

$$\tilde{\nu}/\mathrm{cm^{-1}} = 4.81870 \times 10^{35}(\mu_{\mathrm{He}}/kg)\left(\frac{1}{n_2^2} - \frac{1}{n_1^2}\right)$$

The reduced masses for the ^{4}He and ^{3}He nuclei are

$$\mu = \frac{m_{\mathrm{e}}m_{\mathrm{nuc}}}{m_{\mathrm{e}} + m_{\mathrm{nuc}}}$$

where $m_{\mathrm{nuc}} = 4.00260\ m_{\mathrm{u}}$ for ^{4}He and $3.01603\ m_{\mathrm{u}}$ for ^{3}He, or, in kg

^{4}He $m_{\mathrm{nuc}} = (4.00260\ m_{\mathrm{u}}) \times (1.66054 \times 10^{-27}\ \mathrm{kg}\ m_{\mathrm{u}}^{-1}) = 6.64648 \times 10^{-27}\ \mathrm{kg}$
^{3}He $m_{\mathrm{nuc}} = (3.01603\ m_{\mathrm{u}}) \times (1.66054 \times 10^{-27}\ \mathrm{kg}\ m_{\mathrm{u}}^{-1}) = 5.00824 \times 10^{-27}\ \mathrm{kg}$

The reduced masses are

$$^4\mathrm{He}\ \mu = \frac{(9.10939 \times 10^{-31}\,\mathrm{kg}) \times (6.64648 \times 10^{-27}\,\mathrm{kg})}{(9.10939 \times 10^{-31} + 6.64648 \times 10^{-27})\mathrm{kg}} = 9.10814 \times 10^{-31}\,\mathrm{kg}$$

$$^3\mathrm{He}\ \mu = \frac{(9.10939 \times 10^{-31}\,\mathrm{kg}) \times (5.00824 \times 10^{-27}\,\mathrm{kg})}{(9.10939 \times 10^{-31} + 5.00824 \times 10^{-27})\mathrm{kg}} = 9.10773 \times 10^{-31}\,\mathrm{kg}$$

Finally, the wavenumbers for $n = 3 \to n = 2$ are:

$$^4\mathrm{He}\ \tilde{\nu} = (4.81870 \times 10^{35}) \times (9.10814 \times 10^{-31}) \times (1/4 - 1/9)\mathrm{cm^{-1}} = \boxed{60\,957.4\ \mathrm{cm^{-1}}}$$

$$^3\mathrm{He}\ \tilde{\nu} = (4.81870 \times 10^{35}) \times (9.10773 \times 10^{-31}) \times (1/4 - 1/9)\mathrm{cm^{-1}} = \boxed{60\,954.7\ \mathrm{cm^{-1}}}$$

The wavenumbers for $n = 2 \to n = 1$ are

$$^4\mathrm{He}\ \tilde{\nu} = (4.81870 \times 10^{35}) \times (9.10814 \times 10^{-31}) \times (1/1 - 1/4)\mathrm{cm^{-1}} = \boxed{32\,9170\ \mathrm{cm^{-1}}}$$

$$^3\mathrm{He}\ \tilde{\nu} = (4.81870 \times 10^{35}) \times (9.10773 \times 10^{-31}) \times (1/1 - 1/4)\mathrm{cm^{-1}} = \boxed{32\,9155\ \mathrm{cm^{-1}}}$$

P9.29 (a) Compute the ratios $\upsilon_{\mathrm{star}}/\upsilon$ for all three lines. We are given wavelength data, so we can use:

$$\frac{\upsilon_{\mathrm{star}}}{\upsilon} = \frac{\lambda}{\lambda_{\mathrm{star}}}.$$

The ratios are:

$$\frac{438.392\text{ nm}}{438.882\text{ nm}} = 0.998884, \quad \frac{440.510\text{ nm}}{441.000\text{ nm}} = 0.998889, \text{ and } \frac{441.510\text{ nm}}{442.020\text{ nm}} = 0.998846.$$

The frequencies of the stellar lines are all less than those of the stationary lines, so we infer that the star is $\boxed{\text{receding}}$ from earth. The Doppler effect follows:

$$\nu_{\text{receding}} = \nu f, \quad \text{where } f = \left(\frac{1 - s/c}{1 + s/c}\right)^{1/2}, \text{ so}$$

$$f^2(1 + s/c) = (1 - s/c), \quad (f^2 + 1)s/c = 1 - f^2, \quad s = \frac{1 - f^2}{1 + f^2}c$$

Our average value of f is 0.998873. (Note: the uncertainty is actually greater than the significant figures here imply, and a more careful analysis would treat uncertainty explicitly.) So, the speed of recession with respect to the earth is:

$$s = \left(\frac{1 - 0.997747}{1 + 0.997747}\right)c = \boxed{1.128 \times 10^{-3}\,\text{c}} = \boxed{3.381 \times 10^{5}\text{ m s}^{-1}}$$

(b) One could compute the star's radial velocity with respect to the sun if one knew the earth's speed with respect to the sun along the sun–star vector at the time of the spectral observation. This could be estimated from quantities available through astronomical observation: the earth's orbital velocity times the cosine of the angle between that velocity vector and the earth–star vector at the time of the spectral observation. (The earth–star direction, which is observable by earth-based astronomers, is practically identical to the sun–star direction, which is technically the direction needed.) Alternatively, repeat the experiment half a year later. At that time, the earth's motion with respect to the sun is approximately equal in magnitude and opposite in direction compared to the original experiment. Averaging f values over the two experiments would yield f values in which the earth's motion is effectively averaged out.

P9.31 Electronic configurations of neutral, fourth-period transition atoms in the ground state are summarized in the following table along with observed, positive oxidation states. The most common, positive oxidation states are indicated by unshaded boxes.

Group	3	4	5	6	7	8	9	10	11	12
Oxidation State	Sc	Ti	V	Cr	Mn	Fe	Co	Ni	Cu	Zn
0	$3d4s^2$	$3d^24s^2$	$3d^34s^2$	$3d^54s$	$3d^54s^2$	$3d^64s^2$	$3d^74s^2$	$3d^84s^2$	$3d^{10}4s$	$3d^{10}4s^2$
+1			☺	☺	☺		☺	☺	☺	
+2		☺	☺	☺	☺	☺	☺	☺	☺	☺
+3	☺	☺	☺	☺	☺	☺	☺	☺		
+4		☺	☺	☺	☺	☺	☺	☺		
+5			☺	☺	☺	☺				
+6				☺	☺	☺				
+7					☺					

Toward the middle of the first transition series (Cr, Mn, and Fe) elements exhibit the widest ranges of oxidation states. This phenomenon is related to the availability of both electrons and orbitals favourable for bonding. Elements to the left (Sc and Ti) of the series have few electrons and relatively low effective nuclear charge leaves d orbitals at high energies that are relatively unsuitable for bonding. To the far right (Cu and Zn) effective nuclear charge may be higher but there are few, if any, orbitals available for bonding. Consequently, it is more difficult to produce a range of compounds that promote a wide range of oxidation states for elements at either end of the series. At the middle and right of the series the +2 oxidation state is very commonly observed because normal reactions can provide the requisite ionization energies for the removal of 4s electrons. The readily available +2 and +3 oxidation states of Mn, Fe, and the +1 and +2 oxidation states of Cu make these cations useful in electron-transfer processes occurring in chains of specialized protein within biological cells. The special size and charge of the Zn^{2+} cation makes it useful for the function of some enzymes. The tendency of Fe^{2+} and Cu^{+} to bind oxygen proves very useful in hemoglobin and electron-transport (respiratory) chain, respectively.

10 Molecular structure

Answers to discussion questions

D10.1 Our comparison of the two theories will focus on the manner of construction of the trial wavefunctions for the hydrogen molecule in the simplest versions of both theories. In the valence bond method, the trial function is a linear combination of two simple product wavefunctions, in which one electron resides totally in an atomic orbital on atom A and the other totally in an orbital on atom B. See eqns 10.1 and 10.2, as well as Figure 10.2. There is no contribution to the wavefunction from products in which both electrons reside on either atom A or B, so the valence-bond approach undervalues any ionic contribution to the trial function by totally neglecting such contributions. It is a totally covalent function. The molecular orbital function for the hydrogen molecule is a product of two functions of the form of eqn 10.8, one for each electron, that is

$$\psi = \{A(1) \pm B(1)\}\{A(2) \pm B(2)\} = A(1)A(2) + B(1)B(2) + A(1)B(2) + B(1)A(2)$$

This function gives as much weight to the ionic forms as to the covalent forms. So, the molecular orbital approach greatly overvalues the ionic contributions. At these crude levels of approximation, the valence-bond method gives dissociation energies closer to the experimental values. However, more sophisticated versions of the molecular orbital approach are the methods of choice for obtaining quantitative results on both diatomic and polyatomic molecules. See Sections 10.6–10.8.

D10.3 Both the Pauling and Mulliken methods for measuring the power of atoms in molecules to attract electrons seem to make good chemical sense. If we look at eqn 10.24 (the Pauling scale), we see that if D_0(A–B) were equal to $(\frac{1}{2})\{D_0(\text{A–A}) + D_0(\text{B–B})\}$ the calculated electronegativity difference would be zero, as expected for completely non-polar bonds. Hence, any increased strength of the A–B bond over the average of the A–A and B–B bonds can reasonably be thought of as being due to the polarity of the A–B bond, which in turn is due to the difference in electronegativity of the atoms involved. Therefore, this difference in bond strengths can be used as a measure of electronegativity difference. To obtain numerical values for individual atoms, a reference state (atom) for electronegativity must be established. The value for fluorine is arbitrarily set at 4.0.

The Mulliken scale (eqn 10.25) may be more intuitive than the Pauling scale because we are used to thinking of ionization energies and electron affinities as measures of the electron-attracting powers of atoms. The choice of factor $\frac{1}{2}$, however, is arbitrary, although reasonable, and no more arbitrary than the specific form of eqn 10.24 that defines the Pauling scale.

D10.5 Both the simple and extended Hückel methods parameterize, rather than calculate, the energy integrals that arise in molecular orbital theory. In the simple Hückel method, the overlap integral is also parameterized, whereas in the extended method this integral is evaluated. In the simple method, the energy integrals, α and β, are always considered to be adjustable parameters; their numerical values emerge only at the end of the calculation by comparison to experimental energies. The extended method is less arbitrary: the energy integrals are related to ionization energies of the atoms. The simple method has three other rather drastic approximations, listed in Section 10.6(a) of the text, which eliminate many terms from the secular determinant and make it easier to solve. Ease of solution was important in the early days of quantum chemistry, before the advent of computers, and without the use of these approximations calculations on polyatomic molecules would have been difficult to accomplish.

The simple Hückel method is usually applied only to the calculation of π-electron energies in conjugated organic systems, whereas the extended method is an all-valence electron calculation with a much greater range of applicability. The simple method is based on the assumption of the separability of the σ- and π-electron systems in the molecule. This is a very crude approximation and works best when the energy-level pattern is determined largely by the symmetry of the molecule. See Chapter 11.

D10.7 See *Impact I10.1* for more information. The configurations of the valence electrons in the ground state of these species are found in Figures 10.32, 10.34, and 10.41 of the main text.

N_2	$1\sigma_g^2 1\sigma_u^2 1\pi_u^4 2\sigma_g^2$	$b = 3$	$2S + 1 = 0$
O_2	$1\sigma_g^2 1\sigma_u^2 2\sigma_g^2 1\pi_u^4 1\pi_g^2$	$b = 2$	$2S + 1 = 3$
NO	$1\sigma^2 2\sigma^2 3\sigma^2 1\pi^4 2\pi^1$	$b = 2\frac{1}{2}$	$2S + 1 = 2$

The following figures show the highest occupied molecular orbitals (HOMOs) of each. Shaded vs. unshaded atomic-orbital (AO) lobes represent opposite signs of the wave functions. A relatively large AO represents the major contribution to the molecular orbital (MO).

N_2 2σ MO

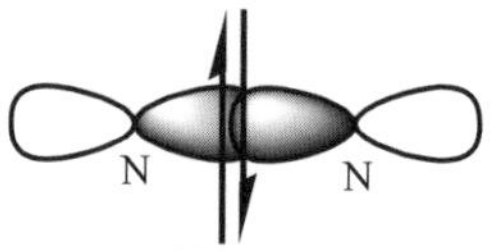

O_2 $1\pi_g$ MO, doubly degenerate

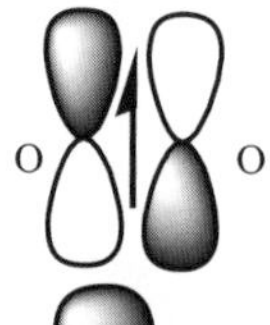

NO 2π

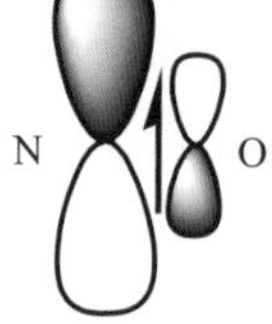

Dinitrogen with a bond order of three and paired electrons in relatively low-energy MOs is very unreactive. Special biological, or industrial, processes are needed to channel energy for promotion of 2σ electrons into high-energy, reactive states. The high-energy $1\pi_g$ lowest unoccupied molecular orbital (LUMO) is not expected to form stable complexes with electron donors.

Molecular nitrogen is very stable in most biological organisms, and as a result the task of converting plentiful atmospheric N_2 to the fixed forms of nitrogen that can be incorporated into proteins is a difficult one. The fact that N_2 possesses no unpaired electrons is itself an obstacle to facile reactivity, and the great strength (large dissociation energy) of the N_2 bond is another obstacle. Molecular orbital theory explains both of these obstacles by assigning N_2 a configuration that gives rise to a high bond order (triple bond) with all electrons paired. (See Figure 10.34 of the text.)

Dioxygen is kinetically stable because of a bond order equal to two and a high effective nuclear charge that causes the MOs to have relatively low energy. But two electrons are in the high-energy $1\pi_g$ HOMO level, which is doubly degenerate. These two electrons are unpaired and can contribute to bonding of dioxygen with other species such as the atomic radicals Fe(II) of hemoglobin and Cu(II) of the electron-transport chain. When sufficient, although not excessively large, energy is available, biological processes can channel an electron into this HOMO to produce the reactive superoxide anion of bond order $1\frac{1}{2}$. As a result, O_2 is very reactive in biological systems in ways that promote function (such as respiration) and in ways that disrupt it (damaging cells).

Although the bond order of nitric oxide is $2\frac{1}{2}$, the nitrogen nucleus has a smaller effective nuclear change than an oxygen atom would have. Thus, the one electron of the 2π HOMO is a high-energy, reactive radical compared to the HOMO of dioxygen. Additionally, the HOMO, being antibonding and predominantly centered on the nitrogen atom, is expected to bond through the nitrogen. Oxidation can result from the loss of the radical electron to form the nitrosyl ion, NO^+, which has a bond order equal to 2. Even though it has a rather high bond order, NO is readily converted to the damagingly reactive peroxynitrite ion ($ONOO^-$) by reaction with O_2^-—without breaking the NO linkage.

D10.9 The Hamiltonians of many-electron atoms and molecules contain terms that involve the interactions of electrons with each other. Any attempt to determine one orbital (i.e. a one-electron wavefunction) requires the distribution of the species' other electrons (i.e. of the orbitals occupied by the other electrons). See *Further information 10.1* for an explicit example involving a Hartree–Fock calculation. Self-consistent field (SCF) procedures address this dilemma by using initial guesses for wavefunctions to put into appropriate Hamiltonians and solving for improved approximations to those wavefunctions based on the initial guesses. The improved wavefunctions then are put into the Hamiltonians, and solution of these improved equations leads to still more improved wavefunctions. These iterative procedures are followed until the wavefunctions converge.

Solutions to exercises

E10.1(a) Let A and B represent H1s atomic orbitals centered on the two different H nuclei, and p_x and p_y two different O2p orbitals centered on the O nucleus. Then, the (unnormalized) spatial portion of wavefunction (for the bonding electrons only) would be

$$\psi = \boxed{\{A(1)p_x(2) + A(2)p_x(1)\} \times \{B(3)p_y(4) + B(4)p_y(3)\}}$$

Note that there are other possible correct answers. For example, any two *different* O2p orbitals may be used: p_x and p_y were arbitrarily selected here. Also, the numbering of the electrons need not be exactly as shown here: two different electron labels must appear in the first linear combination of orbitals, and yet another two different labels in the second linear combination; *within* each linear combination, the two labels must be interchanged.

COMMENT. This simple VB function embodies two perfectly covalent O–H bonds involving the same O nucleus. The bond angle would be 90° because that is the angle that the O2p orbitals make.

Question. Write a VB spatial wavefunction using sp^3 hybrid orbitals on the oxygen.

E10.2(a) Let s represent the H1s atomic orbital and p_z an O2p orbital centered on the O nucleus. Then, the spatial portion of wavefunction (for the bonding electrons only) would be

$$s(1)p_z(2) + s(2)p_z(1)$$

The spatial factor is symmetric with respect to interchange of the electrons' labels, so the spin factor must be antisymmetric:

$$\alpha(1)\beta(2) - \alpha(2)\beta(1).$$

So the total (unnormalized) wavefunction for this bond would be

$$\psi = \boxed{\{s(1)p_z(2) + s(2)p_z(1)\} \times \{\alpha(1)\beta(2) - \alpha(2)\beta(1)\}}$$

COMMENT. This simple VB function embodies a perfectly covalent O–H bond.

E10.3(a) Let A, B, C, and D represent H1s atomic orbitals centered on the four different H nuclei, and h_1, h_2, etc., the four hybrid sp^3 orbitals on C. Then, the (unnormalized) spatial portion of wavefunction (for the bonding electrons only) would be

$$\psi = \{A(1)h_1(2) + A(2)h_1(1)\} \times \{B(3)h_2(4) + B(4)h_2(3)\}$$
$$\times \{C(5)h_3(6) + C(6)h_3(5)\} \times \{D(7)h_4(8) + D(8)h_4(7)\}$$

E10.4(a) $h_3 = s - p_x + p_y - p_z \qquad h_4 = s + p_x - p_y - p_z$

We need to evaluate

$$\int h_3 h_4 \, d\tau = \int (s - p_x + p_y - p_z)(s + p_x - p_y - p_z) \, d\tau$$

We assume that the basis atomic orbitals are normalized and mutually orthogonal. We expand the integrand, noting that all cross terms integrate to zero (because the basis orbitals are orthogonal). The remaining terms integrate to one, yielding

$$\int h_3 h_4 \, d\tau = \int s^2 \, d\tau - \int p_x^2 \, d\tau - \int p_y^2 \, d\tau + \int p_z^2 \, d\tau = 1 - 1 - 1 + 1 = 0$$

E10.5(a) Refer to Figure 10.33 of the text. Place two of the valence electrons in each orbital starting with the lowest-energy orbital, until all valence electrons are used up. Apply Hund's rule to the filling of degenerate orbitals.

(a) Li_2 (2 electrons) $\boxed{1\sigma_g^2, b = 1}$

(b) Be_2 (4 electrons) $\boxed{1\sigma_g^2 1\sigma_u^2, b = 0}$

(c) C_2 (8 electrons) $\boxed{1\sigma_g^2 1\sigma_u^2 1\pi_u^4, b = 2}$

E10.6(a) Note that CO and CN^- are isoelectronic with N_2, so refer to Figure 10.34 of the text for them; note, however, that the σ and π orbitals no longer have u or g symmetry, so they are simply labelled consecutively. For NO, refer to Figure 10.41 of the text.

(a) CO (10 electrons) $\boxed{1\sigma^2 2\sigma^2 1\pi^4 3\sigma^2}$

(b) NO (11 electrons) $\boxed{1\sigma^2 2\sigma^2 3\sigma^2 1\pi^4 2\pi^1}$

(c) CN^- (10 electrons) $\boxed{1\sigma^2 2\sigma^2 1\pi^4 3\sigma^2}$

E10.7(a) B_2 (6 electrons): $1\sigma_g^2 1\sigma_u^2 1\pi_u^2$ $b = 1$
C_2 (8 electrons): $1\sigma_g^2 1\sigma_u^2 1\pi_u^4$ $b = 2$

The bond orders of B_2 and C_2 are, respectively, 1 and 2; so $\boxed{C_2}$ should have the greater bond dissociation enthalpy. The experimental values are approximately 4 and 6 eV, respectively.

E10.8(a) We can use a version of Figure 10.32 of the text, but with the energy levels of F lower than those of Xe as in Figure 10.1 here.

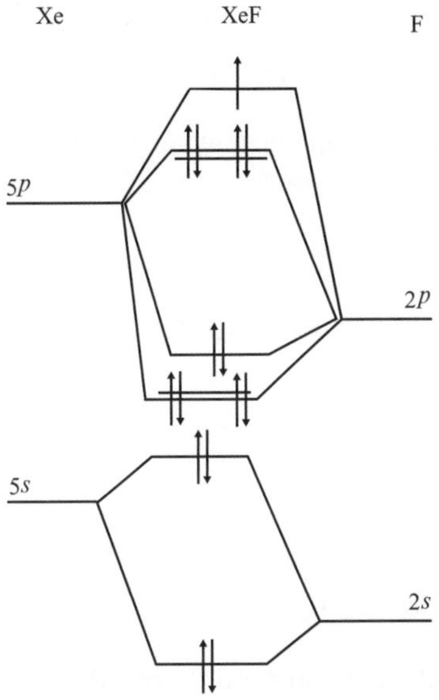

Figure 10.1

For XeF we insert 15 valence electrons. The bond order is increased when XeF^+ is formed from XeF because an electron is removed from an antibonding orbital, therefore XeF^+ will have a shorter bond length than XeF.

E10.9(a) The electron configurations are used to determine the bond orders. Larger bond order corresponds qualitatively to shorter bond length.

The bond orders of NO and N_2 are 2.5 and 3, respectively (*Discussion question 10.7*), hence N_2 should have the shorter bond length. The experimental values are 115 and 110 pm, respectively.

E10.10(a) Call this linear combination of A and B ψ_+. Normalization requires

$$\int \psi_+^2 \, d\tau = 1 = \int (A\sin\theta + B\cos\theta)^2 \, d\tau = \sin^2\theta \int A^2 \, d\tau + \cos^2\theta \int B^2 \, d\tau + 2\sin\theta\cos\theta \int AB \, d\tau$$

(Remember, θ is a parameter, not a variable of integration.) In this expression, the first two integrals are 1 because the atomic orbitals are normalized. The last integral is zero because overlap is zero by hypothesis. Thus, normalization of ψ requires

$$\int \psi_+^2 \, d\tau = 1 = \sin^2\theta + \cos^2\theta,$$

which is a trigonometric identity.

We are looking for a different linear combination of A and B orthogonal to ψ_+. Call the new function $\psi_- = aA + bB$, where we must determine a and b. Thus, we require

$$\int \psi_-\psi_+\,d\tau = 0 = \int (aA + bB)(A\sin\theta + B\cos\theta)d\tau$$
$$= a\sin\theta\int A^2\,d\tau + b\cos\theta\int B^2\,d\tau + a\sin\theta\int AB\,d\tau + b\cos\theta\int BA\,d\tau$$

Using the normalization of the basis functions and their lack of overlap, we have:

$$\int \psi_-\psi_+\,d\tau = 0 = a\sin\theta + b\cos\theta$$

Hence, $\dfrac{b}{a} = -\dfrac{\sin\theta}{\cos\theta}$.

This gives us a ratio of the coefficients. In order to fix the coefficients themselves, we require ψ_- to be normalized. By means of the same steps as we used above for ψ_+, we have

$$\int \psi_-^2\,d\tau = 1 = a^2 + b^2.$$

This would be satisfied if $b = -\sin\theta$ and $a = \cos\theta$. (That is, the coefficients are not only in the ratio of –sine/cosine; they *are* –sine/cosine.) Thus,

$$\psi_- = \boxed{A\cos\theta - B\sin\theta} \text{ is orthogonal to } A\sin\theta + B\cos\theta.$$

COMMENT. Perceptive students will have noticed that A and B were assumed to be real in this solution. The solution is essentially the same if A and B are complex, but with appropriate appearance of complex conjugates.

E10.11(a) We evaluate $\int(\psi_A + \psi_B)(\psi_A - \psi_B)\,d\tau$ and look at the result. If the integral is zero, then they are mutually orthogonal.

$$\int(\psi_A + \psi_B)(\psi_A - \psi_B)\,d\tau = \int(\psi_A^2 - \psi_B^2)\,d\tau = 1 - 1 = \boxed{0}$$

Hence, they are orthogonal.

E10.12(a) Energy is conserved, so when the photon is absorbed, its energy is transferred to the electron. Part of it overcomes the binding energy (ionization energy) and the remainder is manifest as the kinetic energy of the now freed electron.

$$E_{\text{photon}} = I + E_{\text{kinetic}}$$

so $$E_{\text{kinetic}} = E_{\text{photon}} - I = \frac{hc}{\lambda} - I = \frac{(6.626\times10^{-34}\text{ J s})\times(2.998\times10^{8}\text{ m s}^{-1})}{(100\times10^{-9}\text{ m})\times(1.602\times10^{-19}\text{ J eV}^{-1})} - 11.0\text{ eV}$$
$$= \boxed{1.4\text{ eV}} = \boxed{2.2\times10^{-19}\text{ J}}$$

E10.13(a) $E_{\text{photon}} = I + E_{\text{kinetic}}$ [Exercise 10.12(a)]

so $$I = E_{\text{photon}} - E_{\text{kinetic}} = E_{\text{photon}} - \frac{m_e v^2}{2}$$
$$= 21.22\text{ eV} - \frac{(9.11\times10^{-31}\text{ kg})\times(1.90\times10^{6}\text{ m s}^{-1})^2}{2\times(1.602\times10^{-19}\text{ J eV}^{-1})}$$
$$= \boxed{10.9\overline{6}\text{ eV}} = \boxed{1.76\times10^{-18}\text{ J}}$$

E10.14(a) We use results derived in Section 10.5(c) for heteronuclear molecules and no overlap of atomic orbitals. The energies of the molecular orbitals are

$$E_{\pm}=\left(\frac{\alpha_{\rm F}+\alpha_{\rm Xe}}{2}\right)\pm\left(\frac{\alpha_{\rm F}-\alpha_{\rm Xe}}{2}\right)\left\{1+\left(\frac{2\beta}{\alpha_{\rm F}-\alpha_{\rm Xe}}\right)^2\right\}^{1/2} \quad [10.32c]$$

Taking the Coulomb integrals to be equal in magnitude and opposite in sign to the ionization energies, we have

$$E_{\pm}/\mathrm{eV}=\left(\frac{-17.4+-12.1}{2}\right)\pm\left(\frac{-17.4+12.1}{2}\right)\left\{1+\left(\frac{2\times(-1.5)}{-17.4+12.1}\right)^2\right\}^{1/2}$$

so $E_+=\boxed{-17.8\ \mathrm{eV}}$ and $E_-=\boxed{-11.7\ \mathrm{eV}}$.

The coefficients are given by eqn 10.38:

$$c_{\rm F}=\left\{1+\left(\frac{\alpha_{\rm F}-E}{\beta}\right)^2\right\}^{-1/2} \quad \text{and} \quad c_{\rm Xe}=-\left(\frac{\alpha_{\rm F}-E}{\beta}\right)c_{\rm F}$$

So, in the bonding orbital,

$$c_{\rm F}=\left\{1+\left(\frac{-17.4+17.8}{-1.5}\right)^2\right\}^{-1/2}=0.97 \quad \text{and} \quad c_{\rm Xe}=-\left(\frac{-17.4+17.8}{-1.5}\right)\times 0.97=0.25$$

and in the antibonding orbital

$$c_{\rm F}=\left\{1+\left(\frac{-17.4+11.7}{-1.5}\right)^2\right\}^{-1/2}=0.25 \quad \text{and} \quad c_{\rm Xe}=-\left(\frac{-17.4+11.7}{-1.5}\right)\times 0.25=-0.97$$

Thus, the bonding and antibonding orbitals are, respectively

$$\psi_+=\boxed{0.97\chi_{\rm F}+0.25\chi_{\rm Xe}} \quad \text{and} \quad \psi_-=\boxed{0.25\chi_{\rm F}-0.97\chi_{\rm Xe}}$$

E10.15(a) We use results derived in Section 10.5(c) for heteronuclear molecules with overlap of atomic orbitals. The energies of the molecular orbitals are

$$E_{\pm}=\frac{\alpha_{\rm F}+\alpha_{\rm Xe}-2\beta S}{2(1-S^2)}\pm\frac{\{(\alpha_{\rm F}+\alpha_{\rm Xe}-2\beta S)^2-4(1-S^2)(\alpha_{\rm F}\alpha_{\rm Xe}-\beta^2)\}^{1/2}}{2(1-S^2)} \quad [10.32a]$$

$$E_{\pm}/\mathrm{eV}=\frac{-17.4-12.1+2\times1.5\times0.20}{2(1-0.20^2)}$$
$$\pm\frac{\{(-17.4-12.1+2\times1.5\times0.20)^2-4(1-0.20^2)(17.4\times12.1-1.5^2)\}^{1/2}}{2(1-0.20^2)}$$

Thus, taking the lower energy to be that of the bonding orbital, we have

$$E_{\rm bond}=\boxed{-18.1\ \mathrm{eV}} \quad \text{and} \quad E_{\rm anti}=\boxed{-12.0\ \mathrm{eV}}$$

The coefficients are given by eqns 10.36 and 10.34, respectively

$$c_{\rm F}=\left\{1+\left(\frac{\alpha_{\rm F}-E}{\beta-ES}\right)^2-2S\left(\frac{\alpha_{\rm F}-E}{\beta-ES}\right)\right\}^{-1/2} \quad \text{and} \quad c_{\rm Xe}=-\left(\frac{\alpha_{\rm F}-E}{\beta-ES}\right)c_{\rm F}$$

So, in the bonding orbital, $\dfrac{\alpha_F - E}{\beta - ES} = \dfrac{-17.4 + 18.1}{-1.5 + 18.1 \times 0.20} = 0.35\overline{2}$,

$$c_F = (1 + 0.35\overline{2}^2 - 2 \times 0.20 \times 0.35\overline{2})^{-1/2} = 1.01 \quad \text{and} \quad c_{Xe} = -0.35\overline{2} \times 1.01 = -0.36$$

In the antibonding orbital, $\dfrac{\alpha_F - E}{\beta - ES} = \dfrac{-17.4 + 12.0}{-1.5 + 12.0 \times 0.20} = -6.1$,

$$c_F = (1 + 6.1^2 - 2 \times 0.20 \times 6.1)^{-1/2} = 0.16 \quad \text{and} \quad c_{Xe} = -6.1 \times 0.16 = 0.96$$

Thus, the bonding and antibonding orbitals are, respectively,

$$\psi_{bond} = \boxed{1.01\chi_F - 0.36\chi_{Xe}} \quad \text{and} \quad \psi_{anti} = \boxed{0.16\chi_F + 0.96\chi_{Xe}}$$

COMMENT. The alert reader may be concerned over obtaining a coefficient greater than 1. Normalization requires coefficients no greater than 1 when the basis orbitals **do not overlap**. In this exercise, the basis orbitals do overlap. The normalization condition is expressed in eqn 10.35. The reader may verify that the coefficients of both orbitals satisfy that condition.

E10.16(a) The molecular orbitals of the fragments and the molecular orbitals that they form are shown in Figure 10.2.

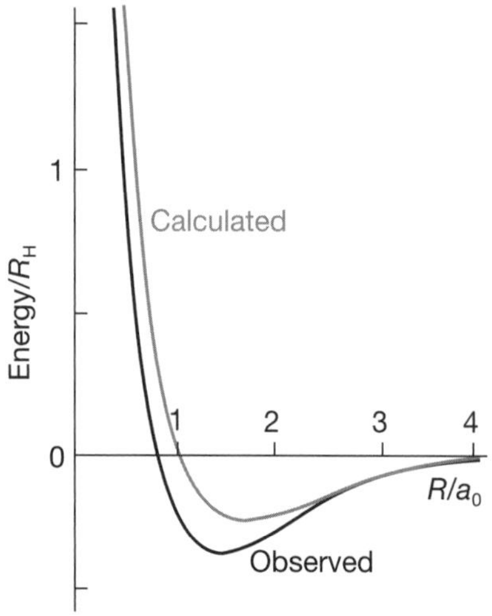

Figure 10.2

COMMENT. Note that the π-bonding orbital must be lower in energy than the σ-antibonding orbital for π-bonding to exist in ethene.

Question. Would the ethene molecule exist if the order of the energies of the π and σ* orbitals were reversed?

E10.17(a) In setting up the secular determinant we use the approximations of Section 10.6

(a) $\begin{vmatrix} \alpha - E & \beta & 0 \\ \beta & \alpha - E & \beta \\ 0 & \beta & \alpha - E \end{vmatrix} = 0$ (b) $\begin{vmatrix} \alpha - E & \beta & \beta \\ \beta & \alpha - E & \beta \\ \beta & \beta & \alpha - E \end{vmatrix} = 0$

The atomic orbital basis is $1s_A$, $1s_B$, $1s_C$ in each case; in linear H_3 we ignore A, C overlap because A and C are not neighbouring atoms; in triangular H_3 we include it because they are. To find the energies, expand the determinants:

(a) $(\alpha - E)\times\{(\alpha - E)(\alpha - E) - \beta^2\} - \beta\times\beta(\alpha - E) = 0,$
$0 = (\alpha - E)\times\{(\alpha - E)(\alpha - E) - 2\beta^2\} = (\alpha - E)(\alpha - E + 2^{1/2}\beta)(\alpha - E - 2^{1/2}\beta)$

The roots (the orbital energies) are $E = \alpha$, $\alpha \pm 2^{1/2}\beta$. The binding energy is the sum of the orbital energies of the occupied orbitals

$$E_{\text{bind}} = 2(\alpha + 2^{1/2}\beta) + \alpha = \boxed{3\alpha + 2^{3/2}\beta}.$$

(b) $(\alpha - E)\times\{(\alpha - E)(\alpha - E) - \beta^2\} - \beta\times\{\beta(\alpha - E) - \beta^2\} + \beta\times\{\beta^2 - \beta(\alpha - E)\} = 0,$
$0 = (\alpha - E)(\alpha - E - \beta)(\alpha - E + \beta) - 2\beta^2(\alpha - E - \beta),$
$0 = (\alpha - E - \beta)(\alpha^2 - 2\alpha E + \alpha\beta + E^2 - \beta E - 2\beta^2).$

At this point, we have a factor linear in E, from which we find a root $E = \alpha - \beta$, and a quadratic factor. One can put the quadratic factor in standard form, remembering that we are solving for E, or one can attempt to factor it further, obtaining

$$0 = (\alpha - E - \beta)(E - \alpha + \beta)(E - \alpha - 2\beta)$$

Thus, we find a second root $E = \alpha - \beta$ (i.e. a doubly degenerate orbital at this energy level) and $E = \alpha + 2\beta$. The binding energy is

$$E_{\text{bind}} = 2(\alpha + 2\beta) + (\alpha - \beta) = \boxed{3\alpha + 3\beta}$$

E10.18(a) We use the molecular orbital energy level diagram in Figure 10.45 of the text. As usual, we fill the orbitals starting from the lowest energy orbital, obeying the Pauli principle and Hund's rule. We then write

(a) $C_6H_6^-$ (7 electrons): $\boxed{a_{2u}^2 e_{1g}^4 e_{2u}^1}$

$$E_\pi = 2(\alpha + 2\beta) + 4(\alpha + \beta) + (\alpha - \beta) = \boxed{7\alpha + 7\beta}$$

(b) $C_6H_6^+$ (5 electrons): $\boxed{a_{2u}^2 e_{1g}^3}$

$$E_\pi = 2(\alpha + 2\beta) + 3(\alpha + \beta) = \boxed{5\alpha + 7\beta}$$

E10.19(a) The structures are numbered to match the row and column numbers shown in the determinants:

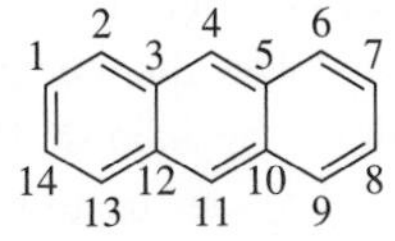

anthracene

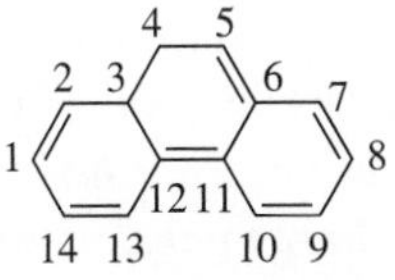

phenanthrene

The secular determinant of anthracene in the Hückel approximation is:

	1	2	3	4	5	6	7	8	9	10	11	12	13	14
1	$\alpha-E$	β	0	0	0	0	0	0	0	0	0	0	0	β
2	β	$\alpha-E$	β	0	0	0	0	0	0	0	0	0	0	0
3	0	β	$\alpha-E$	β	0	0	0	0	0	0	0	β	0	0
4	0	0	β	$\alpha-E$	β	0	0	0	0	0	0	0	0	0
5	0	0	0	β	$\alpha-E$	β	0	0	0	β	0	0	0	0
6	0	0	0	0	β	$\alpha-E$	β	0	0	0	0	0	0	0
7	0	0	0	0	0	β	$\alpha-E$	β	0	0	0	0	0	0
8	0	0	0	0	0	0	β	$\alpha-E$	β	0	0	0	0	0
9	0	0	0	0	0	0	0	β	$\alpha-E$	β	0	0	0	0
10	0	0	0	0	β	0	0	0	β	$\alpha-E$	β	0	0	0
11	0	0	0	0	0	0	0	0	0	β	$\alpha-E$	β	0	0
12	0	0	β	0	0	0	0	0	0	0	β	$\alpha-E$	β	0
13	0	0	0	0	0	0	0	0	0	0	0	β	$\alpha-E$	β
14	β	0	0	0	0	0	0	0	0	0	0	0	β	$\alpha-E$

The secular determinant of phenanthrene in the Hückel approximation is:

	1	2	3	4	5	6	7	8	9	10	11	12	13	14
1	$\alpha-E$	β	0	0	0	0	0	0	0	0	0	0	0	β
2	β	$\alpha-E$	β	0	0	0	0	0	0	0	0	0	0	0
3	0	β	$\alpha-E$	β	0	0	0	0	0	0	0	β	0	0
4	0	0	β	$\alpha-E$	β	0	0	0	0	0	0	0	0	0
5	0	0	0	β	$\alpha-E$	β	0	0	0	0	0	0	0	0
6	0	0	0	0	β	$\alpha-E$	β	0	0	0	β	0	0	0
7	0	0	0	0	0	β	$\alpha-E$	β	0	0	0	0	0	0
8	0	0	0	0	0	0	β	$\alpha-E$	β	0	0	0	0	0
9	0	0	0	0	0	0	0	β	$\alpha-E$	β	0	0	0	0
10	0	0	0	0	0	0	0	0	β	$\alpha-E$	β	0	0	0
11	0	0	0	0	0	β	0	0	0	β	$\alpha-E$	β	0	0
12	0	0	β	0	0	0	0	0	0	0	β	$\alpha-E$	β	0
13	0	0	0	0	0	0	0	0	0	0	0	β	$\alpha-E$	β
14	β	0	0	0	0	0	0	0	0	0	0	0	β	$\alpha-E$

The secular determinants can be diagonalized with the assistance of general-purpose mathematical software. Alternatively, programs specifically designed for Hückel calculations (such as the Simple Huckel Molecular Orbital Theory Calculator at the University of Calgary, http://www.chem.ucalgary.ca/SHMO/ or Hückel software in *Explorations in Physical Chemistry*, 2nd edition by Julio de Paula, Valerie Walters, and Peter Atkins, http://ebooks.bfwpub.com/explorations.php) can be used. As usual, we fill the orbitals starting from the lowest energy orbital, obeying the Pauli principle and Hund's rule.

(a) In anthracene, the energies of the filled orbitals are $\alpha + 2.414\beta$, $\alpha + 2\beta$, $\alpha + 1.414\beta$ (doubly degenerate), $\alpha + \beta$ (doubly degenerate), and $\alpha + 0.414\beta$, so the total π-electron binding energy is $\boxed{14\alpha + 19.314\beta}$.

(b) In phenanthrene, the energies of the filled orbitals are $\alpha + 2.435\beta$, $\alpha + 1.95\beta$, $\alpha + 1.516\beta$, $\alpha + 1.306\beta$, $\alpha + 1.142\beta$, $\alpha + 0.769\beta$, and $\alpha + 0.605\beta$, so the total π-electron binding energy is $\boxed{14\alpha + 19.448\beta}$.

Solutions to problems

Solutions to numerical problems

P10.1 In effect, the question asks us to show graphically that the interference of two waves centred on different locations depends on the both the separation of the centres and the wavelength of the waves (or, as stated in this problem, the parameters k and k', which are inversely proportional to wavelength).

See Figures 10.3(a) and (b). In both figures, the abscissa is x/R, that is x scaled in units of internuclear separation. The positions of the nuclei are shown as dots at $x/R = 0$ (A) and 1 (B). In both graphs, a cosine function with $kR = \pi/2$ is plotted as a dashed curve. This function, then, is $\cos\dfrac{\pi x}{2R}$, and it has a peak at nucleus A.

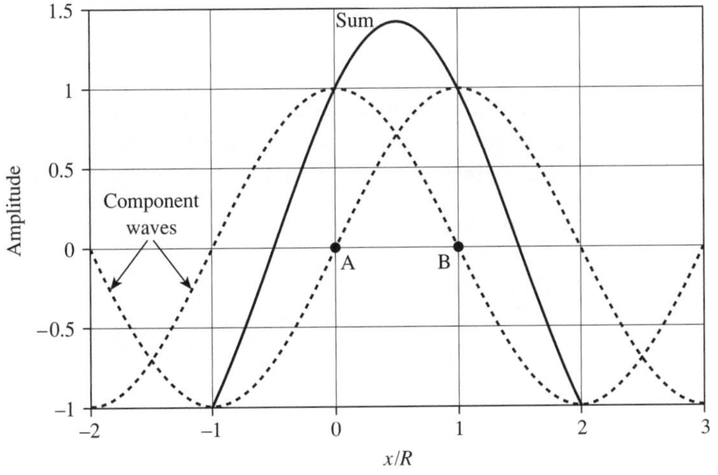

Figure 10.3(a)

In Figure 10.3(a), the other dashed curve is $\cos k'x$, measured from nucleus B, with $k'R = \pi/2$. That is, the second dashed curve has the same wavelength as the first, and it has a peak at nucleus B; its formula is $\cos\dfrac{\pi(x-R)}{2R}$. The solid line is the sum of the two dashed curves, representing the composite wave that results from the interference of the two component cosine functions. The peaks of the two component functions do not coincide; however, in the region between the nuclei, both component functions are reasonably large. Thus, the sum of the components is larger between the nuclei than either component. We call the addition of component waves that have the same sign constructive interference; the components reinforce each other.

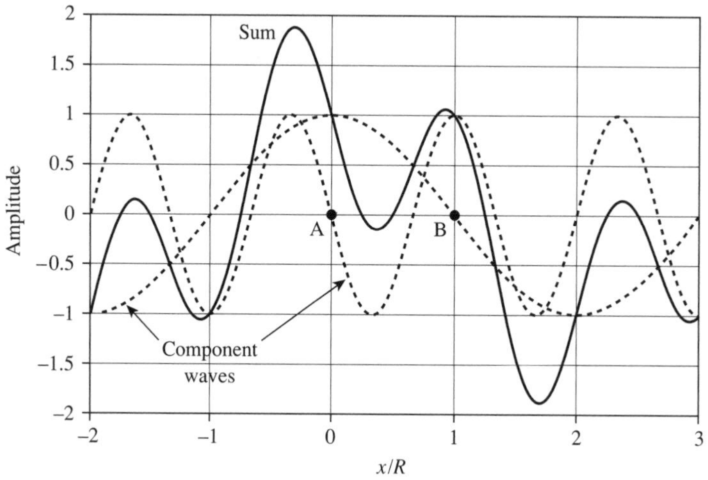

Figure 10.3(b)

In Figure 10.3(b) the second dashed curve is $\cos k'x$, measured from nucleus B, but with $k'R = 3\pi/2$. That is, the second dashed curve still has a peak at nucleus B, but it has a shorter wavelength than the first; its formula is $\cos\dfrac{3\pi(x-R)}{2R}$. Again, the solid line represents the composite wave that results from the interference (sum) of the two component cosine functions. In most of the region between the nuclei, the composite wave has a lower absolute value than either component wave; this is because the component waves have opposite signs in much of this region. We call the addition of component waves that have the opposite signs destructive interference; the components diminish each other.

Question. Does Figure 10.3(b) illustrate **constructive** interference anywhere? If so, where?

P10.3 We require the properly normalized functions

$$\psi_\pm = \left(\frac{1}{2(1\pm S)}\right)^{1/2}(A \pm B)\ [10.8 \text{ and Example 10.1}]$$

We first calculate the overlap integral, S, at $R = 106\ \text{pm} = 2a_0$.

$$S = \left(1 + 2 + \frac{1}{3}(2)^2\right)e^{-2} = 0.586\,[10.13a]$$

Then, $$N_+ = \left(\frac{1}{2(1+S)}\right)^{1/2} = \left(\frac{1}{2(1+0.586)}\right)^{1/2} = 0.561$$

$$N_- = \left(\frac{1}{2(1-S)}\right)^{1/2} = \left(\frac{1}{2(1-0.586)}\right)^{1/2} = 1.09\overline{9}$$

We then calculate with $A = \left(\dfrac{1}{\pi a_0^3}\right)^{1/2} e^{-r_A/a_0}$ and $\psi_\pm = N_\pm\left(\dfrac{1}{\pi a_0^3}\right)^{1/2}\{e^{-r_A/a_0} \pm e^{-r_B/a_0}\}$. [10.9]

Measuring both r_A and r_B from nucleus A, along the internuclear line, we have

$$\psi_\pm = N_\pm\left(\frac{1}{\pi a_0^3}\right)^{1/2}\{e^{-|z|/a_0} \pm e^{-|z-R|/a_0}\}$$

with z measured from A along the axis toward B. We draw up the following table with $R = 106$ pm and $a_0 = 52.9$ pm:

z/pm	−100	−80	−60	−40	−20	0	20	40
$\dfrac{\psi_+}{(1/\pi a_0^3)^{1/2}}$	0.096	0.14	0.20	0.30	0.44	0.64	0.49	0.42
$\dfrac{\psi_-}{(1/\pi a_0^3)^{1/2}}$	0.14	0.21	0.31	0.45	0.65	0.95	0.54	0.20

z/pm (continued)	60	80	100	120	140	160	180	200
$\dfrac{\psi_+}{(1/\pi a_0^3)^{1/2}}$	0.42	0.47	0.59	0.49	0.33	0.23	0.16	0.11
$\dfrac{\psi_-}{(1/\pi a_0^3)^{1/2}}$	−0.11	−0.43	−0.81	−0.73	−0.50	−0.23	−0.23	−0.16

The points are plotted in Figure 10.4.

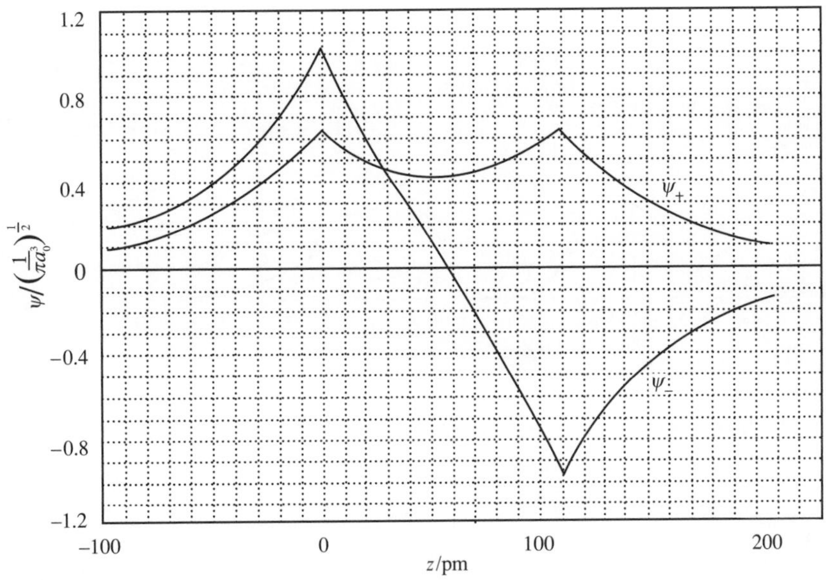

Figure 10.4

P10.5 (a) With spatial dimensions in units (multiples) of a_0, the atomic orbitals of atom A and atom B may be written in the form:

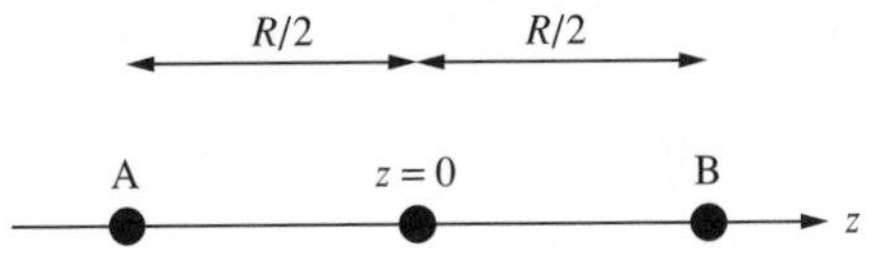

$$p_{z,A} = \frac{1}{4(2\pi)^{1/2}}(z + R/2)e^{-\frac{1}{2}\{x^2+y^2+(z+R/2)^2\}^{1/2}}$$

and $$p_{z,B} = \frac{1}{4(2\pi)^{1/2}}(z - R/2)e^{-\frac{1}{2}\{x^2+y^2+(z-R/2)^2\}^{1/2}}$$

Following eqn 10.8 and Example 10.1, we form LCAO-MOs of the form:

$$\psi_{\sigma^*} = \frac{p_{z,A} + p_{z,B}}{\{2(1+S)\}^{1/2}} \text{ [antibonding] and } \psi_{\sigma} = \frac{p_{z,A} - p_{z,B}}{\{2(1-S)\}^{1/2}} \text{ [bonding]}$$

where $$S = \int_{-\infty}^{\infty}\int_{-\infty}^{\infty}\int_{-\infty}^{\infty} p_{z,A}p_{z,B}\,dx\,dy\,dz \text{ [10.18]}$$

Computations and plots are readily prepared with mathematical software such as Mathcad. Figures 10.5(a) and (b) refer to the σ orbitals formed from the p_z atomic orbitals.

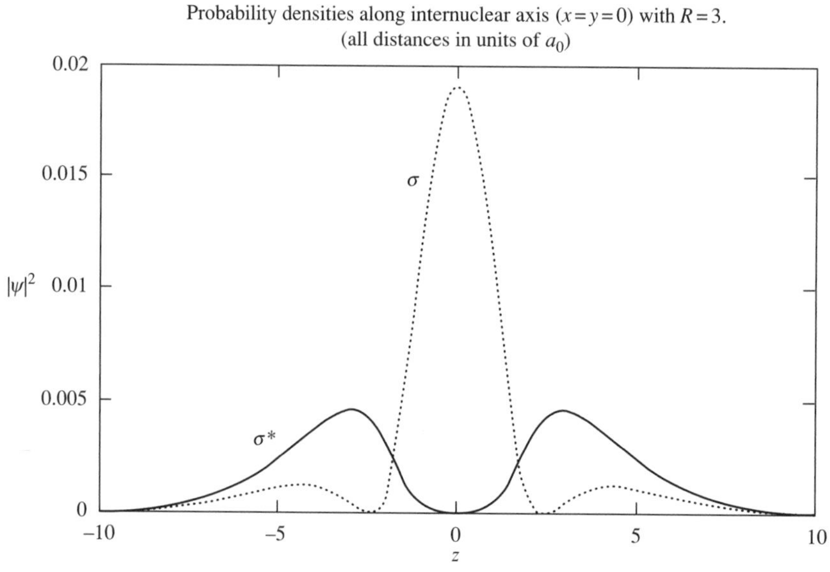

Figure 10.5(a)

(b) With spatial dimensions in units of a_0, the atomic orbitals for the construction of π molecular orbitals are:

$$p_{x,A} = \frac{1}{4(2\pi)^{1/2}}xe^{-\frac{1}{2}\{x^2+y^2+(z+R/2)^2\}^{1/2}}$$

$$p_{x,B} = \frac{1}{4(2\pi)^{1/2}}xe^{-\frac{1}{2}\{x^2+y^2+(z-R/2)^2\}^{1/2}}$$

The π MOs are:

$$\psi_{\pi} = \frac{p_{x,A} + p_{x,B}}{\{2(1+S)\}^{1/2}} \text{ [bonding] and } \psi_{\pi^*} = \frac{p_{x,A} - p_{x,B}}{\{2(1-S)\}^{1/2}} \text{ [antibonding]}$$

where $$S = \int_{-\infty}^{\infty}\int_{-\infty}^{\infty}\int_{-\infty}^{\infty} p_{x,A}p_{x,B}\,dx\,dy\,dz$$

Figure 10.5(c) refers to the π orbital formed from the p_x atomic orbitals.

$R = 3$

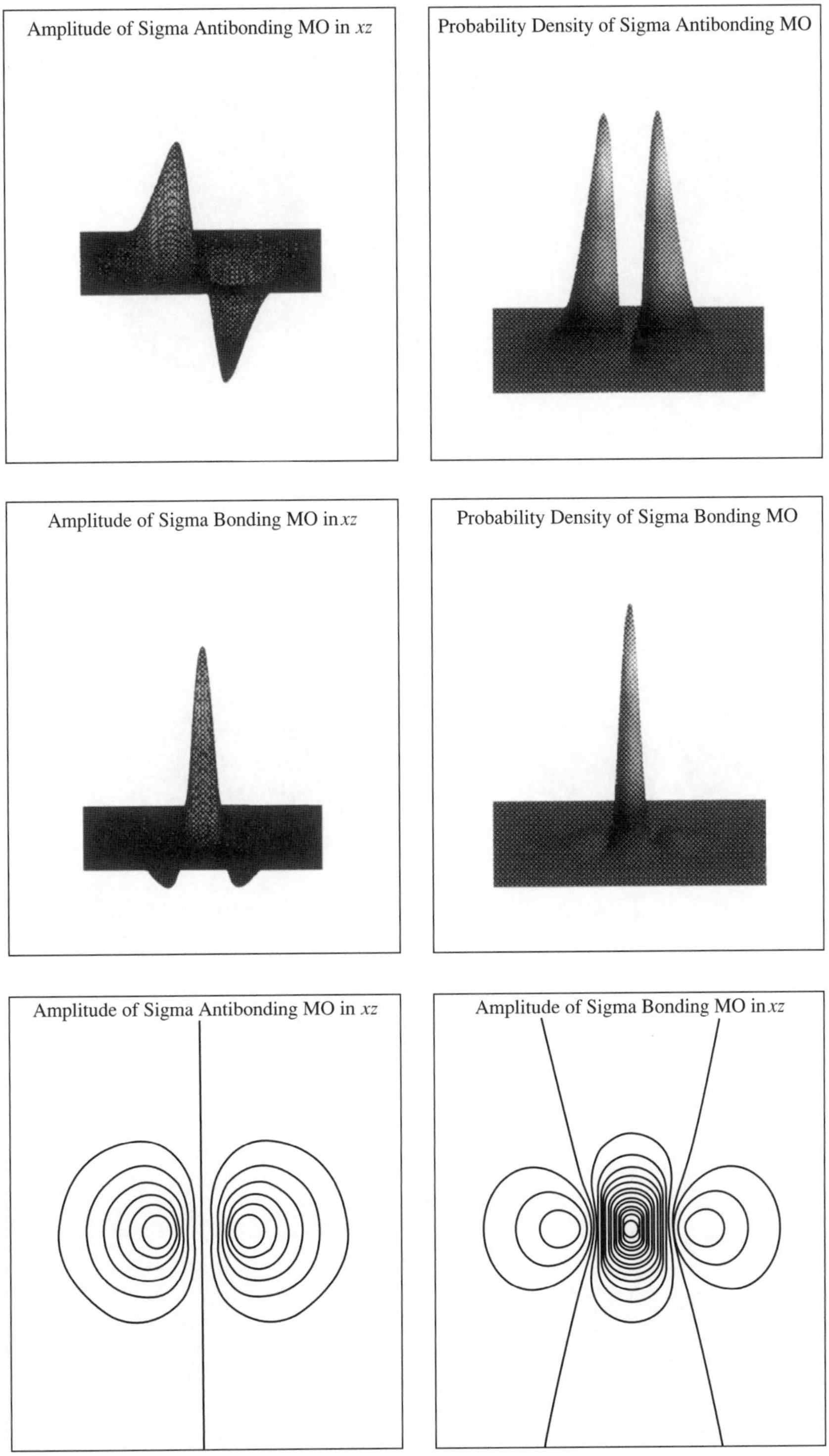
Amplitude of Sigma Antibonding MO in xz
Probability Density of Sigma Antibonding MO
Amplitude of Sigma Bonding MO in xz
Probability Density of Sigma Bonding MO
Amplitude of Sigma Antibonding MO in xz
Amplitude of Sigma Bonding MO in xz

Figure 10.5(b)

$R = 3$

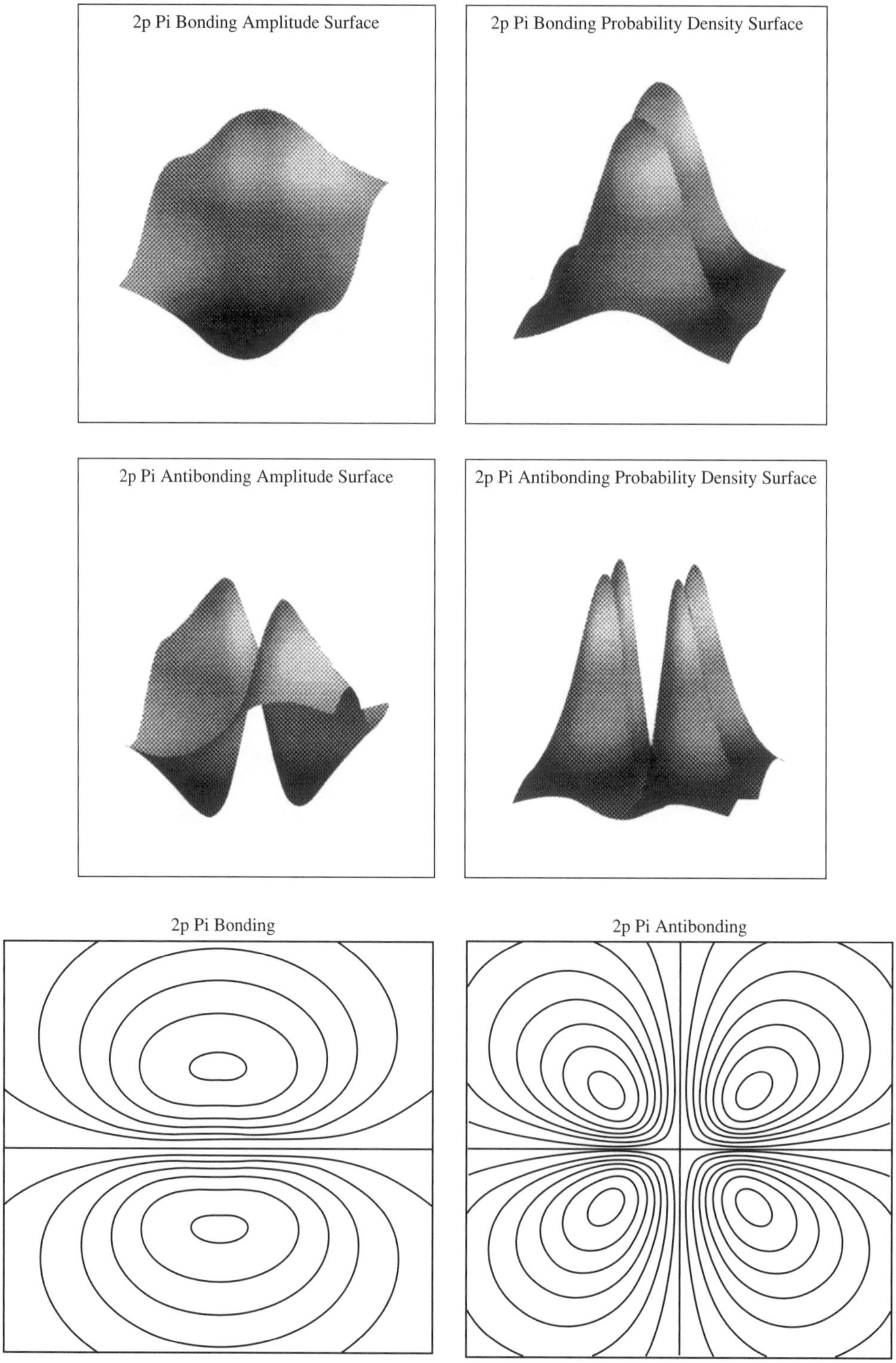
2p Pi Bonding Amplitude Surface
2p Pi Bonding Probability Density Surface
2p Pi Antibonding Amplitude Surface
2p Pi Antibonding Probability Density Surface
2p Pi Bonding
2p Pi Antibonding

Figure 10.5(c)

The plots clearly show the constructive interference that makes a bonding molecular orbital. Nodal planes created by destructive interference are clearly seen in the antibonding molecular orbitals. When calculations and plots are produced for the $R = 10$ case, constructive and destructive interference are seen to be much weaker because of the weak atomic orbital overlap.

P10.7 The electron configuration of F_2 is $1\sigma_g^2 1\sigma_u^{*2} 2\sigma_g^2 1\pi_u^4 1\pi_g^{*4}$. (See Figure 10.33 of the main text; we use an asterisk to denote antibonding orbitals.) That of F_2^- is $1\sigma_g^2 1\sigma_u^{*2} 2\sigma_g^2 1\pi_u^4 1\pi_g^{*4} 2\sigma_u^{*1}$. So F_2^- has one more antibonding electron than does F_2, suggesting a lower bond order ($\frac{1}{2}$ versus 1) and therefore a weaker bond. By definition, a weaker bond has a smaller dissociation energy (hence the difference in D_e). Weaker bonds tend to be longer (hence the difference in R_e) and less stiff (hence the difference in $\tilde{\nu}$, reflecting a difference in the force constant k) than stronger bonds between similar atoms.

P10.9 In the simple Hückel approximation

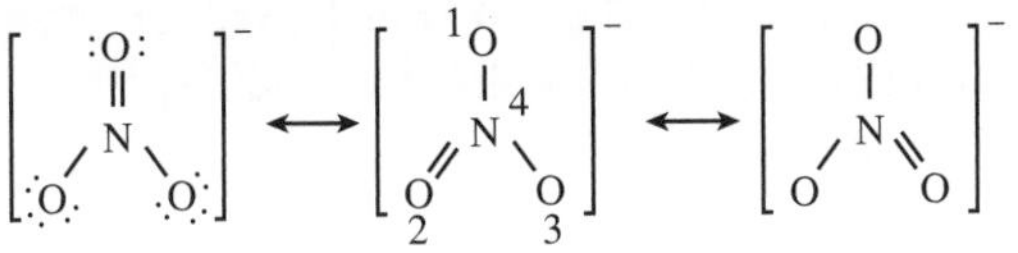

$$\begin{vmatrix} \alpha_O - E & 0 & 0 & \beta \\ 0 & \alpha_O - E & 0 & \beta \\ 0 & 0 & \alpha_O - E & \beta \\ \beta & \beta & \beta & \alpha_N - E \end{vmatrix} = 0$$

$$(E - \alpha_O)^2 \times \{(E - \alpha_O) \times (E - \alpha_N) - 3\beta^2\} = 0$$

Therefore,

$$E - \alpha_O = 0 \text{ (twice)} \quad \text{so} \quad E = \boxed{\alpha_O} \text{ (doubly degenerate)}$$

and $\quad (E - \alpha_O)(E - \alpha_N) - 3\beta^2 = 0 = E^2 - (\alpha_O + \alpha_N)E + \alpha_O\alpha_N - 3\beta^2$

Applying the quadratic equation yields

$$E_\pm = \frac{\alpha_O + \alpha_N \pm \sqrt{(\alpha_O + \alpha_N)^2 - 4(\alpha_O\alpha_N - 3\beta^2)}}{2}$$

$$= \frac{\alpha_O + \alpha_N \pm \sqrt{(\alpha_O - \alpha_N)^2 + 12\beta^2}}{2}$$

$$= \boxed{\frac{1}{2}\left(\alpha_O + \alpha_N \pm (\alpha_O - \alpha_N)\sqrt{1 + \frac{12\beta^2}{(\alpha_O - \alpha_N)^2}}\right)}.$$

These two energies belong to bonding and antibonding orbitals and the degenerate level to a non-bonding orbital. An orbital energy-level diagram is shown in Figure 10.6(a).

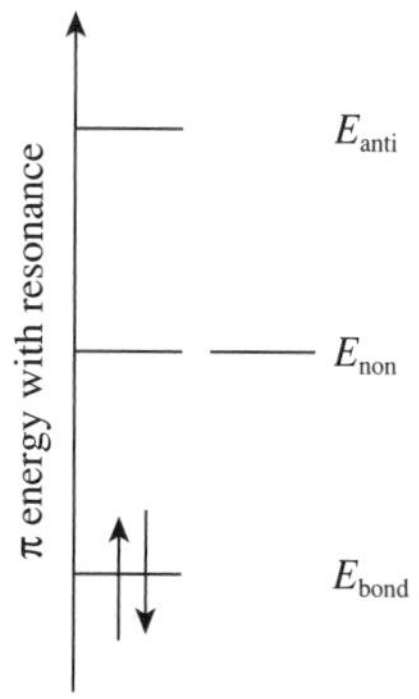

Figure 10.6(a)

To find the delocalization energy, we must compare the energy of the two π electrons in the case with resonance to their energies in the absence of resonance. The latter are derived for just one of the three structures above, i.e. for a structure containing a single localized π bond.

$$\begin{vmatrix} \alpha_O - E & \beta \\ \beta & \alpha_N - E \end{vmatrix} = 0$$

Expanding the determinant and solving for E yields energies of a bonding and antibonding orbital (shown in Figure 10.6(b))

$$(E-\alpha_O)(E-\alpha_N)-\beta^2 = 0 = E^2-(\alpha_O+\alpha_N)E+\alpha_O\alpha_N-\beta^2,$$

so $$E_{local\pm} = \frac{1}{2}\left(\alpha_O+\alpha_N \pm (\alpha_O-\alpha_N)\sqrt{1+\frac{4\beta^2}{(\alpha_O-\alpha_N)^2}}\right)$$

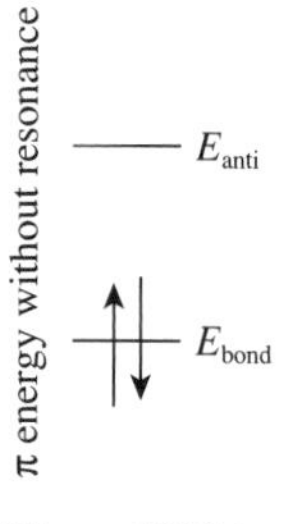

Figure 106(b)

There are two π electrons in the system, so the delocalization energy is

$$2E_+ - 2E_{local+} = \boxed{(\alpha_O-\alpha_N)\left(\sqrt{1+\frac{12\beta^2}{(\alpha_O-\alpha_N)^2}} - \sqrt{1+\frac{4\beta^2}{(\alpha_O-\alpha_N)^2}}\right)}$$

If $12\beta^2 \ll (\alpha_O-\alpha_N)^2$, we can use $(1+x)^{1/2} \approx 1+\frac{x}{2}$, so the delocalization energy is

$$\approx (\alpha_O-\alpha_N)\left(1+\frac{12\beta^2}{2(\alpha_O-\alpha_N)^2} - 1 - \frac{4\beta^2}{2(\alpha_O-\alpha_N)^2}\right) = \boxed{\frac{4\beta^2}{(\alpha_O-\alpha_N)}}$$

P10.11 (a) The transitions occur for photons whose energies are equal to the difference in energy between the highest occupied and lowest unoccupied orbital energies:

$$E_{photon} = E_{LUMO} - E_{HOMO}$$

If N is the number of carbon atoms in these species, then the number of π electrons is also N. These N electrons occupy the first $N/2$ orbitals, so orbital number $N/2$ is the HOMO and orbital number $1+N/2$ is the LUMO. Writing the photon energy in terms of the wavenumber, substituting the given energy expressions with this identification of the HOMO and LUMO gives:

$$hc\tilde{\nu}=\left(\alpha+2\beta\cos\frac{(\frac{1}{2}N+1)\pi}{N+1}\right)-\left(\alpha+2\beta\cos\frac{\frac{1}{2}N\pi}{N+1}\right)=2\beta\left(\cos\frac{(\frac{1}{2}N+1)\pi}{N+1}-\cos\frac{\frac{1}{2}N\pi}{N+1}\right)$$

Solving for β yields

$$\beta=\frac{hc\tilde{\nu}}{2\left(\cos\frac{(\frac{1}{2}N+1)\pi}{N+1}-\cos\frac{\frac{1}{2}N\pi}{N+1}\right)}$$

Draw up the following table:

Species	N	$\tilde{\nu}/\text{cm}^{-1}$	estimated β/eV
C_2H_4	2	61 500	−3.813
C_4H_6	4	46 080	−4.623
C_6H_8	6	39 750	−5.538
C_8H_{10}	8	32 900	−5.873

(b) The total energy of the π electron system is the sum of the energies of occupied orbitals weighted by the number of electrons that occupy them. In C_8H_{10}, each of the first four orbitals are doubly occupied, so

$$E_\pi=2\sum_{k=1}^{4}E_k=2\sum_{k=1}^{4}\left(\alpha+2\beta\cos\frac{k\pi}{9}\right)=8\alpha+4\beta\sum_{k=1}^{4}\cos\frac{k\pi}{9}=8\alpha+9.518\beta$$

The delocalization energy is the difference between this quantity and that of four isolated double bonds:

$$E_{\text{deloc}}=E_\pi-8(\alpha+\beta)=8\alpha+9.518\beta-8(\alpha+\beta)=\boxed{1.518\beta}.$$

Using the estimate of β from part (a) yields $E_{\text{deloc}}=\boxed{8.913\text{ eV}}$.

(c) Draw up the following table, in which the orbital energy decreases as we go down. For the purpose of comparison, we express orbital energies as $(E_k-\alpha)/\beta$. Recall that β is negative (as is α for that matter), so the orbital with the greatest value of $(E_k-\alpha)/\beta$ has the lowest energy.

Orbital	Energy	Coefficients					
	$(E_k-\alpha)/\beta$	1	2	3	4	5	6
6	−1.8019	0.2319	−0.4179	0.5211	−0.5211	0.4179	−0.2319
5	−1.2470	0.4179	−0.5211	0.2319	0.2319	−0.5211	0.4179
4	−0.4450	0.5211	−0.2319	−0.4179	0.4179	0.2319	−0.5211
3	0.4450	0.5211	0.2319	−0.4179	−0.4179	0.2319	0.5211
2	1.2470	0.4179	0.5211	0.2319	−0.2319	−0.5211	−0.4179
1	1.8019	0.2319	0.4179	0.5211	0.5211	0.4179	0.2319

The orbitals are shown schematically in Figure 10.7, with each vertical pair of lobes representing a p orbital on one of the carbons in hexatriene. Shaded lobes represent one sign of the wavefunction (say positive) and unshaded lobes the other sign. Where adjacent atoms have atomic orbitals of the same sign, the resulting molecular orbital is bonding with respect to those atoms; where adjacent atoms have different sign, there is a node between the atoms and the resulting molecular orbital is antibonding with respect to them. The lowest-energy orbital is totally bonding (no nodes between atoms) and the highest-energy orbital totally antibonding (nodes between each adjacent pair). Note that the orbitals have increasing antibonding character as their energy increases. The size of each atomic p orbital is proportional to the magnitude of the coefficient of that orbital in the molecular orbital. So, for example, in orbitals 1 and 6, the largest lobes are in the center of the molecule, so electrons that occupy those orbitals are more likely to be found near the center of the molecule than on the ends. In the ground state of the molecule, there are two electrons in each of orbitals 1, 2, and 3, with the result that the probability of finding a π electron in hexatriene is uniform over the entire molecule.

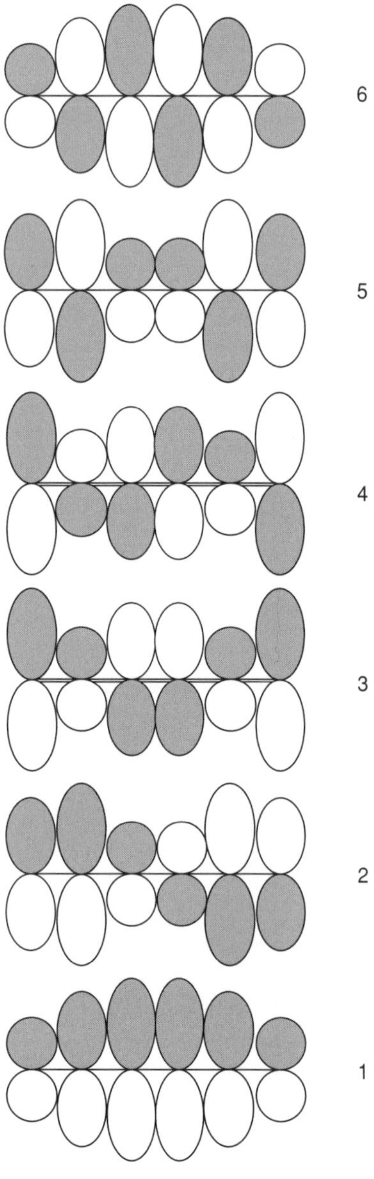

Figure 10.7

P10.13 (a) The table displays computed orbital energies and experimental $\pi^* \leftarrow \pi$ wavenumbers of ethene and the first few conjugated linear polyenes.

Species	E_{LUMO}/eV*	E_{HOMO}/eV*	ΔE/eV*	$\tilde{\nu}$/cm^{-1}
C_2H_4	1.2282	−10.6411	11.8693	61 500
C_4H_6	0.2634	−9.4671	9.7305	46 080
C_6H_8	−0.2494	−8.8993	8.6499	39 750
C_8H_{10}	−0.5568	−8.5767	8.0199	32 900
$C_{10}H_{12}$	−0.7556	−8.3755	7.6199	

*Semi-empirical, PM3 level, PC Spartan Pro™

(b) A plot of the computed energy difference vs. experimental wavenumbers appears in Figure 10.8. The computed points fall on a rather good straight line. Of course a better fit can be obtained to a quadratic and a perfect fit to a cubic polynomial; however, the improvement would be slight and the justification even more slight. The linear least-squares best fit is:

$$\boxed{\Delta E/\text{eV} = 3.3534 + 1.3791 \times 10^{-4}\, \tilde{\nu}/\text{cm}^{-1}} \quad (r^2 = 0.994)$$

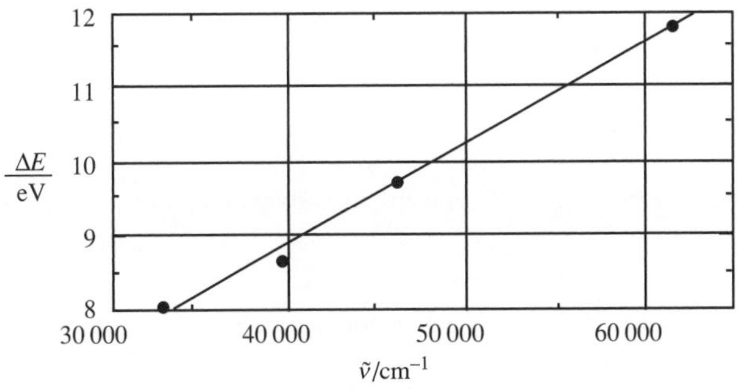

Figure 10.8

(c) Rearrange the fit equation obtained in (b) above:

$$\tilde{\nu}/\text{cm}^{-1} = \frac{\Delta E/\text{eV} - 3.3534}{1.3791 \times 10^{-4}}.$$

So for $C_{10}H_{12}$, we expect a transition at:

$$\tilde{\nu}/\text{cm}^{-1} = \frac{7.6199 - 3.3534}{1.3791 \times 10^{-4}} = \boxed{30\,937\ \text{cm}^{-1}}$$

(d) The fitting procedure is necessary because the orbital energies are only approximate. Remember that an orbital wavefunction is itself an approximation. A semi-empirical computation is a further approximation. If the orbitals were exact, then we would expect the energy difference to be directly proportional to the spectroscopic wavenumbers with the following proportionality:

$$\Delta E = hc\tilde{\nu} = \frac{(6.626 \times 10^{-34}\ \text{J s})(2.998 \times 10^{10}\ \text{cm s}^{-1})\tilde{\nu}}{1.602 \times 10^{-19}\ \text{J/eV}},$$

so $\Delta E/\text{eV} = 1.240 \times 10^{-4}\, \tilde{\nu}/\text{cm}^{-1}$.

Clearly this is different from the fit reported above. A further illustration of why the fitting procedure is necessary can be discerned by comparing the table from part (a) to a corresponding table based on a different computational model, namely Hartree–Fock computations with an STO–3G basis set:

Species	E_{LUMO}/eV*	E_{HOMO}/eV*	ΔE/eV*
C_2H_4	8.9335	−9.1288	18.0623
C_4H_6	6.9667	−7.5167	14.4834
C_6H_8	6.0041	−6.6783	12.6824
C_8H_{10}	5.4488	−6.1811	11.6299
$C_{10}H_{12}$	5.0975	−5.8621	10.9596

**Ab initio*, STO–3G, PC Spartan Pro™

Obviously these energy differences are not the same as the PM3 differences computed above, nor are they energy differences that correspond to the experimental frequencies.

COMMENT. The STO–3G data also fit a straight line. That fit can also be used to estimate the transition in $C_{10}H_{12}$:

$$\tilde{\nu}/\text{cm}^{-1} = \frac{\Delta E/\text{eV} - 3.8311}{2.3045 \times 10^{-4}},$$

so for $C_{10}H_{12}$ we expect a transition at

$$\tilde{\nu}/\text{cm}^{-1} = \frac{10.9596 - 3.8311}{2.3045 \times 10^{-4}} = 30\,933.$$

Even though the computations differed considerably in detail, with the calibration procedure they result in nearly identical predictions.

P10.15 (a) The standard enthalpy of formation ($\Delta_f H^{\ominus}$/kJ mol^{-1}) of ethene and the first few linear polyenes is listed below.

Species	Computed*	Experimental†	% error
C_2H_4	69.580	52.47	32.6
C_4H_6	129.834	108.8 ± 0.79	19.3
		111.9 ± 0.96	16.0
C_6H_8	188.523	168. ± 3	12.2
C_8H_{10}	246.848	295.9‡	16.6

*Semi-empirical, PM3 level, PC Spartan Pro™
†http://webbook.nist.gov/chemistry/
‡Pedley, Naylor, and Kirby, *Thermodynamic Data of Organic Compounds*.

(b) The % error, shown in the table, is defined by:

$$\%\ \text{error} = \frac{\Delta_f H^{\ominus}(\text{calc}) - \Delta_f H^{\ominus}(\text{expt})}{\Delta_f H^{\ominus}(\text{expt})} \times 100\%$$

(c) For all of the molecules, the computed enthalpies of formation exceed the experimental values by much more than the uncertainty in the experimental value. This observation serves to illustrate that molecular modeling software is not a substitute for experimentation when it comes to quantitative measures. It is also worth noting, however, that the experimental uncertainty can vary a great deal. The NIST database lists two different determinations for $\Delta_f H^{\ominus}(C_4H_6)$, and the experimental values differ by more than the uncertainty claimed for each; a critical evaluation of the experimental data is called for. The uncertainty claimed for $\Delta_f H^{\ominus}(C_6H_8)$ is greater still (but still only about 2%). Finally, it should go without saying that not all of the figures reported by the molecular modelling software are physically significant.

Solutions to theoretical problems

P10.17 One approach is to form the explicit forms of the orbitals and find the values of ϕ that maximize them. We need the component unhybridized orbitals, put together from the radial and angular functions listed in Tables 9.1 and 8.2:

$$2s = R_{20}Y_{00} = \frac{1}{8^{1/2}}\left(\frac{Z}{a}\right)^{3/2} \times (2-\rho)e^{-\rho/2} \times \left(\frac{1}{4\pi}\right)^{1/2} = \left(\frac{1}{32\pi}\right)^{1/2}\left(\frac{Z}{a}\right)^{3/2}(2-\rho)e^{-\rho/2}$$

$$2p_x = -\frac{1}{2^{1/2}}R_{21}(Y_{1,+1} - Y_{1,-1})\,[9.22]$$

$$= -\frac{1}{2^{1/2}} \times \frac{1}{24^{1/2}}\left(\frac{Z}{a}\right)^{3/2}\rho e^{-\rho/2} \times \left(\frac{3}{8\pi}\right)^{1/2}\sin\theta(-e^{+i\phi} - e^{-i\phi})$$

$$= \left(\frac{1}{32\pi}\right)^{1/2}\left(\frac{Z}{a}\right)^{3/2}\rho e^{-\rho/2}\sin\theta\cos\phi$$

$$2p_y = \frac{i}{2^{1/2}}R_{21}(Y_{1,+1} + Y_{1,-1})\,[9.22]$$

$$= \frac{i}{2^{1/2}} \times \frac{1}{24^{1/2}}\left(\frac{Z}{a}\right)^{3/2}\rho e^{-\rho/2} \times \left(\frac{3}{8\pi}\right)^{1/2}\sin\theta(-e^{+i\phi} + e^{-i\phi})$$

$$= \left(\frac{1}{32\pi}\right)^{1/2}\left(\frac{Z}{a}\right)^{3/2}\rho e^{-\rho/2}\sin\theta\sin\phi$$

where $\rho = \dfrac{2Zr}{2a} = \dfrac{Zr}{a}$.

In forming each hybrid, we will neglect the factor $\left(\frac{1}{32\pi}\right)^{1/2}\left(\frac{Z}{a}\right)^{3/2}e^{-\rho/2}$ common to each component; an angle-independent multiplicative term cannot influence the angle at which the hybrid is maximal.

Next, form the hybrids, using eqn 10.5:

$$h_1 = s + 2^{1/2}p_y = (2-\rho) + 2^{1/2}(\rho\sin\theta\sin\phi) = 2 + \rho(2^{1/2}\sin\theta\sin\phi - 1)$$

$$h_2 = s + \left(\frac{3}{2}\right)^{1/2}p_x - \left(\frac{1}{2}\right)^{1/2}p_y = (2-\rho) + \left(\frac{3}{2}\right)^{1/2}\rho\sin\theta\cos\phi - \left(\frac{1}{2}\right)^{1/2}\rho\sin\theta\sin\phi$$

$$= 2 + \rho\left(\frac{3^{1/2}\cos\phi - \sin\phi}{2^{1/2}}\sin\theta - 1\right)$$

and $h_3 = s - \left(\frac{3}{2}\right)^{1/2}p_x - \left(\frac{1}{2}\right)^{1/2}p_y = (2-\rho) - \left(\frac{3}{2}\right)^{1/2}\rho\sin\theta\cos\phi - \left(\frac{1}{2}\right)^{1/2}\rho\sin\theta\sin\phi$

$$= 2 - \rho\left(\frac{3^{1/2}\cos\phi + \sin\phi}{2^{1/2}}\sin\theta + 1\right)$$

To find the angle ϕ at which the hybrids have maximum amplitude, differentiate with respect to ϕ. That would work even for h_1, but differentiation is unnecessary. One can see by inspection that the function is maximized when $\sin\phi$ is maximal, namely at $\phi = \pi/2$ (90°). It should come as no surprise that this orbital points along the positive y-axis.

$$\frac{\partial h_2}{\partial \phi} = 0 = \left(\frac{\rho \sin\theta}{2^{1/2}}\right)(-3^{1/2}\sin\phi - \cos\phi)$$

so $$\frac{\sin\phi}{\cos\phi} = -\frac{1}{3^{1/2}} = \tan\phi \quad \text{or} \quad \phi = 5\pi/6 \text{ or } 11\pi/6 \ (150° \text{ or } 330°)$$

Of these, $11\pi/6$ (330°) is a maximum and $5\pi/6$ a minimum. The larger amplitude is at the maximum ($11\pi/6$ or 330°). Finally,

$$\frac{\partial h_3}{\partial \phi} = 0 = -\left(\frac{\rho \sin\theta}{2^{1/2}}\right)(-3^{1/2}\sin\phi + \cos\phi)$$

so $$\frac{\sin\phi}{\cos\phi} = \frac{1}{3^{1/2}} = \tan\phi \quad \text{or} \quad \phi = \pi/6 \text{ or } 7\pi/6 \ (30° \text{ or } 210°)$$

Of these, $7\pi/6$ (210°) is a maximum and corresponds to the larger amplitude. The angles 90°, 210°, and 330° are 120° apart.

P10.19 The energy difference between bonding and antibonding orbitals is

$$\Delta E = E_{2\sigma} - E_{1\sigma} = \left(E_{\rm H} + \frac{j_0}{R} - \frac{j-k}{1-S}\right) - \left(E_{\rm H} + \frac{j_0}{R} - \frac{j+k}{1+S}\right)$$

$$= \left(-\frac{j-k}{1-S}\right) + \left(\frac{j+k}{1+S}\right) = \frac{(-j+k-Sj+Sk)+(j+k-Sj-Sk)}{1-S^2} = \boxed{\frac{2(k-Sj)}{1-S^2}}$$

Use the explicit expressions for S, j, and k in eqn 10.13 to compute ΔE as a function of R, plotted in Figure 10.9.

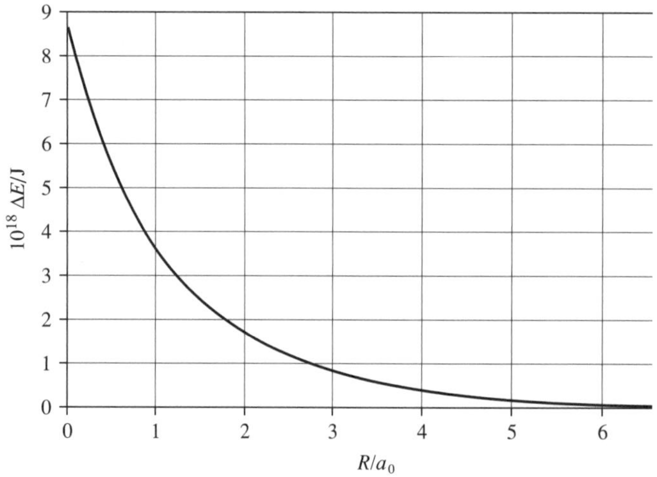

Figure 10.9

P10.21 Refer to *Mathematical Background 6*, which follows Chapter 10, for matrix operations. By hypothesis, $\boldsymbol{M} = a\mathbf{1} + \boldsymbol{O}$, where $\boldsymbol{O}$ has off-diagonal elements. Also by hypothesis, we can find a matrix that diagonalizes $\boldsymbol{O}$; that is, we can find a matrix $\boldsymbol{c}$ such that $\boldsymbol{c}^{-1}\boldsymbol{O}\boldsymbol{c} = \boldsymbol{D}$ [analogous to eqn 10.50], where $\boldsymbol{D}$ is diagonal. We are to show that the original matrix $\boldsymbol{M}$ is diagonalized by the same operation, that is, that $\boldsymbol{c}^{-1}\boldsymbol{M}\boldsymbol{c}$ is also diagonal, but

$$\boldsymbol{c}^{-1}\boldsymbol{M}\boldsymbol{c} = \boldsymbol{c}^{-1}(a\mathbf{1} + \boldsymbol{O})\boldsymbol{c} = a\boldsymbol{c}^{-1}\mathbf{1}\boldsymbol{c} + \boldsymbol{c}^{-1}\boldsymbol{O}\boldsymbol{c} = a\boldsymbol{c}^{-1}\boldsymbol{c} + \boldsymbol{D} = a\mathbf{1} + \boldsymbol{D}$$

In the previous line, we made use of the facts that $\mathbf{1}\boldsymbol{c} = \boldsymbol{c}$ (identity) and $\boldsymbol{c}^{-1}\boldsymbol{c} = \mathbf{1}$ (inverse). Our final result is the sum of two diagonal matrices, so it is also diagonal, which is what was to be proved.

P10.23 Let the two component Gaussian-type orbitals (GTOs) be $e^{-ax_A^2}$ and $e^{-bx_B^2}$, where x_A and x_B are measured from nucleus A and B, respectively. We must express the coordinates with respect to a single origin, however. If we put that origin at nucleus A, then $x_A = x$ and $x_B = x - R$ where R is the internuclear distance. The product of the orbitals, then, is

$$e^{-ax^2}e^{-b(x-R)^2} = e^{-ax^2-b(x-R)^2}$$

We are to show that this product is proportional to a single GTO centered somewhere. That is, the exponent must be shown to have the form $-\zeta(x - x_0)^2 + c$, where ζ, x_0, and c are constants. ($-\zeta(x - x_0)^2$ is, of course, the exponent of a GTO centered at x_0; e^c would then be a proportionality constant between the original product and the new GTO.) So we concentrate on the exponent of our product, first expanding it and then completing the square:

$$-ax^2 - b(x - R)^2 = -(a + b)x^2 + 2bRx - bR^2 = -(a + b)\left(x^2 - \frac{2bRx}{a+b}\right) - bR^2$$

$$= -(a + b)\left(x - \frac{bR}{a+b}\right)^2 + (a + b)\left(\frac{bR}{a+b}\right)^2 - bR^2$$

The constant terms in the exponent can be simplified further, but for our purposes it is sufficient to note that they are independent of x. The new GTO has $\zeta = a + b$, and it is centered at $x_0 = \dfrac{bR}{a+b}$.

Question. Show that this expression reduces to the homonuclear case by letting $b = a$.

P10.25 The Coulomb piece is derived in the Brief illustration

$$\int \chi_A(1)J_a(1)\chi_B(1)\,d\tau_1 = j_0\int \chi_A(1)\int \chi_B(1)\frac{1}{r_{12}}(c_{Aa}\chi_A(2) + c_{Ba}\chi_B(2))(c_{Aa}\chi_A(2) + c_{Ba}\chi_B(2))\,d\tau_1\,d\tau_2$$

$$= j_0c_{Aa}^2\iint \chi_A(1)\chi_B(1)\frac{1}{r_{12}}\chi_A(2)\chi_A(2)\,d\tau_1\,d\tau_2 + j_0c_{Aa}c_{Ba}\iint \chi_A(1)\chi_B(1)\frac{1}{r_{12}}\chi_A(2)\chi_B(2)\,d\tau_1\,d\tau_2$$

$$+ j_0c_{Ba}c_{Aa}\iint \chi_A(1)\chi_B(1)\frac{1}{r_{12}}\chi_B(2)\chi_A(2)\,d\tau_1\,d\tau_2 + j_0c_{Ba}^2\iint \chi_A(1)\chi_B(1)\frac{1}{r_{12}}\chi_B(2)\chi_B(2)\,d\tau_1\,d\tau_2$$

or in abbreviated form

$$\int \chi_A(1)J_a(1)\chi_B(1)d\tau_1 = \boxed{j_0\left\{c_{Aa}^2(AB|AA) + c_{Aa}c_{Ba}(AB|AB) + c_{Ba}c_{Aa}(AB|BA) + c_{Ba}^2(AB|BB)\right\}}$$

Two of the integrals in this expression are equal: $(AB\,|\,AB) = (AB\,|\,BA)$. If the molecule were homonuclear, the other two integrals would be equal to each other, but not to these two.

The exchange piece is

$$\int \chi_A(1)K_a(1)\chi_B(1)\,d\tau_1 = j_0\int \chi_A(1)\int (c_{Aa}\chi_A(1)+c_{Ba}\chi_B(1))\frac{1}{r_{12}}(c_{Aa}\chi_A(2)+c_{Ba}\chi_B(2))\chi_B(2)\,d\tau_1\,d\tau_2$$

$$= j_0c_{Aa}^2\iint \chi_A(1)\chi_A(1)\frac{1}{r_{12}}\chi_A(2)\chi_B(2)\,d\tau_1\,d\tau_2 + j_0c_{Aa}c_{Ba}\iint \chi_A(1)\chi_A(1)\frac{1}{r_{12}}\chi_B(2)\chi_B(2)\,d\tau_1\,d\tau_2$$

$$+ j_0c_{Ba}c_{Aa}\iint \chi_A(1)\chi_B(1)\frac{1}{r_{12}}\chi_A(2)\chi_B(2)\,d\tau_1\,d\tau_2 + j_0c_{Ba}^2\iint \chi_A(1)\chi_B(1)\frac{1}{r_{12}}\chi_B(2)\chi_B(2)\,d\tau_1\,d\tau_2$$

or in abbreviated form

$$\int \chi_A(1)K_a(1)\chi_B(1)d\tau_1 = \boxed{j_0\left\{c_{Aa}^2(AA|AB)+c_{Aa}c_{Ba}(AA|BB)+c_{Ba}c_{Aa}(AB|AB)+c_{Ba}^2(AB|BB)\right\}}$$

None of these integrals in this expression is equal to another. If the molecule were homonuclear, $(AA\,|\,AB)$ would be equal to $(AB\,|\,BB)$. Note that $(AA\,|\,BB) \neq (AB\,|\,AB)$.

Solutions to applications

P10.27 The secular determinant for an N-atom cyclic species H_N has the form

$$\begin{array}{c} \begin{array}{cccccccc} 1 & 2 & 3 & \cdots & \cdots & \cdots & N-1 & N \end{array} \\ \left|\begin{array}{cccccccc} x & 1 & 0 & \cdots & \cdots & \cdots & 0 & 1 \\ 1 & x & 1 & \cdots & \cdots & \cdots & 0 & 0 \\ 0 & 1 & x & 1 & \cdots & \cdots & 0 & 0 \\ 0 & 0 & 1 & x & 1 & \cdots & 0 & 0 \\ \vdots & \vdots & \vdots & \vdots & \vdots & \vdots & \vdots & \vdots \\ \vdots & \vdots & \vdots & \vdots & \vdots & \vdots & \vdots & 1 \\ 1 & 0 & 0 & 0 & 0 & \cdots & 1 & x \end{array}\right| \end{array}$$

where $x = \dfrac{\alpha - E}{\beta}$ or $E = \alpha - \beta x$.

Expanding the determinant, finding the roots of the polynomial, and solving for the total binding energy yields the following table. Keep in mind that $\alpha < 0$ and $\beta < 0$.

Species	Number of e$^-$	Permitted x (roots)	Total binding energy
H_3^+	2	−2, 1 (twice)	$2\alpha + 4\beta$
H_3	3	−2, 1 (twice)	$3\alpha + 3\beta$
H_3^-	4	−2, 1 (twice)	$4\alpha + 2\beta$
H_4	4	−2, 0 (twice), 2	$4\alpha + 4\beta$
H_5^+	4	−2, $(1 - 5^{1/2})/2$ (twice), $(1 + 5^{1/2})/2$ (twice)	$4\alpha + (3 + 5^{1/2})\beta$
		−2, −0.618 (twice), 1.618 (twice)	$4\alpha + 5.236\beta$
H_5	5	−2, $(1 - 5^{1/2})/2$ (twice), $(1 + 5^{1/2})/2$ (twice)	$5\alpha + (5 + 3 \times 5^{1/2})\beta/2$
		−2, −0.618 (twice), 1.618 (twice)	$5\alpha + 5.854\beta$
H_5^-	6	−2, $(1 - 5^{1/2})/2$ (twice), $(1 + 5^{1/2})/2$ (twice)	$6\alpha + (2 + 2 \times 5^{1/2})\beta$
		−2, −0.618 (twice), 1.618 (twice)	$6\alpha + 6.472\beta$
H_6	6	−2, −1 (twice), 1 (twice), 2	$6\alpha + 8\beta$
H_7^+	6	−2, −1.248 (twice), 0.445 (twice), 1.802 (twice)	$6\alpha + 8.992\beta$

Thus, for example, the energy levels in H_5 are

$$\alpha + 2\beta, \alpha + \frac{\beta}{2}(5^{1/2} - 1) \approx \alpha + 0.618\beta \text{ (twice)}, \alpha - \frac{\beta}{2}(1 + 5^{1/2}) \approx \alpha - 1.618\beta \text{ (twice)}$$

and the total binding energy for H_5 is

$$2(\alpha + 2\beta) + 3\left\{\alpha + \frac{\beta}{2}(5^{1/2} - 1)\right\} = 5\alpha + \frac{\beta}{2}(3 \times 5^{1/2} + 5) \approx 5\alpha + 5.854\beta$$

Now we estimate $\Delta_r U$ based on the difference in binding energy between products and reactants in the following reactions:

$$H_4 \rightarrow 2\,H_2 \qquad \Delta_r U = 2(2\alpha + 2\beta) - (4\alpha + 4\beta) = 0$$
$$H_5^+ \rightarrow H_2 + H_3^+ \qquad \Delta_r U = (2\alpha + 2\beta) + (2\alpha + 4\beta) - (4\alpha + 5.236\beta) = 0.764\beta < 0$$

The above $\Delta_r U$ values indicate that H_4 and H_5^+ can fall apart without an energy penalty.

$$H_5^- \rightarrow H_2 + H_3^- \qquad \Delta_r U = (2\alpha + 2\beta) + (4\alpha + 2\beta) - (6\alpha + 6.472\beta) = -2.472\beta > 0$$
$$H_6 \rightarrow 3\,H_2 \qquad \Delta_r U = 3(2\alpha + 2\beta) - (6\alpha + 8\beta) = -2\beta > 0$$
$$H_7^+ \rightarrow 2H_2 + H_3^+ \qquad \Delta_r U = 2(2\alpha + 2\beta) + (2\alpha + 4\beta) - (6\alpha + 8.992\beta) = -0.992\beta > 0$$

The $\Delta_r U$ values for H_5^-, H_6, and H_7^+ suggest that they are stable.

Species	Number of e^-	Satisfies Hückel $4n + 2$ rule?	
		Correct number of e^-?	Stable?
H_4	4	No	No
H_5^+	4	No	No
H_5^-	6	Yes	Yes
H_6	6	Yes	Yes
H_7^+	6	Yes	Yes

Hückel's $4n + 2$ rule successfully predicts the stability of hydrogen rings.

P10.29 This question refers to six 1,4-benzoquinones: the unsubstituted, four methyl-substituted, and a dimethyldimethoxy species. The table below defines the molecules and displays reduction potentials and computed LUMO energies.

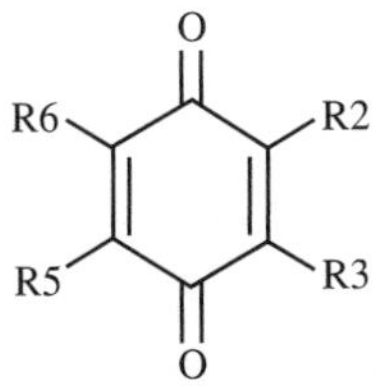

Species	R2	R3	R5	R6	$E^{\ominus}$/V	E_{LUMO}/eV*
1	H	H	H	H	+0.078	−1.706
2	CH_3	H	H	H	+0.023	−1.651
3	CH_3	H	CH_3	H	−0.067	−1.583
4	CH_3	CH_3	CH_3	H	−0.165	−1.371
5	CH_3	CH_3	CH_3	CH_3	−0.260	−1.233
6	CH_3	CH_3	CH_3O	CH_3O		−1.446

*Semi-empirical, PM3 level, PC Spartan Pro™

(a) The LUMO energies for species 1–5 are plotted against standard potentials in Figure 10.10. The figure shows that a linear relationship between the plotted quantities is consistent with these calculations.

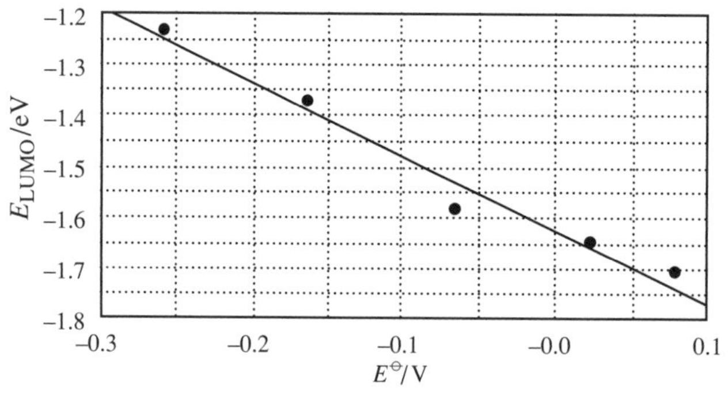

Figure 10.10

(b) The linear least-squares fit from the plot of E_{LUMO} vs. $E^{\ominus}$ is:

$$E_{LUMO}/\text{eV} = -1.621 - 1.435E^{\ominus}/\text{V} \quad (r^2 = 0.927)$$

Solving for $E^{\ominus}$ yields:

$$E^{\ominus}/\text{V} = -(E_{LUMO}/\text{eV} + 1.621)/1.435$$

Substituting the computed LUMO energy for compound 6 (a model of ubiquinone) yields

$$E^{\ominus} = [-(-1.446 + 1.621)/1.435] = \boxed{-0.122\ \text{V}}.$$

(c) The model of plastoquinone defined in the problem is compound 4 in the table above. Its experimental reducing potential is known; however, a comparison to the ubiquinone analogue based on E_{LUMO} ought to use a computed reducing potential:

$$E^{\ominus} = [-(-1.371 + 1.621)/1.435] = \boxed{-0.174\ \text{V}}.$$

The better oxidizing agent is the one that is more easily reduced, the one with the less negative reduction potential. Thus, we would expect compound 6 to be a better oxidizing agent than compound 4, and ubiquinone a better oxidizing agent than plastiquinone.

(d) Respiration oxidizes organic compounds like glucose to carbon dioxide and water; hence oxidizing agents are required. Photosynthesis reduces carbon dioxide and water to glucose; hence reducing agents are required. It stands to reason, then, that the better oxidizing agent, ubiquinone, is employed in oxidizing glucose (i.e. in respiration), while the better reducing agent (that is, the poorer oxidizing agent) is used in reduction, i.e. in photosynthesis. (Note, however, that both species are recycled to their original forms: reduced ubiquinone is oxidized by iron (III) and oxidized plastoquinone is reduced by water.)

11 Molecular symmetry

Answers to discussion questions

D11.1 Within the context of quantum theory and molecular symmetry a **group** is a collection of transformations (R, S, T, etc.) that satisfy these criteria:

1. One of the transformations is the identity (E).
2. For every transformation R, the inverse transformation R^{-1} is included in the collection so that the combination RR^{-1} is equivalent to the identity: $RR^{-1} = E$.
3. The product RS is equivalent to a single member of the collection of transformations.
4. Multiple transformations obey the associative rule: $R(ST) = (RS)T$.

D11.3

Symmetry operation	Symmetry element
Identity, E	The entire object
n-fold rotation	n-fold axis of symmetry, C_n
Reflection	Mirror plane, σ
Inversion	Centre of symmetry, i
n-fold improper rotation	n-fold improper rotation axis, S_n

There are three kinds of mirror planes. The **vertical mirror plane, σ_v**, is parallel to the principal axis while the **horizontal mirror plane, σ_h**, is perpendicular to the principal axis. A mirror plane that bisects the angle between two C_2 axes is called a **dihedral plane, σ_d**. A vertical mirror plane that bisects bonds is also given the σ_d designation.

D11.5 A molecule may be chiral, and therefore optically active, only if it does not possess an axis of improper rotation, S_n. An improper rotation is a rotation followed by a reflection and this combination of operations always converts a right-handed object into a left-handed object and *vice versa*; hence an S_n axis guarantees that a molecule cannot exist in chiral forms. When discussing optical activity, it is helpful to remember that:

(a) the presence of both a C_n and a σ_h is equivalent to an S_n.
(b) $i = S_2$.
(c) $\sigma = S_1$. Thus, a molecule cannot be optically active if it possesses a centre of symmetry or a mirror plane.

D11.7 See Sections 11.4(a)–(d) for the explanation of character table construction and content. Figure 11.1 uses the C_{3v} point group to illustrate structural features of a character table. The central columns are labelled with the symmetry operations, or **symmetry classes**, of the group. Each operation is prefixed with the number of operations in the class, called the **degeneracy of the class**. For example, the degeneracy of the C_3 class is 2, which corresponds to clockwise and counter-clockwise rotations of the principal axis, shown in Figure 11.2, which is a modification of text Figure 11.19. There are three vertical reflection planes in the C_{3v} group; these are also shown in Figure 11.2. The **group order**, h, is the sum of class degeneracies. It is the total number of symmetry operations of the group. The first column of Figure 11.1 shows that the C_{3v} group contains three **symmetry species**. They are the **Mulliken symbols** used to label the irreducible representations of the group. The symmetry species A and B are one-dimensional representations, E is the symbol for a two-dimensional representation, and T is the symbol of a three-dimensional representation. It is useful to remember that E species levels are doubly degenerate, while T species levels are triply degenerate.

Number of symmetry species = number of classes [11.6]

Also, the sum of the squares of the dimension d_i of all symmetry species i equals the group order.

$$\sum_i d_i^2 = h$$

For the C_{3v} group the dimensions are 1, 1, and 2 for the A_1, A_2, and E irreducible representations, respectively. Adding the squares gives $1^2 + 1^2 + 2^2 = 6$ and we see that the sum of the squares equals h.

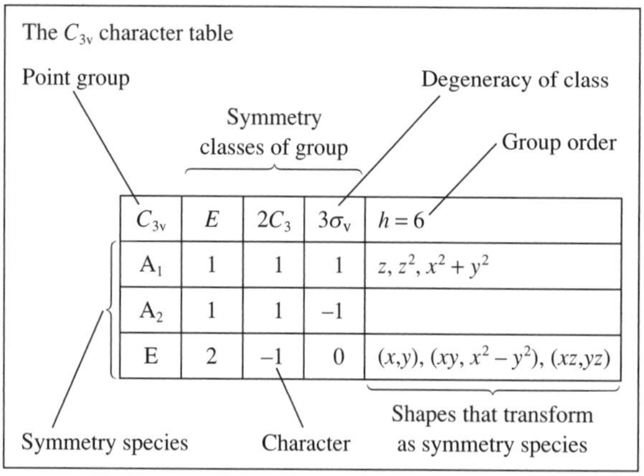

C_{3v}	E	$2C_3$	$3\sigma_v$	$h = 6$
A_1	1	1	1	$z, z^2, x^2 + y^2$
A_2	1	1	−1	
E	2	−1	0	$(x,y), (xy, x^2 - y^2), (xz,yz)$

Figure 11.1

The **characters** below a symmetry class of a point group are the traces (the sum of diagonal elements) of the matrix representations of the class operation(s) for a symmetry species. When the character under the principal axis is +1 for a one-dimensional symmetry species, the Mulliken symbol of the species is A; if the character is −1, the Mulliken symbol of the species is B. All characters of the symmetry species A_1 are +1. Characters of +1 are said to be **symmetric** with respect to the specific operation of the character; characters of −1 are said to be **antisymmetric** with respect to the operation. The subscript 1 or 2 is added to the Mulliken symbol when a species is either symmetric or antisymmetric under a σ_v operation, respectively. The subscripts g and u (gerade and ungerade) indicate that a species is either symmetric or antisymmetric under the i operation. Single primes or double primes are added to the species symbol to indicate symmetry or antisymmetry under the σ_h operation.

The last column of the character table catalogues the symmetry species to which useful functions belong. Any function that is proportional to one of the summarized functions belongs to the same symmetry species. For example, a quick examination of the C_{3v} group reveals that the basic functions z, z^2, and $x^2 + y^2$ belong to the A_1 **irrep** (irreducible representation) so we immediately recognize that the p_z, d_{z^2} and $d_{x^2+y^2}$ atomic orbitals all belong to the A_1 species. The d_{xy} and $d_{x^2-y^2}$ orbitals **jointly** belong to the E symmetry species of the C_{3v} group; they are said to **span** the representation and to be **basis functions** for the representation.

We take a moment to show that x and y jointly span the E irreducible representation of the C_{3v} group. The operations indicated in Figure 11.2 are useful to this effort.

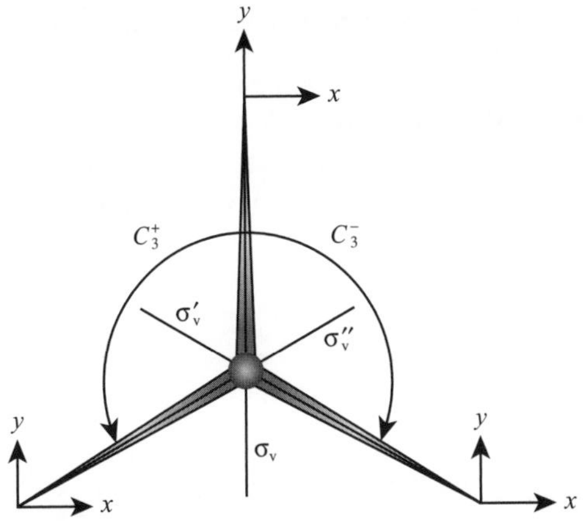

Figure 11.2

By inspection, we deduce that the matrix representation of σ_v is

$$\boldsymbol{D}(\sigma_v) = \begin{pmatrix} -1 & 0 \\ 0 & 1 \end{pmatrix}$$

because this matrix transforms (x,y) to $(-x,y)$ in accord with the requirements of the σ_v reflection shown in Figure 11.2. The trace of this matrix is $-1 + 1 = 0$, which is the character of the σ_v operation of the E symmetry species. The character of the C_3^+ operation is more difficult to deduce. The general form of the two-dimensional rotation of angle ϕ is

$$\boldsymbol{D}(C_\phi) = \begin{pmatrix} \cos\phi & \sin\phi \\ -\sin\phi & \cos\phi \end{pmatrix}$$

so the matrix for a 120° (2π/3 radian) rotation of the principal axis, which is the central z-axis in Figure 11.2, is

$$\boldsymbol{D}(C_3^+) = \begin{pmatrix} \cos(2\pi/3) & \sin(2\pi/3) \\ -\sin(2\pi/3) & \cos(2\pi/3) \end{pmatrix} = \begin{pmatrix} -\frac{1}{2} & \sqrt{3}/2 \\ -\sqrt{3}/2 & -\frac{1}{2} \end{pmatrix}$$

The trace of this matrix is $-\frac{1}{2} - \frac{1}{2} = -1$, which is the character of the C_3 operation of the E symmetry species. Thus, x and y jointly span E.

D11.9 The procedure for the determination of the direct product of symmetry species and its decomposition to a direct sum of irreducible species is illustrated in both Section 11.5(a) and Section 11.6 of the text. We outline the procedure while answering the question: Is $p_x \rightarrow p_y$ an allowed transition for a C_{3v} environment in a field of z-polarized radiation? The answer depends on the z component of the transition dipole moment,

$$\mu_{z,\text{fi}} = -e\int p_y z p_x \, d\tau \quad [11.11]$$

If analysis proves that the transition integral has a non-zero value, we can conclude that the transition is allowed. Group theory says that the integral will be non-zero provided that the direct product of the symmetry species of integrand ($p_y z p_x$) functions either spans the completely symmetrical A_1 species or has a decomposition that contains a component that spans A_1. The group theory analysis involves the following steps.

1. Decide on the symmetry species of the individual functions of the integrand (there are a total of three functions in this example) by reference to the appropriate character table (C_{3v} in this example, see the character table used in D11.7), and write their characters in successive rows in the same order as the table.

Function	E	$2C_3$	$3\sigma_v$	Irrep. of C_{3v}
p_y	2	−1	0	E
z	1	1	1	A_1
p_x	2	−1	0	E

2. The direct product is formed by multiplying the characters in each column.

Function	E	$2C_3$	$3\sigma_v$	Decomposition of the direct product
$p_y z p_x$	4	1	0	$A_1 + A_2 + E$

3. Inspect the direct product produced by step 2 and see if it can be expressed as a direct sum of characters from each column of the group. The integral must be zero if this sum does not contain A_1. For this example we find that $p_y \times z \times p_x = A_1 + A_2 + E$ (i.e. $[4\ 1\ 0] = [1\ 1\ 1] + [1\ 1\ {-1}] + [2\ {-1}\ 0]$) and we conclude that the transition is allowed because the decomposition of the direct product spans A_1. If it proves difficult to find the decomposition by inspection use eqn 11.8b to find the number of times A_1 occurs in the decomposition; if non-zero, the transition is allowed. In this example

$$N_{A_1} = \frac{1}{h}\sum_R \chi^{(A_1)}(R) \times \chi(R) \; [11.8b]$$

$$= \frac{1}{6}\{[1\times 1 \quad 2\times 1 \quad 3\times 1]\times[4 \quad 1 \quad 0]\}$$

[Notice that the degeneracies (1, 2, 3 in this C_{3v} example) of each class must be included in the $\chi^{(A_1)}(R)$ factors.]

$$= \frac{1}{6}\{[1 \quad 2 \quad 3]\times[4 \quad 1 \quad 0]\}$$

$$= \frac{1}{6}\{1\times 4 + 2\times 1 + 3\times 0\} = \frac{1}{6}\{4+2+0\} = 1$$

Since N_{A_1} is non-zero, we once again conclude that the transition is allowed.

Solutions to exercises

E11.1(a) Chloromethane belongs to the point group C_{3v}. The elements, other than the $\boxed{\text{identity } E}$, are a $\boxed{C_3 \text{ axis}}$ and $\boxed{\text{three vertical mirror planes } \sigma_v}$. The symmetry axis passes through the C–Cl nuclei. The mirror planes are defined by the three ClCH planes. The C_3 principal axis and one of the σ_v mirror planes are shown in Figure 11.3.

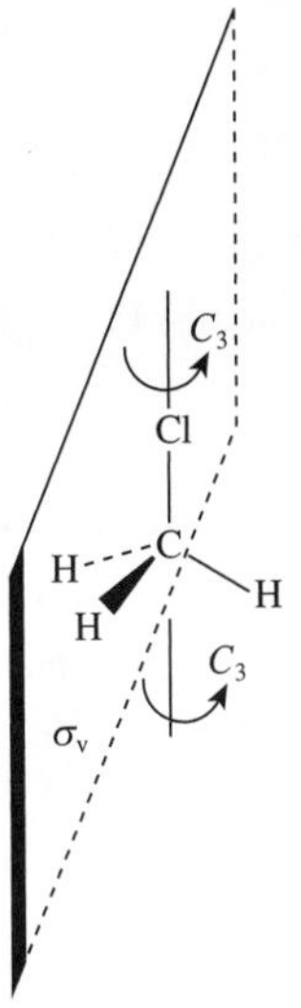

Figure 11.3

E11.2(a) List the symmetry elements of the objects (the principal ones, not necessarily all the implied ones), then use the remarks in Section 11.2, and Figure 11.4 below. Also refer to Figures 11.7 and 11.8 of the text.

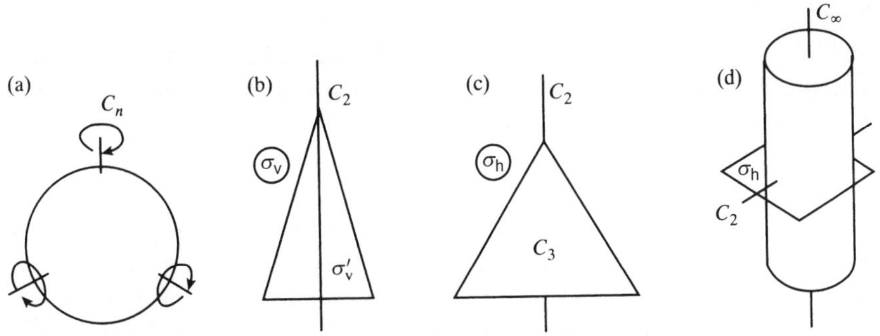

Figure 11.4

(a) Sphere: an infinite number of symmetry axes; therefore $\boxed{R_3}$

(b) Isosceles triangles: E, C_2, σ_v, and σ_v'; therefore $\boxed{C_{2v}}$

(c) Equilateral triangle: $\underbrace{\underbrace{E, C_3, C_2}_{D_3}, \sigma_h}_{\boxed{D_{3h}}}$

(d) Cylinder: E, C_∞, C_2, σ_h; therefore $\boxed{D_{\infty h}}$

E11.3(a) (a) NO_2: $E, C_2, 2\sigma_v$; $\boxed{C_{2v}}$ (b) N_2O: E, C_∞, σ_v; $\boxed{C_{\infty v}}$

(c) $CHCl_3$: $E, C_3, 3\sigma_v$; $\boxed{C_{3v}}$ (d) $CH_2{=}CH_2$: $E, 3C_2, i, \sigma_h$; $\boxed{D_{2h}}$

E11.4(a) (a) *cis*-CHCl=CHCl; $E, C_2, 2\sigma_v$; $\boxed{C_{2v}}$ (b) *trans*-CHCl=CHCl; E, C_2, σ_h, i; $\boxed{C_{2h}}$

E11.5(a) Only molecules belonging to the groups C_n, C_{nv}, and C_s may be polar [Section 11.3(a)]; hence of the molecules listed only $\boxed{\text{(a) pyridine}}$ and $\boxed{\text{(b) nitroethane}}$ are polar.

E11.6(a) NO_2, C_{2v}; N_2O, $C_{\infty v}$; $CHCl_3$, C_{3v}; CH_2CH_2, D_{2h}; *cis*-CHCl=CHCl, C_{2v}; *trans*-CHCl=CHCl, C_{2h}

A molecule cannot be chiral if it has an axis of improper rotation, including disguised or degenerate axes such as an inversion centre (S_2) or a mirror plane (S_1). In Exercises 11.3a and 11.4a, all the molecules have mirror planes, so $\boxed{\text{none}}$ can be chiral.

E11.7(a) Molecules belonging to the point groups D_{2h} or C_{3h} cannot be chiral because, as indicated by their subscripts, they have a $\boxed{\sigma_h}$ mirror plane; see Section 11.3(b). D_{2h} molecules also possess the $\boxed{i}$ symmetry element. In general we must look for an improper rotation axis, perhaps in a disguised form ($S_1 = \sigma$, $S_2 = i$). If the S_n symmetry element is present, the molecule cannot be chiral. However, a molecule may be achiral even though it does not have a centre of inversion. For example, the S_4 and S_8 groups are achiral and optically inactive; although the groups lack i they do have an S_4 and S_8 axis, respectively.

E11.8(a) Since the three C_2 axes are mutually perpendicular, it is convenient to label them as the C_2^z, C_2^y, and C_2^x axes and recognize that they transform the point (x, y, z) as follows:

$$(x, y, z)\boldsymbol{D}(C_2^z) \rightarrow (-x, -y, z)$$
$$(x, y, z)\boldsymbol{D}(C_2^y) \rightarrow (-x, y, -z)$$
$$(x, y, z)\boldsymbol{D}(C_2^z) \rightarrow (x, -y, -z)$$

The group multiplication table is constructed by finding the effect of a second rotational operation on the above expressions. For example, the effect of first applying the C_2^z operation to the point (x, y, z) and subsequently applying the C_2^y operation is

$$(x, y, z)\boldsymbol{D}(C_2^z)\boldsymbol{D}(C_2^y) \rightarrow (-x, -y, z)\boldsymbol{D}(C_2^y) \rightarrow (x, -y, -z)$$

This result is equivalent to the simple application of the C_2^x operation to the point (x, y, z) so we conclude that the group multiplication table has the C_2^x operation at the intersection of the C_2^z column and the C_2^y row. Additionally, the C_2^x operation appears at the intersection of the C_2^y column and the C_2^z row because the operations of symmetry elements commute (i.e. give the same result in either order). With these considerations we construct the following group multiplication table.

Second operation \ First operation	E	C_2^z	C_2^y	C_2^x
E	E	C_2^z	C_2^y	C_2^x
C_2^z	C_2^z	E	C_2^x	C_2^y
C_2^y	C_2^y	C_2^x	E	C_2^z
C_2^x	C_2^x	C_2^y	C_2^z	E

E11.9(a) The p_x orbital spans E of the C_{4v} point group, while z and p_z span A_1. Following the procedure discussed in Section 11.5(a), we write a table of the characters of each function and multiply the rows.

C_{4v}	E	C_2	$2C_4$	$2\sigma_v$	$2\sigma_d$	Symmetry species
p_x	2	−2	0	0	0	E
z	1	1	1	1	1	A_1
p_z	1	1	1	1	1	A_1
$p_x z p_z$	2	−2	0	0	0	E

The characters of the product $p_x z p_z$ are those of E alone, so the integrand does not span A_1. It follows that the integral must be zero.

E11.10(a) For a C_{3v} molecule, x and y span E, while z spans A_1. Thus, the x and y components of the dipole moment [11.11] have transition integrands that span $A_2 \times E \times A_1$ for the $A_1 \to A_2$ transition. By inspection of the C_{3v} character table we find the **decomposition of the direct product** to be $A_2 \times E \times A_1 = E$. The integrand spans E alone. Since it does not span A_1, the x and y components of the transition integral must be zero. The transition integrand for the z component spans $A_2 \times A_1 \times A_1 = A_2$ for the $A_1 \to A_2$ transition. Consequently, the z component of the transition integral must also equal zero and we conclude that the transition is forbidden.

Should these considerations prove confusing, write a table with columns headed by the three components of the electric dipole moment operator, μ.

Component of μ	x	y	z
A_1	1 1 1	1 1 1	1 1 1
$\Gamma(\mu)$	2 −1 0	2 −1 0	1 1 1
A_2	1 1 −1	1 1 −1	1 1 −1
$A_1\Gamma(\mu)A_2$	2 −1 0	2 −1 0	1 1 −1
	E	E	A_2

Since A_1 is not present in any product, the transition dipole moment must be zero.

E11.11(a) We first determine how x and y individually transform under the operations of the C_{4v} group. Using these results we determine how the product xy transforms, from which we extract the symmetry character χ.

Under each operation the functions transform as follows:

E		C_2	$2C_4$	$2\sigma_v$	$2\sigma_d$
x	x	$-x$	y	x	$-y$
y	y	$-y$	$-x$	$-y$	$-x$
xy	xy	xy	$-xy$	$-xy$	xy
χ	1	1	−1	−1	1

From the C_{4v} character table, we see that this set of χ characters belongs of B_2.

E11.12(a) Recall [Section 11.4(d)] that $p_x \propto x$, $p_y \propto y$, $p_z \propto z$, $d_{xy} \propto xy$, $d_{xz} \propto xz$, $d_{yz} \propto yz$, $d_{z^2} \propto z^2$, and $d_{x^2-y^2} \propto x^2 - y^2$. Additionally, when the functions f_1 and f_2 of an overlap integral are bases for irreducible representations of a group, the integral must vanish if they are different symmetry species; if they are the same symmetry species, then the integral may be non-zero [Section 11.5(a)]. Since the combination $p_x(A) - p_x(B)$ of the two O atoms (with x perpendicular to the plane) spans A_2, the orbital on N must span A_2 for a non-zero overlap. Now refer to the C_{2v} character table. The s orbital spans A_1 and the p orbitals of the central N atom span $A_1(p_z)$, $B_1(p_x)$, and $B_2(p_y)$. Therefore, [no orbitals] span A_2, and hence $p_x(A) - p_x(B)$ is a non-bonding combination with the atomic orbitals of N because there is no non-zero overlap. If d orbitals are available, as they are in S of the SO_2 molecule, we could form a molecular orbital with [d_{xy}], which is a basis for A_2.

E11.13(a) Components of the C_{2v} group electric dipole moment transform as the B_1, B_2, and A_1 irreducible species in the $q = x$, y, and z directions, respectively. Transitions are allowed if the integral $\int \psi_f^* q \psi_i \, d\tau$ [11.11] is non-zero, and hence are forbidden unless the direct product of the integrand contains the A_1 species. Since the symmetry species of the ground state (i) is A_1, this requires that the direct product of the symmetry species of ψ_f and q contains the A_1 species. Since $B_1 \times B_1 = A_1$ and $B_2 \times B_2 = A_1$, and $A_1 \times A_1 = A_1$, x-polarized light may cause a transition to a B_1 term, y-polarized light to a B_2 term, and z-polarized light to an A_1 term.

E11.14(a)

C_{4v}, $h = 8$	E	C_2	$2C_4$	$2\sigma_v$	$2\sigma_d$
A_1	1	1	1	1	1
A_2	1	1	1	−1	−1
B_1	1	1	−1	1	−1
B_2	1	1	−1	−1	1
E	2	−2	0	0	0

$$N_n = \frac{1}{h}\sum_R \chi^{(n)}(R)\chi(R) \quad [11.8b], \quad \text{where } \chi(R) = (5,1,1,3,1)$$

$$N_{A_1} = \tfrac{1}{8}\{1(1\times 5) + 1(1\times 1) + 2(1\times 1) + 2(1\times 3) + 2(1\times 1)\} = 2$$

$$N_{A_2} = \tfrac{1}{8}\{1(1\times 5) + 1(1\times 1) + 2(1\times 1) + 2(-1\times 3) + 2(-1\times 1)\} = 0$$

$$N_{B_1} = \tfrac{1}{8}\{1(1\times 5) + 1(1\times 1) + 2(-1\times 1) + 2(1\times 3) + 2(-1\times 1)\} = 1$$

$$N_{B_2} = \tfrac{1}{8}\{1(1\times 5) + 1(1\times 1) + 2(-1\times 1) + 2(-1\times 3) + 2(1\times 1)\} = 0$$

$$N_E = \tfrac{1}{8}\{1(2\times 5) + 1(-2\times 1) + 2(0\times 1) + 2(0\times 3) + 2(0\times 1)\} = 1$$

Thus, this set of basis functions spans [$2A_1 + B_1 + E$].

E11.15(a) (a) The point group of benzene is D_{6h}. In D_{6h} μ spans $E_{1u}(x, y)$ and $A_{2u}(z)$, and the ground term is A_{1g}. Then, using $A_{2u} \times A_{1g} = A_{2u}$, $E_{1u} \times A_{1g} = E_{1u}$, $A_{2u} \times A_{2u} = A_{1g}$, and $E_{1u} \times E_{1u} = A_{1g} + A_{2g} + E_{2g}$, we conclude that the upper term is [either E_{1u} or A_{2u}].

(b) Naphthalene belongs to D_{2h}. In D_{2h} itself, the components span $B_{3u}(x)$, $B_{2u}(y)$, and $B_{1u}(z)$ and the ground term is A_g. Hence, since $A_g \times \Gamma = \Gamma$ in this group, the upper terms are [B_{3u}(x-polarized)], [B_{2u}(y-polarized)], and [B_{1u}(z-polarized)].

E11.16(a) We consider the integral

$$I = \int_{-a}^{a} f_1 f_2 \,d\theta = \int_{-a}^{a} \sin\theta\cos\theta \,d\theta$$

and hence draw up the following table for the effect of operations in the group C_s (see Figure 11.5).

	E	σ_h
$f_1 = \sin\theta$	$\sin\theta$	$-\sin\theta$
$f_2 = \cos\theta$	$\cos\theta$	$\cos\theta$

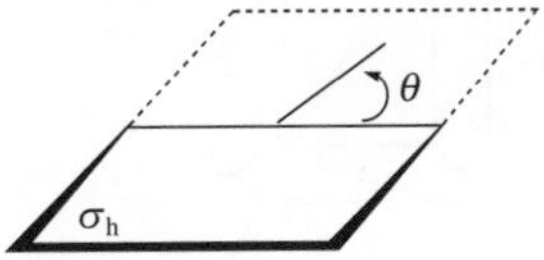

Figure 11.5

In terms of characters:

	E	σ_h	Symmetry species
f_1	1	−1	A″
f_2	1	1	A′
$f_1 f_2$	1	−1	A″

Since the product does not span the totally symmetric species A′, the integral is necessarily zero.

Solutions to problems

Solutions to numerical problems

P11.1 (a) Staggered CH_3CH_3: $E, C_3, C_2, i, 2S_6, 3\sigma_d$; $\boxed{D_{3d}}$ [see Figure 11.6b of the text]

(b) Chair C_6H_{12}: $E, C_3, C_2, i, 2S_6, 3\sigma_d$; $\boxed{D_{3d}}$ Boat C_6H_{12}: $E, C_2, \sigma_v, \sigma_v'$; $\boxed{C_{2v}}$

(c) B_2H_6: $E, C_2, 2C_2', i, \sigma_h, 2\sigma_v'$; $\boxed{D_{2h}}$

(d) $[Co(en)_3]^{3+}$: $E, 2C_3, 3C_2$; $\boxed{D_3}$

(e) Crown S_8: $E, C_4, C_2, 4C_2', 4\sigma_d, 2S_8$; $\boxed{D_{4d}}$

Only the boat C_6H_{12} may be polar, since all the others are D point groups. Only $[Co(en)_3]^{3+}$ belongs to a group without an improper rotation axis ($S_1 = \sigma$), and hence is chiral.

P11.3 Consider Figure 11.6. The effect of σ_h on a point P is to generate $\sigma_h P$, and the effect of C_2 on $\sigma_h P$ is to generate the point $C_2\sigma_h P$. The same point is generated from P by the inversion i, so $C_2\sigma_h P = iP$ for all points P. Hence, $\boxed{C_2\sigma_h = i}$, and i must be a member of the group.

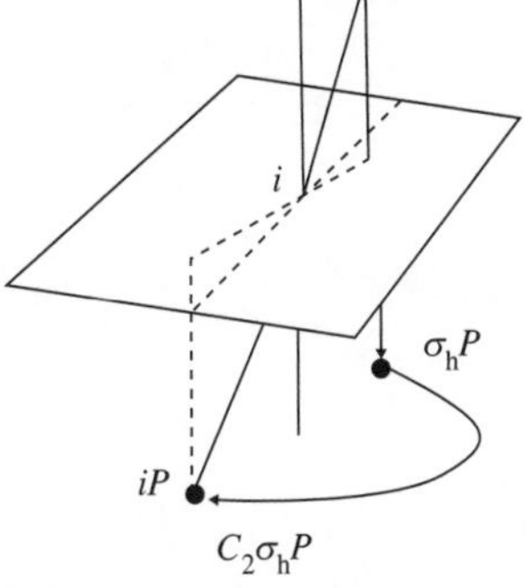

Figure 11.6

P11.5 We examine how the operations of the C_{3v} group affect $l_z = xp_y - yp_x$ when applied to it. The transformation of x, y, and z, and by analogy p_x, p_y, and p_z, are as follows (see Figure 11.7).

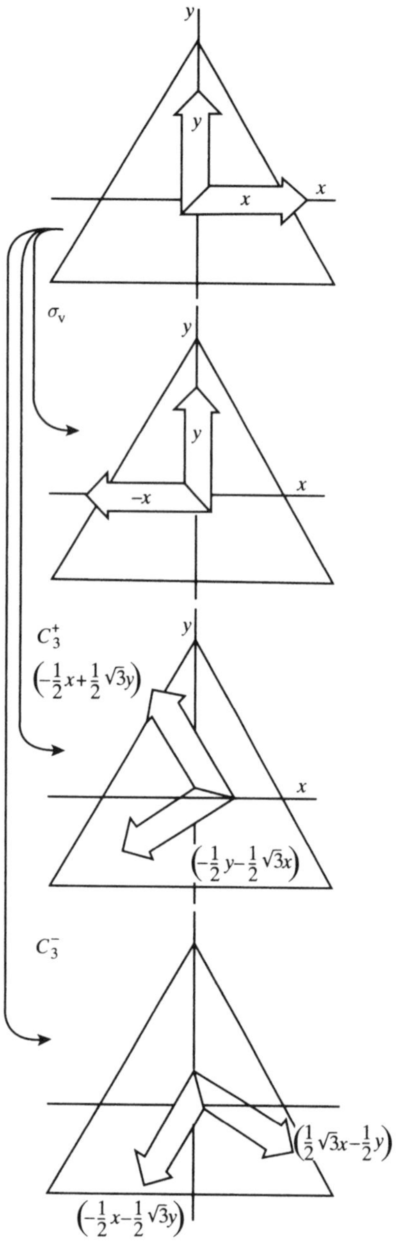

Figure 11.7

$$E(x, y, z) \rightarrow (x, y, z)$$
$$\sigma_v(x, y, z) \rightarrow (-x, y, z)$$
$$\sigma_v'(x, y, z) \rightarrow (x, -y, z)$$
$$\sigma_v''(x, y, z) \rightarrow (x, y, -z)$$
$$C_3^+(x, y, z) \rightarrow (-\tfrac{1}{2}x + \tfrac{1}{2}\sqrt{3}y, -\tfrac{1}{2}\sqrt{3}x - \tfrac{1}{2}y, z)$$
$$C_3^-(x, y, z) \rightarrow (-\tfrac{1}{2}x - \tfrac{1}{2}\sqrt{3}y, \tfrac{1}{2}\sqrt{3}x - \tfrac{1}{2}y, z)$$

The characters of all σ operations are the same, as are those of both C_3 operations (see the C_{3v} character table), hence we need consider only one operation in each class.

$$El_z = xp_y - yp_x = l_z$$

$$\sigma_v l_z = -xp_y + yp_x = -l_z \quad [(x, y, z) \to (-x, y, z)]$$

$$C_3^+ l_z = (-\tfrac{1}{2}x + \tfrac{1}{2}\sqrt{3}y) \times (-\tfrac{1}{2}\sqrt{3}p_x - \tfrac{1}{2}p_y) - (-\tfrac{1}{2}\sqrt{3}x - \tfrac{1}{2}y) \times (-\tfrac{1}{2}p_x + \tfrac{1}{2}\sqrt{3}p_y)$$
$$[(x, y, z) \to (-\tfrac{1}{2}x + \tfrac{1}{2}\sqrt{3}y, -\tfrac{1}{2}\sqrt{3}x - \tfrac{1}{2}y, z)]$$
$$= \tfrac{1}{4}(\sqrt{3}xp_x + xp_y - 3yp_x - \sqrt{3}yp_y - \sqrt{3}xp_x + 3xp_y - yp_x + \sqrt{3}yp_y)$$
$$= xp_y - yp_x = l_z$$

The representatives of E, σ_v, and C_3^+ are therefore all one-dimensional matrices with characters 1, −1, 1, respectively. It follows that l_z is a basis for A_2 (see the C_{3v} character table).

P11.7 The multiplication table, using $\sigma_0 = 1$, is

First	1	σ_x	σ_y	σ_z
Second				
1	1	σ_x	σ_y	σ_z
σ_x	σ_x	1	$i\sigma_z$	$i\sigma_y$
σ_y	σ_y	$-i\sigma_z$	1	$i\sigma_x$
σ_z	σ_z	$i\sigma_y$	$-i\sigma_x$	1

The matrices **do not form a group** since the products $i\sigma_z$, $i\sigma_y$, $i\sigma_x$ and their negatives are not among the four given matrices.

P11.9 (a) In C_{3v} symmetry the H1s orbitals span the same irreducible representations as in NH_3, which is $A_1 + A_1 + E$. There is an additional A_1 orbital because a fourth H atom lies on the C_3 axis. In C_{3v}, the d orbitals span $A_1 + E + E$ (see the final column of the C_{3v} character table). Therefore, **all five d orbitals** may contribute to the bonding.

(b) In C_{2v} symmetry the H1s orbitals span the same irreducible representations as in H_2O, but one 'H_2O' fragment is rotated by 90° with respect to the other. Therefore, whereas in H_2O the H1s orbitals span $A_1 + B_2$ ($H_1 + H_2$, $H_1 - H_2$), in the distorted CH_4 molecule they span $A_1 + B_2 + A_1 + B_1$ ($H_1 + H_2$, $H_1 - H_2$, $H_3 + H_4$, $H_3 - H_4$). In C_{2v} the d orbitals span $2A_1 + B_1 + B_2 + A_2$ (C_{2v} character table]; therefore, **all except $A_2(d_{xy})$** may participate in bonding.

Note: The method used to solve P11.8 also works nicely.

P11.11 (a) We work through the flow diagram in the text (Figure 11.7). We note that this complex with freely rotating CF_3 groups is not linear, it has no C_n axes with $n > 2$. It does have three mutually perpendicular C_2 axes and each has a perpendicular mirror plane. Therefore, the point group is **D_{2h}**.

(b) The plane shown in Figure 11.8 below is a mirror plane so long as the CF_3 groups each have a CF bond in the plane. (i) If the CF_3 groups are staggered, then the Ag–CN axis is an S_2 axis. The Ag–CF_3 axis is also an S_2 axis, which means that the Ag atom is at an inversion centre. There is a C_2

axis perpendicular to the plane of the molecule and the plane of the molecule is a σ_h. So the point group is $\boxed{C_{2h}}$. (ii) If the CF_3 groups are eclipsed, then the axis through the Ag and perpendicular to the plane of the Ag bonds is no longer a C_2 axis; however, the Ag–CN axis is a C_2 axis. There is no σ_h but there are two σ_v planes (the plane shown and the plane perpendicular to it and through the Ag–CN bond). So the point group is $\boxed{C_{2v}}$.

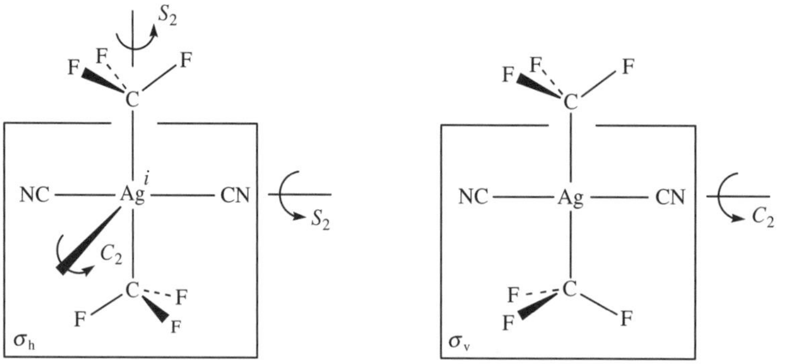

(i) Staggered form, C_{2h} (ii) Eclipsed form, C_{2v}

Figure 11.8

P11.13 (a) C_{2v}. The functions x^2, y^2, and z^2 are invariant under all operations of the group, and so $z(5z^2-3r^2)$ transforms as $z(A_1)$, $y(5y^2-3r^2)$ as $y(B_2)$, $x(5x^2-3r^2)$ as $x(B_1)$, and likewise for $z(x^2-y^2)$, $y(x^2-z^2)$, and $x(z^2-y^2)$. The function xyz transforms as $B_1 \times B_2 \times A_1 = A_2$. Therefore, in group

$$C_{2v},\ f \rightarrow \boxed{2A_1 + A_2 + 2B_1 + 2B_2}$$

(b) C_{3v}. In C_{3v}, z transforms as A_1, and hence so does z^3. From the C_{3v} character table, (x^2-y^2, xy) is a basis for E, and so $(xyz, z(x^2-y^2))$ is a basis for $A_1 \times E = E$. The linear combinations $y(5y^2-3r^2) + 5y(x^2-z^2) \propto y$ and $x(5x^2-3r^2) + 5x(z^2-y^2) \propto x$ are a basis for E. Likewise, the two linear combinations orthogonal to these are another basis for E. Hence, in the group C_{3v}, $f \rightarrow \boxed{A_1 + 3E}$.

(c) T_d. Make the inspired guess that the f orbitals are a basis of dimension $3+3+1$, suggesting the decomposition T + T + A. Is the A representation A_1 or A_2? We see from the character table that the effect of S_4 discriminates between A_1 and A_2. Under S_4, $x \rightarrow y$, $y \rightarrow -x$, $z \rightarrow -z$, and so $xyz \rightarrow xyz$. The character is $\chi = 1$, and so xyz spans A_1. Likewise, $(x^3, y^3, z^3) \rightarrow (y^3, -x^3, -z^3)$ and $\chi = 0+0-1=-1$. Hence, this trio spans T_2. Finally,

$$\{x(z^2-y^2), y(z^2-x^2), z(x^2-y^2)\} \rightarrow \{y(z^2-x^2), -x(z^2-y^2), -z(y^2-x^2)\}$$

resulting in $\chi = 1$, indicating T_1. Therefore, in T_d, $f \rightarrow \boxed{A_1 + T_1 + T_2}$.

(d) O_h. Anticipate an A + T + T decomposition as in the other cubic group. Since x, y, and z all have odd parity, all the irreducible representations will be u. Under S_4, $xyz \rightarrow xyz$ (as in (c)), and so the representation is $\chi = 1$ (see the character table). Under S_4, $(x^3, y^3, z^3) \rightarrow (y^3, -x^3, -z^3)$, as before, and $\chi = -1$, indicating T_{1u}. In the same way, the remaining three functions span T_{2u}. Hence, in O_h, $f \rightarrow \boxed{A_{2u} + T_{1u} + T_{2u}}$.

(The shapes of the orbitals are shown in *Inorganic Chemistry*, 3rd edn, D. F. Shriver, and P. W. Atkins, Oxford University Press and W. H. Freeman & Co (1999))

The f orbitals will cluster into sets according to their irreducible representations.

Thus, (a) $f \rightarrow A_1 + T_1 + T_2$ in T_d symmetry, and there is one non-degenerate orbital and two sets of triply degenerate orbitals. (b) $f \rightarrow A_{2u} + T_{1u} + T_{2u}$, and the pattern of splitting (but not the order of energies) is the same.

P11.15 We begin by drawing up the following table for the C_{2v} group:

	N2s	N2p$_x$	N2p$_y$	N2p$_z$	O2p$_x$	O2p$_y$	O2p$_z$	O′2p$_x$	O′2p$_y$	O′2p$_z$	χ
E	N2s	N2p$_x$	N2p$_y$	N2p$_z$	O2p$_x$	O2p$_y$	O2p$_z$	O′2p$_x$	O′2p$_y$	O′2p$_z$	10
C_2	N2s	−N2p$_x$	−N2p$_y$	N2p$_z$	−O′2p$_x$	−O′2p$_y$	O′2p$_z$	−O2p$_x$	−O2p$_y$	O2p$_z$	0
σ_v	N2s	N2p$_x$	−N2p$_y$	N2p$_z$	O′2p$_x$	−O′2p$_y$	O′2p$_z$	O2p$_x$	−O2p$_y$	O2p$_z$	2
σ_v'	N2s	−N2p$_x$	N2p$_y$	N2p$_z$	−O2p$_x$	O2p$_y$	O2p$_z$	−O′2p$_x$	O′2p$_y$	O′2p$_z$	4

The character set [10, 0, 2, 4] decomposes into $\boxed{4A_1 + 2B_1 + 3B_2 + A_2}$. We then form symmetry-adapted linear combinations as described in Section 11.5(c).

$\psi(A_1) = $ N2s	(column 1)	$\psi(B_1) = $ O2p$_x$ + O′2p$_x$	(column 5)
$\psi(A_1) = $ N2p$_z$	(column 4)	$\psi(B_2) = $ N2p$_y$	(column 3)
$\psi(A_1) = $ O2p$_z$ + O′2p$_z$	(column 7)	$\psi(B_2) = $ O2p$_y$ + O′2p$_y$	(column 6)
$\psi(A_1) = -$O2p$_y$ + O′2p$_y$	(column 9)	$\psi(B_2) = $ O2p$_z$ − O′2p$_z$	(column 7)
$\psi(B_1) = $ N2p$_x$	(column 2)	$\psi(A_2) = $ O2p$_x$ − O′2p$_x$	(column 5)

(The other columns yield the same combinations.)

P11.17 Consider phenanthrene with symmetry elements and carbon atoms as labelled in the structure below.

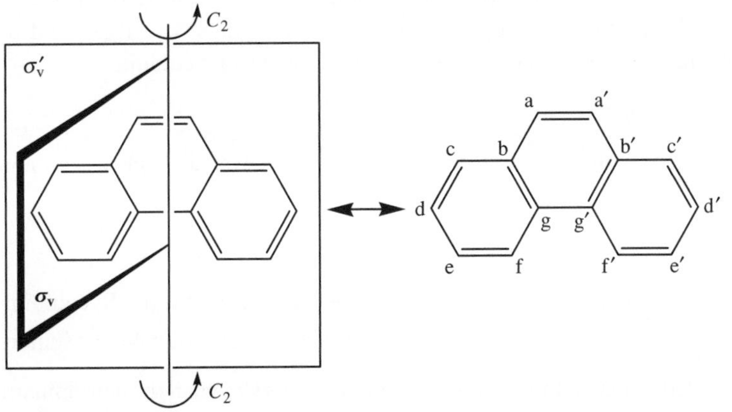

(a) The 2p$_z$ orbitals involved in the π system, which is perpendicular to the σ_v' plane, are the basis we are interested in. To find the irreducible representations spanned by this basis, consider how each basis is transformed under the symmetry operations of the C_{2v} group. To find the character of an operation in this basis, sum the coefficients of the basis terms that are unchanged by the operation.

	a	a′	b	b′	c	c′	d	d′	e	e′	f	f′	g	g′	χ
E	a	a′	b	b′	c	c′	d	d′	e	e′	f	f′	g	g′	14
C_2	−a′	−a	−b′	−b	−c′	−c	−d′	−d	−e′	−e	−f′	−f	−g′	−g	0
σ_v	a′	a	b′	b	c′	c	d′	d	e′	e	f′	f	g′	g	0
σ'_v	−a	−a′	−b	−b′	−c	−c′	−d	−d′	−e	−e′	−f	−f′	−g	−g′	−14

To find the irreducible representations that these orbitals span, multiply the characters in the representation of the orbitals by the characters of the irreducible representations, sum those products, and divide the sum by the order h (= 4) of the group (as in Section 11.5(a)). The table below illustrates the procedure, beginning at the left with the C_{2v} character table.

C_{2v}	E	C_2	σ_v	σ'_v	Product	E	C_2	σ_v	σ'_v	Sum/h
A_1	1	1	1	1		14	0	0	−14	0
A_2	1	1	−1	−1		14	0	0	14	7
B_1	1	−1	1	−1		14	0	0	14	7
B_2	1	−1	−1	1		14	0	0	−14	0

The orbitals span $\boxed{7A_2 + 7B_1}$.

To find symmetry-adapted linear combinations (SALCs), follow the procedure described in Section 11.5(c). Refer to the table above that displays the transformations of the original basis orbitals. To find SALCs of a given symmetry species, take a column of the table, multiply each entry by the character of the species' irreducible representation, sum the terms in the column, and divide by the order of the group. For example, the characters of species A_1 are 1, 1, 1, 1, so the columns to be summed are identical to the columns in the table above. Each column sums to zero, so we conclude that there are no SALCs of A_1 symmetry. (No surprise here: the orbitals span only A_2 and B_1.) An A_2 SALC is obtained by multiplying the characters 1, 1, −1, −1 by the first column:

$$\tfrac{1}{4}(a - a' - a' + a) = \tfrac{1}{2}(a - a').$$

The A_2 combination from the second column is the same. There are seven distinct A_2 combinations in all: $\boxed{\tfrac{1}{2}(a - a'), \tfrac{1}{2}(b - b'), \ldots, \tfrac{1}{2}(g - g')}$.

The B_1 combination from the first column is: $\tfrac{1}{4}(a + a' + a' + a) = \tfrac{1}{2}(a + a')$.

The B_1 combination from the second column is the same. There are seven distinct B_1 combinations in all: $\boxed{\tfrac{1}{2}(a + a'), \tfrac{1}{2}(b + b'), \ldots, \tfrac{1}{2}(g + g')}$. There are no B_2 combinations, as the columns sum to zero.

(b) The structure is labelled to match the row and column numbers shown in the determinant. The Hückel secular determinant of phenanthrene is:

	a	b	c	d	e	f	g	g′	f′	e′	d′	c′	b′	a′
a	$\alpha - E$	β	0	0	0	0	0	0	0	0	0	0	0	β
b	β	$\alpha - E$	β	0	0	0	β	0	0	0	0	0	0	0
c	0	β	$\alpha - E$	β	0	0	0	0	0	0	0	0	0	0
d	0	0	β	$\alpha - E$	β	0	0	0	0	0	0	0	0	0
e	0	0	0	β	$\alpha - E$	β	0	0	0	0	0	0	0	0
f	0	0	0	0	β	$\alpha - E$	β	0	0	0	0	0	0	0
g	0	β	0	0	0	β	$\alpha - E$	β	0	0	0	0	0	0
g′	0	0	0	0	0	0	β	$\alpha - E$	β	0	0	0	β	0
f′	0	0	0	0	0	0	0	β	$\alpha - E$	β	0	0	0	0
e′	0	0	0	0	0	0	0	0	β	$\alpha - E$	β	0	0	0
d′	0	0	0	0	0	0	0	0	0	β	$\alpha - E$	β	0	0
c′	0	0	0	0	0	0	0	0	0	0	β	$\alpha - E$	β	0
b′	0	0	0	0	0	0	0	β	0	0	0	β	$\alpha - E$	β
a′	β	0	0	0	0	0	0	0	0	0	0	0	β	$\alpha - E$

This determinant has the same eigenvalues as in exercise 10.19a(b).

(c) The ground state of the molecule has A_1 symmetry by virtue of the fact that its wavefunction is the product of doubly occupied orbitals, and the product of any two orbitals of the same symmetry has A_1 character. If a transition is to be allowed, the transition dipole must be non-zero, which in turn can only happen if the representation of the product $\Psi_f^* \boldsymbol{\mu} \Psi_i$ includes the totally symmetric species A_1. Consider first transitions to another A_1 wavefunction, in which case we need the product $A_1 \times \mu \times A_1$. Now $A_1 \times A_1 = A_1$, and the only character that returns A_1 when multiplied by A_1 is A_1 itself. The z component of the dipole operator belongs to species A_1, so z-polarized $A_1 \leftarrow A_1$ transitions are allowed. (Note: transitions from the A_1 ground state to an A_1 excited state are transitions from an orbital occupied in the ground state to an excited-state orbital of the same symmetry.) The other possibility is a transition from an orbital of one symmetry (A_2 or B_1) to the other; in that case, the excited-state wavefunction will have symmetry of $A_2B_1 = B_2$ from the two singly occupied orbitals in the excited state. The symmetry of the transition dipole, then, is $A_1 \times \boldsymbol{\mu} \times B_2 = \boldsymbol{\mu}B_2$, and the only species that yields A_1 when multiplied by B_2 is B_2 itself. Now, the y component of the dipole operator belongs to species B_2, so these transitions are also allowed (y-polarized).

P11.19 The p_z orbitals of the fluorine atoms in XeF_4 are shown in Figure 11.9. (The orbital has a positive wavefunction sign in shaded lobes and a negative wavefunction sign in unshaded lobes.) These orbitals may form π molecular orbitals with the xenon atom provided that a non-zero overlap integral exists.

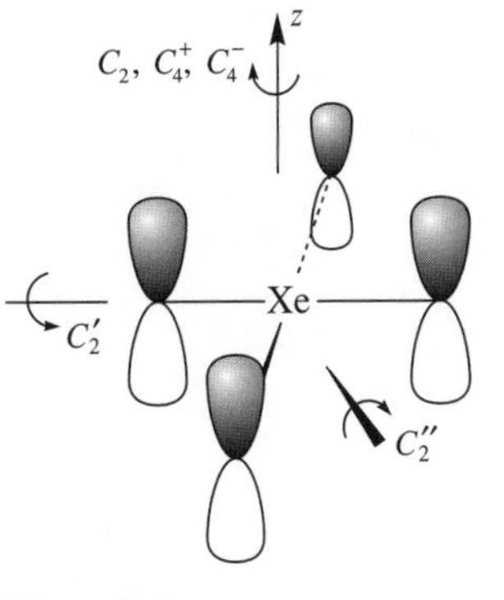

Figure 11.9

To find the symmetry species spanned by the fluorine p_z orbitals, we use a quick rule for determining the character of the basis set under each symmetry operation of the group: count 1 each time a basis function is left unchanged by the operation, because only these functions give a non-zero entry on the diagonal of the matrix representative. In some cases there is a sign change, $(\ldots -f \ldots) \leftarrow (\ldots f \ldots)$; then −1 occurs on the diagonal, and so count −1. The character of the identity is always equal to the dimension of the basis since each function contributes 1 to the trace. Although XeF_4 belongs to the D_{4h} group, we will use the D_4 subgroup for convenience. Here is a summary of the characters exhibited by the fluorine orbitals under each symmetry operation:

D_4 $h = 8$	E	C_2	$2C_4$	$2C_2'$	$2C_2''$
Fluorine p_z orbitals	4	0	0	−2	0

Inspection of the D_4 character table shows that the fluorine p_z orbitals span $A_2 + B_2 + E$. Further inspection of the D_4 character table reveals that p_z belongs to A_2, d_{xy} belongs to B_2, and both the (p_x, p_y) set and the (d_{xz}, d_{yz}) set belong to E. Consequently, only these orbitals of the central atom may possibly have non-zero overlap with the fluorine p_z orbitals. We must now use the procedure of Section 11.5(c) to find the four **symmetry-adapted linear combinations (SALC)** of the fluorine p_z orbitals. Using clockwise labelling of the Fp_z orbitals, we write a table that summarizes the effect of each operation on each orbital. (The symbol A is used to represent the orbital $p_z(A)$ at fluorine atom A, etc.)

D_4	A	B	C	D
E	A	B	C	D
C_2	C	D	A	B
C_4^+	D	A	B	C
C_4^-	B	C	D	A
C_2'(A-C)	−A	−D	−C	−B
C_2'(B-D)	−C	−B	−A	−D
C_2''(A-Xe-B)	−B	−A	−D	−C
C_2''(B-Xe-C)	−D	−C	−B	−A

To generate the B_2 combination, we take the characters for B_2 [1, 1, −1, −1, −1, −1, 1, 1] and multiply column 3 and sum the terms. Then, divide by the order of the group (8):

$$p_1(B_2) = \tfrac{1}{8}\{C + A - B - D + C + A - D - B\} = \tfrac{1}{4}\{p_z(A) - p_z(B) + p_z(C) - p_z(D)\}$$

To generate the A_2 combination, we take the characters for A_2 [1, 1, 1, 1, −1, −1, −1, −1] and multiply column 1:

$$p_2(A_2) = \tfrac{1}{4}\{p_z(A) + p_z(B) + p_z(C) + p_z(D)\}$$

To generate the two E combinations, we take the characters for E [2, −2, 0, 0, 0, 0, 0, 0] and multiply column 1 for one of them and multiply column 2 for the other:

$$p_3(E) = \tfrac{1}{4}\{p_z(A) - p_z(C)\}$$

$$p_4(E) = \tfrac{1}{4}\{p_z(B) - p_z(D)\}$$

Other column multiplications yield these same combinations or zero.

Figure 11.10(a) shows that the $p_z(A_2)$ orbital of the central atom does have non-zero overlap with p_2. Figure 11.10(b) shows that $d_{xz}(E)$ has a non-zero overlap with p_3. However, Figure 11.10(c) shows that, because of the balance between constructive and destructive interference, there is zero overlap between $d_{xy}(B_2)$ and p_1. Thus, there is no orbitial of the central atom that forms a non-zero overlap with the p_1 combination so p_1 is non-bonding. This is confirmed by recognizing that p_1 belongs to the B_{2u} species of D_{4h} while d_{xy} spans B_{2g} and the direct product $B_{2u} \times B_{2g} = A_{1u}$ does not span the completely symmetric A_{1g}; consequently, the overlap integral equals zero.

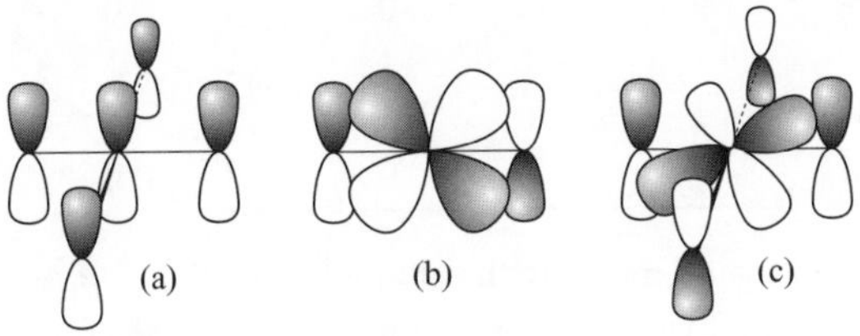

Figure 11.10

Solutions to applications: astrophysics and biology

P11.21 The point group for the square H_4 molecule is D_{4h} with $h = 16$ symmetry species. To find the irreducible representations or symmetry species spanned by four *s* orbitals, we use the methodology of Section 11.5(c).

D_{4h}	E	$2C_4$	C_2	$2C_2'$	$2C_2''$	i	$2S_4$	σ_h	$2\sigma_v$	$2\sigma_d$
Number of unchanged basis members	4	0	0	2	0	0	0	4	2	0

The basis representation is obviously a linear combination of the D_{4h} symmetry species; it is reducible. Only the E, $2C_2'$, σ_h and $2\sigma_v$ symmetry elements contribute (the others have factors of zero) to the number of times symmetry species Γ contributes (N_Γ) to the representation of the basis.

E		$2C_2'$		σ_h		$2\sigma_v$
$N(A_{1g}) = \frac{1}{16}\{4\cdot 1\cdot 1$	+	$2\cdot 2\cdot 1$	+	$4\cdot 1\cdot 1$	+	$2\cdot 2\cdot 1\} = 1$
$N(A_{2g}) = \frac{1}{16}\{4\cdot 1\cdot 1$	+	$2\cdot 2\cdot(-1)$	+	$4\cdot 1\cdot 1$	+	$2\cdot 2\cdot(-1)\} = 0$
$N(B_{1g}) = \frac{1}{16}\{4\cdot 1\cdot 1$	+	$2\cdot 2\cdot 1$	+	$4\cdot 1\cdot 1$	+	$2\cdot 2\cdot 1\} = 1$
$N(B_{2g}) = \frac{1}{16}\{4\cdot 1\cdot 1$	+	$2\cdot 2\cdot(-1)$	+	$4\cdot 1\cdot 1$	+	$2\cdot 2\cdot(-1)\} = 0$
$N(E_g) = \frac{1}{16}\{4\cdot 1\cdot 2$	+	$2\cdot 2\cdot 0$	+	$4\cdot 1\cdot(-2)$	+	$2\cdot 2\cdot 0\} = 0$
$N(A_{1u}) = \frac{1}{16}\{4\cdot 1\cdot 1$	+	$2\cdot 2\cdot 1$	+	$4\cdot 1\cdot(-1)$	+	$2\cdot 2\cdot(-1)\} = 0$
$N(A_{2u}) = \frac{1}{16}\{4\cdot 1\cdot 1$	+	$2\cdot 2\cdot(-1)$	+	$4\cdot 1\cdot(-1)$	+	$2\cdot 2\cdot 1\} = 0$
$N(B_{1u}) = \frac{1}{16}\{4\cdot 1\cdot 1$	+	$2\cdot 2\cdot 1$	+	$4\cdot 1\cdot(-1)$	+	$2\cdot 2\cdot(-1)\} = 0$
$N(B_{2u}) = \frac{1}{16}\{4\cdot 1\cdot 1$	+	$2\cdot 2\cdot(-1)$	+	$4\cdot 1\cdot(-1)$	+	$2\cdot 2\cdot 1\} = 0$
$N(E_u) = \frac{1}{16}\{4\cdot 1\cdot 2$	+	$2\cdot 2\cdot 0$	+	$4\cdot 1\cdot 2$	+	$2\cdot 2\cdot 0\} = 1$

The basis spans $A_{1g} + B_{1g} + E_u$.

P11.23 Can the E_u excited state be reached by a dipole transition from the A_{1g} ground state? Only if the representation of the product $\psi_f^* \mu \psi_i$ includes the totally symmetric species A_{1g}. The z component of the dipole operator belongs to symmetry species A_{2u}, and the x and y components belong to E_u. So the products we must consider are $E_u \times A_{2u} \times A_{1g}$ and $E_u \times E_u \times A_{1g}$. For z-polarized transitions, the relevant characters are:

D_{4h}	E	$2C_4$	C_2	$2C_2'$	$2C_2''$	i	$2S_4$	σ_h	$2\sigma_v$	$2\sigma_d$
E_u	2	0	−2	0	0	−2	0	2	0	0
A_{2u}	1	1	1	−1	−1	−1	−1	−1	1	1
A_{1g}	1	1	1	1	1	1	1	1	1	1
$E_uA_{2u}A_{1g}$	2	0	−2	0	0	2	0	−2	0	0

To see whether $E_u \times A_{2u} \times A_{1g}$ contains A_{1g}, we would multiply the characters of the $E_u \times A_{2u} \times A_{1g}$ by the characters of A_{1g}, sum those products, and divide the sum by the order h of the group; since the characters of A_{1g} are all 1, we can simply sum the characters of $E_u \times A_{2u} \times A_{1g}$. Since they sum to zero, the product $E_u \times A_{2u} \times A_{1g}$ does not contain A_{1g}, and the **z-polarized transition is not allowed**.

For x- or y-polarized transitions:

	E	$2C_4$	C_2	$2C_2'$	$2C_2''$	i	$2S_4$	σ_h	$2\sigma_v$	$2\sigma_d$
E_u	2	0	−2	0	0	−2	0	2	0	0
E_u	2	0	−2	0	0	−2	0	2	0	0
A_{1g}	1	1	1	1	1	1	1	1	1	1
$E_uA_{2u}A_{1g}$	4	0	4	0	0	4	0	4	0	0

Summing the characters of $E_u \times A_{2u} \times A_{1g}$ yields 16, the order of the group. Therefore, the product $E_u \times A_{2u} \times A_{1g}$ does contain A_{1g}, and the **x-, y-polarized transitions are allowed**.

12 Molecular spectroscopy 1: rotational and vibrational spectra

Answers to discussion questions

D12.1 (1) *Doppler broadening*. This contribution to the linewidth is due to the Doppler effect, which shifts the frequency of the radiation emitted or absorbed when the atoms or molecules involved are moving towards or away from the detecting device. Molecules have a wide range of speeds in all directions in a gas and the detected spectral line is the absorption or emission profile arising from all the resulting Doppler shifts. It can be shown that the profile reflects the distribution of molecular velocities parallel to the line of sight, which is a bell-shaped Gaussian curve.

(2) *Lifetime broadening*. The Doppler broadening is significant in gas-phase samples, but lifetime broadening occurs in all states of matter. This kind of broadening is a quantum-mechanical effect related to the uncertainty principle in the form $\delta E \approx \dfrac{\hbar}{\tau}$ and is due to the finite lifetimes, τ, of the states involved in the transition. When τ is finite, the energy of the states is smeared out and hence the transition frequency is broadened.

(3) *Pressure broadening or collisional broadening*. The actual mechanism affecting the lifetime of energy states depends on various processes, one of which is collisional deactivation and another is spontaneous emission. Lowering the pressure can reduce the first of these contributions; the second cannot be changed and results in a natural linewidth.

D12.3 (1) *Rotational Raman spectroscopy*. The gross selection rule is that the molecule must be anisotropically polarizable, which is to say that its polarizability, α, depends on the direction of the electric field relative to the molecule. Non-spherical rotors satisfy this condition, therefore linear and symmetric rotors are rotationally Raman active.

(2) *Vibrational Raman spectroscopy*. The gross selection rule is that the polarizability of the molecule must change as the molecule vibrates. All diatomic molecules satisfy this condition as the molecules swell and contract during a vibration, the control of the nuclei over the electrons varies, and the molecular polarizability changes. Hence, both homonuclear and heteronuclear diatomics are vibrationally Raman active. In polyatomic molecules it is usually quite difficult to judge by inspection whether or not the molecule is anisotropically polarizable; hence group theoretical methods are relied on for judging the Raman activity of the various normal modes of vibration. The procedure is discussed in Section 12.16(b) and demonstrated in the Brief Illustration in that section.

D12.5 The answer to this question depends precisely on what is meant by equilibrium bond length. See the solution to Problem 12.21, where it is demonstrated that the centrifugally distorted bond length r_c is given by the relationship

$$r_c = \frac{r_e}{1 - m_{\text{eff}}\omega^2/k}$$

The angular velocity depends on the quantum number J through the relationship

$$\omega^2 = J(J+1)\hbar^2/m_{\text{eff}}^2 r_e^4;$$

thus, the distortion is greater for higher rotational energy levels. But the equilibrium bond length r_e remains constant, if by that term one means the value of r corresponding to a vibrating non-rotating molecule with $J = 0$. However, if one describes the vibration of the molecule in a higher rotational state as having a new 'equilibrium' distance r_c, the potential energy of vibration will also be different. It is lowered by the amount shown in eqn 12.16, that is, $-\tilde{D}_J J^2(J+1)^2$. A detailed analysis of the combined effects of rotation and vibration is quite complicated. The treatment in Section 12.11 ignores the effects of centrifugal distortion and anharmonicity. Consult advanced treatments for a more thorough discussion.

D12.7 The exclusion rule applies to the benzene molecule because it has a center of symmetry. Consequently, none of the normal modes of vibration of benzene can be both infrared and Raman active. If we wish to characterize all the normal modes we must obtain both kinds of spectra. See the solutions to Exercises 12.27(a) and 12.27(b) for specific illustrations of which modes are infrared active and which are Raman active.

Solutions to exercises

E12.1(a) Polar molecules show a pure rotational absorption spectrum, therefore select the polar molecules based on their well-known structures. Alternatively, determine the point groups of the molecules and use the rule that only molecules belonging to C_n, C_{nv}, and C_s may be polar, and in the case of C_n and C_{nv}, that dipole must lie along the rotation axis. Hence, the polar molecules are

(b) HCl (d) CH_3Cl (e) CH_2Cl_2

Their point group symmetries are

(b) $C_{\infty v}$ (d) C_{3v} (e) C_{2v}

E12.2(a) We select those molecules with an anisotropic polarizability. A practical rule to apply is that spherical rotors do not have anisotropic polarizabilities. Therefore $\boxed{\text{(c) } CH_4 \text{ is inactive}}$. All others are active.

E12.3(a) PH_3 is a symmetric rotor similar to NH_3; we use

$$I_{\parallel} = 2m_H(1-\cos\theta)R^2 \text{ [Table 12.1]}$$
$$= 2 \times 1.0079\, m_u \times 1.66054 \times 10^{-27}\,\text{kg}/m_u \times (1-\cos 93.6°) \times (1.42 \times 10^{-10})^2$$
$$= \boxed{7.173 \times 10^{-47}\ \text{kg m}^2}$$

$\boxed{I_{\parallel} \text{ will not change.}}$ It does not depend on the mass of the P atom. $I_{\perp}$, however, will change; it does depend on the mass of the P atom [Table 12.1].

E12.4(a) $^{35}Cl^{12}CH_3$ is a symmetric rotor; its moments of inertia are given by the formulas [Table 12.1]

$$I_{\parallel} = 2m_{H}(1 - \cos\theta)R^2 \quad \text{and}$$

$$I_{\perp} = m_{H}(1 - \cos\theta)R^2 + \frac{m_H}{m}(m_C + m_{Cl})(1 + 2\cos\theta)R^2 + \frac{m_{Cl}}{m}\{(3m_H + m_C)R'$$

$$+ 6m_H R[\tfrac{1}{3}(1 + 2\cos\theta)]^{1/2}\}R'$$

We substitute the following data into the above formulas and then perform the calculations:

$m_H = 1.0078\, m_u = 1.0078 \times 1.66054 \times 10^{-27}$ kg

$m_{Cl} = 34.9688\, m_u = 34.9688 \times 1.66054 \times 10^{-27}$ kg

$m_C = 12.0000 \times m_u = 12.0000 \times 1.66054 \times 10^{-27}$ kg

$m = 3m_H + m_C + m_{Cl} = 49.9922\, m_u = 49.9922 \times 1.66054 \times 10^{-27}$ kg

$R = 111$ pm $= 1.11 \times 10^{-10}$ m

$R' = 178$ pm $= 1.78 \times 10^{-10}$ m

$\theta = 111°$

Since the factor m_u is common to each term in the formulas, multiplication by its value need not be performed until the end of the calculation.

$$I_{\parallel} = 2 \times 1.0078\, m_u(1 - \cos 111°) \times (1.11 \times 10^{-10}\text{ m})^2$$
$$= \boxed{5.60 \times 10^{-47}\text{ kg m}^2}$$

Substitution of the data in a similar manner into the above formula for $I_{\perp}$ gives

$$\boxed{I_{\perp} = 6.29 \times 10^{-46}\text{ kg m}^2}$$

We use eqns 12.13 to calculate the rotational constants in wavenumbers.

$$\tilde{A} = \frac{\hbar}{4\pi c I_{\parallel}} = 500\text{ m}^{-1} = \boxed{5.00\text{ cm}^{-1}}$$

$$\tilde{B} = \frac{\hbar}{4\pi c I_{\perp}} = 44.5\text{ m}^{-1} = \boxed{0.445\text{ cm}^{-1}}$$

In frequency units we have

$$A = \frac{\hbar}{4\pi I_{\parallel}} = 1.50 \times 10^{11}\text{ s}^{-1} = \boxed{1.50 \times 10^{11}\text{ Hz}}$$

$$B = \frac{\hbar}{4\pi I_{\perp}} = 1.33 \times 10^{10}\text{ s}^{-1} = \boxed{1.33 \times 10^{10}\text{ Hz}}$$

E12.5(a) NO is a linear rotor and we assume there is little centrifugal distortion, hence

$$\tilde{F}(J) = \tilde{B}J(J + 1)\ [12.14]$$

with $\tilde{B} = \dfrac{\hbar}{4\pi c I}$, $I = \mu R^2$ [Table 12.1], and

$$\mu = \frac{m_N m_O}{m_N + m_O} \text{ [nuclide masses from the } \textit{Data Section}\text{]}$$

$$= \left(\frac{(14.003\, m_u) \times (15.995\, m_u)}{(14.003\, m_u) + (15.995\, m_u)}\right) \times (1.6605 \times 10^{-27}\text{ kg } m_u^{-1}) = 1.240 \times 10^{-26}\text{ kg}$$

Then, $I = (1.240 \times 10^{-26}\text{ kg}) \times (1.15 \times 10^{-10}\text{ m})^2 = 1.64\bar{0} \times 10^{-46}\text{ kg m}^2$

and $\tilde{B} = \dfrac{1.0546 \times 10^{-34}\text{ J s}}{(4\pi) \times (2.998 \times 10^8\text{ m s}^{-1}) \times (1.64\bar{0} \times 10^{-46}\text{ kg m}^2)} = 170.\bar{7}\text{ m}^{-1} = 1.70\bar{7}\text{ cm}^{-1}$

The wavenumber of the $J = 4 \leftarrow 3$ transition is

$$\tilde{\nu} = 2\tilde{B}(J+1)[12.21\text{a}] = 8\tilde{B}[J=3] = (8) \times (1.70\bar{7}\text{ cm}^{-1}) = 13.6\text{ cm}^{-1}$$

The frequency is

$$\nu = \tilde{\nu}c = (13.6\bar{5}\text{ cm}^{-1}) \times \left(\frac{10^2\text{ m}^{-1}}{1\text{ cm}^{-1}}\right) \times (2.998 \times 10^8\text{ m s}^{-1}) = \boxed{4.09 \times 10^{11}\text{Hz}}$$

Question. What is the percentage change in these calculated values if centrifugal distortion is included?

E12.6(a) (a) The wavenumber of the transition is related to the rotational constant by

$$hc\tilde{\nu} = \Delta E = hc\tilde{B}[J(J+1) - (J-1)J] = 2hc\tilde{B}J \text{ [12.8, 12.10]}$$

where J refers to the upper state ($J = 3$). The rotational constant is related to molecular structure by

$$\tilde{B} = \frac{\hbar}{4\pi c I} \text{ [12.7]}$$

where I is moment of inertia. Putting these expressions together yields

$$\tilde{\nu} = 2\tilde{B}J = \frac{\hbar J}{2\pi c I} \quad \text{so} \quad I = \frac{\hbar J}{2\pi c\tilde{\nu}} = \frac{(1.0546 \times 10^{-34}\text{ J s}) \times (3)}{2\pi(2.998 \times 10^{10}\text{ cm s}^{-1}) \times (63.56\text{ cm}^{-1})}$$

$$= \boxed{2.642 \times 10^{-47}\text{ kg m}^2}$$

(b) The moment of inertia is related to the bond length by

$$I = m_{\text{eff}}R^2 \quad \text{so} \quad R = \sqrt{\frac{I}{m_{\text{eff}}}}$$

$$m_{\text{eff}}^{-1} = m_H^{-1} + m_{Cl}^{-1} = \frac{(1.0078\, m_u)^{-1} + (34.9688\, m_u)^{-1}}{1.66054 \times 10^{-27}\text{ kg } m_u^{-1}} = 6.1477 \times 10^{26}\text{ kg}^{-1}$$

and $R = \sqrt{(6.1477 \times 10^{26}\text{ kg}^{-1}) \times (2.642 \times 10^{-47}\text{ kg m}^2)} = 1.274 \times 10^{-10}\text{ m} = \boxed{127.4\text{ pm}}$

E12.7(a) If the spacing of lines is constant, the effects of centrifugal distortion are negligible. Hence, we may use for the wavenumbers of the transitions

$$\tilde{F}(J) - \tilde{F}(J-1) = 2\tilde{B}J \text{ [12.9, } J \text{ is upper level]}$$

Since $J = 1, 2, 3, \ldots$, the spacing of the lines is $2\tilde{B}$

$$12.604\ \text{cm}^{-1} = 2\tilde{B}$$

$$\tilde{B} = 6.302\ \text{cm}^{-1} = 6.302 \times 10^{2}\ \text{m}^{-1}$$

$$I = \frac{\hbar}{4\pi c\tilde{B}}\ [\text{Section 12.4}] = m_{\text{eff}}R^2\ [m_{\text{eff}} = \mu \text{ here, see note on good practice}]$$

$$\frac{\hbar}{4\pi c} = \frac{1.0546 \times 10^{-34}\ \text{J s}}{(4\pi) \times (2.9979 \times 10^{8}\ \text{m s}^{-1})} = 2.7993 \times 10^{-44}\ \text{kg m}$$

$$I = \frac{2.7993 \times 10^{-44}\ \text{kg m}}{6.302 \times 10^{2}\ \text{m}^{-1}} = \boxed{4.442 \times 10^{-47}\ \text{kg m}^2}$$

$$m_{\text{eff}} = \frac{m_{\text{Al}}m_{\text{H}}}{m_{\text{Al}} + m_{\text{H}}}$$

$$= \left(\frac{(26.98) \times (1.008)}{(26.98) + (1.008)}\right) m_{\text{u}} \times (1.6605 \times 10^{-27}\ \text{kg}\ m_{\text{u}}^{-1}) = 1.613\overline{6} \times 10^{-27}\ \text{kg}$$

$$R = \left(\frac{I}{m_{\text{eff}}}\right)^{1/2} = \left(\frac{4.442 \times 10^{-47}\ \text{kg m}^2}{1.6136 \times 10^{-27}\ \text{kg}}\right)^{1/2} = 1.659 \times 10^{-10}\ \text{m} = \boxed{165.9\ \text{pm}}$$

E12.8(a) $\tilde{B} = \dfrac{\hbar}{4\pi cI}$ [12.7], implying that $I = \dfrac{\hbar}{4\pi c\tilde{B}}$

Then, with $I = m_{\text{eff}}R^2$, $R = \left(\dfrac{\hbar}{4\pi m_{\text{eff}}c\tilde{B}}\right)^{1/2}$

We use $m_{\text{eff}} = \dfrac{m_1m_2}{m_1 + m_2} = \dfrac{(126.9045) \times (34.9688)}{(126.9045) + (34.9688)} m_{\text{u}} = 27.4146\ m_{\text{u}}$

and hence obtain

$$R = \left(\frac{1.05457 \times 10^{-34}\ \text{J s}}{(4\pi) \times (27.4146) \times (1.66054 \times 10^{-27}\ \text{kg}) \times (2.99792 \times 10^{10}\ \text{cm s}^{-1}) \times (0.1142\ \text{cm}^{-1})}\right)^{1/2}$$

$$= \boxed{232.1\ \text{pm}}$$

E12.9(a) The determination of two unknowns requires data from two independent experiments and the equation that relates the unknowns to the experimental data. In this exercise two independently determined values of B for two isotopically different molecules are used to obtain the moments of inertia of the molecules and from these, by use of the equation for the moment of inertia of linear triatomic rotors (Table 12.1), the interatomic distances R_{HC} and R_{CN} are calculated.

Rotational constants, which are usually expressed in wavenumber (cm^{-1}), are sometimes expressed in frequency units (Hz). The conversion between the two is

$$B/\text{Hz} = c \times \tilde{B}/\text{cm}^{-1}\ [c \text{ in cm s}^{-1}]$$

Thus, $B(\text{in Hz}) = \dfrac{\hbar}{4\pi I}$ and $I = \dfrac{\hbar}{4\pi B}$

Let, $^1H = H$, $^2H = D$, $R_{HC} = R_{DC} = R$, $R_{CN} = R'$. Then,

$$I(\text{HCN}) = \frac{1.05457 \times 10^{-34}\ \text{J s}}{(4\pi) \times (4.4316 \times 10^{10}\ \text{s}^{-1})} = 1.8937 \times 10^{-46}\ \text{kg m}^2$$

$$I(\text{DCN}) = \frac{1.05457 \times 10^{-34}\ \text{J s}}{(4\pi) \times (3.6208 \times 10^{10}\ \text{s}^{-1})} = 2.3178 \times 10^{-46}\ \text{kg m}^2$$

and from Table 12.1 with isotope masses from the *Data Section.*

$$I(\text{HCN}) = m_\text{H}R^2 + m_\text{N}R'^2 - \frac{(m_\text{H}R - m_\text{N}R')^2}{m_\text{H} + m_\text{C} + m_\text{N}}$$

$$I(\text{HCN}) = \left[(1.0078R^2) + (14.0031R'^2) - \left(\frac{(1.0078R - 14.0031R')^2}{1.0078 + 12.0000 + 14.0031}\right)\right]m_\text{u}$$

Multiplying through by $m/m_\text{u} = (m_\text{H} + m_\text{C} + m_\text{N})/m_\text{u} = 27.0109$:

$$27.0109 \times I(\text{HCN}) = \{27.0109 \times (1.0078R^2 + 14.0031R'^2) - (1.0078R - 14.0031R')^2\}m_\text{u}$$

$$\text{or} \left(\frac{27.0109}{1.66054 \times 10^{-27}\ \text{kg}}\right) \times (1.8937 \times 10^{-46}\ \text{kg m}^2) = 3.0804 \times 10^{-18}\ \text{m}^2 \quad \text{(a)}$$

$$= \{27.0109 \times (1.0078R^2 + 14.0031R'^2) - (1.0078R - 14.0031R')^2\}$$

In a similar manner we find for DCN

$$\left(\frac{28.0172}{1.66054 \times 10^{-27}\ \text{kg}}\right) \times (2.3178 \times 10^{-46}\ \text{kg m}^2) = 3.9107 \times 10^{-18}\ \text{m}^2$$

$$= \{28.0172 \times (2.0141R^2 + 14.0031R'^2) - (2.0141R - 14.0031R')^2\} \quad \text{(b)}$$

Thus, there are two simultaneous quadratic equations (a) and (b) to solve for R and R'. These equations are most easily solved by readily available computer programs or by successive approximations. The results are

$$R = 1.065 \times 10^{-10}\ \text{m} = \boxed{106.5\ \text{pm}} \quad \text{and} \quad R' = 1.156 \times 10^{-10}\ \text{m} = \boxed{115.6\ \text{pm}}$$

These values are easily verified by direct substitution into the equations and agree well with the accepted values $R_{HC} = 1.064 \times 10^{-10}$ m and $R_{CN} = 1.156 \times 10^{-10}$ m.

E12.10(a) The Stokes lines appear at

$$\tilde{\nu}(J + 2 \leftarrow J) = \tilde{\nu}_i - 2\tilde{B}(2J + 3)\ [12.25\text{a}] \quad \text{with } J = 0,\ \tilde{\nu} = \tilde{\nu}_i - 6\tilde{B}$$

Since $\tilde{B} = 1.9987\ \text{cm}^{-1}$ (Table 12.2), the Stokes line appears at

$$\tilde{\nu} = (20\,487) - (6) \times (1.9987\ \text{cm}^{-1}) = \boxed{20\,475\ \text{cm}^{-1}}$$

E12.11(a) The separation of lines is $4\tilde{B}$ [Section 12.6, eqns 12.25a and 12.25b], so $\tilde{B} = 0.2438\ \text{cm}^{-1}$. Then we use

$$R = \left(\frac{\hbar}{4\pi m_\text{eff} c\tilde{B}}\right)^{1/2}$$

with $m_{\text{eff}} = \frac{1}{2}m(^{35}\text{Cl}) = (\frac{1}{2}) \times (34.9688\, m_{\text{u}}) = 17.4844\, m_{\text{u}}$

Therefore,

$$R = \left(\frac{1.05457 \times 10^{-34}\ \text{J s}}{(4\pi) \times (17.4844) \times (1.6605 \times 10^{-27}\ \text{kg}) \times (2.9979 \times 10^{10}\ \text{cm s}^{-1}) \times (0.2438\ \text{cm}^{-1})}\right)^{1/2}$$

$$= 1.989 \times 10^{-10}\ \text{m} = \boxed{198.9\ \text{pm}}$$

E12.12(a) The centrifugal distortion constant is given by

$$\tilde{D}_J = \frac{4\tilde{B}^3}{\tilde{\nu}^2}\ [12.17,\ \text{also see Problem 12.21}]$$

$$\tilde{D}_J = \frac{4 \times (6.511\ \text{cm}^{-1})^3}{(2308\ \text{cm}^{-1})^2} = \boxed{2.073 \times 10^{-4}\ \text{cm}^{-1}}$$

$$\tilde{D}_J \propto \tilde{B}^3 \quad \tilde{B} \propto \frac{1}{I} \quad I \propto m_{\text{H}}\ [\text{Table 12.1}]$$

Therefore, $\tilde{D}_J \propto \dfrac{1}{m_{\text{H}}^3}$ and $\dfrac{\tilde{D}_J(^2\text{H})}{\tilde{D}_J(^1\text{H})} = \dfrac{m_{^1\text{H}}^3}{m_{^2\text{H}}^3} = \dfrac{(1.0078\, m_{\text{u}})^3}{(2.0140\, m_{\text{u}})^3} = \boxed{0.1253}$

We have assumed that the internuclear distance remains constant on substitution.

E12.13(a) See eqn 12.22 and Problem 12.26. The most highly populated rotational level is given by

$$J_{\max} \approx \left(\frac{kT}{2hc\tilde{B}}\right)^{1/2} - \frac{1}{2}\ [12.22]$$

For Cl_2 after substituting for the constants this expression becomes

$$J_{\max} \approx \left(\frac{T/\text{K}}{0.7021}\right)^{1/2} - \frac{1}{2}$$

(a) At 25°C = 298.15 K, $J_{\max} \approx \left(\dfrac{298.15/\text{K}}{0.7021}\right)^{1/2} - \dfrac{1}{2} \approx \boxed{20}$

(b) At 100°C = 373.15 K, $J_{\max} \approx \left(\dfrac{373.15/\text{K}}{0.7021}\right)^{1/2} - \dfrac{1}{2} \approx \boxed{23}$

Answers are rounded off to the nearest integer.

E12.14(a) $$\omega = 2\pi\nu = \left(\frac{k_{\text{f}}}{m}\right)^{1/2}$$

$$k_{\text{f}} = 4\pi^2\nu^2 m = 4\pi^2 \times (2.0\ \text{s}^{-1})^2 \times (1.0\ \text{kg}) = 1.6 \times 10^2\ \text{kg s}^{-2} = \boxed{1.6 \times 10^2\ \text{N m}^{-1}}$$

E12.15(a) $$\omega = \left(\frac{k_{\text{f}}}{m_{\text{eff}}}\right)^{1/2}\ [12.33]$$

The fractional difference is

$$\frac{\omega'-\omega}{\omega}=\frac{\left(\frac{k_f}{m'_{\text{eff}}}\right)^{1/2}-\left(\frac{k_f}{m_{\text{eff}}}\right)^{1/2}}{\left(\frac{k_f}{m_{\text{eff}}}\right)^{1/2}}=\frac{\left(\frac{1}{m'_{\text{eff}}}\right)^{1/2}-\left(\frac{1}{m_{\text{eff}}}\right)^{1/2}}{\left(\frac{1}{m_{\text{eff}}}\right)^{1/2}}=\left(\frac{m_{\text{eff}}}{m'_{\text{eff}}}\right)^{1/2}-1$$

$$=\left(\frac{m(^{23}\text{Na})m(^{35}\text{Cl})\{m(^{23}\text{Na})+m(^{37}\text{Cl})\}}{\{m(^{23}\text{Na})+m(^{35}\text{Cl})\}m(^{23}\text{Na})m(^{37}\text{Cl})}\right)^{1/2}-1$$

$$=\left(\frac{m(^{35}\text{Cl})}{m(^{37}\text{Cl})}\times\frac{m(^{23}\text{Na})+m(^{37}\text{Cl})}{m(^{23}\text{Na})+m(^{35}\text{Cl})}\right)^{1/2}-1$$

$$=\left(\frac{34.9688}{36.9651}\times\frac{22.9898+36.9651}{22.9898+34.9688}\right)^{1/2}-1=-0.01077$$

Hence, the difference is 1.077%.

E12.16(a) $$\omega=\left(\frac{k_f}{m_{\text{eff}}}\right)^{1/2}\ [12.33];\quad \omega=2\pi\nu=2\pi\left(\frac{c}{\lambda}\right)=2\pi c\tilde{\nu}$$

Therefore, $k_f=m_{\text{eff}}\omega^2=4\pi^2m_{\text{eff}}c^2\tilde{\nu}^2$, $m_{\text{eff}}=\frac{1}{2}m(^{35}\text{Cl})$

$$=(4\pi^2)\times\left(\frac{34.9688}{2}\right)\times(1.66054\times10^{-27}\ \text{kg})\times[(2.997924\times10^{10}\ \text{cm s}^{-1})\times(564.9\ \text{cm}^{-1})]^2$$

$$=\boxed{328.7\ \text{N m}^{-1}}$$

E12.17(a) We write, with $N'=N$ (upper state) and $N=N$ (lower state),

$$\frac{N'}{N}=\text{e}^{-h\nu/kT}\ [\text{from Boltzmann distribution}]=\text{e}^{-hc\tilde{\nu}/kT}$$

$$\frac{hc\tilde{\nu}}{k}=(1.4388\ \text{cm K})\times(559.7\ \text{cm}^{-1})\ [\text{inside front cover}]=805.3\ \text{K}$$

$$\frac{N(\text{upper})}{N(\text{lower})}=\text{e}^{-805.3\,\text{K}/T}$$

(a) $$\frac{N(\text{upper})}{N(\text{lower})}=\text{e}^{-805.3/298}=\boxed{0.067}\ (1:15)$$ (b) $$\frac{N(\text{upper})}{N(\text{lower})}=\text{e}^{-805.3/500}=\boxed{0.20}\ (1:5)$$

E12.18(a) $$\omega=\left(\frac{k_f}{m_{\text{eff}}}\right)^{1/2}\ [12.33],\ \text{so}\ k_f=m_{\text{eff}}\omega^2=4\pi^2m_{\text{eff}}c^2\tilde{\nu}^2$$

$$m_{\text{eff}}=\frac{m_1m_2}{m_1+m_2}\ [12.32]$$

$$m_{\text{eff}}(\text{H}^{19}\text{F})=\frac{(1.0078)\times(18.9984)}{(1.0078)+(18.9984)}\,m_u=0.9570\,m_u$$

$$m_{\text{eff}}(\text{H}^{35}\text{Cl}) = \frac{(1.0078) \times (34.9688)}{(1.0078) + (34.9688)} m_u = 0.9796\, m_u$$

$$m_{\text{eff}}(\text{H}^{81}\text{Br}) = \frac{(1.0078) \times (80.9163)}{(1.0078) + (80.9163)} m_u = 0.9954\, m_u$$

$$m_{\text{eff}}(\text{H}^{127}\text{I}) = \frac{(1.0078) \times (126.9045)}{(1.0078) + (126.9045)} m_u = 0.9999\, m_u$$

We draw up the following table:

	HF	HCl	HBr	HI
$\tilde{\nu}/\text{cm}^{-1}$	4141.3	2988.9	2649.7	2309.5
m_{eff}/m_u	0.9570	0.9697	0.9954	0.9999
$k/(\text{N m}^{-1})$	$\boxed{967.0}$	$\boxed{515.6}$	$\boxed{411.8}$	$\boxed{314.2}$

Note the order of stiffness HF > HCl > HBr > HI.

Question. Which ratio, $\dfrac{k}{B(A-B)}$ or $\dfrac{\tilde{\nu}}{B(A-B)}$, where B(A – B) are the bond energies of Table 10.3, is the more nearly constant across the series of hydrogen halides? Why?

E12.19(a) Data on three transitions are provided. Only two are necessary to obtain the value of $\tilde{\nu}$ and x_e. The third datum can then be used to check the accuracy of the calculated values.

$$\Delta\tilde{G}(v = 1 \leftarrow 0) = \tilde{\nu} - 2\tilde{\nu}x_e = 1556.22\ \text{cm}^{-1}\ [12.40]$$

$$\Delta\tilde{G}(v = 2 \leftarrow 0) = 2\tilde{\nu} - 6\tilde{\nu}x_e = 3088.28\ \text{cm}^{-1}\ [12.41]$$

Multiply the first equation by 3, then subtract the second

$$\tilde{\nu} = (3) \times (1556.22\ \text{cm}^{-1}) - (3088.28\ \text{cm}^{-1}) = \boxed{1580.38\ \text{cm}^{-1}}$$

Then, from the first equation

$$x_e = \frac{\tilde{\nu} - 1556.22\ \text{cm}^{-1}}{2\tilde{\nu}} = \frac{(1580.38 - 1556.22)\ \text{cm}^{-1}}{(2) \times (1580.38\ \text{cm}^{-1})} = \boxed{7.644 \times 10^{-3}}$$

x_e data are usually reported as $x_e\tilde{\nu}$, which is

$$x_e\tilde{\nu} = 12.08\ \text{cm}^{-1}$$

$$\begin{aligned}\Delta\tilde{G}(v = 3 \leftarrow 0) &= 3\tilde{\nu} - 12\tilde{\nu}x_e \\ &= (3) \times (1580.38\ \text{cm}^{-1}) - (12) \times (12.08\ \text{cm}^{-1}) = 4596.18\ \text{cm}^{-1}\end{aligned}$$

which is very close to the experimental value.

E12.20(a) $\Delta\tilde{G}_{v+1/2} = \tilde{\nu} - 2(v+1)x_e\tilde{\nu}$ [12.40], where $\Delta\tilde{G}_{v+1/2} = \tilde{G}(v+1) - \tilde{G}(v)$

Therefore, since

$$\Delta\tilde{G}_{v+1/2} = (1 - 2x_e)\tilde{\nu} - 2vx_e\tilde{\nu}$$

a plot of $\Delta\tilde{G}_{v+1/2}$ against v should give a straight line, which gives $(1 - 2x_e)\tilde{\nu}$ from the intercept at $v = 0$ and $-2x_e\tilde{\nu}$ from the slope. We draw up the following table:

v	0	1	2	3	4
$(\tilde{\nu})/\text{cm}^{-1}$	1481.86	4367.50	7149.04	9826.48	12399.8
$\Delta\tilde{G}_{v+1/2}/\text{cm}^{-1}$	2885.64	2781.54	2677.44	2573.3$\overline{4}$	

The points are plotted in Figure 12.1. The intercept lies at 2885.6 and the slope is $\dfrac{-312.3}{3} = -104.1$; hence $x_e\tilde{\nu} = 52.1\ \text{cm}^{-1}$.

Since $\tilde{\nu} - 2x_e\tilde{\nu} = 2885.6\ \text{cm}^{-1}$, it follow sthat $\tilde{\nu} = 2989.8\ \text{cm}^{-1}$.

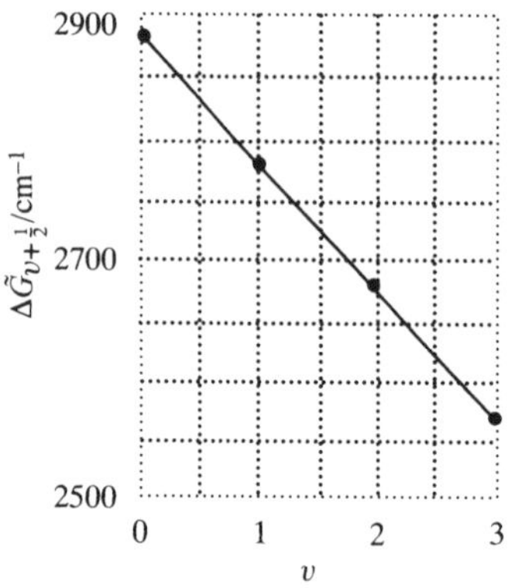

Figure 12.1

The dissociation energy may be obtained by assuming that a Morse potential describes the molecule and that the constant $\tilde{D}_e$ in the expression for the potential is an adequate first approximation for it. Then,

$$\tilde{D}_e = \frac{\tilde{\nu}}{4x_e}\ [12.38] = \frac{\tilde{\nu}^2}{4x_e\tilde{\nu}} = \frac{(2989.8\ \text{cm}^{-1})^2}{(4) \times (52.1\ \text{cm}^{-1})} = 42.9 \times 10^3\ \text{cm}^{-1} \text{ or } 5.32\ \text{eV}$$

However, the depth of the potential well $\tilde{D}_e$ differs from $\tilde{D}_0$, the dissociation energy of the bond, by the zero-point energy; hence

$$\begin{aligned}\tilde{D}_0 &= \tilde{D}_e - \tfrac{1}{2}\tilde{\nu}\ [\text{Fig. } 12.26]\\ &= (42.9 \times 10^3\ \text{cm}^{-1}) - (\tfrac{1}{2}) \times (2989.8\ \text{cm}^{-1})\\ &= 41.4 \times 10^3\ \text{cm}^{-1} = \boxed{5.13\ \text{eV}}\end{aligned}$$

E12.21(a) The dissociation energy may be obtained by assuming that a Morse potential describes the molecule and that the constant $\tilde{D}_e$ in the expression for the potential is an adequate first approximation for it. Then,

$$\tilde{D}_e = \frac{\tilde{\nu}}{4x_e}\ [12.38] \text{ and } x_e = \frac{\tilde{\nu}}{4\tilde{D}_e}$$

However, the depth of the potential well $\tilde{D}_e$ differs from $\tilde{D}_0$, the dissociation energy of the bond, by the zero-point energy; hence

$$\tilde{D}_0 = \tilde{D}_e - \tfrac{1}{2}\tilde{\nu} \text{ [Figure 12.26]}$$

Data in Table 12.2 for D_0 are given in kJ mol^{-1} and the value for $^1H^{19}F$ is 564.4 kJ mol^{-1}. The conversion factor is 83.593 cm^{-1}/kJ mol^{-1}. Hence,

$$\tilde{D}_e = \tilde{D}_0 + \tfrac{1}{2}\tilde{\nu} = 564.4 \text{ kJ mol}^{-1} \times 83.593 \text{ cm}^{-1}/\text{kJ mol}^{-1} + \tfrac{1}{2} \times 4138.32 \text{ cm}^{-1}$$
$$= 49249 \text{ cm}^{-1} = 6.106 \text{ eV}$$

Then, $x_e = \dfrac{\tilde{\nu}}{4\tilde{D}_e} = \dfrac{4138.32 \text{ cm}^{-1}}{4 \times 49249 \text{ cm}^{-1}} = \boxed{0.02101}$

The anharmonicity constant is inversely proportional to the square root of the effective mass of the molecule since it is proportional to $\tilde{\nu}$ [12.34]. $x_e \propto \dfrac{1}{\sqrt{m_{\text{eff}}}}$.

Therefore, $\dfrac{x_e(^2H^{19}F)}{x_e(^1H^{19}F)} = \dfrac{\sqrt{m_{\text{eff}}}(^1H^{19}F)}{\sqrt{m_{\text{eff}}}(^2H^{19}F)} = \dfrac{(0.9570\, m_u)^{1/2}}{(1.8209\, m_u)^{1/2}} = \boxed{0.7250}$

E12.22(a) The R branch obeys the relationship

$$\tilde{\nu}_R(J) = \tilde{\nu} + 2\tilde{B}(J+1) \text{ [12.45c]}$$

Hence, $\tilde{\nu}_R(2) = \tilde{\nu} + 6\tilde{B} = (2648.98) + (6) \times (8.465 \text{ cm}^{-1})$ [Table 12.2] $= \boxed{2699.77 \text{ cm}^{-1}}$

E12.23(a) See the Brief Illustration in Section 12.9. Select those molecules in which a vibration gives rise to a change in dipole moment. It is helpful to write down the structural formulas of the compounds. The infrared-active compounds are

(b) HCl (c) CO_2 (d) H_2O

COMMENT. A more powerful method for determining infrared activity based on symmetry considerations is described in Section 12.16. Also, see Exercises 12.25 and 12.26.

E12.24(a) The number of normal modes of vibration is given by (Section 12.13)

$$N_{\text{vib}} = \begin{Bmatrix} 3N-5 \text{ for linear molecules} \\ 3N-6 \text{ for non-linear molecules} \end{Bmatrix}$$

where N is the number of atoms in the molecule. Hence, since none of these molecules are linear,

(a) 3 (b) 6 (c) 12

COMMENT. Even for moderately sized molecules the number of normal modes of vibration is large and they are usually difficult to visualize.

E12.25(a) See Figures 12.38(H_2O, bent) and 12.37(CO_2, linear) of the text as well as the Brief Illustrations in Section 12.16. Decide which modes correspond to (i) a changing electric dipole moment, (ii) a changing polarizability, and take note of the exclusion rule (Sections 12.15 and 12.16).

(a) Non-linear: all modes both infrared and Raman active.
(b) Linear: the symmetric stretch is infrared inactive but Raman active.

The antisymmetric stretch is infrared active and (by the exclusion rule) Raman inactive. The two bending modes are infrared active and therefore Raman inactive.

E12.26(a) The uniform expansion is depicted in Figure 12.2.

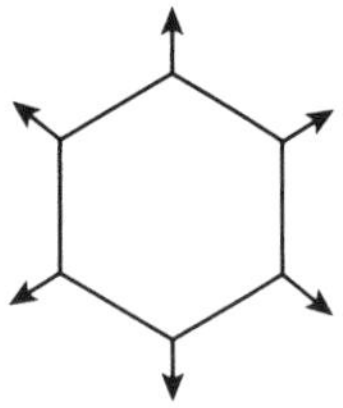

Figure 12.2

Benzene is centrosymmetric, and so the exclusion rule applies (Section 12.15). The mode is infrared inactive (symmetric breathing leaves the molelcular dipole moment unchanged at zero), and therefore the mode may be $\boxed{\text{Raman active}}$ (and is). In group theoretical terms, the breathing mode has symmetry A_{1g} in D_{6h}, which is the point group for benzene, and quadratic forms $x^2 + y^2$ and z^2 have this symmetry (see the character table for C_{6h}, a subgroup of D_{6h}). Hence, the mode is Raman active.

E12.27(a) Use the character table for the group C_{2v} (and see Example 12.6). The rotations span $A_2 + B_1 + B_2$. The translations span $A_1 + B_1 + B_2$. Hence, the normal modes of vibration span the difference, $\boxed{4A_1 + A_2 + 2B_1 + 2B_2}$.

COMMENT. A_1, B_1 and B_2 are infrared active; all modes are Raman active.

Solutions to problems

Solutions to numerical problems

P12.1 Rotational line separations are $2\tilde{B}$ (in wavenumber units), $2B$ (in frequency units), and $(2\tilde{B})^{-1}$ in wavelength units. Hence, the transitions are separated by $\boxed{596\text{ GHz}}$, $\boxed{19.9\text{ cm}^{-1}}$, and $\boxed{0.503\text{ mm}}$.

Ammonia is a symmetric rotor (Section 12.3) and we know that

$$\tilde{B} = \frac{\hbar}{4\pi c I_{\perp}} \text{ [12.13]}$$

and from Table 12.1,

$$I_{\perp} = m_A R^2(1 - \cos\theta) + \left(\frac{m_A m_B}{m}\right) R^2(1 + 2\cos\theta)$$

$m_A = 1.6735 \times 10^{-27}$ kg, $m_B = 2.3252 \times 10^{-26}$ kg, and $m = 2.8273 \times 10^{-26}$ kg with $R = 101.4$ pm and $\theta = 106°78'$, which gives

$$I_\perp = (1.6735 \times 10^{-27}\text{ kg}) \times (101.4 \times 10^{-12}\text{ m})^2 \times (1 - \cos 106°78')$$
$$+ \left(\frac{(1.6735 \times 10^{-27}) \times (2.3252 \times 10^{-26}\text{ kg}^2)}{2.8273 \times 10^{-26}\text{ kg}}\right)$$
$$\times (101.4 \times 10^{-12}\text{ m})^2 \times (1 + 2\cos 106°78')$$
$$= 2.815\overline{8} \times 10^{-47}\text{ kg m}^2$$

Therefore,

$$\tilde{B} = \frac{1.05457 \times 10^{-34}\text{ J s}}{(4\pi) \times (2.9979 \times 10^8\text{ m s}^{-1}) \times (2.815\overline{8} \times 10^{-47}\text{ kg m}^2)} = 994.1\text{ m}^{-1} = \boxed{9.941\text{ cm}^{-1}}$$

which is in accord with the data.

P12.3 Rotation about any axis perpendicular to the C_6 axis may be represented in its essentials by rotation of the pseudolinear molecule in Figure 12.3(a) about the x-axis in the figure.

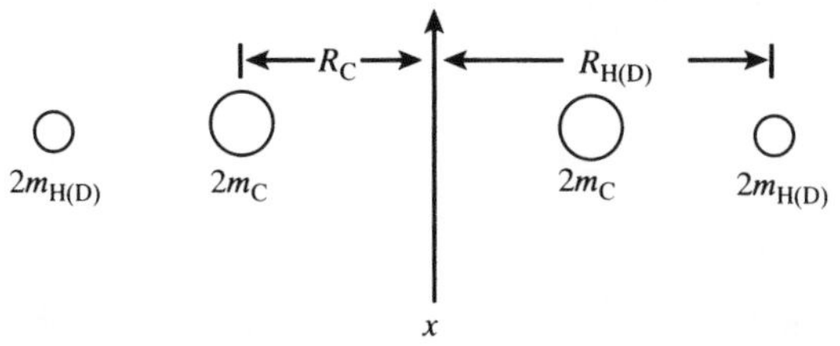

Figure 12.3(a)

The data allow for a determination of R_C and $R_{H(D)}$, which may be decomposed into R_{CC} and $R_{CH(D)}$.

$$I_H = 4m_H R_H^2 + 4m_C R_C^2 = 147.59 \times 10^{-47}\text{ kg m}^2$$
$$I_D = 4m_D R_D^2 + 4m_C R_C^2 = 178.45 \times 10^{-47}\text{ kg m}^2$$

Subtracting I_H from I_D (assume $R_H = R_D$) yields

$$4(m_D - m_H)R_H^2 = 30.86 \times 10^{-47}\text{ kg m}^2$$

$$4(2.01410\, m_u - 1.0078\, m_u) \times (1.66054 \times 10^{-27}\text{ kg } m_u^{-1}) \times (R_H^2) = 30.86 \times 10^{-47}\text{ kg m}^2$$

$$R_H^2 = 4.616\overline{9} \times 10^{-20}\text{ m}^2 \quad R_H = 2.149 \times 10^{-10}\text{ m}$$

$$R_C^2 = \frac{(147.59 \times 10^{-47}\text{ kg m}^2) - (4m_H R_H^2)}{4m_C}$$

$$= \frac{(147.59 \times 10^{-47}\text{ kg m}^2) - (4) \times (1.0078\, m_u) \times (1.66054 \times 10^{-27}\text{ kg } m_u^{-1}) \times (4.616\overline{9} \times 10^{-20}\text{ m}^2)}{(4) \times (12.011\, m_u) \times (1.66054 \times 10^{-27}\text{ kg } m_u^{-1})}$$

$$= 1.4626 \times 10^{-20}\text{ m}^2$$

$$R_C = 1.209 \times 10^{-10}\text{ m}$$

Figure 12.3(b) shows the relationship between R_H, R_C, R_{CC}, and R_{CH}.

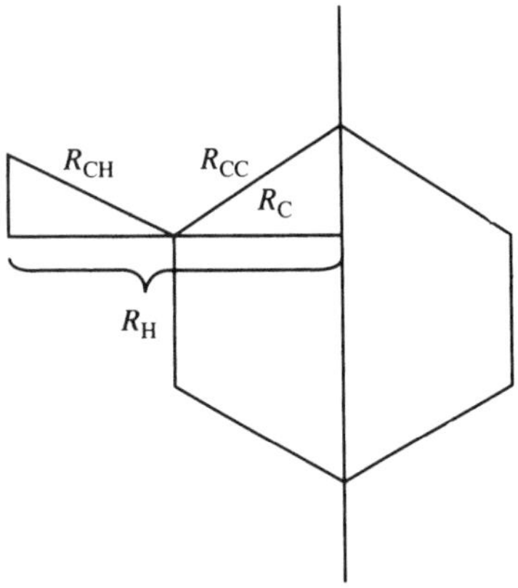

Figure 12.3(b)

$$R_{CC} = \frac{R_C}{\cos 30°} = \frac{1.209 \times 10^{-10}\,\text{m}}{0.8660} = 1.396 \times 10^{-10}\,\text{m} = \boxed{139.6\,\text{pm}}$$

$$R_{CH} = \frac{R_H - R_C}{\cos 30°} = \frac{0.940 \times 10^{-10}}{0.8660} = 1.08\overline{5} \times 10^{-10} = \boxed{108.\overline{5}\,\text{pm}}$$

$$R_{CD} = R_{CH}$$

COMMENT. These values are very close to the interatomic distances quoted by Herzberg in *Electronic Spectra and Electronic Structure of Polyatomic Molecules*, p. 666, which are 139.7 and 108.4 pm, respectively.

P12.5 $\tilde{\nu} = 2\tilde{B}(J+1)\,[12.21\text{a}] = 2\tilde{B}$

Hence, $\tilde{B}(^1\text{HCl}) = 10.4392\,\text{cm}^{-1}$, $\tilde{B}(^2\text{HCl}) = 5.3920\,\text{cm}^{-1}$

$$\tilde{B} = \frac{\hbar}{4\pi cI}\,[12.7] \quad I = m_{\text{eff}}R^2\,[\text{Table 12.1}]$$

$$R^2 = \frac{\hbar}{4\pi c m_{\text{eff}}\tilde{B}} \quad \frac{\hbar}{4\pi c} = 2.79927 \times 10^{-44}\,\text{kg m}$$

$$m_{\text{eff}}(\text{HCl}) = \left(\frac{(1.007825\,m_u) \times (34.96885\,m_u)}{(1.007825\,m_u) + (34.96885\,m_u)}\right) \times (1.66054 \times 10^{-27}\,\text{kg}\,m_u^{-1})$$
$$= 1.62665 \times 10^{-27}\,\text{kg}$$

$$m_{\text{eff}}(\text{DCl}) = \left(\frac{(2.0140\,m_u) \times (34.96885\,m_u)}{(2.0140\,m_u) + (34.96885\,m_u)}\right) \times (1.66054 \times 10^{-27}\,\text{kg}\,m_u^{-1})$$
$$= 3.1622 \times 10^{-27}\,\text{kg}$$

$$R^2(\text{HCl}) = \frac{2.79927 \times 10^{-44}\,\text{kg m}}{(1.62665 \times 10^{-27}\,\text{kg}) \times (1.04392 \times 10^{3}\,\text{m}^{-1})} = 1.64848 \times 10^{-20}\,\text{m}^2$$

$$R(\text{HCl}) = 1.28393 \times 10^{-10}\,\text{m} = \boxed{128.393\,\text{pm}}$$

$$R^2(^2\text{HCl}) = \frac{2.79927 \times 10^{-44}\,\text{kg m}}{(3.1622 \times 10^{-27}\,\text{kg}) \times (5.3920 \times 10^{2}\,\text{m}^{-1})} = 1.6417 \times 10^{-20}\,\text{m}^2$$

$$R(^2\text{HCl}) = 1.2813 \times 10^{-10}\,\text{m} = \boxed{128.13\,\text{pm}}$$

COMMENT. Since the effects of centrifugal distortion have not been taken into account, the number of significant figures in the calculated values of R above should be no greater than four, despite the fact that the data are precise to six figures.

P12.7 From the equation for a linear rotor in Table 12.1 it is possible to show that $Im = m_a m_c(R + R')^2 + m_a m_b R^2 + m_b m_c R'^2$.

$$\text{Thus, } I(^{16}\text{O}^{12}\text{C}^{32}\text{S}) = \left(\frac{m(^{16}\text{O})m(^{32}\text{S})}{m(^{16}\text{O}^{12}\text{C}^{32}\text{S})}\right) \times (R + R')^2 + \left(\frac{m(^{12}\text{C})\{m(^{16}\text{O})R^2 + m(^{32}\text{S})R'^2\}}{m(^{16}\text{O}^{12}\text{C}^{32}\text{S})}\right)$$

$$I(^{16}\text{O}^{12}\text{C}^{34}\text{S}) = \left(\frac{m(^{16}\text{O})m(^{34}\text{S})}{m(^{16}\text{O}^{12}\text{C}^{34}\text{S})}\right) \times (R + R')^2 + \left(\frac{m(^{12}\text{C})\{m(^{16}\text{O})R^2 + m(^{34}\text{S})R'^2\}}{m(^{16}\text{O}^{12}\text{C}^{34}\text{S})}\right)$$

$m(^{16}\text{O}) = 15.9949\,m_u$, $m(^{12}\text{C}) = 12.0000\,m_u$, $m(^{32}\text{S}) = 31.9721\,m_u$, and $m(^{34}\text{S}) = 33.9679\,m_u$. Hence,

$$I(^{16}\text{O}^{12}\text{C}^{32}\text{S})/m_u = (8.5279) \times (R + R')^2 + (0.20011) \times (15.9949R^2 + 31.9721R'^2)$$

$$I(^{16}\text{O}^{12}\text{C}^{32}\text{S})/m_u = (8.7684) \times (R + R')^2 + (0.19366) \times (15.9949R^2 + 33.9679R'^2)$$

The spectral data provides the experimental values of the moments of inertia based on the relationship $\tilde{\nu} = 2\tilde{B}(J+1)$ [12.21a] with $\tilde{B} = \dfrac{\hbar}{4\pi c I}$ [12.7]. These values are set equal to the above equations that are then solved for R and R'. The mean values of I obtained from the data are

$$I(^{16}\text{O}^{12}\text{C}^{32}\text{S}) = 1.37998 \times 10^{-45}\,\text{kg m}^2$$

$$I(^{16}\text{O}^{12}\text{C}^{34}\text{S}) = 1.41460 \times 10^{-45}\,\text{kg m}^2$$

Therefore, after conversion of the atomic mass units to kilograms, the equations we must solve are

$$1.37998 \times 10^{-45}\,\text{m}^2 = (1.4161 \times 10^{-26}) \times (R + R')^2 + (5.3150 \times 10^{-27} R^2) + (1.0624 \times 10^{-26} R'^2)$$

$$1.41460 \times 10^{-45}\,\text{m}^2 = (1.4560 \times 10^{-26}) \times (R + R')^2 + (5.1437 \times 10^{-27} R^2) + (1.0923 \times 10^{-26} R'^2)$$

These two equations may be solved for R and R'. They are tedious to solve by hand, but straightforward. Exercise 12.9(b) illustrates the details of the solution. Readily available mathematical software can be used to quickly give the result. The outcome is $R = \boxed{116.28\,\text{pm}}$ and $R' = \boxed{155.97\,\text{pm}}$. These values may be checked by direct substitution into the equations.

COMMENT. The starting point of this problem is the actual experimental data on spectral line positions. Exercise 12.9(b) is similar to this problem; its starting point is, however, given values of the rotational constants $\tilde{B}$, which were themselves obtained from the spectral line positions. So, the results for R and R' are expected to be essentially identical and they are.

Question. What are the rotational constants calculated from the data on the positions of the absorption lines?

P12.9 The wavenumbers of the transitions with $\Delta v = +1$ are

$$\Delta\tilde{G}_{v+1/2} = \tilde{\nu} - 2(v+1)x_e\tilde{\nu} \text{ [12.40]} \quad \text{and} \quad \tilde{D}_e = \frac{\tilde{\nu}^2}{4x_e\tilde{\nu}} \text{ [12.38]}$$

A plot of $\Delta\tilde{G}_{v+1/2}$ against $v+1$ should give a straight line with intercept $\tilde{\nu}$ at $v+1=0$ and slope $-2x_e\tilde{\nu}$.

Draw up the following table

$v+1$	1	2	3
$\Delta\tilde{G}_{v+1/2}/\text{cm}^{-1}$	284.50	283.00	281.502

The points are plotted in Figure 12.4.

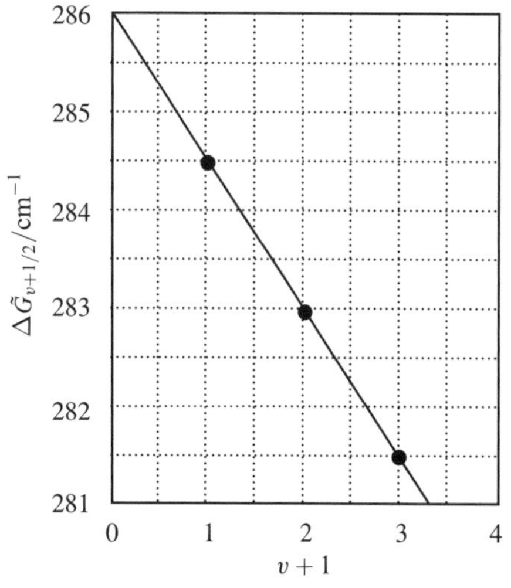

Figure 12.4

The intercept is at 286.0, so $\tilde{\nu} = 286\ \text{cm}^{-1}$. The slope is -1.50, so $x_e\tilde{\nu} = 0.750\ \text{cm}^{-1}$. It follows that

$$\tilde{D}_e = \frac{(286\ \text{cm}^{-1})^2}{(4)\times(0.750\ \text{cm}^{-1})} = 27\ 300\ \text{cm}^{-1}, \quad \text{or} \quad 3.38\ \text{eV}$$

The zero-point level lies at $\boxed{142.81\ \text{cm}^{-1}}$ and so $D_0 = \boxed{3.36\ \text{eV}}$. Since

$$m_{\text{eff}} = \frac{(22.99)\times(126.90)}{(22.99)+(126.90)} m_u = 19.46\overline{4}\ m_u$$

the force constant of the molecule is

$$k = 4\pi^2 m_{\text{eff}} c^2 \tilde{\nu}^2 \text{ [Exercise 12.16(a)]}$$
$$= (4\pi^2) \times (19.46\overline{4}) \times (1.6605 \times 10^{-27} \text{ kg}) \times [(2.998 \times 10^{10} \text{ cm s}^{-1}) \times (286 \text{ cm}^{-1})]^2$$
$$= \boxed{93.8 \text{ N m}^{-1}}$$

P12.11 The set of peaks to the left of center are the P branch, those to the right are the R branch. Within the rigid rotor approximation the two sets are separated by $4\tilde{B}$. The effects of the interactions between vibration and rotation and of centrifugal distortion are least important for transitions with small J values hence the separation between the peaks immediately to the left and right of centre will give good approximate values of $\tilde{B}$ and bond length.

(a) $\tilde{\nu}_Q(J) = \tilde{\nu}$ [12.45b] $= \boxed{2143.26 \text{ cm}^{-1}}$

(b) The zero-point energy is $\frac{1}{2}\tilde{\nu} = 1071.63 \text{ cm}^{-1}$. The molar zero-point energy in J mol^{-1} is

$$N_A hc \times (1071.63 \text{ cm}^{-1}) = N_A hc \times (1.07163 \times 10^5 \text{ m}^{-1})$$
$$= 1.28195 \times 10^4 \text{ J mol}^{-1} = \boxed{12.8195 \text{ kJ mol}^{-1}}$$

(c) $k = 4\pi^2 \mu c^2 \tilde{\nu}^2$

$$\mu(^{12}\text{C}^{16}\text{O}) = \frac{m_\text{C} m_\text{O}}{m_\text{C} + m_\text{O}} = \left(\frac{(12.0000\, m_\text{u}) \times (15.9949\, m_\text{u})}{(12.0000\, m_\text{u}) + (15.9949\, m_\text{u})}\right) \times (1.66054 \times 10^{-27} \text{ kg } m_\text{u}^{-1})$$
$$= 1.13852 \times 10^{-26} \text{ kg}$$

$$k = 4\pi^2 c^2 \times (1.13852 \times 10^{-26} \text{ kg}) \times (2.14326 \times 10^5 \text{ m}^{-1})^2 = \boxed{1.85563 \times 10^3 \text{ N m}^{-1}}$$

(d) $4\tilde{B} \approx 7.655 \text{ cm}^{-1}$

$\tilde{B} \approx \boxed{1.91 \text{ cm}^{-1}}$ [4 significant figures not justified]

(e) $$\tilde{B} = \frac{\hbar}{4\pi c I} \text{ [12.7]} = \frac{\hbar}{4\pi c \mu R^2} \text{ [Table 12.1]}$$

$$R^2 = \frac{\hbar}{4\pi c \mu \tilde{B}} = \frac{\hbar}{(4\pi c) \times (1.13852 \times 10^{-26} \text{ kg}) \times (191 \text{ m}^{-1})} = 1.28\overline{7} \times 10^{-20} \text{ m}^2$$

$R = 1.13 \times 10^{-10} \text{ m} = \boxed{113 \text{ pm}}$

P12.13 $V(R) = hc\tilde{D}_e\{1 - e^{-a(R-R_e)}\}^2$ [12.37]

$$\tilde{\nu} = \frac{\omega}{2\pi c} = 936.8 \text{ cm}^{-1} \quad x_e\tilde{\nu} = 14.15 \text{ cm}^{-1}$$

$$a = \left(\frac{m_{\text{eff}}}{2hc\tilde{D}_e}\right)^{1/2} \omega \quad x_e = \frac{\hbar a^2}{2m_{\text{eff}}\omega} \quad \tilde{D}_e = \frac{\tilde{\nu}}{4x_e}$$

$$m_{\text{eff}}(\text{RbH}) \approx \frac{(1.008) \times (85.47)}{(1.008) + (85.47)} m_\text{u} = 1.654 \times 10^{-27} \text{ kg}$$

$$\tilde{D}_e = \frac{\tilde{\nu}^2}{4x_e\tilde{\nu}} = \frac{(936.8\ \text{cm}^{-1})^2}{(4)\times(14.15\ \text{cm}^{-1})} = 15\ 50\bar{5}\ \text{cm}^{-1}\ (1.92\ \text{eV})$$

$$a = 2\pi\nu\left(\frac{m_{\text{eff}}}{2hc\tilde{D}_e}\right)^{1/2}\ [12.37] = 2\pi c\tilde{\nu}\left(\frac{m_{\text{eff}}}{2hc\tilde{D}_e}\right)^{1/2}$$

$$= (2\pi)\times(2.998\times10^{10}\ \text{cm s}^{-1})\times(936.8\ \text{cm}^{-1})$$

$$\times\left(\frac{1.654\times10^{-27}\ \text{kg}}{(2)\times(15505\ \text{cm}^{-1})\times(6.626\times10^{-34}\ \text{J s})\times(2.998\times10^{10}\ \text{cm s}^{-1})}\right)^{1/2}$$

$$= 9.144\times10^9\ \text{m}^{-1} = 9.144\ \text{nm}^{-1} = \frac{1}{0.1094\ \text{nm}}$$

Therefore, $\dfrac{V(R)}{hc\tilde{D}_e} = \{1 - e^{-(R-R_e)/(0.1094\ \text{nm})}\}^2$

with $R_e = 236.7$ pm. We draw up the following table:

R/pm	50	100	200	300	400	500	600	700	800
$V/(hc\tilde{D}_e)$	20.4	6.20	0.159	0.193	0.601	0.828	0.929	0.971	0.988

These points are plotted in Figure 12.5 as the line labelled $J = 0$.

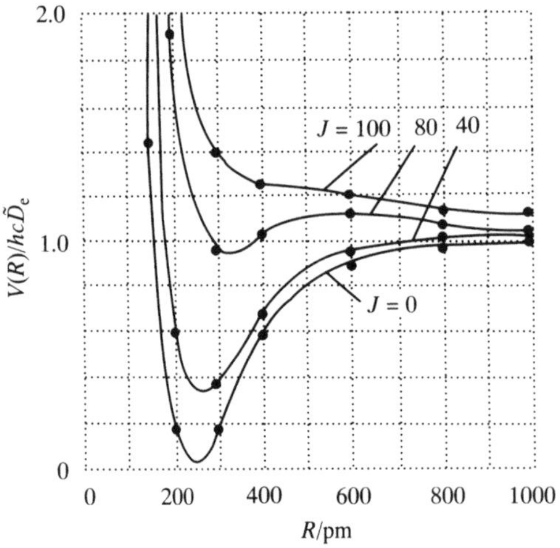

Figure 12.5

For the second part, we note that $\tilde{B} \propto \dfrac{1}{R^2}$ and write

$$V_J^* = V + hc\tilde{B}_e J(J+1)\times\left(\frac{R_e^2}{R^2}\right)$$

with $\tilde{B}_e$ the equilibrium rotational constant, $\tilde{B}_e = 3.020\ \text{cm}^{-1}$.

We then draw up the following table using the values of V calculated above:

R/pm	50	100	200	300	400	600	800	1000
$\frac{R_e}{R}$	4.73	2.37	1.18	0.79	0.59	0.39	0.30	0.24
$\frac{V}{hc\tilde{D}_e}$	20.4	6.20	0.159	0.193	0.601	0.929	0.988	1.000
$\frac{V^*_{40}}{hc\tilde{D}_e}$	27.5	7.99	0.606	0.392	0.713	0.979	1.016	1.016
$\frac{V^*_{80}}{hc\tilde{D}_e}$	48.7	13.3	1.93	0.979	1.043	1.13	1.099	1.069
$\frac{V^*_{100}}{hc\tilde{D}_e}$	64.5	17.2	2.91	1.42	1.29	1.24	1.16	1.11

These points are also plotted in Figure 12.5.

P12.15 (a) Vibrational wavenumbers ($\tilde{\nu}$/cm^{-1}) computed by PC Spartan Pro™ at several levels of theory are tabulated below, along with experimental values.

	A_1	A_1	B_2
Semi-empirical PM3	412	801	896
SCF 6-316G**	592	1359	1569
Density functional	502	1152	1359
Experimental	525	1151	1336

The vibrational modes are shown graphically in Figure 12.6.

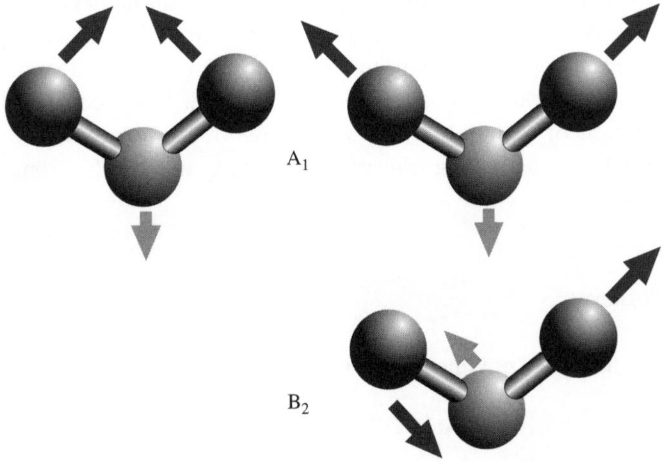

Figure 12.6

(b) The wavenumbers computed by density functional theory agree quite well with experiment. Agreement of the semi-empirical and SCF values with experiment is not so good. In this molecule, experimental wavenumbers can be correlated rather easily to computed vibrational modes even where the experimental and computed wavenumbers disagree substantially. Often, as in this case, computational methods that do a poor job of computing absolute transition wavenumbers still put transitions in proper order by wavenumber. That is, the modelling software systematically overestimates (as in this SCF computation) or underestimates (as in this semi-empirical computation) the wavenumbers, thus keeping them in the correct order. Group theory is another aid in the assignment of transitions: it can classify modes as forbidden, allowed only in particular polarizations, etc. Also, visual examination of the modes of motion can help to classify many modes as predominantly bond stretching, bond bending, or internal rotation; these different modes of vibration can be correlated to quite different ranges of wavenumbers (stretches highest, especially stretches involving hydrogen atoms, and internal rotations lowest).

P12.17 Summarize the six observed vibrations according to their wavenumbers ($\tilde{\nu}/\text{cm}^{-1}$).

IR	870	1370	2869	3417
Raman	877	1408	1435	3407.

(a) If H_2O_2 were linear, it would have $3N - 5 = \boxed{7}$ vibrational modes.

(b) Follow the flow chart in Figure 11.7. Structure **4** is not linear, there is only one C_n axis (a C_2), and there is a σ_h; the point group is $\boxed{C_{2h}}$. Structure **5** is not linear, there is only one C_n axis (a C_2), no σ_h, but two σ_v; the point group is $\boxed{C_{2v}}$. Structure **6** is not linear, there is only one C_n axis (a C_2), no σ_h, no σ_v; the point group is $\boxed{C_2}$.

(c) The exclusion rule applies to structure **2** because it has a center of inversion: no vibrational modes can be both IR and Raman active. So $\boxed{\text{structure } \mathbf{2} \text{ is inconsistent with observation}}$. The vibrational modes of structure **3** span $3A_1 + A_2 + 2B_2$. (The full basis of 12 cartesian coordinates spans $4A_1 + 2A_2 + 2B_1 + 4B_2$; remove translations and rotations.) The C_{2v} character table says that five of these modes are IR active ($3A_1+2B_2$) and all are Raman active. All of the modes of structure **4** are both IR and Raman active. (A look at the character table shows that both symmetry species are IR and Raman active, so determining the symmetry species of the normal modes does not help here.) Both structures **3** and **4** have more active modes than were observed. This is consistent with the observations. After all, group theory can only tell us whether the transition moment *must* be zero by symmetry; it does not tell us whether the transition moment is sufficiently strong to be observed under experimental conditions.

Solutions to theoretical problems

P12.19 Here, we make use of results that have been obtained in Chapter 8 for the average value of $x = R - R_e$ and $x^2 = (R - R_e)^2$.

$$\langle x \rangle = 0 \quad \langle x^2 \rangle = (v + \tfrac{1}{2})\frac{\hbar}{(mk)^{1/2}} \quad [8.32]$$

Consider first $1/\langle R \rangle^2$. We need to evaluate $\langle R \rangle$.

$\langle R\rangle = \langle R_e + x\rangle$ [R_e is a constant] $= \langle R_e\rangle + \langle x\rangle = R_e + 0 = R_e$

$$\boxed{1/\langle R\rangle^2 = 1/R_e^2}$$

Next consider $1/\langle R^2\rangle$. We need to evaluate $\langle R^2\rangle$. $1/\langle R^2\rangle$ is the reciprocal of this quantity.

$$\langle R^2\rangle = \langle (R_e + x)^2\rangle = \langle R_e^2 + 2xR_e + x^2\rangle = \langle R_e^2\rangle + \langle 2xR_e\rangle + \langle x^2\rangle = R_e^2 + 2R_e\langle x\rangle + \langle x^2\rangle = R_e^2 + \langle x^2\rangle$$

$$\boxed{1/\langle R^2\rangle = \frac{1}{R_e^2}\left(\frac{1}{1+\langle x^2\rangle/R_e^2}\right)}$$

Finally consider $\langle 1/R^2\rangle$. We need to evaluate $1/R^2$ and then take the average of this quantity.

$$\frac{1}{R^2} = \frac{1}{(R_e + x)^2} = \frac{1}{R_e^2\left(1+\dfrac{x}{R_e}\right)^2} = \frac{1}{R_e^2}\left(1 - 2\frac{x}{R_e} + 3\frac{x^2}{R_e^2} + \ldots\right)$$

We have expanded $\left(1+\dfrac{x}{R_e}\right)^{-2}$ in a binomial series and dropped terms beyond the second power of $\dfrac{x}{R_e}$. Noting again that $\langle x\rangle = 0$ we finally obtain $\boxed{\left\langle\dfrac{1}{R^2}\right\rangle = \dfrac{1}{R_e^2}\left(1+3\dfrac{\langle x^2\rangle}{R_e^2}\right)}$. Examination of these three results shows that $\left\langle\dfrac{1}{R^2}\right\rangle > \dfrac{1}{\langle R\rangle^2} > \dfrac{1}{\langle R\rangle^2}$.

P12.21 Because the centrifugal force and the restoring force balance,

$k(r_c - r_e) = \mu\omega^2 r_c$, [$\mu \equiv m_{\text{eff}}$ here, see note on good practice, Section 12.8]

we can solve for the distorted bond length as a function of the equilibrium bond length:

$$r_c = \frac{r_e}{1 - \mu\omega^2/k}$$

Classically, then, the energy would be the rotational energy plus the energy of the stretched bond:

$$E = \frac{J^2}{2I} + \frac{k(r_c - r_e)^2}{2} = \frac{J^2}{2I} + \frac{k^2(r_c - r_e)^2}{2k} = \frac{J^2}{2I} + \frac{(\mu\omega^2 r_c)^2}{2k}.$$

How is the energy different form the rigid-rotor energy? Besides the energy of stretching of the bond, the larger moment of inertia alters the strictly rotational piece of the energy. Substitute μr_c^2 for I and substitute for r_c in terms of r_e throughout:

So, $$E = \frac{J^2(1-\mu\omega^2/k)^2}{2\mu r_e^2} + \frac{\mu^2\omega^4 r_e^2}{2k(1-\mu\omega^2/k)^2}.$$

Assuming that $\mu\omega^2/k$ is small (a reasonable assumption for most molecules), we can expand the expression and discard squares or higher powers of $\mu\omega^2/k$:

$$E \approx \frac{J^2(1-2\mu\omega^2/k)}{2\mu r_e^2} + \frac{\mu^2\omega^4 r_e^2}{2k}.$$

(Note that the entire second term has a factor of $\mu\omega^2/k$ even before squaring and expanding the denominator, so we discard all terms of that expansion after the first.) Begin to clean up the expression by using classical definitions of angular momentum:

$$J = I\omega = \mu r^2\omega \quad \text{so} \quad \omega = J/\mu r_e^2,$$

which allows us to substitute expressions involving J for all ω s:

$$E \approx \frac{J^2}{2\mu r_e^2} - \frac{J^4}{\mu^2 r_e^6 k} + \frac{J^4}{2\mu^2 r_e^6 k}.$$

(At the same time, we have expanded the first term, part of which we can now combine with the last term.) Continue to clean up the expression by substituting I/μ for r^2, and then carry the expression over to its quantum-mechanical equivalent by substituting $J(J+1)\hbar^2$ for J^2:

$$E \approx \frac{J^2}{2I} - \frac{J^4\mu}{2I^3k} \Rightarrow E \approx \frac{J(J+1)\hbar^2}{2I} - \frac{J^2(J+1)^2\hbar^4\mu}{2I^3k}.$$

Dividing by hc gives the rotational term, $F(J)$:

$$\tilde{F}(J) \approx \frac{J(J+1)\hbar^2}{2hcI} - \frac{J^2(J+1)^2\hbar^4\mu}{2hcI^3k} = \frac{J(J+1)\hbar}{4\pi cI} - \frac{J^2(J+1)^2\hbar^3\mu}{4\pi cI^3k},$$

where we have used $\hbar = h/2\pi$ to eliminate a common divisor of h. Now use the definition of the rotational constant,

$$\tilde{B} = \frac{\hbar}{4\pi cI} \Rightarrow \tilde{F}(J) \approx J(J+1)\tilde{B} - J^2(J+1)^2\tilde{B}^3\frac{16\pi^2c^2\mu}{k}.$$

Finally, use the relationship between the force constant and vibrational wavenumber:

$$\left(\frac{k}{\mu}\right)^{1/2} = \omega_{\text{vib}} = 2\pi\nu = 2\pi c\tilde{\nu} \quad \text{so} \quad \frac{\mu}{k} = \frac{1}{4\pi^2c^2\tilde{\nu}^2}$$

leaving $\tilde{F}(J) \approx \tilde{B}J(J+1) - \dfrac{4\tilde{B}^3}{\tilde{\nu}^2}J^2(J+1)^2 = \tilde{B}J(J+1) - \tilde{D}J^2(J+1)^2$, where $\boxed{\tilde{D} = \dfrac{4\tilde{B}^3}{\tilde{\nu}^2}}$.

P12.23 In order to plot this potential function define the variable $y \equiv x/a$ and rewrite the potential function as $\upsilon(y) = \dfrac{V(y)}{V_0} = (e^{-1/y^2} - 1)$. Figure 12.7 shows a plot of this function against y.

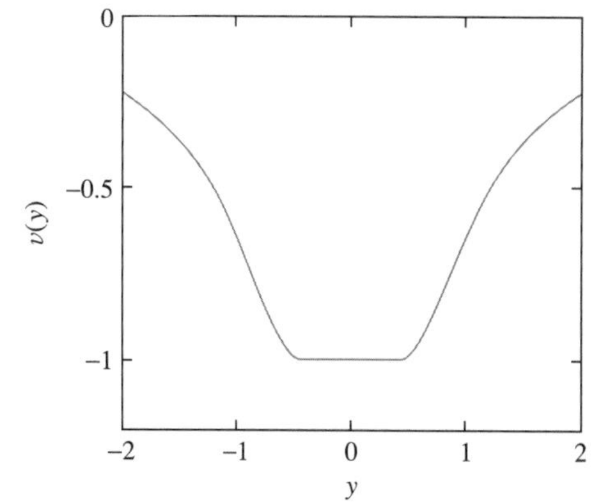

Figure 12.7

Note that for small displacements from $y = 0$ ($x = 0$) the potential energy function is flat (independent of x). Therefore, the first and second derivatives of $V(x)$ are zero and the force constant is zero. There is no restoring force for small displacements from the equilibrium position. The particle $\boxed{\text{cannot undergo simple harmonic motion}}$.

P12.25 Refer to the flow chart in Figure11.7. Yes at the first question (linear?) leads to linear point groups and therefore linear rotors. If the molecule is not linear, then yes at the next question (two or more C_n with $n > 2$?) leads to cubic and icosahedral groups and therefore spherical rotors. If the molecule is not a spherical rotor, yes at the next question leads to symmetric rotors if the highest C_n has $n > 2$; if not, the molecule is an asymmetric rotor.

(a) CH_4: not linear, but more than two $C_n(n > 2)$, so $\boxed{\text{spherical rotor}}$.

(b) CH_3CN: not linear, C_3 (only one of them), so $\boxed{\text{symmetric rotor}}$.

(c) CO_2: linear, so $\boxed{\text{linear rotor}}$.

(d) CH_3OH: not linear, no C_n, so $\boxed{\text{asymmetric rotor}}$.

(e) Benzene: not linear, C_6, but only one high-order axis, so $\boxed{\text{symmetric rotor}}$.

(f) Pyridine: not linear, C_2, is highest rotational axis, so $\boxed{\text{asymmetric rotor}}$.

P12.27 $\tilde{S}(v, J) = (v + \tfrac{1}{2})\tilde{\nu} + \tilde{B}J(J + 1)$ [12.44]

$\Delta\tilde{S}_J^{\mathrm{O}} = \tilde{\nu} - 2\tilde{B}(2J - 1)$ [$\Delta v = 1, \Delta J = -2$]

$\Delta\tilde{S}_J^{\mathrm{S}} = \tilde{\nu} + 2\tilde{B}(2J + 3)$ [$\Delta v = 1, \Delta J = +2$]

The transition of maximum intensity corresponds, approximately, to the transition with the most probable value of J, which was calculated in Problem 12.26:

$$J_{\max} = \left(\frac{kT}{2hc\tilde{B}}\right)^{1/2} - \frac{1}{2}$$

The peak-to-peak separation is then

$$\delta\tilde{\nu} = \Delta\tilde{S}^{\mathrm{S}}_{J_{\max}} - \Delta\tilde{S}^{\mathrm{O}}_{J_{\max}} = 2\tilde{B}(2J_{\max} + 3) - \{-2\tilde{B}(2J_{\max} - 1)\} = 8\tilde{B}(J_{\max} + \tfrac{1}{2})$$

$$= 8\tilde{B}\left(\frac{kT}{2hc\tilde{B}}\right)^{1/2} = \left(\frac{32\tilde{B}kT}{hc}\right)^{1/2}$$

To analyse the data we rearrange the relationship to

$$\tilde{B} = \frac{hc(\delta\tilde{\nu})^2}{32kT}$$

and convert to a bond length using $\tilde{B} = \dfrac{\hbar}{4\pi cI}$, with $I = 2m_x R^2$ (Table 12.1) for a linear rotor. This gives

$$R = \left(\frac{\hbar}{8\pi c m_x B}\right)^{1/2} = \left(\frac{1}{\pi c\delta\tilde{\nu}}\right) \times \left(\frac{2kT}{m_x}\right)^{1/2}$$

We can now draw up the following table:

	$HgCl_2$	$HgBr_2$	HgI_2
T/K	555	565	565
m_x/m_u	35.45	79.1	126.90
$\delta\tilde{\nu}$/cm^{-1}	23.8	15.2	11.4
R/pm	227.6	240.7	253.4

Hence, the three bond lengths are approximately $\boxed{230,\ 240,\ \text{and}\ 250\ \text{pm}}$.

Solutions to applications

P12.29 (a) Refer to Section 13.1. The molar absorption coefficient $\varepsilon(\tilde{\nu})$ is given by

$$\varepsilon(\tilde{\nu}) = \frac{A(\tilde{\nu})}{L[CO_2]} = \frac{RTA(\tilde{\nu})}{Lx_{CO_2}p} \quad [13.4,\ 1.8,\ \text{and}\ 1.15]$$

where $T = 298$ K, $L = 10$ cm, $p = 1$ bar, and $x_{CO_2} = 0.021$.

The absorption band originates with the 001 $\leftarrow$ 000 transition of the antisymmetric stretch vibrational mode at 2349 cm^{-1} (Figure 12.37). The band is very broad because of accompanying rotational transitions and lifetime broadening of each individual absorption (also called collisional broadening or pressure broadening, Section 9.6). The spectra reveals that the Q branch is missing so we conclude that the transition $\boxed{\Delta J = 0 \text{ is forbidden}}$ (Section 12.11) for the $D_{\infty h}$ point group of CO_2. The P branch ($\Delta J = -1$) is evident at lower energies and the R branch ($\Delta J = +1$) is evident at higher energies. See Figures 12.8(a) and (b).

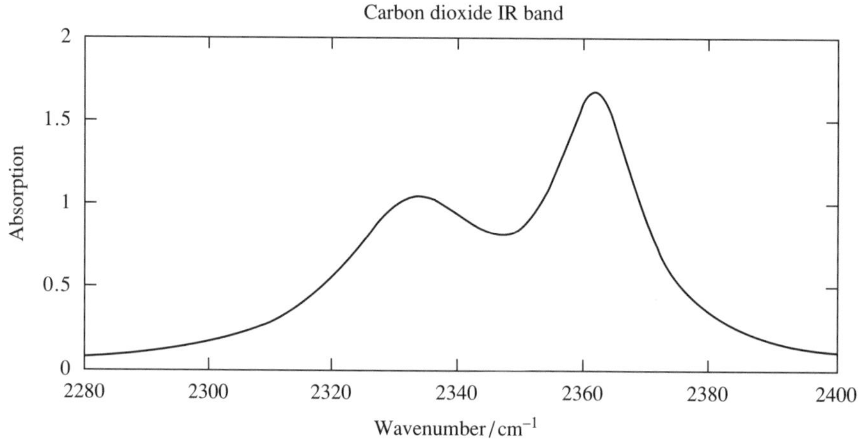

Figure 12.8(a)

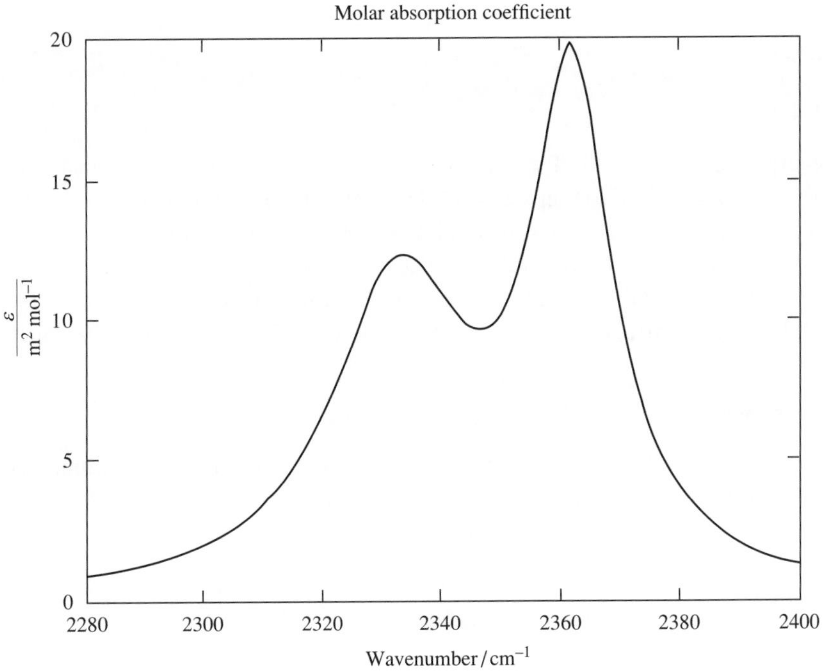

Figure 12.8(b)

(b) $^{16}O{-}^{12}C{-}^{16}O$ has two identical nuclei of zero spin so the CO_2 wavefunction must be symmetric with respect to nuclear interchange and it must obey Bose–Einstein nuclear statistics (Section 12.7). Consequently, J takes on even values only for the $v = 0$ vibrational state and odd values only for the $v = 1$ state. The (v, J) states for this absorption band are $(1, J + 1) \leftarrow (0, J)$ for $J = 0, 2, 4, \ldots$. According to eqn 12.44, the energy of the $(0, J)$ state is

$$\tilde{S}(0, J) = \tfrac{1}{2}\tilde{\nu} + \tilde{B}J(J + 1),$$

where $\tilde{\nu} = 2349\ \mathrm{cm^{-1}}$

$$I = \frac{2M_O R^2}{N_A} = \frac{2(0.01600\ \mathrm{kg\ mol^{-1}})(116.2 \times 10^{-12}\ \mathrm{m})^2}{6.022 \times 10^{23}\ \mathrm{mol^{-1}}}$$
$$= 7.175 \times 10^{-46}\ \mathrm{kg\ m^2}\ [\text{Table 12.1}]$$

$$\tilde{B} = \frac{h}{8\pi^2 cI}\ [12.7]$$
$$= \frac{6.626 \times 10^{-34}\ \mathrm{J\ s}}{8\pi^2(2.998 \times 10^8\ \mathrm{m\ s^{-1}})(7.175 \times 10^{-46}\ \mathrm{kg\ m^2})}$$
$$= 39.02\ \mathrm{m^{-1}} = 0.3902\ \mathrm{cm^{-1}}$$

The transitions of the P and R branches occur at

$$\tilde{\nu}_P = \tilde{\nu} - 2\tilde{B}J\ [12.45a]$$

and

$\tilde{\nu}_R = \tilde{\nu} + 2\tilde{B}(J+1)$ [12.45c]

where $J = 0, 2, 4, 6 \ldots$

The highest energy transition of the P branch is at $\tilde{\nu} - 4\tilde{B}$; the lowest energy transition of the R branch is at $\tilde{\nu} + 2\tilde{B}$. Transitions are separated by $4\tilde{B}(1.5608\ \text{cm}^{-1})$ within each branch. The probability of each transition is proportional to the lower-state population, which we assume to be given by the Boltzman distribution with a degeneracy of $2J+1$. The transition probability is also proportional to both a nuclear degeneracy factor (eqn 12.26) and a transition dipole moment, which is approximately independent of J. The former factors are absorbed into the constant of proportionality.

$$\text{transition probability} \propto (2J+1)e^{-\tilde{S}(0,J)hc/kT}$$

A plot of the right-hand side of this equation against J at 298 K indicates a maximum transition probability at $J_{max} = 16$. We 'normalize' the maximum in the predicted structure, and eliminate the constant of proportionality by examining the transition probability ratio:

$$\frac{\text{transition probability for } J\text{th state}}{\text{transition probability for } J_{max} \text{ state}} = \frac{(2J+1)e^{-\tilde{S}(0,J)hc/kT}}{33e^{-\tilde{S}(0,16)hc/kT}} = \left(\frac{2J+1}{33}\right)e^{-(J^2+J-272)\tilde{B}hc/kT}$$

A plot, Figure 12.8(c), of the above ratio against predicted wavenumbers can be compared to the ratio $A(\tilde{\nu})/A_{max}$ where A_{max} is the observed spectrum maximum (1.677). It shows a fair degree of agreement between the experimental and simple theoretical band shapes.

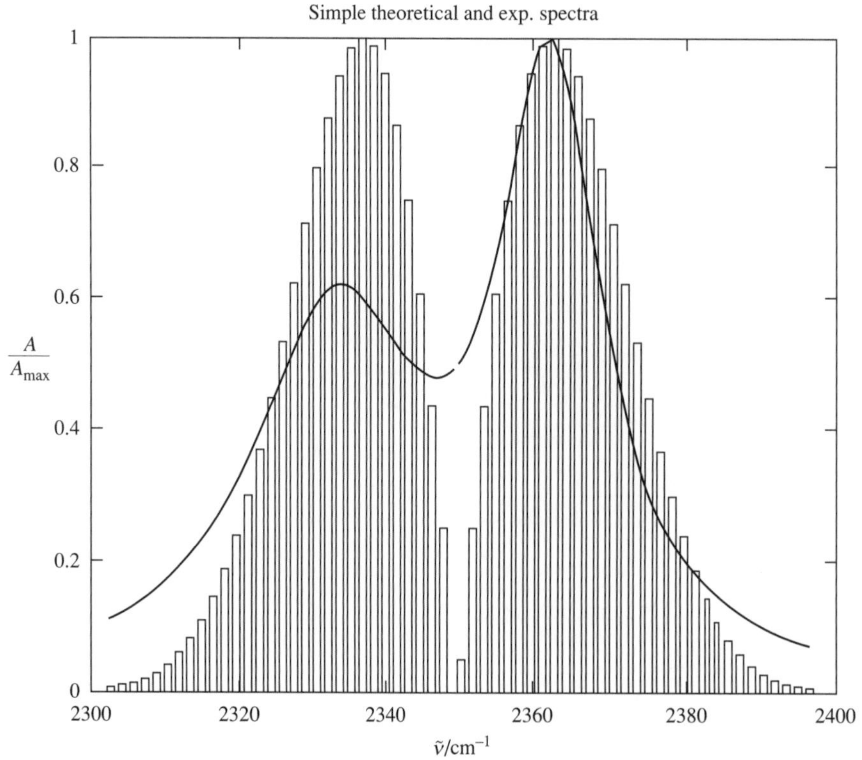

Figure 12.8(c)

(c) Using the equations of *Justification 13.1*, we may write the relationship

$$A = \varepsilon(\tilde{\nu}) \int_0^h [CO_2] \mathrm{d}h$$

The strong absorption of the band suggests that h should not be a very great length and that $[CO_2]$ should be constant between the Earth's surface and h. Consequently, the integration gives

$$A = \varepsilon(\tilde{\nu})[CO_2]h$$

$$= \varepsilon(\tilde{\nu})h\left\{\frac{x_{CO_2}p}{RT}\right\} \quad \text{Dalton's law of partial pressures}$$

p and T are not expected to change much for modest values of h so we estimate that $p = 1$ bar and $T = 288$ K.

$$A = \varepsilon(\tilde{\nu})h\left\{\frac{(3.3\times10^{-4})(1\times10^{5}\,\text{Pa})}{(8.314\,\text{J}\,\cancel{\text{K}^{-1}}\,\text{mol}^{-1})(288\,\cancel{\text{K}})}\right\} = (0.0138\,\text{m}^{-3}\,\text{mol})\varepsilon(\tilde{\nu})h$$

Transmittance $= 10^{-A} = 10^{-(0.0138\,\text{m}^{-3}\,\text{mol})\varepsilon(\tilde{\nu})h}$ [equation after eqn 13.3]

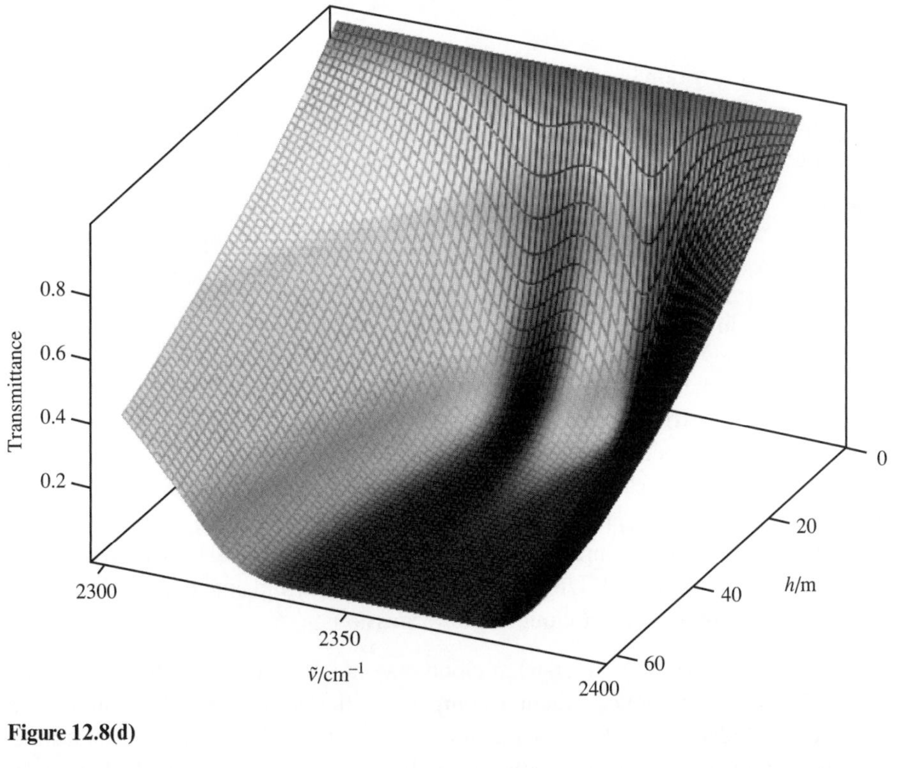

Figure 12.8(d)

The transmittance surface plot, Figure 12.8(d), clearly shows that before a height of about $\boxed{30\text{ m}}$ has been reached all of the Earth's IR radiation in the 2320–2380 cm^{-1} range has been absorbed by atmospheric carbon dioxide.

See C.A. Meserole, F.M. Mulcahy, J. Lutz, and H.A. Yousif, *J. Chem. Ed.*, 74, 316 (1997).

P12.31 $E_J = J(J+1)hc\tilde{B}, \quad g_J = 2J+1$

$$E_1 - E_0 = 2hc\tilde{B} = hc\left(\frac{1}{\lambda_{\text{shorter}}} - \frac{1}{\lambda_{\text{longer}}}\right)$$

$$\tilde{B} = \frac{1}{2}\left(\frac{1}{\lambda_{\text{shorter}}} - \frac{1}{\lambda_{\text{longer}}}\right) = \frac{1}{2}\left(\frac{1}{\lambda_{\text{shorter}}} - \frac{1}{\lambda_{\text{shorter}} + \Delta\lambda}\right)$$

$$= \frac{1}{2}\left(\frac{1}{\lambda_{\text{shorter}}}\right) \times \left(1 - \frac{1}{1 + \frac{\Delta\lambda}{\lambda_{\text{shorter}}}}\right)$$

$$= \frac{1}{2}\left(\frac{1}{387.5\ \text{nm}}\right) \times \left(1 - \frac{1}{1 + \frac{0.061}{387.5}}\right) \times \left(\frac{10^9\ \text{nm}}{10^2\ \text{cm}}\right)$$

$$\tilde{B} = \boxed{2.031\ \text{cm}^{-1}}$$

$$\frac{E_1 - E_0}{k} = \frac{2hc\tilde{B}}{k} = \frac{2(6.626 \times 10^{-34}\ \text{J s}) \times (3.00 \times 10^{10}\ \text{cm s}^{-1}) \times (2.031\ \text{cm}^{-1})}{1.381 \times 10^{-23}\ \text{J K}^{-1}}$$

$$= 5.84\overline{7}\ \text{K}$$

Intensity of $J' \leftarrow J$ absorption line $I_J \propto gJe^{-E_J/kT}$

$$\frac{I_{\lambda_{\text{longer}}}}{I_{I_{\lambda_{\text{shorter}}}}} \simeq \frac{g_1 e^{-E_1/kT}}{g_0 e^{-E_0/kT}} = \frac{g_1}{g_0} e^{-(E_1 - E_0)/kT}$$

solve for T

$$T = \left(\frac{E_1 - E_0}{k}\right) \times \left(\frac{1}{\ln \frac{g_1 I_{\lambda_{\text{shorter}}}}{g_0 I_{\lambda_{\text{longer}}}}}\right) = 5.84\overline{7}\ \text{K}\left(\frac{1}{\ln(3 \times 4)}\right) = \boxed{2.35\ \text{K}}$$

P12.33 Temperature effects. At extremely low temperatures (10 K) only the lowest rotational states are populated. No emission spectrum is expected for the CO in the cloud and star light microwave absorptions by the CO in the cloud are from the lowest rotational states. At higher temperatures additional high-energy lines appear because higher-energy rotational states are populated. Circumstellar clouds may exhibit infrared absorptions due to vibrational excitation as well as electronic transitions in the ultraviolet. Ultraviolet absorptions may indicate the photodissocation of carbon monoxide. High-temperature clouds exhibit emissions.

Density effects. The density of an interstellar cloud may range from one particle to a billion particles per cm^3. This is still very much a vacuum compared to the laboratory high vacuum of a trillion particles per cm^3. Under such extreme vacuum conditions the half-life of any quantum state is expected to be extremely long and absorption lines should be very narrow. At the higher densities the vast size of nebulae obscures distant stars. High densities and high temperatures may create conditions in which emissions stimulate emissions of the same wavelength by molecules. A cascade of stimulated emissions greatly amplifies normally weak lines—the maser phenomena of microwave amplification by stimulated emission of radiation.

Particle velocity effects. Particle velocity can cause Doppler broadening of spectral lines. The effect is extremely small for interstellar clouds at 10 K but is appreciable for clouds near high-temperature stars. Outflows of gas from pulsing stars exhibit a red Doppler shift when moving away at high speed and a blue shift when moving toward us.

There will be many more transitions observable in circumstellar gas than in interstellar gas because many more rotational states will be accessible at the higher temperatures. Higher velocity and density of particles in circumstellar material can be expected to broaden spectral lines compared to those of interstellar material by shortening collisional lifetimes. (Doppler broadening is not likely to be significantly different between circumstellar and interstellar material in the same astronomical neighborhood. The relativistic speeds involved are due to large-scale motions of the expanding universe, compared to which local thermal variations are insignificant.) A temperature of 1000 K is not high enough to significantly populate electronically excited states of CO; such states would have different bond lengths, thereby producing transitions with different rotational constants. Excited vibrational states would be accessible, however, and rovibrational transitions with P and R branches as detailed later in this chapter would be observable in circumstellar but not interstellar material. The rotational constant $\tilde{B}$ for $^{12}C^{16}O$ is 1.691 cm^{-1}. The first excited rotational energy level, $J = 1$, with energy $J(J+1)hc\tilde{B} = 2hc\tilde{B}$, is thermally accessible at about 6 K (based on the rough equation of the rotational energy to thermal energy kT). In interstellar space, only two or three rotational lines would be observable; in circumstellar space (at about 1000 K) the number of transitions would be more like 20.

13 Molecular spectroscopy 2: electronic transitions

Answers to discussion questions

D13.1 The ground electronic configuration of dioxygen, $1\sigma_g^2 1\sigma_u^2 2\sigma_g^2 1\pi_u^4 1\pi_g^2$, is discussed in Section 10.4(d) and the determination of the term symbol, $^3\Sigma_g^-$, is described in Section 13.2(a). The term symbol is Σ to represent a total orbital angular momentum about the internuclear axis of zero. This happens because for every π orbital electron with $\lambda = +1$ there is a π orbital electron with $\lambda = -1$. For example, the two $1\pi_g$ electrons are, according to Hund's rules, in separate degenerate orbitals for which one orbital has $\lambda = +1$ and the other has $\lambda = -1$. Except for the two $1\pi_g$ electrons, electrons have paired α and β spins, which results in zero contribution to the total spin angular momentum. According to Hund's rules, the two $1\pi_g$ electrons (i.e. $1\pi_g^1 1\pi_g^1$) have parallel spins in the ground state. They provide a total spin angular momentum of $S = \frac{1}{2} + \frac{1}{2} = 1$ and a spin multiplicity of $2S + 1 = 3$, which appears as the left superscript 3. The term symbol indicates a gerade total symmetry because electrons are paired in the ungerade molecular orbitals and $u \times u = g$ and the resultant symmetry of electrons in different molecular orbitals must therefore be given by $g \times g = g$. The π orbitals change sign on reflection in the plane that contains the internuclear axis. Consequently, the term symbol has the superscript – to indicate that the molecular wavefunction for O_2 changes sign on reflection in the plane containing the nuclei. To see this, think of O_2 in its ground state as having one electron in $1\pi_{g,x}$, which changes sign under reflection in the yz plane, and the other electron in $1\pi_{g,y}$, which does not change sign under reflection in the same plane, then the overall reflection symmetry is (closed shell) $\times (+) \times (-) = (-)$ and the full term symbol of the ground electronic state of O_2 is $^3\Sigma_g^-$.

D13.3 A band head is the convergence of the frequencies of electronic transitions with increasing rotational quantum number, J. They result from the rotational structure superimposed on the vibrational structure of the electronic energy levels of the diatomic molecule. See Figure 13.11 in the text. To understand how a band head arises, one must examine the equations describing the transition frequencies (eqns 13.11). These equations reveal that convergence can only arise when terms in both $(B' - B)$ and $(B' + B)$ occur in the equation. Since only a term in $(B' - B)$ occurs for the Q branch, no band head can arise for that branch.

D13.5 (a) The transition intensity is proportional to the square of the transition dipole moment. We initially suspect that the transition dipole moment should increase as the length L of the alternating carbon-to-carbon double/single/double/single bond sequence of the polyene is increased and, consequently, the transition intensity should also increase as L increases. To test this hypothesis, consider that the polyene has N π electrons that fill the first $n = N/2$ quantum states of the particle in a one-dimensional box of length $L = Nd = 2nd$, where d is the average carbon-to-carbon bond length.

(The length choice of $L = Nd$ adds half a bond length at each end to the distance between the two end carbon nuclei.) The transition dipole moment of this model is

$$\mu_x = \int_0^L \psi_{n_f} x \psi_{n_i} \mathrm{d}x.$$

We quickly find a selection rule for transitions with the substitution $x = f(x) + L/2$, where $f(x) = x - L/2$ is a function of ungerade symmetry with respect to inversion through the centre of symmetry at $x = L/2$ and we note that the wavefunctions have alternating gerade and ungerade symmetry as n increases. Then,

$$\mu_x = \int_0^L \psi_{n_f}\{f(x) + L/2\}\psi_{n_i}\mathrm{d}x = \int_0^L \psi_{n_f} f(x)\psi_{n_i}\mathrm{d}x + (L/2)\int_0^L \psi_{n_f}\psi_{n_i}\mathrm{d}x.$$

The last integral vanishes because the wavefunctions of different energy levels are orthogonal, which leaves

$$\mu_x = \int_0^L \psi_{n_f} f(x)\psi_{n_i}\mathrm{d}x.$$

Since the integrand factor $f(x)$ has ungerade symmetry, the product $\psi_{n_f}\psi_{n_i}$ of an allowed transition must also have ungerade symmetry so that the total symmetry of the integrand has $\mathrm{u} \times \mathrm{u} = \mathrm{g}$ symmetry and the integral can be non-zero. Thus, the lowest energy transition is $n + 1 \leftarrow n$, where $n = n_i$. It is now convenient to return to the original transition dipole moment integral and substitute the wavefunctions for the final and initial states:

$$\mu_x = \frac{2}{L}\int_0^L \sin\left(\frac{(n+1)\pi x}{L}\right) x \sin\left(\frac{n\pi x}{L}\right)\mathrm{d}x \text{ [8.4b]}.$$

We need not evaluate the integral exactly because the development of a method to improve the intensity of a dye requires only a knowledge of the approximate relationship between μ_x and either n or L (proportional properties for a polyene) so we recognize that $n + 1 \sim n$ for a large polyene and we make the estimate

$$\mu_x \propto \frac{1}{L}\int_0^L x\sin^2\left(\frac{n\pi x}{L}\right)\mathrm{d}x \quad \text{(The integral is found in standard mathematical tables.)}$$

$$\propto L$$

The model confirms the hypothesis that the transition dipole moment and, consequently, the transition intensity is increased by increasing the length of the polyene.

(b) Since $E_n = \dfrac{n^2h^2}{8m_eL^2}$ [8.4a], $\Delta E = \dfrac{(2n+1)h^2}{8m_eL^2}[\Delta n = +1] \sim \dfrac{nh^2}{4m_eL^2}$ for large n.

But, $L = 2nd$ is the length of the chain where d is the average carbon–carbon interatomic distance. Hence,

$$\Delta E \propto \frac{1}{L}.$$

Therefore, the $\boxed{\text{transition moves toward the red as } L \text{ is increased}}$. When white light is used to illuminate the dyed object, the colour absorbed is the **complementary colour** to the reflected, observed colour. Newton's colour wheel, shown in Figure 13.1, usefully displays complementary and observed colours. For example, draw a line from complementary violet through the circle centre to find the observed colour of green-yellow. As the polyene length is increased, the complementary colour progresses from violet to indigo, to blue, to green, etc., while the observed colour progresses from green-yellow, to yellow-orange, to orange-red, to red-violet, to violet-indigo and we say that the $\boxed{\text{apparent colour of the dye shifts towards blue}}$.

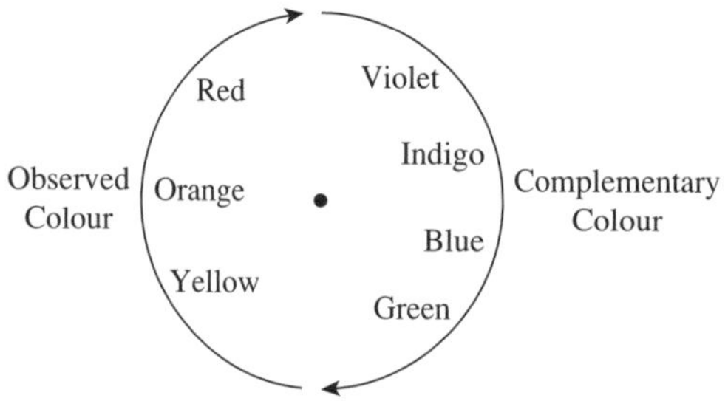

Figure 13.1

D13.7 When the steeply repulsive section of the O_2 potential energy curve for the excited state lies slightly toward the short side of the equilibrium bond length of the ground state and the minimum of the excited state lies to the longer side, a great many excited vibrational states overlap with the lowest-energy vibration of the ground state. The **Franck–Condon factor** is appreciable for many vertical transitions and the absorption band is broad. Furthermore, predissociation to the unbound $^5\Pi_u$ state shortens the lifetime of excited vibrational states. This causes the high-resolution lines of the corresponding vibrational-rotational transitions to be broad through the Heisenberg uncertainty principle $\Delta E \Delta t \geq \hbar/2$.

Solutions to exercises

E13.1(a) The reduction in intensity obeys the Beer–Lambert law introduced in Section 13.1.

$$\log\frac{I}{I_0} = -\log\frac{I_0}{I} = -\varepsilon[\text{J}]L \text{ [13.3 and 13.4]}$$
$$= (-855 \text{ dm}^3 \text{ mol}^{-1} \text{ cm}^{-1}) \times (3.25 \times 10^{-3} \text{ mol dm}^{-3}) \times (0.250 \text{ cm})$$
$$= -0.69\overline{5}$$

Hence, $\frac{I}{I_0} = 10^{-0.69\overline{5}} = 0.20$, and the reduction in intensity is $\boxed{80\%}$.

E13.2(a) $$\log\frac{I}{I_0} = -\log\frac{I_0}{I} = -\varepsilon[\text{J}]L \text{ [13.3, 13.4]}$$

Hence, $\varepsilon = -\frac{1}{[\text{J}]L}\log\frac{I}{I_0} = -\frac{\log(0.201)}{(1.11 \times 10^{-4} \text{ mol dm}^{-3}) \times (1.00 \text{ cm})} = \boxed{6.28 \times 10^3 \text{ dm}^3 \text{ mol}^{-1} \text{ cm}^{-1}}$.

E13.3(a) $\log T = -A = -\varepsilon[\mathrm{J}]L$ [13.2, 13.3, 13.4]

$$[\mathrm{J}] = -\frac{1}{\varepsilon L}\log T = \frac{-\log(1-0.465)}{(286\ \mathrm{dm^3\,mol^{-1}\,cm^{-1}})\times(0.65\ \mathrm{cm})} = \boxed{1.5\ \mathrm{mmol\,dm^{-3}}}$$

E13.4(a) $$\mathcal{A} = \int_{\text{band}} \varepsilon(\tilde{\nu})\,\mathrm{d}\tilde{\nu}\quad [13.5] = \int_{\tilde{\nu}_\mathrm{i}}^{\tilde{\nu}_\mathrm{f}} \varepsilon(\tilde{\nu})\,\mathrm{d}\tilde{\nu}$$

Since $\tilde{\nu} = \lambda^{-1}$ and $\tilde{\nu}/\mathrm{cm^{-1}} = 10^7/(\lambda/\mathrm{nm})$,

$\tilde{\nu}_\mathrm{i}/\mathrm{cm^{-1}} = 10^7/(290) = 3.45\times10^4$,

$\tilde{\nu}_\mathrm{peak}/\mathrm{cm^{-1}} = 10^7/(260) = 3.85\times10^4$, and

$\tilde{\nu}_\mathrm{f}/\mathrm{cm^{-1}} = 10^7/(230) = 4.35\times10^4$.

The positions of the wavenumber end points and peak (max) of the band are schematically presented in Figure 13.2. It is apparent that, because of the relative position of the peak, the molar absorption coefficient is not symmetrically distributed around the peak wavenumber. The distribution is skewed toward higher wavenumbers. However, a reasonable estimate of the area under the curve may be approximated by adding the areas of triangle 1 and triangle 2 shown as dashed lines in the figure.

$$\begin{aligned}\mathcal{A} &= \int_{\tilde{\nu}_\mathrm{i}}^{\tilde{\nu}_\mathrm{f}} \varepsilon(\tilde{\nu})\,\mathrm{d}\tilde{\nu} = \text{area 1} + \text{area 2}\\ &= \tfrac{1}{2}(\tilde{\nu}_\mathrm{peak}-\tilde{\nu}_\mathrm{i})\varepsilon_\mathrm{peak} + \tfrac{1}{2}(\tilde{\nu}_\mathrm{f}-\tilde{\nu}_\mathrm{peak})\varepsilon_\mathrm{peak} = \tfrac{1}{2}(\tilde{\nu}_\mathrm{f}-\tilde{\nu}_\mathrm{i})\varepsilon_\mathrm{peak}\\ &= \tfrac{1}{2}\times(0.90\times10^4\ \mathrm{cm^{-1}})\times(1.21\times10^4\ \mathrm{dm^3\,mol^{-1}\,cm^{-1}})\\ &= \boxed{5.4\times10^7\ \mathrm{dm^3\,mol^{-1}\,cm^{-2}}}\end{aligned}$$

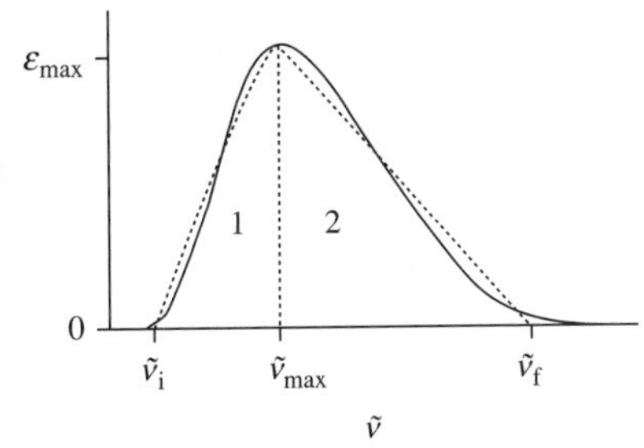

Figure 13.2

E13.5(a) $$\varepsilon = -\frac{1}{[\mathrm{J}]L}\log\frac{I}{I_0}\ [13.3, 13.4]\quad \text{with } L = 0.20\ \mathrm{cm}$$

We use this formula to draw up the following table:

$[\mathrm{Br_2}]/\mathrm{mol\,dm^{-3}}$	0.0010	0.0050	0.0100	0.0500	
I/I_0	0.814	0.356	0.127	3.0×10^{-5}	
$\varepsilon/(\mathrm{dm^3\,mol^{-1}\,cm^{-1}})$	447	449	448	452	mean: $44\overline{9}$

Hence, the molar absorption coefficient is $\varepsilon = \boxed{4.5\times10^2\ \mathrm{dm^3\,mol^{-1}\,cm^{-2}}}$.

E13.6(a) $$\varepsilon = -\frac{1}{[J]L}\log\frac{I}{I_0}\,[13.3,\,13.4] = \frac{-1}{(0.010\ \text{mol dm}^{-3})\times(0.20\ \text{cm})}\log(0.48)$$

$$= \boxed{15\bar{9}\ \text{dm}^3\ \text{mol}^{-1}\ \text{cm}^{-1}}$$

The transmittance in a 4.0 mm cell is

$$T = \frac{I}{I_0} = 10^{-[J]\varepsilon L}\,[13.1, 13.2]$$

$$= 10^{(-0.010\ \text{mol dm}^{-3})\times(15\bar{9}\ \text{dm}^3\ \text{mol}^{-1}\text{cm}^{-1})\times(0.40\ \text{cm})}$$

$$= 10^{-0.63\bar{6}} = 0.23,\ \text{or}\ \boxed{23\%}.$$

E13.7(a) Based on data found in K.S. Shifrin, *Physical Optics of Ocean Water* (American Institute of Physics, 1988), the mean molar absorption coefficient of sea water in the visible is $2.2\times10^{-5}\ \text{cm}^{-1}\ \text{dm}^3\ \text{mol}^{-1}$. We will base our calculations on this value rather than the $6.2\times10^{-3}\ \text{cm}^{-1}\ \text{dm}^3\ \text{mol}^{-1}$ value of very turbid water. We also note that the data shows an extremely low absorption coefficient in the blue, which agrees with the observation the ocean water appears blue in reflected light. The Beer–Lambert law [eqns 13.3 and 13.4] is

$$\log\frac{I}{I_0} = -\varepsilon[J]L, \quad \text{so} \quad L = -\frac{1}{\varepsilon[J]}\log\frac{I}{I_0}$$

For water, $[H_2O] \approx \dfrac{1.00\ \text{kg/dm}^3}{18.02\ \text{g mol}^{-1}} = 55.5\ \text{mol dm}^{-3}$

and $\varepsilon[J] = (2.2\times10^{-5}\ \text{cm}^{-1}\ \text{dm}^3\ \text{mol}^{-1})\times(55.5\ \text{mol dm}^{-3}) = 1.2\times10^{-3}\ \text{cm}^{-1}$, so $\dfrac{1}{\varepsilon[J]} = 8.3\ \text{m}$

Hence, $L/\text{m} = -8.3\times\log\dfrac{I}{I_0}$

(a) $\dfrac{I}{I_0} = 0.50,\ L = -8.3\ \text{m}\times\log(0.50) = \boxed{2.5\ \text{m}}$

(b) $\dfrac{I}{I_0} = 0.1,\ L = -8.3\ \text{m}\times\log(0.10) = \boxed{8.3\ \text{m}}$

E13.8(a) We will make the assumption that the absorption curve can be approximated by an isosceles triangle, as shown in Figure 13.3. The base of the absorption band triangle is twice the half-width at half-height, $\Delta\tilde{\nu}_{1/2}$, because the small triangle of base $\Delta\tilde{\nu}_{1/2}$ is similar to the absorption band triangle but has half the height. Thus, the area of the absorption band, which equals the integrated absorption coefficient [13.5], is given by

$$\mathcal{A} = \tfrac{1}{2}\times\varepsilon_{\text{max}}\times2\Delta\tilde{\nu}_{1/2}\ [\text{area} = \tfrac{1}{2}\times\text{height}\times\text{base}]$$
$$= \varepsilon_{\text{max}}\Delta\tilde{\nu}_{1/2} = (5000\ \text{cm}^{-1})\times\varepsilon_{\text{max}}$$

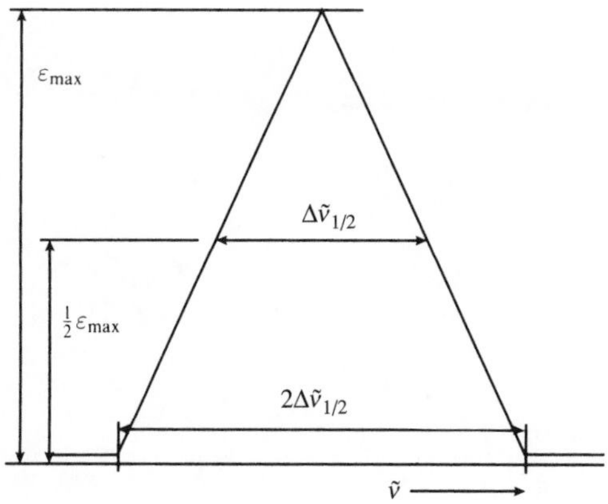

Figure 13.3

(a) $\mathcal{A} = (5000\ \text{cm}^{-1}) \times (1 \times 10^4\ \text{dm}^3\ \text{mol}^{-1}\ \text{cm}^{-1}) = \boxed{5 \times 10^7\ \text{dm}^3\ \text{mol}^{-1}\ \text{cm}^{-2}}$

(b) $\mathcal{A} = (5000\ \text{cm}^{-1}) \times (5 \times 10^2\ \text{dm}^3\ \text{mol}^{-1}\ \text{cm}^{-1}) = \boxed{2.5 \times 10^6\ \text{dm}^3\ \text{mol}^{-1}\ \text{cm}^{-2}}$

E13.9(a) The left superscript of the dihydrogen excited state $^3\Pi_u$ is the value of $2S + 1 = 3$, so $S = 1$, which means that the two electrons of H_2 are parallel ($S = s_1 + s_2 = \frac{1}{2} + \frac{1}{2} = 1$) in this excited state. The symbol Π indicates that the total orbital angular momentum around the molecular axis is $|\Lambda| = 1$. Since Λ is the sum of the individual electron orbital angular momentum quantum numbers around the molecular axis, we see that one of the unpaired electrons must be in a σ orbital ($\lambda = 0$) and the other electron must be in a π orbital ($\lambda = 1$). The excited state has ungerade overall parity. Since a σ bonding orbital has gerade symmetry, while a π bonding orbital has ungerade symmetry and $g \times u = u$, we deduce that one possible electron configuration is $\boxed{1\sigma_g^1 1\pi_u^1}$ (see Figures 10.23 and 10.32).

E13.10(a) The $1\sigma_g^2 1\sigma_u^2 1\pi_u^3 1\pi_g^1$ valence configuration has two unpaired electrons so $S = s_1 + s_2 = \frac{1}{2} + \frac{1}{2} = 1$ and the spin multiplicity is given by $2S + 1 = 2(1) + 1 = \boxed{3}$. Because $u \times u = g$ and $g \times g = g$, the net parity of two electrons paired in an orbital is always gerade. Consequently, the overall parity is found by multiplying the parity of unpaired electrons. For this configuration, $u \times g = \boxed{u}$.

E13.11(a) The electronic spectrum selection rules concerned with changes in angular momentum are (Section 13.2(b)): $\Delta\Lambda = 0, \pm 1 \quad \Delta S = 0 \quad \Delta\Sigma = 0 \quad \Delta\Omega = 0, \pm 1$, where $\Omega = \Lambda + \Sigma$. Λ gives the total orbital angular momentum about the internuclear axis and Σ gives the total spin angular momentum about the internuclear axis. The $\pm$ superscript selection rule for reflection in the plane along the internuclear axis is $+\leftrightarrow+$ or $-\leftrightarrow-$ (i.e. $+\leftrightarrow-$ is forbidden). The **Laporte selection rule** states that for a centrosymmetric molecule (those with a centre of inversion) the only allowed transitions are transitions that are accompanied by a change of parity: $u\leftrightarrow g$.

(a) The changes in the transition $^2\Pi \leftrightarrow {}^2\Pi$ are $\Delta\Lambda = 0$, $\Delta S = 0$, $\Delta\Sigma = 0$, and $\Delta\Omega = 0$ so the transition is $\boxed{\text{allowed}}$.

(b) The changes in the transition $^1\Sigma \leftrightarrow {}^1\Sigma$ are $\Delta\Lambda = 0$, $\Delta S = 0$, $\Delta\Sigma = 0$, and $\Delta\Omega = 0$ so the transition is $\boxed{\text{allowed}}$.

(c) The changes in the transition $\Sigma \leftrightarrow \Delta$ are $\Delta\Lambda = 2$ so the transition is $\boxed{\text{forbidden}}$.

(d) The transition $\Sigma^+ \leftrightarrow \Sigma^-$ is $\boxed{\text{forbidden}}$ because $+ \not\leftrightarrow -$.

(e) The transition $\Sigma^+ \leftrightarrow \Sigma^+$ is $\boxed{\text{allowed}}$ because $\Delta\Lambda = 0$ and $+ \leftrightarrow +$.

E13.12(a) We begin by evaluating the normalization constants N_0 and $N_{v'}$.

$$N_0^2 = \frac{1}{\int_{-\infty}^{\infty} e^{-2ax^2}\,dx} = \left(\frac{2a}{\pi}\right)^{1/2} \text{(standard integral)}; \; N_0 = \left(\frac{2a}{\pi}\right)^{1/4}.$$

Likewise, $N_{0'}^2 = \dfrac{1}{\int_{-\infty}^{\infty} e^{-2b(x-x_0)^2}\,dx} = \left(\dfrac{2b}{\pi}\right)^{1/2}; \; N_{0'} = \left(\dfrac{2b}{\pi}\right)^{1/4}.$

Furthermore, we can easily check that

$$ax^2 + b(x-x_0)^2 = z^2 + \frac{ab}{a+b}x_0^2, \text{ where } z = (a+b)^{1/2}x - \frac{b}{(a+b)^{1/2}}x_0 \text{ and } dx = \frac{1}{(a+b)^{1/2}}dz.$$

Then, the vibration overlap integral between the vibrational wavefunction in the upper (0′) and lower electronic states (0) is:

$$S(0',0) = \langle 0'|0\rangle = N_0 N_{0'}\int_{-\infty}^{\infty} e^{-ax^2}e^{-b(x-x_0)^2}\,dx = N_0 N_{0'}\int_{-\infty}^{\infty} e^{-\{ax^2+b(x-x_0)^2\}}\,dx$$

$$= \frac{N_0 N_{0'}}{(a+b)^{1/2}}\int_{-\infty}^{\infty} e^{-\left\{z^2+\frac{ab}{a+b}x_0^2\right\}}\,dz = \frac{N_0 N_{0'}}{(a+b)^{1/2}} e^{-\frac{ab}{a+b}x_0^2}\int_{-\infty}^{\infty} e^{-z^2}\,dz = N_0 N_{0'}\left(\frac{\pi}{a+b}\right)^{1/2} e^{-\frac{ab}{a+b}x_0^2}$$

$$= \left(\frac{2a}{\pi}\right)^{1/4}\left(\frac{2b}{\pi}\right)^{1/4}\left(\frac{\pi}{a+b}\right)^{1/2} e^{-\frac{ab}{a+b}x_0^2} = (4ab)^{1/4}\left(\frac{1}{a+b}\right)^{1/2} e^{-\frac{ab}{a+b}x_0^2}$$

For the case $b = a/2$, this simplifies to

$$S(0',0) = \frac{2}{(3\sqrt{2})^{1/2}} e^{-ax_0^2/3}.$$

The Franck–Condon factor is

$$|S(0',0)|^2 = \boxed{\frac{2\sqrt{2}}{3} e^{-2ax_0^2/3}}.$$

E13.13(a) The rotational constant of the excited state is $\tilde{B}' = 0.3101\ \text{cm}^{-1}$, while the rotational constant for the ground state is $\tilde{B} = 0.3540\ \text{cm}^{-1}$. Since $\tilde{B}' < \tilde{B}$, the bond length of the excited state is longer than the equilibrium bond length and the $\boxed{\text{R branch has a band head}}$ (text Figure 13.11a).

R branch $(\Delta J = +1)$: $\Delta\tilde{\nu}_R(J) = \tilde{\nu}_R(J) - \tilde{\nu} = (\tilde{B}' + \tilde{B})(J+1) + (\tilde{B}' - \tilde{B})(J+1)^2$ [13.11c]

To find the J value at which the head occurs, we need only compute a table of $\Delta\tilde{\nu}_R(J)$ values from low to high J. The point at which $\Delta\tilde{\nu}_R(J)$ is greatest is the branch head.

J	0	1	2	3	4	5	6	7	8	9
$\Delta\tilde{\nu}_R(J)/\mathrm{cm}^{-1}$	0.6202	1.1526	1.5972	1.9540	2.2230	2.4042	2.4976	2.5032	2.4210	2.2510

As indicated by the table, $\Delta\tilde{\nu}_R(J)$ is a maximum (i.e. the R branch head) when $\boxed{J = 7}$.

An alternative method involves deriving an equation that must be satisfied when $J = J_{\text{head}}$. We start by recognizing that eqn 13.11c indicates that $\Delta\tilde{\nu}_R(J)$ is a quadratic function of J and, in fact, the quadratic shape of the $\tilde{\nu}_R$ against J curve is called the Fortrat parabola. This means that we can derive an equation for J_{head} by finding the maximum of the Fortrat parabola: $\mathrm{d}\Delta\tilde{\nu}_R/\mathrm{d}J = 0$ when $J = J_{\text{head}}$.

$$\frac{\mathrm{d}\Delta\tilde{\nu}_R}{\mathrm{d}J} = \frac{\mathrm{d}}{\mathrm{d}J}\{(\tilde{B}' + \tilde{B})(J+1) - |(\tilde{B}' - \tilde{B})|(J+1)^2\} \text{ (Note: } \tilde{B}' - \tilde{B} = -|\tilde{B}' - \tilde{B}| \text{ because } \tilde{B}' < \tilde{B}.)$$

$$= +(\tilde{B}' + \tilde{B}) - 2|(\tilde{B}' - \tilde{B})|(J+1)$$

$$(\tilde{B}' + \tilde{B}) - 2|(\tilde{B}' - \tilde{B})|(J_{\text{head}} + 1) = 0$$

$$J_{\text{head}} = \frac{(\tilde{B}' + \tilde{B})}{2|(\tilde{B}' - \tilde{B})|} - 1 = \frac{(0.3101 + 0.3540)}{2|0.3101 - 0.3540|} - 1 = 6.56 = 7$$

E13.14(a) The R branch has a band head so we immediately conclude both that the rotational constant of the exited state, $\tilde{B}'$, is less than the rotational constant of the ground state, $\tilde{B}$, and that $\boxed{\text{the bond length of the excited state is longer than the equilibrium bond length}}$. Exercise E13.13a demonstrates that the J value of the band head, J_{head}, is the closest integer to $\frac{1}{2}(\tilde{B}' + \tilde{B})/|(\tilde{B}' - \tilde{B})| - 1$. Thus, if we are only given that $J_{\text{head}} = 1$ and $\tilde{B} = 60.80\ \mathrm{cm}^{-1}$, we know only that

$$0.5 < \tfrac{1}{2}(\tilde{B}' + \tilde{B})/(\tilde{B} - \tilde{B}') - 1 < 1.5$$

because the fractional value of a $\frac{1}{2}(\tilde{B}' + \tilde{B})/(\tilde{B}' - \tilde{B}) - 1$ calculation must be rounded-off to give the integer value J_{head}. Algebraic manipulation of the inequality yields

$$\frac{\{1 + 2(0.5)\}\tilde{B}}{\{3 + 2(0.5)\}} < \tilde{B}' < \frac{\{1 + 2(1.5)\}\tilde{B}}{\{3 + 2(1.5)\}}$$

$$\frac{\tilde{B}}{2} < \tilde{B}' < \frac{2\tilde{B}}{3}$$

$$\boxed{30.4\ \mathrm{cm}^{-1} < \tilde{B}' < 40.5\ \mathrm{cm}^{-1}}$$

Here is an alternative solution that gives the same answer with insight into the band head concept. At the head of an R band, $\tilde{\nu}_{J_{\text{head}}} > \tilde{\nu}_{J_{\text{head}}-1}$, where $\tilde{\nu}_{J_{\text{head}}-1}$ is the transition $J_{\text{head}} \leftarrow J = J_{\text{head}} - 1$. Substitution of eqn 13.11c into this inequality yields the relationship $\tilde{B}' > J_{\text{head}}\tilde{B}/(J_{\text{head}} + 1)$. Similarly, $\tilde{\nu}_{J_{\text{head}}} > \tilde{\nu}_{J_{\text{head}}+1}$ where $\tilde{\nu}_{J_{\text{head}}+1}$ is the transition $J_{\text{head}} + 2 \leftarrow J = J_{\text{head}} + 1$. Substitution of eqn 13.11c into this inequality yields the relationship $\tilde{B}' < (J_{\text{head}} + 1)\tilde{B}/(J_{\text{head}} + 2)$. Consequently,

$$J_{\text{head}}\tilde{B}/(J_{\text{head}} + 1) < \tilde{B}' < (J_{\text{head}} + 1)\tilde{B}/(J_{\text{head}} + 2).$$

E13.15(a) π electrons in polyenes may be considered as particles in a one-dimensional box. Applying the Pauli exclusion principle, the N conjugated electrons will fill the levels, two electrons at a time, up to the level $n = N/2$. Since N is also the number of alkene carbon atoms. Nd is the length of the box, which includes an extra half bond length on each end of the polyene, with d the carbon–carbon interatomic distance. Hence,

$$E_n = \frac{n^2h^2}{8mN^2d^2} \text{ [8.4a]}$$

For the lowest-energy transition ($\Delta n = +1$):

$$\Delta E = h\nu = \frac{hc}{\lambda} = E_{N/2+1} - E_{N/2} = \frac{(N+1)h^2}{8mN^2d^2}$$

Therefore, the larger N, the larger λ. Hence, the absorption at 243 nm is due to the diene and that at 192 nm to the butene.

Question. How accurate is the formula derived above in predicting the wavelengths of the absorption maxima in these two compounds?

An alternative analysis uses simple Hückel theory of π molecular orbitals (Section 10.6). The substituted butene is modelled with the ethene π energies of text Figure 10.42, which shows a $\pi_{LUMO} \leftarrow \pi_{HOMO}$ transition energy of -2β (β is negative). The substituted 2,4-hexadiene is modelled with the butadiene π energies of text Figure 10.43, which shows a $\pi_{LUMO} \leftarrow \pi_{HOMO}$ transition energy of -1.24β. Thus, the transition energy of the conjugated diene is lower and has a longer wavelength. This is an example of the general principle (see Table 10.5) that the difference between neighbouring energy levels becomes smaller as N becomes larger.

E13.16(a) The weak absorption at 30,000 cm^{-1} (330 nm) is typical of a carbonyl chromophore of an enol. The assignment is $\pi^* \leftarrow n$, where a non-bonding electron comes from one of the two lone pairs of the oxygen valence. The two lone pairs of oxygen are in sp^2 hybrid orbitals, which define the xy plane that contains the σ bond of the carbonyl. The π^* molecular orbital is perpendicular to this plane. There is little overlap between the n and π^* orbitals, thereby producing a low value for the dipole transition integral and a low molar absorption coefficient.

The strong absorption at 46 950 cm^{-1} (213 nm) has the $\pi^* \leftarrow \pi$ assignment. The conjugation of the π bonds of the ethenic chromophore and the carbonyl chromophore causes this transition to be shifted to lower energies w/r/t both the $\pi^* \leftarrow \pi$ transition of ethene (165 nm) and the $\pi^* \leftarrow \pi$ transition of propanone (190 nm). This shift can be understood in terms of the simple Hückel theory of π molecular orbitals using the butadiene π energy model shown in text Figure 10.43 and Figure 13.4 below. Figure 13.4 demonstrates a broad principle: the difference between neighbouring energy levels becomes smaller as the number of adjacent, overlapping orbitals becomes larger.

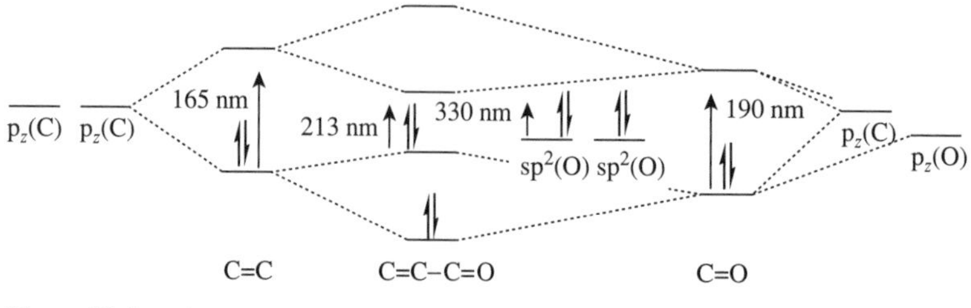

Figure 13.4

E13.17(a) The transition wavenumber is $\tilde{\nu} = \dfrac{1}{\lambda} = \dfrac{1}{700\ \text{nm}} = 14 \times 10^3\ \text{cm}^{-1}$.

Water molecules are weak ligand field splitters, so we expect the d^5 electrons of Fe^{3+} to have the $t_{2g}^3e_g^2$ high-spin ground-state configuration in the octahedral $[Fe(H_2O)_6]^{3+}$ complex. The d orbital electron spins are expected to be parallel with $S = 5/2$ and $2S + 1 = 6$ by Hund's maximum multiplicity rule. We also expect that $P > \Delta_O$, where P is the energy of repulsion for pairing two electrons in an orbital. A d–d transition to the $t_{2g}^4e_g^1$ octahedral excited state is expected to be both spin and parity forbidden and, therefore, have a very small molar absorption coefficient. This transition releases the energy Δ_O and requires the energy P needed to pair two electrons within a t_{2g} orbital. Thus, $\tilde{\nu} = P - \Delta_O$ and $\boxed{\Delta_O = P - \tilde{\nu}}$. Using the typical value $P \sim 28 \times 10^3\ \text{cm}^{-1}$ yields the estimate $\Delta_O \sim \boxed{14 \times 10^3\ \text{cm}^{-1}}$. See F.A. Cotton and G. Wilkinson, *Advanced Inorganic Chemistry*, 4th edn, p. 646, Wiley-Interscience Publishers, New York (1980) for electron-pairing energies.

E13.18(a) (a) Vibrational energy spacings of the $\boxed{\text{lower}}$ state are determined by the spacing of the peaks of fluorescence spectrum A of benzophenone $\boxed{\tilde{\nu} \approx 1800\ \text{cm}^{-1}}$.

(b) The peaks of A give no information about the spacing of the upper vibrational levels (without a detailed analysis of the intensities of the lines).

E13.19(a) Only an integral number of half-wavelengths fit into the cavity. These are the **resonant modes**.

$\lambda = 2L/n$ [13.18] where n is an integer and L is the length of the cavity.

The resonant frequencies are given by $\nu = c/\lambda = nc/2L$. The lowest-energy resonant modes ($n = 1$) in a 30 cm cavity are $\boxed{\lambda = 60\ \text{cm}\ (\nu = 500\ \text{MHz})}$.

E13.20(a) Referring to Example 13.2, we have

$P_{\text{peak}} = E_{\text{pulse}}/t_{\text{pulse}}$ and $P_{\text{average}} = E_{\text{total}}/t = E_{\text{pulse}} \times \nu_{\text{repetition}}$,
where $\nu_{\text{repetition}}$ is the pulse repetition rate.

$$t_{\text{pulse}} = E_{\text{pulse}}/P_{\text{peak}} = \frac{0.10\ \text{mJ}}{5.0\ \text{MW}} = \boxed{20\ \text{ps}}$$

$$\nu_{\text{repetition}} = P_{\text{average}}/E_{\text{pulse}} = \frac{7.0\ \text{kW}}{0.10\ \text{mJ}} = \boxed{70\ \text{MHz}}$$

E13.21(a) This Mathcad worksheet simulates the output of a mode-locked laser. The radiation intensity is shown in *Justification 13.5* to be proportional to the function $f(t, N)$ of the worksheet. The plots demonstrate that the superposition of a great many modes creates very narrow spikes separated by $t = 2L/c$.

$L := 30\ \text{cm}$ $\quad c := 299792458\ \text{m s}^{-1}$ $\quad ns := 10^{-9}\ \text{s}$

$t_{max} := 5 \cdot 2 \cdot L \cdot c^{-1}$ $\quad t_{max} := 1.001 \times 10^{-8}\ \text{s}$

$i_{max} := 1000$ $\quad i := 1..i_{max}$ $\quad t_i := \dfrac{i \cdot t_{max}}{i_{max}}$

$$f(t, N) := \left(\frac{\sin(N \cdot \pi c \cdot t \cdot 2^{-1} \cdot L^{-1})}{\sin(\pi c \cdot t \cdot 2^{-1} \cdot L^{-1})} \right)^2$$

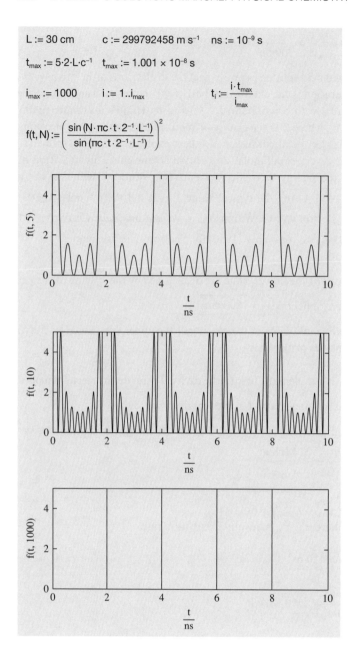

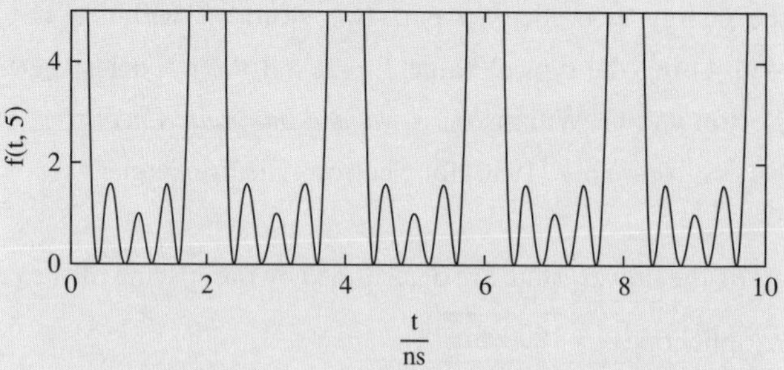

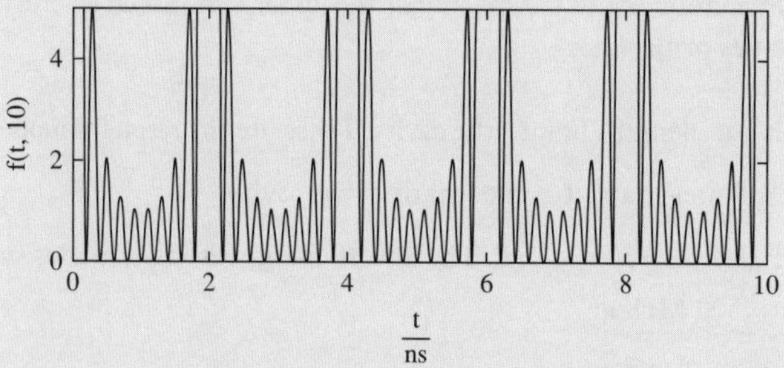

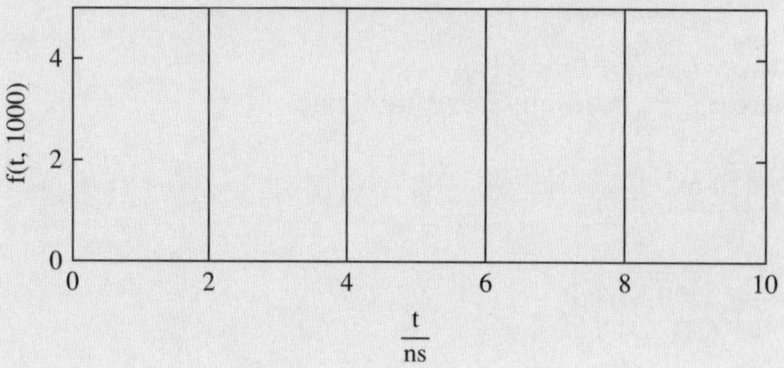

Solutions to problems

Solutions to numerical problems

P13.1 The potential energy curves for the $X^3\Sigma_g^-$ and $B^3\Sigma_u^-$ electronic states of O_2 are represented schematically in Figure 13.5 along with the notation used to represent the energy separation of this problem. Curves for the other electronic states of O_2 are not shown. Ignoring rotational structure and anharmonicity we may write

$$\tilde{\nu}_{00} \approx T_e + \tfrac{1}{2}(\tilde{\nu}' - \tilde{\nu}) = 6.175\ \text{eV} \times \left(\frac{8065.5\ \text{cm}^{-1}}{1\ \text{eV}}\right) + \tfrac{1}{2}(700 - 1580)\ \text{cm}^{-1} \approx \boxed{49\ 36\bar{4}\ \text{cm}^{-1}}$$

COMMENT. Note that the selection rule $\Delta v = \pm 1$ does not apply to vibrational transitions between different electronic states.

Question. What is the percentage change in $\tilde{\nu}_{00}$ if the anharmonicity constants $x_e\tilde{\nu}$, 12.0730 cm^{-1} and 8.002 cm^{-1} for the ground and excited states, respectively, are included in the analysis?

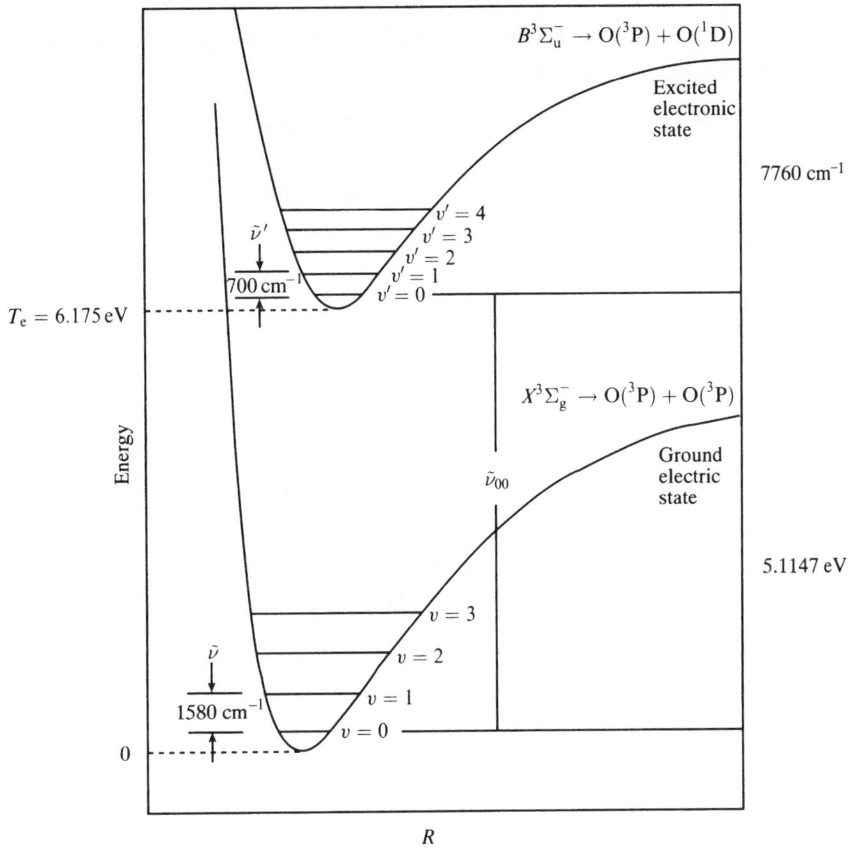

Figure 13.5

P13.3 Text Figure 13.27 provides the conceptual framework for the analysis. Initially, we cannot decide whether the dissociation products, atomic iodine and bromine, are produced in their ground atomic states or excited states. But we note that the two convergence limits are separated by an amount of energy exactly equal to the excitation energy of the bromine atom: 18 345–14 660 cm^{-1} = 3685 cm^{-1}. Consequently, dissociation at 14 660 cm^{-1} must yield a bromine atom in its ground state. Therefore, the possibilities for the dissociation energy are 14 660 cm^{-1} or 14 660–7598 cm^{-1} = 7062 cm^{-1}, depending on whether the iodine atom is produced in its ground or excited electronic state, respectively.

In order to decide which of these two possibilities is correct we can set up the following Born–Haber cycle.

(1) $IBr(g) \rightarrow \frac{1}{2} I_2(s) + \frac{1}{2} Br_2(l)$ $\quad \Delta H_1^\ominus = -\Delta_f H^\ominus(IBr, g)$

(2) $\frac{1}{2} I_2(s) \rightarrow \frac{1}{2} I_2(g)$ $\quad \Delta H_2^\ominus = \frac{1}{2}\Delta_{sub} H^\ominus(I_2, s)$

(3) $\frac{1}{2} Br_2(l) \rightarrow \frac{1}{2} Br_2(g)$ $\quad \Delta H_3^\ominus = \frac{1}{2}\Delta_{vap} H^\ominus(Br_2, l)$

(4) $\frac{1}{2} I_2(g) \rightarrow I(g)$ $\quad \Delta H_4^\ominus = \frac{1}{2}\Delta_{diss} H(I-I)$

(5) $\frac{1}{2} Br_2(g) \rightarrow Br(g)$ $\quad \Delta H_5^\ominus = \frac{1}{2}\Delta_{diss} H(Br-Br)$

$IBr(g) \rightarrow I(g) + Br(g)$ $\quad \Delta_{diss} H^\ominus(I-Br) = \Delta H_1^\ominus + \Delta H_2^\ominus + \Delta H_3^\ominus + \Delta H_4^\ominus + \Delta H_5^\ominus$

The value $\Delta_f H^\ominus(IBr, g) = +40.79\ \mathrm{kJ\ mol^{-1}}$ is provided in the problem. We use values found in the text appendix to calculate the values of $\Delta_{diss}H(I-I)$ and $\Delta_{diss}H(Br-Br)$.

$$\Delta_{diss}H(I-I) = 2 \times \Delta_f H^\ominus(I, g) - \Delta_f H^\ominus(I_2, g) = \{2 \times 106.84 - 62.44\}\ \mathrm{kJ\ mol^{-1}} = 151.24\ \mathrm{kJ\ mol^{-1}}$$
$$\Delta_{diss}H(Br-Br) = 2 \times \Delta_f H^\ominus(Br, g) - \Delta_f H^\ominus(Br_2, g) = \{2 \times 111.88 - 30.907\}\ \mathrm{kJ\ mol^{-1}} = 192.85\ \mathrm{kJ\ mol^{-1}}$$

Values for $\Delta_{sub}H^\ominus(I_2, s) = \Delta_f H^\ominus(I_2, g)$ and $\Delta_{vap}H^\ominus(Br_2, l) = \Delta_f H^\ominus(Br_2, g)$ are also provided by the appendix of the text. Thus,

$$\begin{aligned}\Delta_{diss}H^\ominus(I-Br) &= \Delta H_1^\ominus + \Delta H_2^\ominus + \Delta H_3^\ominus + \Delta H_4^\ominus + \Delta H_5^\ominus \\ &= -\Delta_f H^\ominus(IBr, g) - \tfrac{1}{2}\Delta_{sub}H^\ominus(I_2, s) + \tfrac{1}{2}\Delta_{vap}H^\ominus(Br_2, l) \\ &\quad + \tfrac{1}{2}\Delta_{diss}H(I-I) + \tfrac{1}{2}\Delta_{diss}H(Br-Br) \\ &= \{-40.79 + \tfrac{1}{2} \times 62.44 + \tfrac{1}{2} \times 30.907 + \tfrac{1}{2} \times 151.24 + \tfrac{1}{2} \times 192.85\}\ \mathrm{kJ\ mol^{-1}} \\ &= 177.93\ \mathrm{kJ\ mol^{-1}} \qquad \text{[Data table]}\end{aligned}$$

$$\Delta_{diss}H^\ominus(I-Br)/N_A hc = 14\,874\ \mathrm{cm^{-1}}$$

Comparison to the possibilities 14 660 cm^{-1} and 7062 cm^{-1} shows that it is the former that is the correct dissociation energy so we conclude that the photo dissociation yields the ground state of both the iodine and bromine atoms.

P13.5 The spectrum gives the peak and half-height points:

$$\varepsilon_{peak} = 250\ \mathrm{dm^3\ mol^{-1}\ cm^{-1}},\ \lambda_{peak} = 284\ \mathrm{nm}\ (\tilde{\nu} = 35\,200\ \mathrm{cm^{-1}})$$
$$\varepsilon_{1/2} = 125\ \mathrm{dm^3\ mol^{-1}\ cm^{-1}},\ \lambda_{1/2} = 305\ \mathrm{nm}\ (\tilde{\nu} = 32\,800\ \mathrm{cm^{-1}})\ \text{and}\ \lambda_{1/2} = 265\ \mathrm{nm}\ (\tilde{\nu} = 37\,700\ \mathrm{cm^{-1}})$$

We estimate that the wavenumber band has a normal Gaussian shape.

$$\varepsilon = \varepsilon_{max} e^{-(\tilde{\nu}-\tilde{\nu}_{peak})^2/a^2}$$

where a is a constant related to the half-width

$$\Delta\tilde{\nu}_{1/2} = (37\,700 - 32\,800)\ \mathrm{cm^{-1}} = 4900\ \mathrm{cm^{-1}}$$

$$\mathcal{A} = \int_{band} \varepsilon(\tilde{\nu})\, d\tilde{\nu}\ [13.5] = \varepsilon_{max} \int_{-\infty}^{\infty} e^{-(\tilde{\nu}-\tilde{\nu}_{peak})^2/a^2}\, d\tilde{\nu} = \varepsilon_{max} a\sqrt{\pi} \quad \text{(standard integral)}$$

The relationship between the half-width and a is found by evaluation of the lineshape at $\varepsilon(\tilde{\nu}_{1/2}) = \varepsilon_{max}/2$:

$$\varepsilon_{max}/2 = \varepsilon_{max} e^{-(\tilde{\nu}_{1/2}-\tilde{\nu}_{peak})^2/a^2}$$
$$\ln(1/2) = -(\tilde{\nu}_{1/2} - \tilde{\nu}_{peak})^2/a^2$$

$$a^2 = \frac{(\tilde{\nu}_{1/2} - \tilde{\nu}_{\text{peak}})^2}{\ln(2)} = \frac{(\Delta\tilde{\nu}_{1/2}/2)^2}{\ln(2)}$$

$$a = \frac{\Delta\tilde{\nu}_{1/2}}{2\sqrt{\ln 2}}$$

Thus,

$$\mathcal{A} = \boxed{\tfrac{1}{2}\Delta\tilde{\nu}_{1/2}\varepsilon_{\max}\sqrt{\pi/\ln(2)}} = 1.0645\,\Delta\tilde{\nu}_{1/2}\varepsilon_{\max}$$

$$\mathcal{A} = \tfrac{1}{2}(4900\text{ cm}^{-1}) \times (250\text{ dm}^3\text{ mol}^{-1}\text{ cm}^{-1})\sqrt{\pi/\ln(2)} = \boxed{1.3\overline{0} \times 10^6\text{ dm}^3\text{ mol}^{-1}\text{ cm}^{-2}}$$

Since the dipole moment components transform as $A_1(z)$, $B_1(x)$, and $B_2(y)$, excitations from A_1 to A_1, B_1, and B_2 terms are allowed.

P13.7 Essential concepts associated with the vibration transition structure of an electronic absorption band are shown in text Figure 13.26 and Figure 13.6 below. The $X^3\Sigma_g^-$ ground state of O_2 dissociates into two $O(^3P)$ atoms while the $B^3\Sigma_u^-$ excited state dissociates into a $O(^3P)$ atom and a $O(^1D)$ atom. The electron absorption band for the $B^3\Sigma_u^- \leftarrow X^3\Sigma_g^-$ transition is characterized by the vibrational states ($\upsilon = 0, 1, 2, \ldots, \upsilon_{\max}$) of the excited state because a great majority of the ground-state molecules are in the lowest vibrational state. Thus, the vibrational transitions are $0 \leftarrow 0$, $1 \leftarrow 0$, $2 \leftarrow 0$, $\ldots$, $\upsilon \leftarrow 0$. It is important to recognize that the vibrational transitions near the dissociation limit are difficult to accurately identify among noise of comparable magnitude. To begin the data analysis, define $\Delta\tilde{\nu}_\upsilon$ to be the difference between the wavenumber for the $\upsilon + 1 \leftarrow 0$ transition and the wavenumber for the $\upsilon \leftarrow 0$ transition:

$$\Delta\tilde{\nu}_\upsilon = \tilde{\nu}_{\upsilon+1} - \tilde{\nu}_\upsilon$$

Examination of Figure 13.6 will confirm that $D_0(B^3\Sigma_u^-)$ equals the sum of all $\Delta\tilde{\nu}_\upsilon$, many of which must be determined with a Birge–Sponer extrapolation, plus a small energy needed to dissociate an excited molecule in the $\upsilon_{\max}$ vibrational state. We take the latter energy to be given by the fractional quantum number obtained in the Birge–Sponer extrapolation.

$$D_0(B^3\Sigma_u^-) = \sum_{\upsilon=0} \Delta\tilde{\nu}_\upsilon$$

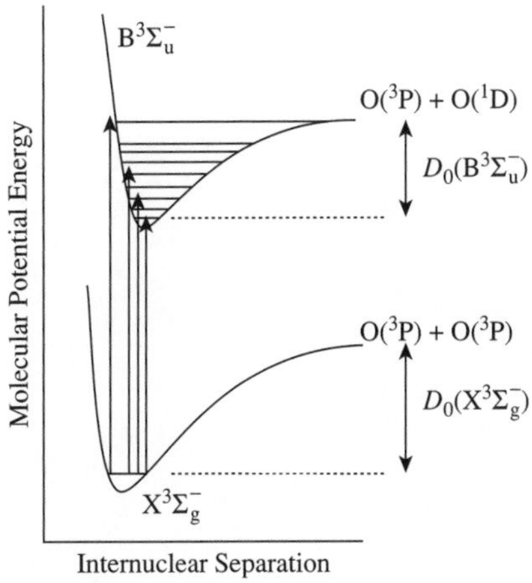

Figure 13.6

We draw a table of $\Delta\tilde{\nu}_\upsilon$ computations, prepare a plot of $\Delta\tilde{\nu}_\upsilon$ against υ, and calculate a quadratic regression fit to the curve, which provides the Birge–Sponer extrapolation to $\Delta\tilde{\nu}_\upsilon = 0$.

υ	0	1	2	3	4	5	6	7
$\tilde{\nu}_\upsilon$/cm^{-1}	50 062.6	50 725.4	51 369.0	51 988.6	52 579.0	53 143.4	53 679.6	54 177.0
$\Delta\tilde{\nu}_\upsilon$/cm^{-1}	662.8	643.6	619.6	590.4	564.4	536.2	497.4	464.8

υ	8	9	10	11	12	13	14
$\tilde{\nu}_\upsilon$/cm^{-1}	54 641.8	55 078.2	55 460.0	558 03.1	56 107.3	56 360.3	56 570.6
$\Delta\tilde{\nu}_\upsilon$/cm^{-1}	436.4	381.8	343.1	304.2	253.0	210.3	

The quadratic fit to the $\Delta\tilde{\nu}_\upsilon$ against υ plot is shown in Figure 13.7 and the fit $y(\upsilon)$ provides a Birge–Sponer extrapolation, which indicates that $\upsilon_{\text{max}} = 16$ and $\upsilon_{\text{dissociation limit}} = 16.8376$. Thus,

$$D_0(\text{B}^3\Sigma_\text{u}^-) = \sum_{\upsilon=0}^{15} \Delta\tilde{\nu}_\upsilon + 0.8376 \times \Delta\tilde{\nu}_{16}$$

where the values of $\Delta\tilde{\nu}_\upsilon$ are computed with the fit $y(\upsilon)$. Doing the sum, we find that

$$\begin{aligned} D_0(\text{B}^3\Sigma_\text{u}^-) &= 6767.908\ \text{cm}^{-1} + 0.8376 \times (48.0992\ \text{cm}^{-1}) \\ &= \boxed{6808.2\ \text{cm}^{-1} \text{ or } 0.84411\ \text{eV}} \end{aligned}$$

Examination of Figure 13.6 also indicates that

$$D_0(\text{X}^3\Sigma_\text{g}^-) + E_{\text{O}(^1\text{D})\leftarrow\text{O}(^3\text{P})} = \tilde{\nu}_0 + D_0(\text{B}^3\Sigma_\text{u}^-)$$

where $E_{\text{O}(^1\text{D})\leftarrow\text{O}(^3\text{P})}$ is 190 kJ mol^{-1} or 1.97 and $\tilde{\nu}_0$ (i.e. $\tilde{\nu}_{00}$) equals 50 062.6 cm^{-1} or 6.20697 eV.

$$\begin{aligned} D_0(\text{X}^3\Sigma_\text{g}^-) &= \tilde{\nu}_0 + D_0(\text{B}^3\Sigma_\text{u}^-) - E_{\text{O}(^1\text{D})\leftarrow\text{O}(^3\text{P})} \\ &= (6.20697 + 0.84411 - 1.97)\ \text{eV} = \boxed{5.08\ \text{eV}} \end{aligned}$$

This value of the ground-state dissociation energy is in complete agreement with the accepted value.

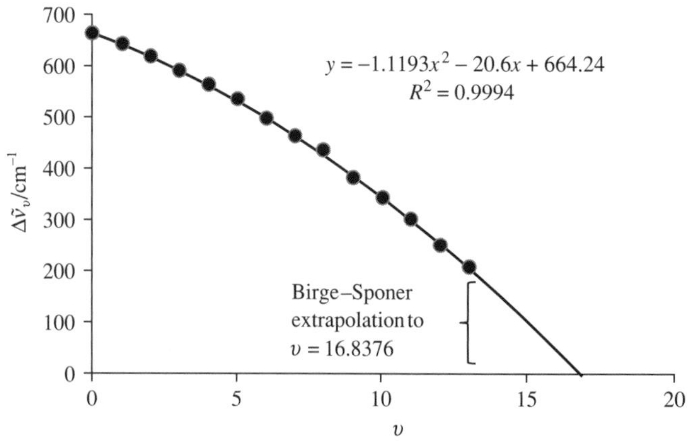

Figure 13.7

P13.9 Draw up a table like the following:

Hydrocarbon	$h\upsilon_{max}$/eV	E_{HOMO}/eV*
Benzene	4.184	−9.7506
Biphenyl	3.654	−8.9169
Naphthalene	3.452	−8.8352
Phenanthrene	3.288	−8.7397
Pyrene	2.989	−8.2489
Anthracene	2.890	−8.2477

*Semi-empirical, PM3 level, PC Spartan Pro™

The energy of charge transfer to I_2, $h\upsilon_{max}$, is plotted against E_{HOMO} in Figure 13.8. There is a clear correlation between the energy of charge transfer and the HOMO of hydrocarbon electron donor: as E_{HOMO} increases, the energy of charge transfer decreases. The correlation appears to be linear and the correlation coefficient ($R = 0.9855$) indicates that about 98.6% of the variation is explained by the linear correlation. This supports the hypothesis that for π-donor hydrocarbons $h\upsilon_{max} = E_{LUMO}(I_2) - E_{HOMO}(\pi)$.

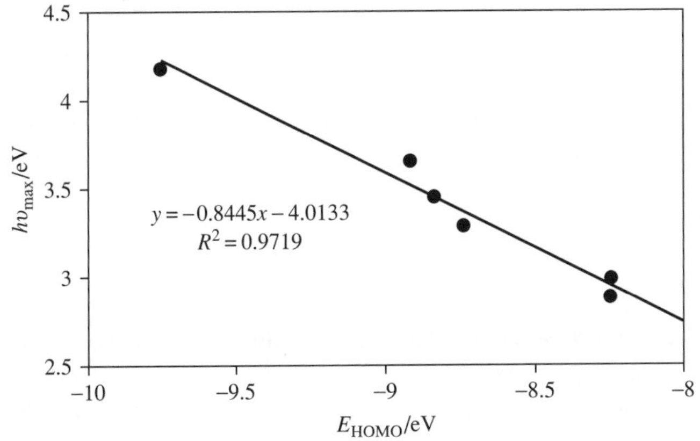

Figure 13.8

P13.11 The anthracene vapour fluorescence spectrum gives the vibrational splitting of the lower state. The wavelengths stated correspond to the wavenumbers 22 730, 24 390, 25 640, and 27 030 cm^{-1}, indicating spacings of 1660, 1250, and 1390 cm^{-1}. The absorption spectrum spacing gives the separation of the vibrational levels of the upper state. The wavenumbers of the absorption peaks are 27 800, 29 000, 30 300, and 32 800 cm^{-1}. The vibrational spacings are therefore 1200, 1300, and 2500 cm^{-1}. The data is compatible with the deactivation of the excited-state vibrational modes before spontaneous emission returns the molecule to the ground electronic state. This produces a fluorescence band of lower energy than the absorption band. Furthermore, while the absorption band has a vibrational progression that depends on vibrational modes of the excited state, the fluorescence band has a vibrational progression that depends on vibrational modes of the ground state. The absorption and fluorescence spectra are not mirror images.

P13.13 The laser is delivering photons of energy:

$$E = h\upsilon = \frac{hc}{\lambda} = \frac{(6.626 \times 10^{-34}\ \text{J s}) \times (2.998 \times 10^{8}\ \text{m s}^{-1})}{488 \times 10^{-9}\ \text{m}} = 4.07 \times 10^{-19}\ \text{J}$$

Since the laser is putting out 1.0 mJ of these photons every second, the rate of photon emission is

$$r = \frac{1.0 \times 10^{-3}\ \text{J s}^{-1}}{4.07 \times 10^{-19}\ \text{J}} = 2.5 \times 10^{15}\ \text{s}^{-1}$$

The time it takes the laser to deliver 10^6 photons (and therefore the time the dye remains fluorescent) is

$$t = \frac{10^6}{2.5 \times 10^{15}\ \text{s}^{-1}} = \boxed{4 \times 10^{-10}\ \text{s or } 0.4\ \text{ns}}$$

Solutions to theoretical problems

P13.15 Suppose that non-absorbing species B (pyridine in this problem) is progressively added to a solution of light-absorbing species A (I_2 in this problem). Furthermore, suppose that A and B form a complex of light-absorbing species AB. Suppose that all absorbance measurements are made at the equilibrium $A + B \rightleftharpoons AB$. By the conservation of mass the sum of the equilibrium concentrations of the absorbing species, $[A]_e + [AB]_e$, is a constant for all additions of B. That is, the sum of concentrations of all species that contain A must equal the concentration of A present before any B is added, $[A]_0$. The absorbance at any wavelength is

$$\begin{aligned} A_\lambda &= \varepsilon_{A,\lambda}L[A]_e + \varepsilon_{AB,\lambda}L[AB]_e \\ &= \varepsilon_{A,\lambda}L([A]_0 - [AB]_e) + \varepsilon_{AB,\lambda}L[AB]_e \\ &= \varepsilon_{A,\lambda}L[A]_0 + (\varepsilon_{AB,\lambda} - \varepsilon_{A,\lambda})L[AB]_e \end{aligned}$$

The first term to the right in the above equation is a constant for all additions of B; the second term is generally not constant and, consequently, A_λ generally varies with the addition of B. However, at any wavelength for which $\varepsilon_{AB,\lambda} = \varepsilon_{A,\lambda}$ the second term vanishes, making A_λ a constant at all additions of B. The wavelength at which this happens is called the **isosbestic point**. It is characterized by the equilibrium between absorbing species and $\varepsilon_{AB,\lambda} = \varepsilon_{A,\lambda}$.

P13.17 We need to establish whether the transition dipole moments

$$\mu_{fi} = \int \psi_f^* \mu \psi_i\, d\tau \text{ [9.25 and 11.11]}$$

connecting the states 1 and 2 and the states 1 and 3 are zero or nonzero. The particle in a box wavefunctions are $\psi_n = (2/L)^{1/2}\sin(n\pi x/L)$ [8.4b].

Thus, $$\mu_{2,1} \propto \int \sin\left(\frac{2\pi x}{L}\right) x \sin\left(\frac{\pi x}{L}\right) dx \propto \int x\left[\cos\left(\frac{\pi x}{L}\right) - \cos\left(\frac{3\pi x}{L}\right)\right] dx$$

and $$\mu_{3,1} \propto \int \sin\left(\frac{3\pi x}{L}\right) x \sin\left(\frac{\pi x}{L}\right) dx \propto \int x\left[\cos\left(\frac{2\pi x}{L}\right) - \cos\left(\frac{4\pi x}{L}\right)\right] dx$$

having used $\sin\alpha \sin\beta = \frac{1}{2}\cos(\alpha-\beta) - \frac{1}{2}\cos(\alpha+\beta)$. Both of these integrals can be evaluated using the standard form

$$\int x(\cos ax)\,dx = \frac{1}{a^2}\cos ax + \frac{x}{a}\sin ax$$

$$\int_0^L x\cos\left(\frac{\pi x}{L}\right)dx = \frac{1}{(\pi/L)^2}\cos\left(\frac{\pi x}{L}\right)\Bigg|_0^L + \frac{x}{(\pi/L)}\sin\left(\frac{\pi x}{L}\right)\Bigg|_0^L = -2\left(\frac{L}{\pi}\right)^2 \neq 0$$

$$\int_0^L x\cos\left(\frac{3\pi x}{L}\right)dx = \frac{1}{(3\pi/L)^2}\cos\left(\frac{3\pi x}{L}\right)\Bigg|_0^L + \frac{x}{(3\pi/L)}\sin\left(\frac{3\pi x}{L}\right)\Bigg|_0^L = -2\left(\frac{L}{3\pi}\right)^2 \neq 0$$

Thus, $\mu_{2,1} \neq 0$

In a similar manner, $\mu_{3,1} = 0$.

COMMENT. A general formula for μ_{fi} applicable to all possible particle in a box transitions may be derived. The result is (n = f, m = i)

$$\mu_{nm} = -\frac{eL}{\pi^2}\left[\frac{\cos(n-m)\pi - 1}{(n-m)^2} - \frac{\cos(n+m)\pi - 1}{(n+m)^2}\right]$$

Question. For m and n both even or both odd numbers, $\mu_{nm} = 0$; if one is even and the other odd, $\mu_{nm} \neq 0$. Can you establish the general relationship for μ_{nm} above?

P13.19 $$S = \left\{1 + \frac{R}{a_0} + \frac{1}{3}\left(\frac{R}{a_0}\right)^2\right\}e^{-R/a_0}\ [10.19]$$

$\mu = -eSR$ [given]

$\mu/(-ea_0) = (R/a_0) \times S$

$$= (R/a_0) \times \left[1 + \frac{R}{a_0} + \frac{1}{3}\left(\frac{R}{a_0}\right)^2\right]e^{-R/a_0}$$

We then draw up the following table:

R/a_0	0	1	2	3	4	5	6	7	8
$\mu/(-ea_0)$	0	0.858	1.173	1.046	0.757	0.483	0.283	0.155	0.081

These points are plotted in Figure 13.9. The dipole-transition moment rises from zero at $R = 0$ to a peak at $R \sim 2.1a_0$, after which it declines to zero at infinity. The transition moment is zero at $R = 0$ because the initial state ($1s_A$) and the final state ($1s_A$) are identical and, consequently, there is no dipole moment and no transition. As $R \to \infty$, the electron is confined to a single atom because its wavefunction does not extend to the other and, consequently, no transition is possible. The oscillator strength is proportional to $|\mu|^2$.

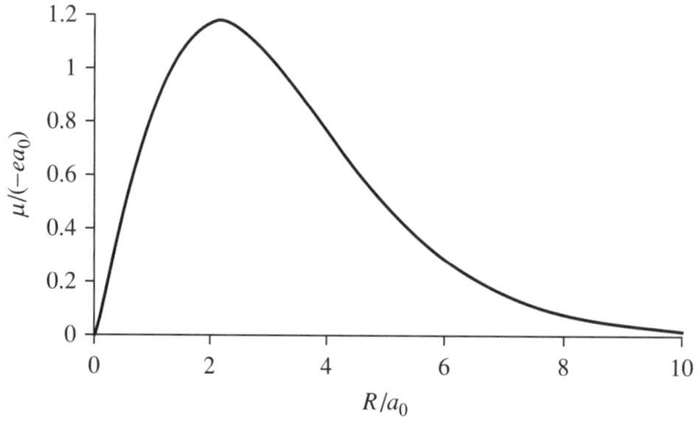

Figure 13.9

Solutions to applications: biochemistry, environmental science, and astrophysics

P13.21 An **isosbestic wavenumber** is a point in the spectrum at which two solution components have identical molar absorption coefficients (see Problem 13.15 for an example application). It is very rare to observe a situation in which three solution components share an isosbestic point because of the very low probability that three components could have identical molar absorption coefficients at a particular wavenumber. The spectrum of text Figure 13.43 shows three isosbestic points, evidenced by the common points of the spectra under different concentrations of CNS^-. Consequently, we infer that only two absorbing species ($Her(CNS)_8$ and $Her(OH)_8$) are present.

P13.23 The following table summarizes AM1 calculations (an extended Hückel method) of the LUMO–HOMO separation in the 11-*cis* (Figure 13.10) and 11-*trans* molecule (10) model of retinal. The −46.0° torsional angle between the first two alternate double bonds indicates that they are not coplanar. In contrast, the C11C12C13C14 torsion angle shows that the C11C12 double bond is close to coplanar with neighbouring double bonds. The aromatic character of the alternating π-bond system is evidenced by contrasting the computed bond lengths at a single bond away from the π-system (C1–C2), a double bond (C11–C12), and a single bond between doubles (C12–C13) within the Lewis structure. We see a typical single bond length, a slightly elongated double bond length, and a bond length that is intermediate between a single and a double, respectively. The latter lengths are characteristic of aromaticity. The HOMO and LUMO of 11-*cis* are shown in Figure 13.10.

Conformation	11-*trans* (10)	11-*cis* (10)
$\Delta_f H^\ominus$/kJ mol^{-1}	725.07	738.1
E_{LUMO}/eV	-5.142	-5.138
E_{HOMO}/eV	−10.770	−10.888
ΔE/eV	(a) 5.628	(b) 5.750
λ/nm	(a) 220.3	(b) 215.6
C5C6C7C8 torsion angle/°	−44.5	−46.0
C11C12C13C14 torsion angle/°	179.7	−165.5
C1–C2/pm	153.2	153.2
C11–C12/pm	137.3	136.7
C12–C13/pm	142.0	142.1

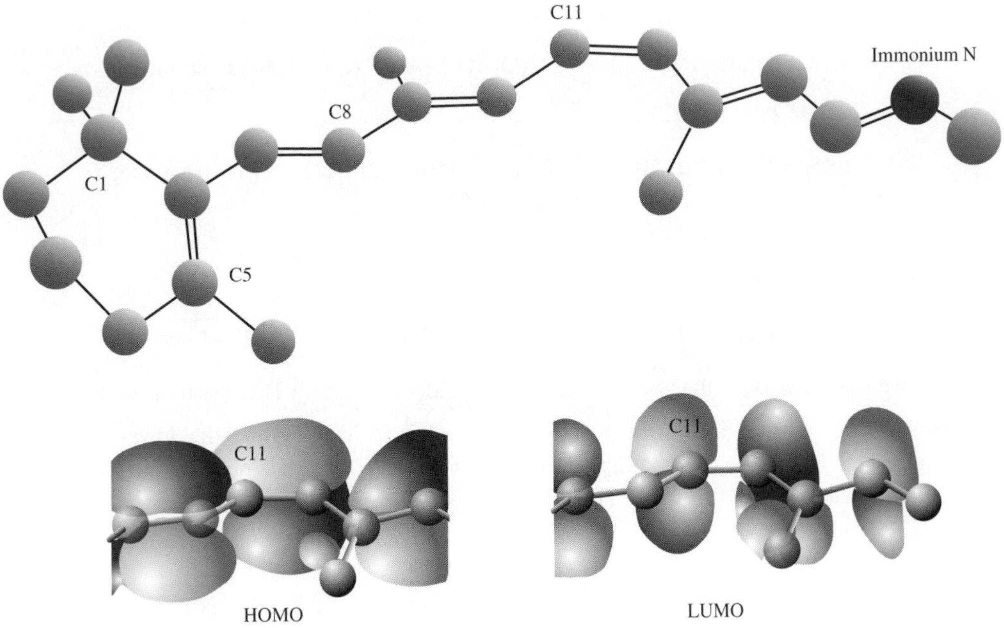

Figure 13.10

(c) The lowest $\pi^* \leftarrow \pi$ transition occurs in the ultraviolet with the 11-*cis* transition at higher energy (higher frequency, lower wavelength). It is apparent that important interactions between retinal and a surrounding opsin molecule are responsible for reducing the transition energy to the observed strong absorption in the 400–600 nm visible range.

P13.25 The Dobson unit (DU) is the height of a pure ozone column defined by the factor 10^{-3} cm DU^{-1} with the ozone at 1 atm and 0°C. The ozone concentration under the conditions of the Dobson unit is

$$[O_3] = \frac{n}{V} = \frac{p}{RT} = \frac{1\ \text{atm}}{(0.0820574\ \text{dm}^3\ \text{atm mol}^{-1}\ \text{K}^{-1}) \times (273.15\ \text{K})} = 4.4615 \times 10^{-2}\ \text{mol dm}^{-3}$$

The absorbance of the ozone column is given by

$$\begin{aligned} A &= \varepsilon[O_3]h_{\text{column}}\ [13.4] \\ &= (476\ \text{dm}^3\ \text{mol}^{-1}\ \text{cm}^{-1}) \times (4.4615 \times 10^{-2}\ \text{mol dm}^{-3}) \times h_{\text{column}} = (21.2\bar{4}\ \text{cm}^{-1}) \times h_{\text{column}} \end{aligned}$$

For 300 DU: $A = (21.2\bar{4}\ \text{cm}^{-1}) \times (300\ \text{DU}) \times (10^{-3}\ \text{cm DU}^{-1}) = \boxed{6.37}$

For 100 DU: $A = (21.2\bar{4}\ \text{cm}^{-1}) \times (100\ \text{DU}) \times (10^{-3}\ \text{cm DU}^{-1}) = \boxed{2.12}$

P13.27 Process 1: $Cl_2 + O_2 \rightarrow Cl_2O_2$(structure 1) $\Delta_f H^\ominus$(structure 1) = +133 kJ mol^{-1}

Process 2: Cl_2O_2(structure 2) $\rightarrow Cl + OClO^+ + e^-$ $\Delta_{r2} H^\ominus$(structure 2) = 10.95 eV = +1057 kJ mol^{-1}

$\Delta_f H^\ominus(OClO^+) = +1096$ kJ mol^{-1}

$\Delta_f H^\ominus(e^-) = 0$

Cl_2O_2(structure 2) $\rightarrow Cl_2O_2^+$(structure 2) + e^- $\Delta_{I2} H^\ominus$(structure 2) = 11.05 eV = +1066 kJ mol^{-1}

The relationship between $\Delta_{r2}H^{\ominus}$(structure 2) and formation enthalpies is

$$\Delta_{r2}H^{\ominus}(\text{structure 2}) = \Delta_f H^{\ominus}(\text{Cl}) + \Delta_f H^{\ominus}(\text{OClO}^+) + \Delta_f H^{\ominus}(\text{e}^-) - \Delta_f H^{\ominus}(\text{structure 2})$$

Solving for $\Delta_f H^{\ominus}$(structure 2) gives

$$\begin{aligned}\Delta_f H^{\ominus}(\text{structure 2}) &= \Delta_f H^{\ominus}(\text{Cl}) + \Delta_f H^{\ominus}(\text{OClO}^+) + \Delta_f H^{\ominus}(\text{e}^-) - \Delta_{r2}H^{\ominus}(\text{structure 2}) \\ &= (121.68 + 1096 + 0 - 1057)\ \text{kJ mol}^{-1} \\ &= +161\ \text{kJ mol}^{-1}\end{aligned}$$

Thus, we find the experimental result

$$\Delta_f H^{\ominus}(\text{structure 2}) - \Delta_f H^{\ominus}(\text{structure 1}) = (161-133)\ \text{kJ mol}^{-1} = \boxed{+28\ \text{kJ mol}^{-1}}$$

and we conclude that $\boxed{\text{the two structures are not identical}}$. This is consistent with the quantum-mechanical computations, which say that ClOOCl is likely to be the lowest-energy isomer. Experimentally, we see that the Cl_2O_2 structure of process 2, which is not ClOOCl, is not very much greater in energy than the lowest-energy isomer.

14 Molecular spectroscopy 3: magnetic resonance

Answers to discussion questions

D14.1 Resonance is a mechanical phenomenon in which energy is exchanged between two systems that have the same natural frequency. An example of classical resonance is the exchange of energy between two similar pendulums connected by a light spring. When one pendulum is set in motion the other pendulum starts to oscillate. Energy is transferred from the first pendulum to the second. The transfer of energy is most efficient when the two pendulums have the same natural frequency. The pendulums are then said to be in resonance. Quantum-mechanical resonance has some of the same characteristics. An important case is when one system is the electromagnetic field and the other is an atom or molecule. In spectroscopy, resonance denotes the strong coupling and efficient energy transfer that occurs when the frequency of the electromagnetic field just matches the transition frequency associated with the energy separation between states of the atomic or molecular system. In principle all spectroscopic techniques are resonance techniques, but the term is most commonly applied to NMR and EPR. In NMR and EPR either the energy levels of the system can be adjusted to achieve resonance with the frequency of the electromagnetic field or the frequency of the field can be adjusted to match the energy separation of the states of the atom or molecule.

D14.3 Examination of eqn 14.17 implies that in an NMR spectrometer operating at a fixed frequency, the external magnetic field required to fulfill the resonance condition is given by

$$\mathcal{B}_0 = \frac{2\pi\nu_L}{\gamma(1-\sigma)}$$

Thus, a positive value of the shielding constant, σ, shifts the resonance field to 'high field' and a negative value of σ shifts the resonance field to 'low field'. Conversely, in a spectrometer at fixed external magnetic field, positive values of σ shift the resonant frequency to lower values and negative values of σ shift it to higher values. Chemical shifts are also reported in terms of the parameter, δ, called the chemical shift and defined by eqn 14.18. σ and δ are independent of the external magnetic field, but the chemical shift in frequency units, $\nu - \nu_0$, is not. See the Brief Illustration in Section 14.5(a).

D14.5 See Section 14.6(e) for a detailed discussion of chemical and magnetic equivalence as applied to NMR and the distinction between them. Here, we will summarize the basic concepts. Two nuclei are chemically equivalent if they are related by a symmetry operation of the molecule. Symmetrically equivalent nuclei will have the same resonance frequency, i.e. the same chemical shift. Examples are the protons in benzene and the protons *meta*- to each other (H-2, H-6 and H-3, H-5) in *para*-nitrophenol. In benzene the protons are related by a C_6 operation (as well as others) and in *para*-nitrophenol the protons are related by a plane of symmetry and a C_2 operation. Two nuclei are magnetically equivalent if in addition to being chemically equivalent they have identical spin–spin interactions with all other magnetic nuclei in the molecule. Examples are CH_2CF_2 and 1,2,3-trichlorobenzene. Chemical equivalence does not imply magnetic equivalence. In the case of *para*-nitrophenol, the protons H-2 and H-6, although chemically equivalent, are not magnetically equivalent because the coupling of H-2 to H-3 is different from the coupling of H-6 to H-3.

D14.7 Both spin–lattice and spin–spin relaxation are caused by fluctuating magnetic and electric fields at the nucleus in question and these fields result from the random thermal motions present in the solution or other form of matter. These random motions can be a result of a number of processes and it is hard to summarize all that could be important. In theory every known nuclear interaction coupled with every type of motion can contribute to relaxation and detailed treatments can be exceedingly complex. However, they all depend on the magnetogyric ratio of the atom in question and the magnetogyric ratio of the proton is much larger than that of ^{13}C. Hence, the interaction of the proton with fluctuating local magnetic fields caused by the presence of neighbouring magnetic nuclei will be greater, and the relaxation will be quicker, corresponding to a shorter relaxation time for protons. Another consideration is the structure of compounds containing carbon and hydrogen. Typically, the C atoms are in the interior of the molecule bonded to other C atoms, 99% of which are non-magnetic, so the primary relaxation effects are due to bonded protons. Protons are on the outside of the molecule and are subject to many more interactions and hence faster relaxation.

D14.9 The ESR spectra of a spin probe, such as the di-*tert*-butyl nitroxide radical, broaden with restricted motion of the probe. This suggests that the width of spectral lines may correlate with the depth to which a probe may enter into a biopolymer crevice. Deep crevices are expected to severely restrict probe motion and broaden the spectral lines. Additionally, the splitting and center of ESR spectra of an oriented sample can provide information about the shape of the biopolymer-probe environment because the probe ESR signal is anisotropic and depends on the orientation of the probe with respect to the external magnetic field. Oriented biopolymers occur in lipid membranes and in muscle fibres.

Solutions to exercises

E14.1(a) $$\nu_L = -\frac{\gamma_e \mathcal{B}_0}{2\pi} \text{ [14.9, note the negative sign]} = \frac{1.761\times10^{11}\,\text{s}^{-1}\,\text{T}^{-1}\times 1.0\,\text{T}}{2\pi} = 2.8\times10^{10}\,\text{s}^{-1} = \boxed{28\ \text{GHz}}$$

COMMENT. γ_e is a *negative* quantity, but frequencies are thought of as positive quantities, hence the negative sign in eqn 14.9.

E14.2(a) The relationship between angular velocity and angular displacement is

$$\omega = 2\pi\nu = \frac{\Delta\theta}{t} \text{ [See any general physics text.]}$$

In this case, ν is the Larmor frequency, ν_L, calculated in Exercise 14.1(a), and $\Delta\theta$ is $\pi/2$. Solving for t we obtain

$$t = \frac{\Delta\theta}{2\pi\nu_L} = \frac{1}{4\nu_L} = \frac{1}{4 \times 2.8 \times 10^{10}\,\text{s}^{-1}} = \boxed{8.9 \times 10^{-12}\,\text{s}}$$

E14.3(a) The resonance frequency is equal to the Larmor frequency of the proton and is given by

$$\nu = \nu_L = \frac{\gamma\mathcal{B}_0}{2\pi}\,[14.9] \text{ with } \gamma = \frac{g_I\mu_N}{\hbar}\,[14.11]$$

$$\text{hence, } \nu = \frac{g_I\mu_N\mathcal{B}_0}{h} = \frac{(5.5857) \times (5.0508 \times 10^{-27}\,\text{J T}^{-1}) \times (14.1\,\text{T})}{6.626 \times 10^{-34}\,\text{J s}} = \boxed{600\text{ MHz}}$$

E14.4(a) The energy level separation is

$$\Delta E = h\nu, \quad \text{where } \nu = \frac{\gamma\mathcal{B}_0}{2\pi}\,[14.9]$$

So,

$$\nu = \frac{(6.73 \times 10^7\,\text{T}^{-1}\,\text{s}^{-1}) \times 14.4\,\text{T}}{2\pi} = 1.54 \times 10^8\,\text{s}^{-1}$$

$$= 1.54 \times 10^8\,\text{Hz} = \boxed{154\text{ MHz}}$$

E14.5(a) As shown in Exercise 14.3(a) a 600-MHz NMR spectrometer operates in a magnetic field of 14.1 T.

For the proton

$$\Delta E = \gamma\hbar\mathcal{B}_0 = h\nu_L = h\nu \text{ at resonance}$$
$$= (6.626 \times 10^{-34}\,\text{Js}) \times (6.00 \times 10^8\,\text{s}^{-1}) = \boxed{3.98 \times 10^{-25}\,\text{J}}$$

A 600-MHz NMR spectrometer means 600 MHz is the resonance frequency for protons for which the magnetic field is 14.1 T. In high-field NMR it is the field not the frequency that is fixed, so for the deuteron

$$\nu = \frac{g_I\mu_N\mathcal{B}_0}{h}\,[\text{Exercise 14.3(a)}]$$
$$= \frac{(0.8575) \times (5.051 \times 10^{-27}\,\text{J T}^{-1}) \times (14.1\,\text{T})}{6.626 \times 10^{-34}\,\text{J s}} = 9.22 \times 10^7\,\text{Hz} = 92.2\text{ MHz}$$

$$\Delta E = h\nu = (6.626 \times 10^{-34}\,\text{J s}) \times (9.21\bar{6} \times 10^7\,\text{s}^{-1}) = \boxed{6.11 \times 10^{-26}\,\text{J}}$$

Thus, the separation in energy is $\boxed{\text{larger for the proton.}}$

E14.6(a) In all cases the selection rule $\Delta m_I = \pm 1$ is applied, hence

$$\mathcal{B}_0 = \frac{h\nu}{g_I\mu_N} = \frac{6.626 \times 10^{-34}\,\text{J Hz}^{-1}}{5.0508 \times 10^{-27}\,\text{J T}^{-1}} \times \frac{\nu}{g_I}$$
$$= (1.3119 \times 10^{-7}) \times \frac{(\nu/\text{Hz})}{g_I}\,\text{T} = (0.13119) \times \frac{(\nu/\text{MHz})}{g_I}\,\text{T}$$

We can draw up the following table

$\mathcal{B}_0$/T	(a)^{1}H	(b)^{2}H	(c)^{13}C
g_I	5.5857	0.85745	1.4046
(i) 250 MHz	5.87	38.3	23.4
(ii) 500 MHz	11.7	76.6	46.8

COMMENT. Magnetic fields above 23 T have not yet been obtained for use in NMR spectrometers. As discussed in the solution to Exercise 14.5(a), it is the field, not the frequency, that is fixed in high-field NMR spectrometers. Thus, an NMR spectrometer that is called a 500-MHz spectrometer refers to the resonance frequency for protons and has a magnetic field fixed at 11.7 T.

Question. What are the resonance frequencies of these nuclei in 250-MHz and 500-MHz spectrometers? See Exercise 14.5(a) part (b).

E14.7(a) The ground state has

$$m_I + \tfrac{1}{2} = \alpha \text{ spin},\ m_I = -\tfrac{1}{2} = \beta \text{ spin}$$

Hence, with

$$\delta N = N_\alpha - N_\beta$$

$$\frac{\delta N}{N} = \frac{N_\alpha - N_\beta}{N_\alpha + N_\beta} = \frac{N_\alpha - N_\alpha e^{-\Delta E/kT}}{N_\alpha + N_\alpha e^{-\Delta E/kT}} \quad [\textit{Justification}\ 14.1]$$

$$= \frac{1 - e^{-\Delta E/kT}}{1 + e^{-\Delta E/kT}} \approx \frac{1-(1-\Delta E/kT)}{1+1} \approx \frac{\Delta E}{2kT} = \frac{g_I \mu_N \mathcal{B}_0}{2kT} \quad [\text{for } \Delta E \ll kT]$$

That is, $\dfrac{\delta N}{N} \approx \dfrac{g_I \mu_N \mathcal{B}_0}{2kT} = \dfrac{(5.5857)\times(5.0508\times10^{-27}\ \text{J T}^{-1})\times(\mathcal{B}_0)}{(2)\times(1.38066\times10^{-23}\ \text{J T}^{-1})\times(298\ \text{K})} \approx 3.43\times10^{-6}\mathcal{B}_0/\text{T}$

(a) $\mathcal{B}_0 = 0.3\ \text{T},\quad \delta N/N = \boxed{1\times10^{-6}}$

(b) $\mathcal{B}_0 = 1.5\ \text{T},\quad \delta N/N = \boxed{5.1\times10^{-6}}$

(c) $\mathcal{B}_0 = 10\ \text{T},\quad \delta N/N = \boxed{3.4\times10^{-5}}$

E14.8(a) We adapt eqn 14.28 to the case of the magnetic field of the electron.

$$\mathcal{B}_e = \frac{\gamma_e \hbar \mu_0}{4\pi R^3}(1 - 3\cos^2\theta)\, m_S$$

We use $m_S = \tfrac{1}{2}$ and $R = 100$ pm $= 1.00\times10^{-10}$ m and substitute for the constants in the expression. We obtain

$$\mathcal{B}_e = \frac{-1.761\times10^{11}\ \text{s}^{-1}\,\text{T}^{-1}\times1.055\times10^{-34}\ \text{J s}\times12.57\times10^{-7}\ \text{T}^2\,\text{J}^{-1}\,\text{m}^3}{8\pi(1.00\times10^{-10}\ \text{m})^3}(1-3\cos^2\theta)$$

$$\mathcal{B}_e = -0.929\ \text{T}\times(1-3\cos^2\theta)$$

(a) $\theta = 0°,\ \mathcal{B}_e = \boxed{1.86\ \text{T}}$ (b) $\theta = 90°,\ \mathcal{B}_e = \boxed{0}$

E14.9(a) $\delta N \approx \dfrac{Ng_I\mu_N\mathcal{B}_0}{2kT}$ [Exercise 14.7(a)] $= \dfrac{Nh\nu}{2kT}$

Thus, $\delta N \propto \nu$

$$\frac{\delta N(800\ \text{MHz})}{\delta N(60\ \text{MHz})} = \frac{800\ \text{MHz}}{60\ \text{MHz}} = \boxed{13}$$

This ratio is not dependent on the nuclide, as long as the approximation $\Delta E \ll kT$ holds [Exercise 14.7(**a**)].

E14.10(a) $\mathcal{B}_{loc} = (1-\sigma)\mathcal{B}_0$ [14.16]

$$|\Delta\mathcal{B}_{loc}| = |(\Delta\sigma)|\,\mathcal{B}_0 \approx |[\delta(CH_3) - \delta(CHO)]| \times 10^{-6}\,\mathcal{B}_0 \left[|\Delta\sigma| \approx \left|\frac{\nu - \nu^\circ}{\nu^\circ}\right|\right]$$

$$= |(2.20 - 9.80)| \times 10^{-6}\,\mathcal{B}_0 = 7.60 \times 10^{-6}\,\mathcal{B}_0$$

(a) $\mathcal{B}_0 = 1.5\ \text{T}$, $|\Delta\mathcal{B}_{loc}| = 7.60 \times 10^{-6} \times 1.5\ \text{T} = \boxed{11\,\mu\text{T}}$

(b) $\mathcal{B}_0 = 15\ \text{T}$, $|\Delta\mathcal{B}_{loc}| = \boxed{110\,\mu\text{T}}$

E14.11(a) $\nu - \nu^\circ = \nu^\circ\delta \times 10^{-6}$ [14.18]

$$\begin{aligned}|\Delta\nu| &\equiv (\nu - \nu^\circ)(CHO) - (\nu - \nu^\circ)(CH_3)\\ &= \nu(CHO) - \nu(CH_3)\\ &= \nu^\circ[\delta(CHO) - \delta(CH_3)] \times 10^{-6}\\ &= (9.80 - 2.20) \times 10^{-6}\nu^\circ = 7.60 \times 10^{-6}\nu^\circ\end{aligned}$$

(a) $\nu^\circ = 250\ \text{MHz}$, $|\Delta\nu| = 7.60 \times 10^{-6} \times 250\ \text{MHz} = 1.90\ \text{kHz}$

(b) $\nu^\circ = 500\ \text{MHz}$, $|\Delta\nu| = 3.80\ \text{MHz}$

(a) The spectrum is shown in Figure 14.1 with the value of $|\Delta\nu|$ as calculated above.

(b) When the frequency is changed to 500 MHz, the $|\Delta\nu|$ changes to 3.80 kHz. The fine structure (the splitting within groups) remains the same as spin–spin splitting is unaffected by the strength of the applied field. However, the intensity of the lines increases by a factor of 2 because $\delta N/N \propto \nu$ (Exercise 14.9(a)).

The observed splitting pattern is that of an AX_3 (or A_3X) species, the spectrum of which is described in Section 14.6.

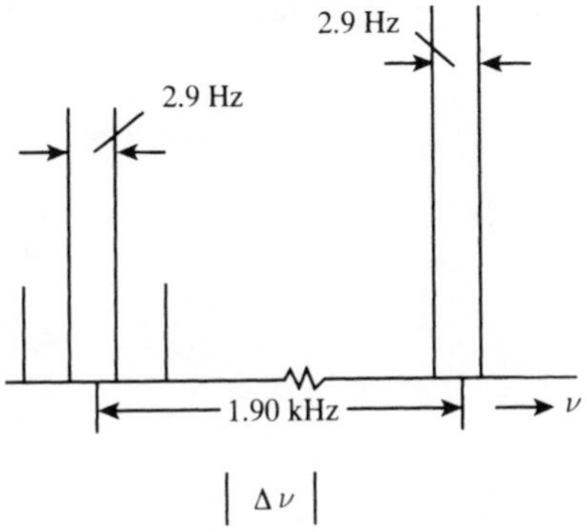

Figure 14.1

E14.12(a) See Section 14.6(b), Example 14.1 and Figures 14.16–14.19 for the approach to the solution to this exercise. Also, see Example 14.2 and Figures 14.55 and 14.56. That latter example and those figures are applied specifically to EPR spectra, but the process of determining the intensity pattern in the fine structure of an NMR spectrum is the same. See the table below for the version of Pascal's triangle for up to four spin-$\frac{3}{2}$ nuclei. Each number in the table is the sum of the four ($I=\frac{3}{2}$, $2I+1=4$) numbers above it (two to the right and two to the left).

												1												
									1		1		1		1									
						1		2		3		4		3		2		1						
			1		3		6		10		12		12		10		6		3		1			
1		4		10		20		31		40		44		40		31		20		10		4		1

E14.13(a) $\tau \approx \frac{\sqrt{2}}{\pi\Delta\nu}$ [14.29, with $\delta\nu$ written as $\Delta\nu$]

$\Delta\nu = \nu^{\circ}(\delta' - \delta)\times 10^{-6}$ [Exercise 14.11(a)]

$$\tau \approx \frac{\sqrt{2}}{\pi\nu^{\circ}(\delta' - \delta)\times 10^{-6}}$$

Then,

$$\approx \frac{\sqrt{2}}{(\pi)\times 250\times 10^{6}\,\text{Hz}\times(5.2-4.0)\times 10^{-6}}$$

$$\approx 1.5\times 10^{-3}\,\text{s}$$

Therefore, the signals merge when the lifetime of each isomer is less than about 1.5 ms, corresponding to a conversion rate of about $\boxed{6.7\times 10^{2}\,\text{s}^{-1}}$.

E14.14(a) The four equivalent ^{19}F nuclei ($I=\frac{1}{2}$) give a single line. However, the ^{10}B nucleus ($I = 3$, 19.6% abundant) splits this line into $2\times 3+1 = 7$ lines and the ^{11}B nucleus ($I=\frac{3}{2}$, 80.4% abundant) into $2\times\frac{3}{2}+1=4$ lines. The splitting arising from the ^{11}B nucleus will be larger than that arising from the ^{10}B nucleus (since its magnetic moment is larger, by a factor of 1.49, Table 14.2). Moreover, the total intensity of the four lines due to the ^{11}B nuclei will be greater (by a factor of $80.4/19.6\approx 4$ than the total intensity of the seven lines due to the ^{10}B nuclei. The individual line intensities will be in the ratio $\frac{7}{4}\times 4 = 7$ ($\frac{4}{7}$ the number of lines and about four times as abundant). The spectrum is sketched in Figure 14.2.

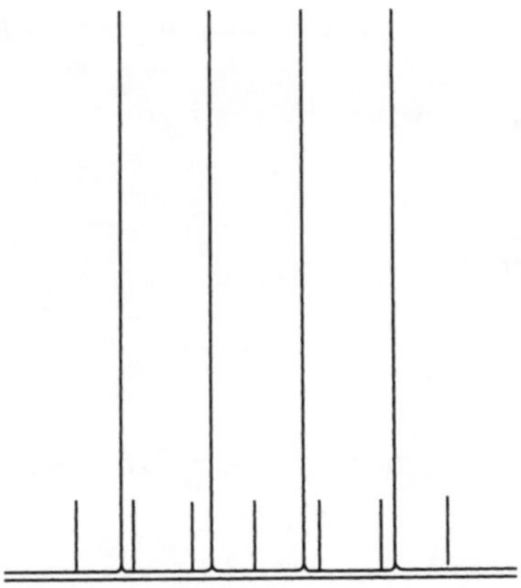

Figure 14.2

E14.15(a) The A, M, and X resonances lie in distinctively different groups. The A resonance is split into a 1:2:1 triplet by the M nuclei, and each line of that triplet is split into a 1:4:6:4:1 quintet by the X nuclei (with $J_{AM} > J_{AX}$). The M resonance is split into a 1:3:3:1 quartet by the A nuclei and each line is split into a quintet by the X nuclei (with $J_{AM} > J_{MX}$). The X resonance is split into a quartet by the A nuclei and then each line is split into a triplet by the M nuclei (with $J_{AX} > J_{MX}$). The spectrum is sketched in Figure 14.3.

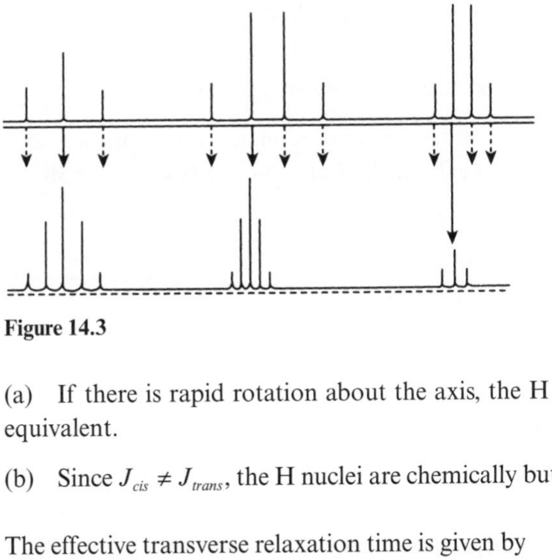

Figure 14.3

E14.16(a) (a) If there is rapid rotation about the axis, the H nuclei are both chemically and magnetically equivalent.

(b) Since $J_{cis} \neq J_{trans}$, the H nuclei are chemically but not magnetically equivalent.

E14.17(a) The effective transverse relaxation time is given by

$$T_2^* = \frac{1}{\pi\Delta\nu_{1/2}}\,[14.34] = \frac{1}{\pi \times 1.5\ \mathrm{s^{-1}}} = \boxed{0.21\ \mathrm{s}}$$

E14.18(a) The maximum enhancement is given by

$$\eta = \frac{\gamma_{^1\mathrm{H}}}{2 \times \gamma_{^{31}\mathrm{P}}}\,[14.37] = \frac{26.752 \times 10^7\ \mathrm{T^{-1}\,s^{-1}}}{2 \times 10.840 \times 10^7\ \mathrm{T^{-1}\,s^{-1}}}\,[\text{Table 14.2}] = \boxed{1.234}$$

E14.19(a) Analogous to precession of the magnetization vector in the laboratory frame due to the presence of $\mathcal{B}_0$, that is

$$\nu_L = \frac{\gamma \mathcal{B}_0}{2\pi} [14.9],$$

there is a precession in the rotating frame, due to the presence of $\mathcal{B}_1$, namely

$$\nu_L = \frac{\gamma \mathcal{B}_1}{2\pi} \quad \text{or} \quad \omega_1 = \gamma \mathcal{B}_1 \quad [\omega = 2\pi\nu]$$

Since ω is an angular frequency, the angle through which the magnetization vector rotates is

$$\theta = \gamma \mathcal{B}_1 t = \frac{g_I \mu_N}{\hbar} \mathcal{B}_1 t$$

and $\mathcal{B}_1 = \dfrac{\theta\hbar}{g_I \mu_N t} = \dfrac{(\frac{\pi}{2}) \times (1.055 \times 10^{-34}\text{ J s})}{(5.586) \times (5.051 \times 10^{-27}\text{ J T}^{-1}) \times (1.0 \times 10^{-5}\text{ s})} = \boxed{5.9 \times 10^{-4}\text{ T}}$

A 180° pulse requires $2 \times 10\ \mu\text{s} = \boxed{20\ \mu\text{s}}$.

E14.20(a) (a) $\mathcal{B}_0 = \dfrac{h\nu}{g_I \mu_N} = \dfrac{(6.626 \times 10^{-34}\text{ J Hz}^{-1}) \times (9 \times 10^9\text{ Hz})}{(5.5857) \times (5.051 \times 10^{-27}\text{ J T}^{-1})} = \boxed{2 \times 10^2\text{ T}}$

(b) $\mathcal{B}_0 = \dfrac{h\nu}{g_e \mu_B} = \dfrac{(6.626 \times 10^{-34}\text{ J Hz}^{-1}) \times (300 \times 10^6\text{ Hz})}{(2.0023) \times (9.274 \times 10^{-24}\text{ J T}^{-1})} = \boxed{10\text{ mT}}$

COMMENT. Because of the sizes of these magnetic fields neither experiment seems feasible.

Question. What frequencies are required to observe electron resonance in the magnetic field of a 300-MHz NMR magnet and nuclear resonance in the field of a 9-GHz ($g = 2.00$) ESR magnet? Are these experiments feasible?

E14.21(a) $$g = \frac{h\nu}{\mu_B \mathcal{B}_0} [14.39]$$

We shall often need the value

$$\frac{h}{\mu_B} = \frac{6.62608 \times 10^{-34}\text{ J Hz}^{-1}}{9.27401 \times 10^{-24}\text{ J T}^{-1}} = 7.14478 \times 10^{-11}\text{ T Hz}^{-1}$$

Then, in this case

$$g = \frac{(7.14478 \times 10^{-11}\text{ T Hz}^{-1}) \times (9.2231 \times 10^9\text{ Hz})}{329.12 \times 10^{-3}\text{ T}} = \boxed{2.0022}$$

E14.22(a) $$a = \mathcal{B}(\text{line } 3) - \mathcal{B}(\text{line } 2) = \mathcal{B}(\text{line } 2) - \mathcal{B}(\text{line } 1)$$

$$\left.\begin{aligned} \mathcal{B}_3 - \mathcal{B}_2 &= (334.8 - 332.5)\text{mT} = 2.3\text{ mT} \\ \mathcal{B}_2 - \mathcal{B}_1 &= (332.5 - 330.2)\text{mT} = 2.3\text{ mT} \end{aligned}\right\} a = \boxed{2.3\text{ mT}}$$

Use the centre line to calculate g:

$$g = \frac{h\nu}{\mu_B \mathcal{B}_0} = (7.14478 \times 10^{-11}\,\text{T Hz}^{-1}) \times \frac{9.319 \times 10^9\,\text{Hz}}{332.5 \times 10^{-3}\,\text{T}} = \boxed{2.002\bar{5}}$$

E14.23(a) The centre of the spectrum will occur at 332.5 mT. Proton 1 splits the line into two components with separation 2.0 mT and hence at 332.5 ± 1.0 mT. Proton 2 splits these two hyperfine lines into two, each with separation 2.6 mT, and hence the lines occur at 332.5 ± 1.0 ± 1.3 mT. The spectrum therefore consists of four lines of equal intensity at the fields 330.2 mT, 332.2 mT, 332.8 mT, 334.8 mT.

E14.24(a) We construct Figure 14.4(a) for $\cdot CH_3$ and Figure 14.4(b) for $\cdot CD_3$. The predicted intensity distribution is determined by counting the number of overlapping lines of equal intensity from which the hyperfine line is constructed.

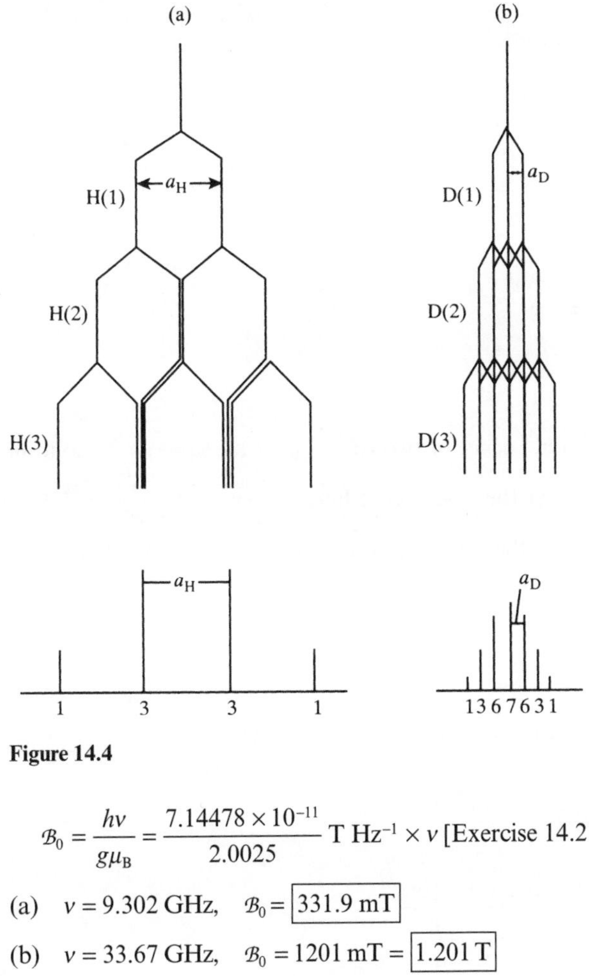

Figure 14.4

E14.25(a) $$\mathcal{B}_0 = \frac{h\nu}{g\mu_B} = \frac{7.14478 \times 10^{-11}}{2.0025}\,\text{T Hz}^{-1} \times \nu\,[\text{Exercise 14.21(a)}] = 35.68\,\text{mT} \times (\nu/\text{GHz})$$

(a) $\nu = 9.302$ GHz, $\mathcal{B}_0 = \boxed{331.9\,\text{mT}}$

(b) $\nu = 33.67$ GHz, $\mathcal{B}_0 = 1201\,\text{mT} = \boxed{1.201\,\text{T}}$

E14.26(a) Since the number of hyperfine lines arising from a nucleus of spin I is $2I+1$, we solve $2I+1=4$ and find that $\boxed{I=\frac{3}{2}}$.

COMMENT. Four lines of equal intensity could also arise from two inequivalent nuclei with $I=\frac{1}{2}$.

E14.27(a) The X nucleus produces six lines of equal intensity. The pair of H nuclei in XH_2 split each of these lines into a 1 : 2 : 1 triplet (Figure 14.5(a)). The pair of D nuclei ($I=1$) in XD_2 split each line into a 1 : 2 : 3 : 2 : 1 quintet (Figure 14.5(b)). The total number of hyperfine lines observed is then $6\times 3=18$ in XH_2 and $6\times 5=30$ in XD_2.

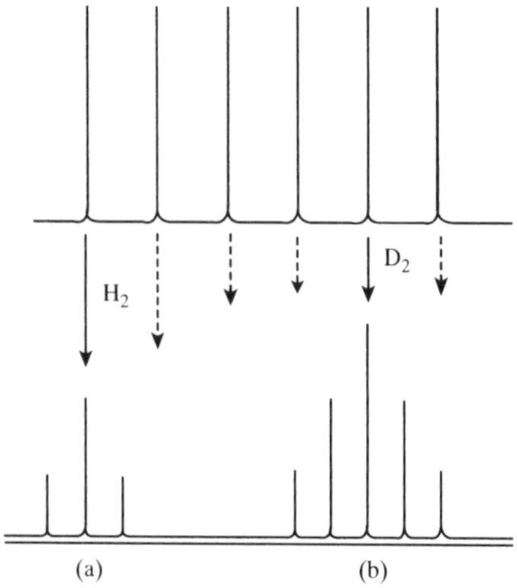

Figure 14.5

E14.28(a) We assume that the condition for the coalescence of two EPR lines is the same as the condition for the coalescence of two NMR lines, that is, the coalescence time is given by $\tau=\dfrac{\sqrt{2}}{\pi\delta\nu}$ [14.29]. Then, we need to calculate $\delta\nu$ corresponding to the frequency difference of the two EPR resonances, $g_{\parallel}$ and $g_{\perp}$. We have $h\nu=g\mu_B\mathcal{B}_0$ [14.39]. Solve for ν.

$$\nu=\frac{\mu_B}{h}\mathcal{B}_0 g\text{, and then}$$

$$\delta\nu=\frac{\mu_B}{h}\mathcal{B}_0(\delta g)=\frac{9.274\times10^{-24}\ \text{J T}^{-1}}{6.626\times10^{-34}\ \text{J s}}\times0.30\ \text{T}\times(2.032-2.012)=8.4\times10^{7}\ \text{s}^{-1}$$

$$\tau=\frac{\sqrt{2}}{\pi\delta\nu}=\frac{\sqrt{2}}{\pi\times8.4\times10^{7}\ \text{s}^{-1}}=5.36\times10^{-9}\ \text{s}$$

The tumbling rate is the inverse of this time, $\boxed{1.9\times10^{8}\ \text{s}^{-1}}$.

Solutions to problems

Solutions to numerical problems

P14.1 $g_I = -3.8260$ (Table 14.2)

$$\mathcal{B}_0 = \frac{h\nu}{g_I\mu_N} = \frac{(6.626\times10^{-34}\,\text{J Hz}^{-1})\times\nu}{(-)(3.8260)\times(5.0508\times10^{-27}\,\text{J T}^{-1})} = 3.429\times10^{-8}(\nu/\text{Hz})\text{T}$$

Therefore, with $\nu = 300$ MHz,

$$\mathcal{B}_0 = (3.429\times10^{-8})\times(300\times10^{6}\,\text{T}) = \boxed{10.3\,\text{T}}$$

$$\frac{\delta N}{N} \approx \frac{g_I\mu_N\mathcal{B}_0}{2kT}\ \text{[Exercise 14.7(a)]}$$

$$= \frac{(-3.8260)\times(5.0508\times10^{-27}\,\text{J T}^{-1})\times(10.3\,\text{T})}{(2)\times(1.381\times10^{-23}\,\text{J K}^{-1})\times(298\,\text{K})} = \boxed{2.42\times10^{-5}}$$

Since $g_I < 0$ (as for an electron, the magnetic moment is antiparallel to its spin), the $\boxed{\beta}$ state $\boxed{(m_I = -\tfrac{1}{2})}$ lies lower.

P14.3 The envelopes of maxima and minima of the curve are determined by T_2 through eqn 14.30, but the time interval between the maxima of this decaying curve corresponds to the reciprocal of the frequency difference $\Delta\nu$ between the pulse frequency ν° and the Larmor frequency ν_L, that is $\Delta\nu = |\nu^{\circ} - \nu_L|$

$$\Delta\nu = \frac{1}{0.10\,\text{s}} = 10\,\text{s}^{-1} = 10\,\text{Hz}$$

Therefore, the Larmor frequency is $\boxed{300\times10^{6}\pm10\,\text{Hz}}$.

According to eqns 14.30 and 14.32 the intensity of the maxima in the FID curve decays exponentially as e^{-t/T_2} Therefore, T_2 corresponds to the time at which the intensity has been reduced to $\frac{1}{e}$ of the original value. In the text figure, this corresponds to a time slightly before the fourth maximum has occurred, or about $\boxed{0.29\,\text{s}}$.

P14.5 The three rotational conformations of $F_2BrC-CBrCl_2$ are shown in Figure 14.6. In conformation I, the two F atoms are equivalent. However, in conformations II and III they are non-equivalent. At low temperature, the molecular residence time in conformation I is longer (because this conformation has the lowest repulsive energy of the large bromine atoms) than that of conformations II and III, which have equal residence times. With its longer residence time, we expect that the NMR signal intensity of conformation I should be stronger and we can conclude that it is the low-temperature singlet. It is a singlet because equivalent atoms do not have detectable spin–spin couplings.

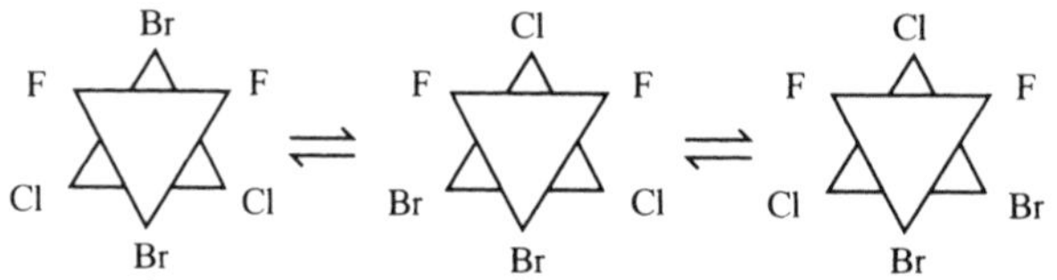

Figure 14.6

The fluorines of conformations II and III are non-equivalent, so their coupling is observed at low temperature. Fluorine has a nuclear spin of $\frac{1}{2}$, so we expect a doublet for each fluorine. These are observed with strong geminal coupling of 160 Hz. As temperature increases, the rate of rotation between II and III increases and the two fluorines become equivalent in these conformations, but remain distinct from I. The doublets collapse to singlets. With a further temperature increase to −30°C, and above, the rate of rotation about the C–C bond becomes so rapid that the residence times of the three conformations become equal. The very short residence times produce an average NMR signal that is a singlet and the fluorines appear totally equivalent.

The peaks for conformations II and III in Figure 14.60 show both spin–spin coupling and differences in chemical shift. The spin–spin splitting is 160 Hz. The difference in chemical shift can be estimated from the separation between the doublet centres, Δ:

$$\Delta = (J^2 + \delta\nu^2)^{1/2}$$

Δ is estimated from the figure to be 210 Hz. This yields for $\delta\nu$, the chemical shift,

$$\begin{aligned}\delta\nu &= (\Delta^2 - J^2)^{1/2}\\ &= (210^2 - 160^2)^{1/2}\,\text{Hz} \approx 140\,\text{Hz}\end{aligned}$$

Collapse to a single line will occur when the rate of interconversion satisfies

$$k \approx \frac{1}{\tau} \approx \frac{\pi\delta\nu}{\sqrt{2}}\,[14.29]$$

$$k = \frac{\pi \times 140\,\text{s}^{-1}}{\sqrt{2}} \approx \boxed{3 \times 10^2\,\text{s}^{-1}}$$

The relative intensities, I, of the lines at −80°C can be used to estimate the energy difference ($E_{\text{II}} - E_{\text{I}}$) between conformation I and conformations II and III. We assume that the relative intensities of the lines are proportional to the populations of conformers and that these populations follow the Boltzmann distribution (Chapters 2 and 15). Then,

$$\frac{I_{\text{I}}}{I_{\text{II}}} = \frac{e^{-E_{\text{I}}/RT}}{e^{-E_{\text{II}}/RT}} = e^{(E_{\text{II}} - E_{\text{I}})/RT}$$

$$E_{\text{II}} - E_{\text{I}} = RT\ln\left(\frac{I_{\text{I}}}{I_{\text{II}}}\right) = 8.314\,\text{J K}^{-1}\,\text{mol}^{-1} \times (273 - 80)\,\text{K}\ln(10)$$

$$= 3.7 \times 10^3\,\text{J mol}^{-1} = \boxed{3.7\,\text{kJ mol}^{-1}}$$

This energy difference is not, however, the rotational energy barrier between the rotational isomers. The latter can be estimated from the rate of interconversion between the isomers as a function of temperature. That rate of interconversion is roughly $3 \times 10^2\,\text{s}^{-1}$ at −30°C. At −60°C, as estimated from the line width at that temperature (see the Brief Illustration in Section 9.6(b)), it is roughly $\frac{1}{3}$ of that value, or ~ $1.0\times10^2\ \mathrm{s^{-1}}$. Assuming that the rate of interconversion satisfies an Arrhenius type of behaviour, $k \propto e^{-E_a/RT}$, where E_a is the rotational energy barrier,

$$\frac{k(-30\,°\mathrm{C})}{k(-60\,°\mathrm{C})} = 3 = e^{\left\{-\frac{E_a}{R}\left(\frac{1}{243\,\mathrm{K}} - \frac{1}{213\,\mathrm{K}}\right)\right\}}$$

$$E_a = \frac{R\ln 3}{\left(\frac{1}{213\,\mathrm{K}} - \frac{1}{243\,\mathrm{K}}\right)} = 1.6\times10^4\ \mathrm{J\,mol^{-1}} = \boxed{16\ \mathrm{kJ\,mol^{-1}}}$$

This value is typical of the rotational barriers observed in compounds of this kind.

P14.7 (a) The Karplus equation [14.27] for $^3J_{HH}$ is a linear equation in $\cos\phi$ and $\cos 2\phi$. The experimentally determined equation for $^3J_{SnSn}$ is a linear equation in $^3J_{HH}$. In general, if $F(f)$ is linear in f, and if $f(x)$ is linear in x, then $F(x)$ is linear. So we expect $^3J_{SnSn}$ to be linear in $\cos\phi$ and $\cos 2\phi$. This is demonstrated in (b).

(b) $^3J_{SnSn}/\mathrm{Hz} = 78.86(^3J_{HH}/\mathrm{Hz}) + 27.84$

Inserting the Karplus equation for $^3J_{HH}$ we obtain

$^3J_{SnSn}/\mathrm{Hz} = 78.86\{A + B\cos\phi + C\cos 2\phi\} + 27.84$. Using $A = 7$, $B = -1$, and $C = 5$, we obtain

$$^3J_{SnSn}/\mathrm{Hz} = \boxed{580 - 79\cos\phi + 395\cos 2\phi}$$

The plot of $^3J_{SnSn}$ is shown in Figure 14.7.

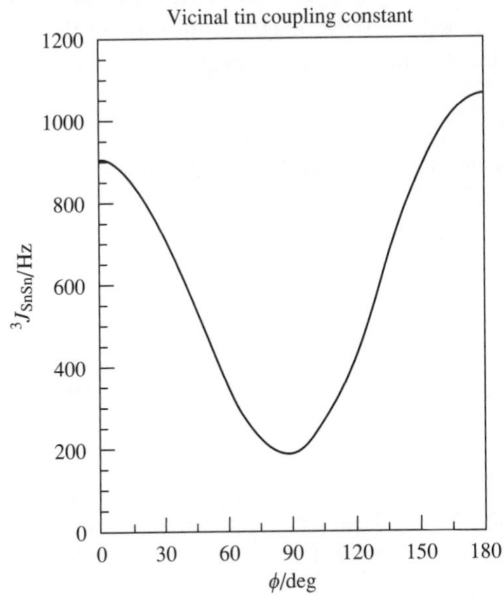

Figure 14.7

(c) A staggered configuration (Figure 14.8) with the $SnMe_3$ groups *trans* to each other is the preferred configuration. The $SnMe_3$ repulsions are then at a minimum.

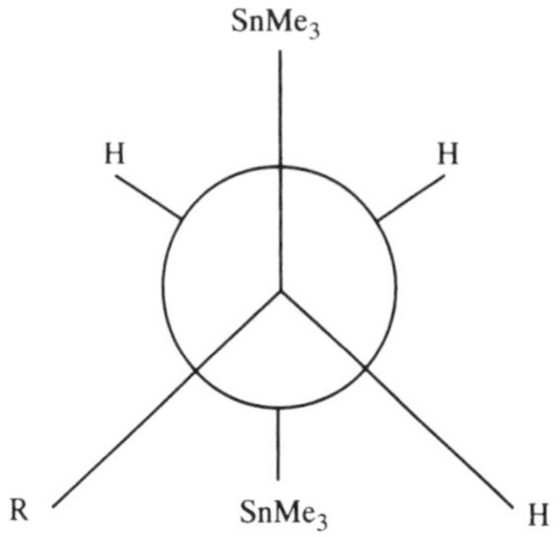

Figure 14.8

P14.9 $$\mathcal{B}_{\text{nuc}} = -\frac{\gamma\hbar\mu_0 m_I}{4\pi R^3}(1-3\cos^2\theta)\,[14.29] = \frac{g_I\mu_N\mu_0}{4\pi R^3}\left[m_I = +\frac{1}{2}, \theta = 0, \gamma\hbar = g_I\mu_N\right]$$

which rearranges to

$$R = \left(\frac{g_I\mu_N\mu_0}{4\pi\mathcal{B}_{\text{nuc}}}\right)^{1/3} = \left(\frac{(5.5857)\times(5.0508\times10^{-27}\ \text{JT}^{-1})\times(4\pi\times10^{-7}\ \text{T}^2\,\text{J}^{-1}\,\text{m}^3)}{(4\pi)\times(0.715\times10^{-3}\ \text{T})}\right)^{1/3}$$

$$= (3.946\times10^{-30}\ \text{m}^3)^{1/3} = \boxed{158\ \text{pm}}$$

P14.11 Refer to the figure in the solution to Exercise 14.24a. The width of the CH_3 spectrum is $3a_{\text{H}} = \boxed{6.9\ \text{mT}}$. The width of the CD_3 spectrum is $6a_{\text{D}}$. It seems reasonable to assume, since the hyperfine interaction is an interaction of the magnetic moments of the nuclei with the magnetic moment of the electron, that the strength of the interactions is proportional to the nuclear moments.

$$\mu = g_I\mu_N I \quad \text{or} \quad \mu_z = g_I\mu_N m_I\ [14.11, 14.10]$$

and thus nuclear magnetic moments are proportional to the nuclear g values, hence

$$a_{\text{D}} \approx \frac{0.85745}{5.5857}\times a_{\text{H}} = 0.1535a_{\text{H}} = 0.35\ \text{mT}$$

Therefore, the overall width is $6a_{\text{D}} = \boxed{2.1\ \text{mT}}$.

P14.13 We write $P(\text{N2s}) = \dfrac{5.7\ \text{mT}}{55.2\ \text{mT}} = \boxed{0.10}$ (10% of its time)

$$P(\text{N2p}_z) = \frac{1.3\ \text{mT}}{4.8\ \text{mT}} = \boxed{0.27}\ \text{(27\% of its time)}$$

The total probability is

(a) $P(\text{N}) = 0.10 + 0.27 = \boxed{0.37}$ (37% of its time).

(b) $P(\text{O}) = 1 - P(\text{N}) = \boxed{0.63}$ (63% of its time).

The hybridization ratio is

$$\frac{P(\text{N2p})}{P(\text{N2s})} = \frac{0.27}{0.10} = \boxed{2.7}$$

The unpaired electron therefore occupies an orbital that resembles an sp^3 hybrid on N, in accord with the radical's non-linear shape.

Solutions to theoretical problems

P14.15 Use eqn 14.22 and see the Brief Illustration.

$$\sigma_d = \frac{e^2\mu_0}{12\pi m_e}\left\langle\frac{1}{r}\right\rangle [14.22]$$

In order to obtain σ_d for the 1s and 2s orbitals of hydrogenic atoms of atomic number Z we need to evaluate $\left\langle\frac{1}{r}\right\rangle$ for these orbitals. This has already been done in the solution to Problem 9.19 in this manual.

(a) $\langle r^{-1}\rangle_{1s} = \boxed{\frac{Z}{a_0}}$ [P9.19]; therefore, $(\sigma_d)_{1s} = \boxed{\frac{e^2\mu_0 Z}{12\pi m_e a_0} = 1.78\times10^{-5}\,Z}$.

(b) $\langle r^{-1}\rangle_{2s} = \boxed{\frac{Z}{4a_0}}$ [P9.19]; therefore, $(\sigma_d)_{2s} = \boxed{\frac{e^2\mu_0 Z}{48\pi m_e a_0} = 4.45\times10^{-6}\,Z}$.

P14.17 Equation 14.28 may be written

$$\mathcal{B}_{\text{nuc}} = k(1-3\cos^2\theta)$$

where k is a constant independent of angle. Thus,

$$\langle\mathcal{B}_{\text{nuc}}\rangle \propto \int_0^{\pi}(1-3\cos^2\theta)\sin\theta\,d\theta\int_0^{2\pi}d\phi$$

$$\propto \int_1^{-1}(1-3x^2)dx\times2\pi\;[x=\cos\theta,\ dx=-\sin\theta\,d\theta]$$

$$\propto (x-x^3)\Big|_1^{-1} = 0$$

P14.19 The shape of spectral line $I(\omega)$ is related to the free induction decay signal $S(t)$ by

$$I(\omega) = a\,\text{Re}\int_0^{\infty}S(t)e^{i\omega t}\,dt,$$

where A is a constant and Re means take the real part of what follows. Calculate the line shape corresponding to an oscillating, decaying function

$$S(t) = \cos\omega_0 t\,e^{-t/\tau}$$

$$I(\omega) = a\,\mathrm{Re}\int_0^\infty S(t)\mathrm{e}^{\mathrm{i}\omega t}\,\mathrm{d}t$$

$$= a\,\mathrm{Re}\int_0^\infty \cos\omega_0 t\mathrm{e}^{-t/\tau+\mathrm{i}\omega t}\,\mathrm{d}t$$

$$= \frac{1}{2}a\,\mathrm{Re}\int_0^\infty (\mathrm{e}^{-\mathrm{i}\omega_0 t} + \mathrm{e}^{\mathrm{i}\omega_0 t})\mathrm{e}^{-t/\tau+\mathrm{i}\omega t}\,\mathrm{d}t$$

$$= \frac{1}{2}a\,\mathrm{Re}\int_0^\infty \{\mathrm{e}^{\mathrm{i}(\omega_0+\omega+\mathrm{i}/\tau)t} + \mathrm{e}^{-\mathrm{i}(\omega_0-\omega-\mathrm{i}/\tau)t}\}\mathrm{d}t$$

$$= -\frac{1}{2}a\,\mathrm{Re}\left[\frac{1}{\mathrm{i}(\omega_0+\omega+\mathrm{i}/\tau)} - \frac{1}{\mathrm{i}(\omega_0-\omega-\mathrm{i}/\tau)}\right]$$

when ω and ω_0 are similar to magnetic resonance frequencies (or higher), only the second term in brackets is significant $\left(\text{because } \frac{1}{(\omega_0+\omega)} \ll 1, \text{ but } \frac{1}{(\omega_0-\omega)} \text{ may be large if } \omega \approx \omega_0\right)$. Therefore,

$$I(\omega) \approx \frac{1}{2}a\,\mathrm{Re}\frac{1}{\mathrm{i}(\omega_0-\omega)+1/\tau}$$

$$= \frac{1}{2}a\,\mathrm{Re}\frac{-\mathrm{i}(\omega_0-\omega)+1/\tau}{(\omega_0-\omega)^2+1/\tau^2} = \frac{1}{2}a\frac{1/\tau}{(\omega_0-\omega)^2+1/\tau^2}$$

$$= \boxed{\left(\frac{1}{2}\right)\frac{a\tau}{1+(\omega_0-\omega)^2\tau^2}}$$

which is a Lorentzian line centred on ω_0, of amplitude $\frac{1}{2}A\tau$ and width $\frac{2}{\tau}$ at half-height.

P14.21 (a) Let $S(t) = \mathrm{e}^{-|t|/\tau}$. The signal in the time domain is proportional to $S(t)$. We want the signal $I(\omega)$ in the frequency domain, ω. $I(\omega)$ is the Fourier transform of $S(t)$.

$$I(\omega,\tau) = \frac{2(1/\tau)}{(1/\tau)^2+\omega^2} \quad \text{[See MB7.2]}$$

The maximum in this function occurs at $\omega = 0$ and is $I_{\max} = 2\tau$. We now solve for the value of ω, ($\omega_{1/2}$), when $I = \frac{1}{2}I_{\max} = \tau$. This yields $\boxed{\omega_{1/2} = \frac{1}{\tau}}$. This is the half-width (linewidth) of the signal at half-height.

(b) Let $S(t) = \mathrm{e}^{-t^2/\tau^2}$. We proceed as in part (a). We need the Fourier transform of $S(t)$. The Fourier transform of this function is most easily performed with a computer algebra system (CAS) such as MathCad®. Also, see Self-test MB7.2, which examines a form of this function. The result is

$$I(\omega,\tau) = (\pi\tau^2)^{1/2}\,\mathrm{e}^{-(\omega\tau/2)^2}$$

The maximum in this function occurs when $\omega = 0$, giving $I_{\max} = (\pi\tau^2)^{1/2}$. We now solve for the value of ω, ($\omega_{1/2}$), when $I = \frac{1}{2}, I_{\max} = \frac{1}{2}(\pi\tau^2)^{1/2}$.

$\frac{1}{2}(\pi\tau^2)^{1/2} = (\pi\tau^2)^{1/2}\, e^{-(\omega_{1/2}\tau/2)^2}$ or $\left(\frac{\omega_{1/2}\tau}{2}\right)^2 = \ln 2$. This yields

$$\boxed{\omega_{1/2} = 2(\ln 2)^{1/2}\left(\frac{1}{\tau}\right)}$$

Solutions to applications

P14.23 At, say, room temperature, the tumbling rate of benzene, the small molecule, in a mobile solvent, may be close to the Larmor frequency, and hence its spin-lattice relaxation time will be short. As the temperature increases, the tumbling rate may increase well beyond the Larmor frequency, resulting in an increased spin–lattice relaxation time.

For the large oligopeptide at room temperature, the tumbling rate may be well below the Larmor frequency, but with increasing temperature it will approach the Larmor frequency due to the increased thermal motion of the molecule combined with the decreased viscosity of the solvent. Therefore, the spin-lattice relaxation time may decrease.

P14.25 (a) The first figure displays spin densities computed by molecular modelling software (*ab initio*, density functional theory, Gaussian 98™).

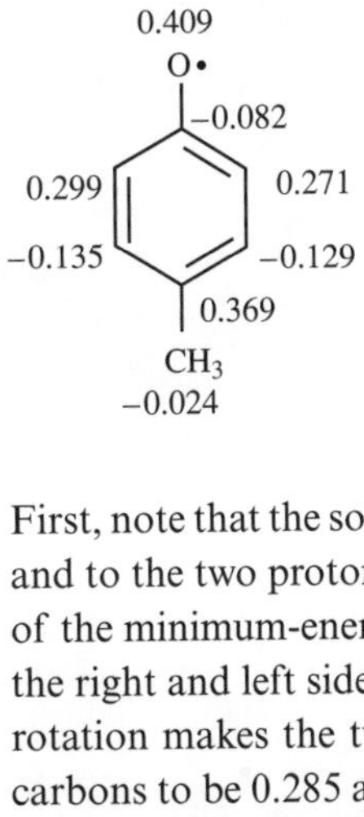

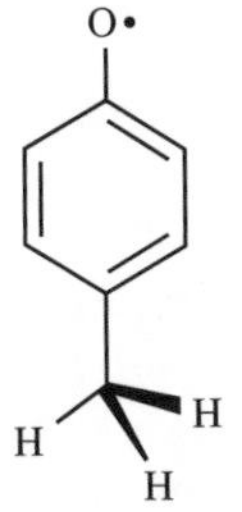

First, note that the software assigned slightly different values to the two protons *ortho* to the oxygen and to the two protons *meta* to the oxygen. This is undoubtedly a computational artifact, a result of the minimum-energy structure having one methyl proton in the plane of the ring, which makes the right and left sides of the ring slightly non-equivalent (see second figure). In fact, fast internal rotation makes the two halves of the ring equivalent. We will take the spin density at the *ortho* carbons to be 0.285 and those of the *meta* carbons to be −0.132. Predict the form of the spectrum by using the McConnell equation 14.42 for the splittings. The two *ortho* protons give rise to a 1:2:1 triplets with splitting $0.285 \times 2.25\ \text{mT} = 0.64\ \text{mT}$; these will in turn be split by the two *meta* protons into 1:2:1 triplets with splitting

$$0.132 \times 2.25\ \text{mT} = 0.297\ \text{mT}.$$

Finally, these lines will be seen to be further split by the three methyl protons into 1:3:3:1 quartets with splittings 1.045 mT. Note that the McConnel relationship cannot be applied to calculate these latter splittings, but the software generates them directly from calculated spin densities on the methyl hydrogens. The computed splittings agree well with experiment at the *ortho* positions (0.60 mT) and at the methyl hydrogens (1.19 mT), but less well at the *meta* positions (0.145 mT).

P14.27 We use $\nu = \dfrac{\gamma_N \mathcal{B}_{loc}}{2\pi} = \dfrac{\gamma_N}{2\pi}(1-\sigma)\mathcal{B}_0$ [14.17]

where $\mathcal{B}_0$ is the applied field.

Because shielding constants are quite small (a few parts per million) compared to 1, we may write for the purposes of this calculation

$$\nu = \frac{\gamma_N \mathcal{B}_0}{2\pi}$$

$$\nu_L - \nu_R = 100\ \text{Hz} = \frac{\gamma_N}{2\pi}(\mathcal{B}_L - \mathcal{B}_R)$$

$$\mathcal{B}_L - \mathcal{B}_R = \frac{2\pi \times 100\ \text{s}^{-1}}{\gamma_N}$$

$$= \frac{2\pi \times 100\ \text{s}^{-1}}{26.752 \times 10^7\ \text{T}^{-1}\,\text{s}^{-1}} = 2.35 \times 10^{-6}\ \text{T}$$

$$= 2.35\ \mu\text{T}$$

The field gradient required is then

$$\frac{2.35\ \mu\text{T}}{0.08\ \text{m}} = \boxed{29\ \mu\text{T m}^{-1}}$$

Note that knowledge of the spectrometer frequency, applied field, and the numerical value of the chemical shift (because constant) is not required.

15 Statistical thermodynamics 1: the concepts

Answers to discussion questions

D15.1 Consider the value of the partition function at the extremes of temperature. The limit of q as T approaches zero is simply g_0, the degeneracy of the ground state. As T approaches infinity, each term in the sum is simply the degeneracy of the energy level. If the number of levels is infinite, the partition function is infinite as well. In some special cases where we can effectively limit the number of states, the upper limit of the partition function is just the total number of states. In general, we see that the molecular partition function gives an indication of the average number of states thermally accessible to a molecule at the temperature of the system.

D15.3 The **population** of a state is the number of molecules of a sample that are *in* that state, on average. The population of a state is the number of molecules in the sample times the probability of the state. The **configuration** of a system is a list of populations in order of the energy of the corresponding states. For example, $\{N-3, 2, 1, 0, \ldots\}$ is a possible configuration of a system of N molecules in which all but three molecules are in the ground state, two are in the next lowest state, one in the next state, etc. The **weight** of a configuration is the number of ways a given configuration can be comprised. In our example, the single molecule in the second excited state could be any of the system's N molecules, so there are N ways to arrange that state alone. The weight of the configuration $\{N_0, N_1, N_2, \ldots\}$ is

$$\mathcal{W} = \frac{N!}{N_0!N_1!N_2!\cdots} \quad [15.1]$$

When N is large (as it is for any macroscopic sample), the most probable configuration is so much more probable than all other possible configurations that it is the system's dominant configuration. See Section 15.1.

D15.5 Distinct 'states' differ in one or more observable physical quantity such as energy or angular momentum. Thus, distinct states may have the same energy (may belong to the same 'energy level') if they differ in some other quantity; these are degenerate states. Each energy level has at least one state—more for degenerate levels. The partition function is a sort of count of **states**; it is a sum over all **states** of each state's Boltzmann factor, namely $e^{-\beta\varepsilon_i}$. Because Boltzmann factors depend only on energy, the sum may practically be done as a sum over energy levels, as long as the degeneracy is included. That is, the sum over states within a given energy level is a sum of identical Boltzmann factors, equal to the degeneracy of the level times the Boltzmann factor.

D15.7 Identical particles can be regarded as distinguishable when they are localized as in a crystal lattice where we can assign a set of coordinates to each particle. Strictly speaking, it is the lattice site that carries the set of coordinates, but as long as the particle is fixed to the site, it too can be considered distinguishable. Identical particles that are free to move and intermingle with each other are indistinguishable because their trajectories cannot be followed, even in principle. In fact, it would be more accurate to say that such particles do not have trajectories, according to quantum mechanics.

Solutions to exercises

E15.1(a) The weight is given by

$$\mathcal{W} = \frac{N!}{N_0!N_1!N_2!\cdots} = \frac{16!}{0!1!2!3!8!0!0!0!0!2!}$$

This can be simplified by removing the common factor of 8! from the numerator and denominator and noting that $0! = 1! = 1$:

$$\mathcal{W} = \frac{16 \times 15 \times 14 \times 13 \times 12 \times 11 \times 10 \times 9}{2 \times (3 \times 2) \times 2} = \boxed{21\,621\,600} = 2.16216 \times 10^7$$

E15.2(a) Apply eqn 15.7 to each level. For two non-degenerate levels,

$$\frac{N_2}{N_1} = \frac{e^{-\beta\varepsilon_2}}{q} \times \frac{q}{e^{-\beta\varepsilon_1}} = e^{-\beta(\varepsilon_2-\varepsilon_1)} = e^{-\beta\Delta\varepsilon} = e^{-\Delta\varepsilon/kT}$$

Hence, as $T \to \infty$, $\frac{N_2}{N_1} = e^{-0} = \boxed{1}$. That is, the two levels would become equally populated.

E15.3(a) For two non-degenerate levels,

$$\frac{N_2}{N_1} = e^{-\Delta\varepsilon/kT} \text{ [Exercise 15.2(a)]}$$

$$\text{so,} \quad \ln\frac{N_2}{N_1} = -\frac{\Delta\varepsilon}{kT} \quad \text{and} \quad T = -\frac{\Delta\varepsilon}{k\ln\frac{N_2}{N_1}}$$

$$\text{Thus,} \quad T = -\frac{6.626 \times 10^{-34}\ \text{J s} \times 2.998 \times 10^{10}\ \text{cm s}^{-1} \times 400\ \text{cm}^{-1}}{1.381 \times 10^{-23}\ \text{J K}^{-1} \times \ln(1/3)} = \boxed{524\ \text{K}}$$

E15.4(a) For two non-degenerate levels,

$$\frac{N_2}{N_1} = e^{-\Delta\varepsilon/kT} \text{ [Exercise 15.2(a)]}$$

so, assuming that other states (if any) are negligibly populated,

$$\ln\frac{N_2}{N_1} = -\frac{\Delta\varepsilon}{kT} \quad \text{and} \quad T = -\frac{\Delta\varepsilon}{k\ln\frac{N_2}{N_1}}$$

$$\text{Thus,}\quad T=-\frac{6.626\times10^{-34}\ \text{J s}\times2.998\times10^{10}\ \text{cm s}^{-1}\times540\ \text{cm}^{-1}}{1.381\times10^{-23}\ \text{J K}^{-1}\times\ln(10/90)}=\boxed{35\bar{4}\ \text{K}}$$

E15.5(a) (a) The thermal wavelength is

$$\Lambda=\frac{h}{(2\pi mkT)^{1/2}}\ [15.19]$$

We need the molecular mass, not the molar mass:

$$m=\frac{150\times10^{-3}\ \text{kg mol}^{-1}}{6.022\times10^{23}\ \text{mol}^{-1}}=2.49\times10^{-25}\ \text{kg}$$

$$\text{So}\quad \Lambda=\frac{6.626\times10^{-34}\ \text{J s}}{(2\pi\times2.49\times10^{-25}\ \text{kg}\times1.381\times10^{-23}\ \text{J K}^{-1}\times T)^{1/2}}=\frac{1.43\times10^{-10}\ \text{m}}{(T/\text{K})^{1/2}}$$

(i) $$T=300\ \text{K}: \Lambda=\frac{1.43\times10^{-10}\ \text{m}}{(300)^{1/2}}=\boxed{8.23\times10^{-12}\ \text{m}}=\boxed{8.23\ \text{pm}}$$

(ii) $$T=3000\ \text{K}: \Lambda=\frac{1.43\times10^{-10}\ \text{m}}{(3000)^{1/2}}=\boxed{2.60\times10^{-12}\ \text{m}}=\boxed{2.60\ \text{pm}}$$

(b) The translational partition function is

$$q^{\text{T}}=\frac{V}{\Lambda^3}\ [15.19]$$

(i) $$T=300\ \text{K}: q^{\text{T}}=\frac{1.00\ \text{cm}^3}{(8.23\times10^{-12}\ \text{m})^3}=\frac{(1.00\times10^{-2}\ \text{m})^3}{(8.23\times10^{-12}\ \text{m})^3}=\boxed{1.79\times10^{27}}$$

(ii) $$T=3000\ \text{K}: q^{\text{T}}=\frac{(1.00\times10^{-2}\ \text{m})^3}{(2.60\times10^{-12}\ \text{m})^3}=\boxed{5.67\times10^{28}}$$

E15.6(a) $$q^{\text{T}}=\frac{V}{\Lambda^3}\ [15.19],\ \text{implying that}\ \frac{q}{q'}=\left(\frac{\Lambda'}{\Lambda}\right)^3$$

$$\text{However, as}\ \Lambda\propto\frac{1}{m^{1/2}},\ \frac{q}{q'}=\left(\frac{m}{m'}\right)^{3/2}$$

Therefore, $\frac{q}{q'}=2^{3/2}=\boxed{2.83}$ because D_2 is twice as massive as H_2.

E15.7(a) $$S=k\ln\mathcal{W}\ [15.27]$$

Therefore,

$$\left(\frac{\partial S}{\partial U}\right)_V=\frac{k}{\mathcal{W}}\left(\frac{\partial\mathcal{W}}{\partial U}\right)_V$$

$$\text{so}\quad\left(\frac{\partial\mathcal{W}}{\partial U}\right)_V=\frac{\mathcal{W}}{k}\left(\frac{\partial S}{\partial U}\right)_V$$

But from eqn 3.48

$$\left(\frac{\partial U}{\partial S}\right)_V = T$$

So, $$\left(\frac{\partial S}{\partial U}\right)_V = \frac{1}{T}$$

then $$\left(\frac{\partial \mathcal{W}}{\partial U}\right)_V = \frac{\mathcal{W}}{k}\left(\frac{1}{T}\right)$$

Therefore,

$$\frac{\Delta \mathcal{W}}{\mathcal{W}} \approx \frac{\Delta U}{kT} = \frac{100\times 10^{3}\,\text{J}}{(1.381\times 10^{-23}\,\text{J K}^{-1})\times 298\,\text{K}} = \boxed{2.4\times 10^{25}}$$

E15.8(a) The high-temperature expression for the rotational partition function of a linear molecule is

$$q^{\text{R}} = \frac{kT}{\sigma hc\tilde{B}}\,[16.15\text{b}],\quad \tilde{B} = \frac{\hbar}{4\pi cI}\,[12.7],\quad I = \mu R^2\,[\text{Table 12.1}]$$

Hence, $$q = \frac{8\pi^2 kTI}{\sigma h^2} = \frac{8\pi^2 kT\mu R^2}{\sigma h^2}$$

For O_2, $\mu = \frac{1}{2}m(\text{O}) = \frac{1}{2}\times 16.00m_{\text{u}} = 8.00m_{\text{u}}$, and $\sigma = 2$; therefore

$$q = \frac{(8\pi^2)\times(1.381\times 10^{-23}\,\text{J K}^{-1})\times(300\,\text{K})\times(8.00\times 1.6605\times 10^{-27}\,\text{kg})\times(1.2075\times 10^{-10}\,\text{m})^2}{(2)\times(6.626\times 10^{-34}\,\text{J s})^2}$$

$$= \boxed{72.2}$$

E15.9(a) The high-temperature expression for the rotational partition function of a non-linear molecule is

$$q^{\text{R}} = \frac{1}{\sigma}\left(\frac{kT}{hc}\right)^{3/2}\left(\frac{\pi}{\tilde{A}\tilde{B}\tilde{C}}\right)^{1/2}\quad [16.14\text{b with }\sigma]$$

Combining the universal constants yields

$$\left(\frac{k}{hc}\right)^{3/2}\pi^{1/2} = \left(\frac{1.381\times 10^{-23}\,\text{J K}^{-1}}{6.626\times 10^{-34}\,\text{J s}\times 2.998\times 10^{10}\,\text{cm s}^{-1}}\right)^{3/2}\pi^{1/2} = 1.027\,\text{K}^{-3/2}\,\text{cm}^{-3/2}$$

Thus, $$q^{\text{R}} = \frac{1.027}{\sigma}\frac{(T/\text{K})^{3/2}}{(\tilde{A}\tilde{B}\tilde{C}/\text{cm}^{-3})^{1/2}} = \frac{1.027\times(T/\text{K})^{3/2}}{(3.1752\times 0.3951\times 0.3505)^{1/2}} = 1.549\times(T/\text{K})^{3/2}$$

(a) At 25°C, $q^{\text{R}} = 1.549\times(298)^{3/2} = \boxed{7.97\times 10^{3}}$

(b) At 100°C, $q^{\text{R}} = 1.549\times(373)^{3/2} = \boxed{1.12\times 10^{4}}$

E15.10(a) The rotational partition function of a non-symmetrical linear molecule is

$$q^{\text{R}} = \sum_J (2J+1)\text{e}^{-hc\tilde{B}J(J+1)/kT}\quad [16.13\text{ with }\beta = 1/kT]$$

Use $\dfrac{hc\tilde{B}}{k} = \dfrac{6.626\times10^{-34}\ \text{J s}\times2.998\times10^{10}\ \text{cm s}^{-1}\times1.931\ \text{cm}^{-1}}{1.381\times10^{-23}\ \text{J K}^{-1}} = 2.778\ \text{K}$,

so $q^{\text{R}} = \sum_J (2J+1)\text{e}^{-2.778\ \text{K}\times J(J+1)/T}$

Use a spreadsheet or other mathematical software to evaluate the terms of the sum and to sum the terms until they converge. The high-temperature expression is

$$q^{\text{R}} = \frac{kT}{hc\tilde{B}} = \frac{T}{2.778\ \text{K}}$$

The explicit and high-temperature expressions are compared in Figure 15.1. The difference between the two expressions is very nearly constant after the first few degrees. That difference drops to 5% of the explicit sum at $\boxed{18\ \text{K}}$. As both expressions rise, their absolute difference becomes relatively smaller.

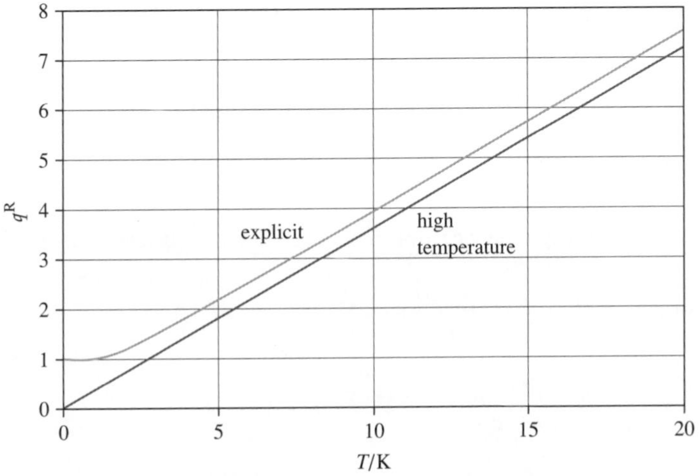

Figure 15.1

E15.11(a) The rotational partition function of a spherical rotor molecule, ignoring nuclear statistics, is

$$q^{\text{R}} = \sum_J g_J \text{e}^{-\varepsilon_J^{\text{R}}/kT}\ [15.9] = \sum_J (2J+1)^2\text{e}^{-hc\tilde{B}J(J+1)/kT}\ [12.8\ \&\ \text{Section 12.4(d)}]$$

Use $\dfrac{hc\tilde{B}}{k} = \dfrac{6.626\times10^{-34}\ \text{J s}\times2.998\times10^{10}\ \text{cm s}^{-1}\times5.241\ \text{cm}^{-1}}{1.381\times10^{-23}\ \text{J K}^{-1}} = 7.539\ \text{K}$

so $q^{\text{R}} = \sum_J (2J+1)^2\text{e}^{-7.539\ \text{K}\times J(J+1)/T}$

Use a spreadsheet or other mathematical software to evaluate the terms of the sum and to sum the terms until they converge. The high-temperature expression is eqn 16.14b, neglecting σ and with $\tilde{A} = \tilde{B} = \tilde{C}$:

$$q^{\text{R}} = \pi^{1/2}\left(\frac{kT}{hc\tilde{B}}\right)^{3/2} = \pi^{1/2}\left(\frac{T}{7.539\ \text{K}}\right)^{3/2}$$

The explicit and high-temperature expressions are compared in Figure 15.2. The difference between the two expressions actually grows (albeit rather slowly) after the first few degrees. Because both expressions grow faster than does the difference between them, the relative difference drops; it reaches 5% of the explicit sum at $\boxed{37\ \text{K}}$.

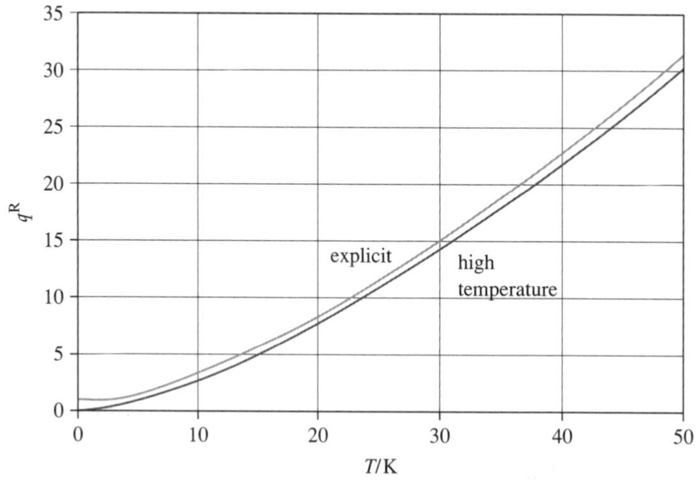

Figure 15.2

E15.12(a) The rotational partition function of a symmetric rotor molecule, ignoring nuclear statistics, is

$$q^{\text{R}} = \sum_{J,K} g_{J,K} \text{e}^{-\varepsilon^{\text{R}}_{J,K}/kT} \,[15.9] = \sum_{J=0} (2J+1)\text{e}^{-hc\tilde{B}J(J+1)/kT}\left(1 + 2\sum_{K=1}^{J} \text{e}^{-hc(\tilde{A}-\tilde{B})K^2/kT}\right) [12.12]$$

Use $\dfrac{hc\tilde{B}}{k} = \dfrac{6.626\times10^{-34}\ \text{J s}\times 2.998\times10^{10}\ \text{cm s}^{-1}\times 0.443\ \text{cm}^{-1}}{1.381\times10^{-23}\ \text{J K}^{-1}} = 0.637\ \text{K},$

and $\dfrac{hc(\tilde{A}-\tilde{B})}{k} = \dfrac{6.626\times10^{-34}\ \text{J s}\times 2.998\times10^{10}\ \text{cm s}^{-1}\times (5.097-0.443)\ \text{cm}^{-1}}{1.381\times10^{-23}\ \text{J K}^{-1}} = 6.694\ \text{K}$

so $q^{\text{R}} = \displaystyle\sum_{J=0} (2J+1)\text{e}^{-0.637\ \text{K}\times J(J+1)/T}\left(1 + 2\sum_{K=1}^{J} \text{e}^{-6.694\ \text{K}\times K^2/T}\right)$

Write a brief computer program or use other mathematical software to evaluate the terms of the sum and to sum the terms until they converge. Nested sums are straightforward to program in languages such as BASIC or FORTRAN, whereas in spreadsheets they are more unwieldy. Compare the results of the direct sum with the high-temperature expression, eqn 16.14b, with $\tilde{B} = \tilde{C}$:

$$q^{\text{R}} = \left(\frac{\pi}{\tilde{A}}\right)^{1/2}\left(\frac{kT}{hc}\right)^{3/2}\frac{1}{\tilde{B}}.$$

The explicit and high-temperature expressions are compared in Figure 15.3. The difference between the two expressions actually grows (albeit rather slowly) after the first few degrees. Because both expressions grow faster than does the difference between them, the relative difference drops; it reaches 5% of the explicit sum at $\boxed{4.5\ \text{K}}$.

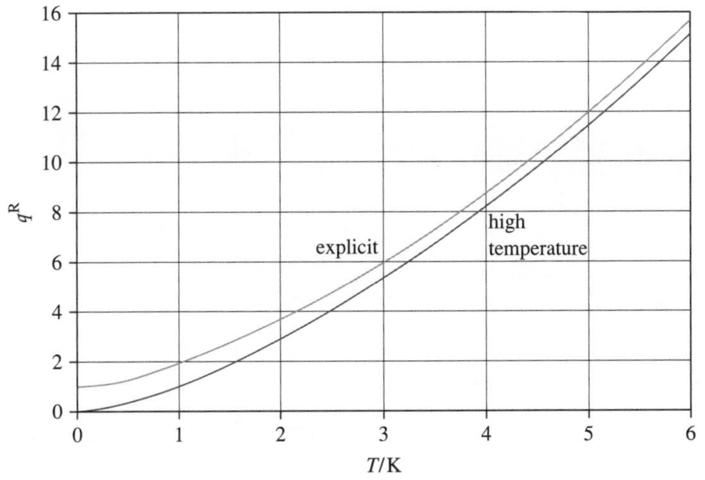

Figure 15.3

E15.13(a) The symmetry number is the order of the rotational subgroup of the group to which a molecule belongs (except for linear molecules, for which $\sigma = 2$ if the molecule has inversion symmetry and 1 otherwise).

(a) CO: full group $C_{\infty v}$; subgroup C_1; hence $\sigma = \boxed{1}$

(b) O_2: full group $D_{\infty h}$; subgroup C_2; $\sigma = \boxed{2}$

(c) H_2S: full group C_{2v}; subgroup C_2; $\sigma = \boxed{2}$

(d) SiH_4: full group T_d; subgroup T; $\sigma = \boxed{12}$

(e) $CHCl_3$: full group C_{3v}; subgroup C_3; $\sigma = \boxed{3}$

E15.14(a) Ethene has four indistinguishable atoms that can be interchanged by rotations, so $\sigma = 4$. The rotational partition function of a non-linear molecule is [Exercise 15.9(a)]

$$q^R = \frac{1.027}{\sigma}\frac{(T/\mathrm{K})^{3/2}}{(\tilde{A}\tilde{B}\tilde{C}/\mathrm{cm}^{-3})^{1/2}} = \frac{1.027 \times (298.15)^{3/2}}{4 \times (4.828 \times 1.0012 \times 0.8282)^{1/2}} = \boxed{660.6}$$

E15.15(a) The partition function for a mode of molecular vibration is

$$q^V = \sum_{\upsilon} e^{-\upsilon hc\tilde{\nu}/kT} = \frac{1}{1 - e^{-hc\tilde{\nu}/kT}} \quad [16.19 \text{ with } \beta = 1/kT]$$

Use $$\frac{hc\tilde{\nu}}{k} = \frac{6.626 \times 10^{-34}\,\mathrm{J\,s} \times 2.998 \times 10^{10}\,\mathrm{cm\,s^{-1}} \times 323.2\,\mathrm{cm^{-1}}}{1.381 \times 10^{-23}\,\mathrm{J\,K^{-1}}} = 464.9\,\mathrm{K},$$

so $$q^V = \sum_{\upsilon} e^{-\upsilon hc\tilde{\nu}/kT} = \frac{1}{1 - e^{-464.9\,\mathrm{K}/T}}$$

The high-temperature expression is

$$q^V = \frac{kT}{hc\tilde{\nu}} = \frac{T}{464.9\,\mathrm{K}} \quad [16.21]$$

The explicit and high-temperature expressions are compared in Figure 15.4. The high-temperature expression reaches 95% of the explicit sum at $\boxed{4500\,\mathrm{K}}$.

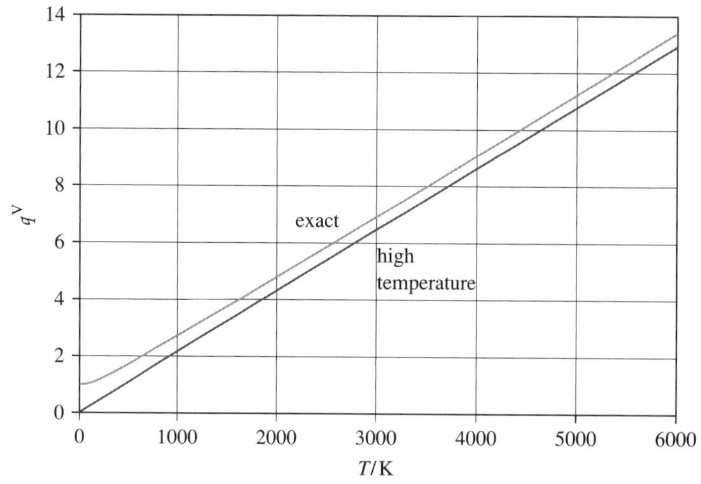

Figure 15.4

E15.16(a) The partition function for a mode of molecular vibration is

$$q^{V}_{\text{mode}} = \sum_{v} e^{-vhc\tilde{\nu}/kT} = \frac{1}{1-e^{-hc\tilde{\nu}/kT}} \quad [16.19 \text{ with } \beta = 1/kT]$$

and the overall vibrational partition function is the product of the partition functions of the individual modes. (See Example 16.3.) We draw up the following table:

mode	1	2	3	4
$\tilde{\nu}/\text{cm}^{-1}$	658	397	397	1535
$hc\tilde{\nu}/kT$	1.893	1.142	1.142	4.416
q^{V}_{mode}	1.177	1.469	1.469	1.012

The overall vibrational partition function is

$$q^{V} = 1.177 \times 1.469 \times 1.469 \times 1.012 = \boxed{2.571}$$

E15.17(a) The partition function for a mode of molecular vibration is

$$q^{V}_{\text{mode}} = \sum_{v} e^{-vhc\tilde{\nu}/kT} = \frac{1}{1-e^{-hc\tilde{\nu}/kT}} \quad [16.19 \text{ with } \beta = 1/kT]$$

and the overall vibrational partition function is the product of the partition functions of the individual modes. (See Example 16.3.) We draw up the following table, including the degeneracy of each level:

mode	1	2	3	4
$\tilde{\nu}/\text{cm}^{-1}$	459	217	776	314
g_{mode}	1	2	3	3
$hc\tilde{\nu}/kT$	1.320	0.624	2.232	0.903
q^{V}_{mode}	1.364	2.15	1.120	1.681

The overall vibrational partition function is

$$q^{\mathrm{V}} = 1.364 \times 2.15^2 \times 1.120^3 \times 1.681^3 = \boxed{42.3}$$

E15.18(a) $$q = \sum_{\text{levels}} g_j \mathrm{e}^{-\beta\varepsilon_j} \,[15.9] = \sum_{\text{levels}} g_j \mathrm{e}^{-hc\tilde{\nu}_j/kT} = 3 + \mathrm{e}^{-hc\tilde{\nu}_1/kT} + 3\mathrm{e}^{-hc\tilde{\nu}_2/kT}$$

where $$\frac{hc\tilde{\nu}_j}{kT} = \frac{6.626 \times 10^{-34}\ \mathrm{J\ s} \times 2.998 \times 10^{10}\ \mathrm{cm\ s^{-1}} \times \tilde{\nu}_j}{1.381 \times 10^{-23}\ \mathrm{J\ K^{-1}} \times 1900\ \mathrm{K}} = 7.571 \times 10^{-4} \times (\tilde{\nu}_j/\mathrm{cm^{-1}})$$

Therefore,

$$q = 3 + \mathrm{e}^{-7.571\times10^{-4}\times3500} + 3\mathrm{e}^{-7.571\times10^{-4}\times4700} = 3 + 0.0707 + 3 \times 0.0285 = \boxed{3.1561}$$

E15.19(a) $$U_{\mathrm{m}} - U_{\mathrm{m}}(0) = -\frac{N_{\mathrm{A}}}{q}\left(\frac{\mathrm{d}q}{\mathrm{d}\beta}\right)_V \quad [15.24\text{a for 1 mol}]$$

Use $$\frac{hc\tilde{\nu}_j}{kT} = hc\tilde{\nu}_j\beta = 7.571 \times 10^{-4} \times (\tilde{\nu}_j/\mathrm{cm^{-1}}) \text{ [Exercise 15.18(a)]}$$

and $$q = 3 + \mathrm{e}^{-hc\tilde{\nu}_1\beta} + 3\mathrm{e}^{-hc\tilde{\nu}_2\beta} = 3.1561 \text{ [Exercise 15.18(a)]}$$

Thus, $$U_{\mathrm{m}} - U_{\mathrm{m}}(0) = -\frac{N_{\mathrm{A}}hc}{q}(\tilde{\nu}_1\mathrm{e}^{-hc\tilde{\nu}_1\beta} + 3\tilde{\nu}_2\mathrm{e}^{-hc\tilde{\nu}_2\beta})$$

$$= \frac{6.022 \times 10^{23}\ \mathrm{mol^{-1}} \times 6.626 \times 10^{-34}\ \mathrm{J\ s} \times 2.998 \times 10^{10}\ \mathrm{cm\ s^{-1}}}{3.1561}$$

$$\times (3500\ \mathrm{cm^{-1}} \times \mathrm{e}^{-7.571\times10^{-4}\times3500} + 3 \times 4700\ \mathrm{cm^{-1}} \times \mathrm{e}^{-7.571\times10^{-4}\times4700})$$

$$= \boxed{+2.46 \times 10^3\ \mathrm{J}} = \boxed{+2.46\ \mathrm{kJ}}$$

E15.20(a) The partition function is

$$q = \sum_i \mathrm{e}^{-\beta\varepsilon_i} = \boxed{1 + \mathrm{e}^{-2\mu_{\mathrm{B}}\beta\mathcal{B}}} \text{ [energies measured from lower state]}$$

The mean energy per electron is

$$\langle\varepsilon\rangle = \frac{E(T)}{N} = -\frac{1}{q}\frac{\mathrm{d}q}{\mathrm{d}\beta} = \boxed{\frac{2\mu_{\mathrm{B}}\mathcal{B}\mathrm{e}^{-2\mu_{\mathrm{B}}\beta\mathcal{B}}}{1 + \mathrm{e}^{-2\mu_{\mathrm{B}}\beta\mathcal{B}}}} \quad [15.22]$$

Alternatively, if we measure energy with respect to zero magnetic field (rather than to ground-state energy), then

$$\langle\varepsilon\rangle = \varepsilon_{\mathrm{gs}} - \frac{1}{q}\frac{\mathrm{d}q}{\mathrm{d}\beta} = \boxed{-\mu_{\mathrm{B}}\mathcal{B} + \frac{2\mu_{\mathrm{B}}\mathcal{B}\,\mathrm{e}^{-2\mu_{\mathrm{B}}\beta\mathcal{B}}}{1 + \mathrm{e}^{-2\mu_{\mathrm{B}}\beta\mathcal{B}}}}$$

We write $x = 2\mu_{\mathrm{B}}\beta\mathcal{B}$, so the partition function becomes

$$q = 1 + \mathrm{e}^{-x}$$

and the mean energy, scaled by the energy separation, becomes

$$\frac{\langle\varepsilon\rangle}{2\mu_B\mathcal{B}} = -\frac{1}{2} + \frac{e^{-x}}{1+e^{-x}} = -\frac{1}{2} + \frac{1}{e^{x}+1}$$

The functions are plotted in Figures 15.5(a) and (b). The effect of increasing the magnetic field is to concentrate population into the lower level.

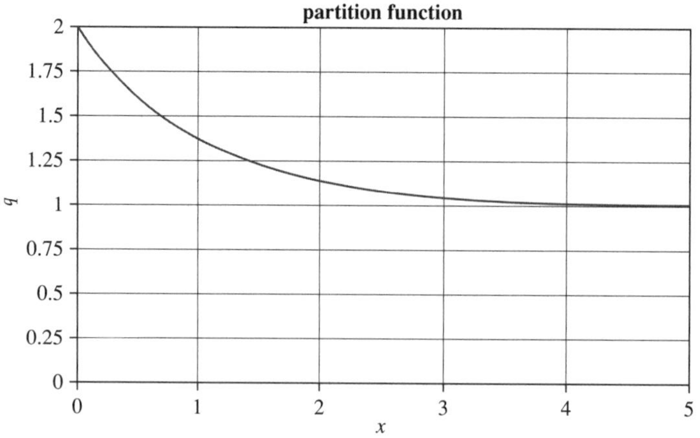

Figure 15.5(a)

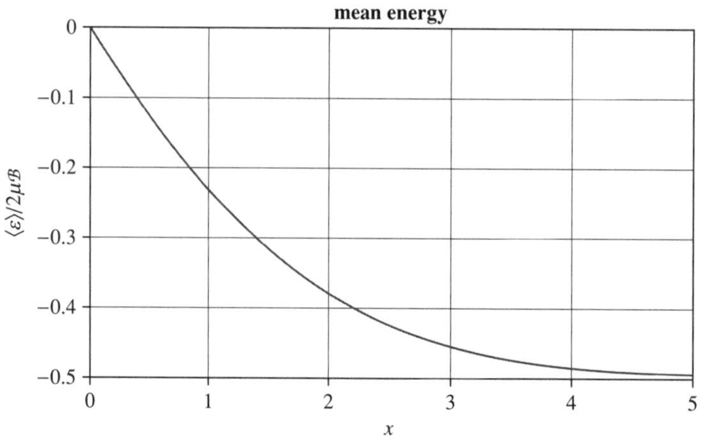

Figure 15.5(b)

The relative populations are

$$\frac{N_+}{N_-} = e^{-\beta\Delta\varepsilon}\,[\text{Exercise 15.2(a)}] = e^{-x}$$

$$x = 2\mu_B\beta\mathcal{B} = \frac{2\mu_B\mathcal{B}}{kT} = \frac{2\times 9.274\times 10^{-24}\,\text{J T}^{-1}\times 1.0\,\text{T}}{(1.381\times 10^{-23}\,\text{J K}^{-1})T} = \frac{1.343}{T/\text{K}}$$

(a) At $T = 4$ K, $\dfrac{N_+}{N_-} = e^{-1.343/4} = \boxed{0.71}$ (b) At $T = 298$ K, $\dfrac{N_+}{N_-} = e^{-1.343/298} = \boxed{0.996}$

E15.21(a) The energy separation is $\varepsilon = k \times 10\ \text{K}$

(a) $\dfrac{n_1}{n_0} = \text{e}^{-\beta(\varepsilon_1-\varepsilon_0)}$ [Exercise 15.2(a)] $= \text{e}^{-\beta\varepsilon} = \text{e}^{-10/(T/\text{K})}$

(1) $T = 1.0\ \text{K};\quad \dfrac{n_1}{n_0} = \text{e}^{-10} = \boxed{5 \times 10^{-5}}$

(2) $T = 10\ \text{K};\quad \dfrac{n_1}{n_0} = \text{e}^{-1.0} = \boxed{0.4}$

(3) $T = 100\ \text{K};\quad \dfrac{n_1}{n_0} = \text{e}^{-0.100} = \boxed{0.905}$

(b) $q = \sum_j g_j \text{e}^{-\varepsilon_j/kT} = \text{e}^0 + \text{e}^{-1.0} = 1.3\overline{68} = \boxed{1.4}$

(c) $E(T) = -\dfrac{N}{q}\dfrac{\text{d}q}{\text{d}\beta}$ [15.22]

$$E_\text{m}(T) = -\frac{N_\text{A}}{q}\{-(10\ \text{K}) \times k\text{e}^{-(10\ \text{K}\times k\beta)}\} = \frac{(10\ \text{K}) \times R}{1 + \text{e}^{(10\ \text{K}\times k\beta)}} = \frac{(10\ \text{K}) \times R}{1 + \text{e}^{10/(T/\text{K})}}$$

At $T = 10$ K, $E_\text{m} = \dfrac{(10\ \text{K}) \times R}{1 + \text{e}} = \dfrac{(10\ \text{K}) \times (8.3145\ \text{J K}^{-1}\,\text{mol}^{-1})}{3.718} = 22.\overline{36}\ \text{J mol}^{-1} = \boxed{22\ \text{J mol}^{-1}}$

(d) $C_V = \left(\dfrac{\partial U}{\partial T}\right)_V = \dfrac{\text{d}E}{\text{d}T}\quad \dfrac{\text{d}}{\text{d}T} = -\dfrac{1}{kT^2}\dfrac{\text{d}}{\text{d}\beta}$

$$C_{V,\text{m}} = -\frac{1}{kT^2}\frac{\text{d}}{\text{d}\beta}\left(\frac{(10\ \text{K}) \times R}{1 + \text{e}^{(10\ \text{K}\times k\beta)}}\right) = R\left(\frac{\text{e}}{(1+\text{e})^2}\right) = 0.19\overline{7}R = \boxed{1.6\ \text{J K}^{-1}\,\text{mol}^{-1}}$$

(e) $S = \dfrac{U - U(0)}{T} + Nk \ln q$ [15.28] $= \dfrac{E}{T} + nR \ln q$

$$S_\text{m} = \left(\frac{22.\overline{36}\ \text{J mol}^{-1}}{10\ \text{K}}\right) + (R \ln 1.3\overline{68}) = \boxed{4.8\ \text{J K}^{-1}\,\text{mol}^{-1}}$$

E15.22(a) $\dfrac{n_1}{n_0} = \text{e}^{-\beta\Delta\varepsilon} = \text{e}^{-hc\tilde{\nu}/kT} = \dfrac{1}{\text{e}}$

Taking the natural logarithm of both sides yields

$$-1 = \frac{-hc\tilde{\nu}}{kT},$$

so $T = \dfrac{hc\tilde{\nu}}{k} = \dfrac{(6.626 \times 10^{-34}\ \text{J s}) \times (2.998 \times 10^{10}\ \text{cm s}^{-1}) \times (2991\ \text{cm}^{-1})}{1.381 \times 10^{-23}\ \text{J K}^{-1}} = \boxed{4303\ \text{K}},$

where we used $\tilde{\nu} = 2991\ \text{cm}^{-1}$ from Table 12.2, assuming $^1\text{H}^{35}\text{Cl}$.

COMMENT. Vibrational energy level separations are large compared to kT at room temperature, which is 207 cm^{-1} at 298 K. Thus, high temperatures are required to achieve substantial population in excited vibrational states.

Question. If thermal decomposition of HCl occurs when 1% of HCl molecules find themselves in a vibrational state of energy corresponding to the bond dissociation energy (431 kJ mol^{-1}), what temperature is required? Assume $\tilde{\nu}$ is constant at 2991 cm^{-1} and do not take the result too seriously.

E15.23(a) The Sackur–Tetrode equation gives the statistical entropy; for 1 mole at standard pressure:

$$S_m^\ominus = R\ln\left(\frac{e^{5/2}kT}{p^\ominus \Lambda^3}\right)\,[15.39b], \quad \text{where} \quad \Lambda = \frac{h}{(2\pi mkT)^{1/2}}\,[15.19]$$

$$\Lambda = \frac{6.626\times10^{-34}\,\text{J s}}{\{2\pi(20.18)(1.6605\times10^{-27}\,\text{kg})(1.381\times10^{-23}\,\text{J K}^{-1})T\}^{1/2}} = \frac{3.886\times10^{-10}\,\text{m}}{(T/\text{K})^{1/2}}$$

$$S_m^\ominus = R\ln\left\{\frac{e^{5/2}(1.381\times10^{-23}\,\text{J K}^{-1})T}{(1\times10^5\,\text{Pa})(3.886\times10^{-10}\,\text{m})^3}\times\left(\frac{T}{\text{K}}\right)^{3/2}\right\} = R\ln\{28.67\times(T/\text{K})^{5/2}\}$$

(a) At 200 K, $S_m^\ominus = (8.3145\,\text{J K}^{-1}\,\text{mol}^{-1})\times\ln\{28.67\times(200)^{5/2}\} = \boxed{138\,\text{J K}^{-1}\,\text{mol}^{-1}}$

(b) At 298.15 K, $S_m^\ominus = (8.3145\,\text{J K}^{-1}\,\text{mol}^{-1})\times\ln\{28.67\times(298.15)^{5/2}\} = \boxed{146\,\text{J K}^{-1}\,\text{mol}^{-1}}$

E15.24(a) The entropy in terms of partition function is

$$S(T) = \frac{U(T)-U(0)}{T} + Nk\ln q\,[15.28] = -\frac{N}{T}\left(\frac{\partial \ln q}{\partial \beta}\right)_V + Nk\ln q\,[15.24b]$$

$$q = \frac{1}{1-e^{-\beta\varepsilon}}\,[\text{Example 15.2}] = \frac{1}{1-e^{-hc\beta\tilde{\nu}}}$$

$$\text{and} \quad \left(\frac{\partial \ln q}{\partial \beta}\right)_V = (1-e^{-hc\beta\tilde{\nu}})\times\frac{-hc\tilde{\nu}e^{-hc\beta\tilde{\nu}}}{(1-e^{-hc\beta\tilde{\nu}})^2} = -\frac{hc\tilde{\nu}e^{-hc\beta\tilde{\nu}}}{1-e^{-hc\beta\tilde{\nu}}} = -\frac{hc\tilde{\nu}}{e^{hc\beta\tilde{\nu}}-1}$$

$$hc\beta\tilde{\nu} = \frac{(6.626\times10^{-34}\,\text{J s})\times(2.998\times10^{10}\,\text{cm s}^{-1})\times(560\,\text{cm}^{-1})}{(1.381\times10^{-23}\,\text{J K}^{-1})\times(500\,\text{K})} = 1.611$$

Therefore, $q = \dfrac{1}{1-e^{-1.611}} = 1.249$

$$\text{and} \quad \left(\frac{\partial \ln q}{\partial \beta}\right)_V = -\frac{(6.626\times10^{-34}\,\text{J s})\times(2.998\times10^{10}\,\text{cm s}^{-1})\times(560\,\text{cm}^{-1})}{e^{1.611}-1}$$

$$= -2.775\times10^{-21}\,\text{J}$$

$$\text{Hence, } S_m = (6.022\times10^{23}\,\text{mol}^{-1})\times\left(\frac{2.775\times10^{-21}\,\text{J}}{500\,\text{K}} + (1.381\times10^{-23}\,\text{J K}^{-1})\ln 1.249\right)$$

$$= \boxed{5.20\,\text{J K}^{-1}\,\text{mol}^{-1}}$$

E15.25(a) Inclusion of a factor of $1/N!$ is necessary when considering indistinguishable particles. Because of their translational freedom, gases are collections of indistinguishable particles. The factor, then, must be included in calculations on (a) $\boxed{\text{He gas}}$, (b) $\boxed{\text{CO gas}}$, and (d) $\boxed{H_2O\ \text{vapour}}$.

Solutions to problems

Solutions to numerical problems

P15.1 Listing all possible configurations for a 20-particle system would be very time consuming and tedious indeed; however, listing representative configurations for a given total energy is manageable if done systematically. 'Representative' means listing only one configuration that has a given weight.

For example, consider systems that have a total energy of 10ε, where ε is the separation between adjacent energy levels. There are many distinct configurations that have 17 particles in the ground state and one particle in each of three different states, including configurations in which the singly occupied levels are (ε, 2ε, 7ε), (2ε, 3ε, 5ε), and (ε, 4ε, 5ε). Any one of these configurations is 'representative' of all of them, however, because they all have the same weight, namely

$$\mathcal{W} = \frac{N!}{N_0!N_1!N_2!\cdots}[13.1] = \frac{20!}{17!1!1!1!} = 20 \times 19 \times 18 = 6840$$

so it is sufficient to enumerate just one of the configurations whose occupancy numbers are 17, 1, 1, and 1 (to make sure that one such configuration exists consistent with the desired total energy). A systematic way of keeping track of representative configurations is to lower maximum occupancy numbers. The next set of occupancy numbers to look at would be 16, 4; then 16, 3, 1; 16, 2, 2; 16, 2, 1, 1; and 16, 1, 1, 1, 1. We would eliminate 16, 4 because no such occupancy numbers yield a total energy of 10ε.

Set up a spreadsheet to generate one of each kind of representative configuration, while keeping the total number of particles constant at 20 and total energy constant (at 10ε for the moment). Again, the most systematic way to do this is to give the highest occupancy numbers to the lowest-energy available states. (This rule also generates the most 'exponential' configurations.) For example, out of several configurations corresponding to occupancy numbers of 16, 3, and 1, the one used is $N_0 = 16$, $N_1 = 3$, and $N_7 = 1$.

Next examine systems of total energy 10ε, 15ε, and 20ε. The most probable configurations and corresponding weights are:

10ε $\{N_0 = 12, N_1 = 6, N_2 = 2, N_3 = 0, \ldots\}$ $\mathcal{W} = \dfrac{20!}{12!6!2!} = 3\,527\,160$

15ε $\{N_0 = 10, N_1 = 6, N_2 = 3, N_3 = 1, N_4 = 0, \ldots\}$ $\mathcal{W} = \dfrac{20!}{10!6!3!1!} = 155\,195\,040$

20ε $\{N_0 = 10, N_1 = 4, N_2 = 3, N_3 = 2, N_4 = 1, N_5 = 0, \ldots\}$ $\mathcal{W} = \dfrac{20!}{10!4!3!2!1!} = 2\,327\,925\,600$

and $\{N_0 = 9, N_1 = 6, N_2 = 2, N_3 = 2, N_4 = 1, N_5 = 0, \ldots\}$ $\mathcal{W} = \dfrac{20!}{9!6!2!2!1!} = 2\,327\,925\,600$

As the total energy increases, the most probable configuration has more occupied levels and occupancy of higher-energy levels.

The Boltzmann distribution would predict the following relative probabilities for equally spaced energy levels above the ground state:

$$\frac{p_j}{p_0} = \mathrm{e}^{-\beta j\varepsilon} \quad \text{so} \quad \ln\frac{p_j}{p_0} = -\beta j\varepsilon$$

Thus, a plot of the natural log of relative probability vs. the ordinal number of the energy level (j) should be a straight line whose slope is $-\beta\varepsilon$. Using the non-zero occupancy numbers from the configuration, the plots are indeed roughly linear. Furthermore, the value of β decreases with increasing total energy, corresponding to an increase in temperature.

P15.3 Although He is a liquid at these temperatures ($T_b = 4.22$ K), we will test it as if it were a perfect gas with no interaction potential:

$$p_I = \frac{N_I}{N} = \frac{g_I \mathrm{e}^{-\beta\varepsilon_I}}{q} \text{ [15.7, with degeneracy included]}$$

$$q = \frac{V}{\Lambda^3} = \frac{V}{h^3}(2\pi mkT)^{3/2}\,[15.19] = \frac{V}{h^3}(2\pi kTm_\mathrm{u})^{3/2}\left(\frac{m}{m_\mathrm{u}}\right)^{3/2}$$

$$= \frac{(1.0\times10^{-2}\,\mathrm{m})^3 \times \{2\pi\times(1.381\times10^{-23}\,\mathrm{J\,K^{-1}})\times(1.6605\times10^{-27}\,\mathrm{kg})\}^{3/2}}{(6.626\times10^{-34}\,\mathrm{J\,s})^3}\times\left(\frac{mT}{m_\mathrm{u}}\right)^{3/2}$$

$$= (1.8\overline{8}\times10^{20})\times\left(\frac{m}{m_\mathrm{u}}\times\frac{T}{\mathrm{K}}\right)^{3/2}$$

The energy levels, measured with respect to the ground state, are

$$\varepsilon_I = \frac{h^2}{8mL^2}(n_1^2+n_2^2+n_3^2)\,[8.11\mathrm{b}] - \varepsilon_0 = \frac{h^2}{8mL^2}(n_1^2+n_2^2+n_3^2-3)$$

because the non-degenerate ground state has, $n_1 = n_2 = n_3 = 1$.

The first excited state is triply degenerate:

$$\left.\begin{array}{ll} n_1 = n_2 = 1; & n_3 = 2 \\ n_1 = n_3 = 1; & n_2 = 2 \\ n_2 = n_3 = 1; & n_1 = 2 \end{array}\right\} n_1^2+n_2^2+n_3^2 = 6;\ g = 3$$

So, $\varepsilon_1 = \dfrac{3h^2}{8mL^2}$

and $$\beta\varepsilon = \left(\frac{1}{kT}\right)\times\left(\frac{3h^2}{8mL^2}\right)$$

$$= \frac{3(6.626\times10^{-34}\,\mathrm{J\,s})^2}{8(1.381\times10^{-23}\,\mathrm{J\,K^{-1}})\times(1.0\times10^{-2}\,\mathrm{m})^2\times(1.6605\times10^{-27}\,\mathrm{kg})}\times\frac{m_\mathrm{u}}{mT}$$

$$= (7.1\overline{7}\times10^{-14})\times\frac{m_\mathrm{u}}{m}\times\frac{\mathrm{K}}{T}$$

$$p_1 = \frac{g_1\mathrm{e}^{-\beta\varepsilon_1}}{q} = \frac{3\mathrm{e}^{-\left((7.1\overline{7}\times10^{-14})\times\frac{m_\mathrm{u}}{m}\times\frac{\mathrm{K}}{T}\right)}}{1.8\overline{8}\times10^{20}}\times\left(\frac{m_\mathrm{u}}{m}\times\frac{\mathrm{K}}{T}\right)^{3/2}$$

Isotope	m/m_u	T/K	p_1	Occupancy $= p_1N = 10^{22}p_1$
^{4}He	4.003	0.0010	6.3×10^{-17}	6.3×10^5
		2.0	7.0×10^{-22}	7
		4.0	2.5×10^{-22}	2
^{3}He	3.016	0.0010	9.6×10^{-17}	9.6×10^5
		2.0	1.1×10^{-21}	11
		4.0	3.8×10^{-22}	4

These results may at first seem to contradict the expected common sense result that the populations of excited states increase as the temperature increases; however, the energy separations of these states is so small that even a slight increase in temperature promotes the particles from low-lying levels such as the first excited one to much higher quantum levels.

P15.5 $q = \dfrac{V}{\Lambda^3}$, $\Lambda = \dfrac{h}{(2\pi mkT)^{1/2}}$ [15.19]

and hence

$$T = \left(\frac{h^2}{2\pi mk}\right) \times \left(\frac{q}{V}\right)^{2/3}$$

$$= \left(\frac{(6.626 \times 10^{-34}\ \text{J s})^2}{2\pi \times (39.95) \times (1.6605 \times 10^{27}\ \text{kg}) \times (1.381 \times 10^{-23}\ \text{J K}^{-1})}\right) \times \left(\frac{10}{(0.01\ \text{m})^3}\right)^{2/3}$$

$$= \boxed{3.5 \times 10^{-15}\ \text{K}}\ \text{[a very low temperature]}$$

The exact partition function in one dimension is

$$q = \sum_{n=1}^{\infty} e^{-(n^2-1)h^2\beta/8mL^2}$$

For an Ar atom in a cubic box of side 1.0 cm,

$$\frac{h^2\beta}{8mL^2} = \frac{(6.626 \times 10^{-34}\ \text{J s})^2}{8 \times (39.95) \times (1.661 \times 10^{-27}\ \text{kg}) \times (1.381 \times 10^{-23}\ \text{J K}^{-1}) \times (3.5 \times 10^{-15}\ \text{K}) \times (0.010\ \text{m})^2}$$

$$= 0.17\bar{1}$$

Then, $q = \sum_{n=1}^{\infty} e^{-0.17\bar{1}(n^2-1)} = 1.00 + 0.60 + 0.25 + 0.08 + 0.02 + \cdots = 1.95$

The partition function for motion in three dimensions is therefore $q = (1.95)^3 = \boxed{7.41}$.

COMMENT. Temperatures as low as 3.5×10^{-15} K have never been achieved. However, a temperature of 1×10^{-10} K has been attained by adiabatic nuclear demagnetization.

Question. Does the integral approximation apply at 1×10^{-10} K?

P15.7 (a) $q = \sum_I g_I e^{-\beta\varepsilon_I}$ [15.9] $= \sum_I g_I e^{-hc\beta\tilde{\nu}_I}$

We use $hc\beta = \dfrac{6.626 \times 10^{-34}\ \text{J s} \times 2.998 \times 10^{10}\ \text{cm s}^{-1}}{1.381 \times 10^{-23}\ \text{J K}^{-1} \times 298\ \text{K}} = \dfrac{1}{207\ \text{cm}^{-1}}$ at 298 K and $hc\beta = \dfrac{1}{3476\ \text{cm}^{-1}}$ at 5000 K. Therefore,

(i) $q = 5 + e^{-4707/207} + 3e^{-4751/207} + 5e^{-10559/207}$

$= (5) + (1.3 \times 10^{-10}) + (3.2 \times 10^{-10}) + (3.5 \times 10^{-22}) = \boxed{5.00}$

(ii) $q = 5 + e^{-4707/3476} + 3e^{-4751/3476} + 5e^{-10559/3476}$

$= (5) + (0.26) + (0.76) + (0.24) = \boxed{6.26}$

(b) $p_I = \dfrac{g_I e^{-\beta\varepsilon_I}}{q} = \dfrac{g_I e^{-hc\beta\tilde{\nu}_I}}{q}$ [15.7, with degeneracy included]

Therefore, $p_0 = \dfrac{5}{q} = \boxed{1.00}$ at 298 K and $\boxed{0.80}$ at 5000 K.

$p_2 = \dfrac{3e^{-4751/207}}{5.00} = \boxed{6.58 \times 10^{-11}}$ at 298 K $\qquad p_2 = \dfrac{3e^{-4751/3476}}{6.26} = \boxed{0.122}$ at 5000 K

(c) The statistical entropy is

$$S(T) = \frac{U(T) - U(0)}{T} + Nk \ln q\ [15.28] = -\frac{N}{qT}\left(\frac{\partial q}{\partial \beta}\right)_V + Nk \ln q\ [15.24a]$$

where $\left(\dfrac{\partial q}{\partial \beta}\right)_V = \sum_I g_I \left(\dfrac{\partial e^{-hc\beta\tilde{\nu}_I}}{\partial \beta}\right)_V = -hc\sum_I \tilde{\nu}_I g_I e^{-hc\beta\tilde{\nu}_I}$

(i) At 298 K, $\left(\dfrac{\partial q}{\partial \beta}\right)_V = -(6.626 \times 10^{-34}\ \text{J s})(2.998 \times 10^{10}\ \text{cm s}^{-1})$
$\times (4707 \times e^{-4707/207} + 4751 \times 3e^{-4751/207} + 10559 \times 5e^{-10559/207})\ \text{cm}$
$= -4.38 \times 10^{-29}\ \text{J}$

so $S(T) = 6.022 \times 10^{23}\ \text{mol}^{-1} \times \left(\dfrac{4.38 \times 10^{-29}\ \text{J}}{5 \times 298\ \text{K}} + 1.381 \times 10^{-23}\ \text{J K}^{-1} \times \ln 5\right) = \boxed{13.4\ \text{J K mol}^{-1}}$

(ii) At 5000 K, $\left(\dfrac{\partial q}{\partial \beta}\right)_V = -(6.626 \times 10^{-34}\ \text{J s})(2.998 \times 10^{10}\ \text{cm s}^{-1})$
$\times (4707 \times e^{-4707/3476} + 4751 \times 3e^{-4751/3476} + 10559 \times 5e^{-10559/3476})\ \text{cm}$
$= -1.47 \times 10^{-19}\ \text{J}$

so $S(T) = 6.022 \times 10^{23}\ \text{mol}^{-1} \times \left(\dfrac{1.47 \times 10^{-19}\ \text{J}}{6.26 \times 5000\ \text{K}} + 1.381 \times 10^{-23}\ \text{J K}^{-1} \times \ln 6.26\right) = \boxed{18.1\ \text{J K mol}^{-1}}$

P15.9 $q = \sum_I g_I e^{-\beta\varepsilon_I}\ [15.9] = \sum_I g_I e^{-hc\beta\tilde{\nu}_I}$

$p_I = \dfrac{g_I e^{-\beta\varepsilon_I}}{q} = \dfrac{g_I e^{-hc\beta\tilde{\nu}_I}}{q}$ [15.7, with degeneracy included]

Here, $hc\beta\tilde{\nu} = \dfrac{hc\tilde{\nu}}{kT} = \dfrac{6.626 \times 10^{-34}\ \text{J s} \times 2.998 \times 10^{10}\ \text{cm s}^{-1} \times 121.1\ \text{cm}^{-1}}{1.381 \times 10^{-23}\ \text{J K}^{-1} \times (T/\text{K})} = \dfrac{174.2}{T/\text{K}}$

We measure energies from the lower states and write

$$q = 2 + 2e^{-hc\beta\tilde{\nu}} = 2 + 2e^{-174.2/(T/\mathrm{K})}$$

This function is plotted in Figure 15.6.

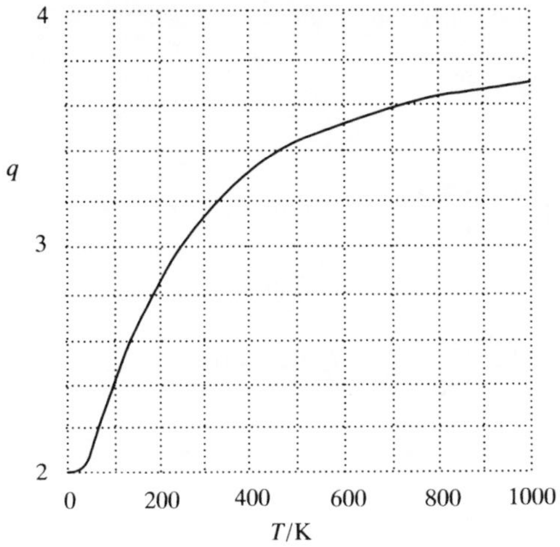

Figure 15.6

(a) At 300 K

$$p_0 = \frac{2}{q} = \frac{1}{1+e^{-174.2/300}} = \boxed{0.641}$$

and $p_1 = 1 - p_0 = \boxed{0.359}$

(b) The electronic contribution to the mean molecular energy is

$$\langle \varepsilon \rangle = \frac{E(T)}{N} = -\frac{1}{q}\frac{dq}{d\beta}\,[15.22] = \frac{2hc\tilde{\nu}e^{-hc\beta\tilde{\nu}}}{q}$$

$$= \frac{6.626\times10^{-34}\ \mathrm{J\ s}\times2.998\times10^{10}\ \mathrm{cm\ s^{-1}}\times121.1\ \mathrm{cm^{-1}}\times e^{-174.2/300}}{1+e^{-174.2/300}} = \boxed{8.63\times10^{-22}\ \mathrm{J}}$$

which corresponds to $\boxed{0.520\ \mathrm{kJ\ mol^{-1}}}$.

(c) The statistical entropy is

$$S(T) = \frac{U(T)-U(0)}{T} + Nk\ln q\,[15.28] = \frac{E(T)}{T} + Nk\ln q$$

(i) At 300 K,

$$S(T) = 6.022\times10^{23}\ \mathrm{mol^{-1}}\left(\frac{8.63\times10^{-22}\ \mathrm{J}}{300\ \mathrm{K}} + 1.381\times10^{-23}\ \mathrm{J\ K^{-1}}\times\ln(2+2e^{-174.2/300})\right)$$

$$= \boxed{11.2\ \mathrm{J\ K\ mol^{-1}}}$$

(ii) At 500 K,

$$\langle \varepsilon \rangle = \frac{6.626\times10^{-34}\ \mathrm{J\ s}\times2.998\times10^{10}\ \mathrm{cm\ s^{-1}}\times121.1\ \mathrm{cm^{-1}}\times e^{-174.2/500}}{1+e^{-174.2/500}} = \boxed{9.95\times10^{-22}\ \mathrm{J}}$$

$$S(T) = 6.022 \times 10^{23}\ \text{mol}^{-1}\left(\frac{9.95 \times 10^{-22}\ \text{J}}{500\ \text{K}} + 1.381 \times 10^{-23}\ \text{J K}^{-1} \times \ln(2 + 2\text{e}^{-174.2/500})\right)$$

$$= \boxed{11.4\ \text{J K mol}^{-1}}$$

P15.11 The partition function is

$$q = \sum_i \text{e}^{-\beta\varepsilon_i}[15.8] = \sum_\upsilon \text{e}^{-hc\tilde{\nu}_\upsilon\beta} = \sum_\upsilon \text{e}^{-hc\tilde{\nu}_\upsilon/kT}$$

(a) At 100 K, $\dfrac{hc\tilde{\nu}_\upsilon}{kT} = \dfrac{6.626 \times 10^{-34}\ \text{J s} \times 2.998 \times 10^{10}\ \text{cm s}^{-1} \times \tilde{\nu}_\upsilon}{1.381 \times 10^{-23}\ \text{J K}^{-1} \times (100\ \text{K})} = 1.44 \times 10^{-2} \times \dfrac{\tilde{\nu}_\upsilon}{\text{cm}^{-1}}$,

so $q = 1 + \text{e}^{-0.0144\times213.30} + \text{e}^{-0.0144\times425.39} + \text{e}^{-0.0144\times636.27} + \text{e}^{-0.0144\times845.93} = \boxed{1.049}$

(b) At 298 K, $\dfrac{hc\tilde{\nu}_\upsilon}{kT} = \dfrac{6.626 \times 10^{-34}\ \text{J s} \times 2.998 \times 10^{10}\ \text{cm s}^{-1} \times \tilde{\nu}_\upsilon}{1.381 \times 10^{-23}\ \text{J K}^{-1} \times (298\ \text{K})} = 4.83 \times 10^{-3} \times \dfrac{\tilde{\nu}_\upsilon}{\text{cm}^{-1}}$,

so $q = 1 + \text{e}^{-4.83\times10^{-3}\times213.30} + \text{e}^{-4.83\times10^{-3}\times425.39} + \text{e}^{-4.83\times10^{-3}\times636.27} + \text{e}^{-4.83\times10^{-3}\times845.93} = \boxed{1.548}$

The fractions of molecules at the various levels are

$$p_\upsilon = \frac{\text{e}^{-\beta\varepsilon_\upsilon}}{q}[15.7] = \frac{\text{e}^{-hc\tilde{\nu}_\upsilon/kT}}{q}$$

So, $p_0 = \dfrac{1}{q} = \text{(a)}\ \boxed{0.953},\quad \text{(b)}\ \boxed{0.645}$

$$p_1 = \frac{\text{e}^{-hc\tilde{\nu}_1/kT}}{q} = \text{(a)}\ \boxed{0.044},\quad \text{(b)}\ \boxed{0.230}$$

$$p_2 = \frac{e^{-hc\tilde{\nu}_2/kT}}{q} = \text{(a)}\ \boxed{0.002},\quad \text{(b)}\ \boxed{0.083}$$

The vibrational contribution to the mean molar energy is

$$E = -\frac{N}{q}\frac{\text{d}q}{\text{d}\beta}[15.22] = \frac{Nhc}{q}\sum_\upsilon \tilde{\nu}_\upsilon \text{e}^{-hc\tilde{\nu}_\upsilon\beta} = \frac{Nhc}{q}\sum_\upsilon \tilde{\nu}_\upsilon \text{e}^{-hc\tilde{\nu}_\upsilon/kT}$$

(a) At 100 K, $\displaystyle\sum_\upsilon \frac{\tilde{\nu}_\upsilon}{\text{cm}^{-1}}\text{e}^{-hc\tilde{\nu}_\upsilon/kT} = 213.30 \times \text{e}^{-0.0144\times213.30} + 425.39 \times \text{e}^{-0.0144\times425.39}$

$$+ 636.27 \times \text{e}^{-0.0144\times636.27} + 845.93 \times \text{e}^{-0.0144\times845.93} = 10.9$$

and $E = \dfrac{6.022 \times 10^{23}\ \text{mol}^{-1} \times 6.626 \times 10^{-34}\ \text{J s} \times 2.998 \times 10^{10}\ \text{cm s}^{-1}}{1.049} \times 10.9\ \text{cm}^{-1} = \boxed{125\ \text{J mol}^{-1}}$

(b) At 298 K, $\displaystyle\sum_\upsilon \frac{\tilde{\nu}_\upsilon}{\text{cm}^{-1}}\text{e}^{-hc\tilde{\nu}_\upsilon/kT} = 213.30 \times \text{e}^{-4.83\times10^{-3}\times213.30} + 425.39 \times \text{e}^{-4.83\times10^{-3}\times425.39}$

$$+ 636.27 \times \text{e}^{-4.83\times10^{-3}\times636.27} + 845.93 \times \text{e}^{-4.83\times10^{-3}\times845.93} = 175$$

and $E = \dfrac{6.022 \times 10^{23}\ \text{mol}^{-1} \times 6.626 \times 10^{-34}\ \text{J s} \times 2.998 \times 10^{10}\ \text{cm s}^{-1}}{1.548} \times 175\ \text{cm}^{-1}$

$$= \boxed{1.35 \times 10^3\ \text{J mol}^{-1}} = \boxed{1.35\ \text{kJ mol}^{-1}}$$

The statistical entropy is

$$S(T) = \frac{U(T) - U(0)}{T} + Nk \ln q \,[15.28] = \frac{E}{T} + Nk \ln q$$

(a) At 100 K,

$$S = \frac{125\ \text{J}}{100\ \text{K}} + 6.022 \times 10^{23}\ \text{mol}^{-1} \times 1.381 \times 10^{-23}\ \text{J K}^{-1} \times \ln(1.049) = \boxed{1.64\ \text{J K mol}^{-1}}$$

(b) At 298 K,

$$S = \frac{125\ \text{J}}{298\ \text{K}} + 6.022 \times 10^{23}\ \text{mol}^{-1} \times 1.381 \times 10^{-23}\ \text{J K}^{-1} \times \ln(1.548) = \boxed{8.16\ \text{J K mol}^{-1}}$$

COMMENT. Eqn 15.12 gives a closed-form expression for the vibrational partition function, based on an infinite ladder of harmonic oscillator states. The explicit sum will deviate slightly from this number because an actual molecule has a finite number of slightly anharmonic vibrational states below the bond dissociation energy.

Solutions to theoretical problems

P15.13 Refer to the equations that precede eqn 15.2 for the expression for $\ln \mathcal{W}$:

$$\ln \mathcal{W} = \ln N! - \sum_i \ln N_i!$$

Substitute the full version of Stirling's approximation:

$$\begin{aligned}\ln \mathcal{W} &= \ln\{(2\pi)^{1/2} N^{N+1/2} \mathrm{e}^{-N}\} - \sum_j \ln\{(2\pi)^{1/2} N_j^{N_j+1/2} \mathrm{e}^{-N_j}\} \\ &= \tfrac{1}{2}\ln(2\pi) + (N + \tfrac{1}{2})\ln N - N - \sum_j \{\tfrac{1}{2}\ln(2\pi) + (N_j + \tfrac{1}{2})\ln N_j - N_j\} \\ &= \tfrac{1}{2}\ln(2\pi) + (N + \tfrac{1}{2})\ln N - \sum_j \{\tfrac{1}{2}\ln(2\pi) + (N_j + \tfrac{1}{2})\ln N_j\}.\end{aligned}$$

With this expression in hand, we turn to *Further Information 15.1*, to the derivation of the Boltzmann distribution. There, the method of undetermined multipliers still yields

$$0 = \frac{\partial \ln \mathcal{W}}{\partial N_i} + \alpha - \beta\varepsilon_i \,[15.42]$$

Evaluating the derivative is similar with the more accurate expression for $\mathcal{W}$ as with the more approximate form. First, derivatives of the 2π terms vanish, leaving

$$\frac{\partial \ln \mathcal{W}}{\partial N_i} = \frac{\partial\{(N + \frac{1}{2})\ln N\}}{\partial N_i} - \sum_j \frac{\partial\{(N_j + \frac{1}{2})\ln N_j\}}{\partial N_i}$$

The first term is

$$\begin{aligned}\frac{\partial\{(N + \frac{1}{2})\ln N\}}{\partial N_i} &= \left(\frac{\partial(N + \frac{1}{2})}{\partial N_i}\right)\ln N + (N + \tfrac{1}{2})\left(\frac{\partial \ln N}{\partial N_i}\right) \\ &= \ln N + \left(\frac{N + \frac{1}{2}}{N}\right)\left(\frac{\partial N}{\partial N_i}\right) = \ln N + \left(\frac{N + \frac{1}{2}}{N}\right).\end{aligned}$$

This differs from the derivation in *Further Information 15.1* only insomuch as $\dfrac{N+\frac{1}{2}}{N}$ differs from 1. In macroscopic samples, N is so large for this difference to be utterly negligible. Similarly

$$\sum_j \frac{\partial\{(N_j+\frac{1}{2})\ln N_j\}}{\partial N_i} = \ln N_i + \left(\frac{N_i+\frac{1}{2}}{N_i}\right)$$

Again, in macroscopic samples, the last term differs from unity only negligibly *for any states that have a reasonable probability of occupation*. For those samples, the remainder of the derivation is exactly as in *Further Information 15.1*. Any deviations are limited to states of exceedingly low occupancy—the extreme tail of the distribution.

P15.15 We draw up the following table:

0	ε	2ε	3ε	4ε	5ε	6ε	7ε	8ε	9ε	$\mathcal{W}$
8	0	0	0	0	0	0	0	0	1	9
7	1	0	0	0	0	0	0	1	0	72
7	0	1	0	0	0	0	1	0	0	72
7	0	0	1	0	0	1	0	0	0	72
7	0	0	0	1	1	0	0	0	0	72
6	2	0	0	0	0	0	1	0	0	252
6	0	2	0	0	1	0	0	0	0	252
6	0	0	3	0	0	0	0	0	0	84
6	1	0	0	2	0	0	0	0	0	252
6	1	1	0	0	0	1	0	0	0	504
6	1	0	1	0	1	0	0	0	0	504
6	0	1	1	1	0	0	0	0	0	504
5	3	0	0	0	0	1	0	0	0	504
5	0	3	1	0	0	0	0	0	0	504
5	2	1	0	0	1	0	0	0	0	1512
5	2	0	1	1	0	0	0	0	0	1512
5	1	2	0	1	0	0	0	0	0	1512
5	1	1	2	0	0	0	0	0	0	1512
4	4	0	0	0	1	0	0	0	0	630
4	3	1	0	1	0	0	0	0	0	2520
4	3	0	2	0	0	0	0	0	0	1260
4	2	2	1	0	0	0	0	0	0	3780
3	5	0	0	1	0	0	0	0	0	504
3	4	1	1	0	0	0	0	0	0	2520
2	6	0	1	0	0	0	0	0	0	252
2	5	2	0	0	0	0	0	0	0	756
1	7	1	0	0	0	0	0	0	0	72
0	9	0	0	0	0	0	0	0	0	1

For a configuration to resemble an exponential decay, populations of successively higher states must be no greater than the previous state. In this entire list, the only configuration that meets this criterion is $\boxed{\{4,2,2,1,0,0,0,0,0,0\}}$. This is the most likely configuration:

$$\mathcal{W} = \frac{N!}{N_0!N_1!N_2!\cdots}[15.1] = \frac{9!}{4!2!2!1!} = \boxed{3780}$$

P15.17 (a) $q = \sum_I g_I e^{-\beta\varepsilon_I} = 1 + 3e^{-\beta\varepsilon} = \boxed{1 + 3e^{-\varepsilon/kT}}$

At $T = \dfrac{\varepsilon}{k}$, $q = 1 + 3e^{-1} = 2.104$

(b) $$E_m(T) = -\frac{N_A}{q}\frac{dq}{d\beta}[15.22] = \frac{N_A}{q}(3\varepsilon e^{-\beta\varepsilon}) = \frac{N_A}{q}(3kTe^{-1}) = \frac{3RT}{2.104\,e} = \boxed{0.5245\,RT}$$

A numerical value cannot be obtained for the energy without specific knowledge of the temperature, but that is not required for the heat capacity or the entropy.

$$C_V = \left(\frac{\partial U_m}{\partial T}\right)_V = \left(\frac{\partial E_m}{\partial T}\right)_V$$

Since $$\frac{d}{dT} = \frac{d\beta}{dT}\times\frac{d}{d\beta} = -\frac{1}{kT^2}\frac{d}{d\beta} = -k\beta^2\frac{d}{d\beta},$$

$$C_V = -k\beta^2\left(\frac{\partial E}{\partial \beta}\right)_V = -k\beta^2(3\varepsilon N_A)\frac{\partial}{\partial\beta}\left(\frac{e^{-\beta\varepsilon}}{q}\right) = -k\beta^2(3\varepsilon N_A)\frac{\partial}{\partial\beta}\left(\frac{e^{-\beta\varepsilon}}{1+3e^{-\beta\varepsilon}}\right)$$

$$= -k\beta^2(3\varepsilon N_A)\left[\frac{(1+3e^{-\beta\varepsilon})\times(-\varepsilon)e^{-\beta\varepsilon} - e^{-\beta\varepsilon}(-3\varepsilon e^{-\beta\varepsilon})}{(1+3e^{-\beta\varepsilon})^2}\right]$$

$$= -k\beta^2(3\varepsilon N_A)\left[\frac{-\varepsilon e^{-\beta\varepsilon} - 3\varepsilon e^{-2\beta\varepsilon} + 3\varepsilon e^{-2\beta\varepsilon}}{(1+3e^{-\beta\varepsilon})^2}\right] = -k\beta^2(3\varepsilon N_A)\left[\frac{-\varepsilon e^{-\beta\varepsilon}}{(1+3e^{-\beta\varepsilon})^2}\right]$$

$$= \frac{3R\varepsilon^2 e^{-\beta\varepsilon}}{(kT)^2(1+3e^{-\beta\varepsilon})^2}$$

For $$\varepsilon = kT,\ C_V = \frac{3Re^{-1}}{(1+3e^{-1})^2} = \frac{3R}{e\left(1+\frac{3}{e}\right)^2} = \boxed{2.074\ \mathrm{J\,K^{-1}\,mol^{-1}}}$$

Note that taking the derivative of $0.5245RT$ with respect to T does not give the correct answer. That is because $0.5245RT$ does not give the temperature dependence of E, but rather the value of E at a particular temperature:

$$\frac{\partial}{\partial T}(0.5245RT) = 0.5245R = 4.361\ \mathrm{J\,K^{-1}\,mol^{-1}}$$

and this is not the correct value.

The calculation of S does not require taking another derivative, so we can use $E = 0.5245RT$:

$$S_m = \frac{E}{T} + R\ln q = 0.5245R + R\ln(2.104) = \boxed{10.55\ \mathrm{J\,K^{-1}\,mol^{-1}}}.$$

P15.19 $dU = TdS - pdV$ [3.46] and $dA = dU - d(TS)$ [3.33] $= SdT - pdV$,

so $p = -\left(\frac{\partial A}{\partial V}\right)_{T,N}$

Now, let us express A in terms of partition functions,

$$A = U - TS = U(0) - \left(\frac{\partial \ln Q}{\partial \beta}\right)_V - T\left(\frac{U(T) - U(0)}{T} + k \ln Q\right) [15.34, 15.35] = -kT \ln Q$$

Hence, $p = kT\left(\frac{\partial \ln Q}{\partial V}\right)_{T,N} = kT\left(\frac{\partial \ln(q^N/N!)}{\partial V}\right)_{T,N} = kT\left(\frac{\partial(N\ln q - \ln N!)}{\partial V}\right)_{T,N} = NkT\left(\frac{\partial \ln q}{\partial V}\right)_{T,N}$

$$= NkT\left(\frac{\partial \ln(V/\Lambda^3)}{\partial V}\right)_{T,N} = NkT\left(\frac{\partial(\ln V - \ln \Lambda^3)}{\partial V}\right)_{T,N} = NkT\left(\frac{\partial \ln V}{\partial V}\right)_{T,N}$$

$$= \frac{NkT}{V} = \frac{nRT}{V} \quad \text{or} \quad \boxed{pV = nRT}$$

P15.21 The expressions obtained in Problem 15.20 are

$$U(T) - U(0) = \frac{\varepsilon N e^{-\varepsilon/kT}}{1 + e^{-\varepsilon/kT}}$$

for the internal energy relative to that at zero temperature and

$$S_m(T) = \frac{\varepsilon R e^{-\varepsilon/kT}}{kT(1 + e^{-\varepsilon/kT})} + R\ln(1 + e^{-\varepsilon/kT})$$

for the entropy. Figure 15.7 is a plot of $\frac{U - U(0)}{\varepsilon N}$ vs. $\frac{S_m}{R}$.

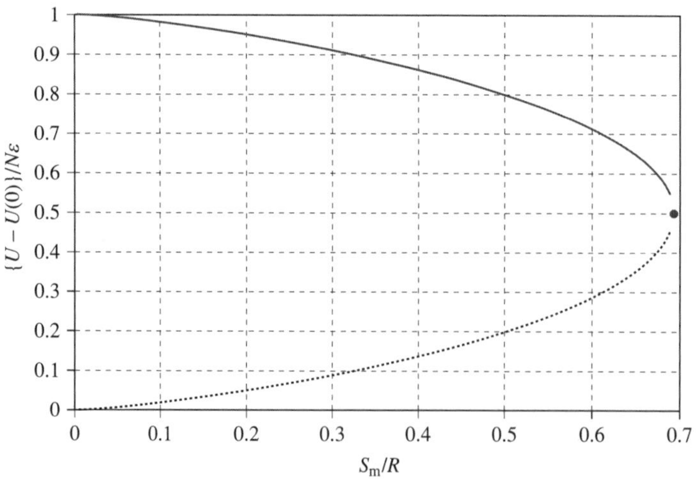

Figure 15.7

The solid curve (upper portion of the graph) corresponds to negative temperatures; note that $\frac{\partial U}{\partial S} < 0$ over this range. The dashed curve (lower portion) corresponds to positive temperatures; note that $\frac{\partial U}{\partial S} > 0$ over this range. The isolated point corresponds to $\beta = 0$; at that point, the slope passes from very large and negative to very large and positive.

Solutions to applications

P15.23 If the atmosphere were at equilibrium, then the Boltzmann distribution would apply, so the relative populations per unit volume would be

$$\frac{N_i}{N_j} = \frac{e^{-\beta\varepsilon_i}}{e^{-\beta\varepsilon_j}} = e^{-(\varepsilon_j-\varepsilon_i)/kT}$$

What distinguishes the states and energies of molecules in a planet's gravitational field is the distance r from the centre of the planet. The energy is gravitational, measured from the ground-state energy

$$\varepsilon(r) = V(r) - V(r_0) = -GMm\left(\frac{1}{r} - \frac{1}{r_0}\right)$$

Note that the ground-state energy is literally the energy at the ground—or more precisely at the lowest point of the atmosphere, r_0

so $$\frac{N(r)}{N(r_0)} = e^{-\{V(r)-V(r_0)\}/kT}$$

Far from the planet, we have

$$\lim_{r\to\infty}\varepsilon(r) = -GMm\left(0 - \frac{1}{r_0}\right) = \frac{GMm}{r_0}$$

and $$\lim_{r\to\infty}\frac{N(r)}{N(r_0)} = e^{-GMm/r_0kT}$$

Hence, if the atmosphere were at equilibrium, as one ventured further from the planet, the concentration of molecules would tend toward a non-zero fraction of the concentration at the surface. This is obviously not the current distribution for planetary atmospheres where the corresponding limit is zero. Consequently, we may conclude that no planet's atmosphere, including Earth's, is at equilibrium.

16 Statistical thermodynamics 2: applications

Answers to discussion questions

D16.1 An approximation involved in the derivation of all of these expressions is the assumption that the contributions from the different modes of motion are separable. The expression $q^{R} = kT/hc\tilde{B}$ is the high-temperature approximation to the rotational partition function for non-symmetrical linear rotors. The expression $q^{V} = kT/hc\tilde{\nu}$ is the high-temperature form of the partition function for one vibrational mode of the molecule in the harmonic approximation. The expression $q^{E} = g^{E}$ for the electronic partition function applies at normal temperatures to atoms and molecules with no low-lying excited electronic energy levels.

D16.3 Residual entropy is due to the presence of some disorder in the system even at $T = 0$. It is observed in systems where there is very little energy difference—or none—between alternative arrangements of the molecules at very low temperatures. Consequently, the molecules cannot lock into a preferred orderly arrangement and some disorder persists.

D16.5 Identical particles can be regarded as distinguishable when they are localized, as in a crystal lattice, where we can assign a set of coordinates to each particle. Strictly speaking, it is the lattice site that carries the set of coordinates, but as long as the particle is fixed to the site, it too can be considered distinguishable.

D16.7 See *Justification 16.3* for a derivation of the general expression (eqn 16.52b) for the equilibrium constant in terms of the partition functions and difference in molar energy, $\Delta_r E_0$, of the products and reactants in a chemical reaction. The partition functions are functions of temperature and the ratio of partition functions in eqn 16.52b will therefore vary with temperature. However, the most direct effect of temperature on the equilibrium constant is through the exponential term $e^{-\Delta_r E_0/RT}$. The manner in which both factors affect the magnitudes of the equilibrium constant and its variation with temperature is described in detail for a simple $R \rightleftharpoons P$ gas-phase equilibrium in Section 16.8(c) and *Justification 16.4*.

Solutions to exercises

E16.1(a) The thermal wavelength is given by $\Lambda = \dfrac{h}{(2\pi mkT)^{1/2}}$ [15.19, 16.12]. When numerical values for h, k, and π are substituted in this expression it becomes

$$\Lambda/\text{pm} = \frac{1749}{(T/\text{K})^{1/2}(M/\text{g mol}^{-1})^{1/2}}$$

At 298.15 K, with $M = 44.01\ \text{g mol}^{-1}$ for CO_2, we obtain

$$\Lambda = \frac{1749}{(298.15 \times 44.01)^{1/2}}\ \text{pm} = \boxed{15.27\ \text{pm}} = 15.27 \times 10^{-12}\ \text{m}$$

E16.2(a) See Exercise 16.1(a).

$$q^{\text{T}} = \frac{V}{\Lambda^3}\,[16.12] = \frac{1.0\ \text{cm}^3 \times \dfrac{1\ \text{m}^3}{10^6\ \text{cm}^3}}{(15.27 \times 10^{-12}\ \text{m})^3} = \boxed{2.8 \times 10^{26}}$$

E16.3(a) $C_{V,\text{m}} = \frac{1}{2}(3 + \nu_{\text{R}}^* + 2\nu_{\text{V}}^*)R$ [16.35]

with a mode active if $T > \theta_{\text{M}}$.

(a) $\nu_{\text{R}}^* = 2,\ \nu_{\text{V}}^* \approx 1$; hence $C_{V,\text{m}} = \frac{1}{2}(3 + 2 + 2 \times 1)R = \boxed{\tfrac{7}{2}R}$ [experimental $= 3.4R$]

Note that I_2 has quite a low vibrational wavenumber, so

$$\theta_{\text{V}} = \frac{hc\tilde{\nu}}{k} = \frac{(6.626 \times 10^{-34}\ \text{J s})(2.998 \times 10^{10}\ \text{cm s}^{-1})(214\ \text{cm}^{-1})}{1.381 \times 10^{-23}\ \text{J K}^{-1}} = 308\ \text{K}$$

The temperature specified for this exercise is less than this, but only slightly. Looking at Figure 16.11 in the text suggests that the mode is closer to active than inactive when the temperature is even close to the vibrational temperature.

(b) $\nu_{\text{R}}^* = 3,\ \nu_{\text{V}}^* \approx 0$; hence $C_{V,\text{m}} = \frac{1}{2}(3 + 3 + 0)R = \boxed{3R}$ [experimental $= 3.2R$]

(c) $\nu_{\text{R}}^* = 3,\ \nu_{\text{V}}^* \approx 4$; hence $C_{V,\text{m}} = \frac{1}{2}(3 + 3 + 2 \times 4)R = \boxed{7R}$ [experimental $= 8.8R$]

Note that data from the book by Herzberg (*Molecular Spectra and Molecular Structure II*) show low vibrational wavenumbers for four modes of benzene, and we have included these above. There are 26 more vibrational modes we have neglected, taking them to be inactive. Slight activity from these modes accounts for the difference of about $1.8R$ between the experimental value and our estimate.

E16.4(a) The rotational temperature is given by

$$\theta_{\text{R}} = \frac{hc\tilde{B}}{k}\,[\text{Section 16.2(b)}],\quad \frac{hc}{k} = 1.4388\ \text{cm K},\ \theta_{\text{R}} = 1.4388\ \text{cm K} \times \tilde{B}$$

$$\theta_{\text{R}} = 1.4388\ \text{cm K} \times 10.593\ \text{cm}^{-1}\ [\text{Table 12.2}] = \boxed{15.24\ \text{K}}$$

E16.5(a) The rotational partition function of a linear molecule at high temperature (298 K) is [see Section 16.2(b)]

$$q^{\text{R}} = \frac{kT}{\sigma hc\tilde{B}} = \frac{T}{\sigma\theta_{\text{R}}}\qquad \theta_{\text{R}} = \frac{hc\tilde{B}}{k}$$

Substituting values for the constants this may also be written as

$$q^{\mathrm{R}} = \frac{0.6950}{\sigma} \times \frac{T/\mathrm{K}}{(\tilde{B}/\mathrm{cm}^{-1})} = \frac{(0.6950) \times (T/\mathrm{K})}{10.59} [\sigma = 1] = 0.06563(T/\mathrm{K})$$

(a) $q^{\mathrm{R}} = (0.06563) \times (298) = \boxed{19.6}$ (b) $q^{\mathrm{R}} = (0.06563) \times (523) = \boxed{34.3}$.

E16.6(a) Look for the rotational subgroup of the molecule (the group of the molecule composed only of the identity and the rotational elements, and assess its order).

(a) CO: full group $C_{\infty v}$; subgroup C_1; hence $\sigma = \boxed{1}$

(b) O_2: full group $D_{\infty h}$; subgroup C_2; $\sigma = \boxed{2}$

(c) H_2S: full group C_{2v}; subgroup C_2; $\sigma = \boxed{2}$

(d) SiH_4: full group T_d; subgroup T; $\sigma = \boxed{12}$

(e) $CHCl_3$: full group C_{3v}; subgroup C_3; $\sigma = \boxed{3}$

E16.7(a) For non-linear molecules at high temperature, $q^{\mathrm{R}} = \frac{1}{\sigma}\left(\frac{kT}{hc}\right)^{3/2}\left(\frac{\pi}{\tilde{A}\tilde{B}\tilde{C}}\right)^{1/2}$ [Section 16.2b]

After substituting values for the constants this becomes

$$q^{\mathrm{R}} = \frac{1.0270}{\sigma} \frac{(T/\mathrm{K})^{3/2}}{(\tilde{A}\tilde{B}\tilde{C}/\mathrm{cm}^{-3})^{1/2}}$$

$$= \frac{1.0270 \times 298^{3/2}}{(2) \times (27.878 \times 14.509 \times 9.287)^{1/2}} [\sigma = 2 \text{ from Table 16.2}] = \boxed{43.1}$$

The high-temperature approximation is valid if $T > \theta_{\mathrm{R}}$, where

$$\theta_{\mathrm{R}} = \frac{hc(\tilde{A}\tilde{B}\tilde{C})^{1/3}}{k}$$

$$= \frac{(6.626 \times 10^{-34}\ \mathrm{J\ s}) \times (2.998 \times 10^{10}\ \mathrm{cm\ s^{-1}}) \times [(27.878) \times (14.509) \times (9.287)\ \mathrm{cm}^{-3}]^{1/3}}{1.38 \times 10^{-23}\ \mathrm{J\ K^{-1}}}$$

$$= \boxed{22.36\ \mathrm{K}}.$$

E16.8(a) $q^{\mathrm{R}} = 43.1$ [Exercise 16.7(a)]

All the rotational modes of water are fully active at 25°C [Exercise 16.7(a)]; therefore

$$U_{\mathrm{m}}^{\mathrm{R}} - U_{\mathrm{m}}^{\mathrm{R}}(0) = E^{\mathrm{R}} = \tfrac{3}{2}RT$$

$$S_{\mathrm{m}}^{\mathrm{R}} = \frac{E^{\mathrm{R}}}{T} + R \ln q^{\mathrm{R}}$$

$$= \tfrac{3}{2}R + R \ln 43.1 = \boxed{43.76\ \mathrm{J\ K^{-1}\ mol^{-1}}}$$

COMMENT. Division of q^{R} by $N_{\mathrm{A}}!$ is not required for the internal contributions; internal motions may be thought of as localized (distinguishable). It is the overall canonical partition function, which is a product of internal and external contributions, that is divided by $N_{\mathrm{A}}!$

E16.9(a) (a) For a spherical rotor (Section 12.5(a))

$$E_J = hc\tilde{B}J(J+1)\ [12.8],$$

where $\tilde{B} = 5.2412\ \text{cm}^{-1}$ for CH_4 and the degeneracy is $g(J) = (2J+1)^2$. Hence,

$$q^{\text{R}} \approx \frac{1}{\sigma}\sum_J (2J+1)^2 e^{-\beta hc\tilde{B}J(J+1)} \text{ [16.13 corrected by symmetry number and degeneracy factor]}$$

The relevant parameters are

$$hc\tilde{B}\beta = \frac{(1.4388\ \text{K}) \times (5.2412)}{T} = \frac{7.5410}{T/\text{K}}, \quad \sigma = 12,$$

so $q^{\text{R}} = \dfrac{1}{12}\sum_J (2J+1)^2 e^{-7.5410J(J+1)/(T/\text{K})}$

At 298 K,

$$q^{\text{R}} = \frac{1}{12}(1.0000 + 8.5561 + 21.480 + 36.173 + \cdots) = \frac{1}{12} \times 443.427 = \boxed{36.95}$$

The sum converges after 20 terms. Similarly, at 500 K

$$q^{\text{R}} = \frac{1}{12}(1.0000 + 8.7326 + 22.8370 + 40.8880 + \cdots) = \frac{1}{12} \times 960.96 = \boxed{80.08}$$

The sum converged after 24 terms.

(Note that the results are still approximate because the symmetry number is a valid corrector only at high temperatures. To get exact values of q^{R} we should do a detailed analysis of the rotational states allowed by the Pauli principle.)

(b) For non-linear molecules at high temperature, $q^{\text{R}} = \dfrac{1}{\sigma}\left(\dfrac{kT}{hc}\right)^{3/2}\left(\dfrac{\pi}{\tilde{A}\tilde{B}\tilde{C}}\right)^{1/2}$ [Section 16.2b]

After substituting values for the constants this becomes, with $\tilde{A} = \tilde{B} = \tilde{C}$, $\tilde{A}\tilde{B}\tilde{C} = \tilde{B}^3$,

$$q^{\text{R}} = \frac{1.0270}{\sigma}\frac{(T/\text{K})^{3/2}}{(\tilde{B}/\text{cm}^{-1})^{3/2}} = \frac{1.0270}{12}\frac{(T/\text{K})^{3/2}}{(5.2412)^{3/2}} = 7.133 \times 10^{-3} \times (T/\text{K})^{3/2}$$

At 298 K, $q = 7.133 \times 10^{-3} \times 298^{3/2} = \boxed{36.7}$

At 500 K, $q = 7.133 \times 10^{-3} \times 500^{3/2} = \boxed{79.7}$

The difference in this case is small.

E16.10(a) $q^{\text{R}} = \dfrac{kT}{\sigma hc\tilde{B}}$ [16.15b], $\tilde{B} = \dfrac{\hbar}{4\pi cI}$ [12.7], $I = \mu R^2$ [Table 12.1]

Hence, $q^{\text{R}} = \dfrac{8\pi^2 kTI}{\sigma h^2} = \dfrac{8\pi^2 kT\mu R^2}{\sigma h^2}$

For O_2, $\mu = \frac{1}{2}m(\text{O}) = \frac{1}{2} \times 16.00\ m_u = 8.00\ m_u$, and $\sigma = 2$; therefore

$$q^{\text{R}} = \frac{(8\pi^2) \times (1.381 \times 10^{-23}\ \text{J K}^{-1}) \times (300\ \text{K}) \times (8.00 \times 1.6605 \times 10^{-27}\ \text{kg}) \times (1.2075 \times 10^{-10}\ \text{m})^2}{(2) \times (6.626 \times 10^{-34}\ \text{J s})^2}$$

$$= \boxed{72.5}$$

E16.11(a) Assuming that all rotational modes are active we can draw up the following table for $C_{V,\text{m}}$, $C_{p,\text{m}}$, and γ with and without active vibrational modes.

	$C_{V,\text{m}}$	$C_{p,\text{m}}$	γ	Exptl	
$NH_3(\nu_V^* = 0)$	$3R$	$4R$	1.33	1.31	closer
$NH_3(\nu_V^* = 6)$	$9R$	$10R$	1.11		
$CH_4(\nu_V^* = 0)$	$3R$	$4R$	1.33	1.31	closer
$CH_4(\nu_V^* = 9)$	$12R$	$13R$	1.08		

The experimental values are obtained from Tables 2.6 and 2.8 assuming $C_{p,\text{m}} = C_{V,\text{m}} + R$. It is clear from the comparison in the above table that the vibrational modes are not active. This is confirmed by the experimental vibrational wavenumbers (see Herzberg, *Molecular Spectra and Molecular Spectra II*), all of which are much greater than kT at 298 K.

E16.12(a) $$C_{V,\text{m}}/R = f, \quad f = \left(\frac{\theta_V}{T}\right)^2 \times \left(\frac{\text{e}^{-\theta_V/2T}}{1 - \text{e}^{-\theta_V/T}}\right)^2 \ [16.34];\ \theta_V = \frac{hc\tilde{\nu}}{k}$$

We write $x = \frac{\theta_V}{T}$; then $C_{V,\text{m}}/R = \frac{x^2\text{e}^{-x}}{(1 - \text{e}^{-x})^2}$

This function is plotted in Figure 16.1. For the acetylene (ethyne) calculation, use the expression above for each mode. We draw up the following table using $kT/hc = 207\ \text{cm}^{-1}$ at 298 K and $348\ \text{cm}^{-1}$ at 500 K, and $\theta_V/T = hc\tilde{\nu}/kT$:

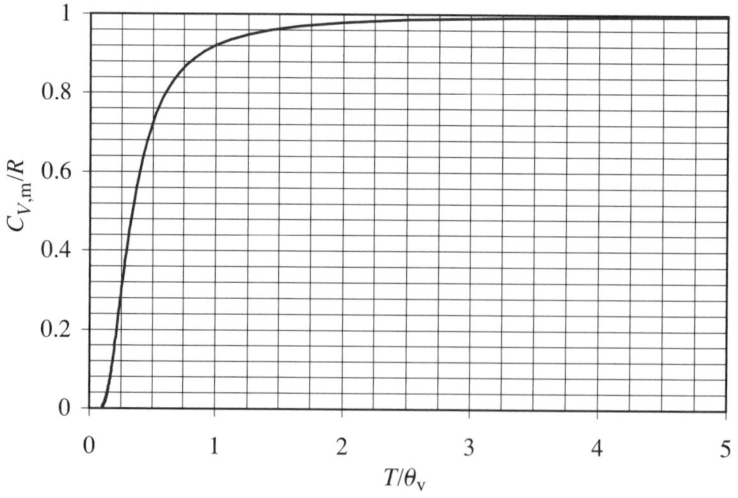

Figure 16.1

	x		$C_{V,m}/R$	
$\tilde{\nu}/\mathrm{cm^{-1}}$	298 K	500 K	298 K	500 K
612	2.96	1.76	0.505	0.777
612	2.96	1.76	0.505	0.777
729	3.52	2.09	0.389	0.704
729	3.52	2.09	0.389	0.704
1974	9.54	5.67	0.007	0.112
3287	15.88	9.45	3.2×10^{-5}	0.007
3374	16.30	9.70	2.2×10^{-5}	0.006

The heat capacity of the molecule is the sum of these contributions, namely

(a) $1.796R = \boxed{14.93\ \mathrm{J\,K^{-1}\,mol^{-1}}}$ at 298 K and

(b) $3.086R = \boxed{25.65\ \mathrm{J\,K^{-1}\,mol^{-1}}}$ at 500 K.

E16.13(a) In each case the contribution to G is given by

$$G - G(0) = -nRT\ln q \text{ [16.9; also see Comment to Exercise 16.8(a)]}$$

Therefore, we first evaluate q^{R} and q^{V}.

$$q^{\mathrm{R}} = \frac{kT}{\sigma hc\tilde{B}} = \frac{T}{\sigma\theta_{\mathrm{R}}} \qquad \theta_{\mathrm{R}} = \frac{hc\tilde{B}}{k}$$

After substituting values for the constants this becomes

$$q^{\mathrm{R}} = \frac{0.6950}{\sigma}\,\frac{T/\mathrm{K}}{\tilde{B}/\mathrm{cm^{-1}}}[\sigma = 2] = \frac{0.6950\times(298)}{(2)\times(0.3902)} = 265$$

The vibrational partition function is a product of three terms of the form of eqn 16.19, one for each vibrational mode

$$q^{\mathrm{V}} = \left(\frac{1}{1-\mathrm{e}^{-a}}\right)\times\left(\frac{1}{1-\mathrm{e}^{-b}}\right)^2\times\left(\frac{1}{1-\mathrm{e}^{-c}}\right)\text{[16.19]}$$

with

$$a = \frac{(1.4388)\times(1388.2)}{298} = 6.70\bar{2}$$

$$b = \frac{(1.4388)\times(667.4)}{298} = 3.22\bar{2}$$

$$c = \frac{(1.4388)\times(2349.2)}{298} = 11.3\bar{4}$$

Hence,

$$q^{\mathrm{V}} = \frac{1}{1-\mathrm{e}^{-6.702}}\times\left(\frac{1}{1-\mathrm{e}^{-3.222}}\right)^2\times\frac{1}{1-\mathrm{e}^{-11.34}} = 1.08\bar{6}$$

Therefore, the rotational contribution to the molar Gibbs energy is

$$-RT \ln q^{\mathrm{R}} = -8.314\ \mathrm{J\ K^{-1}\ mol^{-1}} \times 298\ \mathrm{K} \times \ln 265$$
$$= \boxed{-13.8\ \mathrm{kJ\ mol^{-1}}}$$

and the vibrational contribution is

$$-RT \ln q^{\mathrm{V}} = -8.314\ \mathrm{J\ K^{-1}\ mol^{-1}} \times 298\ \mathrm{K} \times \ln 1.08\overline{6}$$
$$= \boxed{-0.20\ \mathrm{kJ\ mol^{-1}}}$$

E16.14(a) We assume based on the information given that Cl can effectively be treated as a two-level system. The partition function is then

$$q = \sum_j g_j \mathrm{e}^{-\beta\varepsilon_j}$$

with degeneracies $g_j = 2J + 1$ [J here is the subscript on the term symbol]

so $q = 4 + 2\mathrm{e}^{-\beta\varepsilon}[g(^2\mathrm{P}_{3/2}) = 4,\ g(^2\mathrm{P}_{1/2}) = 2]$

(a) $q = 4 + 2\mathrm{e}^{-\frac{hc \times 881\ \mathrm{cm^{-1}}}{kT}} = 4 + 2\mathrm{e}^{-\frac{1.4388\ \mathrm{cm\ K} \times 881\ \mathrm{cm^{-1}}}{500\ \mathrm{K}}} = \boxed{4.158}$

(b) $q = 4 + 2\mathrm{e}^{-\frac{hc \times 881\ \mathrm{cm^{-1}}}{kT}} = 4 + 2\mathrm{e}^{-\frac{1.4388\ \mathrm{cm\ K} \times 881\ \mathrm{cm^{-1}}}{900\ \mathrm{K}}} = \boxed{4.489}$

E16.15(a) The partition function is

$$q = \sum_j g_j \mathrm{e}^{-\beta\varepsilon_j}$$

with degeneracies $g_j = 2J + 1$ [J here is the subscript on the term symbol]

so $q = 4 + 2\mathrm{e}^{-\beta\varepsilon}[g(^2\mathrm{P}_{3/2}) = 4,\ g(^2\mathrm{P}_{1/2}) = 2]$

$$U - U(0) = -\frac{N}{q}\left(\frac{\partial q}{\partial \beta}\right)_V = \frac{N\varepsilon \mathrm{e}^{-\beta\varepsilon}}{2 + \mathrm{e}^{-\beta\varepsilon}}$$

$$C_V = \left(\frac{\partial U}{\partial T}\right)_V = -k\beta^2\left(\frac{\partial U}{\partial \beta}\right)_V = \frac{2R(\varepsilon\beta)^2 \mathrm{e}^{-\beta\varepsilon}}{(2 + \mathrm{e}^{-\beta\varepsilon})^2}[N = N_{\mathrm{A}}]$$

(a) Therefore, since at 500 K $\beta\varepsilon = 2.53\overline{5}$

$$C_{V,\mathrm{m}}/R = \frac{(2) \times (2.53\overline{5})^2 \times (\mathrm{e}^{-2.53\overline{5}})}{(2 + \mathrm{e}^{-2.53\overline{5}})^2} = \boxed{0.236}$$

(b) At 900 K, when $\beta\varepsilon = 1.408$,

$$C_{V,\mathrm{m}}/R = \frac{(2) \times (1.408)^2 \times (\mathrm{e}^{-1.408})}{(2 + \mathrm{e}^{-1.408})^2} = \boxed{0.193}$$

COMMENT. $C_{V,\mathrm{m}}$ is smaller at 900 K than at 500 K, for then the temperature is higher than the peak in the 'two-level' heat capacity curve.

E16.16(a) See the solution to Exercise 16.15(a).

(a) At 500 K

$$\beta\varepsilon = \frac{(1.4388\ \text{cm K}) \times (881\ \text{cm}^{-1})}{500\ \text{K}} = 2.53\overline{5}$$

Therefore, the contribution to G_m is

$$G_m - G_m(0) = -RT \ln q\ [16.9, \text{also see the Comment to Exercise 16.8(a)}]$$

$$-RT \ln q = -(8.314\ \text{J K}^{-1}\ \text{mol}^{-1}) \times (500\ \text{K}) \times \ln(4 + 2 \times e^{-2.535})$$
$$= -(8.314\ \text{J K}^{-1}\ \text{mol}^{-1}) \times (500\ \text{K}) \times \ln(4 + 0.158) = \boxed{-6.42\ \text{kJ mol}^{-1}}$$

(b) At 900 K

$$\beta\varepsilon = \frac{(1.4388\ \text{cm K}) \times (881\ \text{cm}^{-1})}{900\ \text{K}} = 1.40\overline{8}$$

Therefore, the contribution to G_m is

$$G_m - G_m(0) = -RT \ln q$$

$$-RT \ln q = -(8.314\ \text{J K}^{-1}\ \text{mol}^{-1}) \times (900\ \text{K}) \times \ln(4 + 2 \times e^{-1.408})$$
$$= -(8.314\ \text{J K}^{-1}\ \text{mol}^{-1}) \times (900\ \text{K}) \times \ln(4 + 0.489) = \boxed{-14.0\ \text{kJ mol}^{-1}}$$

E16.17(a) We assume that the upper nine of the $(2 \times \frac{9}{2} + 1) = 10$ spin-orbit states of the ion lie at an energy much greater than kT at 1 K; hence, since the spin degeneracy of Co^{2+} is 4 (the ion is a spin quartet), $q = 4$. The contribution to the entropy is

$$R \ln q = (8.314\ \text{J K}^{-1}\ \text{mol}^{-1}) \times (\ln 4) = \boxed{11.5\ \text{J K}^{-1}\ \text{mol}^{-1}}$$

E16.18(a) The Mayer f-function is $f = e^{-\beta E_p} - 1$. For the hard-sphere potential of eqn 16.43 we have:

$r \le \sigma \quad E_p = \infty \quad f = -1$

$r > \sigma \quad E_p = 0 \quad f = 0$

We plot the f-function against $x = r/\sigma$ for x between 0 and 2. It rises steeply at $x = r/\sigma = 1$. See Figure 16.2.

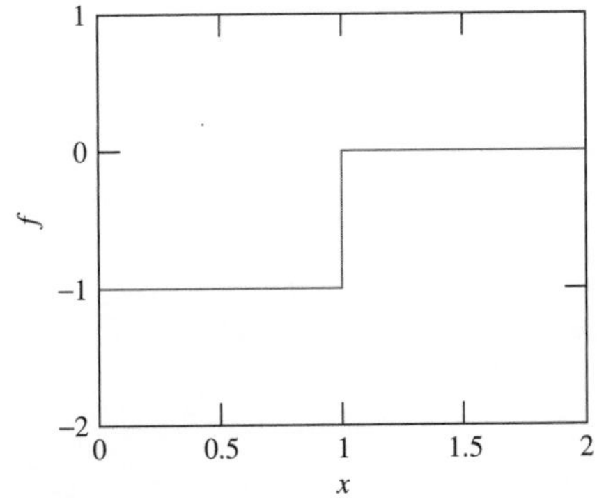

Figure 16.2

$x = r/\sigma$

E16.19(a) In each case $S_m = R \ln s$ [16.50b]. Therefore,

(a) $S_m = R \ln 3 = 8.3145\ \mathrm{J\ K^{-1}\ mol^{-1}} \times \ln 3 = \boxed{9.13\ \mathrm{J\ K^{-1}\ mol^{-1}}}$

(b) $S_m = R \ln 5 = 8.3145\ \mathrm{J\ K^{-1}\ mol^{-1}} \times \ln 5 = \boxed{13.4\ \mathrm{J\ K^{-1}\ mol^{-1}}}$

(c) $S_m = R \ln 6 = 8.3145\ \mathrm{J\ K^{-1}\ mol^{-1}} \times \ln 6 = \boxed{14.9\ \mathrm{J\ K^{-1}\ mol^{-1}}}$

E16.20(a) We use eqn 16.54a with $X = I$, $X_2 = I_2$, $\Delta_r E_0 = D_0$.

$$\tilde{D}_0 = \tilde{D}_e - \frac{1}{2}\tilde{\nu} = 1.5422\ \mathrm{eV} \times \frac{8065.5\ \mathrm{cm^{-1}}}{1\ \mathrm{eV}} - 107.18\ \mathrm{cm^{-1}} = 1.2331 \times 10^4\ \mathrm{cm^{-1}} = 1.475 \times 10^5\ \mathrm{J\ mol^{-1}}$$

$$K = \left(\frac{(q^{\ominus}_{\mathrm{I,m}})^2}{q^{\ominus}_{\mathrm{I_2,m}} N_A}\right) e^{-\Delta_r E_0/RT}\ [16.54a]$$

$$q^{\ominus}_{\mathrm{I,m}} = q^{T\ominus}_m(\mathrm{I}) q^E(\mathrm{I}), \quad q^E(\mathrm{I}) = 4$$

$$q^{\ominus}_{\mathrm{I_2,m}} = q^{T\ominus}_m(\mathrm{I_2}) q^R(\mathrm{I_2}) q^V(\mathrm{I_2}) q^E(\mathrm{I_2}), \quad q^E(\mathrm{I_2}) = 1$$

$$\frac{q^{T\ominus}_m(\mathrm{I_2})}{N_A} = \frac{kT}{p^{\ominus}\Lambda^3} \qquad \Lambda = \frac{h}{(2\pi mkT)^{1/2}}$$

After substituting the values of the constants we obtain

$$\frac{q^{T\ominus}_m(\mathrm{I_2})}{N_A} = 2.561 \times 10^{-2} (T/\mathrm{K})^{5/2} \times (M/\mathrm{g\ mol^{-1}})^{3/2}$$

$$\frac{q^{T\ominus}_m(\mathrm{I_2})}{N_A} = 2.561 \times 10^{-2} \times 1000^{5/2} \times 253.8^{3/2} = 3.27 \times 10^9$$

$$\frac{q^{T\ominus}_m(\mathrm{I})}{N_A} = 2.561 \times 10^{-2} \times 1000^{5/2} \times 126.9^{3/2} = 1.16 \times 10^9$$

$$q^R(\mathrm{I_2}) = \frac{0.6950}{\sigma} \times \frac{T/\mathrm{K}}{\tilde{B}/\mathrm{cm^{-1}}}\ [\text{See Exercise 16.13(a)}] = \frac{1}{2} \times 0.6950 \times \frac{1000}{0.0373} = 931\overline{6}$$

$$q^V(\mathrm{I_2}) = \frac{1}{1 - e^{-a}}, \quad a = 1.4388 \frac{\tilde{\nu}/\mathrm{cm^{-1}}}{T/\mathrm{K}} = \frac{1}{1 - e^{-1.4388 \times 214.36/1000}} = 3.77$$

$$K = \frac{(1.16 \times 10^9 \times 4)^2 e^{-17.741}}{(3.27 \times 10^9) \times (9316) \times (3.77)} = \boxed{3.70 \times 10^{-3}}$$

Solutions to problems

Solutions to numerical problems

P16.1 $$q^E = \sum_j g_j e^{-\beta\varepsilon_j} = 2 + 2e^{-\beta\varepsilon}, \quad \varepsilon = \Delta\varepsilon = 121.1\ \mathrm{cm^{-1}}$$

$$U_m - U_m(0) = -\frac{N_A}{q^E}\left(\frac{\partial q^E}{\partial \beta}\right)_V = \frac{N_A \varepsilon e^{-\beta\varepsilon}}{q^E}$$

$$C_{V,m} = -k\beta^2 \left(\frac{\partial U_m}{\partial \beta}\right)_V\ [16.31a]$$

Let $x = \beta\varepsilon$, then $\mathrm{d}\beta = \dfrac{1}{\varepsilon}\mathrm{d}x$

$$C_{V,\mathrm{m}} = -k\left(\frac{x}{\varepsilon}\right)^2 \varepsilon \frac{\partial}{\partial x}\left(\frac{N_\mathrm{A}\varepsilon \mathrm{e}^{-x}}{1+\mathrm{e}^{-x}}\right) = -N_\mathrm{A} k x^2 \times \frac{\partial}{\partial x}\left(\frac{\mathrm{e}^{-x}}{1+\mathrm{e}^{-x}}\right) = R\left(\frac{x^2\mathrm{e}^{-x}}{(1+\mathrm{e}^{-x})^2}\right)$$

Therefore,

$$C_{V,\mathrm{m}}/R = \frac{x^2\mathrm{e}^{-x}}{(1+\mathrm{e}^{-x})^2}, \quad x = \beta\varepsilon$$

We then draw up the following table:

T/K	50	298	500
(kT/hc)/mol^{-1}	34.8	207	348
x	3.48	0.585	0.348
$C_{V,\mathrm{m}}/R$	0.351	0.079	0.029
$C_{V,\mathrm{m}}$/(J K^{-1} mol^{-1})	2.91	0.654	0.244

In order to plot the heat capacity against T we first evaluate $x = \beta\varepsilon = \dfrac{\varepsilon}{k}\left(\dfrac{1}{T}\right)$. After substituting values for ε and k we find $x = \dfrac{174.23}{T/\mathrm{K}}$. $C_{V,\mathrm{m}}/R \equiv c(T)$ is evaluated in the following Mathcad® worksheet and plotted against T in Figure 16.3.

T := 1, 2 .. 500

$$c(T) := \frac{\left[\left(\frac{174.23}{T}\right)^2 \cdot e^{\frac{-174.23}{T}}\right]}{\left(1+e^{\frac{-174.23}{T}}\right)^2}$$

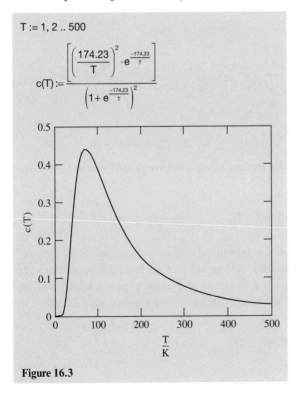

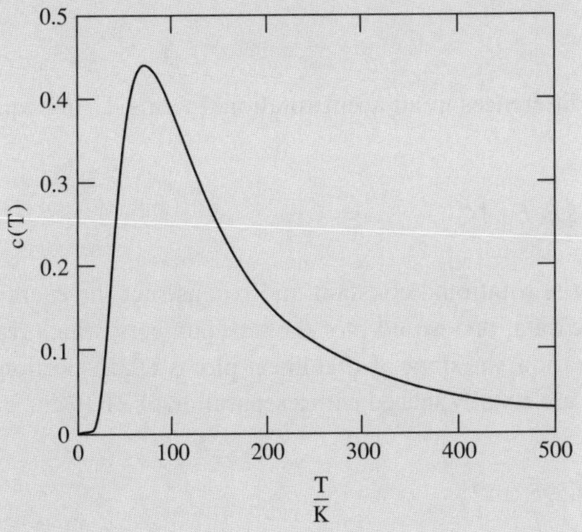

Figure 16.3

COMMENT. Note that the double degeneracies do not affect the results because the two factors of 2 in q cancel when U is formed. In the range of temperature specified, the electronic contribution to the heat capacity decreases with increasing temperature.

P16.3 The energy expression for a particle on a ring is

$$E = \frac{\hbar^2 m_l^2}{2I} \quad [8.38a]$$

Therefore

$$q = \sum_{m=-\infty}^{\infty} e^{-m_l^2\hbar^2/2IkT} = \sum_{m=-\infty}^{\infty} e^{-\beta m_l^2\hbar^2/2I}$$

The summation may be approximated by an integration

$$q \approx \frac{1}{\sigma}\int_{-\infty}^{\infty} e^{-m_l^2\hbar^2/2IkT}\,dm_l = \frac{1}{\sigma}\left(\frac{2IkT}{\hbar^2}\right)^{1/2}\int_{-\infty}^{\infty} e^{-x^2}dx \approx \frac{1}{\sigma}\left(\frac{2\pi IkT}{\hbar^2}\right)^{1/2} = \frac{1}{\sigma}\left(\frac{2\pi I}{\hbar^2\beta}\right)^{1/2}$$

$$U - U(0) = -N\frac{\partial \ln q}{\partial\beta} = -N\frac{\partial}{\partial\beta}\ln\frac{1}{\sigma}\left(\frac{2\pi I}{\hbar^2\beta}\right)^{1/2} = \frac{N}{2\beta} = \frac{1}{2}NkT = \frac{1}{2}RT \quad (N = N_A)$$

$$C_{V,m} = \left(\frac{\partial U_m}{\partial T}\right)_V = \frac{1}{2}R = \boxed{4.2\ \mathrm{J\,K^{-1}\,mol^{-1}}}$$

$$S_m = \frac{U_m - U_m(0)}{T} + R\ln q$$

$$= \frac{1}{2}R + R\ln\frac{1}{\sigma}\left(\frac{2\pi IkT}{\hbar^2}\right)^{1/2}$$

$$= \frac{1}{2}R + R\ln\frac{1}{3}\left(\frac{(2\pi)\times(5.341\times10^{-47}\ \mathrm{kg\,m^2})\times(1.381\times10^{-23}\ \mathrm{J\,K^{-1}})\times(298)}{(1.055\times10^{-34}\ \mathrm{J\,s})^2}\right)^{1/2}$$

$$= \frac{1}{2}R + 1.31R = 1.81R, \text{ or } \boxed{15\ \mathrm{J\,K^{-1}\,mol^{-1}}}$$

P16.5 The absorption lines are the values of differences in adjacent rotational terms. Using eqns 12.8, 12.9, and 12.10, we have

$$\tilde{F}(J+1) - \tilde{F}(J) = \frac{E(J+1) - E(J)}{hc} = 2\tilde{B}(J+1)$$

for $J = 0, 1, \ldots$. Therefore, we can find the rotational constant and reconstruct the energy levels from the data. To make use of all of the data, one would plot the wavenumbers, which represent $\tilde{F}(J+1) - \tilde{F}(J)$, vs. J; from the above equation, the slope of that linear plot is $2\tilde{B}$. Inspection of the data shows that the lines in the spectrum are equally spaced with a separation of 21.19 $\mathrm{cm^{-1}}$, so that is the slope:

$$\text{slope} = 21.19\ \mathrm{cm^{-1}} = 2\tilde{B} \quad \text{so} \quad \tilde{B} = 10.59\overline{5}\ \mathrm{cm^{-1}}$$

The partition function is

$$q = \sum_{J=0}^{\infty}(2J+1)e^{-\beta E(J)}, \quad \text{where} \quad E(J) = hc\tilde{B}J(J+1)\,[12.8]$$

and the factor of $2J+1$ is the degeneracy of the energy levels.

$$\text{At } 25°\text{C}, \; hc\tilde{B}\beta = \frac{hc\tilde{B}}{kT} = \frac{6.626\times10^{-34}\text{ J s}\times 2.998\times10^{10}\text{ cm s}^{-1}\times 10.59\bar{5}\text{ cm}^{-1}}{1.381\times10^{-23}\text{ J K}^{-1}\times 298.15\text{ K}} = 0.05112$$

$$\begin{aligned} q &= \sum_{J=0}^{\infty}(2J+1)e^{-0.05112J(J+1)} \\ &= 1 + 3e^{-0.05112\times1\times2} + 5e^{-0.05112\times2\times3} + 7e^{-0.05112\times3\times4} + \cdots \\ &= 1 + 2.708 + 3.679 + 3.791 + 3.238 + \cdots = \boxed{19.90}. \end{aligned}$$

P16.7 The molar entropy is given by [*Justification 15.6*]

$$S_m^{\ominus} = \frac{U_m - U_m(0)}{T} + R\left(\ln\frac{q_m}{N_A} + 1\right), \quad \text{where} \quad U_m - U_m(0) = -N_A\left(\frac{\partial \ln q}{\partial \beta}\right)_V$$

where $\dfrac{q_m}{N_A} = \dfrac{q_m^T}{N_A} q^R q^V q^E$

The energy term $U_m - U_m(0)$ works out to be

$$U_m - U_m(0) = N_A[\langle\varepsilon^T\rangle + \langle\varepsilon^R\rangle + \langle\varepsilon^V\rangle + \langle\varepsilon^E\rangle]$$

Translation:

$$\frac{q_m^{T\ominus}}{N_A} = \frac{kT}{p^{\ominus}\Lambda^3} \qquad \Lambda = \frac{h}{(2\pi mkT)^{1/2}}$$

After substituting the values of the constants we obtain

$$\begin{aligned} \frac{q_m^{T\ominus}}{N_A} &= 2.561\times10^{-2}(T/\text{K})^{5/2}\times(M/\text{g mol}^{-1})^{3/2} \\ &= 2.561\times10^{-2}\times(298)^{5/2}\times(38.00)^{3/2} = 9.20\times10^{6} \end{aligned}$$

and $\langle\varepsilon^T\rangle = \frac{3}{2}kT$

Rotation of a linear molecule:

$$q^R = \frac{0.6950}{\sigma}\times\frac{T/\text{K}}{\tilde{B}/\text{cm}^{-1}} \text{ [See Exercise 16.13(a)]}$$

The rotational constant is

$$\begin{aligned} \tilde{B} &= \frac{\hbar}{4\pi cI} = \frac{\hbar}{4\pi c\mu R^2} \\ &= \frac{(1.0546\times10^{-34}\text{ J s})\times(6.022\times10^{23}\text{ mol}^{-1})}{4\pi(2.998\times10^{10}\text{ cm s}^{-1})\times(\frac{1}{2}\times19.00\times10^{-3}\text{ kg mol}^{-1})\times(190.0\times10^{-12}\text{ m})^2} \\ &= 0.4915\text{ cm}^{-1} \end{aligned}$$

so $q^{\mathrm{R}} = \dfrac{0.6950}{2} \times \dfrac{298}{0.4915} = 210.\overline{7}$

Also, $\langle \varepsilon^{\mathrm{R}} \rangle = kT$

Vibration:

$$q^{\mathrm{V}} = \frac{1}{1-\mathrm{e}^{-hc\tilde{\nu}/kT}} = \frac{1}{1-\exp\left(\dfrac{-1.4388(\tilde{\nu}/\mathrm{cm}^{-1})}{T/\mathrm{K}}\right)} = \frac{1}{1-\exp\left(\dfrac{-1.4388(450.0)}{298}\right)}$$

$$= 1.129$$

$$\langle \varepsilon^{\mathrm{V}} \rangle = \frac{hc\tilde{\nu}}{\mathrm{e}^{hc\tilde{\nu}/kT}-1} = \frac{(6.626\times10^{-34}\ \mathrm{J\ s})\times(2.998\times10^{10}\ \mathrm{cm\ s^{-1}})\times(450.0\ \mathrm{cm^{-1}})}{\exp\left(\dfrac{1.4388(450.0)}{298}\right)-1}$$

$$= 1.149\times10^{-21}\ \mathrm{J}$$

The Boltzmann factor for the lowest-lying excited electronic state is

$$\exp\left(\frac{-(1.609\ \mathrm{eV})\times(1.602\times10^{-19}\ \mathrm{J\ eV^{-1}})}{(1.381\times10^{-23}\ \mathrm{J\ K^{-1}})\times(298\ \mathrm{K})}\right) = 6\times10^{-28}$$

so we may take q^{E} to equal the degeneracy of the ground state, namely 2 and $\langle \varepsilon^{\mathrm{E}} \rangle$ to be zero. Putting it all together yields

$$\frac{U_{\mathrm{m}}-U_{\mathrm{m}}(0)}{T} = \frac{N_{\mathrm{A}}}{T}\left(\tfrac{3}{2}kT + kT + 1.149\times10^{-21}\ \mathrm{J}\right) = \tfrac{5}{2}R + \frac{N_{\mathrm{A}}(1.149\times10^{-21}\ \mathrm{J})}{T}$$

$$= (2.5)\times(8.3145\ \mathrm{J\ mol^{-1}\ K^{-1}}) + \frac{(6.022\times10^{23}\ \mathrm{mol^{-1}})\times(1.149\times10^{-21}\ \mathrm{J})}{298\ \mathrm{K}}$$

$$= 23.11\ \mathrm{J\ mol^{-1}\ K^{-1}}$$

$$R\left(\ln\frac{q_{\mathrm{m}}}{N_{\mathrm{A}}}-1\right) = (8.3145\ \mathrm{J\ mol^{-1}\ K^{-1}})\times\{\ln[(9.20\times10^{6})\times(210.7)\times(1.129)\times(2)]-1\}$$

$$= 176.3\ \mathrm{J\ mol^{-1}\ K^{-1}} \quad \text{and} \quad S_{\mathrm{m}}^{\ominus} = \boxed{199.4\ \mathrm{J\ mol^{-1}\ K^{-1}}}$$

P16.9 (a) The probability distribution of rotational energy levels is the Boltzmann factor of each level, weighted by the degeneracy, over the partition function

$$p_J^{\mathrm{R}}(T) = \frac{g(J)\mathrm{e}^{-\varepsilon_J/kT}}{q^{\mathrm{R}}} = \frac{(2J+1)\mathrm{e}^{-hc\tilde{B}J(J+1)/kT}}{\sum_{J=0}(2J+1)\mathrm{e}^{-hc\tilde{B}J(J+1)/kT}}\ [16.13, 15.9]$$

It is conveniently plotted against J at several temperatures using mathematical software. This distribution at 100 K is shown in Figure 16.4 as both a bar plot and a line plot.

The plots show that higher rotational states become more heavily populated at higher temperature. Even at 100 K the most populated state has 4 quanta of rotational energy; it is elevated to 13 quanta at 1000 K.

Values of the vibrational state probability distribution,

$$p_v^{\mathrm{V}}(T) = \frac{\mathrm{e}^{-\varepsilon_J/kT}}{q^{\mathrm{V}}} = \mathrm{e}^{-vhc\tilde{\nu}/kT}(1-\mathrm{e}^{-hc\tilde{\nu}/kT})\ [16.19]$$

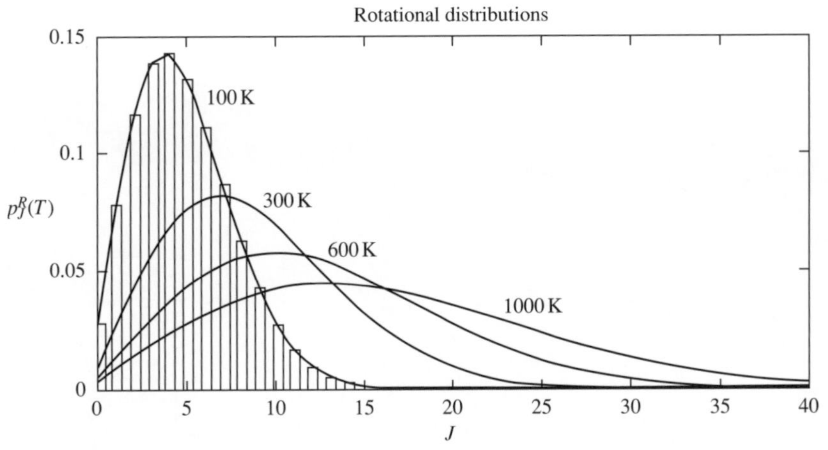

Figure 16.4(a)

are conveniently tabulated against v at several temperatures. Computations may be discontinued when values drop below some small number like 10^{-7}.

	$p_v^{V}(T)$			
v	100 K	300 K	600 K	1000 K
0	1	1	0.095	0.956
1	2.77×10^{-14}	3.02×10^{-5}	5.47×10^{-3}	0.042
2		9.15×10^{-10}	3.01×10^{-5}	1.86×10^{-3}
3			1.65×10^{-7}	8.19×10^{-5}
4				3.61×10^{-6}
5				1.59×10^{-7}

Only the state $v = 0$ is appreciably populated below 1000 K and even at 1000 K only 4% of the molecules have 1 quanta of vibrational energy.

(b) The classical (equipartition) rotational partition function is

$$q^{\rm R}_{\rm classical}(T) = \frac{kT}{hc\tilde{B}} = \frac{T}{\theta_{\rm R}} \quad [16.15b]$$

where $\theta_{\rm R}$ is the rotational temperature. We would expect the partition function to be well approximated by this expression for temperatures much greater than the rotational temperature.

$$\theta_{\rm R} = \frac{hc\tilde{B}}{k} = \frac{(6.626 \times 10^{-34}\ {\rm J\ s}) \times (2.998 \times 10^{10}\ {\rm cm\ s^{-1}}) \times (1.931\ {\rm cm^{-1}})}{1.381 \times 10^{-23}\ {\rm J\ K^{-1}}}$$

$$\theta_{\rm R} = 2.779\ {\rm K}$$

In fact $\theta_{\rm R} \ll T$ for all temperatures of interest in this problem (100 K or more). Agreement between the classical expression and the explicit sum is indeed good, as Figure 16.4(b) confirms. The figure displays the percentage deviation $(q^{\rm R}_{\rm classical} - q^{\rm R})100/q^{\rm R}$. The maximum deviation is about –0.9% at 100 K and the magnitude decreases with increasing temperature.

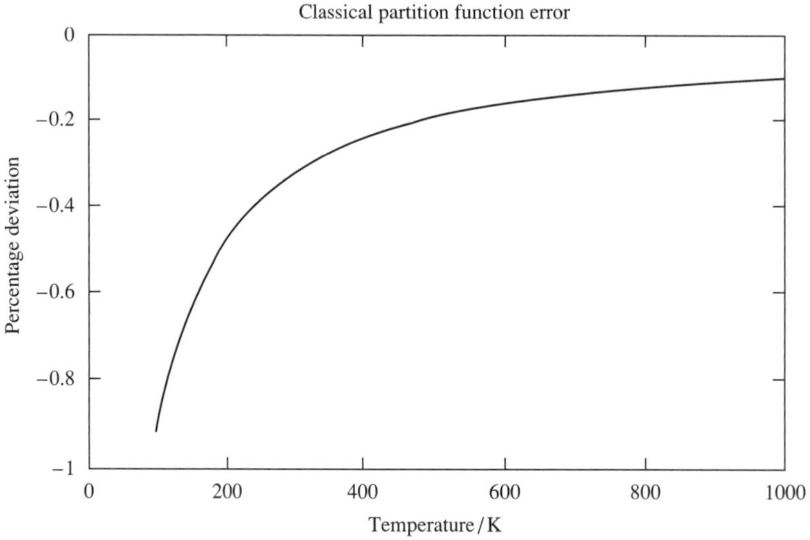

Figure 16.4(b)

(c) The translational, rotational, and vibrational contributions to the total energy are specified by eqns 16.25b, 16.26b, and 16.28, respectively. As molar quantities, they are:

$$U^{\mathrm{T}} = \tfrac{3}{2}RT, \quad U^{\mathrm{R}} = RT, \quad U^{\mathrm{V}} = \frac{N_{\mathrm{A}}hc\tilde{\nu}}{e^{hc\tilde{\nu}/kT} - 1}$$

The contributions to the difference in energy from its 100 K value are $\Delta U^{\mathrm{T}}(T) = U^{\mathrm{T}}(T) - U^{\mathrm{T}}(100\ \mathrm{K})$, etc. Figure 16.4(c) shows the individual contributions to $\Delta U(T)$. Translational motion contributes 50% more than the rotational motion because it has three quadratic degrees of freedom compared to two quadratic degrees of freedom for rotation. Very little change occurs in the vibrational energy because very high temperatures are required to populate $v = 1, 2, \ldots$ states (see part (a)).

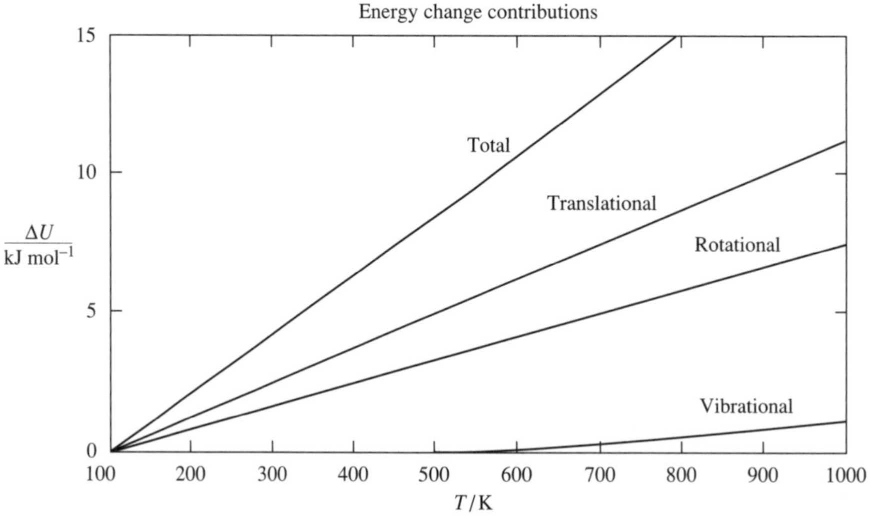

Figure 16.4(c)

$$C_{V,\mathrm{m}}(T) = \left(\frac{\partial U(T)}{\partial T}\right)_V = \left(\frac{\partial}{\partial T}\right)_V (U^{\mathrm{T}} + U^{\mathrm{R}} + U^{\mathrm{V}})$$

$$= \frac{3}{2}R + R + \frac{\mathrm{d}U^{\mathrm{V}}}{\mathrm{d}T} = \frac{5}{2}R + \frac{\mathrm{d}U^{\mathrm{V}}}{\mathrm{d}T}$$

The derivative $\mathrm{d}U^{\mathrm{V}}/\mathrm{d}T$ may be evaluated numerically with numerical software (we advise exploration of the technique) or it may be evaluated analytically using eqn 16.34:

$$C^{\mathrm{V}}_{V,\mathrm{m}} = \frac{\mathrm{d}U^{\mathrm{V}}}{\mathrm{d}T} = R\left\{\frac{\theta_{\mathrm{V}}}{T}\left(\frac{\mathrm{e}^{-\theta_{\mathrm{V}}/2T}}{1-\mathrm{e}^{-\theta_{\mathrm{V}}/T}}\right)\right\}^2$$

where $\theta_{\mathrm{V}} = hc\tilde{\nu}/k = 3122$ K. Figure 16.4(d) shows the ratio of the vibrational contribution to the sum of translational and rotational contributions. Below 300 K, vibrational motion makes a small, perhaps negligible, contribution to the heat capacity. The contribution is about 10% at 700 K and grows with increasing temperature.

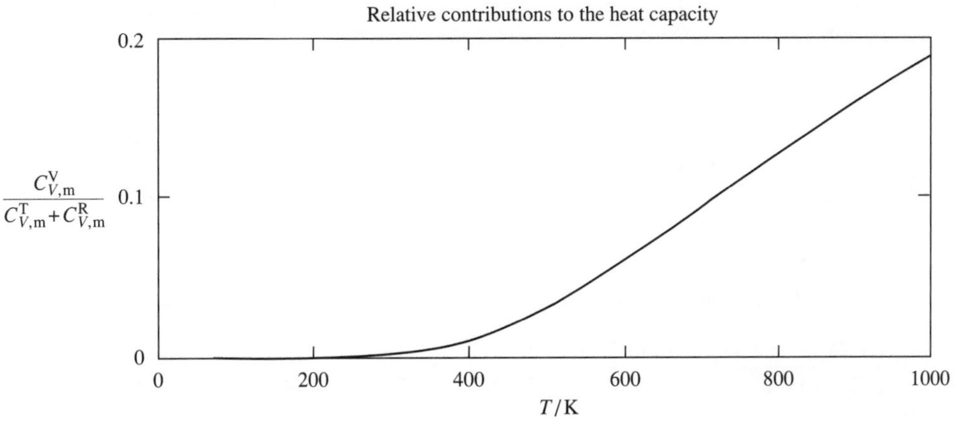

Figure 16.4(d)

The change with temperature of molar entropy may be evaluated by numerical integration with mathematical software.

$$\Delta S(T) = S(T) - S(100\ \mathrm{K}) = \int_{100\ \mathrm{K}}^{T} \frac{C_{p,\mathrm{m}}(T)\mathrm{d}T}{T}\ [3.22]$$

$$= \int_{100\ \mathrm{K}}^{T} \frac{C_{V,\mathrm{m}}(T) + R}{T}\mathrm{d}T\ [2.47]$$

$$= \int_{100\ \mathrm{K}}^{T} \frac{\frac{7}{2}R + C^{\mathrm{V}}_{V,\mathrm{m}}(T)}{T}\mathrm{d}T.$$

$$\Delta S(T) = \underbrace{\frac{7}{2}R\ln\left(\frac{T}{100\ \mathrm{K}}\right)}_{\Delta S^{\mathrm{T+R}}(T)} + \underbrace{\int_{100\ \mathrm{K}}^{T}\frac{C^{\mathrm{V}}_{V,\mathrm{m}}(T)}{T}\mathrm{d}T}_{\Delta S^{\mathrm{V}}(T)}$$

Figure 16.4(e) shows the ratio of the vibrational contribution to the sum of translational and rotational contributions. Even at the highest temperature the vibrational contribution to the entropy change is less than 2.5% of the contributions from translational and rotational motion. The vibrational contribution is negligible at low temperature.

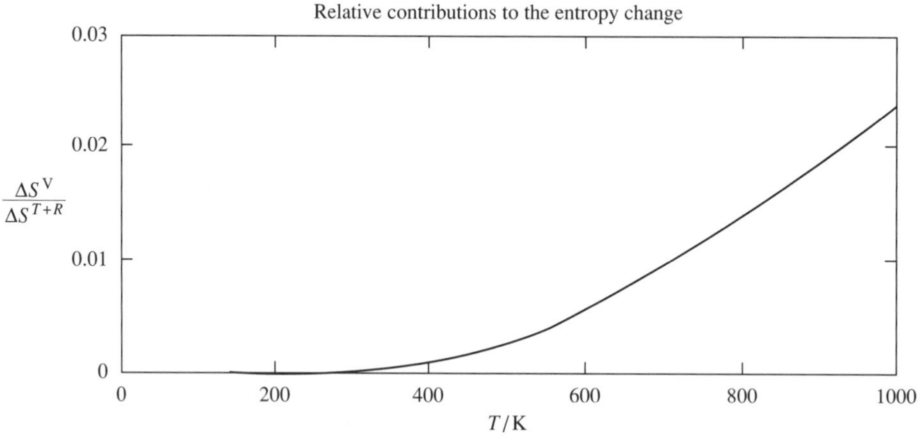

Figure 16.4(e)

P16.11 $$K = \frac{q^{\ominus}_{m}(CHD_3)q^{\ominus}_{m}(DCl)}{q^{\ominus}_{m}(CD_4)q^{\ominus}_{m}(HCl)} e^{-\beta\Delta_r E_0}$$ [16.52, with $\Delta_r E_0$ here defined as the molecular, not molar, energy difference; N_A factors cancel]

Use partition function expressions from checklists of key equations, Chapters 15 and 16. The ratio of translational partition functions is

$$\frac{q^{T}_{m}(CHD_3)q^{T}_{m}(DCl)}{q^{T}_{m}(CD_4)q^{T}_{m}(HCl)} = \left(\frac{M(CHD_3)M(DCl)}{M(CD_4)M(HCl)}\right)^{3/2} = \left(\frac{19.06 \times 37.46}{20.07 \times 36.46}\right)^{3/2} = 0.964$$

The ratio of rotational partition functions is

$$\frac{q^{R}(CHD_3)q^{R}(DCl)}{q^{R}(CD_4)q^{R}(HCl)} = \frac{\sigma(CD_4)}{\sigma(CHD_3)}\frac{(\tilde{B}(CD_4)/cm^{-1})^{3/2}\tilde{B}(HCl)/cm^{-1}}{(\tilde{A}(CHD_3)\tilde{B}(CHD_3)^2/cm^{-3})^{1/2}\tilde{B}(DCl)/cm^{-1}}$$

$$= \frac{12}{3} \times \frac{2.63^{3/2} \times 10.59}{(2.63 \times 3.28^2)^{1/2} \times 5.445} = 6.24$$

The ratio of vibrational partition functions (called Q for convenience below) is

$$Q = \frac{q^{V}(CHD_3)q^{V}(DCl)}{q^{V}(CD_4)q^{V}(HCl)} = \frac{q(2993)q(2142)q(1003)^3 q(1291)^2 q(1036)^2 q(2145)}{q(2109)q(1092)^2 q(2259)^3 q(996)^3 q(2991)}$$

where $q(x) = \dfrac{1}{1 - e^{-1.4388x/(T/K)}}$

We also require $\Delta_r E_0$, which is equal to the difference in zero point energies

$$\frac{\Delta_r E_0}{hc} = \tfrac{1}{2}\{(2993 + 2142 + 3 \times 1003 + 2 \times 1291 + 2 \times 1036 + 2145)$$
$$- (2109 + 2 \times 1092 + 3 \times 2259 + 3 \times 996 + 2991)\}\ cm^{-1}$$
$$= -1053\ cm^{-1}$$

So, the exponent in the energy term is

$$-\beta\Delta_r E_0 = -\frac{\Delta_r E_0}{kT} = -\frac{hc}{k} \times \frac{\Delta_r E_0}{hc} \times \frac{1}{T} = -\frac{1.4388 \times (-1053)}{T/\text{K}} = +\frac{1515}{T/\text{K}}$$

Hence,

$$K = 0.964 \times 6.24 \times Q\text{e}^{+1515/(T/\text{K})} = 6.02Q\text{e}^{+1515/(T/\text{K})}$$

We can now evaluate K (on a computer) and obtain the following values

T/K	300	400	500	600	700	800	900	1000
K	945	273	132	83	61	49	42	37

The values of K are plotted in Figure 16.5.

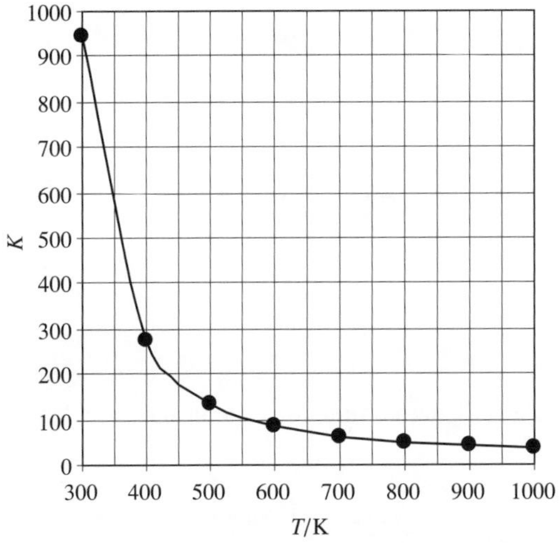

Figure 16.5

Solutions to theoretical problems

P16.13 A Sackur–Tetrode type of equation describes the translational entropy of the gas. Here,

$$q^{\text{T}} = q_x^{\text{T}} q_y^{\text{T}} \quad \text{with} \quad q_x^{\text{T}} = \left(\frac{2\pi m}{\beta h^2}\right)^{1/2} X \text{ [15.15]}$$

where X is the length of the surface. Therefore,

$$q^{\text{T}} = \left(\frac{2\pi m}{\beta h^2}\right) XY = \frac{2\pi m\sigma}{\beta h^2}, \quad \sigma = XY$$

$$U_{\text{m}} - U_{\text{m}}(0) = -\frac{N_{\text{A}}}{q}\left(\frac{\partial q}{\partial \beta}\right) = RT \text{ [or by equipartition]}$$

$$S_m = \frac{U_m - U_m(0)}{T} + R(\ln q_m - \ln N_A + 1)\left[q_m = \frac{q}{n}\right]$$

$$= R + R\ln\left(\frac{eq_m}{N_A}\right) = R\ln\left(\frac{e^2 q_m}{N_A}\right)$$

$$= \boxed{R\ln\left(\frac{2\pi e^2 m\sigma_m}{h^2 N_A \beta}\right)}\left[\sigma_m = \frac{\sigma}{n}\right]$$

Call this molar entropy of the mobile two-dimensional film S_{m2}. The molar entropy of condensation is the difference between this entropy and that of a (three-dimensional) gas:

$$\Delta S_m = S_{m2} - S_{m3}.$$

The three-dimensional value is given by the Sackur–Tetrode equation

$$S_m = R\ln\left\{e^{5/2}\left(\frac{2\pi m}{h^2\beta}\right)^{3/2}\frac{V_m}{N_A}\right\}$$

So $\Delta S_m = R\ln\dfrac{e^2(2\pi m/h^2\beta)\times(\sigma_m/N_A)}{e^{5/2}(2\pi m/h^2\beta)^{3/2}\times(V_m/N_A)} = \boxed{R\ln\left\{\left(\dfrac{\sigma_m}{V_m}\right)\times\left(\dfrac{h^2\beta}{2\pi me}\right)^{1/2}\right\}}$

P16.15 Begin with the partition function of an oscillator (checklist of key equations, Chapter 16)

$$q^V = \frac{1}{1-e^{-x}}, \quad x = \frac{\theta_V}{T} = hc\tilde{\nu}\beta = \hbar\omega\beta$$

Expressions for internal energy and other thermodynamic functions are in the checklists of key equations, Chapters 15 and 16.

$$U - U(0) = -\frac{N}{q}\left(\frac{\partial q}{\partial \beta}\right)_V = -N(1-e^{-x})\frac{d}{d\beta}(1-e^{-x})^{-1} = \frac{N\hbar\omega e^{-x}}{1-e^{-x}} = \boxed{\frac{N\hbar\omega}{e^x - 1}}$$

$$C_V = \left(\frac{\partial U}{\partial T}\right)_V = -k\beta^2\frac{\partial U}{\partial \beta}[16.31a] = -k\beta^2\hbar\omega\frac{\partial U}{\partial x}$$

$$= k(\beta\hbar\omega)^2 N\left\{\frac{e^x}{(e^x-1)^2}\right\} = \boxed{kN\left\{\frac{x^2 e^x}{(e^x-1)^2}\right\}}$$

$$H - H(0) = U - U(0)\,[q \text{ is independent of } V] = \boxed{\frac{N\hbar\omega}{e^x - 1}}$$

$$S = \frac{U - U(0)}{T} + nR\ln q = \frac{Nkxe^{-x}}{1-e^{-x}} - Nk\ln(1-e^{-x})$$

$$= \boxed{Nk\left(\frac{x}{e^x - 1} - \ln(1-e^{-x})\right)}$$

$$A - A(0) = G - G(0) = -nRT\ln q = \boxed{NkT\ln(1-e^{-x})}$$

The functions are plotted in Figure 16.6.

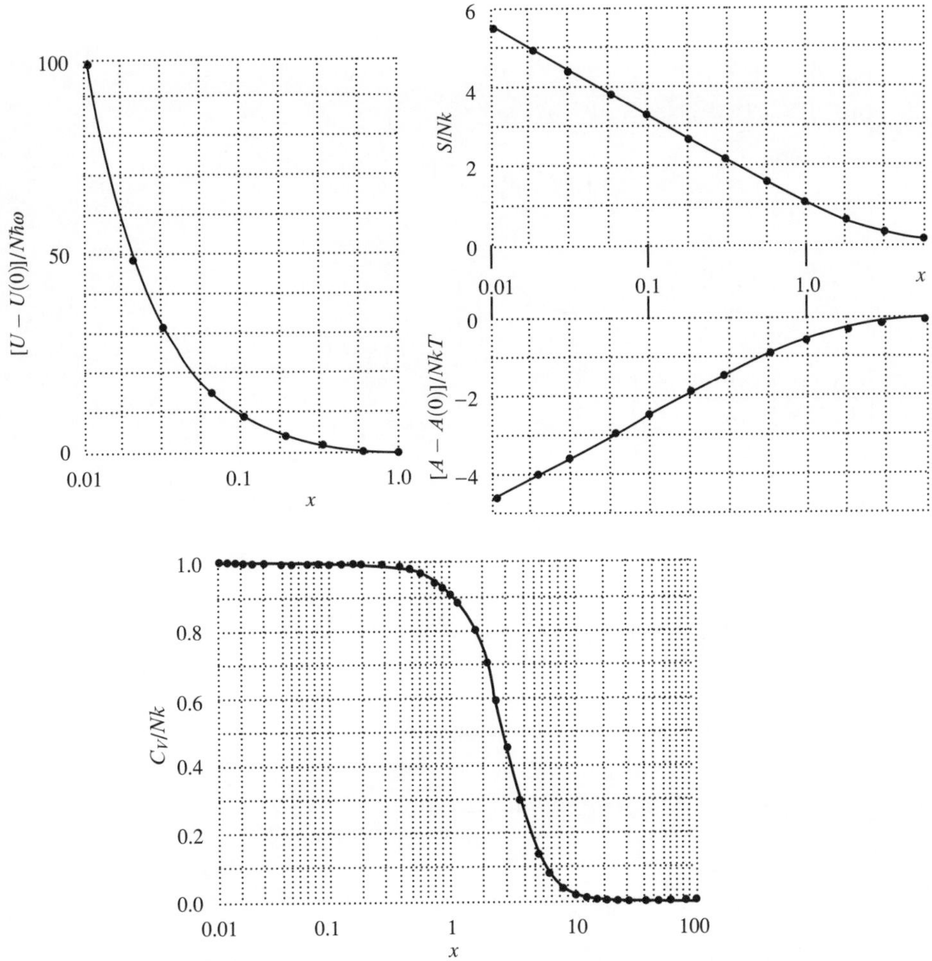

Figure 16.6

P16.17 We use as given

$$\mu = \frac{V_m}{C_{p,m}}(\alpha T - 1) \quad \text{and} \quad \mu_T = -C_{p,m}\mu = -V_m(\alpha T - 1)$$

We substitute $\alpha = \frac{1}{V_m}\left(\frac{\partial V_m}{\partial T}\right)_p$ and obtain $\mu_T = V_m - \left(\frac{\partial V_m}{\partial T}\right)_p T$. We express V_m in terms of the second virial coefficient by using the virial expansion in powers of p, namely $Z = \frac{pV_m}{RT} = 1 + \frac{B(T)}{RT}p + \cdots$ [1.19]. In the limit as $p \to 0$, we need not retain terms beyond the first power of p. Then $V_m = \frac{RT}{p} + B(T)$ and $\left(\frac{\partial V_m}{\partial T}\right)_p = \frac{R}{p} + \frac{dB(T)}{dT}$. Substituting into the expression for μ_T we obtain the required formula:

$$\boxed{\mu_T = B(T) - T\frac{dB(T)}{dT}.}$$

P16.19 (a) $U - U(0) = -\dfrac{N}{q}\dfrac{\partial q}{\partial \beta} = \dfrac{N}{q}\sum \varepsilon_j e^{-\beta\varepsilon_j} = \dfrac{NkT}{q}\dot{q} = \boxed{nRT\left(\dfrac{\dot{q}}{q}\right)}$

$$C_V = \left(\frac{\partial U}{\partial T}\right)_V = \frac{\partial \beta}{\partial T}\left(\frac{\partial U}{\partial \beta}\right)_V = -\frac{1}{kT^2}\frac{\partial}{\partial \beta}\left(\frac{N}{q}\sum_j \varepsilon_j e^{-\beta\varepsilon_j}\right)$$

$$= \left(\frac{N}{kT^2}\right)\times\left[\frac{1}{q}\sum_j \varepsilon_j^2 e^{-\beta\varepsilon_j} + \frac{1}{q^2}\left(\frac{\partial q}{\partial \beta}\right)\sum_j \varepsilon_j e^{-\beta\varepsilon_j}\right]$$

$$= \left(\frac{N}{kT^2}\right)\times\left[\frac{1}{q}\sum_j \varepsilon_j^2 e^{-\beta\varepsilon_j} - \frac{1}{q^2}\left(\sum_j \varepsilon_j e^{-\beta\varepsilon_j}\right)^2\right]$$

$$= \left(\frac{N}{kT^2}\right)\times\left[\frac{k^2T^2\ddot{q}}{q} - \frac{k^2T^2}{q^2}\dot{q}^2\right]$$

$$= \boxed{nR\left\{\frac{\ddot{q}}{q} - \left(\frac{\dot{q}}{q}\right)^2\right\}}$$

$$S = \frac{U - U(0)}{T} + nR\ln\left(\frac{q}{N} + 1\right) = \boxed{nR\left(\frac{\dot{q}}{q} + \ln\frac{eq}{N}\right)}$$

(b) At 5000 K, $\dfrac{kT}{hc} = 3475\ \text{cm}^{-1}$. We form the sums

$$q = \sum_j g e^{-\beta\varepsilon_j} = 1 + e^{-21850/3475} + 3e^{-21870/3475} + \cdots = 1.0167$$

$$\dot{q} = \sum_j \frac{\varepsilon_j}{kT} e^{-\beta\varepsilon_j} = \frac{hc}{kT}\sum_j \tilde{\nu}_j e^{-\beta\varepsilon_j}$$

$$= \left(\frac{1}{3475}\right)\times\{0 + 21850\, e^{-21850/3475} + 3\times 21870\, e^{-21870/3475} + \cdots\} = 0.1057$$

$$\ddot{q} = \sum_j \left(\frac{\varepsilon_j}{kT}\right)^2 e^{-\beta\varepsilon_j} = \left(\frac{hc}{kT}\right)^2 \sum_j \tilde{\nu}_j^2 e^{-\beta\varepsilon_j}$$

$$= \left(\frac{1}{3475}\right)^2 \times\{0 + 21850^2\, e^{-21850/3475} + 3\times 21870^2\, e^{-21870/3475} + \cdots\} = 0.6719$$

The electronic contribution to the molar constant-volume heat capacity is

$$C_{V,\text{m}} = R\left\{\frac{\ddot{q}}{q} - \left(\frac{\dot{q}}{q}\right)^2\right\}$$

$$= 8.314\ \text{J K}^{-1}\,\text{mol}^{-1}\times\left\{\frac{0.6719}{1.0167} - \left(\frac{0.1057}{1.0167}\right)^2\right\} = \boxed{5.41\ \text{J K}^{-1}\,\text{mol}^{-1}}$$

P16.21 The derivation of

$$C_V = \frac{kN\beta^2}{2}\zeta(\beta)$$

given in P16.20 is completely general. That is, it makes no use of the fact that the energies and degeneracies in question were those of a linear rotor. The derivation and therefore the result can apply equally well to a non-linear rotor, to electronic energy levels, or to the vibrational energy levels involved of this problem.

To evaluate contributions of individual excitations to the heat capacity, we re-write $\zeta(\beta)$ in notation associated with vibrational energy levels

$$\zeta(\beta) = \frac{1}{q^2}\sum_{\upsilon,\upsilon'}\{\varepsilon(\upsilon)-\varepsilon(\upsilon')\}^2 g(\upsilon)g(\upsilon')\mathrm{e}^{-\beta[\varepsilon(\upsilon)+\varepsilon(\upsilon')]} = \frac{1}{q^2}\sum_{\upsilon,\upsilon'}\{\varepsilon(\upsilon)-\varepsilon(\upsilon')\}^2 \mathrm{e}^{-\beta[\varepsilon(\upsilon)+\varepsilon(\upsilon')]}$$

where the levels are non-degenerate, or at least are treated as such because vibrational modes are treated one by one. The energy levels are

$$\varepsilon(\upsilon) = hc\tilde{\nu}\upsilon = \theta_V k\upsilon \quad \text{so} \quad \beta\varepsilon(\upsilon) = \theta_V\upsilon/T.$$

The total heat capacity and the contributions of several transitions are plotted in Figure 16.7. For vibration, one can compute q and the total C_V/R analytically, using expressions from the checklists of key equations in Chapters 15 and 16.

$$q^V = \frac{1}{1-\mathrm{e}^{-\theta_V/T}}\ [16.19] \quad \text{and} \quad \frac{C_{V,\mathrm{m}}}{R} = \left(\frac{\theta_V}{T}\right)^2 \frac{\mathrm{e}^{-\theta_V/T}}{(1-\mathrm{e}^{-\theta_V/T})^2}\ [16.34]$$

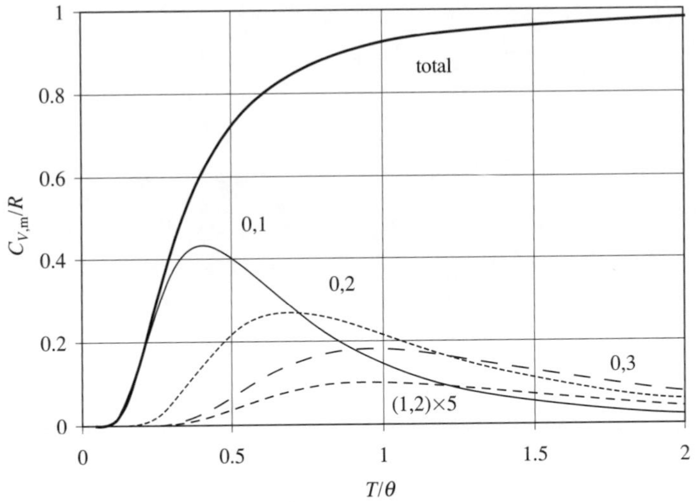

Figure 16.7

P16.23 The solution to this problem requires the use of mathematical software. Our solution is provided in the following MathCad® worksheet.

$k := 1.381 \cdot 10^{-23}\ J \cdot K^{-1}$ $N_A := 6.022 \cdot 10^{23} \cdot mol^{-1}$

The radial distribution function as a function of the unitless or 'reduced' or 'red' variable $r_{red} = r/d = r/\sigma$ where σ is the Lennard–Jones potential separation at which the pair potential equals zero is:

$$g(r_{red}) := 1 + \cos(4 \cdot r_{red} - 4) \cdot e^{-(r_{red}-1)} \quad \text{where} \quad r_{red} = \frac{r}{d} = \frac{r}{r_{min}} = \frac{r}{\sigma} > 1$$

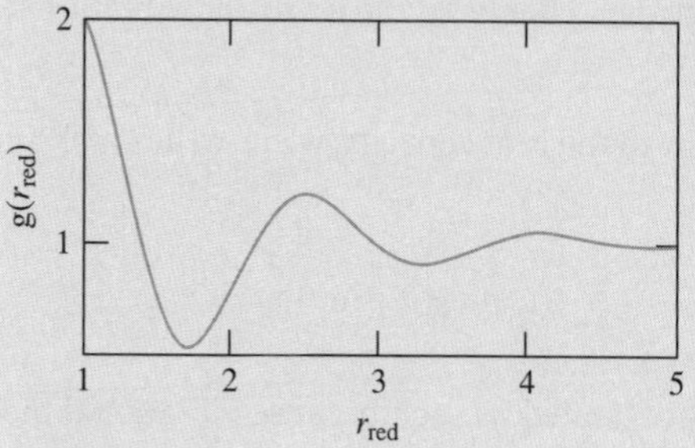

Figure 16.8(a)

This function, plotted in Figure 16.8(a), clearly resembles the form of Figure 16.14 of the text.

The Lennard-Jones potential in the unitless form $V_{red} = E_p/\varepsilon$ is

$$V_{red}(r_{red}) := 4(r_{red}^{-12} - r_{red}^{-6})$$

The virial in the unitless form $v_{2r} = v_2/\varepsilon = r_r dV_r/dr_r$ [16.49a] is

$$v_{2red}(r_{red}) := r_{red} \cdot \left(\frac{d}{dr_{red}} V_{red}(r_{red})\right)$$

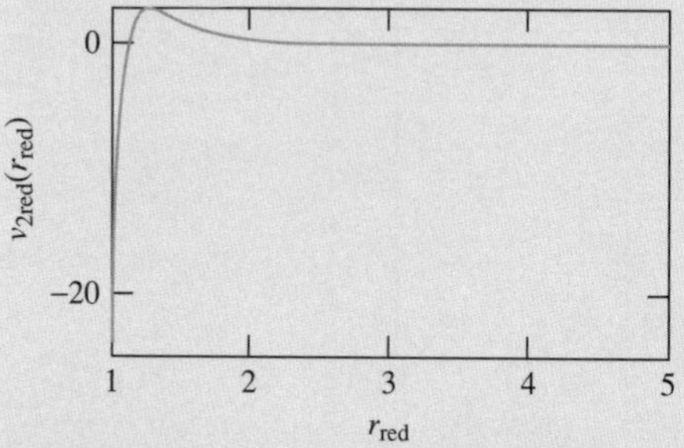

Figure 16.8(b)

The force between the molecular pair is attractive when the virial is positive and Figure 16.8(b) shows that the attraction is greatest at $r_{red} = 1.244$.

The internal pressure is related to the pair distribution function by the expression:

$$p = \frac{n \cdot R \cdot T}{V} - 2\frac{\pi}{3}\left(\frac{N}{V}\right)^2 \int_{\sigma}^{\infty} g(r) \cdot v_2(r) \cdot r^2 dr \ [16.49b]$$

or in terms of the unitless variables

$$p = \frac{n \cdot R \cdot T}{V} - 2\frac{\pi}{3} \cdot \varepsilon \cdot \sigma^3 \left(\frac{N}{V}\right)^2 \int_{I}^{\infty} g(r_{re}) \cdot v_{2red}(r_{red}) \cdot r_{red}^2 dr_{red}$$

Since $nRT/V = \rho kT$ where $\rho = N/V$,

$$\frac{p}{\rho \cdot k \cdot T} = 1 - \frac{2\pi \cdot \varepsilon \cdot \sigma^3 \cdot \rho}{3k \cdot T} \int_l^{\infty} g(r_{red}) \cdot v_{2red}(r_{red}) \cdot r_{red}^2 \, dr_{red}$$

Or, using the unitless pressure $p_{red} = p/\rho kT$ and the unitless temperature $T_{red} = kT/\varepsilon$ and the unitless density $\rho_{red} = \rho\sigma^3$, we have

$$p_{red}(\rho_{red}, T_{red}) := 1 - \frac{2\pi \cdot \rho_{red}}{3T_{red}} \int_l^{\infty} g(r_{red}) \cdot v_{2red}(r_{red}) \cdot r_{red}^2 \, dr_{red}$$

This expression contains no matter specific parameter. Consequently, it is an example of the principle of corresponding states, which states that in the appropriate dimensionless reduced variables all types of gases and fluids, no matter what their composition might be, are described by the same equation of state. Isotherms of p_{red} against ρ_{red}, shown in Figure 16.8(c), are linear in ρ_{red}. Notice that at very large values of T_{red} the equation becomes the perfect gas law (as expected) because the second term on the right becomes insignificantly small. Also, it is apparent that this equation of state is very incorrect at high number densities ($\rho_{red} > 0.4$ for $T_{red} = 1.35$) because p_{red} becomes negative.

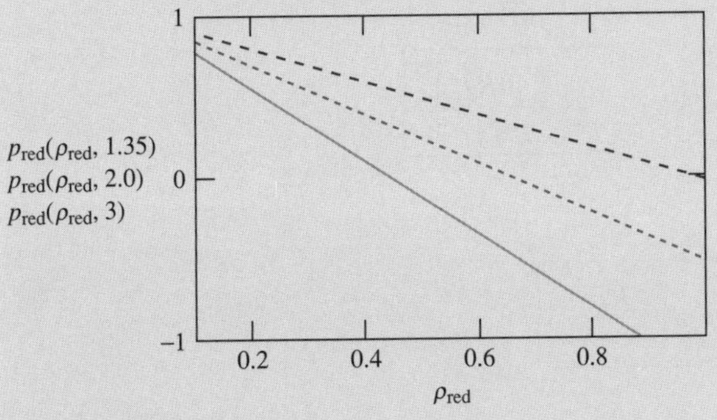

Figure 16.8(c)

The principle of corresponding states is an approximation, so it is instructive to compare experimental values with the value predicted by the above expression. Consider argon at the critical point. From tables in the appendix and the above definitions we find:

$\varepsilon := (111.84 \cdot \mathrm{K}) \cdot \mathrm{k} \quad \sigma := 362.3 \cdot 10^{-12} \cdot \mathrm{m}$

$V_c := 75.25 \cdot \mathrm{cm}^3 \cdot \mathrm{mol}^{-1} \quad T_c := 150.72 \cdot \mathrm{K} \quad p_c := 48.00 \cdot \mathrm{atm}$

$\rho_c := \frac{N_A}{V_c} \quad \rho_C = 8.003 \times 10^{27}\ \mathrm{m}^{-3}$

$T_{red} := \frac{k \cdot T_c}{\varepsilon} \quad T_{red} = 1.348$

$\rho_{red} := \rho_c \cdot \sigma^3 \quad \rho_{red} = 0.381$

Thus, the experimental values at the critical point give the ratio

$\frac{p_c}{\rho_c \cdot k \cdot T_c} = 0.292$

while the ratio given by the equation of state with the pair distribution function gives

$$p_{\text{red}}(\rho_{\text{red}}, 1.348) = 0.138$$

It is apparent that the equation of state based on the simple pair distribution function that is provided in this problem and the simple Lennard-Jones pair potential gives a very poor prediction of the $p/\rho kT$ ratio at the critical point. It may, however, give better results at lower number densities. The discrepancy probably arises with the deficiency of the pair potential and the assumption that total potential is the sum of pairwise potential additivity.

P16.25 $$c_s = \left(\frac{\gamma RT}{M}\right)^{1/2}, \quad \gamma = \frac{C_{p,\text{m}}}{C_{V,\text{m}}}, \quad C_{p,\text{m}} = C_{V,\text{m}} + R$$

(a) $C_{V,\text{m}} = \tfrac{1}{2}R(3 + \nu_{\text{R}}^* + 2\nu_{\text{V}}^*) = \tfrac{1}{2}R(3+2) = \tfrac{5}{2}R$

$$C_{p,\text{m}} = \tfrac{5}{2}R + R = \tfrac{7}{2}R$$

$$\gamma = \tfrac{7}{5} = 1.40; \quad \text{hence} \quad \boxed{c_s = \left(\frac{1.40RT}{M}\right)^{1/2}}$$

(b) $C_{V,\text{m}} = \tfrac{1}{2}R(3+2) = \tfrac{5}{2}R, \quad \gamma = 1.40, \quad \boxed{c_s = \left(\frac{1.40RT}{M}\right)^{1/2}}$

(c) $C_{V,\text{m}} = \tfrac{1}{2}R(3+3) = 3R$

$$C_{p,\text{m}} = 3R + R = 4R, \quad \gamma = \tfrac{4}{3}, \quad \boxed{c_s = \left(\frac{4RT}{3M}\right)^{1/2}}$$

For air, $M \approx 29\ \text{g mol}^{-1}$, $T \approx 298\ \text{K}$, $\gamma = 1.40$

$$c_s = \left(\frac{(1.40)\times(2.48\ \text{kJ mol}^{-1})}{29\times10^{-3}\ \text{kg mol}^{-1}}\right)^{1/2} = \boxed{350\ \text{m s}^{-1}}$$

Solutions to applications

P16.27 (a) The heat capacity is

$$C_V = -k\beta^2\left(\frac{\partial E}{\partial \beta}\right)_V \quad [16.31\text{a}]$$

First, express E as a function of β:

$$E = \frac{N\varepsilon e^{-\beta\varepsilon}}{1+e^{-\beta\varepsilon}}$$

Hence, $$\frac{C_V}{-k\beta^2} = \left(\frac{\partial E}{\partial \beta}\right)_V = \frac{1}{1+e^{-\beta\varepsilon}} \times (-N\varepsilon^2 e^{-\beta\varepsilon}) - \frac{N\varepsilon}{(1+e^{-\beta\varepsilon})^2} \times (-\varepsilon e^{-\beta\varepsilon})$$

Collecting terms over a common denominator yields

$$C_V = \frac{kN\beta^2\varepsilon^2 e^{-\beta\varepsilon}}{(1+e^{-\beta\varepsilon})^2}(1+e^{-\beta\varepsilon}-1) = \frac{kN\beta^2\varepsilon^2 e^{-2\beta\varepsilon}}{(1+e^{-\beta\varepsilon})^2} = \frac{kN(1/kT)^2\varepsilon^2 e^{-2\varepsilon/kT}}{(1+e^{-\varepsilon/kT})^2}$$

Change the expression to molar rather than molecular quantities:

$$N = N_A, \quad R = N_A k, \quad \text{and} \quad \varepsilon/k = \varepsilon_m/R,$$

so, $$C_{V,m} = \frac{R(\varepsilon_m/RT)^2 e^{-2\varepsilon_m/RT}}{(1+e^{-\varepsilon_m/RT})^2}$$

(b) It is convenient to plot $C_{V,m}$ (in units of R) as a function of x, where $x = kT/\varepsilon = RT/\varepsilon_m$. See Figure 16.9

$$C_{V,m} = \frac{Re^{-2/x}}{x^2\left(1+e^{-1/x}\right)^2} \qquad c(x) \equiv \frac{C_{V,m}}{R}$$

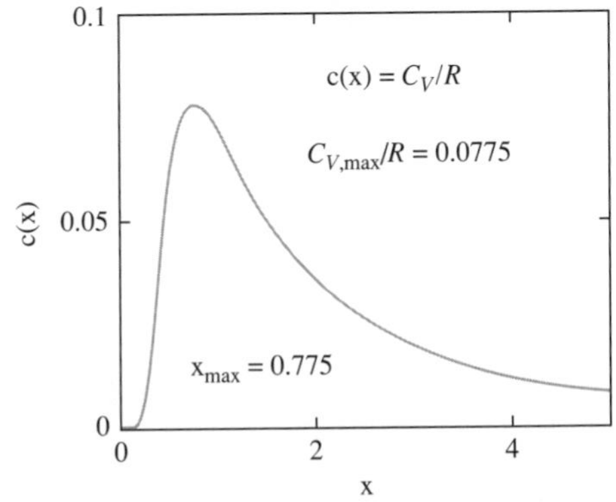

Figure 16.9

(c) The above plot indicates a maximum heat capacity at about 0.08 R at a value for x of about 0.8. The X–Y trace feature of mathematical software may be used to find a more accurate value for x_{max} of 0.775 and for $c(x_{max})$ of 0.775. A formula for the maximum is determined by the criterion that $dC_{V,m}/dx = 0$ at the maximum.

$$\frac{d(C_{V,m}/R)}{dx} = \frac{d}{dx}\left\{\frac{e^{-2/x}}{x^2(1+e^{-1/x})^2}\right\} = \frac{2e^{-2/x}}{x^4(1+e^{-1/x})^3}\{1 - x - xe^{-1/x}\}$$

Thus, $C_{V,m}$ is a maximum when $x = x_{max}$ satisfies the equation

$$1 - x_{max} - x_{max}e^{-1/x_{max}} = 0$$

This is a transcendental equation so it is necessary to solve for x_{max} with a numerical method. x_{max} may be numerically determined with the numeric solver application of the modern scientific calculator. The Given/Find solve block of Mathcad can also be used and the following presents a Mathcad solution.

$x := .7$ (Estimate for using solve block.)

Given $\left(1 - x - x \cdot e^{\frac{-1}{x}}\right) = 0$

Find(x) = 0.782

This represents the best value of $x_{\max}$. $\boxed{T_{\max} = \dfrac{\varepsilon_m x_{\max}}{R}}$ with $x_{\max}$ determined as above.

P16.29 The standard molar Gibbs energy is given by

$$G_m^\ominus - G_m^\ominus(0) = RT\ln\frac{q_m^\ominus}{N_A} \quad \text{where } \frac{q_m^\ominus}{N_A} = \frac{q_m^{T\ominus}}{N_A} q^R q^V q^E \text{ [16.51]}$$

Translation: (see the solutions to Exercises 16.13(a) and 16.20(a) and Problem 16.7 for all partition function expressions)

$$\frac{q_m^{T\ominus}}{N_A} = 2.561\times10^{-2}(T/K)^{5/2}(M/\text{g mol}^{-1})^{3/2}$$

$$= 2.561\times10^{-2}\times(200.0)^{5/2}\times(102.9)^{3/2} = 1.512\times10^{7}$$

Rotation of a non-linear molecule

$$q^R = \frac{1}{\sigma}\left(\frac{kT}{hc}\right)^{3/2}\left(\frac{\pi}{\tilde{A}\tilde{B}\tilde{C}}\right)^{1/2} = \frac{1.0270}{\sigma}\times\frac{(T/K)^{3/2}}{(\tilde{A}\tilde{B}\tilde{C}/\text{cm}^{-3})^{1/2}}$$

$$= \frac{1.0270}{2}\times\frac{[(200.0)\times(2.998\times10^{10}\text{ cm s}^{-1})]^{3/2}}{[(13109.4)\times(2409.8)\times(2139.7)\times(10^{6}\text{ s}^{-1})^{3}/\text{cm}^{-3}]^{1/2}} = 2.900\times10^{4}$$

Vibration

$$q_1^V = \frac{1}{1-\exp\left(\frac{-1.4388(\tilde{\nu}/\text{cm}^{-1})}{T/K}\right)} = \frac{1}{1-\exp\left(\frac{-1.4388(753)}{200.0}\right)} = 1.004$$

$$q_2^V = \frac{1}{1-\exp\left(\frac{-1.4388(542)}{200.0}\right)} = 1.021$$

$$q_3^V = \frac{1}{1-\exp\left(\frac{-1.4388(310)}{200.0}\right)} = 1.120$$

$$q_4^V = \frac{1}{1-\exp\left(\frac{-1.4388(127)}{200.0}\right)} = 1.670$$

$$q_5^V = \frac{1}{1-\exp\left(\frac{-1.4388(646)}{200.0}\right)} = 1.010$$

$$q_6^V = \frac{1}{1-\exp\left(\frac{-1.4388(419)}{200.0}\right)} = 1.052$$

$$q^V = \prod_{i=1}^{6} q_i^V = 2.037$$

Putting it all together yields

$$G_m^\ominus - G_m^\ominus(0) = (8.3145\text{ J mol}^{-1}\text{ K}^{-1})\times(200.0\text{ K})\times\ln[(1.512\times10^{7})\times(2.900\times10^{4})\times(2.037)\times(1)]$$

$$G_m^\ominus - G_m^\ominus(0) = 4.576\times10^{4}\text{ J mol}^{-1} = \boxed{45.76\text{ kJ mol}^{-1}}$$

17 Molecular interactions

Answers to discussion questions

D17.1 A molecule with a permanent separation of electric charge has a **permanent dipole moment** and is said to be a **polar molecule**. In molecules containing atoms of differing electronegativity, the bonding electrons may be displaced in such a way as to produce a net separation of charge. Separation of charge may also arise from a difference in atomic radii of the bonded atoms. The separation of charges in the bonds is usually, although not always, in the direction of the more electronegative atom but depends on the precise bonding and nuclear arrangement in the molecule, as described in Section 17.1. A heteronuclear diatomic molecule necessarily has a dipole moment if there is a difference in electronegativity between the atoms, but the situation in polyatomic molecules is more complex. A polyatomic molecule has a permanent dipole moment only if it belongs to the symmetry point groups C_n, C_{nv}, or C_s, as discussed in Section 11.3(a) of the text.

An external electric field can distort the electron density in both polar and non-polar molecules. This results in an **induced dipole moment** that is proportional to the field and the constant of proportionality is called the **polarizability**. Molecules with small HOMO−LUMO gaps typically have large polarizabilities.

D17.3 Dipole moments are not measured directly, but are calculated from a measurement of the **relative permittivity**, $\varepsilon_r = \varepsilon/\varepsilon_0$ [17.13, dielectric constant], of the medium as measured by comparing the capacitance of a capacitor with and without the sample present using $\varepsilon_r = C/C_0$. Equations 17.14 (the **Debye equation**) and 17.15,

$$\frac{\varepsilon_r - 1}{\varepsilon_r + 2} = \frac{\rho P_m}{M} \; [17.14], \quad \text{where} \quad P_m = \frac{N_A}{3\varepsilon_0}\left(\alpha + \frac{\mu^2}{3kT}\right) \; [17.15]$$

imply that the dipole moment can be determined from a measurement of ε_r as a function of temperature. This approach is illustrated in Example 17.2. In another method, the relative permittivity of a solution of the polar molecule is measured as a function of concentration. The calculation is again based on the Debye equation, but in a modified form. The values obtained by this method are accurate only to about 10%. A third method is based on the relationship between relative permittivity and refractive index, eqn 17.17, and thus reduces to a measurement of the refractive index. Accurate values of the dipole moments of gaseous molecules can be obtained from the Stark effect in their microwave spectra.

D17.5 See Figure 17.4 of the text for typical charge arrays corresponding to electric multipoles.

The interaction potential between a point charge Q_2 (monopole, $n = 1$) and any of the multipoles ($m = 2$ or 3 or …) is given by eqn 17.19 as $V \propto \dfrac{1}{r^m}$, where r is the separation distance between Q_2 and the multipole. This is a steeper potential energy decrease with r than that observed for the Coulombic interaction between two point charges: $V \propto \dfrac{1}{r}$. The steeper decline originates in the case for which $r \gg l$, where l is the separation of charge within the multipole because, as r becomes relatively large, the array of charges in the multipole appears to blend together into neutrality, causing lower-order interaction terms to cancel. For example, the dipole terms within the monopole–quadrupole ($m = 3$) interaction potential cancel, leaving only a $1/r^3$ term when $r \gg l$.

We use the linear quadrupole charge arrangement shown in Figure 17.1 to show this cancellation of lower-order terms. Since we are interested in the case $x = l/r \ll 1$, the following Taylor series expansions are useful substitutions:

$$(1+x)^{-1} = 1 - x + x^2 - x^3 + \cdots \quad \text{and} \quad (1-x)^{-1} = 1 + x + x^2 + x^3 + \cdots$$

Begin by adding the terms for the Coulomb potential interaction between the charge array of the quadrupole and the monopole Q_2, substitute $x = l/r$, and perform Taylor series expansions on the functions of x:

$$4\pi\varepsilon_0 V = \frac{Q_1Q_2}{r+l} - \frac{2Q_1Q_2}{r} + \frac{Q_1Q_2}{r-l} = \frac{Q_1Q_2}{r}\left\{\frac{1}{1+x} - 2 + \frac{1}{1-x}\right\}$$

$$= \frac{Q_1Q_2}{r}\{\cancel{1} \cancel{-x} + x^2 - x^3 + x^4 + \cdots \cancel{-2} \cancel{+1} \cancel{+x} + x^2 + x^3 + x^4 \cdots\}$$

$$= \frac{2x^2Q_1Q_2}{r}\{1 + x^2 + x^4 + \cdots\}$$

The higher-order terms within the polynomial are negligibly small compared to 1 in the case for which $x = l/r \ll 1$, leaving the simple expression:

$$V = \frac{2x^2Q_1Q_2}{4\pi\varepsilon_0 r} = \frac{l^2Q_1Q_2}{2\pi\varepsilon_0 r^3} \quad \text{or} \quad V \propto \frac{1}{r^3}.$$

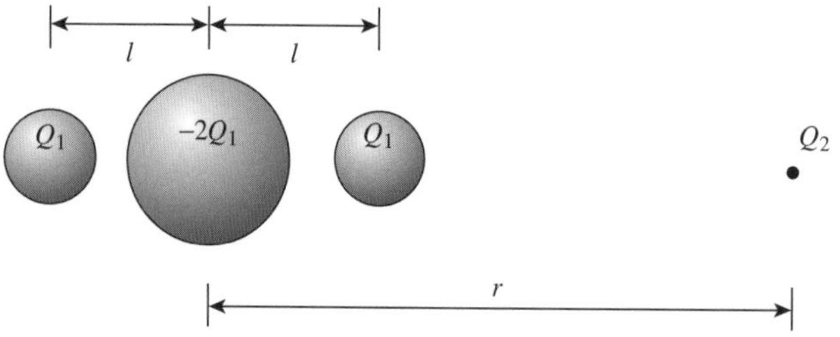

Figure 17.1

D17.7 A **hydrogen bond** (···) is an attractive interaction between two species that arises from a link of the form A − H ··· B , where A and B are highly electronegative elements (usually nitrogen, oxygen, or fluorine) and B possesses a lone pair of electrons. It is a contact-like attraction that requires AH to touch B. Experimental evidences supports a linear or near-linear structural arrangement and a bond strength of about 20 kJ mol^{-1}. The hydrogen bond strength is considerably weaker than a covalent bond but it is larger than, and dominates, other intermolecular attractions such as dipole–dipole attractions. Its formation can be understood in terms of either the (a) electrostatic interaction model or (b) with molecular orbital calculations (Chapter 10).

(a) A and B, being highly electronegative, are viewed as having partial negative charges (δ^-) in the electrostatic interaction model of the hydrogen bond. Hydrogen, being less electronegative than A, is viewed as having a partial positive (δ^+). The linear structure maximizes the electrostatic attraction between H and B:

$$\overset{\delta^-}{\mathrm{A}}\text{——}\overset{\delta^+}{\mathrm{H}}\text{---------}:\overset{\delta^-}{\mathrm{B}}$$

This model is conceptually very useful. However, it is impossible to exactly calculate the interaction strength with this model because the partial atomic charges cannot be precisely defined. There is no way to define which fraction of the electrons of the AB covalent bond should be assigned to one or the other nucleus.

(b) *Ab initio* quantum calculations are needed in order to explore questions about the linear structure, the role of the lone pair, the shape of the potential energy surface, and the extent to which the hydrogen bond has covalent σ bond character. Yes, the hydrogen bond appears to have some σ bond character. This was initially suggested by Linus Pauling in the 1930s and more recent experiments with Compton scattering of X-rays and NMR techniques indicate that the covalent character may provide as much as 20% of the hydrogen-bond strength. A three-centre molecular orbital model provides a degree of insight. A linear combination of an appropriate σ orbital on A, the 1s hydrogen orbital, and an appropriate orbital for the lone pair on B yields a total of three molecular orbitals. One of the MOs is bonding, one is almost non-bonding, and the third is antibonding (see text Figure 17.8). Both bonding MO and the almost non-bonding orbital are occupied by two electrons (the σ bonding electrons of A–H and the lone pair of B). The antibonding MO is empty. Thus, depending on the precise location of the almost non-bonding orbital, the non-bonding orbital may lower the total energy and account for the hydrogen bond.

D17.9 As discussed in Section 3.9 and depicted in text Figure 3.20, when the temperature of a thermodynamically stable, pure gaseous substance declines at constant pressure there comes a temperature at which the chemical potential of the liquid phase ($\mu_{\text{pure liquid}} = G_{\text{m,pure liquid}}$) is no longer greater than the chemical potential of the gas phase ($\mu_{\text{pure gas}} = G_{\text{m,pure gas}}$). That temperature, the boiling point, is part of the equilibrium vapour pressure curve for which $\mu_{\text{pure liquid}} = \mu_{\text{pure gas}}$ and the pressure is called the **vapour pressure of the liquid**. The relationship between the vapour pressure p and the boiling point T, the liquid–vapour phase boundary, indicates that a plot of $\ln p$ against $1/T$ is linear with a slope equal to $-\Delta_{\text{vap}}H/R$:

$$p = p_{\text{ref}}\mathrm{e}^{-\chi}, \quad \text{where } \chi = \frac{\Delta_{\text{vap}}H}{R}\left(\frac{1}{T} - \frac{1}{T_{\text{ref}}}\right)$$

[4.12, perfect gas, constant $\Delta_{\text{vap}}H$, experimental reference point (p_{ref}, T_{ref})]

Conditions below the boiling point are $\mu_{\text{pure liquid}} < \mu_{\text{pure gas}}$, the liquid is the thermodynamically stable state, and condensation is spontaneous. Above the boiling point $\mu_{\text{pure liquid}} > \mu_{\text{pure gas}}$, the gas is the thermodynamically stable state, and vaporization is spontaneous.

If an insoluble, inert gas is used to apply additional pressure ΔP on the liquid with the result that $P = p + \Delta P$, the vapour pressure becomes

$$p = p^* e^{V_m(l)\Delta P/RT} \; [4.4]$$

where p^* is the vapour pressure of the liquid in the absence of an additional pressure. The positive exponential factor indicates that the vapour pressure increases when the pressure acting on the condensed phase is increased.

The above discussion relates to bulk, macroscopic samples of matter in which surface effects are negligible. However, the condensation process can be explored further by considering the formation of small spherical droplets of liquid within the gas phase. The curvature of the droplet surface of radius r and surface tension γ plays a significant role because it gives rise to a positive pressure differential $\Delta P = 2\gamma/r$ between the droplet bulk and the gas phase. Equation 4.4 becomes the **Kelvin equation**:

$$p = p^* e^{2\gamma V_m(l)/rRT} \; [17.51]$$

The relationship indicates that droplets with smaller values of r have greater vapour pressure. This presents a severe problem for the formation of liquid droplets from a pure gas state that has been cooled below its boiling point. We can imagine a swarm of gas-phase molecules congregating into what would certainly be a very, very small droplet at best. The droplet, being exceedingly small, has an enhanced vapour pressure. Therefore, instead of growing it evaporates, even though $\mu_{\text{pure liquid}} < \mu_{\text{pure gas}}$. This effect kinetically stabilizes the gas phase, while liquid formation remains spontaneous. The rate of condensation is extremely slow and the direct mechanism of condensation is hindered. Thus, rapid formation of liquid–gas equilibrium depends on the presence of minute dust particles or other kinds of foreign matter to which the gas molecules can stick. These serve as **nucleation centres** for condensation by having larger radii with reasonably low vapour pressure.

Solutions to exercises

E17.1(a) A molecule with a centre of symmetry may not be polar but molecules belonging to the groups C_n, C_{nv}, and C_s may be polar (Section 11.3). The C atom of CIF_3 is approximately sp^3 hybridized, which causes the molecule to belong to the C_{3v} point group. The highly electronegative F atoms cause the C–F bonds to be very polar and the average dipole of the three very polar C–F bonds is unbalanced by the less polar C–I bond, therefore $\boxed{CIF_3}$ is polar.

Ozone is a bent molecule that belongs to the C_{2v} point group. Lewis resonance structures for the molecule show a central atom with a double bond to an end oxygen atom and a single bond to oxygen at the other end. The central oxygen has a formal charge of +1 while the double-bonded end oxygen has a formal charge of zero. The single-bonded end oxygen has a −1 formal charge. The average position of the −1 formal charge of the two resonance structure predicts a small negative charge that lies half-way between the extremities and a fraction of a bond length away from the central oxygen, which is expected to have a small positive charge. Consequently, $\boxed{O_3}$ is polar.

Hydrogen peroxide belongs to the C_2 point group, with each carbon atom being approximately sp^3 hybridized. The H–O–O bond angles are 96.87°. The H–O–O plane of one hydrogen atom is at a 93.85° angle with the O–O–H plane of the other hydrogen atom. Consequently, the dipole moments of the two polar O–H bonds do not cancel and the $\boxed{H_2O_2}$ molecule is polar.

E17.2(a) $\mu_{res} = (\mu_1^2 + \mu_2^2 + 2\mu_1\mu_2 \cos\theta)^{1/2}$ [17.2a]

$$= [(1.5)^2 + (0.80)^2 + (2) \times (1.5) \times (0.80) \times (\cos 109.5°)]^{1/2} \text{D} = \boxed{1.4 \text{ D}}$$

E17.3(a) The dipole moment is the vector sum shown in Figure 17.2.

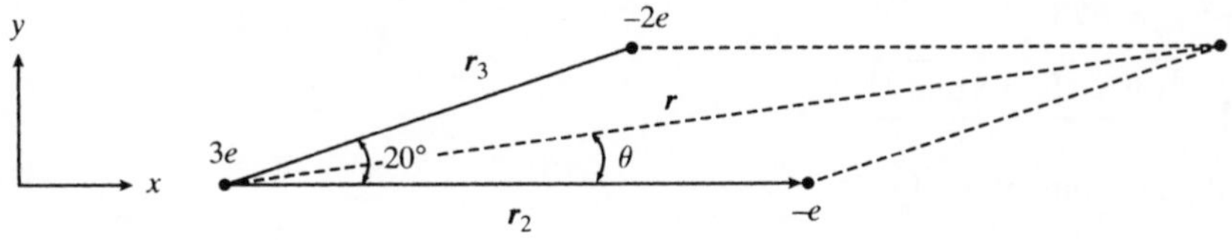

Figure 17.2

$$\mu = \sum_i Q_i \mathbf{r}_i = 3e(0) - e\mathbf{r}_2 - 2e\mathbf{r}_3, \quad \text{where} \quad r_2 = \mathbf{i}x_2 \quad \text{and} \quad r_3 = \mathbf{i}x_3 + \mathbf{j}y_3$$

$$x_2 = +0.32 \text{ nm}$$

$$x_3 = r_3 \cos 20° = (+0.23 \text{ nm}) \times (0.94\bar{0}) = 0.21\bar{6} \text{ nm}$$

$$y_3 = r_3 \sin 20° = (+0.23 \text{ nm}) \times (0.34\bar{2}) = 0.078\bar{7} \text{ nm}$$

The components of the vector sum are the sums of the components.

$$\mu_x = -ex_2 - 2ex_3 = -e \times \{(0.32) + 2 \times (0.21\bar{6})\} \text{ nm} = -e \times (0.75\bar{2} \text{ nm})$$

$$\mu_y = -2ey_3 = -2e \times (0.078\bar{7}) = -e \times (0.15\bar{7} \text{ nm})$$

$$\mu = (\mu_x^2 + \mu_y^2)^{1/2} \text{ [17.3b]}$$
$$= e \times \{(0.75\bar{2} \text{ nm})^2 + (0.15\bar{7} \text{ nm})^2\}^{1/2} = (1.602 \times 10^{-19} \text{ C}) \times (0.76\bar{8} \times 10^{-9} \text{ m})$$
$$= (1.2\bar{3} \times 10^{-28} \text{ C m}) \times \left(\frac{1 \text{ D}}{3.33564 \times 10^{-30} \text{ C m}}\right) = \boxed{37 \text{ D}}$$

The angle that μ makes with x-axis is given by

$$\cos\theta = \frac{|\mu_x|}{\mu} = \frac{0.75\bar{2}}{0.76\bar{8}} \quad \text{so} \quad \theta = \cos^{-1}\left(\frac{0.75\bar{2}}{0.76\bar{8}}\right) = \boxed{11.\bar{7}°}$$

E17.4(a) The O–H bond length of a water molecule is 95.85 pm and the Li^+ cation is 100 pm from the dipole centre. Because these lengths are comparable, a calculation based on the assumption that the water dipole acts like a point dipole with a dipole length much shorter than the dipole–ion distance is unlikely to provide an accurate value of the dipole–ion interaction energy. However, such a calculation does provide an 'order-of-magnitude' estimate. The minimum value of the dipole–ion interaction occurs with the dipole pointing towards the cation.

$$V_{min} \sim -\frac{\mu_{H_2O} Q_{Li^+}}{4\pi\varepsilon_0 r^2} \text{ [17.18]} = -\frac{\mu_{H_2O} e}{4\pi\varepsilon_0 r^2}$$

$$\sim -\frac{(1.85 \text{ D}) \times (3.336 \times 10^{-30} \text{ C m D}^{-1}) \times (1.602 \times 10^{-19} \text{ C})}{(1.113 \times 10^{-10} \text{ J}^{-1} \text{ C}^2 \text{ m}^{-1}) \times (100 \times 10^{-12} \text{ m})^2}$$

$$\sim -8.88 \times 10^{-19} \text{ J}$$

The interaction potential becomes a maximum on flipping the dipole. This effectively changes the sign of the dipole in the previous calculation, giving

$$V_{max} \sim 8.88 \times 10^{-19}\ \mathrm{J}$$

The work w required to flip the dipole is the difference $V_{max} - V_{min}$.

$$w \sim V_{max} - V_{min} = 1.78 \times 10^{-18}\ \mathrm{J}$$

$$w_m = wN_A \sim \boxed{1.07 \times 10^3\ \mathrm{kJ\ mol^{-1}}}$$

E17.5(a) The induced dipole moment of H_2O is

$$\mu^*_{H_2O} = \alpha\mathcal{E}\ [17.4] = 4\pi\varepsilon_0\alpha'_{H_2O}\mathcal{E}\ [17.5]$$
$$= 4\pi \times (8.854 \times 10^{-12}\ \mathrm{J^{-1}\,C^2\,m^{-1}}) \times (1.48 \times 10^{-30}\ \mathrm{m^3}) \times (1.0 \times 10^5\ \mathrm{V\ m^{-1}})\ [1\ \mathrm{J} = 1\ \mathrm{C\ V}]$$
$$= (1.6\bar{5} \times 10^{-35}\ \mathrm{C\ m}) \times \left(\frac{1\ \mathrm{D}}{3.33564 \times 10^{-30}\ \mathrm{C\ m}}\right)$$
$$= \boxed{5.0\ \mu\mathrm{D}}$$

Thus, we see that the induced dipole is much smaller than the permanent dipole moment of H_2O (1.85 D).

E17.6(a) Polarizability α, dipole moment μ, and molar polarization P_m are related by

$$P_m = \left(\frac{N_A}{3\varepsilon_0}\right) \times \left(\alpha + \frac{\mu^2}{3kT}\right) [17.15]$$

In order to solve for α, it is first necessary to obtain μ from the temperature variation of P_m.

$$\alpha + \frac{\mu^2}{3kT} = \frac{3\varepsilon_0 P_m}{N_A}$$

Therefore, $\left(\frac{\mu^2}{3k}\right) \times \left(\frac{1}{T} - \frac{1}{T'}\right) = \left(\frac{3\varepsilon_0}{N_A}\right) \times (P_m - P'_m)\ [P_m \text{ at } T,\ P'_m \text{ at } T']$

and hence,

$$\mu^2 = \frac{9\varepsilon_0 k \times (P_m - P'_m)}{N_A \times \left(\frac{1}{T} - \frac{1}{T'}\right)}$$
$$= \frac{9 \times (8.854 \times 10^{-12}\ \mathrm{J^{-1}\,C^2\,m^{-1}}) \times (1.381 \times 10^{-23}\ \mathrm{J\ K^{-1}}) \times (70.62 - 62.47) \times 10^{-6}\ \mathrm{m^3\ mol^{-1}}}{(6.022 \times 10^{23}\ \mathrm{mol^{-1}}) \times \left(\frac{1}{351.0\ \mathrm{K}} - \frac{1}{423.2\ \mathrm{K}}\right)}$$
$$= 3.06\bar{4} \times 10^{-59}\ \mathrm{C^2\,m^2}$$

$$\mu = 5.54 \times 10^{-30}\ \mathrm{C\ m} \times \left(\frac{1\ \mathrm{D}}{3.33564 \times 10^{-30}\ \mathrm{C\ m}}\right) = \boxed{1.66\ \mathrm{D}}$$

$$\alpha = \frac{3\varepsilon_0 P_m}{N_A} - \frac{\mu^2}{3kT}$$

$$= \frac{3 \times (8.854 \times 10^{-12}\ \mathrm{J^{-1}\,C^2\,m^{-1}}) \times (70.62 \times 10^{-6}\ \mathrm{m^3\,mol^{-1}})}{6.022 \times 10^{23}\ \mathrm{mol^{-1}}}$$

$$- \frac{3.06\overline{4} \times 10^{-59}\ \mathrm{C^2\,m^2}}{3 \times (1.381 \times 10^{-23}\ \mathrm{J\,K^{-1}}) \times (351.0\ \mathrm{K})}$$

$$= \boxed{1.01 \times 10^{-39}\ \mathrm{J^{-1}\,C^2\,m^2}}$$

Corresponding to $\alpha' = \frac{\alpha}{4\pi\varepsilon_0}$ [17.5] $= \boxed{9.06 \times 10^{-30}\ \mathrm{m^3}}$

E17.7(a) $M_{ClF_3} = 92.45\ \mathrm{g\,mol^{-1}}$

$$\varepsilon_r - 1 = \frac{\rho P_m}{M} \times (\varepsilon_r + 2)\ [17.14]$$

$$\left(1 - \frac{\rho P_m}{M}\right)\varepsilon_r = 1 + \frac{2\rho P_m}{M}$$

$$\varepsilon_r = \frac{M + 2\rho P_m}{M - \rho P_m}$$

$$= \frac{92.45\ \mathrm{g\,mol^{-1}} + 2 \times (1.89\ \mathrm{g\,cm^{-3}}) \times (27.18\ \mathrm{cm^3\,mol^{-1}})}{92.45\ \mathrm{g\,mol^{-1}} - (1.89\ \mathrm{g\,cm^{-3}}) \times (27.18\ \mathrm{cm^3\,mol^{-1}})}$$

$$= \boxed{4.75}$$

E17.8(a) $n_r = (\varepsilon_r)^{1/2}$ [17.17] and $\frac{\varepsilon_r - 1}{\varepsilon_r + 2} = \frac{\rho N_A \alpha}{3M\varepsilon_0}$ [17.16]

Therefore,

$$\alpha = \frac{3M\varepsilon_0}{\rho N_A}\left(\frac{n_r^2 - 1}{n_r^2 + 2}\right)$$

$$= \frac{3 \times (267.83\ \mathrm{g\,mol^{-1}}) \times (8.854 \times 10^{-12}\ \mathrm{J^{-1}\,C^2\,m^{-1}})}{(3.32 \times 10^6\ \mathrm{g\,m^{-3}}) \times (6.022 \times 10^{23}\ \mathrm{mol^{-1}})} \times \left(\frac{1.732^2 - 1}{1.732^2 + 2}\right)$$

$$= \boxed{1.42 \times 10^{-39}\ \mathrm{J^{-1}\,C^2\,m^2}}$$

E17.9(a) $\alpha' = \frac{\alpha}{4\pi\varepsilon_0}$ [17.5]

$$\alpha = 4\pi\varepsilon_0\alpha' = (1.11265 \times 10^{-10}\ \mathrm{J^{-1}\,C^2\,m^{-1}}) \times (1.5 \times 10^{-30}\ \mathrm{m^3}) = 1.6\overline{7} \times 10^{-40}\ \mathrm{J^{-1}\,C^2\,m^2}$$

Let

$$C = \frac{\rho N_A \alpha}{3M\varepsilon_0} = \frac{(0.99707 \times 10^6\ \mathrm{g\,m^{-3}}) \times (6.022 \times 10^{23}\ \mathrm{mol^{-1}}) \times (1.6\overline{7} \times 10^{-40}\ \mathrm{J^{-1}\,C^2\,m^2})}{3 \times (18.02\ \mathrm{g\,mol^{-1}}) \times (8.85419 \times 10^{-12}\ \mathrm{J^{-1}\,C^2\,m^{-1}})} = 0.20\overline{9}$$

and solve the Clausius–Mossotti equation [17.16] for ε_r with which we calculate the refractive index.

$$\varepsilon_r = \frac{1+2C}{1-C}$$

$$= \frac{1+2\times(0.20\overline{9})}{1-0.20\overline{9}}$$

$$= 1.7\overline{9}$$

$$n_r = \varepsilon_r^{1/2}\,[17.17] = (1.7\overline{9})^{1/2} = \boxed{1.34}$$

This result agrees with the experimental value of 1.33 because it lies within the uncertainty of the refractive index measurement.

E17.10(a) $\mu = 1.57\text{ D} = 5.23\overline{7}\times10^{-30}\text{ C m}$ for chlorobenzene (112.55 g mol^{-1})

$$\alpha = 4\pi\varepsilon_0\alpha'\,[17.5] = (1.11265\times10^{-10}\text{ J}^{-1}\text{ C}^2\text{ m}^{-1})\times(1.23\times10^{-29}\text{ m}^3) = 1.36\overline{9}\times10^{-39}\text{ J}^{-1}\text{ C}^2\text{ m}^2$$

$$P_m = \frac{N_A}{3\varepsilon_0}\left(\alpha + \frac{\mu^2}{3kT}\right)[17.15]$$

$$= \frac{6.022\times10^{23}\text{ mol}^{-1}}{3\times(8.85419\times10^{-12}\text{ J}^{-1}\text{ C}^2\text{ m}^{-1})}\left((1.36\overline{9}\times10^{-39}\text{ J}^{-1}\text{ C}^2\text{ m}^2) + \frac{(5.23\overline{7}\times10^{-30}\text{ C m})^2}{3\times(1.3807\times10^{-23}\text{ J K}^{-1})\times(298.15\text{ K})}\right)$$

$$= 8.13\overline{8}\times10^{-5}\text{ m}^3\text{ mol}^{-1}$$

Let

$$C = \frac{\rho P_m}{M} = \frac{(1.173\times10^6\text{ g m}^{-3})\times(8.13\overline{8}\times10^{-5}\text{ m}^3\text{ mol}^{-1})}{112.55\text{ g mol}^{-1}}$$

$$= 0.848\overline{1}$$

and solve the Debye equation [17.14] for ε_r.

$$\frac{\varepsilon_r - 1}{\varepsilon_r + 2} = C\ [17.14\text{, the Debye equation}]$$

$$\varepsilon_r = \frac{1+2C}{1-C}$$

$$= \frac{1+2\times(0.848\overline{1})}{1-0.848\overline{1}}$$

$$= \boxed{17.7}$$

E17.11(a) $$V_{\text{London}} = -\frac{3(\alpha'_{\text{He}} I_{\text{He}})^2}{2(I_{\text{He}} + I_{\text{He}})r^6}\,[17.25] = -\frac{3(\alpha'_{\text{He}})^2 I_{\text{He}}}{4r^6}$$

$$= -\frac{3 \times (0.20 \times 10^{-30}\ \text{m}^3)^2 \times (2372.3\ \text{kJ mol}^{-1})}{4 \times (1.0 \times 10^{-9}\ \text{m})^6}$$

$$= \boxed{0.071\ \text{J mol}^{-1}}$$

E17.12(a) Using the partial charge presented in Table 17.2, we estimate the partial charge on each hydrogen atom of a water molecule to be $Q_H = \delta e$, where $\delta = 0.42$. The electroneutrality of an H_2O molecule implies that the estimated partial charge on the oxygen atom is $Q_O = -2\delta e$. With a hydrogen bond length of 170 pm, the point-charge model of the hydrogen bond in a vacuum estimates the potential of interaction to be

$$V = \frac{Q_H Q_O}{4\pi\varepsilon_0 r} = -\frac{2(\delta e)^2}{4\pi\varepsilon_0 r}\,[17.12a]$$

$$= -\frac{2(0.42 \times 1.60 \times 10^{-19}\ \text{C})^2}{4\pi(8.85 \times 10^{-12}\ \text{J}^{-1}\,\text{C}^2\,\text{m}^{-1}) \times (170 \times 10^{-12}\ \text{m})} = -4.8 \times 10^{-19}\ \text{J}$$

The molar energy required to break these bonds is

$$E_m = -N_A V = -(6.022 \times 10^{23}\ \text{mol}^{-1}) \times (-4.8 \times 10^{-19}\ \text{J}) = \boxed{28\overline{9}\ \text{kJ mol}^{-1}}$$

E17.13(a) $$V_m = \frac{M}{\rho} = \frac{18.02\ \text{g mol}^{-1}}{0.9982\ \text{g cm}^{-3}} = 18.05\ \text{cm}^3$$

$$p = p^* e^{2\gamma V_m(\text{l})/rRT}\ [17.51,\ \text{the Kelvin equation}]$$

$$= (2.3\ \text{kPa}) \times \exp\left\{\frac{2 \times \left(72.75 \times 10^{-3}\ \text{N m}^{-1}\right) \times (18.05 \times 10^{-6}\ \text{m}^3\,\text{mol}^{-1})}{(10 \times 10^{-9}\ \text{m}) \times (8.3145\ \text{J K}^{-1}\,\text{mol}^{-1}) \times (293\ \text{K})}\right\}$$

$$= \boxed{2.6\ \text{kPa}}$$

E17.14(a) $$\gamma = \tfrac{1}{2}\rho g r h\,[17.40]$$
$$= \tfrac{1}{2} \times (998.2\ \text{kg m}^{-3}) \times (9.80665\ \text{m s}^{-2}) \times (0.300 \times 10^{-3}\ \text{m}) \times (4.96 \times 10^{-2}\ \text{m})$$
$$= 0.0728\overline{3}\ \text{kg s}^{-2} = \boxed{72.8\ \text{mN m}^{-1}}$$

This value is in good agreement with Table 17.5.

E17.15(a) $$p_{\text{in}} - p_{\text{out}} = \frac{2\gamma}{r}\,[17.38,\ \text{the Laplace equation}]$$

$$= \frac{2 \times (72.75 \times 10^{-3}\ \text{N m}^{-1})}{200 \times 10^{-9}\ \text{m}}\,[\text{Table 17.5}]$$

$$= 7.28 \times 10^5\ \text{N m}^{-2} = \boxed{728\ \text{kPa}}$$

Pressure differentials for small droplets are quite large.

Solutions to problems

Solutions to numerical problems

P17.1 The positive end of the dipole, which bisects the hydrogen atom positions, will lie closer to the (negative) anion. The electric field generated by a dipole is

$$\mathcal{E} = \frac{\mu}{2\pi\varepsilon_0 r^3}\,[17.21]$$

$$= \frac{(1.85\ \text{D}) \times (3.34 \times 10^{-30}\ \text{C m D}^{-1})}{2\pi \times (8.854 \times 10^{-12}\ \text{J}^{-1}\,\text{C}^2\,\text{m}^{-1}) \times r^3} = \frac{1.11 \times 10^{-19}\ \text{V m}^{-1}}{(r/\text{m})^3} = \frac{1.11 \times 10^{8}\ \text{V m}^{-1}}{(r/\text{nm})^3}$$

(a) $\mathcal{E} = \boxed{1.1 \times 10^{8}\ \text{V m}^{-1}}$ when $r = 1.0$ nm

(b) $\mathcal{E} = \dfrac{1.11 \times 10^{8}\ \text{V m}^{-1}}{0.3^3} = \boxed{4 \times 10^{9}\ \text{V m}^{-1}}$ for $r = 0.3$ nm

(c) $\mathcal{E} = \dfrac{1.11 \times 10^{8}\ \text{V m}^{-1}}{30^3} = \boxed{4\ \text{kV m}^{-1}}$ for $r = 30$ nm.

P17.3 The point charge model can be used to estimate the magnitude of the electric dipole moment of hydrogen peroxide as a function of ϕ (defined in Figure 17.3b as a view down the z-axis of the O–O bond). Each hydrogen atom has a partial charge of δ; each oxygen atom has a partial charge of $-\delta$. The dipole moment magnitude is

$$\mu = (\mu \cdot \mu)^{1/2} = (\mu_x^2 + \mu_y^2 + \mu_z^2)^{1/2}$$

where

$$\mu_x = \sum_{\text{J}} Q_{\text{J}} x_{\text{J}} = \delta \times \{x_{\text{H}_1} - x_{\text{O}_1} - x_{\text{O}_2} + x_{\text{H}_2}\}\,[17.3\text{a,b}],\ \text{etc.}$$

We will use the Cartesian coordinate system defined in Figure 17.3a. The bond lengths are $l_{\text{OH}} = 97$ pm and $l_{\text{OO}} = 149$ pm. We also use the ratio $l_{\text{ratio}} = l_{\text{OO}}/l_{\text{OH}} = 1.54$ and calculate μ in units of δl_{OH} so that it is unnecessary to estimate the magnitude of δ (although an estimate is found in Table 17.2). The O–O–H bond angle, θ, may be estimated as 90° but we will use the experimental value of 100°. The computations of μ_x, μ_y, and μ_z require the coordinates of each atom; those of H_1 and the oxygen atoms are shown in Figure 17.3a.

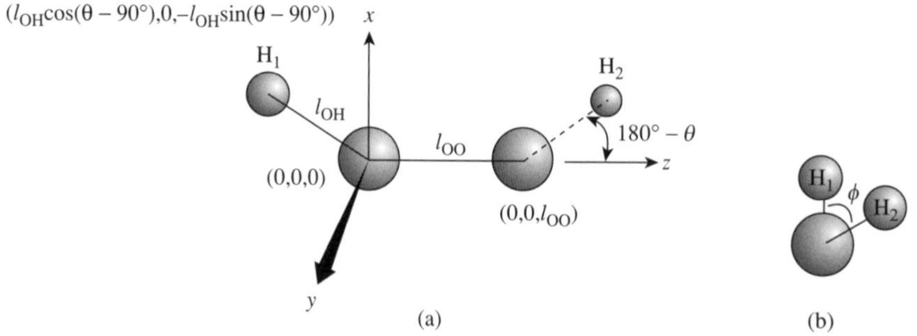

Figure 17.3

The coordinates of H_2 can be determined by analogy to the relationships between Cartesian coordinates and spherical polar coordinates. They are:

$$x = l_{OH}\sin(180° - \theta)\cos\phi$$
$$y = l_{OH}\sin(180° - \theta)\sin\phi$$
$$z = l_{OO} + l_{OH}\cos(180° - \theta)$$

Substitution of variables into eqn. 17.3b yields

$$\begin{aligned}(\mu/\delta l_{OH})^2 &= (\mu_x/\delta l_{OH})^2 + (\mu_y/\delta l_{OH})^2 + (\mu_z/\delta l_{OH})^2 \\ &= \{\cos(10°) + \sin(80°)\cos\phi\}^2 + \{\sin(80°)\sin\phi\}^2 + \{-\sin(10°) - l_{ratio} + l_{ratio} + \cos(80°)\}^2 \\ &= \{\cos(10°) + \sin(80°)\cos\phi\}^2 + \{\sin(80°)\sin\phi\}^2 + \{-\sin(10°) + \cos(80°)\}^2\end{aligned}$$

We now draw a table to calculate $(\mu/\delta l_{OH})^2$ in ϕ increments of 15° and, subsequently, calculate $\mu/\delta l_{OH}$ values at each ϕ. Figure 17.4 is a plot of the variation. As expected, there the dipole is a maximum of almost twice the single O–H bond dipole when the hydrogen atoms are eclipsed and it is zero when they have a gauche conformation.

ϕ/deg	ϕ/radians	sq($\mu/\delta l$)	$\mu/\delta l$
0	0	3.879385	1.969616
15	0.261799	3.813292	1.952765
30	0.523599	3.619516	1.902502
45	0.785398	3.311262	1.819687
60	1.047198	2.909539	1.705737

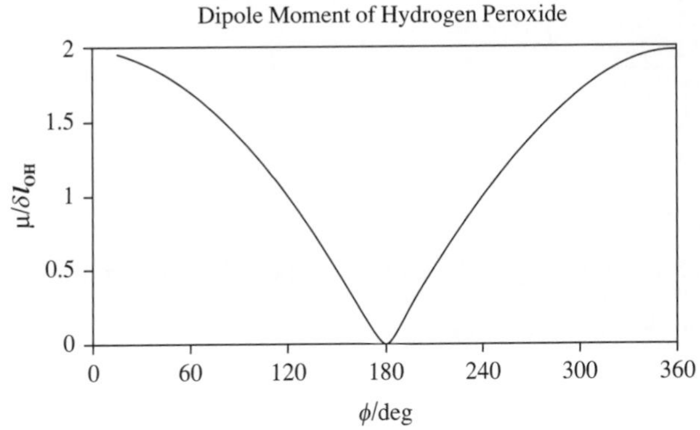

Figure 17.4

P17.5 The equations relating dipole moment and polarizability volume to the experimental quantities ε_r and ρ are

$$P_m = \left(\frac{M}{\rho}\right) \times \left(\frac{\varepsilon_r - 1}{\varepsilon_r + 2}\right) [17.14] \quad \text{and} \quad P_m = \frac{4\pi}{3}N_A\alpha' + \frac{N_A\mu^2}{9\varepsilon_0 kT} [17.15, \text{ with } \alpha = 4\pi\varepsilon_0\alpha']$$

Therefore, we draw up the following table (with $M = 119.4\ \text{g mol}^{-1}$):

$\theta/°\text{C}$	–80	–70	–60	–40	–20	0	20
T/K	193	203	213	233	253	273	293
$\dfrac{1000}{T/\text{K}}$	5.18	4.93	4.69	4.29	3.95	3.66	3.41
ε_r	3.1	3.1	7.0	6.5	6.0	5.5	5.0
$\dfrac{\varepsilon_r - 1}{\varepsilon_r + 2}$	0.41	0.41	0.67	0.65	0.63	0.60	0.57
$\rho/\text{g cm}^{-3}$	1.65	1.64	1.64	1.61	1.57	1.53	1.50
$P_m/(\text{cm}^3\ \text{mol}^{-1})$	29.8	29.9	48.5	48.0	47.5	56.8	45.4

P_m is plotted against $\dfrac{1}{T}$ in Figure 17.5.

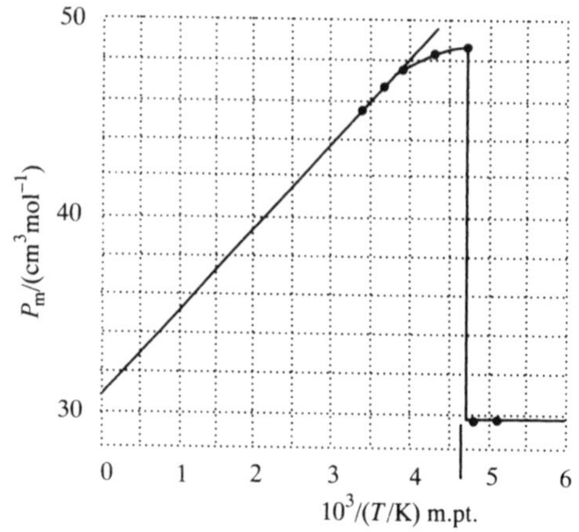

Figure 17.5

The (dangerously unreliable) intercept is $\approx 30\ \text{cm}^3\ \text{mol}^{-1}$ and the slope below the melting point is such that $\dfrac{dP_m}{d(1/T)} \approx 4.5 \times 10^3\ \text{cm}^3\ \text{mol}^{-1}\ \text{K}$. It follows that

$$\alpha' = \frac{3 \times (30\ \text{cm}^3\ \text{mol}^{-1})}{4\pi \times (6.022 \times 10^{23}\ \text{mol}^{-1})} = \boxed{1.2 \times 10^{-23}\ \text{cm}^3}$$

$$\mu^2 = \frac{9\varepsilon_0 k}{N_A}\frac{dP_m}{d(1/T)}$$

$$= \left\{\frac{9 \times (8.85419 \times 10^{-12}\ \text{J}^{-1}\ \text{C}^2\ \text{m}^{-1}) \times (1.3807 \times 10^{-23}\ \text{J K}^{-1})}{6.022 \times 10^{23}\ \text{mol}^{-1}}\right\} \times (4.5 \times 10^{-3}\ \text{m}^3\ \text{mol}^{-1}\ \text{K})$$

$$= 8.2 \times 10^{-30}\ \text{C}^2\ \text{m}^2$$

$$\mu = (8.2 \times 10^{-60}\ \text{C}^2\ \text{m}^2)^{1/2} \times \left(\frac{1\ \text{D}}{3.33564 \times 10^{-30}\ \text{C m}}\right) = \boxed{0.86\ \text{D}}$$

A sharp decrease in P_m occurs at the freezing point of chloroform (−64°C), indicating that the dipole reorientation term no longer contributes. Note that P_m for the solid corresponds to the extrapolated, dipole-free, value of P_m, so the extrapolation is less hazardous than it looks.

P17.7 $$P_m = \frac{4\pi}{3}N_A\alpha' + \frac{N_A\mu^2}{9\varepsilon_0 kT}\,[17.15, \text{ with } \alpha = 4\pi\varepsilon_0\alpha']$$

Eqn 17.15 indicates that a plot of P_m against $1/T$ should be linear with a slope, $\frac{dP_m}{d(1/T)}$, equal to $\frac{N_A\mu^2}{9\varepsilon_0 k}$ and a $1/T = 0$ intercept that equals $\frac{4\pi}{3}N_A\alpha'$. We therefore draw up the following table and prepare a plot of P_m against $1/T$. If it is linear, we perform a linear least squares regression fit of the plot so as to acquire the slope and intercept from which we calculate α' and μ. A suitable plot is shown in Figure 17.6.

T/K	292.2	309.0	333.0	387.0	413.0	446.0
$\frac{1000}{T/\text{K}}$	3.42	3.24	3.00	2.58	2.42	2.24
$P_m/(\text{cm}^3\,\text{mol}^{-1})$	57.57	55.01	51.22	44.99	42.51	39.59

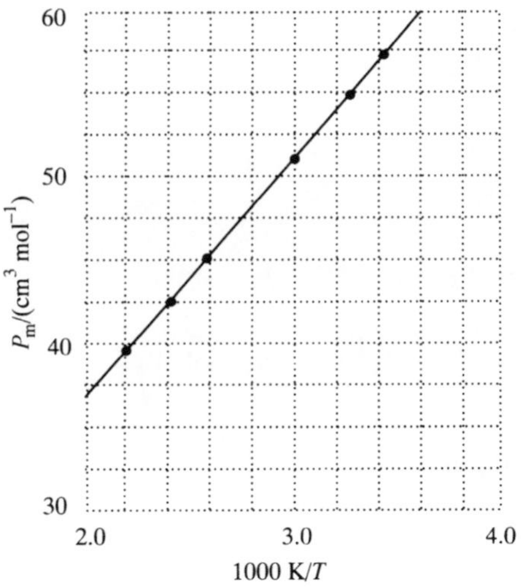

Figure 17.6

The plot of P_m against $1/T$ is linear with a regression fit that gives an intercept of 5.65 cm³ mol⁻¹ (not shown in the figure), and the slope is such that $dP_m/d(1/T) = 1.52 \times 10^4\ \text{cm}^3\ \text{mol}^{-1}\ \text{K}$. It follows that

$$\alpha' = \frac{3P_m(\text{at intercept})}{4\pi N_A} = \frac{3 \times (5.65\ \text{cm}^3\ \text{mol}^{-1})}{4\pi \times (6.022 \times 10^{23}\ \text{mol}^{-1})} = \boxed{2.24 \times 10^{-24}\ \text{cm}^3}$$

$$\mu^2 = \frac{9\varepsilon_0 k}{N_A}\frac{dP_m}{d(1/T)}$$

$$= \left\{\frac{9 \times (8.85419 \times 10^{-12}\ \text{J}^{-1}\,\text{C}^2\,\text{m}^{-1}) \times (1.3807 \times 10^{-23}\ \text{J K}^{-1})}{6.022 \times 10^{23}\ \text{mol}^{-1}}\right\} \times (1.52 \times 10^{-2}\ \text{m}^3\ \text{mol}^{-1}\ \text{K})$$

$$= 2.78 \times 10^{-59}\ \text{C}^2\,\text{m}^2$$

$$\mu = (2.78 \times 10^{-59}\,\mathrm{C^2\,m^2})^{1/2} \times \left(\frac{1\,\mathrm{D}}{3.33564 \times 10^{-30}\,\mathrm{C\,m}}\right) = \boxed{1.58\,\mathrm{D}}$$

The high-frequency contribution to the molar polarization, P'_m, at 273 K may be calculated from the refractive index

$$P'_\mathrm{m} = \left(\frac{M}{\rho}\right) \times \left(\frac{\varepsilon_\mathrm{r} - 1}{\varepsilon_\mathrm{r} + 2}\right) [17.14] = \left(\frac{M}{\rho}\right) \times \left(\frac{n_\mathrm{r}^2 - 1}{n_\mathrm{r}^2 + 2}\right)$$

Assuming that ammonia under these conditions (1.00 atm pressure assumed) can be considered a perfect gas, we have

$$\rho = \frac{PM}{RT} \text{ and } \frac{M}{\rho} = \frac{RT}{p} = \frac{(82.06\,\mathrm{cm^3\,atm\,K^{-1}\,mol^{-1}}) \times (273\,\mathrm{K})}{1.00\,\mathrm{atm}}$$
$$= 2.24 \times 10^4\,\mathrm{cm^3\,mol^{-1}}$$

Then, $P'_\mathrm{m} = 2.24 \times 10^4\,\mathrm{cm^3\,mol^{-1}} \times \left\{\frac{(1.000379)^2 - 1}{(1.000379)^2 + 2}\right\} = \boxed{5.66\,\mathrm{cm^3\,mol^{-1}}}$.

If we assume that the high-frequency contribution to P_m remains the same at 292.2 K then we have

$$\frac{N_\mathrm{A}\mu^2}{9\varepsilon_0 kT} = P_\mathrm{m} - P'_\mathrm{m} = (57.57 - 5.66)\,\mathrm{cm^3\,mol^{-1}} = 51.91\,\mathrm{cm^3\,mol^{-1}}$$
$$= 5.191 \times 10^{-5}\,\mathrm{m^3\,mol^{-1}}$$

Solving for μ we have

$$\mu = \left(\frac{9\varepsilon_0 k}{N_\mathrm{A}}\right)^{1/2} T^{1/2}(P_\mathrm{m} - P'_\mathrm{m})^{1/2} = 4.275 \times 10^{-29}(\mathrm{C^2\,mol\,K^{-1}\,m^{-1}})^{1/2} T^{1/2}(P_\mathrm{m} - P'_\mathrm{m})^{1/2}$$
$$= \{4.275 \times 10^{-29}\,(\mathrm{C^2\,mol\,K^{-1}\,m^{-1}})^{1/2}\} \times (292.2\,\mathrm{K})^{1/2} \times (5.191 \times 10^{-5}\,\mathrm{m^3\,mol^{-1}})^{1/2}$$
$$= 5.26 \times 10^{-30}\,\mathrm{C\,m} = 1.58\,\mathrm{D}$$

The agreement between the two computational methods is exact.

P17.9 (a) $V = 4\varepsilon\left\{\left(\frac{r_0}{r}\right)^{12} - \left(\frac{r_0}{r}\right)^{6}\right\}$ [17.33, Lennard-Jones potential]

The depth of the well in Joules is $\varepsilon = hc\tilde{D}_\mathrm{e} = \boxed{1.51 \times 10^{-23}\,\mathrm{J}}$.

The distance at which the potential is zero is given by

$$r_\mathrm{e} = 2^{1/6} r_0, \quad \text{so} \quad r_0 = r_\mathrm{e} 2^{-1/6} = 2^{-1/6} \times (297\,\mathrm{pm}) = \boxed{265\,\mathrm{pm}}.$$

A plot of the Lennard-Jones potential is shown in Figure 17.7.

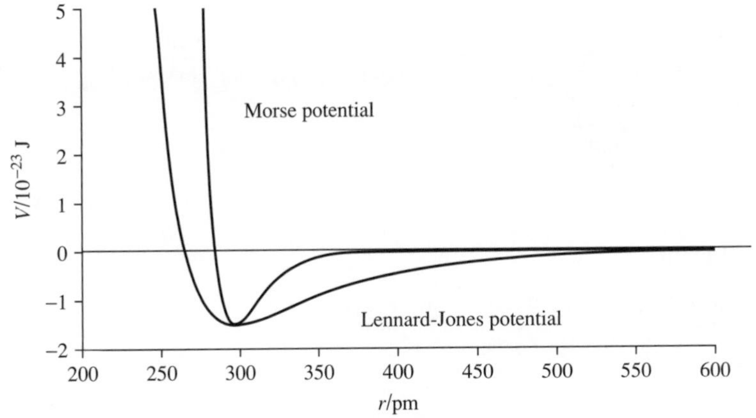

Figure 17.7

(b) $V = hc\tilde{D}_e\{1 - e^{-a\times(r-r_e)}\}^2 - hc\tilde{D}_e$ [12.37, Morse potential energy]

The constant $-hc\tilde{D}_e$ has been added to the above expression for the Morse potential so that it has a minimum of $-hc\tilde{D}_e$ at $r = r_e$. This makes it easy to compare the Morse potential with the Lennard-Jones potential of part (a), which also has a minimum of $-hc\tilde{D}_e$ at $r = r_e$. Plots of both potentials are shown in Figure 17.7, with $a = 0.0579\ \text{pm}^{-1}$ for the Morse potential. The Morse potential has a much steeper repulsive component at separations lower than r_e, while the Lennard-Jones potential shows a longer-range attractive component.

P17.11 $$P_m = \frac{M}{\rho} \times \left(\frac{\varepsilon_r - 1}{\varepsilon_r + 2}\right) [17.14] \quad \text{and} \quad P_m = \frac{N_A}{3\varepsilon_0}\left(\alpha + \frac{\mu^2}{3kT}\right) [17.15]$$

The dipole moment μ contributes to the molar polarization of eqn 17.15 in the low-frequency limit of the applied electric field where molecular rotation can orientate quickly enough to follow the change in direction of the field. This corresponds to field oscillations of less than about 10^{12} Hz, a typical rotational frequency, which is the range of microwaves and radio waves. In the high-frequency limit of 10^{14} Hz or higher (infrared, visible, and ultraviolet) molecules cannot orientate quickly enough to follow the change in direction of the field and the dipole moment does not contribute to eqn 17.15. The limiting values of the molar polarization, calculated below, should be compared to the values reported and discussed in P17.6.

Low-frequency limit of methanol:

$$P_m = \frac{N_A}{3\varepsilon_0}\left(\alpha + \frac{\mu^2}{3kT}\right) [17.15]$$

$$= \frac{6.022 \times 10^{23}\ \text{mol}^{-1}}{3 \times (8.85419 \times 10^{-12}\ \text{J}^{-1}\,\text{C}^2\,\text{m}^{-1})}$$

$$\times \left(3.59 \times 10^{-40}\ \text{J}^{-1}\,\text{C}^2\,\text{m}^2 + \frac{(5.70 \times 10^{-30}\ \text{C m})^2}{3 \times (1.3807 \times 10^{-23}\ \text{J K}^{-1}) \times (293.15\ \text{K})}\right)$$

$$= \boxed{68.8\ \text{cm}^3\,\text{mol}^{-1}}$$

Solving eqn 17.14 for ε_r gives

$$\varepsilon_r = \frac{1+\dfrac{2\rho P_m}{M}}{1-\dfrac{\rho P_m}{M}} = \frac{M+2\rho P_m}{M-\rho P_m} = \frac{32.0\ \text{g mol}^{-1} + 2\times(0.7914\ \text{g cm}^{-3})\times(68.8\ \text{cm}^3\ \text{mol}^{-1})}{32.0\ \text{g mol}^{-1} - (0.7914\ \text{g cm}^{-3})\times(68.8\ \text{cm}^3\ \text{mol}^{-1})}$$

$$= \boxed{4.40}$$

$$n_r = \varepsilon_r^{1/2}\ [17.17] = (4.40)^{1/2} = \boxed{2.10}$$

High-frequency limit of methanol:

$$P_m = \frac{N_A\alpha}{3\varepsilon_0} = \frac{(6.022\times10^{23}\ \text{mol}^{-1})\times(3.59\times10^{-40}\ \text{J}^{-1}\ \text{C}^2\ \text{m})}{3\times(8.85419\times10^{-12}\ \text{J}^{-1}\ \text{C}^2\ \text{m}^{-1})}$$

$$= \boxed{8.14\ \text{cm}^3\ \text{mol}^{-1}}$$

Solving eqn 17.14 for ε_r gives

$$\varepsilon_r = \frac{M+2\rho P_m}{M-\rho P_m} = \frac{32.0\ \text{g mol}^{-1} + 2\times(0.7914\ \text{g cm}^{-3})\times(8.14\ \text{cm}^3\ \text{mol}^{-1})}{32.0\ \text{g mol}^{-1} - (0.7914\ \text{g cm}^{-3})\times(8.14\ \text{cm}^3\ \text{mol}^{-1})}$$

$$= \boxed{1.76}$$

$$n_r = \varepsilon_r^{1/2}\ [17.17] = (1.76)^{1/2} = \boxed{1.33}$$

Solutions to theoretical problems

P17.13 (a) The two linear quadrupoles are shown in Figure 17.8 with a collinear configuration.

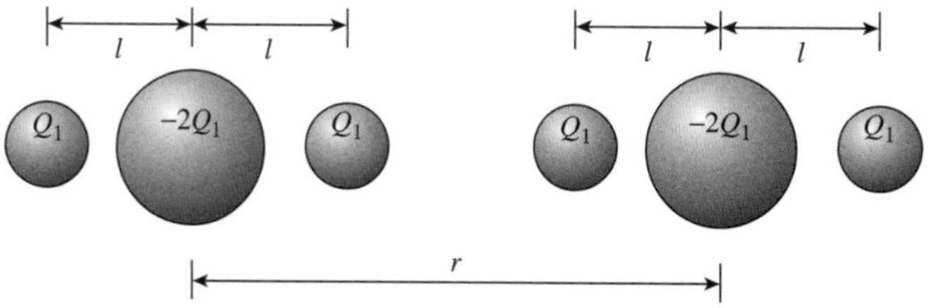

Figure 17.8

The total potential energy of the interaction between the quadrupoles is:

$$4\pi\varepsilon_0 V = \frac{Q_1^2}{r} - \frac{2Q_1^2}{r+l} + \frac{Q_1^2}{r+2l} - \frac{2Q_1^2}{r-l} + \frac{4Q_1^2}{r} - \frac{2Q_1^2}{r+l} + \frac{Q_1^2}{r-2l} - \frac{2Q_1^2}{r-l} + \frac{Q_1^2}{r}$$

$$\frac{4\pi\varepsilon_0 rV}{Q_1^2} = 6 - \frac{4}{1+x} - \frac{4}{1-x} + \frac{1}{1+2x} + \frac{1}{1-2x}, \quad \text{where} \quad x = \frac{l}{r}$$

With the point quadrupole condition that $x \ll 1$ the last four terms in the above expression can be expanded with the Taylor series:

$$(1+z)^{-1} = 1 - z + z^2 - z^3 + z^4 - \cdots \quad \text{and} \quad (1-z)^{-1} = 1 + z + z^2 + z^3 + z^4 + \cdots$$

where z is either $2x$ or x.

$$\begin{aligned}\frac{4\pi\varepsilon_0 rV}{Q_1^2} &= 6 - 4\{1 - x + x^2 - x^3 + x^4 - \cdots + 1 + x + x^2 + x^3 + x^4 + \cdots\} \\ &\quad + \{1 - (2x) + (2x)^2 - (2x)^3 + (2x)^4 - \cdots\} + \{1 + (2x) + (2x)^2 + (2x)^3 + (2x)^4 + \cdots\} \\ &= -8x^2 - 8x^4 + 8x^2 + 32x^4 + \cdots \\ &= 24x^4 + \text{higher-order terms}\end{aligned}$$

In the limit of small x values the higher-order terms are negligibly small, leaving

$$V = \frac{6x^4Q_1^2}{\pi\varepsilon_0 r} = \boxed{\frac{6l^4Q_1^2}{\pi\varepsilon_0 r^5}}$$

Thus, $V \propto \frac{1}{r^5}$ for the quadrupole–quadrupole interaction, in agreement with eqn 17.19.

(b) The two linear quadrupoles are shown in Figure 17.9 with a parallel configuration.

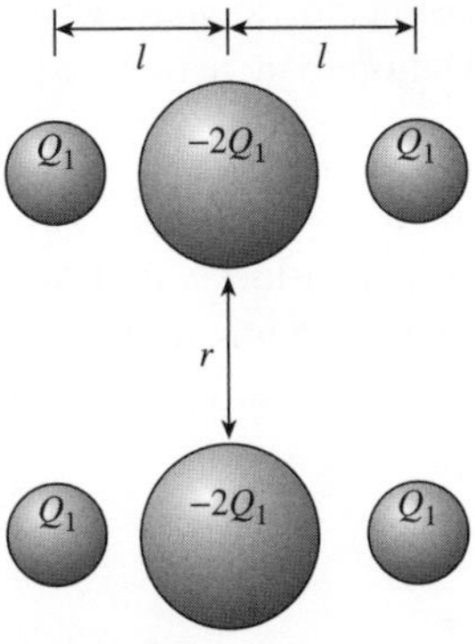

Figure 17.9

In addition to the distance r between some of the point charges in one quadrupole and point charges in the other, the Pythagorean theorem provides the distances $(r^2 + l^2)^{1/2}$ and $(r^2 + 4l^2)^{1/2}$. The total potential energy of the interaction between the quadrupoles is:

$$\begin{aligned}4\pi\varepsilon_0 V &= \frac{Q_1^2}{r} - \frac{2Q_1^2}{(r^2+l^2)^{1/2}} + \frac{Q_1^2}{(r^2+4l^2)^{1/2}} - \frac{2Q_1^2}{(r^2+l^2)^{1/2}} + \frac{4Q_1^2}{r} \\ &\quad - \frac{2Q_1^2}{(r^2+l^2)^{1/2}} + \frac{Q_1^2}{(r^2+4l^2)^{1/2}} - \frac{2Q_1^2}{(r^2+l^2)^{1/2}} + \frac{Q_1^2}{r}\end{aligned}$$

$$\frac{2\pi\varepsilon_0 rV}{Q_1^2} = 3 - \frac{4}{(1+x^2)^{1/2}} + \frac{1}{(1+4x^2)^{1/2}}, \quad \text{where} \quad x = \frac{l}{r}$$

With the point quadrupole condition that $x \ll 1$, the last two terms in the above expression can be expanded with the Taylor series:

$$\begin{aligned}(1+z)^{-1/2} &= 1 - \tfrac{1}{2}z + \tfrac{1}{2}\tfrac{3}{4}z^2 - \tfrac{1}{2}\tfrac{3}{4}\tfrac{5}{6}z^3 + \tfrac{1}{2}\tfrac{3}{4}\tfrac{5}{6}\tfrac{7}{8}z^4 - \cdots \\ &= 1 - \tfrac{1}{2}z + \tfrac{3}{8}z^2 - \tfrac{15}{48}z^3 + \tfrac{105}{384}z^4 - \cdots\end{aligned}$$

where z is either x^2 or $4x^2$.

$$\frac{2\pi\varepsilon_0 rV}{Q_1^2} = 3 - 4\{1 - \tfrac{1}{2}x^2 + \tfrac{3}{8}x^4 - \tfrac{15}{48}x^6 + \tfrac{105}{384}x^8 - \cdots\} + \{1 - 2x^2 + 6x^4 - 20x^6 + 70x^8 - \cdots\}$$
$$= -\tfrac{9}{2}x^4 + \text{higher-order terms}$$

In the limit of small x values the higher-order terms are negligibly small, leaving

$$V = -\frac{9x^4q_1^2}{4\pi\varepsilon_0 r} = \boxed{-\frac{9l^4q_1^2}{4\pi\varepsilon_0 r^5}}$$

Thus, $V \propto \dfrac{1}{r^5}$ for the quadrupole–quadrupole interaction, in agreement with eqn 17.9.

P17.15 Individual acetic acid molecules have a non-zero dipole moment but the two dipoles of the dimer are exactly opposed and cancel. At low temperature a significant fraction of molecules are a part of a dimer, which means that their individual dipole moments will not be observed. However, as temperature is increased, hydrogen bonds of the dimer are broken, releasing individual molecules and the apparent dipole moment increases. This causes the molar polarization to increase by eqn 17.15 and the relative permittivity to increase by eqn 17.14.

Isothermal dilution in benzene solution (a decrease in the preparation concentration c_{A_2}, where A_2 is acetic acid dimer) has an effect that is analogous to heating. To see this, consider equilibrium constant K for dimer dissociation and solve for the dissociation constant α. The equilibrium concentrations of A_2 and A are $(1-\alpha)c_{A_2}$ and $2\alpha c_{A_2}$, respectively. Thus, the equilibrium constant expression for

$$A_2(\text{benzene}) \rightarrow 2\,A(\text{benzene})$$

is

$$K = [A]^2/[A_2] = (2\alpha c_{A_2})^2/\{(1-\alpha)c_{A_2}\}$$
$$4c_{A_2}\alpha^2 + K\alpha - K = 0$$
$$\alpha = \tfrac{1}{8}\{-x + \sqrt{x^2 + 16x}\} \quad \text{with} \quad x = K/c_{A_2} \quad \text{[quadratic equation solution]}$$

A plot of α against x shows that α increases as x increases. Thus, a decrease in c_{A_2} by isothermal dilution, which is equivalent to an increase in x, causes an increase in [A], an increase in the apparent dipole moment, an increase in the molar polarization, and an **increase in the relative permittivity**.

P17.17 We want to construct the exponential-6 potential so that $V(r_0) = 0$ and the depth of the potential well is $-|\varepsilon|$. Consequently, we write

$$V(r) = A\varepsilon\left[Be^{-r/r_0} - \left(\frac{r_0}{r}\right)^6\right]$$

where the constants A and B are chosen to satisfy the conditions. For example, examine $V(r_0)$ to find an expression for B:

$$V(r_0) = A\varepsilon\left[Be^{-r_0/r_0} - \left(\frac{r_0}{r_0}\right)^6\right] = A\varepsilon[Be^{-1} - 1] = 0$$

So, $B = \mathrm{e}$ and the potential becomes $V(r) = A\varepsilon\left[\mathrm{e}^{1-r/r_0} - \left(\frac{r_0}{r}\right)^6\right]$.

The repulsive exponential term, the attractive term, and the potential (with $A\varepsilon = -1$) are sketched in Figure 17.10. The point at which the potential is a minimum is labelled as $x_e = r_e/r_0$.

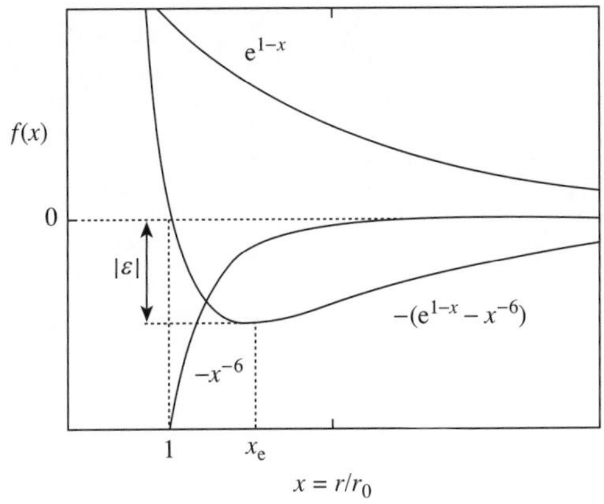

Figure 17.10

Let $x = r/r_0$, then $V(x) = A\varepsilon\left[\mathrm{e}^{1-x} - \frac{1}{x^6}\right]$ and the potential minimum occurs when $\left.\frac{\mathrm{d}V(x)}{\mathrm{d}x}\right|_{x=x_e} = 0$.

$$\frac{\mathrm{d}V(x)}{\mathrm{d}x} = A\varepsilon\left[-\mathrm{e}^{1-x} + \frac{6}{x^7}\right]$$

$$\left.\frac{\mathrm{d}V(x)}{\mathrm{d}x}\right|_{x=x_e} = A\varepsilon\left[-\mathrm{e}^{1-x_e} + \frac{6}{x_e^7}\right] = 0$$

Thus, the solution of the transcendental equation $\mathrm{e}^{1-x_e} = \frac{6}{x_e^7}$ gives the value x_e at which V is a minimum.

x_e may be found as the intersection of the curves e^{1-x} and $6/x^7$ or it may be found using the numeric solver of a scientific calculator. Here is a short Mathcad worksheet solution for x_e.

$x_e := 1$	Estimate of x_e for Given/Find solve block. ($x_e = r_e/r_0$)	
Given	$\mathrm{e}^{1-x_e} = 6x_e^{-7}$	$x_e :=$ Find(x_e)
$x_e = 1.3598$		

Thus, $\boxed{r_e = 1.3598\ r_0}$.

Parameter A is determined with the criterion that $V(r_e) = \varepsilon$:

$$V(x_e) = \varepsilon = A\varepsilon\left[e^{1-x_e} - \frac{1}{x_e^6}\right]$$

$$A = \left[e^{1-x_e} - \frac{1}{x_e^6}\right]^{-1} = \left[e^{1-1.3598} - \frac{1}{1.3598^6}\right]^{-1}$$

$$\boxed{A = 1.8531}$$

P17.19 Refer to Figure 17.11(a).

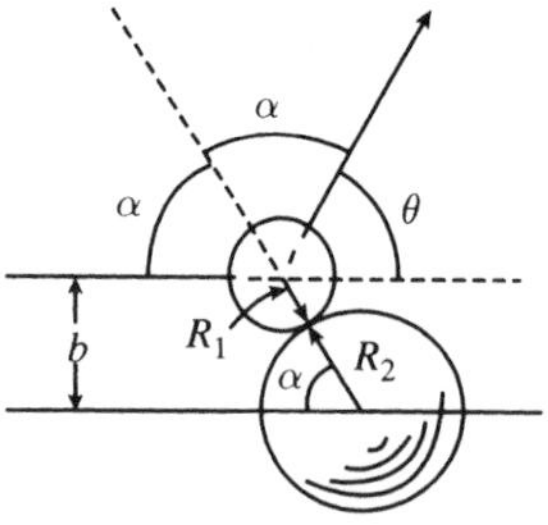

Figure 17.11(a)

The scattering angle is $\theta = \pi - 2\alpha$ if specular reflection occurs in the collision (angle of impact equal to angle of departure from the surface). For $b \le R_1 + R_2$, $\sin\alpha = \dfrac{b}{R_1 + R_2}$.

$$\theta = \begin{cases} \pi - 2\arcsin\left(\dfrac{b}{R_1 + R_2}\right) & b \le R_1 + R_2 \\ 0 & b > R_1 + R_2 \end{cases}$$

The function is plotted in Figure 17.11(b).

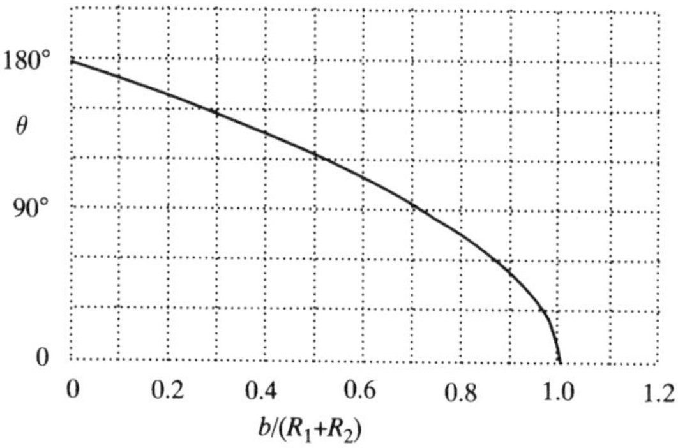

Figure 17.11(b)

Solutions to applications: biochemistry

P17.21 The interaction is a dipole–induced-dipole interaction. The energy is given by eqn 17.24:

$$V = -\frac{\mu_1^2 \alpha_2'}{4\pi\varepsilon_0 r^6} = -\frac{[(2.7\text{ D})(3.336\times10^{-30}\text{ C m D}^{-1})]^2(1.04\times10^{-29}\text{ m}^3)}{4\pi(8.854\times10^{-12}\text{ J}^{-1}\text{ C}^2\text{ m}^{-1})(4.0\times10^{-9}\text{ m})^6}$$

$$V = \boxed{-1.8\times10^{-27}\text{ J} = -1.1\times10^{-3}\text{ J mol}^{-1}}.$$

COMMENT. This value seems exceedingly small. The distance suggested in the problem may be too large compared to typical values. A distance of 0.40 nm yields $V = -1.1$ kJ mol^{-1}.

P17.23 (a) The left-side molecule in Figure 17.12 is methyladenine. Note that we have taken the liberty of placing the methyl group in the position that would be occupied by a sugar in RNA and DNA. The wavefunction, structure, and atomic electrostatic charges (shown in the figure) calculation was performed with Spartan '06™ using a Hartree–Fock procedure with a 6-31G* basis set. The atomic electrostatic charge (ESP) numerical method generates charges that reproduce the electrostatic field from the entire wavefunction. The right-side molecule in Figure 17.12 is methylthymine.

(b) The two molecules will hydrogen bond into a stable dimer in an orientation for which hydrogen bonding is linear, maximized, and steric hindrance is avoided. We expect hydrogen bonds of the type N–H···O andN–H···N with the N and O atoms having large negative electrostatic charges and the H atoms having large positive charges. These atoms are evident in the figure.

(c) Figure 17.12 shows one of three arrangements of hydrogen bonding between the two molecules. Another can be drawn by rotating methylthymine over the top of methyladenine and a third involves rotation to the bottom. The dashed lines show the alignments of two strong hydrogen bonds between the molecules.

(d) The A-to-T base pairing shown in Figure 17.12 has the largest charges in the most favourable positions for strong hydrogen bonding. Also the N-to-O distance of one hydrogen bond equals the N-to-N distance of the other, a favourable feature in RNA and DNA polymers where this pairing and alignment is observed naturally.

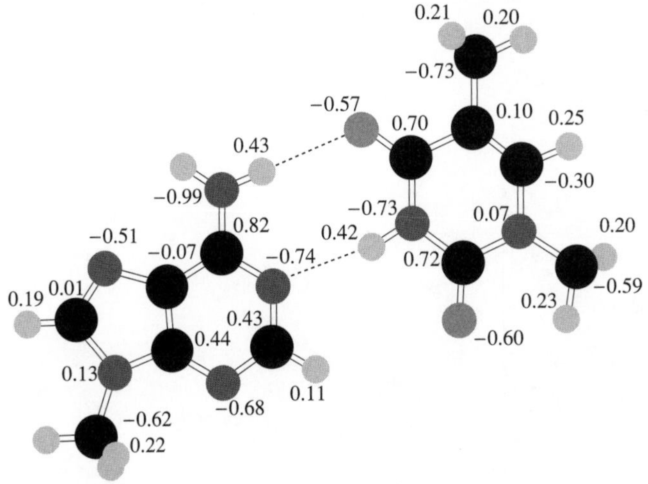

Figure 17.12

(e) The favourable orientation for hydrogen bonding between methylguanine (left-side molecule) and methylcytosine (right-side molecule) is shown in Figure 17.13. Large counter charges align within distances, resulting in three strong hydrogen bonds.

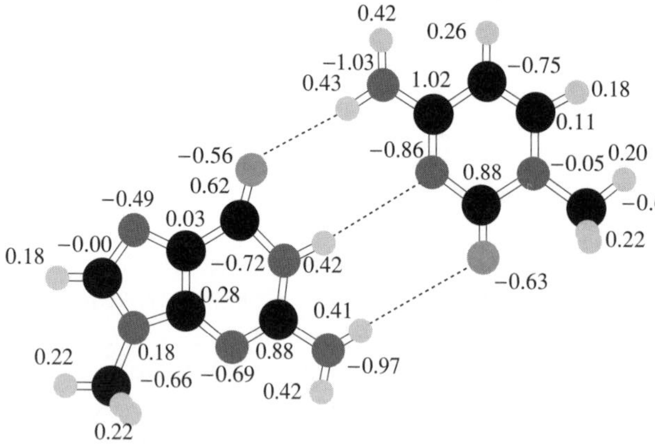

Figure 17.13

P17.25 (a) The hydrocarbons in questions form a homologous series. They are straight-chain alkanes of the formula C_nH_{2n+2}, or R–H, where R = C_nH_{2n+1}. Draw up the following table:

n	1	2	3	4	5
π	0.5	1.0	1.5	2.0	2.5

The relationship here is evident by inspection: $\pi = n/2$, so we predict for the seven-carbon hydrocarbon in question:

$$\pi = 7/2 = \boxed{3.5}$$

(b) The plot shown in Figure 17.14 is consistent with a linear relationship, for $R^2 = 0.997$ is close to unity. The best linear fit is:

$$\log K_1 = -1.95 - 1.49\pi,$$

so $\boxed{\text{slope} = -1.49}$ and $\boxed{\text{intercept} = -1.95}$.

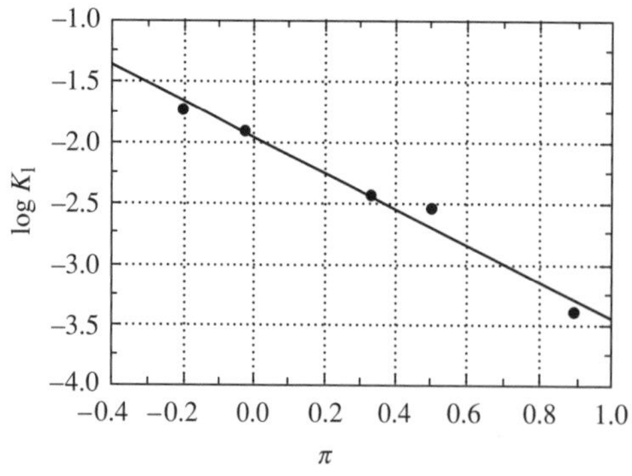

Figure 17.14

(c) If we know π for the substituent R = H, then we can use the linear SAR just derived. Our best estimate of π can be obtained by considering the zero-carbon 'alkane' H_2, whose radical H ought to have a hydrophobicity constant $\pi = 0/2 = 0$. This value yields

$$\log K_1 = -1.95 - 1.49(0) = -1.95 \quad \text{so} \quad K_1 = 10^{-1.95} = \boxed{1.12 \times 10^{-2}}$$

Note: The assumption that R = H is part of the homologous series of straight-chain alkanes is a reasonable but questionable one.

18 Materials 1: macromolecules and self-assembly

Answers to discussion questions

D18.1 **Contour length, R_c**: the length of the macromolecule measured along its backbone, the length of all its monomer units placed end to end. This is the stretched-out length of the macromolecule with bond angles maintained within the monomer units and 180° angles at unit links. It is proportional to the number of monomer units, N, and to the length of each unit (eqn 18.3).

Root mean square separation, R_{rms}: one measure of the average separation of the ends of a random coil. It is the square root of the mean value of R^2, where R is the separation of the two ends of the coil. R_{rms} is proportional to $N^{1/2}$ and the length of each unit (eqn 18.4).

Radius of gyration, R_g: the radius of a thin hollow spherical shell of the same mass and moment of inertia as the macromolecule: $R_g^2 = I/m_{tot}$. In general, it is not easy to visualize this distance geometrically. However, for the simple case of a molecule consisting of an unconstrained chain of identical atoms this quantity can be visualized as the root mean square distance of the atoms from the centre of mass. It also depends on $N^{1/2}$, but is smaller than the root mean square separation by a factor of $(1/6)^{1/2}$ (eqn 18.6).

D18.3 The formation of micelles is favoured by the interaction between hydrocarbon tails and is opposed by charge repulsion of the polar groups that are placed close together at the micelle surface. As salt concentration is increased, the repulsion of head groups is reduced because their charges are partly shielded by the ions of the salt. This favours micelle formation, causing the micelles to be larger and the critical micelle concentration to be smaller.

D18.5 Polymers are unlike small molecules in that all small molecules of the same species have nearly identical masses with variations caused by a distribution of isotopes. Polymers, however, vary widely in mass because they can vary in the number of monomeric units they contain. Depending on how a polymer mixture is synthesized and purified, it is entirely possible for one macromolecule to contain 1000 monomer units and another 1500. We call a polymer sample **polydisperse** if there is a large variation in mass among the molecules of the sample; conversely, a sample is **monodisperse** if its range of masses is narrow.

The various experimental methods for measuring the average molar mass of a polymer mixture (such as mass spectrometry, light scattering, osmometry, and viscosity) yield similar results for a monodisperse polymer solution, but the measurements can show great variation for a polydisperse polymer mixture. The differences in averages are in the weighting factors. The weighting factor for

$\overline{M_n}$ is the number of molecules that have a particular mass and the weighting factor in $\overline{M_w}$ is the mass fraction of a sample that has a particular mass. Different measurement techniques yield different weighting factors because they are sensitive to different factors. The intensity of a mass spectrometry peak, for instance, is proportional to the number of molecules of a given mass. Some techniques, like light scattering, are more sensitive to the size (volume) and shape of particles, and some, like sedimentation, are more sensitive to the mass. Discussions in the text reveal, however, that the measurements capture a complicated function of size, shape, mass, and number. Osmometry, measuring a colligative property, is sensitive to the number of molecules N_i that have molar mass M_i. Consequently, average osmotic properties depend on the number average molar mass. Light scattering depends on molecular size and shape, which indirectly depend on mass, so weight average molar mass becomes important. Other mass averages become important when the technique is sensitive to intermolecular attractions and repulsions, molecular entanglements, and gravitational and centrifuge effects.

Solutions to exercises

E18.1(a) $R_{rms} = N^{1/2}l$ [18.4] $= (700)^{1/2} \times (0.90 \text{ nm}) = \boxed{27 \text{ nm}}$

E18.2(a) The repeating monomer unit of polyethylene is (—CH_2—CH_2—), which has a molar mass of 28 g mol^{-1}. The number of repeating units, N, is therefore

$$N = \frac{M_{polymer}}{M_{monomer}} = \frac{280\,000 \text{ g mol}^{-1}}{28 \text{ g mol}^{-1}} = 1.00 \times 10^4$$

$l = 2R(\text{C—C})$ [add half a bond length on either side of monomer]

$$R_c = Nl \text{ [18.3]} = 2 \times (1.00 \times 10^4) \times (154 \text{ pm}) = 3.08 \times 10^6 \text{ pm} = \boxed{3.08 \text{ μm}}$$

$$R_{rms} = N^{1/2}l = 2 \times (1.00 \times 10^4)^{1/2} \times (154 \text{ pm}) = 3.08 \times 10^4 \text{ pm} = \boxed{30.8 \text{ nm}}$$

E18.3(a) For a random coil, the radius of gyration is

$$R_g = \left(\frac{N}{6}\right)^{1/2} l \text{ [18.6]} \quad \text{so} \quad N = 6\left(\frac{R_g}{l}\right)^2 = 6 \times \left(\frac{7.3 \text{ nm}}{0.154 \text{ nm}}\right)^2 = \boxed{1.4 \times 10^4}$$

E18.4(a) The repeating monomer unit of polyethylene is (—CH_2—CH_2—), which has a molar mass of 28 g mol^{-1}. The number of repeating units, N, is therefore

$$N = \frac{M_{polymer}}{M_{monomer}} = \frac{65\,000 \text{ g mol}^{-1}}{28 \text{ g mol}^{-1}} = 2.3\overline{2} \times 10^3$$

$l = 2R(\text{C—C}) = 2 \times 154 \text{ pm} = 308 \text{ pm}$ [add half a bond length on either side of monomer]

In units of l the polymer ends are separated by the distance

$$n = \frac{10 \times 10^{-9} \text{ m}}{l} = \frac{10 \times 10^{-9} \text{ m}}{308 \times 10^{-12} \text{ m}} = 32.\overline{5}$$

Thus,

$$P=\left(\frac{2}{\pi N}\right)^{1/2}\mathrm{e}^{-n^2/2N}\,[18.1]=\left(\frac{2}{\pi\times 2.3\bar{2}\times 10^3}\right)^{1/2}\mathrm{e}^{-(32.\bar{5})^2/\{2\times(2.3\bar{2}\times 10^3)\}}=\boxed{0.017}$$

E18.5(a) We obtain (see Exercise E18.4a) $N = 2.3\bar{2}\times 10^3$ and $l = 308$ pm. Thus,

$$a=\left(\frac{3}{2Nl^2}\right)^{1/2}[18.2]=\left(\frac{3}{2\times(2.3\bar{2}\times 10^3)\times(308\times 10^{-12}\text{ m})^2}\right)^{1/2}=8.2\bar{6}\times 10^7\text{ m}^{-1}$$

The 10.00 nm to 10.10 nm range of distances between the polymer ends is very small so we estimate that the distribution function $f(r)$ is the constant given by $f(r) = f(10.05\text{ nm})$. The probability that the polymer ends are in this range is

$$P=f(r)\Delta r=4\pi\left(\frac{a}{\pi^{1/2}}\right)^3 r^2\mathrm{e}^{-a^2r^2}\Delta r\,[18.2]$$

$$=4\pi\left(\frac{8.2\bar{6}\times 10^7\text{ m}^{-1}}{\pi^{1/2}}\right)^3\times(10.05\times 10^{-9}\text{ m})^2\times \mathrm{e}^{-(8.2\bar{6}\times 10^7\text{ m}^{-1})^2(10.05\times 10^{-9}\text{ m})^2}\times(10.1-10.00)\times 10^{-9}\text{ m}$$

$$=\boxed{6.4\times 10^{-3}}$$

E18.6(a) We obtain (see Exercise E18.4a) $N = 2.3\bar{2}\times 10^3$ and $l = 308$ pm. In units of l the polymer ends are moved apart by the distance

$$n=\frac{1.0\times 10^{-9}\text{ m}}{l}=\frac{1.0\times 10^{-9}\text{ m}}{308\times 10^{-12}\text{ m}}=3.2\bar{5}$$

Thus, the molar change in conformational entropy arising from the stretch is

$$\Delta S=-\tfrac{1}{2}RN\ln\{(1+\nu)^{1+\nu}(1-\nu)^{1-\nu}\}\,[18.7]\quad\text{with}\quad \nu=n/N=3.2\bar{5}/2.3\bar{2}\times 10^3=1.4\bar{0}\times 10^{-3}$$

$$=-\tfrac{1}{2}\times(8.3145\text{ J mol}^{-1}\text{ K}^{-1})\times(2.3\bar{2}\times 10^3)\times\ln\{(1.0014\bar{0})^{1.0014\bar{0}}\times(0.99860)^{0.99860}\}$$

$$=\boxed{-19\text{ mJ mol}^{-1}\text{ K}^{-1}}$$

E18.7(a) The radius of gyration for a constrained chain that has successive individual bonds constrained to a single cone of angle $\theta = 109.5°$ (i.e. $\cos\theta = -\tfrac{1}{3}$) is

$$R_{\text{g,constrained coil}}=R_{\text{g,random coil}}F\quad\text{with}\quad F=\left(\frac{1-\cos\theta}{1+\cos\theta}\right)^{1/2}[18.8]=\left(\frac{1+\frac{1}{3}}{1-\frac{1}{3}}\right)^{1/2}=\sqrt{2}$$

The percentage change in the radius of gyration on application of the constraint is

$$\left(\frac{R_{\text{g,constrained coil}}-R_{\text{g,random coil}}}{R_{\text{g,random coil}}}\right)\times 100\%=\left(\frac{R_{\text{g,constrained coil}}}{R_{\text{g,random coil}}}-1\right)\times 100\%=(F-1)\times 100\%$$

$$=(\sqrt{2}-1)\times 100\%$$

$$=\boxed{+41.42\%}$$

The percentage change in the volume on application of the constraint is

$$\left(\frac{V_{\text{constrained coil}} - V_{\text{random coil}}}{V_{\text{random coil}}}\right) \times 100\% = \left(\frac{R^3_{\text{g,constrained coil}} - R^3_{\text{g,random coil}}}{R^3_{\text{g,random coil}}}\right) \times 100\%$$

$$= \left\{\left(\frac{R_{\text{g,constrained coil}}}{R_{\text{g,random coil}}}\right)^3 - 1\right\} \times 100\% = (F^3 - 1) \times 100\%$$

$$= (2^{3/2} - 1) \times 100\%$$

$$= \boxed{+182.8\%}$$

E18.8(a) In analogy to eqn 18.11 we have

$$R_{\text{g}} = R_{\text{g,random coil}} F = \left(\frac{N}{6}\right)^{1/2} lF, \quad \text{where} \quad F = \left(\frac{2l_p}{l} - 1\right)^{1/2}$$

Thus, the percentage increase in the radius of gyration when the polymer persistence length is changed from l to $0.050R_c$ with constant contour length is given by the expression

$$\left(\frac{R_{\text{g}} - R_{\text{g,random coil}}}{R_{\text{g,random coil}}}\right) \times 100\% = \left(\frac{R_{\text{g}}}{R_{\text{g,random coil}}} - 1\right) \times 100\% = (F - 1) \times 100\%$$

$$= \left\{\left(\frac{2l_p}{l} - 1\right)^{1/2} - 1\right\} \times 100\%$$

$$= \left\{\left(\frac{2 \times (0.05R_c)}{l} - 1\right)^{1/2} - 1\right\} \times 100\%$$

$$= \left\{\left(\frac{2 \times (0.05 \times Nl)}{l} - 1\right)^{1/2} - 1\right\} \times 100\%$$

$$= \{(0.10N - 1)^{1/2} - 1\} \times 100\%$$

$$= \boxed{+895\% \text{ when } N = 1000}$$

The percentage change in the volume as determined by the radius of gyration is

$$\left(\frac{V - V_{\text{g,random coil}}}{V_{\text{g,random coil}}}\right) \times 100\% = \left(\frac{R^3_{\text{g}} - R^3_{\text{g,random coil}}}{R^3_{\text{g,random coil}}}\right) \times 100\% = (F^3 - 1) \times 100\%$$

$$= \left\{\left(\frac{2l_p}{l} - 1\right)^{3/2} - 1\right\} \times 100\% = \left\{\left(\frac{2 \times (0.05R_c)}{l} - 1\right)^{3/2} - 1\right\} \times 100\%$$

$$= \left\{\left(\frac{2 \times (0.05 \times Nl)}{l} - 1\right)^{3/2} - 1\right\} \times 100\%$$

$$= \{(0.10N - 1)^{3/2} - 1\} \times 100\%$$

$$= \boxed{+9.84 \times 10^4\% \text{ when } N = 1000}$$

E18.9(a) In analogy to eqn 18.11 we have

$$R_g = R_{g,\text{random coil}}F = \left(\frac{N}{6}\right)^{1/2} lF, \quad \text{where} \quad F = \left(\frac{2l_p}{l} - 1\right)^{1/2}$$

Solving for l_p gives

$$l_p = \frac{l}{2}\left\{\frac{6R_g^2}{Nl^2} + 1\right\}$$

$$= \frac{150\ \text{pm}}{2}\left\{\frac{6 \times (2100\ \text{pm})^2}{1000 \times (150\ \text{pm})^2} + 1\right\} = \boxed{1.3 \times 10^4\ \text{pm}}$$

E18.10(a) Each polyethene chain bond has a length $l = 154$ pm $= 0.154$ nm and the number N chain bonds is the polymer molar mass divided by the molar mass of the repeating CH_2 unit.

$$N = \frac{M_{\text{polymer}}}{M_{CH_2}} = \frac{65\,000\ \text{g mol}^{-1}}{14\ \text{g mol}^{-1}} = 4.6\overline{4} \times 10^3$$

The restoring force of a freely jointed chain is

$$\mathcal{F} = \frac{kT}{2l}\ln\left(\frac{1+\nu}{1-\nu}\right) \qquad \nu = n/N \text{ [18.12a and } \textit{Justification 18.3}\text{]}$$

n is the displacement from equilibrium in units of l so

$$\nu = \frac{n}{N} = \frac{(1.0\ \text{nm})/(0.154\ \text{nm})}{4.6\overline{4} \times 10^3} = 1.4 \times 10^{-3}$$

Since $\nu \ll 1$, we use the simplified form of eqn 18.12a.

$$\mathcal{F} = \frac{nkT}{Nl} \text{ [18.12b]}$$

$$= \frac{(6.4\overline{9}) \times (1.381 \times 10^{-23}\ \text{J K}^{-1}) \times (293.15\ \text{K})}{(4.6\overline{4} \times 10^3) \times (154 \times 10^{-12}\ \text{m})} = \boxed{3.7 \times 10^{-14}\ \text{N}}$$

E18.11(a) Equal amounts imply equal numbers of molecules. Hence, the number average is (eqn 18.19)

$$\overline{M_n} = \frac{N_1M_1 + N_2M_2}{N} = \frac{n_1M_1 + n_2M_2}{n} = \tfrac{1}{2}(M_1 + M_2)\ [n_1 = n_2 = \tfrac{1}{2}n]$$

$$= \tfrac{1}{2}(62 + 78)\ \text{kg mol}^{-1} = \boxed{70\ \text{kg mol}^{-1}}$$

and the weight average is (eqns 18.20a, 18.20b)

$$\overline{M_w} = \frac{m_1M_1 + m_2M_2}{m} = \frac{n_1M_1^2 + n_2M_2^2}{n_1M_1 + n_2M_2} = \frac{M_1^2 + M_2^2}{M_1 + M_2}\ [n_1 = n_2]$$

$$= \frac{62^2 + 78^2}{62 + 78}\ \text{kg mol}^{-1} = \boxed{71\ \text{kg mol}^{-1}}$$

E18.12(a) (a) Osmometry gives the number average molar mass, so

$$\overline{M_n} = \frac{N_1M_1 + N_2M_2}{N_1 + N_2}[18.19] = \frac{n_1M_1 + n_2M_2}{n_1 + n_2} = \frac{\left(\frac{m_1}{M_1}\right)M_1 + \left(\frac{m_2}{M_2}\right)M_2}{\left(\frac{m_1}{M_1}\right) + \left(\frac{m_2}{M_2}\right)} = \frac{m_1 + m_2}{\left(\frac{m_1}{M_1}\right) + \left(\frac{m_2}{M_2}\right)}$$

$$= \frac{100\text{ g}}{\left(\frac{30\text{ g}}{30\text{ kg mol}^{-1}}\right) + \left(\frac{70\text{ g}}{15\text{ kg mol}^{-1}}\right)}[\text{assume 100 g of solution}] = \boxed{18\text{ kg mol}^{-1}}$$

(b) Light scattering gives the mass average molar mass, so

$$\overline{M_w} = \frac{m_1M_1 + m_2M_2}{m_1 + m_2} = \frac{(30)\times(30) + (70)\times(15)}{100}\text{ kg mol}^{-1} = \boxed{20\text{ kg mol}^{-1}}$$

E18.13(a) Since the two spherical particles have different radii but identical densities, we immediately recognize that they have identical buoyancy and, therefore, their effective masses are proportional to the particle mass. The solute particle mass equals its volume multiplied by its density ρ_s. Therefore, the particle mass is proportional to the cube of the particle radius:

$$m_{\text{eff}} \propto a^3\rho_s$$

According to eqn 18.31, the Stoke's frictional coefficient is proportional to a and eqn 18.28 indicates that the sedimentation rate s is proportional to the effective mass and inversely proportional to the frictional coefficient. Thus,

$$s \propto \frac{a^3\rho_s}{a}$$

$$\propto a^2\rho_s$$

The constant of proportionality cancels when taking ratios so the relative rates of sedimentation of the two different particles is

$$\frac{s_2}{s_1} = \frac{a_2^2\rho_2}{a_1^2\rho_1} = 10^2 = \boxed{100} \text{ because } \rho_1 = \rho_2$$

E18.14(a) $M = \overline{M_n}$ and eqn 18.30 gives

$$M = \frac{SfN_A}{b}[18.30] = \frac{SfN_A}{1 - \rho v_s}[18.27]$$

Assuming that the solution density equals that of water at 298 K (0.9969 g cm^{-3}) and substitution of the Stokes–Einstein relationship, $f = kT/D$ [20.51], gives

$$M = \frac{SRT}{(1 - \rho v_s)D}$$

$$= \frac{(4.48\text{ Sv})\times(10^{-13}\text{ s Sv}^{-1})(8.3145\text{ J mol}^{-1}\text{ K}^{-1})\times(298.15\text{ K})}{\{1 - (996.9\text{ kg m}^{-3})\times(0.749\times10^{-3}\text{ m}^3\text{ kg}^{-1})\}\times(6.9\times10^{-11}\text{ m}^2\text{ s}^{-1})} = \boxed{64\text{ kg mol}^{-1}}$$

E18.15(a) The net force acting upon the settling particle equals zero because of the balance between the gravitation pull, $m_{\text{eff}}g$, and the frictional force, fs, where s is the drift speed and the frictional coefficient is given by Stokes's relationship $f = 6\pi a\eta$ [18.31] for a particle of radius a. Thus,

$$fs = m_{\text{eff}}g$$

$$s = \frac{m_{\text{eff}}g}{f} = \frac{bmg}{f} = \frac{(1-\rho/\rho_s)\times\{(\tfrac{4}{3}\pi a^3)\times\rho_s\}g}{6\pi a\eta} \quad [18.27 \text{ and } 18.31]$$

$$= \frac{2(\rho_s-\rho)a^2g}{9\eta}$$

$$= \frac{2\times(1750\ \text{kg m}^{-3} - 1000\ \text{kg m}^{-3})\times(20\times10^{-6}\ \text{m})^2\times(9.8067\ \text{m s}^{-2})}{9\times(8.9\times10^{-4}\ \text{kg m}^{-1}\,\text{s}^{-1})}$$

$$= \boxed{0.73\ \text{mm s}^{-1}}$$

E18.16(a) $$\overline{M}_n = \frac{SfN_A}{b}\ [18.30] = \frac{SfN_A}{1-\rho v_s}\ [18.27]$$

Substitution of the Stokes–Einstein relationship, $f = kT/D$ [20.51], gives

$$\overline{M}_n = \frac{SRT}{(1-\rho v_s)D} = \frac{(3.2\ \text{Sv})\times(10^{-13}\ \text{s Sv}^{-1})(8.3145\ \text{J mol}^{-1}\,\text{K}^{-1})\times(293.15\ \text{K})}{\{1-(1.06\ \text{g cm}^{-3})\times(0.656\ \text{cm}^3\,\text{g}^{-1})\}\times(8.3\times10^{-11}\ \text{m}^2\,\text{s}^{-1})}$$

$$= \boxed{31\ \text{kg mol}^{-1}}$$

E18.17(a) The number of solute molecules with potential energy E is proportional to $e^{-E/kT}$, hence

$$c \propto N \propto e^{-E/kT}, \quad \text{where} \quad E = \tfrac{1}{2}m_{\text{eff}}r^2\omega^2$$

Therefore, $c \propto e^{Mb\omega^2r^2/2RT}$ [$m_{\text{eff}} = bm$, $M = mN_A$] and

$$\ln c = \text{const.} + \frac{Mb\omega^2r^2}{2RT} \quad [b = 1-\rho v_s]$$

This expression indicates that the slope of a plot of ln c against r^2 is equal to $Mb\omega^2/2RT$. Therefore,

$$M = \frac{2RT\times\text{slope}}{b\omega^2} = \frac{2\times(8.3145\ \text{J K}^{-1}\,\text{mol}^{-1})\times(300\ \text{K})\times(729\times10^4\ \text{m}^{-2})}{(1-0.997\times0.61)\times\{(50000)\times(2\pi/60\ \text{s})\}^2}$$

$$= \boxed{3.4\times10^3\ \text{kg mol}^{-1}}$$

Solutions to problems

Solutions to numerical problems

P18.1 The glass transition temperature T_g is the temperature at which internal bond rotations freeze. In effect, the easier such rotations are, the lower T_g. Internal rotations are more difficult for polymers that have bulky side chains than for polymers without such chains because the side chains of neighbouring molecules can impede each others' motion. Of the four polymers in this problem, polystyrene has the largest side chain (phenyl) and the largest T_g. The chlorine atoms in poly(vinyl chloride) interfere with each other's motion more than the smaller hydrogen atoms that hang from the carbon backbone of polyethylene. Poly(oxymethylene), like polyethylene, has only hydrogen atoms protruding from its backbone; however, poly(oxymethylene) has fewer hydrogen protrusions and a still lower T_g than polyethylene.

P18.3 R_g is the radius of rotation of a point mass that has the same mass m and moment of inertia I as the object of interest. For an object that has a continuum of mass within its macroscopic boundaries:

$I = mR_g^2$, where $I = \int \rho R^2 \mathrm{d}\tau$ and R is the distance from the axis of rotation

For a homogeneous object $\rho = m/V$ so

$R_g^2 = V^{-1}\int R^2 \mathrm{d}\tau$ (the integrand is the square distance from the axis of rotation)

(a) For a solid sphere of radius a: $V = \frac{4}{3}\pi a^3$, $\mathrm{d}\tau = r^2 \sin\theta \mathrm{d}r\,\mathrm{d}\theta\,\mathrm{d}\phi$, and $R^2 = x^2 + y^2$.

$$R_g^2 V = \int_{\phi=0}^{2\pi}\int_{\theta=0}^{\pi}\int_{r=0}^{a} (x^2+y^2)r^2 \sin\theta\,\mathrm{d}r\,\mathrm{d}\theta\,\mathrm{d}\phi$$

$$= \int_{\phi=0}^{2\pi}\int_{\theta=0}^{\pi}\int_{r=0}^{a} (r^2\sin^2\theta\cos^2\phi + r^2\sin^2\theta\sin^2\phi)r^2 \sin\theta\,\mathrm{d}r\,\mathrm{d}\theta\,\mathrm{d}\phi$$

$$= \int_{\phi=0}^{2\pi}\int_{\theta=0}^{\pi}\int_{r=0}^{a} r^4\sin^3\theta\,\mathrm{d}r\,\mathrm{d}\theta\,\mathrm{d}\phi = \left\{\int_0^{2\pi}\mathrm{d}\phi\right\}\left\{\int_0^{\pi}\sin^3\theta\,\mathrm{d}\theta\right\}\left\{\int_0^{a} r^4\,\mathrm{d}r\right\}$$

$$= 2\pi \times \{-\cos\theta + \tfrac{1}{3}\cos^3\theta\}\Big|_{\theta=0}^{\theta=\pi} \times \left[\frac{r^5}{5}\right]_{r=0}^{r=a} = \frac{8\pi a^5}{15}$$

$$R_g^2 = \frac{8\pi a^5}{15V} = \frac{8\pi a^5}{15\times\left(\frac{4}{3}\pi a^3\right)} = \tfrac{2}{5}a^2$$

$$R_g = \boxed{\sqrt{\tfrac{2}{5}}\,a}$$

COMMENT. A common error involves using r^2 in place of the squared distance from the axis of rotation, which is actually equal to x^2+y^2 not to $x^2+y^2+z^2=r^2$. The common error gives the result $\sqrt{\frac{3}{5}}\,a$.

For a spherical macromolecule, the specific volume is:

$$\upsilon_s = \frac{V}{m} = \frac{4\pi a^3}{3}\times\frac{N_A}{M} \quad \text{so } a = \left(\frac{3\upsilon_s M}{4\pi N_A}\right)^{1/3}$$

Therefore,

$$R_g = \left(\frac{2}{5}\right)^{1/2}\times\left(\frac{3\upsilon_s M}{4\pi N_A}\right)^{1/3} = \left(\frac{2}{5}\right)^{1/2}\times\left(\frac{(3\upsilon_s/\mathrm{cm^3\,g^{-1}})\times \mathrm{cm^3\,g^{-1}}\times(M/\mathrm{g\,mol^{-1}})\times \mathrm{g\,mol^{-1}}}{4\pi\times(6.022\times10^{23}\,\mathrm{mol^{-1}})}\right)^{1/3}$$

$$= (4.6460\times10^{-9}\,\mathrm{cm})\times\{(\upsilon_s/\mathrm{cm^3\,g^{-1}})\times(M/\mathrm{g\,mol^{-1}})\}^{1/3}$$

$$R_g/\mathrm{nm} = \boxed{0.046460\times\{(\upsilon_s/\mathrm{cm^3\,g^{-1}})\times(M/\mathrm{g\,mol^{-1}})\}^{1/3}}$$

COMMENT. The common error, discussed above, gives

$0.056902 \times \{(v_s/\text{cm}^3\,\text{g}^{-1}) \times (M/\text{g mol}^{-1})\}^{1/3}$.

When $M = 100$ kg mol^{-1} and $v_s = 0.750$ cm^3 g^{-1},

$$R_g = 0.046460 \times \{(0.750) \times (100 \times 10^3)\}^{1/3}\,\text{nm}$$

$$= \boxed{1.96\,\text{nm}}$$

(b) For a solid rod of radius a and length l there are two moments of inertia. One about the axis of its length ($R_{g,\|}$, $R = r$) and another about the axis that is perpendicular to its length ($R_{g,\perp}$, $R = z$).

$$R_{g,\|}^2 = (\pi a^2 l)^{-1} \int r^2\,d\tau = (\pi a^2 l)^{-1} \int_{r=0}^{a} \int_{z=0}^{l} \int_{\theta=0}^{2\pi} r^3\,d\theta\,dz\,dr \quad \text{[cylindrical coordinates]}$$

$$= (\pi a^2 l)^{-1} 2\pi l \int_0^a r^3\,dr = \frac{2}{a^2}\left[\frac{r^4}{4}\right]_{r=0}^{r=a} = \tfrac{1}{2}a^2$$

$$R_{g,\|} = \boxed{\sqrt{\tfrac{1}{2}}\,a}$$

$$R_{g,\perp}^2 = (\pi a^2 l)^{-1} \int z^2\,d\tau = (\pi a^2 l)^{-1} \int_{r=0}^{a} \int_{z=-l/2}^{l/2} \int_{\theta=0}^{2\pi} z^2 r\,d\theta\,dz\,dr \quad \text{[cylindrical coordinates]}$$

$$= \tfrac{1}{12} l^2$$

$$R_{g,\perp} = \boxed{\sqrt{\tfrac{1}{12}}\,l}$$

For a rod-like macromolecule, the specific volume is:

$$v_s = \frac{V}{m} = \pi a^2 l \times \frac{N_A}{M} \quad \text{so } l = \frac{v_s M}{\pi a^2 N_A}$$

Therefore,

$$R_{g,\|} = \sqrt{\tfrac{1}{2}} \times (0.50\,\text{nm}) = \boxed{0.35\,\text{nm}}$$

$$R_{g,\perp} = \sqrt{\tfrac{1}{12}}\left(\frac{v_s M}{\pi a^2 N_A}\right)$$

$$= \sqrt{\tfrac{1}{12}}\left\{\frac{(0.750 \times 10^{-6}\,\text{m}^3\,\text{g}^{-1}) \times (100 \times 10^3\,\text{g mol}^{-1})}{\pi \times (0.50 \times 10^{-9}\,\text{m})^2 \times (6.022 \times 10^{23}\,\text{mol}^{-1})}\right\}$$

$$= \boxed{46\,\text{nm}}$$

P18.5 $$[\eta] = \lim_{c \to 0}\left(\frac{\eta/\eta_0 - 1}{c}\right) \text{[18.35]}$$

We see that the y-intercept of a plot of the right-hand side against c, extrapolated to $c = 0$, gives $[\eta]$. We begin by constructing the following table using $\eta_0 = 0.985$ g m^{-1} s^{-1}.

c/g dm^{-3}	1.32	2.89	5.73	9.17
$\left(\frac{\eta/\eta_0 - 1}{c}\right)\Big/$dm^3 g^{-1}	0.0731	0.0755	0.0771	0.0825

The points are plotted in Figure 18.1. The least-squares intercept is at 0.0716, so $[\eta] = \boxed{0.0716\ \text{dm}^3\ \text{g}^{-1}}$

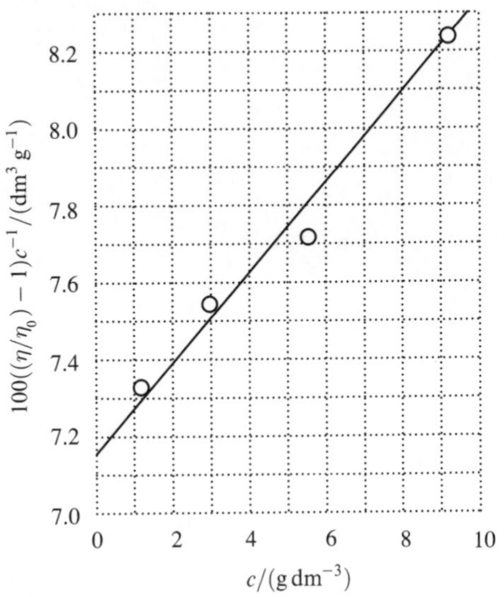

Figure 18.1

P18.7 We follow the procedure of Example 18.5. Also, compare to Problems 18.5 and 18.6

$$[\eta] = \lim_{c\to 0}\left(\frac{\eta/\eta_0 - 1}{c}\right) [18.35] \quad \text{and} \quad [\eta] = K\overline{M}_v^{\,a}\ [18.37]$$

with K and a from Table 18.4. We draw up the following table using $\eta_0 = 0.647 \times 10^{-3}$ kg m^{-1} s^{-1}, plot the points, and perform a linear regression fit as shown in Figure 18.2.

100 c/g cm^{-3}	0	0.2	0.4	0.6	0.8	1.0
$\eta/10^{-3}$ kg m^{-1} s^{-1}	0.647	0.690	0.733	0.777	0.821	0.865
$\left(\frac{\eta/\eta_0 - 1}{c}\right)\Big/$100 cm^3 g^{-1}	—	$0.332\overline{3}$	$0.332\overline{3}$	$0.334\overline{9}$	$0.336\overline{2}$	$0.336\overline{9}$

The intercept of the linear regression fit gives $[\eta]$. Hence,

$$[\eta] = (0.330\overline{6}) \times (100\ \text{cm}^3\ \text{g}^{-1}) = 33.0\overline{6}\ \text{cm}^3\ \text{g}^{-1}$$

and

$$\overline{M}_v = \left(\frac{[\eta]}{K}\right)^{1/a} \text{g mol}^{-1} = \left(\frac{33.0\overline{6}\ \text{cm}^3\ \text{g}^{-1}}{8.3 \times 10^{-2}\ \text{cm}^3\ \text{g}^{-1}}\right)^{1/0.50} \text{g mol}^{-1} = \boxed{1.6 \times 10^5\ \text{g mol}^{-1}}$$

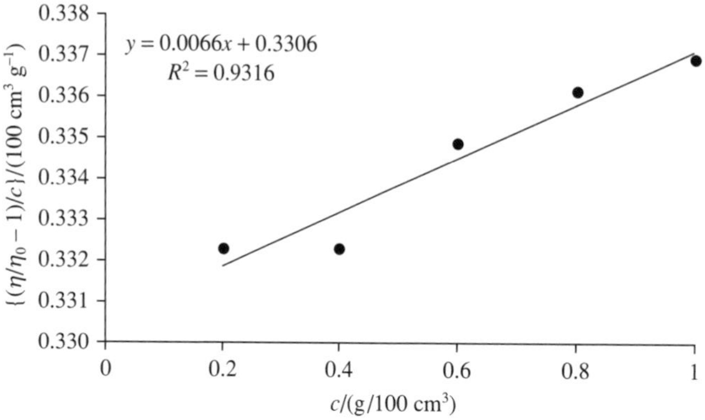

Figure 18.2

P18.9 (a) The empirical Mark–Kuhn–Houwink–Sakurada equation [18.38] is

$$[\eta] = K\overline{M}_{\rm v}^{\,a}$$

or, explicitly showing units,

$$\frac{[\eta]}{\text{cm}^3\,\text{g}^{-1}} = \left(\frac{K}{\text{cm}^3\,\text{g}^{-1}}\right) \times \left(\frac{\overline{M}_{\rm v}}{\text{g mol}^{-1}}\right)^a = \left(\frac{10^{3a}\,K}{\text{cm}^3\,\text{g}^{-1}}\right) \times \left(\frac{\overline{M}_{\rm v}}{\text{kg mol}^{-1}}\right)^a$$

We fit the data to the above equation and obtain K and a from a regression power fitting procedure. The plot is shown in Figure 18.3 along with the regression power fit. The fit indicates that $\boxed{a = 0.71}$, while

$$\frac{10^{3a}\,K}{\text{cm}^3\,\text{g}^{-1}} = 1.67$$

$$K = 1.67 \times 10^{-3\times(0.71)}\ \text{cm}^3\,\text{g}^{-1} = \boxed{1.2 \times 10^{-2}\ \text{cm}^3\,\text{g}^{-1}}$$

(b) The values of a and K found in part (a) are comparable to those of polystyrene in benzene reported in Table 18.4. The difference may be due to a polymer-swelling effect that makes polystyrene more mobile in benzene than in tetrahydrofuran. This causes the specific viscosity to be slightly smaller in benzene, thereby, lowering the value of K.

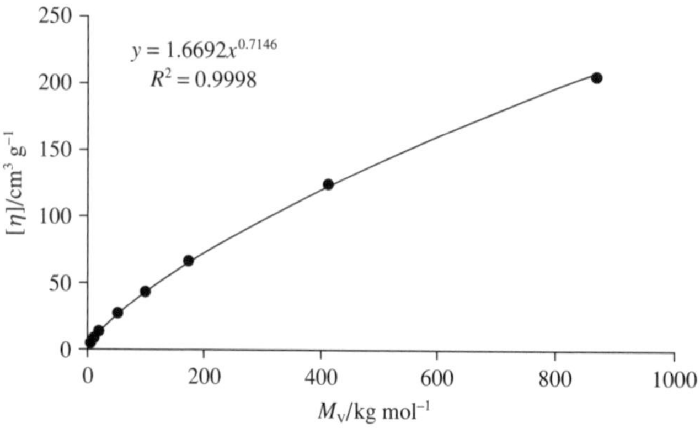

Figure 18.3

P18.11 $$\frac{h}{c} = \frac{RT}{\rho g \overline{M}_n} + \frac{BRT}{\rho g \overline{M}_n^{\,2}} \times c \text{ [see Section 5.5(e), Example 5.4; } M = \overline{M}_n]$$

A plot of h/c against c has an intercept equal to $RT/\rho g\overline{M}_n$ at $c = 0$ and the slope at the intercept equals $BRT/\rho g\overline{M}_n^{\,2}$. Thus, $\overline{M}_n = RT/(\rho g \times \textit{intercept})$ and $B = \textit{slope} \times \rho g\overline{M}_n^{\,2}/RT$. We draw up the following table and prepare the requisite plot shown in Figure 18.4.

$c/(\mathrm{g}/100\ \mathrm{cm}^3)$	0.200	0.400	0.600	0.800	1.00
h/cm	0.48	1.12	1.86	2.76	3.88
$\frac{h}{c}\Big/100\ \mathrm{cm}^4\ \mathrm{g}^{-1}$	2.4	2.80	3.10	3.45	3.88

The plot of Figure 18.4 appears to be linear with a linear regression fit that gives an intercept of $204.\bar{3}\ \mathrm{cm}^4\ \mathrm{g}^{-1}$ and a slope of $1.80\bar{5} \times 10^4\ \mathrm{cm}^7\ \mathrm{g}^{-2}$ (i.e. $1.80\bar{5} \times 10^{-4}\ \mathrm{m}^7\ \mathrm{kg}^{-2}$). It follows that

$$\overline{M}_n = \frac{(8.3145\ \mathrm{J\ K^{-1}\ mol^{-1}}) \times (298\ \mathrm{K})}{(798\ \mathrm{kg\ m^{-3}}) \times (9.8067\ \mathrm{m\ s^{-2}}) \times (204.\bar{3} \times 10^{-5}\ \mathrm{m^4\ kg^{-1}})} = \boxed{155\ \mathrm{kg\ mol^{-1}}}$$

$$B = \frac{(798\ \mathrm{kg\ m^{-3}}) \times (9.8067\ \mathrm{m\ s^{-2}}) \times (155\ \mathrm{kg\ mol^{-1}})^2 \times (1.80\bar{5} \times 10^{-4}\ \mathrm{m^7\ kg^{-2}})}{(8.3145\ \mathrm{J\ K^{-1}\ mol^{-1}}) \times (298\ \mathrm{K})} = \boxed{13.7\ \mathrm{m^3\ mol^{-1}}}$$

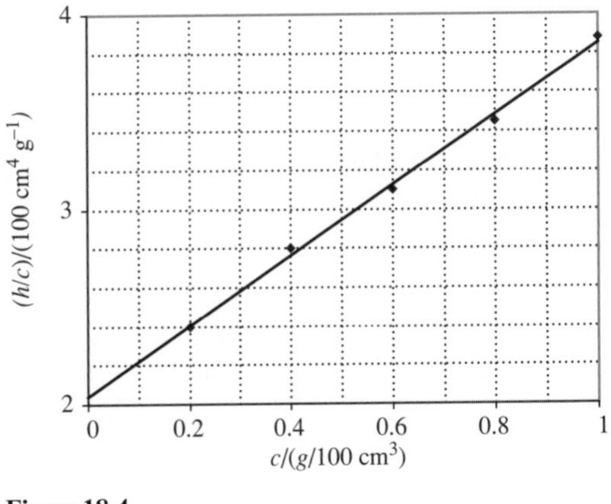

Figure 18.4

Solutions to theoretical problems

P18.13 See the discussion of radius of gyration in Section 18.2(a). For a random coil, $R_g \propto N^{1/2} \propto M^{1/2}$. For a rigid rod, the radius of gyration is proportional to the length of the rod, which is in turn proportional to the number of polymer units, N, and therefore also proportional to M. Therefore, poly(γ-benzyl-L-glutamate) in formamide is rod-like, whereas polystyrene in butanone is a random coil.

P18.15 The probability that the ends of a three-dimensional freely jointed chain lie in the range r to $r + \mathrm{d}r$ is $f(r)\mathrm{d}r$, where

$$f(r) = 4\pi\left(\frac{a}{\pi^{1/2}}\right)^3 r^2 \mathrm{e}^{-a^2r^2} \qquad a = \left(\frac{3}{2Nl^2}\right)^{1/2} \quad [18.2]$$

The mean nth power of the end-to-end separation is

$$\langle r^n \rangle = \int_0^\infty r^n f(r)\mathrm{d}r$$

(a) The square of the root mean square separation of the chain ends is

$$R_{\mathrm{rms}}^2 = \langle r^2 \rangle = \int_0^\infty r^2 f(r)\mathrm{d}r$$

$$= 4\pi\left(\frac{a}{\pi^{1/2}}\right)^3 \int_0^\infty r^4 \mathrm{e}^{-a^2r^2}\mathrm{d}r$$

$$= 4\pi\left(\frac{a}{\pi^{1/2}}\right)^3 \frac{\Gamma\left(\frac{5}{2}\right)}{2a^5}, \quad \text{where } \Gamma\left(\tfrac{5}{2}\right) = \tfrac{3}{4}\pi^{1/2} \text{ [standard integral]}$$

$$= \frac{3}{2a^2} = \frac{3}{2}\left(\frac{2Nl^2}{3}\right) = Nl^2$$

$$R_{\mathrm{rms}} = \langle r^2 \rangle^{1/2} = \boxed{N^{1/2}l}$$

Note: The general form of the above standard integral is

$$\int_0^\infty r^m \mathrm{e}^{-a^2r^2}\mathrm{d}r = \frac{\Gamma[(m+1)/2]}{2a^{m+1}}, \quad \text{where } \Gamma \text{ is the gamma function found in mathematics handbooks}$$

When $N = 4000$ and $l = 154$ pm, $R_{\mathrm{rms}} = 4000^{1/2} \times 154\ \mathrm{pm} = \boxed{9.74\ \mathrm{nm}}$

(b) The mean separation is

$$R_{\mathrm{mean}} = \langle r \rangle = \int_0^\infty r f(r)\mathrm{d}r$$

$$= 4\pi\left(\frac{a}{\pi^{1/2}}\right)^3 \int_0^\infty r^3 \mathrm{e}^{-a^2r^2}\mathrm{d}r$$

$$= 4\pi\left(\frac{a}{\pi^{1/2}}\right)^3 \frac{\Gamma(2)}{2a^4} \quad \text{where } \Gamma(2) = 1 \text{ [standard integral]}$$

$$= \frac{2}{\pi^{1/2}a} = \boxed{\left(\frac{8N}{3\pi}\right)^{1/2} l}$$

When $N = 4000$ and $l = 154$ pm, $R_{\mathrm{mean}} = \left(\frac{8 \times 4000}{3\pi}\right)^{1/2} 154\ \mathrm{pm} = \boxed{8.98\ \mathrm{nm}}$

(c) The most probable separation, R^*, is the value of r for which f is a maximum, so set $\left.\frac{\mathrm{d}f}{\mathrm{d}r}\right]_{r=R^*} = 0$ and solve for R^*.

$$\frac{\mathrm{d}f}{\mathrm{d}r} = 4\pi\left(\frac{a}{\pi^{1/2}}\right)^3 (2r - 2a^2r^3)\mathrm{e}^{-a^2r^2}$$

Evaluation of the slope of f at $r = R^*$ shows that it is zero when

$$1 - a^2R^{*2} = 0$$

$$R^* = \frac{1}{a} = \boxed{\left(\tfrac{2}{3}N\right)^{1/2} l}$$

When $N = 4000$ and $l = 154$ pm, $R^* = \left(\tfrac{2}{3} \times 4000\right)^{1/2} \times 154 \text{ pm} = \boxed{7.95 \text{ nm}}$

P18.17 There are two standard sums found in mathematics handbooks that are used in the following derivation. They are:

$$\sum_{j=1}^{i} j = \tfrac{1}{2}i(i+1) \quad \text{and} \quad \sum_{j=1}^{i} j^2 = \tfrac{1}{6}i(i+1)(2i+1)$$

We begin with the following definition of the radius of gyration:

$$2N^2R_g^2 = \sum_{i=1}^{N}\sum_{j=1}^{N}\langle R_{ij}^2\rangle$$

where $\langle R_{ij}^2\rangle$ is the mean over all conformations of the separation between atoms i and j in a chain of identical atoms linked with the bond length l. In fact, $\langle R_{ij}^2\rangle$ is identical to the end separation of a chain of $|i-j|$ atoms and, consequently, $\langle R_{ij}^2\rangle = |i-j|l^2$ [18.4].

$$2N^2R_g^2/l^2 = \sum_{i=1}^{N}\sum_{j=1}^{N}|i-j|$$

We now break the inner summation of absolute values into two separate summations of positive terms.

$$2N^2R_g^2/l^2 = \sum_{i=1}^{N}\left\{\sum_{j=1}^{i}(i-j) + \sum_{j=i+1}^{N}(j-i)\right\}$$

$$= \sum_{i=1}^{N}\left\{i\sum_{j=1}^{i}1 - \sum_{j=1}^{i}j + \sum_{j=i+1}^{N}j - i\sum_{j=i+1}^{N}1\right\}$$

$$= \sum_{i=1}^{N}\left\{i^2 - \tfrac{1}{2}i(i+1) + \sum_{j=1}^{N}j - \sum_{j=1}^{i}j - i(N-i)\right\}$$

$$= \sum_{i=1}^{N}\{i^2 - \tfrac{1}{2}i(i+1) + \tfrac{1}{2}N(N+1) - \tfrac{1}{2}i(i+1) - i(N-i)\}$$

$$= \sum_{i=1}^{N}\{i^2 - i(N+1) + \tfrac{1}{2}N(N+1)\}$$

$$= \sum_{i=1}^{N}i^2 - (N+1)\sum_{i=1}^{N}i + \tfrac{1}{2}N(N+1)\sum_{i=1}^{N}1$$

$$= \tfrac{1}{6}N(N+1)(2N+1) - \tfrac{1}{2}N(N+1)^2 + \tfrac{1}{2}N^2(N+1)$$

Since $N \gg 1$, $N+1$ effectively equals N.

$$2N^2R_g^2/l^2 = \tfrac{1}{3}N^3 - \tfrac{1}{2}N^3 + \tfrac{1}{2}N^3 = \tfrac{1}{3}N^3$$

$R_g^2 = \tfrac{1}{6}Nl^2$ or $\boxed{R_g = \sqrt{\tfrac{N}{6}}\,l}$, which confirms eqn 18.6

P18.19 A simple procedure is to generate numbers in the range 1 to 8, and to step north for a 1 or 2, east for 3 or 4, south for 5 or 6, and west for 7 or 8 on a uniform grid. One such walk is shown in Figure 18.5. Roughly, the mean and most probable separations would appear to vary as $N^{1/2}$.

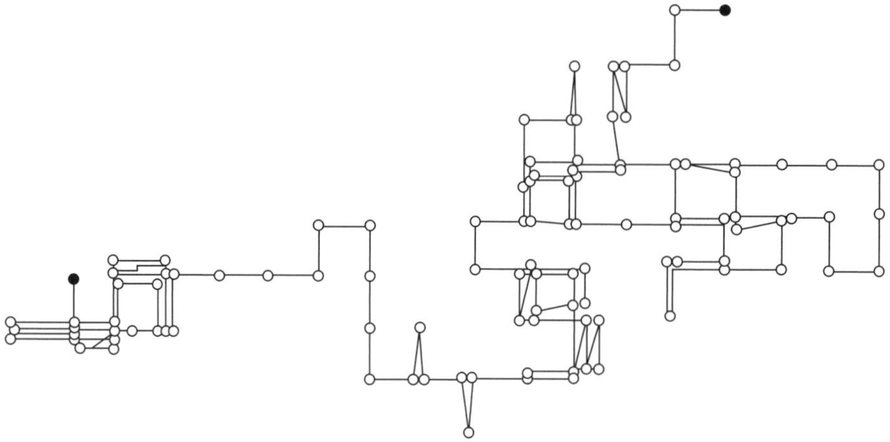

Figure 18.5

P18.21 With concentration c in g dm^{-3} the osmotic pressure equation can be written in the form

$$\Pi = RT\left\{\frac{c}{M} + B \times \left(\frac{c}{M}\right)^2 + \cdots\right\}$$

where B is the osmotic virial coefficient in dm^3 mol^{-1}. The osmotic virial coefficient arises largely from the effect of excluded volume. If we imagine a solution of a macromolecule being built by the successive addition of macromolecules of effective radius a to the solvent, each one being excluded by the ones that preceded it, then B is the excluded volume per mole of molecules. The volume of a molecule is $v_{\text{mol}} = \frac{4}{3}\pi a^3$ but the excluded volume is determined by the smallest distance possible between centres of two molecules, which is $2a$, so the excluded volume is $\frac{4}{3}\pi(2a)^3 = 8v_{\text{mol}}$ for a pair of molecules. The volume excluded per molecule is half of this volume or $4v_{\text{mol}}$. Thus, for an effective radius of $a = \gamma R_g = 0.85R_g$ the osmotic virial coefficient is

$$B = 4N_A v_{\text{mol}} = \frac{16\pi}{3}N_A a^3 = \frac{16\pi}{3}N_A \gamma^3 R_g^3$$

(a) For a freely jointed chain: $R_g = \left(\frac{N}{6}\right)^{1/2} l$ [18.6]

$$B = \boxed{\frac{16\pi}{3}N_A(\gamma l)^3\left(\frac{N}{6}\right)^{3/2}}$$

$$= \frac{16\pi}{3} \times (6.022 \times 10^{23}\ \text{mol}^{-1})\{0.85 \times (154 \times 10^{-12}\ \text{m})\}^3 \times \left(\frac{4000}{6}\right)^{3/2}$$

$$= \boxed{0.38\ \text{m}^3\ \text{mol}^{-1}}$$

(b) For a chain with tetrahedral bond angles: $R_g = \left(\frac{N}{3}\right)^{1/2} l$ [18.9]

$$B = \boxed{\frac{16\pi}{3} N_A (\gamma l)^3 \left(\frac{N}{3}\right)^{3/2}} = \frac{16\pi}{3} \times (6.022 \times 10^{23}\ \text{mol}^{-1})\{0.85 \times (154 \times 10^{-12}\ \text{m})\}^3 \times \left(\frac{4000}{3}\right)^{3/2}$$

$$= \boxed{1.1\ \text{m}^3\ \text{mol}^{-1}}$$

P18.23 Given that $G = U - TS - tl$ and $dU = TdS + tdl$, we take the differential, obtaining

$$dG = dU - TdS - SdT - ldt - tdl$$

$$= TdS + tdl - TdS - SdT - ldt - tdl = \boxed{-SdT - ldt}$$

Since $A = U - TS$, we have $A = G + tl$, so

$$dA = dG + tdl + ldt = -SdT - ldt + tdl + ldt = \boxed{-SdT + tdl}$$

Since dA and dG are both exact differentials

$$\left(\frac{\partial S}{\partial l}\right)_T = -\left(\frac{\partial t}{\partial T}\right)_l \quad \text{and} \quad \left(\frac{\partial S}{\partial t}\right)_T = \left(\frac{\partial l}{\partial T}\right)_t$$

Since $dU = TdS + tdl$ [given],

$$\left(\frac{\partial U}{\partial l}\right)_T = T\left(\frac{\partial S}{\partial l}\right)_T + t = \boxed{-T\left(\frac{\partial t}{\partial T}\right)_l + t} \text{ [Maxwell relation above]}$$

Solutions to applications: biochemistry and technology

P18.25 Molecular mechanics computations with the AMBER force field using the HyperChem package are reported below. The value of the total potential energy will vary between different force fields, as will the shape of the potential energy surface. The local energy minimum at $\phi = -179.6°$ and $\psi = -4.1°$ is found to have a potential energy equal to 28.64 kJ mol^{-1} when R = H. This value is used as a reference to calculate energy differences (ΔE) on the potential energy surface. ΔE values give the relative stability of different conformations with higher values indicating energetically unstable conformations. Similarly, ΔE values were calculated with respect to the local energy minimum at $\phi = -152.3°$ and $\psi = 163.2°$ when R = CH_3.

	Initial		Optimized			
	ϕ/°	ψ/°	ϕ/°	ψ/°	E/(kJ mol^{-1})	ΔE/(kJ mol^{-1})
(a) R = H	75	–65	–176.0	8.3	28.765	0.126
	180	180	180	180	32.156	3.515
	65	35	–179.6	–4.1	28.639	0.000
(b) R = CH_3	75	–65	54.5	19.7	46.338	7.531
	180	180	–152.3	163.2	38.807	0.000
	65	35	52.9	24.1	46.250	7.443

The computations were set up by using the software's 'model build' command, that is, initially setting default values for bond lengths and angles except for the specified initial values of ϕ and ψ. Care must be taken to build the proper chirality at the central carbon when R = CH_3. Then, the constraints were removed, and the entire structure was allowed to relax to a minimum energy. Not all of the initial conformations relaxed to the same final conformation. The different final conformations appear to represent local energy minima. It ought not to be surprising that there are several such minima in even a short peptide chain that contains several nearly free internal rotations. It is instructive to compare the all *trans* ($\phi = \psi = 180°$) initial conformation in the R = H and R = CH_3 cases. In the former, neither angle changes, but the resulting structure is not the lowest-energy structure. In the latter, the methyl group appears to push the planes of the peptide link away from each other (albeit not far) due to steric effects; however, the resulting energy is lower than that of the other conformations examined.

Two of the initial conformations of each molecule converge to the same energy minimum. These energy wells are rather broad and the exact angle at which the computation stops within the minimum depends on details of convergence criteria used in the iterative methodology of the software as well as details of the force field. Both sets of computations also found a second local energy minimum.

An alternative method for studying the energy dependence on ϕ and ψ involves a method like that specified above but with the AMBER computation performed at fixed values of both angles. Figure 18.6 summarizes a set of computations with $-180° < \phi < 180°$ and $\psi = 90°$. To characterize the energy surface, one would carry out similar calculations for several values of ψ.

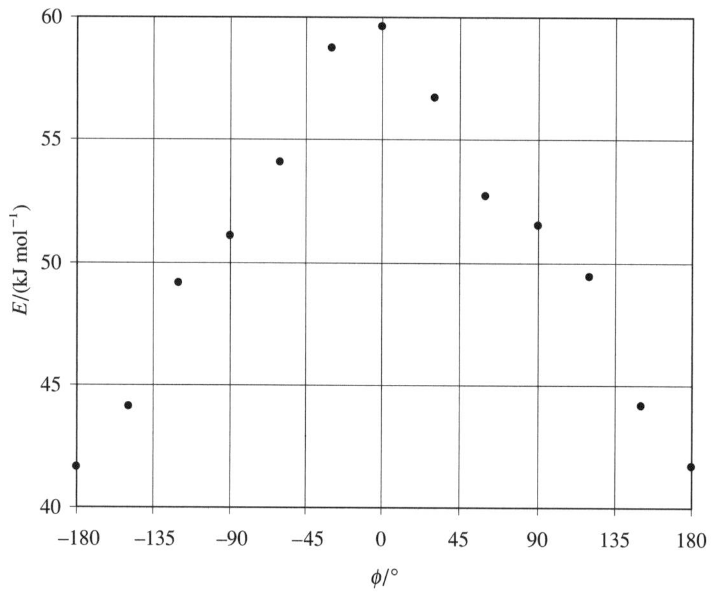

Figure 18.6

P18.27 As demonstrated in P18.3 the radius of gyration of a solid sphere is given by:

(i) $R_g/\text{nm} = 0.046460 \times \{(\upsilon_s/\text{cm}^3\,\text{g}^{-1}) \times (M/\text{g mol}^{-1})\}^{1/3}$

but you may use

(ii) $R_g/\text{nm} = 0.056902 \times \{(v_s/\text{cm}^3\ \text{g}^{-1}) \times (M/\text{g mol}^{-1})\}^{1/3}$

because either of these very similar equations serves as a simple, rough model of a more or less spherical macromolecule. These equations are not expected to do more than coarsely estimate the radius of gyration. If the calculated value approximates the observed value, that is evidence for a spherical-like macromolecule. We present calculated values of both eqns (i) and (ii) in the following table for your convenience only. Comparison of the calculated radius of gyration with the observed suggests that serum albumin and bushy stunt virus resemble solid globular spheres, but DNA does not.

	$M/(\text{g mol}^{-1})$	$v_s/(\text{cm}^3\ \text{g}^{-1})$	$(R_g/\text{nm})_{\text{calc i}}$	$(R_g/\text{nm})_{\text{calc ii}}$	$(R_g/\text{nm})_{\text{expt}}$
Serum albumin	66×10^3	0.752	1.71	2.09	2.98
Bushy stunt virus	10.6×10^6	0.741	9.24	11.3	12.0
DNA	4×10^6	0.556	6.06	7.43	117.0

The large value of the DNA radius of gyration suggests that it may be rod-like. To explore this hypothesis we use the equation for $R_{g,\perp}$ of a solid rod (see P18.3(b)):

$$R_{g,\perp} = \sqrt{\tfrac{1}{12}}\left(\frac{v_s M}{\pi a^2 N_A}\right) \text{[P18.3(b)]}$$

Using DNA data to calculate a according to the rod-like model gives

$$a^2 = \sqrt{\tfrac{1}{12}}\left(\frac{v_s M}{\pi N_A R_{g,\perp}}\right)$$

$$= \sqrt{\tfrac{1}{12}}\left(\frac{(0.556 \times 10^{-6}\ \text{m}^3\ \text{g}^{-1}) \times (4 \times 10^6\ \text{g mol}^{-1})}{\pi N_A (117.0 \times 10^{-9}\ \text{m})}\right) = 2.9 \times 10^{-18}\ \text{m}^2$$

$$a = 1.7\ \text{nm}$$

This is not too terribly different from the known radius of the DNA double strand (~1 nm) so we cannot reject the rod-like hypothesis.

P18.29 Rearrange eqn 18.33 yields

$$\ln c = \text{const.} + \frac{\overline{M_w} b\omega^2 r^2}{2RT}$$

so a plot of $\ln c$ against r^2 should be a straight line of slope $\overline{M_w} b\omega^2/2RT$ from which we find that $\overline{M_w} = \dfrac{2RT \times slope}{b\omega^2}$. We construct the following table and prepare the plot shown in Figure 18.7.

r/cm	5.0	5.1	5.2	5.3	5.4
$c/(\text{mg cm}^{-3})$	0.536	0.284	0.148	0.077	0.039
$r^2/(\text{cm}^2)$	25.0	26.0	27.0	28.1	29.2
$\ln(c/\text{mg cm}^{-3})$	−0.624	−1.259	−1.911	−2.564	−3.244

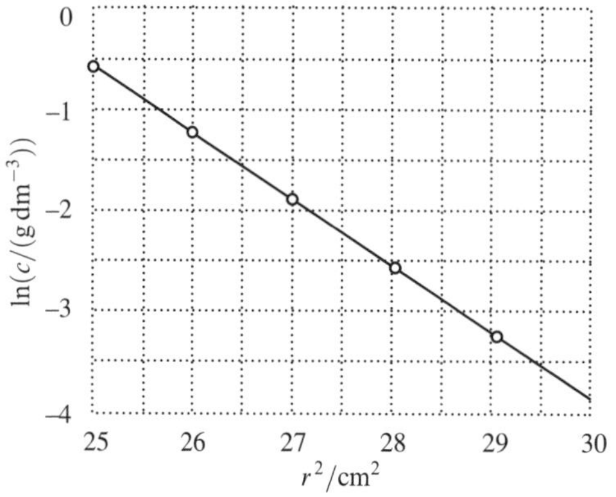

Figure 18.7

The plot of Figure 18.7 is seen to be linear and a linear regression fit of the data gives a slope equal to $-0.623\ \text{cm}^{-2}$.

Therefore,

$$\overline{M_w} = \frac{2RT \times \text{slope}}{b\omega^2} = \frac{2RT \times \text{slope}}{(1-\rho v_s)\omega^2}\ [18.27]$$

$$= \frac{2 \times (8.3145\ \text{J K}^{-1}\ \text{mol}^{-1}) \times (293\ \text{K}) \times (-0.623 \times 10^4\ \text{m}^{-2})}{\{1-(1.001\ \text{g cm}^{-3}) \times (1.112\ \text{cm}^3\ \text{g}^{-1})\} \times (2\pi \times 322\ \text{s}^{-1})^2} = \boxed{65.6\ \text{kg mol}^{-1}}$$

P18.31 $S = \dfrac{s}{r\omega^2}$ [18.29]

Since $s = \dfrac{\mathrm{d}r}{\mathrm{d}t}$, $\dfrac{s}{r} = \dfrac{1}{r}\dfrac{\mathrm{d}r}{\mathrm{d}t} = \dfrac{\mathrm{d}\ln r}{\mathrm{d}t}$

and if we plot ln r against t, the slope gives S through

$$S = \frac{1}{\omega^2}\frac{\mathrm{d}\ln r}{\mathrm{d}t}$$

We prepare the following table and plot the points as shown in Figure 18.8.

t/s	0	300	600	900	1200	1500	1800
r/cm	6.127	6.153	6.179	6.206	6.232	6.258	6.284
ln(r/cm)	1.813	1.817	1.821	1.826	1.830	1.834	1.838

Figure 18.8 shows that the plot is linear with a regression slope of $1.48 \times 10^{-5}\ \text{s}^{-1}$. Thus,

$$S = \frac{(1.48 \times 10^{-5}\ \text{s}^{-1})}{\left(2\pi \times \dfrac{50 \times 10^3}{60\ \text{s}}\right)^2} = 5.40 \times 10^{-13}\ \text{s} \quad \text{or} \quad \boxed{5.40\ \text{Sv}}$$

$$\overline{M_n} = \frac{SfN_A}{b}\ [18.30] = \frac{SfN_A}{1-\rho v_s}\ [18.27]$$

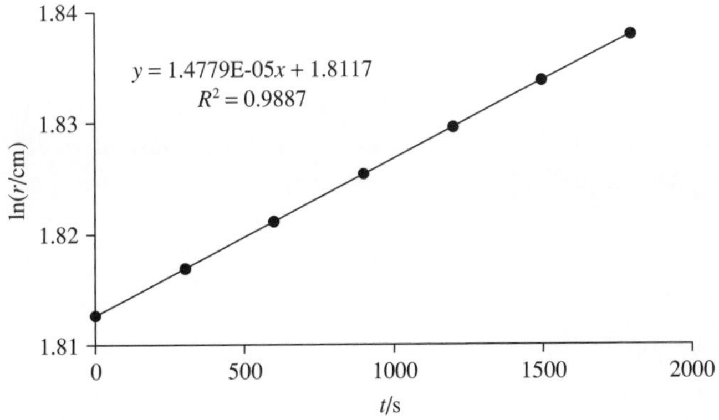

Figure 18.8

Substitution of the Stokes–Einstein relationship, $f = kT/D$ [20.51], gives

$$\overline{M}_n = \frac{SRT}{(1-\rho v_s)D}$$

$$= \frac{(5.40\ \text{Sv}) \times (10^{-13}\ \text{s Sv}^{-1})(8.3145\ \text{J mol}^{-1}\ \text{K}^{-1}) \times (293.15\ \text{K})}{\{1-(998.1\ \text{kg m}^{-3}) \times (0.728 \times 10^{-3}\ \text{m}^3\ \text{kg}^{-1})\} \times (7.62 \times 10^{-11}\ \text{m}^2\ \text{s}^{-1})}$$

$$= \boxed{63.2\ \text{kg mol}^{-1}}$$

Information about the molecular geometry is inferred by calculation of the ratio of the actual frictional coefficient, f, of the macromolecule to that of the frictional coefficient, f_0, of a sphere of the same volume. Interpolation between the ratios given in Table 18.3 provides the dimensions of the molecular ellipsoid.

$$f = \frac{kT}{D} = \frac{(1.381 \times 10^{-23}\ \text{J K}^{-1}) \times (293\ \text{K})}{7.62 \times 10^{-11}\ \text{m}^2\ \text{s}^{-1}} = 5.31 \times 10^{-11}\ \text{kg s}^{-1}$$

$$V_m = (0.728\ \text{cm}^3\ \text{g}^{-1}) \times (63.2 \times 10^3\ \text{g mol}^{-1}) = 46.0 \times 10^3\ \text{cm}^3\ \text{mol}^{-1} = 4.60 \times 10^{-2}\ \text{m}^3\ \text{mol}^{-1}.$$

Then, the effective sphere radius is

$$a = \left(\frac{3V_m}{4\pi N_A}\right)^{1/3} = \left(\frac{3 \times (4.60 \times 10^{-2}\ \text{m}^3\ \text{mol}^{-1})}{4\pi \times (6.022 \times 10^{23}\ \text{mol}^{-1})}\right)^{1/3} = 2.63\ \text{nm}$$

$$f_0 = 6\pi a\eta = 6\pi \times (2.63 \times 10^{-9}\ \text{m}) \times (1.00 \times 10^{-3}\ \text{kg m}^{-1}\ \text{s}^{-1}) = 4.96 \times 10^{-11}\ \text{kg s}^{-1}$$

Thus,

$$\frac{f}{f_0} = \frac{5.31}{4.96} = 1.07$$

Interpolation for the Table 18.3 value of 1.07 reveals that the a/b ellipsoidal ratio of major–minor axes is about 2.3 for both the prolate and oblate ellipsoids. So, it appears that the protein is an ellipsoid with $a/b = 2.3$. We cannot, however, determine whether it is a prolate or an oblate ellipsoid.

P18.33 (a) $S/k = \ln(W)$ and $\Delta S/k = \ln(W_{\text{circular}}) - \ln(W_{\text{ideal chain}})$, where k is the Boltzmann constant.

W_{circular} is the configuration weight of the DNA molecule that has joined ends ($n = 0$), while $W_{\text{ideal chain}}$ is the configuration weight for the molecule chain for which no segment has a constraint and two possible configurations (right-pointing and left-pointing, see text *Further information 18.1(a)*). $W_{\text{ideal chain}}$ for a molecule of N segments equals 2^N.

$$W_{\text{circular}} = \frac{N!}{(N/2)!(N/2)!}$$

$$\begin{aligned}\ln(W_{\text{circular}}) &= \ln(N!) - 2\ln\{(N/2)!\}\\ &= \ln(2\pi)^{1/2} + \left(N + \frac{1}{2}\right)\ln(N) - N\\ &\quad - 2\left\{\ln(2\pi)^{1/2} + \left(\frac{N}{2} + \frac{1}{2}\right)\ln\left(\frac{N}{2}\right) - \frac{N}{2}\right\} \quad \text{(Stirling's approx.)}\\ &= \ln\left\{2^N\left(\frac{2}{\pi N}\right)^{1/2}\right\}\end{aligned}$$

$$\Delta S/k = \ln\left\{2^N\left(\frac{2}{\pi N}\right)^{1/2}\right\} - \ln(2^N)$$

$$\Delta S/k = \ln\left(\frac{2}{\pi N}\right)^{1/2}$$

Since $2/\pi N < 1$, entropy decreases in forming the closed circular (cc)DNA. The following graph, Figure 18.9, shows the dependence of ΔS upon N.

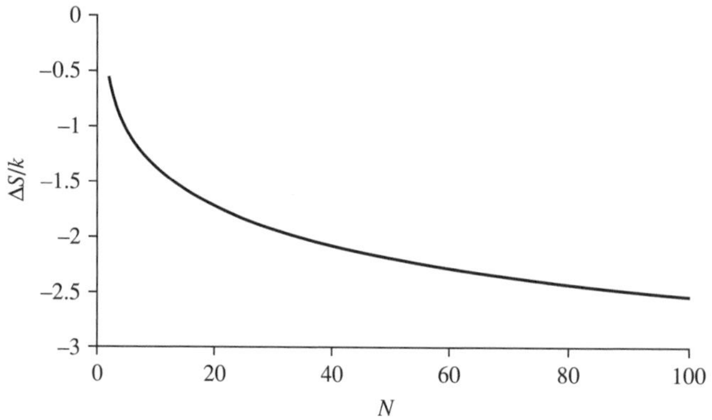

Figure 18.9

(b) (i) A continuous, normalized Gaussian function, which is also called the normal distribution or the bell-curve, has the form: $f(x) = (1/2\pi\sigma^2)^{1/2}e^{(x-\langle x\rangle)^2/2\sigma^2}$, where $-\infty < x < \infty$. $\langle x\rangle$ is the mean value of x and σ is the standard deviation. It can be shown that $\sigma^2 = \langle x^2\rangle - \langle x\rangle^2$.

The discrete energy distribution for a twisted ccDNA molecule is:

$$p_i = e^{-\beta\varepsilon_i}/q = e^{-\beta\kappa i^2}/q$$

where κ is an empirical constant and $i = 0, \pm1, \pm2, \ldots$ Because it has an exponent in i^2, which is comparable to x^2 in the above Gaussian function, the energy distribution has the form of a Gaussian function with a maximum value that is centred on $i = 0$. The standard deviation of the discrete Gaussian distribution is found by comparing the two equations. It is $\sigma = (1/2\beta\kappa)^{1/2} = (kT/2\kappa)^{1/2}$.

(ii) The following MathCad worksheet plots (Figure 18.10) the energy distribution at several values of the unitless temperature $T_{ratio} = 1/\beta\kappa = kT/\kappa$. Bar plots (histograms), shown in Figure 18.11, are appropriate for discrete distributions in which the argument takes on specific values only ($i = 0, \pm1, \pm2, \pm3, \ldots$). Even though the argument is not defined for non-integer values of i, Figure 18.11

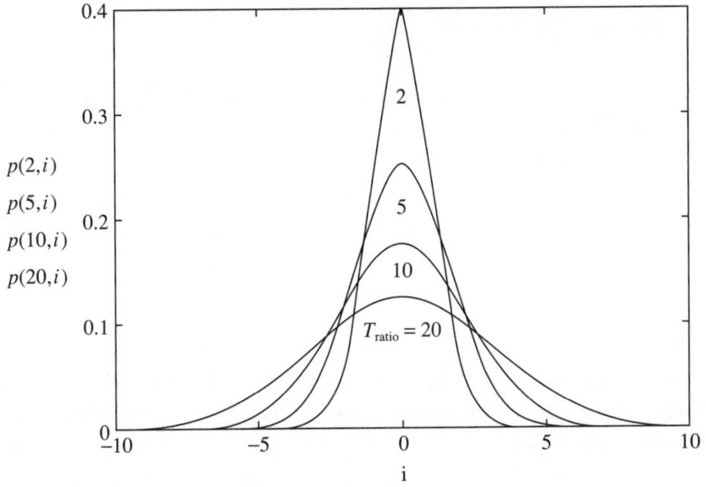

Figure 18.10

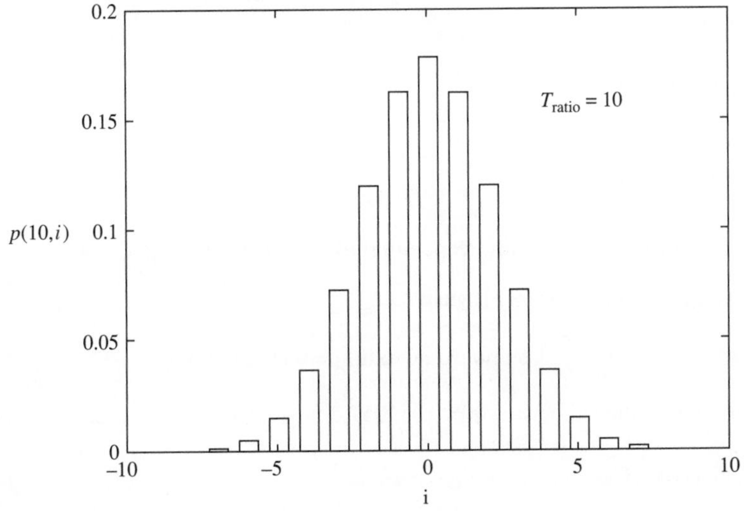

Figure 18.11

presents line plots with the understanding of discrete values. This reduces the visual confusion of overlapping bars from multiple plots. The plots show that at higher temperatures there are fewer molecules in the lowest energy state $i = 0$ and a greater number of molecules in the high-energy states.

MathCad plotting program:

$$N_{max} := 10 \qquad i := -N_{max} \ldots N_{max}$$

$$q(T_{ratio}) := \sum_{i=-75}^{75} e^{-i^2/T_{ratio}} \qquad p(T_{ratio}, i) := \frac{e^{-i^2/T_{ratio}}}{q(T_{ratio})}$$

(iii) Suppose that the ratio of kT to κ is 4. The desired probabilities are easily provided by the above MathCad program. They are:

$p(4, 0) = 0.282 \qquad p(4, 1) = 0.22$

$p(4, 5) = 5.446 \times 10^{-4} \qquad p(4, 10) = 3.918 \times 10^{-12}$

P18.35 (a) The increase in transition temperatures with the hydrophobic chain length is a result of the increased strength of the van der Waals interaction between long saturated portions of the chains that can interlock well with each other. The introduction of double bonds in the chains can affect the interlocking of the parallel chains by putting kinks in the chains, thereby decreasing the strength of the van der Waals interactions between chains. Double bonds can be either *cis* or *trans*. Only *cis* double bonds produce a kink, but most fatty acids are the *cis*-isomer. So, we expect that the transition temperatures will decrease in rough proportion to the number of C=C bonds.

(b) The addition of cholesterol is expected to increase the temperature of the transition from the liquid crystalline state to the liquid state by altering the conformations of the hydrocarbon chains. Cholesterol stabilizes extended chain conformations of adjacent hydrocarbon sections by van der Waals interactions relative to the coiled conformations that predominate when cholesterol is absent. The extended chains can pack better than coiled arrangements. However, the transition temperature for the melting of the solid crystalline state into the liquid-crystalline form is probably decreased on addition of cholesterol; its presence prevents the hydrophobic chains from freezing into a solid array by disrupting their packing. This will also spread the melting point over a range of temperatures.

P18.37 (a) $\dfrac{\eta}{\eta^*} = \dfrac{t}{t^*} \approx 1 + [\eta]c + k'[\eta]^2 c^2$

Define $F = \dfrac{t/t^* - 1}{c} = [\eta] + k'[\eta^2]c$

A linear regression of F against c yields an intercept equal to $[\eta]$ and a slope equal to $k'[\eta]^2$.

(1) In toluene: linear regression ($R = 0.99954$) yields

$[\eta] = 0.08\overline{566}\ \text{dm}^3\,\text{g}^{-1} = \boxed{0.086\ \text{dm}^3\,\text{g}^{-1}}$; standard deviation $= 0.00020\ \text{dm}^3\,\text{g}^{-1}$

$k'[\eta]^2 = 0.002\overline{688}\ \text{dm}^6\,\text{g}^{-2}$; standard deviation $= 0.000057\ \text{dm}^6\,\text{g}^{-2}$

$$k' = \frac{0.002\overline{688}\ \text{dm}^6\,\text{g}^{-2}}{\left(0.08\overline{566}\ \text{dm}^3\,\text{g}^{-1}\right)^2} = \boxed{0.37}$$

(2) In cyclohexane: linear regression ($R = 0.98198$) yields

$$[\eta] = 0.041\overline{50}\ \mathrm{dm^3\,g^{-1}} = \boxed{0.042\ \mathrm{dm^3\,g^{-1}}};\ \text{standard deviation} = 0.00018\ \mathrm{dm^3\,g^{-1}}$$

$$k'[\eta]^2 = 0.006\overline{001}\ \mathrm{dm^6\,g^{-2}};\ \text{standard deviation} = 0.000116\ \mathrm{dm^6\,g^{-2}}$$

$$k' = \frac{0.0006\overline{001}\ \mathrm{dm^6\,g^{-2}}}{(0.041\overline{50}\ \mathrm{dm^3\,g^{-1}})^2} = \boxed{0.35}$$

(b) $[\eta] = K\overline{M}_\mathrm{v}^{\,a}$ or $\overline{M}_\mathrm{v} = \left(\dfrac{[\eta]}{K}\right)^{1/a}$

(1) In toluene

$$\overline{M}_\mathrm{v} = \left(\frac{0.085\overline{66}\ \mathrm{dm^2\,g^{-1}}}{1.15\times10^{-5}\ \mathrm{dm^3\,g^{-1}}}\right)^{(1/0.72)}\ \mathrm{g\,mol^{-1}} = \boxed{2.4\times10^2\ \mathrm{kg\,mol^{-1}}}$$

(2) In cyclohexane

$$\overline{M}_\mathrm{v} = \left(\frac{0.041\overline{50}\ \mathrm{dm^3\,g^{-1}}}{8.2\times10^{-5}\ \mathrm{dm^3\,g^{-1}}}\right)^{(1/(1/2))}\ \mathrm{g\,mol^{-1}} = \boxed{2.6\times10^2\ \mathrm{kg\,mol^{-1}}}$$

(c) $[\eta]/(\mathrm{dm^3\,g^{-1}}) = \phi(r_\mathrm{rms}/\mathrm{m})^3/M,\ \phi = 2.84\times10^{26}$

$$r_\mathrm{rms} = \left(\frac{[\eta]M}{\phi}\right)^{1/3}\ \mathrm{m},\quad \text{where}\quad r_\mathrm{rms} = \langle r^2\rangle^{1/2}$$

(1) In toluene

$$r_\mathrm{rms} = \left(\frac{0.085\overline{66}\times2.3\overline{9}\times10^5}{2.84\times10^{26}}\right)^{1/3}\ \mathrm{m} = \boxed{42\ \mathrm{nm}}$$

(2) In cyclohexane

$$r_\mathrm{rms} = \left(\frac{0.041\overline{50}\times2.5\overline{6}\times10^5}{2.84\times10^{26}}\right)^{1/3}\ \mathrm{m} = \boxed{33\ \mathrm{nm}}$$

(d) $M(\text{styrene}) = 104\ \mathrm{g\,mol^{-1}}$

Average number of monomeric units, $\langle n\rangle$ is

$$\langle n\rangle = \frac{\overline{M}_\mathrm{v}}{M(\text{styrene})}$$

(1) In toluene

$$\langle n\rangle = \frac{2.3\overline{9}\times10^5\ \mathrm{g\,mol^{-1}}}{104\ \mathrm{g\,mol^{-1}}} = \boxed{2.3\times10^3}$$

(2) In cyclohexane

$$\langle n\rangle = \frac{2.5\overline{6}\times10^5\ \mathrm{g\,mol^{-1}}}{104\ \mathrm{g\,mol^{-1}}} = \boxed{2.5\times10^3}$$

(e) Consider the geometry in Figure 18.12.

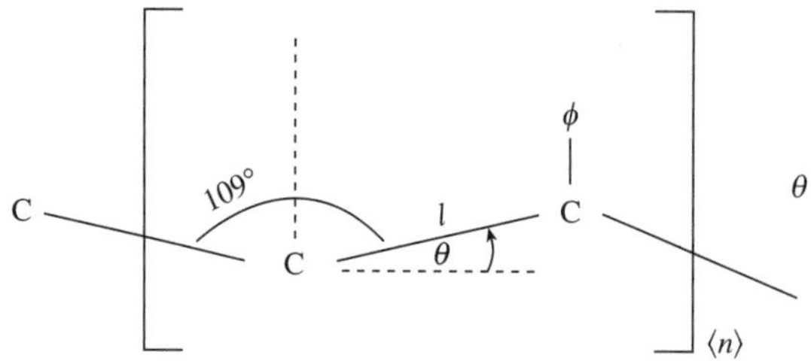

Figure 18.12

For a polymer molecule consisting of $\langle n \rangle$ monomers, the maximum molecular length, L_{max}, is

$$L_{\text{max}} = 2l\langle n \rangle \cos\theta = 2(0.154\text{ nm})\langle n \rangle \cos 35° = (0.2507\text{ nm})\langle n \rangle$$

(1) In toluene

$$L_{\text{max}} = (0.2507\text{ nm}) \times (2.3\bar{0} \times 10^3) = \boxed{5.8 \times 10^2\text{ nm}}$$

(2) In cyclohexane

$$L_{\text{max}} = (0.2507\text{ nm}) \times (2.4\bar{6} \times 10^3) = \boxed{6.2 \times 10^2\text{ nm}}$$

(f) $R_g = \left(\dfrac{\langle n \rangle}{3}\right)^{1/2} l = (0.0889\text{ nm})\langle n \rangle^{1/2}$

Kirkwood–Riseman: $r_{\text{rms}}^{\text{KR}} = \left(\dfrac{[\eta]M}{\phi}\right)^{1/3} = \left(\dfrac{[\eta]M}{2.84 \times 10^{26}}\right)^{1/3}$

Constrained tetrahedral chain: $r_{\text{rms}} = (2\langle n \rangle)^{1/2}\, l$ [18.9]

or unconstrained chain: $r_{\text{rms}} = \langle n \rangle^{1/2} l$ [18.4]

Solvent	$\langle n \rangle$	R_g/nm	$r_{\text{rms}}^{\text{KR}}$/nm	r_{rms}/nm
Toluene	$2.3\bar{0} \times 10^3$	$\boxed{4.3}$	42	$\boxed{10.\bar{4}\text{ or }7.4}$
Cyclohexane	2.5×10^3	$\boxed{4.4}$	33	$\boxed{10.\bar{8}\text{ or }7.6}$

(g) The manufacturer's claim that the polystyrene beads have an average molar mass of 250 kg mol^{-1} is consistent with the values of $\overline{M}_v$ found in part (b). Differences may be due to the use of a different solvent by the manufacturer or to the use of a non-viscosity technique for the molar mass measurement. The manufacturer's claim appears to be valid.

19 Materials 2: solids

Answers to discussion questions

D19.1 A **space lattice** is the three-dimensional structural pattern formed by lattice points representing the locations of motifs that may be atoms, molecules, or groups of atoms, molecules, or ions within a crystal. All points of the space lattice have identical environments and they define the crystal structure. The **unit cell** is an imaginary parallelepiped from which the entire crystal structure can be generated, without leaving gaps, using translations of the unit cell alone. Each unit cell is defined in terms of lattice points. The smallest possible unit cell is called the **primitive unit cell**. **Non-primitive unit cells** may exhibit lattice points within the cell, at the cell centre, on cell faces, or on cell edges.

D19.3 We can use the Debye–Scherrer powder diffraction method, follow the procedure of Sections 19.3(b)–(d), and in particular look for systematic absences in the diffraction patterns. We can proceed through the following sequence:

1. Measure distances of the lines in the diffraction pattern from the centre.
2. From the known radius of the camera, convert the distances to angles.
3. Calculate $\sin^2\theta$.
4. Find the common factor $A = \lambda^2/4a^2$ in $\sin^2\theta = (\lambda^2/4a^2)(h^2 + k^2 + l^2)$.
5. Index the lines using $\sin^2\theta/A = h^2 + k^2 + l^2$.
6. Look for the systematic absences in (hkl). For body-centred cubic (bcc), diffraction lines corresponding to $h + k + l$ that are odd will be absent (see text Figure 19.23). For face-centred cubic (fcc), only lines for which h, k, and l are either all even or all odd will be present, others will be absent.
7. Solve $A = \lambda^2/4a^2$ for a.

D19.5 There are two, large-angle, meridional reflection spots in the X-ray diffraction pattern obtained from parallel fibres of B-DNA (text Figure 19.42). They are reflections off the relatively close planes (340 pm) of nucleotide bases that run perpendicular to the axis of the molecule (text Figure 19.44). Between these meridional spots is an X-shaped distribution of reflection spots that originates from planes of longer periodicity (3400 pm; text Figure 19.43). The pitch p of the helix, the distance d between parallel, reflective planes, and the angle α from the meridional of the spots of the X-pattern are shown in Figure 19.1. The X-rays approach the helix on a path almost perpendicular to the molecular axis. Regions of high electron density act like reflective planes with each coil of the helix providing one plane that reflects at a clockwise angle and another that reflects at a counter-clockwise angle. A set of parallel planes that reflect the X-rays at the angle α is shown in Figure 19.1; one of the set of parallel planes that reflect the X-rays at the angle $-\alpha$ is also shown.

Reflections from each set of parallel planes constructively interfere to produce one of the legs of the crossed spot patterns seen at the centre of text Figure 19.42. Constructive interference requires that $d = p\cos\alpha$. The sequence of spots outward along a leg corresponds to the first-, second-,... order diffraction ($n = 1, 2, \ldots$ where by Bragg's law $n\lambda = 2d\sin\theta$ [19.4]). Examination of Figure 19.1 reveals that $\boxed{\alpha \text{ increases as the pitch increases}}$.

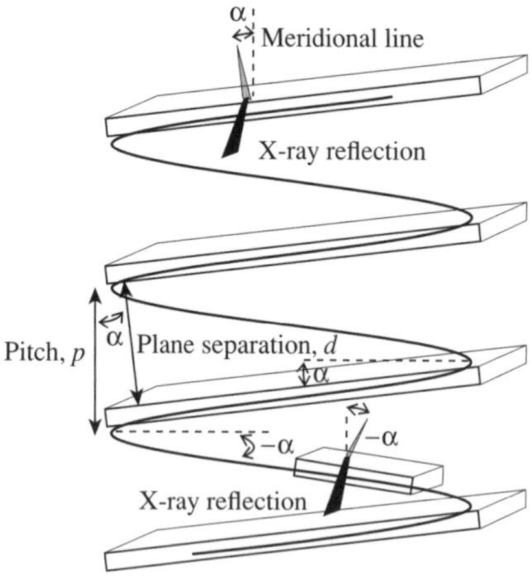

Figure 19.1

D19.7 The **structure factor** F_{hkl} is the sum over all j atoms of terms each of which has a scattering factor f_j:

$$F_{hkl} = \sum_j f_j \mathrm{e}^{\mathrm{i}\phi_{hkl}(j)}, \quad \text{where} \quad \phi_{hkl}(j) = 2\pi(hx_j + ky_j + lz_j)\ [19.7]$$

The importance of the structure factor to the X-ray crystallographic method of structure determination is its **Fourier synthesis** relationship to the electron density distribution, $\rho(r)$, within the crystal:

$$\rho(\mathbf{r}) = \frac{1}{V}\sum_{hkl} F_{hkl}\mathrm{e}^{-2\pi\mathrm{i}(hx+ky+lz)}\ [19.8]$$

Eqn 19.8 reveals that, if the structure factors for the lattice planes can be measured, the electron density distribution can be calculated by performing the indicated sum. Therein lays the **phase problem**. Measurement detectors yield only the intensity of scattered radiation, which is proportional to $|F_{hkl}|^2$, and give no direct information about F_{hkl}. To see this, consider the structure factor form $F_{hkl} = |F_{hkl}|\mathrm{e}^{\mathrm{i}\alpha_{hkl}}$, where α_{hkl} is the phase of the hkl reflection plane. Then,

$$|F_{hkl}|^2 = \{|F_{hkl}|\mathrm{e}^{\mathrm{i}\alpha_{hkl}}\}^*\{|F_{hkl}|\mathrm{e}^{\mathrm{i}\alpha_{hkl}}\} = |F_{hkl}| \times |F_{hkl}|\mathrm{e}^{-\mathrm{i}\alpha_{hkl}}\mathrm{e}^{\mathrm{i}\alpha_{hkl}} = |F_{hkl}| \times |F_{hkl}|$$

and we see that all information about the phase is lost in an intensity measurement. It seems impossible to perform the sum of eqn 19.8 since we do not have the important factor $\mathrm{e}^{\mathrm{i}\alpha_{hkl}}$. Crystallographers have developed numerous methods to resolve the phase problem. In the **Patterson synthesis**, X-ray diffraction spot intensities are used to acquire separation and relative orientations of atom pairs. Another method uses the dominance of heavy-atom scattering to deduce phase. **Heavy-atom replacement** may be necessary for this type of application. **Direct methods** dominate modern X-ray diffraction analysis. These methods use statistical techniques, and the considerable computational capacity of the modern computer, to compute the probabilities that the phases have a particular value.

D19.9 Unit cells of ionic solids must be electrically neutral. Consequently, it is impossible to achieve the 12-coordinate close-packed ionic structure. The most efficient use of space in an ionic solid is the (8,8)-coordinate **caesium-chloride structure** in which each cation is surrounded by eight anions (text Figure 19.32) and each anion is surrounded by eight cations. Ionic compounds typically exhibit the caesium-chloride structure when the **radius ratio**, $\gamma = r_{smaller}/r_{larger}$ [19.12], is greater than 0.732. Using the radii of Table 19.3 we find that for CsCl $\gamma = 167/181 = 0.923$. In this structure, an ion of one charge occupies the centre of a cubic unit cell with eight counter ions at its corners. The caesium-chloride structure may be viewed as an expanded cubic P array of anions in which cations occupy cubic holes. Similarly, the structure may be viewed as an expanded cubic P array of cations in which anions occupy cubic holes.

When the radii of ions differ to the extent that $0.414 < \gamma < 0.732$ the (6,6)-coordinate **rock-salt structure** typified by NaCl is observed. Illustrated in text Figure 19.33, the structure can be pictured as consisting of two interpenetrating slightly expanded cubic F (fcc) arrays, one composed of cations and the other of anions. Emphasizing the expanded fcc array of bulky Cl^- anions, we see that the cations occupy octahedral holes as each cation is surrounded by six anion nearest neighbours in an octahedral arrangement. Because a similar arrangement exists for the anions within the expanded fcc array of Na^+ cations, the lattice structure is said to be (6,6) coordinated.

See S.-M. Ho and B. E. Douglas, *Journal of Chemical Education* **46**, 208, 1969, for additional discussion and diagrams.

D19.11 The **Fermi–Dirac distribution** is a version of the Boltzmann distribution that takes into account the effect of the Pauli exclusion principle. It can therefore be used to calculate the population, P, of a state of given energy in a many-electron system at a temperature T:

$$P = \frac{1}{e^{(E-\mu)/kT} + 1} \quad [19.23]$$

In this expression, μ is the **Fermi energy**, or **chemical potential**, the energy of the level for which $P = \frac{1}{2}$. The Fermi energy should be distinguished from the Fermi level, which is the energy of the highest occupied state at $T = 0$ (see text Figure 19.51). For energies well above μ, the population resembles a Boltzmann distribution, decaying exponentially with increasing energy. The higher the temperature, the longer the exponential tail.

From thermodynamics we know that $\mathrm{d}U = -p\,\mathrm{d}V + T\mathrm{d}S + \mu\,\mathrm{d}n$ for a one-component system. This may also be written $\mathrm{d}U = -p\,\mathrm{d}V + T\mathrm{d}S + \mu\,\mathrm{d}N$, and this μ is the chemical potential per particle that appears in the F–D distribution law. The term in $\mathrm{d}U$ containing μ is the chemical work and gives the change in internal energy with change in the number of particles. Thus, μ has a wider significance than its interpretation as a partial molar Gibbs energy and it is not surprising that it occurs in the F–D expression in comparison to the energy of the particle. The Helmholtz energy, A, and μ are related through $\mathrm{d}A = -p\,\mathrm{d}V + S\mathrm{d}T + \mu\,\mathrm{d}N$, and so μ also gives the change in the Helmholtz energy with change in number of particles. To fully understand how the chemical potential μ enters into the F–D expression for P, we must examine its derivation, which makes use of the relationship between μ and A and of that between A and the partition function for F–D particles.

D19.13 The most obvious difference is that there is no magnetic analogue of electric charge; hence, there are no magnetic 'ions.' Both electric and magnetic moments exist and these can be either permanent or induced. Induced magnetic moments in the entire sample can be either parallel or antiparallel to the applied field producing them (paramagnetic or diamagnetic moments), whereas in the electric case they are always parallel. Magnetization, $\mathcal{M}$, is the analog of polarization, $\mathcal{P}$. Although both magnetization and induced dipole moment are proportional to the fields producing them, they are not analogous quantities, neither are volume magnetic susceptibility, χ, and electric polarizability, α. The magnetic quantities refer to the sample as a whole, the electric quantities to the molecules. Molar magnetic susceptibility is analogous to molar polarization, as can be seen by comparing eqns 19.32 and 17.15, and magnetizability is analogous to electric polarizability.

Solutions to exercises

E19.1(a) There are four equivalent lattice points in the fcc unit cell. One way of choosing them is shown by the positions of the Cl^- ions in either Figure 19.2 below or in text Figure 19.22. The three lattice points equivalent to $(\frac{1}{2}, 0, 0)$ are $\boxed{(1, \frac{1}{2}, 0)}$, $\boxed{(1, 0, \frac{1}{2})}$, and $\boxed{(\frac{1}{2}, \frac{1}{2}, \frac{1}{2})}$. Figure 19.2 shows the location of the atoms in the fcc unit cell of NaCl. The tinted circles are Na^+; the open circles are Cl^-.

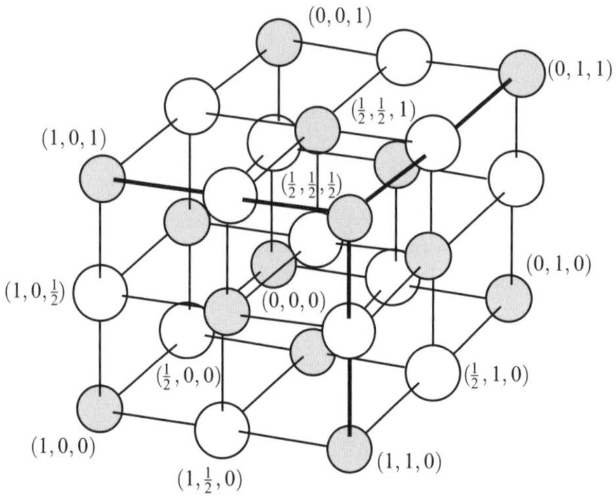

Figure 19.2

COMMENT. The positions of the other Cl^- ions in Figure 19.2 do not correspond to lattice points of the unit cell shown, as they are generated by full unit cell translations, and hence belong to neighbouring unit cells.

Question. What Na^+ positions define the unit cell of NaCl in Figure 19.2? What lattice points are equivalent to (0, 0, 0)?

E19.2(a) The planes are sketched in Figure 19.3. Expressed in multiples of the unit cell distances the planes are labelled (2, 3, 2) and (2, 2, ∞). Their Miller indices are the reciprocals of these multiples with all fractions cleared, thus

$$(2, 3, 2) \rightarrow (\tfrac{1}{2}, \tfrac{1}{3}, \tfrac{1}{2}) \rightarrow (3, 2, 3) \quad \text{[multiply by the lowest common denominator 6]}$$

$$(2, 2, \infty) \rightarrow (\tfrac{1}{2}, \tfrac{1}{2}, 0) \rightarrow (1, 1, 0) \quad \text{[multiply by the lowest common denominator 2]}$$

Dropping the commas, the planes are written $\boxed{(323) \text{ and } (110)}$.

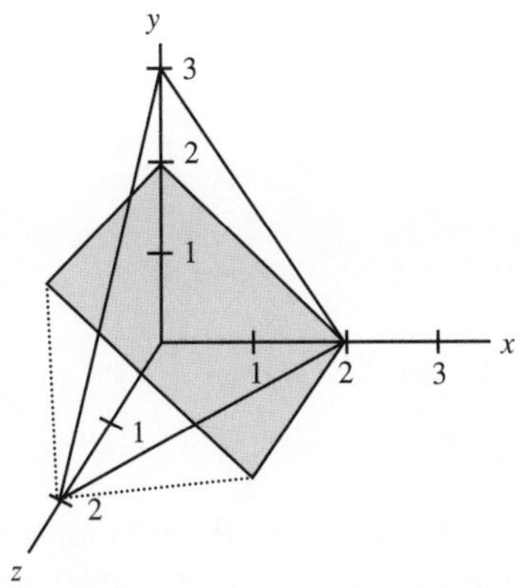

Figure 19.3

E19.3(a) The distance between planes in a cubic lattice is $d_{hkl} = \dfrac{a}{(h^2+k^2+l^2)^{1/2}}$ [19.2].

$$d_{111} = \frac{432\ \text{pm}}{(1^2+1^2+1^2)^{1/2}} = \boxed{249\ \text{pm}}$$

$$d_{211} = \frac{432\ \text{pm}}{(2^2+1^2+1^2)^{1/2}} = \boxed{176\ \text{pm}}$$

$$d_{100} = \frac{432\ \text{pm}}{(1^2+0^2+0^2)^{1/2}} = \boxed{432\ \text{pm}}$$

E19.4(a) $\lambda = 2d\sin\theta$ [19.5] $= 2\times(99.3\ \text{pm})\times\sin 20.85° = \boxed{70.7\ \text{pm}}$

E19.5(a) Refer to Figure 19.23 of the text. Systematic absences in the bcc lattice correspond to $h+k+l=$ odd. Hence, the first three lines are from planes (110), (200), and (211).

$$\sin\theta_{hkl} = \frac{\lambda}{2d_{hkl}}\ [19.5] \quad \text{and} \quad d_{hkl} = \frac{a}{(h^2+k^2+l^2)^{1/2}}\ [19.2], \text{ so}$$

$$\sin\theta_{hkl} = (h^2+k^2+l^2)^{1/2}\times\left(\frac{\lambda}{2a}\right)$$

In a bcc unit cell, the body diagonal of the cube is $4R$ where R is the atomic radius. The relationship of the side of the unit cell to R is therefore (using the Pythagorean theorem twice)

$$(4R)^2 = a^2 + 2a^2 = 3a^2 \quad \text{or} \quad a = \frac{4R}{3^{1/2}}$$

This can be seen from Figure 19.4.

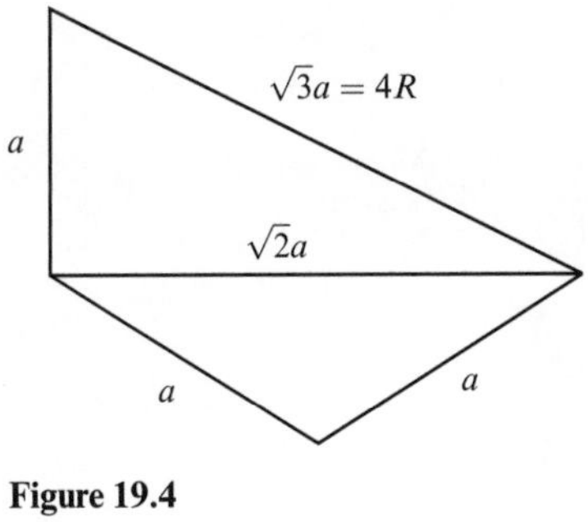

Figure 19.4

$$a = \frac{4 \times 126\text{ pm}}{3^{1/2}} = 291\text{ pm}$$

$$\frac{\lambda}{2a} = \frac{58\text{ pm}}{2 \times (291\text{ pm})} = 0.099\overline{7}$$

$$\sin\theta_{110} = \sqrt{2} \times (0.099\overline{7}) = 0.14\overline{1} \qquad 2\theta_{110} = \boxed{16°}$$

$$\sin\theta_{200} = 2 \times (0.099\overline{7}) = 0.19\overline{9} \qquad 2\theta_{200} = \boxed{23°}$$

$$\sin\theta_{211} = \sqrt{6} \times (0.099\overline{7}) = 0.24\overline{4} \qquad 2\theta_{211} = \boxed{28°}$$

E19.6(a) In a circular camera, the distance between adjacent lines is $D = R\Delta(2\theta)$, where R is the radius of the camera (distance from sample to film) and θ is the diffraction angle in radians. Combining these quantities with the Bragg law ($\lambda = 2d\sin\theta$ [19.5], relating the glancing angle to the wavelength and separation of planes), we get

$$D = 2R\Delta\theta = 2R\Delta\left(\sin^{-1}\frac{\lambda}{2d}\right)$$

$$= 2(5.74\text{ cm}) \times \left(\sin^{-1}\frac{154.051\text{ pm}}{2(77.8\text{ pm})} - \sin^{-1}\frac{154.433\text{ pm}}{2(77.8\text{ pm})}\right) = \boxed{0.214\text{ cm}}$$

E19.7(a) *Justification 19.1* demonstrates that the scattering factor in the forward direction equals the number of electrons in the atom or simple ion. Consequently, $\boxed{f_{\text{Br}^-} = 36}$.

E19.8(a) A tetragonal unit cell, as shown in Figure 19.8 of the text, has $a = b \neq c$. Therefore,

$$V = (651\text{ pm}) \times (651\text{ pm}) \times (934\text{ pm}) = 3.96 \times 10^{-28}\text{ m}^3 = \boxed{0.396\text{ nm}^3}$$

E19.9(a)

$$\rho = \frac{\text{mass of unit cell}}{\text{volume of unit cell}} = \frac{m}{V}$$

$$m = nM = \frac{N}{N_A}M \text{ [}N\text{ is the number of formula units per unit cell]}$$

Then, $\rho = \dfrac{NM}{VN_A}$

and $N = \dfrac{\rho V N_A}{M}$

$$= \frac{(3.9 \times 10^6\text{ g m}^{-3}) \times (634) \times (784) \times (516 \times 10^{-36}\text{ m}^3) \times (6.022 \times 10^{23}\text{ mol}^{-1})}{154.77\text{ g mol}^{-1}} = 3.9$$

Therefore, $\boxed{N = 4}$ and the true calculated density (in the absence of defects) is

$$\rho = \frac{4 \times (154.77\text{ g mol}^{-1})}{(634) \times (784) \times (516 \times 10^{-30}\text{ cm}^3) \times (6.022 \times 10^{23}\text{ mol}^{-1})} = \boxed{4.01\text{ g cm}^{-3}}$$

E19.10(a)

$$d_{hkl} = \left[\left(\frac{h}{a}\right)^2 + \left(\frac{k}{b}\right)^2 + \left(\frac{l}{c}\right)^2\right]^{-1/2} \quad [19.3]$$

$$d_{411} = \left[\left(\frac{4}{812}\right)^2 + \left(\frac{1}{947}\right)^2 + \left(\frac{1}{637}\right)^2\right]^{-1/2} \text{pm} = \boxed{190\ \text{pm}}$$

E19.11(a) Since the reflection at 32.6° is (220), we know that

$$d_{220} = \frac{\lambda}{2\sin\theta}[19.5] = \frac{154\ \text{pm}}{2 \times \sin 32.6°} = 143\ \text{pm}$$

and hence, since $d_{220} = \dfrac{a}{(2^2+2^2)^{1/2}}$ [19.1] $= \dfrac{a}{8^{1/2}}$ for a cubic unit cell, it follows that

$$a = (8^{1/2}) \times (143\ \text{pm}) = 404\ \text{pm}$$

The indices of the other reflections are obtained from

$$(h^2 + k^2 + l^2) = \left(\frac{a}{d_{hkl}}\right)^2 [19.2] = \left(\frac{2a\sin\theta}{\lambda}\right)^2 [19.5]$$

We draw up the following table.

$\theta/°$	$\left(\dfrac{2a\sin\theta}{\lambda}\right)^2$	$h^2 + k^2 + l^2$	(hkl)	a/pm
19.4	3.04	3	$\boxed{(111)}$	402
22.5	4.03	4	$\boxed{(200)}$	402
32.6	7.99	8	(220)	404
39.4	11.09	11	$\boxed{(311)}$	402

By comparing the reflections found in the table with the summary of text Figure 19.23, we conclude that the substance has a $\boxed{\text{cubic F}}$ (fcc) unit cell. The final column values are obtained from

$$a = \left(\frac{\lambda}{2\sin\theta}\right) \times (h^2 + k^2 + l^2)^{1/2} \text{ and averaged to } a = 402\ \text{pm}.$$

E19.12(a)

$$\theta_{hkl} = \arcsin\frac{\lambda}{2d_{hkl}}[\text{from eqn 19.5}] = \arcsin\left\{\frac{\lambda}{2}\left[\left(\frac{h}{a}\right)^2 + \left(\frac{k}{b}\right)^2 + \left(\frac{l}{c}\right)^2\right]^{1/2}\right\}[\text{from eqn 19.3}]$$

$$\theta_{100} = \arcsin\left\{\frac{154}{2}\left[\left(\frac{1}{542}\right)^2 + \left(\frac{0}{917}\right)^2 + \left(\frac{0}{645}\right)^2\right]^{1/2}\right\} = \boxed{8.17°}$$

$$\theta_{010} = \arcsin\left\{\frac{154}{2}\left[\left(\frac{0}{542}\right)^2 + \left(\frac{1}{917}\right)^2 + \left(\frac{0}{645}\right)^2\right]^{1/2}\right\} = \boxed{4.82°}$$

$$\theta_{111} = \arcsin\left\{\frac{154}{2}\left[\left(\frac{1}{542}\right)^2 + \left(\frac{1}{917}\right)^2 + \left(\frac{1}{645}\right)^2\right]^{1/2}\right\} = \boxed{11.75°}$$

E19.13(a) From the text discussion of systematic absences (Section 19.3 and Figure 19.23 of the text) we can conclude that the unit cell is $\boxed{\text{face-centred cubic}}$.

E19.14(a) $F_{hkl} = \sum_j f_j e^{2\pi i(hx_j+ky_j+lz_j)}$ [19.7] with $f_j = \frac{1}{8}f$ (each atom is shared by eight cells).

Therefore, $F_{hkl} = \frac{1}{8}f\{1 + e^{2\pi ih} + e^{2\pi ik} + e^{2\pi il} + e^{2\pi i(h+k)} + e^{2\pi i(h+l)} + e^{2\pi i(k+l)} + e^{2\pi i(h+k+l)}\}$

However, all the exponential terms equal 1 since, h, k, and l are all integers and

$$e^{i\theta} = \cos\theta + i\sin\theta\ [\theta = 2\pi h, 2\pi k, \ldots] = \cos\theta = 1$$

Therefore, $F_{hkl} = \boxed{f}$ for the primitive cubic unit cell.

E19.15(a) The electron density is given by

$$\rho(r) = \frac{1}{V}\sum_{hkl} F_{hkl}e^{-2\pi i(hx+ky+lz)}\ [19.8]$$

The component along the x direction is

$$\rho(x) = \frac{1}{V}\sum_h F_h e^{-2\pi ihx}.$$

Using the data of this problem, we sum from $h = -9$ to $+9$ and use the relationship $F_h = F_{|h|}$. The following Mathcad worksheet computation of $\rho(0.5)$ uses unit volume. Figure 19.5 shows $\rho(x)$ with x in units of a. High electron density in indicated at the centre of a unit cell edge.

$$F := \begin{pmatrix} 10 \\ -10 \\ 8 \\ -8 \\ 6 \\ -6 \\ 4 \\ -4 \\ 2 \\ -2 \end{pmatrix} \qquad P(x) := \sum_{h=-9}^{9} \left(F_{|h|} \cdot e^{-2\cdot\pi\cdot i\cdot h\cdot x}\right)$$

$$\rho(.5) = 110$$

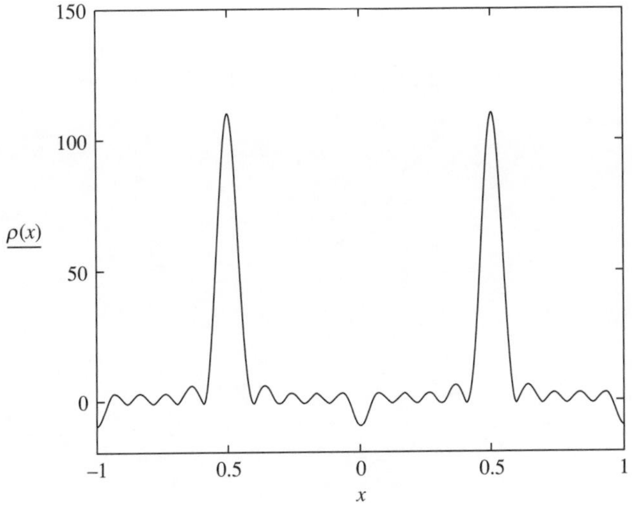

Figure 19.5

E19.16(a) Using the information of Exercise 19.15a, the Mathcad worksheet computation of $P(1.0)$ is performed with eqn 19.9.

$$F := \begin{pmatrix} 10 \\ -10 \\ 8 \\ -8 \\ 6 \\ -6 \\ 4 \\ -4 \\ 2 \\ -2 \end{pmatrix} \qquad P(x) := \sum_{h=-9}^{9} \left(F_{|h|} \cdot e^{-2\cdot\pi\cdot i\cdot h\cdot x}\right) \qquad P(1.0) = 780$$

The Patterson synthesis $P(x)$ of Figure 19.6 with x in units of a shows that atoms represented by this data are separated by 1 a unit along the x-axis.

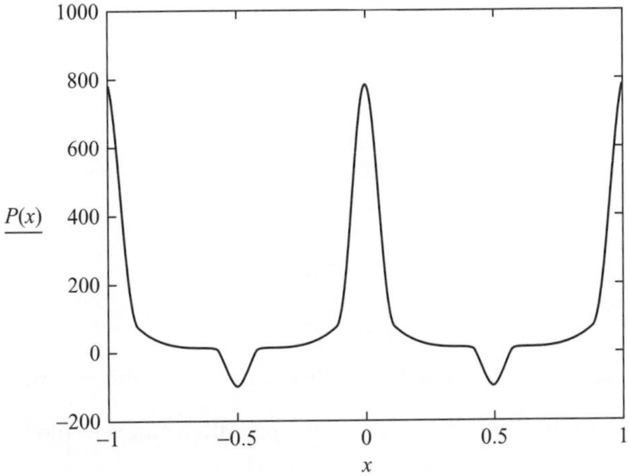

Figure 19.6

E19.17(a) Draw points corresponding to the vectors joining each pair of atoms. Heavier atoms give more intense contribution than light atoms. Remember that there are two vectors joining any pair of atoms ($\overleftarrow{AB}$ and $\overrightarrow{AB}$; don't forget the AA zero vectors for the centre point of the diagram. See Figure 19.7 for BF_3.

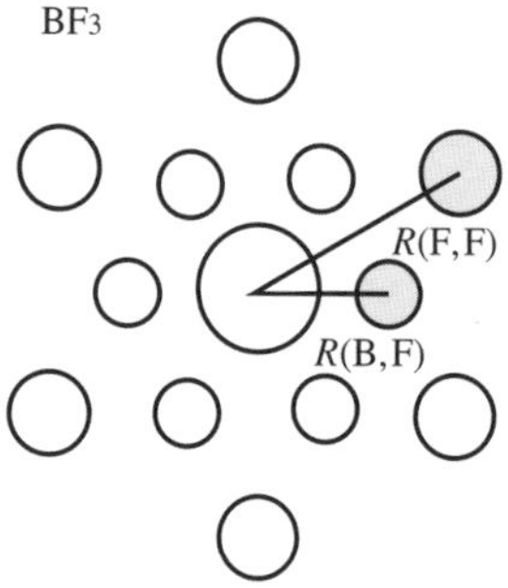

Figure 19.7

E19.18(a) By the de Broglie relationship: $\lambda = \dfrac{h}{p} = \dfrac{h}{mv}$

Hence, $v = \dfrac{h}{m\lambda} = \dfrac{6.626\times10^{-34}\ \text{J s}}{(1.675\times10^{-27}\ \text{kg})\times(50\times10^{-12}\ \text{m})} = \boxed{7.9\ \text{km s}^{-1}}$

E19.19(a) As discussed in text Example 19.4 the wavelength of a thermal neutron is

$$\lambda = \frac{h}{(mkT)^{1/2}} = \frac{6.626\times10^{-34}\ \text{J s}}{[(1.675\times10^{-27}\ \text{kg})\times(1.381\times10^{-23}\ \text{J K}^{-1})\times(300\ \text{K})]^{1/2}} = \boxed{252\ \text{pm}}$$

E19.20(a) The hatched area in Figure 19.8 is $h\times 2R = 3^{1/2}R\times 2R = 2\sqrt{3}R^2$, where $h = 2R\cos 30°$. The net number of cylinders in a hatched area is 1, and the area of the cylinder's base is πR^2. The volume of the prism (of which the hatched area is the base) is $2\sqrt{3}R^2L$, and the volume occupied by the cylinders is $\pi R^2 L$. Hence, the packing fraction is

$$f = \frac{\pi R^2 L}{2\sqrt{3}R^2 L} = \frac{\pi}{2\sqrt{3}} = \boxed{0.9069}$$

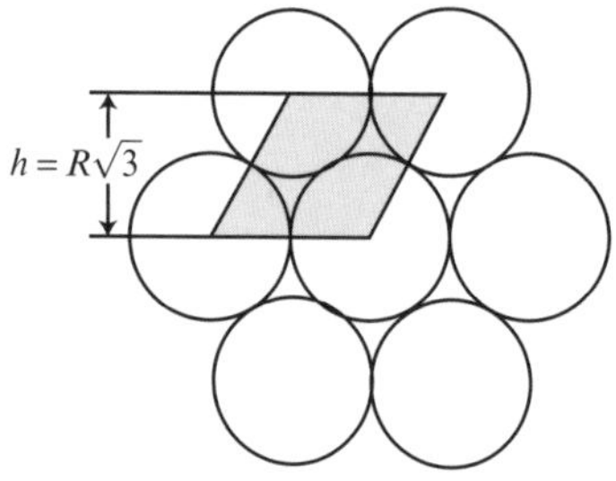

Figure 19.8

E19.21(a) $f = \dfrac{NV_a}{V_c}$, where N is the number of atoms in each unit cell, V_a their individual volumes, and V_c the volume of the unit cell itself. Refer to Figure 19.9 for a view of the primitive unit cell, the bcc unit cell, and the fcc unit cell.

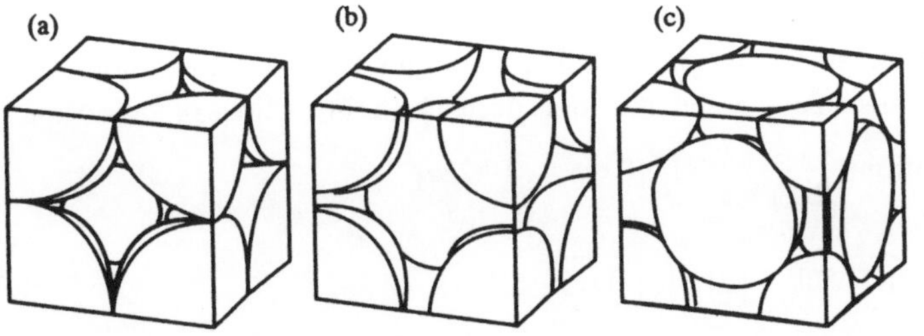

Figure 19.9

(a) For a primitive unit cell: $N = 1$, $V_a = \frac{4}{3}\pi R^3$, and $V_c = (2R)^3$

$$f = \frac{\left(\frac{4}{3}\pi R^3\right)}{(2R)^3} = \frac{\pi}{6} = \boxed{0.5236}$$

(b) For the bcc unit cell: $N = 2$, $V_a = \frac{4}{3}\pi R^3$, and $V_c = \left(\frac{4R}{\sqrt{3}}\right)^3$ [body diagonal of a unit cube is $4R$]

$$f = \frac{2 \times \frac{4}{3}\pi R^3}{\left(\frac{4R}{\sqrt{3}}\right)^3} = \frac{\pi\sqrt{3}}{8} = \boxed{0.6802}$$

(c) For the fcc unit cell: $N = 4$, $V_a = \frac{4}{3}\pi R^3$, and $V_c = (2\sqrt{2}R)^3$

$$f = \frac{4 \times \frac{4}{3}\pi R^3}{(2\sqrt{2}R)^3} = \frac{\pi}{3\sqrt{2}} = \boxed{0.7405}$$

E19.22(a) Consider the sixfold coordination shown in Figure 19.10, where we take the smallest distance between the larger hard spheres to be $2R$ and the closest approach of a large and small hard sphere to be $R + r$. This lattice gives the minimum value of the **radius ratio**, $\gamma = r/R$ [19.12]. By the Pythagorean theorem

$$(R + r)^2 = 2R^2 \quad \text{or} \quad \left(1 + \frac{r}{R}\right)^2 = 2$$

$$\gamma = \sqrt{2} - 1 = \boxed{0.41421}$$

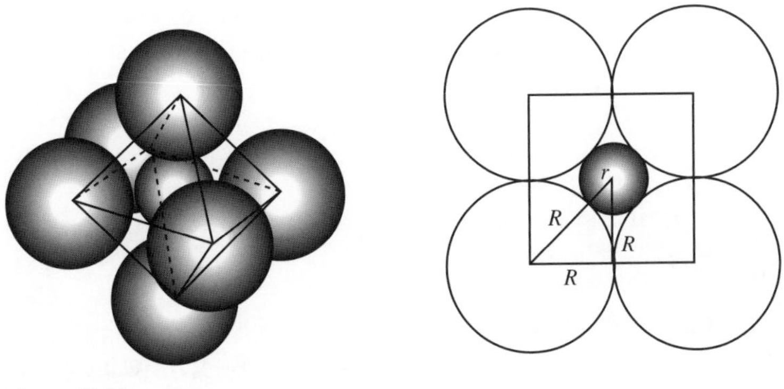

Figure 19.10

E19.23(a) The radius ratios determined in Exercises 19.22a and 19.22b correspond to the smallest value of the radius of the interior cation, since any smaller value would tend to bring the anions closer and increase their interionic repulsion and at the same time decrease the attractions of cation and anion.

(a) For sixfold coordination: $\dfrac{r_{\text{cation}}}{R_{\text{anion}}} = 0.414$ [Exercise 19.22a]

$$r_{\text{cation}} = (0.414) \times (140\ \text{pm})\ [\text{Table 19.3}] = \boxed{58.0\ \text{pm}}$$

(b) For eightfold coordination: $\dfrac{r_{\text{cation}}}{R_{\text{anion}}} = 0.732$ [Exercise 19.22b]

$$r_{\text{cation}} = (0.732) \times (142\ \text{pm}) = \boxed{104\ \text{pm}}\ [\textit{CRC Handbook of Chemistry and Physics}]$$

E19.24(a) The volume change is a result of two counteracting factors: different packing fractions f and different radii r. Let V and v be the unit cell volume per atom and the atomic hard-sphere volume, respectively. $v \propto r^3$ so the ratio of unit cell volumes is given by

$$\frac{V_{\text{bcc}}}{V_{\text{hcp}}} = \frac{f_{\text{hcp}}}{f_{\text{bcc}}} \times \frac{v_{\text{bcc}}}{v_{\text{hcp}}} = \frac{f_{\text{hcp}}}{f_{\text{bcc}}} \times \frac{r_{\text{bcc}}^3}{r_{\text{hcp}}^3}$$

Since the packing fractions are

$f_{\text{hcp}} = 0.7405$ and $f_{\text{bcc}} = 0.6802$ [*Justification 19.3* and Exercise 19.21a],

the unit cell volume ratio for titanium is

$$\frac{V_{\text{bcc}}}{V_{\text{hcp}}} = \frac{0.7405}{0.6802} \times \frac{(142.5)^3}{(145.8)^3} = 1.016$$

Hence, there is an $\boxed{\text{expansion}}$ of 1.6% when titanium transforms from hexagonal close-packed (hcp) to bcc.

E19.25(a) The lattice enthalpy is the difference in enthalpy between an ionic solid and the corresponding isolated ions. In this exercise, it is the enthalpy corresponding to the process

$$CaO(s) \rightarrow Ca^{2+}(g) + O^{2-}(g)$$

The standard lattice enthalpy can be computed from the standard enthalpies given in the exercise by considering the formation of CaO(s) from its elements as occurring through the following steps: sublimation of Ca(s), removing two electrons from Ca(g), atomization of O_2(g), two-electron attachment to O(g), and formation of CaO(s) lattice from gaseous ions. The formation reaction of CaO(s) is

$$Ca(s) + \tfrac{1}{2}O_2(g) \rightarrow CaO(s)$$

$$\Delta_f H^\ominus(\text{CaO,s}) = \Delta_{\text{sub}} H^\ominus(\text{Ca,s}) + \Delta_{\text{ion}} H^\ominus(\text{Ca,g}) + \tfrac{1}{2}\Delta_{\text{bond diss}} H^\ominus(\text{O}_2\text{,g}) + \Delta_{\text{eg}} H^\ominus(\text{O,g}) + \Delta_{\text{eg}} H^\ominus(\text{O}^-\text{,g}) - \Delta_L H^\ominus(\text{CaO,s})$$

So the lattice enthalpy is

$$\Delta_L H^\ominus(\text{CaO,s}) = \Delta_{\text{sub}} H^\ominus(\text{Ca,s}) + \Delta_{\text{ion}} H^\ominus(\text{Ca,g}) + \tfrac{1}{2}\Delta_{\text{bond diss}} H^\ominus(\text{O}_2\text{,g}) + \Delta_{\text{eg}} H^\ominus(\text{O,g}) + \Delta_{\text{eg}} H^\ominus(\text{O}^-\text{,g}) - \Delta_f H^\ominus(\text{CaO,s})$$

$$\Delta_L H^\ominus(\text{CaO,s}) = [178 + 1735 + \tfrac{1}{2}(497) - 141 + 844 + 635]\ \text{kJ mol}^{-1} = \boxed{3500\ \text{kJ mol}^{-1}}$$

E19.26(a) Young's modulus: $E = \dfrac{\text{normal stress}}{\text{normal strain}}$ [19.16a] $= 1.2\ \text{GPa} = 1.2 \times 10^9\ \text{kg m}^{-1}\ \text{s}^{-2}$

We solve this relationship for the normal strain after calculating the normal stress from the data provided.

normal stress = force per unit area = F/A

normal stress = relative elongation = $\Delta L/L$

$$\Delta L/L = \frac{F/A}{E} = \frac{mg/A}{E} = \frac{mg}{AE} = \frac{mg}{\pi(d/2)^2 E}$$

$$= \frac{1.0\ \text{kg} \times (9.81\ \text{m s}^{-2})}{\pi(1.0 \times 10^{-3}\ \text{m}/2)^2 \times (1.2 \times 10^9\ \text{kg m}^{-1}\ \text{s}^{-2})}$$

$$= \boxed{0.010} \text{ or about } 1.0\% \text{ elongation}$$

E19.27(a) Poisson's ratio: $v_\text{p} = \dfrac{\text{transverse strain}}{\text{normal strain}}$ [19.17] $= 0.45$

We note that the transverse strain is usually a contraction and that it is usually evenly distributed in both transverse directions. That is, if $(\Delta L/L)_z$ is the normal strain, then the transverse strains, $(\Delta L/L)_x$ and $(\Delta L/L)_y$, are equal. In this case of a 1.0% uniaxial stress:

$$\left(\frac{\Delta L}{L}\right)_z = +0.010, \quad \left(\frac{\Delta L}{L}\right)_x = \left(\frac{\Delta L}{L}\right)_y = -0.010 \times 0.45 = -0.0045 \text{ [a contraction of widths]}$$

Application of the stress to 1 cm^3 cube of polyethene results in a volume equal to

$$(1 - 0.0045)^2 \times (1 + 0.010) \times 1\ \text{cm}^3 = 1.00093\ \text{cm}^3$$

The change in volume is $\boxed{9.3 \times 10^{-4}\ \text{cm}^3}$.

E19.28(a) n-type; the dopant, arsenic, belongs to Group 15, whereas germanium belongs to Group 14.

E19.29(a)

$$E_\text{g} = h\nu_\text{min} = \frac{hc}{\lambda_\text{max}} = \frac{(6.626 \times 10^{-34}\ \text{J s})(2.998 \times 10^8\ \text{m s}^{-1})}{350 \times 10^{-9}\ \text{m}}\left(\frac{1\ \text{eV}}{1.602 \times 10^{-19}\ \text{J}}\right) = \boxed{3.54\ \text{eV}}$$

E19.30(a)

$\mu = g_\text{e}\{S(S+1)\}^{1/2}\mu_\text{B}$ [19.34 with S in place of s]

Therefore, since $\mu = 3.81\mu_\text{B}$ and $g_\text{e} \approx 2$,

$$S(S+1) = (\tfrac{1}{4}) \times (3.81)^2 = 3.63, \quad \text{implying that} \quad S = 1.47 = \tfrac{3}{2}$$

which indicates that Cr^{3+} has $\boxed{\text{three unpaired electrons}}$ in $CrCl_3$.

E19.31(a) $\chi_\text{m} = \chi V_\text{m}$ [19.28] $= \chi M/\rho = (-7.2 \times 10^{-7}) \times (78.11\ \text{g mol}^{-1})/(0.879\ \text{g cm}^{-3}) = \boxed{-6.4 \times 10^{-5}\ \text{cm}^3\ \text{mol}^{-1}}$

E19.32(a) The molar susceptibility is given by

$$\chi_\text{m} = \frac{N_\text{A} g_\text{e}^2 \mu_0 \mu_\text{B}^2 S(S+1)}{3kT} \text{ [19.35]}, \quad \text{so} \quad S(S+1) = \frac{3kT\chi_\text{m}}{N_\text{A} g_\text{e}^2 \mu_0 \mu_\text{B}^2}$$

$$S(S+1) = \frac{3(1.381\times10^{-23}\,\mathrm{J\,K^{-1}})\times(294.53\,\mathrm{K})\times(0.1463\times10^{-6}\,\mathrm{m^3\,mol^{-1}})}{(6.022\times10^{23}\,\mathrm{mol^{-1}})\times(2.0023)^2\times(4\pi\times10^{-7}\,\mathrm{T^2\,J^{-1}\,m^3})\times(9.27\times10^{-24}\,\mathrm{J\,T^{-1}})^2} = 6.847$$

so,

$$S^2+S-6.847=0 \quad \text{and} \quad S=\frac{-1+\sqrt{1+4(6.847)}}{2}=2.164$$

corresponding to 4.328 effective unpaired spins. The theoretical number is $\boxed{5}$ corresponding to the $3d^5$ electronic configuration of Mn^{2+}.

COMMENT. The discrepancy between the two values is accounted for by an antiferromagnetic interaction between the spins that alters χ_m from the form of eqn 19.35.

E19.33(a) $$\chi_m = (6.3001\times10^{-6})\times\left(\frac{S(S+1)}{T/\mathrm{K}}\,\mathrm{m^3\,mol^{-1}}\right) \text{[See Brief Illustration, Section 19.11(b)]}$$

Since Cu(II) is a d^9 species, it has one unpaired spin, and so $S = s = \frac{1}{2}$. Therefore,

$$\chi_m = \frac{(6.3001\times10^{-6})\times\left(\frac{1}{2}\right)\times\left(\frac{3}{2}\right)}{298}\,\mathrm{m^3\,mol^{-1}} = \boxed{+1.6\times10^{-8}\,\mathrm{m^3\,mol^{-1}}}$$

E19.34(a) The relationship between critical temperature and critical magnetic field is given by

$$\mathcal{H}_c(T)=\mathcal{H}_c(0)\left(1-\frac{T^2}{T_c^2}\right) \text{[19.36]}$$

Solving for T gives the critical temperature for a given magnetic field:

$$T=T_c\left(1-\frac{\mathcal{H}_c(T)}{\mathcal{H}_c(0)}\right)^{1/2}=(7.19\,\mathrm{K})\times\left(1-\frac{20\,\mathrm{kA\,m^{-1}}}{63.9\,\mathrm{kA\,m^{-1}}}\right)^{1/2}=\boxed{6.0\,\mathrm{K}}.$$

Solutions to problems

Solutions to numerical problems

P19.1 Having a (100) reflection, Pd must have a primitive cubic unit cell (see text Figure 19.23) and eqn 19.2 describes distance between reflection planes.

$$\lambda = 2d_{hkl}\sin\theta_{hkl}=\frac{2a\sin\theta_{hkl}}{(h^2+k^2+l^2)^{1/2}} \text{[19.5 and 19.2]} = 2a\sin 6.0° = 0.209a$$

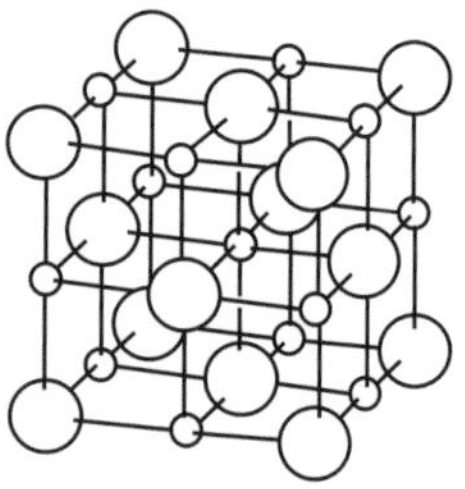

Figure 19.11

In an NaCl unit cell (Figure 19.11) the number of formula units is 4 (each corner ion is shared by 8 cells, each edge ion by 4, and each face ion by 2). Therefore,

$$\rho = \frac{NM}{VN_A} = \frac{4M}{a^3N_A}, \quad \text{implying that} \quad a = \left(\frac{4M}{\rho N_A}\right)^{1/3}$$

$$a = \left(\frac{(4)\times(58.44\text{ g mol}^{-1})}{(2.17\times10^6\text{ g m}^{-3})\times(6.022\times10^{23}\text{ mol}^{-1})}\right)^{1/3} = 563.\overline{5}\text{ pm}$$

and hence, $\lambda = (0.209)\times(563.\overline{5}\text{ pm}) = \boxed{118\text{ pm}}$

P19.3 For the three given reflections

$$\sin 19.076° = 0.32682 \qquad \sin 22.171° = 0.37737 \qquad \sin 32.256° = 0.53370$$

For cubic lattices $\sin\theta_{hkl} = \dfrac{\lambda(h^2+k^2+l^2)^{1/2}}{2a}$ [19.5 with 19.2]

First, consider the possibility of primitive cubic unit cell; the first three reflections are (100), (110), and (200). (See Figure 19.23 of the text.)

$$\frac{\sin\theta_{100}}{\sin\theta_{110}} = \frac{1}{\sqrt{2}} \neq \frac{0.32682}{0.37737}\text{ [not simple cubic]}$$

Consider next the possibility of body-centred cubic; the first three reflections are (110), (200), and (211).

$$\frac{\sin\theta_{110}}{\sin\theta_{200}} = \frac{\sqrt{2}}{\sqrt{4}} = \frac{1}{\sqrt{2}} \neq \frac{0.32682}{0.37737}\text{ (not bcc)}$$

Consider finally face-centred cubic; the first three reflections are (111), (200) and (220)

$$\frac{\sin\theta_{111}}{\sin\theta_{200}} = \frac{\sqrt{3}}{\sqrt{4}} = 0.86603$$

which compares very favourably to $\dfrac{0.32682}{0.37737} = 0.86605$. Therefore, the lattice is $\boxed{\text{face-centred cubic}}$.

This conclusion may easily be confirmed in the same manner using the second and third reflections.

$$a = \frac{\lambda}{2\sin\theta}(h^2+k^2+l^2)^{1/2} = \frac{\lambda}{2\sin\theta_{111}}(1^2+1^2+1^2)^{1/2}$$

$$= \left(\frac{154.18\text{ pm}}{(2)\times(0.32682)}\right)\times\sqrt{3} = \boxed{408.55\text{ pm}}$$

There are 4 atoms per fcc unit cell ($\frac{1}{8}$ for each corner plus $\frac{1}{2}$ for each face) and the volume of a cubic cell is a^3. Thus,

$$\rho = \frac{NM}{N_AV} = \frac{4\times(107.87\text{ g mol}^{-1})}{(6.0221\times10^{23}\text{ mol}^{-1})\times(4.0855\times10^{-8}\text{ cm})^3} = \boxed{10.507\text{ g cm}^{-3}}$$

This compares favourably to the value for silver listed in the Data section.

P19.5 $\theta_{111}(100\text{ K}) = 22.0403°$, $\theta_{111}(300\text{ K}) = 21.9664°$

$\sin\theta_{111}(100\text{ K}) = 0.37526$, $\sin\theta_{111}(300\text{ K}) = 0.37406$

$$\frac{\sin\theta(300\text{ K})}{\sin\theta(100\text{ K})} = 0.99681 = \frac{a(100\text{ K})}{a(300\text{ K})} \text{ [see Problem 19.4]}$$

$$a(300\text{ K}) = \frac{\lambda\sqrt{3}}{2\sin\theta_{111}} = \frac{(154.0562\text{ pm})\times\sqrt{3}}{(2)\times(0.37406)} = 356.67\text{ pm}$$

$$a(100\text{ K}) = (0.99681)\times(356.67\text{ pm}) = 355.53\text{ pm}$$

$$\frac{\delta a}{a} = \frac{356.67 - 355.53}{355.53} = 3.206\times10^{-3}$$

$$\frac{\delta V}{V} = \frac{356.67^3 - 355.53^3}{355.53^3} = 9.650\times10^{-3}$$

$$\alpha_{\text{volume}} = \frac{1}{V}\frac{\delta V}{\delta T} = \frac{9.560\times10^{-3}}{200\text{ K}} = \boxed{4.8\times10^{-5}\text{ K}^{-1}}$$

$$\alpha_{\text{linear}} = \frac{1}{a}\frac{\delta a}{\delta T} = \frac{3.206\times10^{-3}}{200\text{ K}} = \boxed{1.6\times10^{-5}\text{ K}^{-1}}$$

P19.7 $V = abc\sin\beta$

and the information given tells us that $a = 1.377b$, $c = 1.436b$, and $\beta = 122°49'$; hence

$$V = (1.377)\times(1.436b^3)\sin 122°49' = 1.662b^3$$

Since $\rho = \dfrac{NM}{VN_A} = \dfrac{2M}{1.662b^3N_A}$ we find that

$$b = \left(\frac{2M}{1.662\rho N_A}\right)^{1/3}$$

$$= \left(\frac{(2)\times(128.18\text{ g mol}^{-1})}{(1.662)\times(1.152\times10^6\text{ g m}^{-3})\times(6.022\times10^{23}\text{ mol}^{-1})}\right)^{1/3}$$

$$= 605.8\text{ pm}$$

Therefore, $a = \boxed{834\text{ pm}}$, $a = \boxed{606\text{ pm}}$, $a = \boxed{870\text{ pm}}$

P19.9 The volume per unit cell is

$$V = abc = (3.6881\text{ nm})\times(0.9402\text{ nm})\times(1.7652\text{ nm}) = 6.121\text{ nm}^3 = 6.121\times10^{-21}\text{ cm}^3$$

The mass per unit cell is eight times the mass of the formula unit, $[N(C_4H_9)_4][Ru(N)(S_2C_6H_4)_2]$, for which the molar mass is

$$M = \{101.07 + 2(14.007) + 28(12.011) + 44(1.008) + 4(32.066)\}\text{ g mol}^{-1}$$
$$= 638.01\text{ g mol}^{-1}$$

The density is

$$\rho = \frac{m}{V} = \frac{8M}{N_A V} = \frac{8 \times (638.01\ \text{g mol}^{-1})}{(6.022 \times 10^{23}\ \text{mol}^{-1}) \times (6.121 \times 10^{-21}\ \text{cm}^3)}$$

$$= \boxed{1.385\ \text{g cm}^{-3}}$$

The osmium analogue has a molar mass of 727.1 g mol^{-1}. If the volume of the crystal changes negligibly with the substitution, then the densities of the complexes are in proportion to their molar masses:

$$\rho_{\text{Os}} = \frac{727.1}{638.01}(1.385\ \text{g cm}^{-3}) = \boxed{1.578\ \text{g cm}^{-3}}$$

P19.11 Examining the powder diffraction pattern of text Figure 19.68, we see that neither tungsten (W) nor copper (Cu) is a primitive cubic unit lattice because the distinctive closeness of lines five (the (210) line) and six (the (211) line) with respect to the separation of lines six and seven (the (220) line) is not present. Thus, we must check the possibility of bcc and fcc lattices. Carefully measuring the distance of each line from the centre spot of the diffraction pattern, d_c, we acquire the data of the following table.

i^{th} line	1	2	3	4	5	6	7	8
W, d_c/mm	9.2	13.1	17.5	19.6	22.6	25.9	29.5	34.3
Cu, d_c/mm	9.9	12.0	17.0	20.6	22.0	25.2	31.4	33.3

In a circular camera, $cd_c = (\text{camera circumference}) \times \dfrac{2\theta}{2\pi} = 2R\theta$, where R is the radius of the camera (distance from sample to film, see Figure 19.18) and θ is the diffraction angle in radians. The constant c is a scaling constant that accounts for Figure 19.68 having a scaling that is possibly different from the original film. We eliminate this scaling factor by taking a ratio of measured values to the value of the first line. Then the ratio for the ith line is related to the glancing angles by $d_{c,i}/d_{c,1} = \theta_i/\theta_1 \equiv Y_{\text{exp},i}$. This glancing angle ratio can also be expressed as a function of the Miller indices and the cubic unit cell dimension a.

$$\theta_{hkl,i} = \arcsin\left(\frac{\lambda}{2d_{hkl,i}}\right)[19.5] = \arcsin\left(\frac{\lambda(h_i^2 + k_i^2 + l_i^2)^{1/2}}{2a}\right)[19.2]$$

$$Y_{\text{fit},i} \equiv \frac{\theta_{hkl,i}}{\theta_{hkl,1}} = \frac{\arcsin\left(\dfrac{\lambda(h_i^2 + k_i^2 + l_i^2)^{1/2}}{2a}\right)}{\arcsin\left(\dfrac{\lambda(h_1^2 + k_1^2 + l_1^2)^{1/2}}{2a}\right)}$$

This relationship for the angle ratio is called Y_{fit} because a fitting procedure must be used to adjust both the Miller indices and the value of a so that the sum of the squares of errors function, $SSE = \sum_i^{\text{all data}} (Y_{\text{exp}} - Y_{\text{fit}})_i^2$, is minimized. We will perform the fit with Mathcad as its Given/Minerr solve block is very convenient for the minimization of SSE.

(a) Here is a Mathcad worksheet for performing the tungsten data fit with the body-centred unit cell assumption. The Miller indices for bcc are taken from text Figure 19.23.

Measured distance of diffraction line from central spot of Figure 19.69(a) and Miller indices of body-centred cubic lattice:

$$d_c := \begin{pmatrix} 9.2 \\ 13.1 \\ 17.5 \\ 19.6 \\ 22.6 \\ 25.9 \\ 29.5 \\ 34.3 \end{pmatrix} \text{mm} \quad h := \begin{pmatrix} 1 \\ 2 \\ 2 \\ 2 \\ 3 \\ 2 \\ 3 \\ 4 \end{pmatrix} \quad k := \begin{pmatrix} 1 \\ 0 \\ 1 \\ 2 \\ 1 \\ 2 \\ 2 \\ 0 \end{pmatrix} \quad l := \begin{pmatrix} 0 \\ 0 \\ 1 \\ 0 \\ 0 \\ 2 \\ 1 \\ 0 \end{pmatrix}$$

Transformation of measured distances to form used in analysis:

$$i := 1..8 \qquad Yexp_i := \frac{d_{c_i}}{d_{c_1}}$$

Setup for SSE:

$$Yfit(x, y, z, a) := \left| \frac{\operatorname{asin}\left[\dfrac{154(x^2+y^2+z^2)^{1/2}}{2 \cdot a}\right]}{\operatorname{asin}\left[\dfrac{154[(h_1)^2+(k_1)^2+(l_1)^2]^{1/2}}{2 \cdot a}\right]} \right|$$

$$SSE(a) := \sum_{i=1}^{8} (Yexp_i - Yfit(h_i, k_i, l_i, a))^2$$

Initial guess of value for parameter *a* and Given/Minerr solve block: $a := 500$

Given $SSE(a) = 0$ $a := Minerr(a)$

Dimension of unit cell in pm with bcc assumption: $a = 320.901$

Standard deviation of fit: $\left(\dfrac{SSE(a)}{8-1}\right)^{1/2} = 0.039$

The standard deviation of the fit for tungsten is very small in the bcc model. To test the fcc model, simply change the Miller indices in the above worksheet to the fcc indices shown in text Figure 19.23. The standard deviation of the fcc model is found to be 0.246, which is about six times larger than the bcc fit. Thus, we confidently conclude that $\boxed{\text{tungsten has the bcc unit cell with } a = 321 \text{ pm}}$. Figure 19.12 with $r = R$ indicates that the atomic radius is given by

$$r = \tfrac{1}{4}\sqrt{3}a = \tfrac{1}{4}\sqrt{3} \times (321 \text{ pm}) = \boxed{139 \text{ pm}}$$

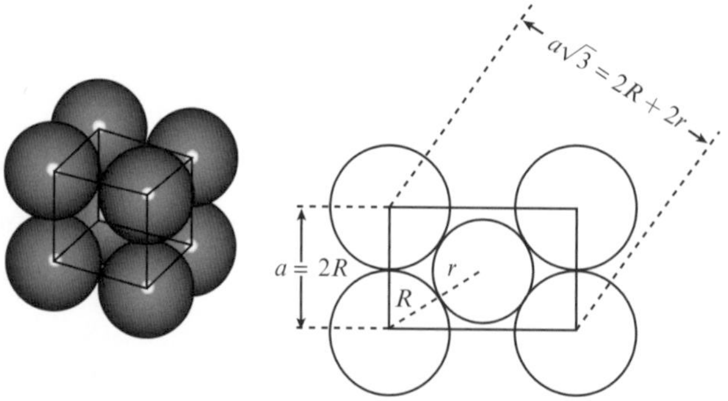

Figure 19.12

(b) Here is a Mathcad worksheet for performing the copper data fit with the face-centred unit cell assumption. The Miller indices for fcc are taken from text Figure 19.23.

Measured distance of diffraction line from central spot of Figure 19.69(a) and Miller indices of body-centred cubic lattice:

$$d_c := \begin{pmatrix} 9.9 \\ 12.0 \\ 17.0 \\ 20.6 \\ 22.0 \\ 25.2 \\ 31.4 \\ 33.3 \end{pmatrix} \text{mm} \quad h := \begin{pmatrix} 1 \\ 2 \\ 2 \\ 3 \\ 2 \\ 4 \\ 3 \\ 4 \end{pmatrix} \quad k := \begin{pmatrix} 1 \\ 0 \\ 2 \\ 1 \\ 2 \\ 0 \\ 3 \\ 2 \end{pmatrix} \quad l := \begin{pmatrix} 1 \\ 0 \\ 0 \\ 1 \\ 2 \\ 0 \\ 1 \\ 0 \end{pmatrix}$$

Transformation of measured distances to form used in analysis:

$$i := 1..\,8 \qquad Yexp_i := \frac{d_{c_i}}{d_{c_1}}$$

Setup for SSE:

$$\mathbf{Yfit}(x, y, z, a) := \left| \frac{\text{asin}\left[\dfrac{154(x^2+y^2+z^2)^{1/2}}{2\cdot a}\right]}{\text{asin}\left[\dfrac{154[(h_1)^2+(k_1)^2+(l_1)^2]^{1/2}}{2\cdot a}\right]} \right|$$

$$\mathbf{SSE}(a) := \sum_{i=1}^{8}(Yexp_1 - Yfit(h_i, k_i, l_i, a))^2$$

Initial guess of value for parameter a and Given/Minerr solve block: $a := 400$

Given SSE(a) = 0 $a :=$ Minerr(a)

Dimension of unit cell in pm with fcc assumption: $a = 361.66$

Standard deviation of fit: $\left(\dfrac{SSE(a)}{8-1}\right)^{1/2} = 0.063$

The standard deviation of the fit for copper is very small in the fcc model. To test the bcc model, simply change the Miller indices in the above worksheet to the bcc indices shown in text Figure 19.23. The standard deviation of the bcc model is found to be 0.146, which is about twice as large as the fcc fit. Thus, we confidently conclude that $\boxed{\text{copper has the fcc unit cell with } a = 362 \text{ pm}}$. The face diagonal of an fcc cell is related to the edge length a by

$(4r)^2 = a^2 + a^2$, where r is the atomic radius

Thus, $r = \frac{1}{4}\sqrt{2}\, a = \frac{1}{4}\sqrt{2} \times (362 \text{ pm}) = \boxed{128 \text{ pm}}$

P19.13 $$\rho = \frac{m\,(\text{unit cell})}{V\,(\text{unit cell})} = \frac{2 \times M(\text{CH}_2\text{CH}_2)/N_A}{abc}$$

$$\rho = \frac{(2) \times (28.05 \text{ g mol}^{-1})}{(6.022 \times 10^{23} \text{ mol}^{-1}) \times [(740 \times 493 \times 253) \times 10^{-36}] \text{ m}^3} = 1.01 \times 10^6 \text{ g m}^{-3} = \boxed{1.01 \text{ g cm}^{-3}}$$

P19.15 The problem asks for an estimate of $\Delta_f H^\ominus(\text{CaCl,s})$. A Born–Haber cycle would envision formation of CaCl(s) from its elements as sublimation of Ca(s), ionization of Ca(g), atomization of $Cl_2(g)$, electron gain of Cl(g), and formation of CaCl(s) from gaseous ions. Therefore

$$\Delta_f H^\ominus(\text{CaCl,s}) = \Delta_{\text{sub}}H^\ominus(\text{Ca,s}) + \Delta_{\text{ion}}H^\ominus(\text{Ca,g}) + \Delta_f H^\ominus(\text{Cl,g}) + \Delta_{\text{eg}}H^\ominus(\text{Cl,g}) - \Delta_L H^\ominus(\text{CaCl,s})$$

Before we can estimate the lattice enthalpy of CaCl, we select a lattice with the aid of the radius ratio rule. The ionic radius for Cl^- is 181 pm; use the ionic radius of K^+ (138 pm) for Ca^+.

$$\gamma = \frac{138\text{ pm}}{181\text{ pm}} = 0.762$$

suggesting the CsCl structure. We can interpret the Born–Mayer equation (eqn. 19.15) as giving the negative of the lattice enthalpy.

$$\Delta_L H^\ominus \approx \frac{A|z_A z_B|N_A e^2}{4\pi\varepsilon_0 d}\left(1 - \frac{d^*}{d}\right), \quad \text{where } d^* \text{ is taken to be 34.5 pm (common choice).}$$

The distance d is $d = (138 + 181)\text{ pm} = 319\text{ pm}$. The Madelung constant is found in Table 19.4.

$$\Delta_L H^\ominus \approx \frac{1.763 \times |(1)(-1)| \times (6.022 \times 10^{23}\text{ mol}^{-1}) \times (1.602 \times 10^{-19}\text{ C})^2}{4\pi(8.854 \times 10^{-12}\text{ J}^{-1}\text{C}^2\text{ m}^{-1}) \times (319 \times 10^{-12}\text{ m})}\left(1 - \frac{34.5\text{ pm}}{319\text{ pm}}\right)$$

$$\approx 6.85 \times 10^5\text{ J mol}^{-1} = 685\text{ kJ mol}^{-1}$$

The enthalpy of formation, then, is

$$\Delta_f H^\ominus(\text{CaCl,s}) \approx [176 + 589.7 + 121.7 - 348.7 - 685]\text{ kJ mol}^{-1} = \boxed{-146\text{ kJ mol}^{-1}}.$$

Although formation of CaCl(s) from its elements is exothermic, formation of $CaCl_2(s)$ is still more favoured energetically. Consider the disproportionation reaction 2 CaCl(s) → Ca(s) + $CaCl_2(s)$ for which

$$\Delta H^\ominus = \Delta_f H^\ominus(\text{Ca,s}) + \Delta_f H^\ominus(\text{CaCl}_2\text{,s}) - 2\Delta_f H^\ominus(\text{CaCl,s})$$
$$\approx [0 - 795.8 - 2(-146)]\text{ kJ mol}^{-1}$$
$$\approx -504\text{ kJ mol}^{-1}$$

and the thermodynamic instability of CaCl(s) toward disproportionation to Ca(s) and $CaCl_2(s)$ becomes apparent.

Note: Using the tabulated ionic radius of Ca (i.e. that of Ca^{2+}) would be less valid than using the atomic radius of a neighbouring monovalent ion, for the problem asks about a hypothetical compound of monovalent calcium. Predictions with the smaller Ca^{2+} radius (100 pm) differ from those listed above but the conclusion remains the same: the expected structure changes to rock-salt, the lattice enthalpy to 758 kJ mol^{-1}, $\Delta_f H^\ominus(\text{CaCl}_2\text{,s})$ to −219 kJ mol^{-1} and the disproportionation enthalpy to −358 kJ mol^{-1}.

P19.17 The molar magnetic susceptibility is given by

$$\chi_m = \frac{N_A g_e^2 \mu_0 \mu_B^2 S(S+1)}{3kT} [19.35] = (6.3001 \times 10^{-6}) \times \frac{S(S+1)}{T/K} \text{ m}^3\text{mol}^{-1}$$

[See Brief Illustration, Section 19.11(b)]

$$\text{For } S = 2, \chi_m = \frac{(6.3001 \times 10^{-6}) \times (2) \times (2+1)}{298} \text{ m}^3 \text{ mol}^{-1} = \boxed{0.127 \times 10^{-6} \text{ m}^3 \text{ mol}^{-1}}$$

$$\text{For } S = 3, \chi_m = \frac{(6.3001 \times 10^{-6}) \times (3) \times (3+1)}{298} \text{ m}^3 \text{ mol}^{-1} = \boxed{0.254 \times 10^{-6} \text{ m}^3 \text{ mol}^{-1}}$$

$$\text{For } S = 4, \chi_m = \frac{(6.3001 \times 10^{-6}) \times (4) \times (4+1)}{298} \text{ m}^3 \text{ mol}^{-1} = \boxed{0.423 \times 10^{-6} \text{ m}^3 \text{ mol}^{-1}}$$

Instead of a single value of S we use an average weighted by the Boltzmann factor

$$\exp\left(\frac{-50 \times 10^3 \text{ J mol}^{-1}}{(8.3145 \text{ J mol}^{-1} \text{K}^{-1}) \times (298 \text{ K})}\right) = 1.7 \times 10^{-9}$$

Thus, the $S = 2$ and $S = 4$ forms are present in negligible quantities compared to the $S = 3$ form. The compound's susceptibility, then, is that of the $S = 3$ form, namely $\boxed{0.254 \text{ cm}^3 \text{ mol}^{-1}}$.

Solutions to theoretical problems

P19.19 Consider for simplicity the two-dimensional lattice and planes shown in Figure 19.13.

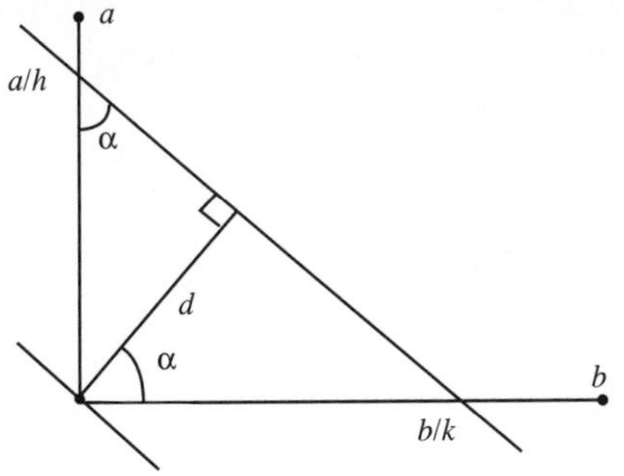

Figure 19.13

The (hk) plane cut the a- and b-axes at a/h and b/k, and we have

$$\sin\alpha = \frac{d}{(a/h)} = \frac{hd}{a}, \quad \cos\alpha = \frac{d}{(b/k)} = \frac{kd}{b}$$

Then, since $\sin^2\alpha + \cos^2\alpha = 1$, we can write

$$\left(\frac{hd}{a}\right)^2 + \left(\frac{kd}{b}\right)^2 = 1$$

and therefore

$$\frac{1}{d^2} = \left(\frac{h}{a}\right)^2 + \left(\frac{k}{b}\right)^2$$

The same argument extends by analogy (or further trigonometry) to three dimensions, to give

$$\boxed{\frac{1}{d^2} = \left(\frac{h}{a}\right)^2 + \left(\frac{k}{b}\right)^2 + \left(\frac{l}{c}\right)^2} \quad [19.3]$$

P19.21 The scattering factor is given by

$$f = 4\pi\int_0^\infty \rho(r)\frac{\sin kr}{kr}r^2\mathrm{d}r \ [19.6], \quad \text{where} \quad k = \frac{4\pi}{\lambda}\sin\theta$$

and the integral may be divided into segments such as

$$f = 4\pi\int_0^R \rho(r)\frac{\sin kr}{kr}r^2\mathrm{d}r + 4\pi\int_R^\infty \rho(r)\frac{\sin kr}{kr}r^2\mathrm{d}r$$

In the case for which $\rho = 3Z/4\pi R^3$ when $0 \le r \le R$, and $\rho = 0$ when $r > R$, the second integral vanishes, leaving

$$f = \frac{3Z}{R^3}\int_0^R \frac{\sin kr}{kr}r^2\mathrm{d}r = \frac{3Z}{R^3k}\int_0^R \sin(kr)\,r\mathrm{d}r = \frac{3Z}{R^3k}\left[\frac{\sin kr}{k^2} - \frac{r\cos kr}{k}\right]_{r=0}^{r=R} = \frac{3Z}{(kR)^2}\left\{\frac{\sin kR}{kR} - \cos kR\right\}$$

This shows that the scattering factor is proportional to the atomic number Z, which is illustrated in a plot of f against $\sin(\theta)/\lambda$ in Figure 19.14. As expected, $f = Z$ in the forward direction.

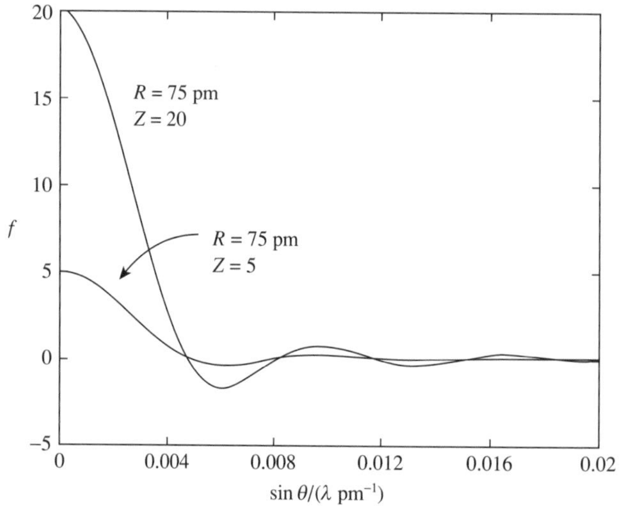

Figure 19.14

Figure 19.15 is a plot of the scattering factor for several R values at constant Z. As R increases, the scattering factor shifts to the forward direction.

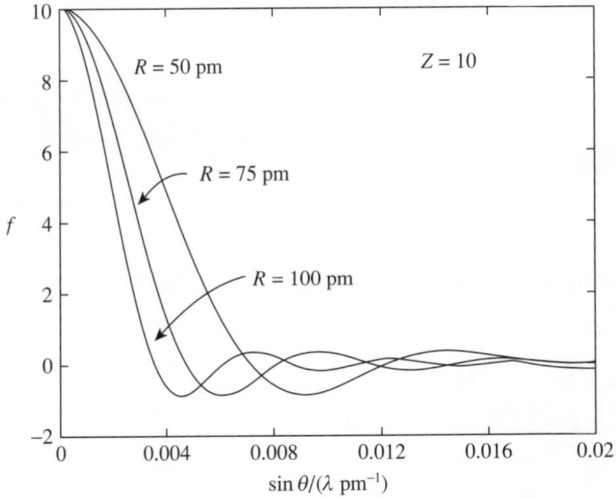

Figure 19.15

P19.23 Replacing an exponentially decreasing 1s wavefunction with a Gaussian function effectively replaces the cusp of the exponential decay near $r = 0$ with a broader, flatter function, while simultaneously decreasing the long tail of the exponential. Physically, this is analogous to contraction of the orbital by increasing the nuclear charge Z. In Problem 19.22 an increase in Z was found to cause a shift of the scattering factor away from the forward direction. To confirm this hypothesis, we first fit the $1s(Z = 1)$ exponential $\exp(-\rho/2)$ with the $\exp(-br^2)$ Gaussian by choosing the constant b so as to minimize the sum of the squared errors (SSE) in the range $0 \le r \le 2a$. Doing this in a Mathcad worksheet yields the following b value.

$N := 40 \quad j := 0..N \quad r_{max} := 2 \cdot a \quad r_j := \frac{j \cdot r_{max}}{N}$

$SSE(b) := \sum_j \left[e^{\frac{-\rho(r_j,1)}{2}} - e^{-b \cdot (r_j)^2} \right]^2 \quad$ Estimate of b: $b := 1 \cdot pm^{-2}$

Given $\quad SSE(b) = 0 \quad b := Minerr(b)$

$b = 0.00035 pm^{-2}$

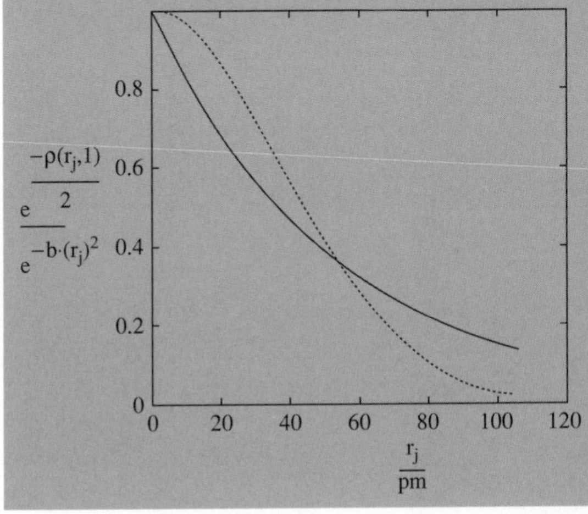

The absence of both the cusp and long tail in the Gaussian is clearly apparent in the worksheet plot.

For the one-electron hydrogen-like atom $4\pi\rho(r) = R_{1s}(r)^2$ [9.18] or $4\pi\rho(r) = g(r)^2$ and the scattering factor is given by

$$f(r, Z) = 4\pi \int_0^{\infty} \rho(r, Z)\frac{\sin(4\pi xr)}{4\pi xr} r^2\,\mathrm{d}r \text{ [19.6]}, \quad \text{where} \quad x = \frac{\sin\theta}{\lambda}$$

The next section of the worksheet normalizes the Gaussian $g(r)$ and defines the Gaussian scattering factor f_g.

$$N := \left[\frac{1}{\int_{0\cdot\mathrm{pm}}^{50\cdot a} e^{-b\cdot(r)^2}\cdot e^{-b\cdot(r)^2}\cdot r^2\,\mathrm{dr}}\right]^{1/2} \qquad N = 0.00649\mathrm{pm}^{\frac{-3}{2}}$$

$$g(r) := N\cdot e^{-b\cdot(r)^2} \qquad f_g(x) := \int_{0\cdot\mathrm{pm}}^{100\cdot a} \frac{g(r)^2\cdot\sin(4\cdot\pi\cdot x\cdot r)\cdot r}{4\cdot\pi\cdot x}\,\mathrm{dr}$$

The worksheet setup for the 1s radial wavefunction (Table 9.1) and its scattering factor f follows.

$$\mathrm{pm} := 10^{-12}\cdot\mathrm{m} \qquad a := 52.91772108\mathrm{pm}$$

$$\rho(r, Z) := \frac{2\cdot Z\cdot r}{a} \qquad R_{1s}(r, Z) := 2\cdot\left(\frac{Z}{a}\right)^{3/2}\cdot e^{\frac{-\rho(r,Z)}{2}}$$

$$f(x, Z) := \int_{0\cdot\mathrm{pm}}^{100\cdot a} \frac{R_{1s}(r, Z)^2\cdot\sin(4\cdot\pi\cdot x\cdot r)\cdot r}{4\cdot\pi\cdot x}\,\mathrm{dr}$$

The subsequent plot of $f(x, 1)$ and $f_g(x)$ confirms that replacement of the 1s wavefunction by a Gaussian shifts the scattering factor away from the forward direction.

$$N := 200 \qquad i = 0..N \qquad x_{max} := 0.01\cdot\mathrm{pm}^{-1} \qquad x_i := \frac{i\cdot x_{max}}{N}$$

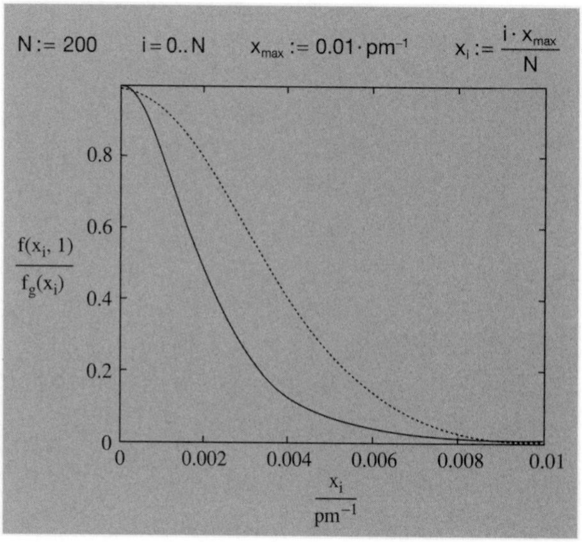

P19.25 The four values of $hx + ky + lz$ that occur in the exponential functions of F have the values 0, 5/2, 3, and 7/2, and so

$$F_{114} = f_I \sum_j e^{i\phi_{114}(j)} [19.7] = f_I \sum_j e^{2i\pi(x_j + y_j + 4z_j)} = f_I\{1 + e^{5i\pi} + e^{6i\pi} + e^{7i\pi}\} = f_I\{1 - 1 + 1 - 1\} = \boxed{0}$$

P19.27 (a) $E_P = -A \times \dfrac{|z_A z_B| N_A e^2}{4\pi\varepsilon_0 d}$ [19.13] and $E_P^* = N_A C' e^{-d/d^*}$ [19.14]

The sum of these two expressions gives the total potential energy and the minimum in this quantity is obtained by differentiation with respect to d.

$$\frac{d(E_P + E_P^*)}{dd} = 0 = \frac{A|z_A z_B| N_A e^2}{4\pi\varepsilon_0 d^2} - N_A C'\left(\frac{1}{d^*}\right) e^{-d/d^*}$$

N_A can be eliminated and we can write

$$C' e^{-d/d^*} = \frac{A|z_A z_B| e^2}{4\pi\varepsilon_0 d}\left(\frac{d^*}{d}\right)$$

Then, after substitution into eqn 19.14 we obtain

$$E_{P,\min} = -A\frac{|z_A z_B| N_A e^2}{4\pi\varepsilon_0 d} + AN_A\frac{|z_A z_B| e^2}{4\pi\varepsilon_0 d}\left(\frac{d^*}{d}\right) = -AN_A\left(\frac{|z_A z_B| e^2}{4\pi\varepsilon_0 d}\right)\left(1 - \frac{d^*}{d}\right)$$

which is eqn 19.15.

P19.29 (a) The density of energy levels is:

$$\rho(E) = \frac{dk}{dE} = \left(\frac{dE}{dk}\right)^{-1}$$

where $\dfrac{dE}{dk} = \dfrac{d}{dk}\left(\alpha + 2\beta\cos\dfrac{k\pi}{N+1}\right)[19.21] = -\dfrac{2\pi\beta}{N+1}\sin\dfrac{k\pi}{N+1}$

so $\rho(E) = -\dfrac{N+1}{2\pi\beta}\left(\sin\dfrac{k\pi}{N+1}\right)^{-1}$

Unlike the expression just derived, the relationship the problem asks us to derive has no trigonometric functions and it contains E and α within a square root. This comparison suggests that the trigonometric identity $\sin^2\theta + \cos^2\theta = 1$ will be of use here. Let $\theta = k\pi/(N+1)$; then

$$\sin\theta = (1 - \cos^2\theta)^{1/2}$$

however, $\cos\theta$ is related to the energy

$$E = \alpha + 2\beta\cos\theta \quad \text{so} \quad \cos\theta = \frac{E - \alpha}{2\beta}$$

and $\sin\theta = \left[1 - \left(\dfrac{E-\alpha}{2\beta}\right)^2\right]^{1/2}$

Finally, $\boxed{\rho(E) = -\dfrac{(N+1)/2\pi\beta}{\left[1-\left(\dfrac{E-\alpha}{2\beta}\right)^2\right]^{1/2}}}$

(b) The denominator of this expression vanishes as the energy approaches $\alpha \pm 2\beta$. Near those limits, $E - \alpha$ becomes $\pm 2\beta$, making the quantity under the square root zero, and ρ(E) approach infinity.

P19.31 Figure 19.16 depicts the single transition of the monomer with the transition dipole moment μ_{mon}. It also defines the excited states and transitions of the dimer.

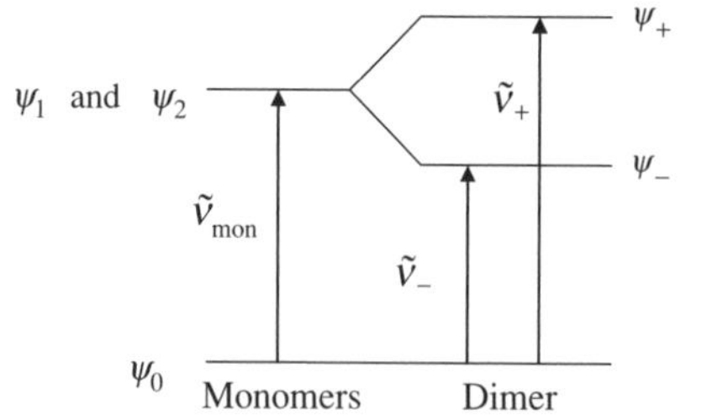

Figure 19.16

$\mu_{+\text{or}-} = \langle\psi_{+\text{or}-}|\mu|\psi_0\rangle = \langle c_{+\text{or}-,1}\psi_1 + c_{+\text{or}-,2}\psi_2|\mu|\psi_0\rangle = c_{+\text{or}-,1}\langle\psi_1|\mu|\psi_0\rangle + c_{+\text{or}-,2}\langle\psi_2|\mu|\psi_0\rangle$
But $\langle\psi_1|\mu|\psi_0\rangle = \langle\psi_2|\mu|\psi_0\rangle = \mu_{\text{mon}}$, so
$\mu_{+\text{or}-} = (c_{+\text{or}-,1} + c_{+\text{or}-,2})\mu_{\text{mon}}$

P19.33 If a substance responds non-linearly to an electric field E, then it induces a dipole moment:

$$\mu = \alpha E + \beta E^2$$

If the electric field is oscillating at two frequencies, we can write the electric field as

$$E = E_1 \cos\omega_1 t + E_2 \cos\omega_2 t$$

and the non-linear response as

$$\beta E^2 = \beta(E_1 \cos\omega_1 t + E_2 \cos\omega_2 t)^2$$

$$\beta E^2 = \beta(E_1^2 \cos^2\omega_1 t + E_2^2 \cos^2\omega_2 t + 2E_1E_2 \cos\omega_1 t \cos\omega_2 t)$$

Application of trigonometric identities allows a product of cosines to be re-written as a sum:

$$\cos A \cos B = \tfrac{1}{2}\cos(A-B) + \tfrac{1}{2}\cos(A+B)$$

Using this result (a special case of which applies to the $\cos^2$ terms), yields:

$$\beta E^2 = \tfrac{1}{2}\beta[E_1^2(1+\cos 2\omega_1 t) + E^2(1+\cos 2\omega_2 t) + 2E_1E_2(\cos(\omega_1+\omega_2)t + \cos(\omega_1-\omega_2)t]$$

This expression includes responses at twice the original frequencies as well as at the sum and difference frequencies.

P19.35 $N_2O_4(g) \overset{K}{\rightleftharpoons} 2\, NO_2(g)$

$(1-\alpha)n \qquad 2\alpha n \quad$ amounts

$\dfrac{1-\alpha}{1+\alpha} \qquad \dfrac{2\alpha}{1+\alpha} \quad$ mole fractions

$\left(\dfrac{1-\alpha}{1+\alpha}\right)p \qquad \left(\dfrac{2\alpha}{1+\alpha}\right)p \quad$ partial pressures $[p \equiv p/p^{\ominus}$ here]

$$K = \frac{(2\alpha/1+\alpha)^2 p}{(1-\alpha/1+\alpha)} = \frac{4\alpha^2}{1-\alpha^2}p$$

Now solve for α.

$$\alpha^2 = \frac{K}{4p+K}, \quad \alpha = \left(\frac{K}{4p+K}\right)^{1/2}$$

The degree of dimerization is $d = 1-\alpha = 1-\left(\dfrac{K}{4p+K}\right)^{1/2} = \boxed{1-\left(\dfrac{1}{4(p/K)+1}\right)^{1/2}}$.

The susceptibility varies in proportion to $\alpha = 1 - d$. As pressure increases, α decreases, and the susceptibility $\boxed{\text{decreases}}$.

To determine the effect of temperature we need $\Delta_r H \approx \Delta_r H^{\ominus}$ for the reaction above.

$$\Delta_r H^{\ominus} = 2 \times (33.18\ \text{kJ mol}^{-1}) - 9.16\ \text{kJ mol}^{-1} = +57.2\ \text{kJ mol}^{-1}$$

A positive $\Delta_r H^{\ominus}$ indicates that $NO_2(g)$ is favoured as the temperature increases; hence the susceptibility $\boxed{\text{increases}}$ with temperature.

Solutions to applications: biochemistry and nanoscience

P19.37 The density of a face-centred cubic crystal is $4m/V$ where m is the mass of the unit hung on each lattice point and V is the volume of the unit cell. (The 4 comes from the fact that each of the cell's 8 vertices is shared by 8 cells, and each of the cell's 6 faces is shared by 2 cells.)

So, $\rho = \dfrac{4m}{a^3} = \dfrac{4M}{N_A a^3}$ and $M = \frac{1}{4}\rho N_A a^3$

$$M = \tfrac{1}{4}(1.287\ \text{g cm}^{-3}) \times (6.022 \times 10^{23}\ \text{mol}^{-1}) \times (12.3 \times 10^{-7}\ \text{cm})^3 = \boxed{3.61 \times 10^5\ \text{g mol}^{-1}}$$

P19.39 Tans and coworkers (S.J. Tans *et al.*, *Nature*, **393**, 49 (1998)) have draped a semiconducting carbon nanotube (CNT) over metal (gold in Figure 19.17) electrodes that are 400 nm apart atop a silicon surface coated with silicon dioxide. A bias voltage between the electrodes provides the source and drain of the molecular field-effect transistor (FET). The silicon serves as a gate electrode and the thin silicon oxide layer (at least 100 nm thick) insulates the gate from the CNT circuit. By adjusting the magnitude of an electric field applied to the gate, current flow across the CNT may be turned on and off.

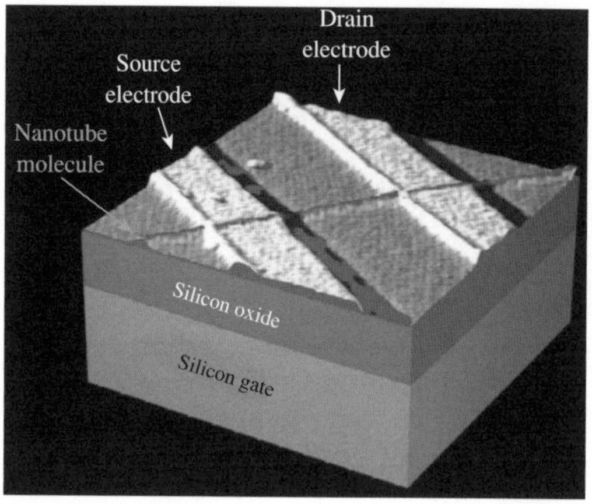

Figure 19.17

Wind and coworkers (S.J. Wind *et al.*, *Applied Physics Letters*, **80**(20, May 20), 3817 (2002)) have designed (Figure 19.18) a CNT-FET of improved current carrying capability. The gate electrode is above the conduction channel and separated from the channel by a thin oxide dielectric. In this manner the CNT-to-air contact is eliminated, an arrangement that prevents the circuit from acting like a p-type transistor. This arrangement also reduces the gate oxide thickness to about 15 nm, allowing for much smaller gate voltages and a steeper subthreshold slope, which is a measure of how well a transistor turns on or off.

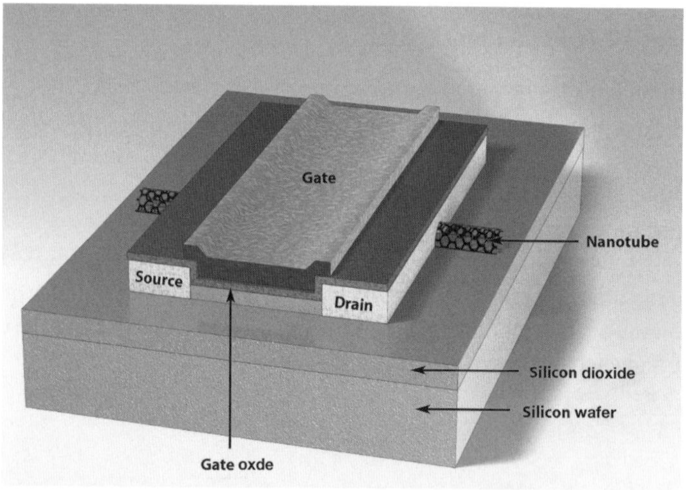

Figure 19.18
(http://researchweb.watson.ibm.com/resources/news/20020520_nanotubes.shtml)

A single-electron transistor (SET) has been prepared by Cees Dekker and coworkers (*Science*, **293**, 76, (2001)) with a CNT. The SET is prepared by putting two bends in a CNT with the tip of an AFM (Figure 19.19). Bending causes two buckles that, at a distance of 20 nm, serves as a conductance barrier. When an appropriate voltage is applied to the gate below the barrier, electrons tunnel one at a time across the barrier.

Figure 19.19

Weitz *et al.* (*Physica Status Solidi* (b) **243**, 13, 3394 (2006)) report on the construction of a single-wall CNT using a silane-based organic self-assembled monolayer (SAM) as a gate dielectric on top of a highly doped silicon wafer. The organic SAM is made of 18-phenoxyoctadecyltrichlorosilane. This ultrathin layer (Figure 19.20) ensures strong gate coupling and therefore low operation voltages. Single-electron transistors (SETs) were obtained from individual metallic SWCNTs. Field-effect transistors made from individual semiconducting SWCNTs operate with gate-source voltages of −2 V show good saturation, small hysteresis (200 mV), and a low subthreshold swing (290 mV/dec).

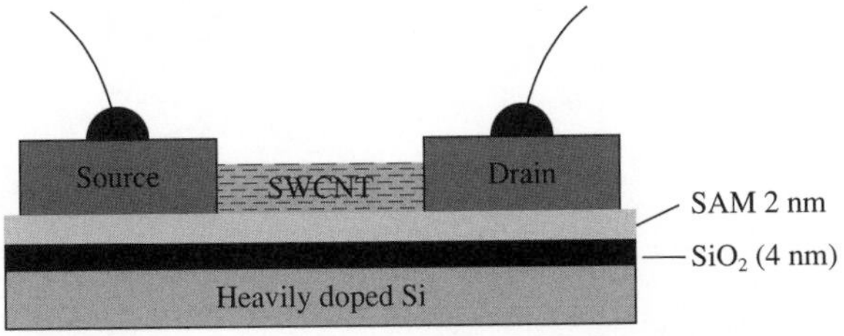

Figure 19.20

John Rodgers and researchers at the University of Illinois have reported a technique for producing near-perfect alignment of CNT transistors (Figure 19.21). The array is prepared by patterning thin strips of an iron catalyst on quartz crystals and then growing nanometer-wide CNTs along those strips using conventional carbon vapour deposition. The quartz crystal aligns the nanotubes. Then the researchers make transistors by depositing source, drain, and gate electrodes using conventional photolithography. Transistors made with about 2000 nanotubes can carry currents of one ampere, which is several orders of magnitude larger than the current possible with single nanotubes. The research group has also developed a technique for transferring the nanotube arrays onto any substrate, including silicon, plastic, and glass. See Coskun Kocabas, Seong Jun Kang, Taner Ozel, Moonsub Shim, and John A. Rogers, *Journal of Physical Chemistry C* **2007**, *111*, 17879, Improved synthesis of aligned arrays of single-walled carbon nanotubes and their implementation in thin film type transistors.

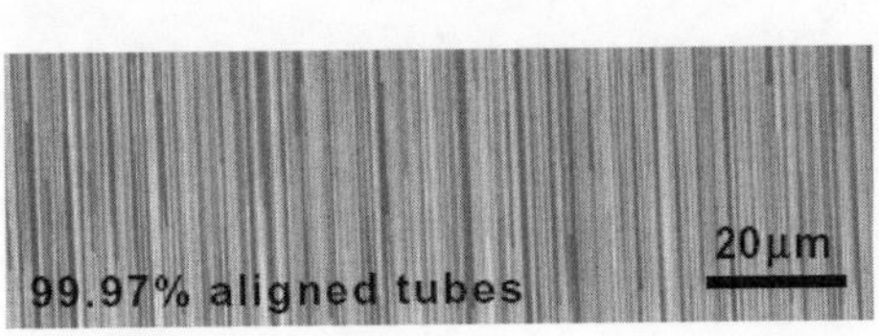

Figure 19.21

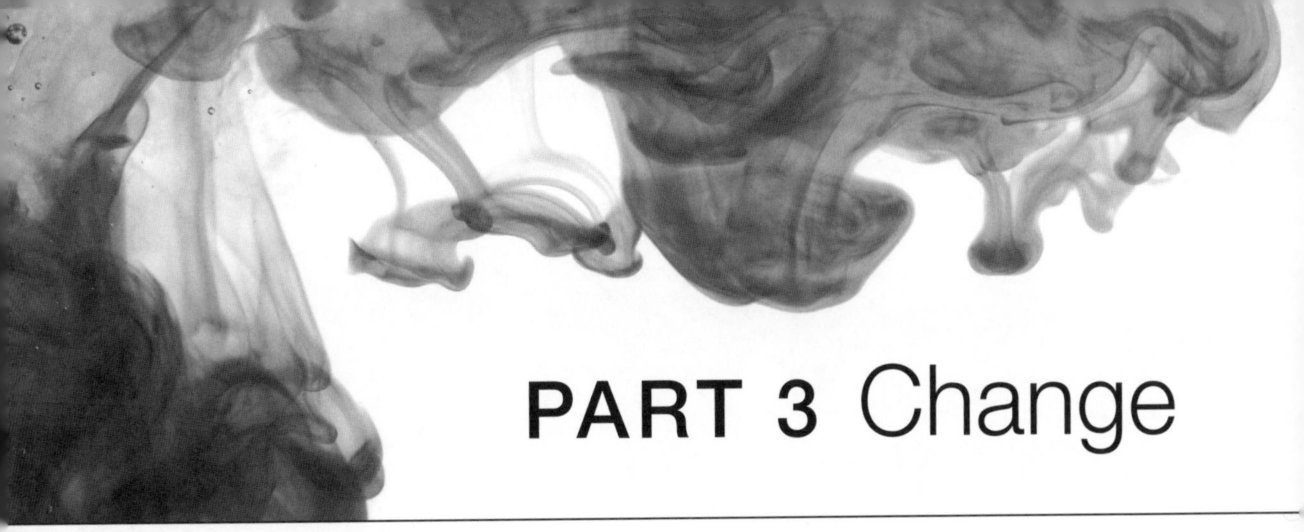

PART 3 Change

20 Molecules in motion

Answers to discussion questions

D20.1 **Diffusion** is the non-convective migration of particles (molecules) down a concentration gradient. Diffusion can be interpreted at the molecular level as being the result of the random jostling of the molecules in a fluid. The motion of the molecules is the result of a series of short jumps in random directions, a so-called **random walk**.

In the random walk model of diffusion, although a molecule may take many steps in a given time, it has only a small probability of being found from its starting point because some of the steps lead it away from the starting point but others lead it back. As a result, the net distance travelled increases only as the square root of the time. There is no net flow of molecules unless there is a concentration gradient in the fluid, also there are just as many molecules moving in one direction as another. The rate at which the molecules spread out is proportional to the concentration gradient. The constant of proportionality is called the **diffusion coefficient, *D***.

On the molecular level in a gas, **thermal conduction** occurs because of random molecular motions in the presence of a temperature gradient. Across any plane in the gas, there is a net flux of energy from the high-temperature side because molecules coming from that side carry a higher average energy per molecule across the plane than those coming from the low-temperature side. In solids, the situation is more complex as energy transport occurs through quantized elastic waves (phonons) and, in metals, also by electrons. Conduction in liquids can occur by all the mechanisms mentioned.

At the molecular (ionic) level, **electrical conduction** in an electrolytic solution is the net non-convective migration of ions in any given direction. When a gradient in electrical potential exists in a conductivity cell there will be a greater flow of positive ions in the direction of the negative electrode than in the direction of the positive electrode, hence there is a net flow of positive charge toward the region of low electrical potential. Likewise, a net flow of negative ions in the direction of the positive electrode will occur. In metals, only negatively charged electrons contribute to the current.

To see the connection between the flux of momentum and the viscosity, consider a fluid in a state of **Newtonian, laminar flow**, which can be imagined as steady flow in which the fluid moves past a surface in parallel layers of different velocities (Figure 20.10 of the text). The layer next to the wall of the vessel is stationary, and the velocity of successive layers varies linearly with distance, z, from the wall. Molecules ceaselessly move between the layers and bring with them the x component of linear momentum they possessed in their original layer. A layer is retarded by molecules arriving from a more slowly moving layer because they have a low momentum in the x direction. A layer is accelerated by molecules arriving from a more rapidly moving layer. We interpret the net retarding effect as the fluid's viscosity.

D20.3 According to the **Grotthuss mechanism**, there is an effective motion of a proton that involves the rearrangement of bonds in a group of water molecules. However, the actual mechanism is still highly contentious. Attention now focuses on the $H_9O_4^+$ unit, in which the nearly trigonal planar H_3O^+ ion is linked to three strongly solvating H_2O molecules. This cluster of atoms is itself hydrated, but the hydrogen bonds in the secondary sphere are weaker than in the primary sphere. It is envisaged that the rate-determining step is the cleavage of one of the weaker hydrogen bonds of this secondary sphere (Figure 20.1(a)). After this bond cleavage has taken place, and the released molecule has rotated through a few degrees (a process that takes about 1 ps), there is a rapid adjustment of bond lengths and angles in the remaining cluster to form a $H_5O_2^+$ cation of structure $H_2O\cdots H^+\cdots OH_2$ (Figure 20.1(b)). Shortly after this reorganization has occurred, a new $H_9O_4^+$ cluster forms as other molecules rotate into a position where they can become members of a secondary hydration sphere, but now the positive charge is located one molecule to the right of its initial location (Figure 20.1(c)). According to this model, there is no coordinated motion of a proton along a chain of molecules, simply a very rapid hopping between neighbouring sites, with a low activation energy. The model is consistent with the observation that the molar conductivity of protons increases as the pressure is raised, for increasing pressure ruptures the hydrogen bonds in water.

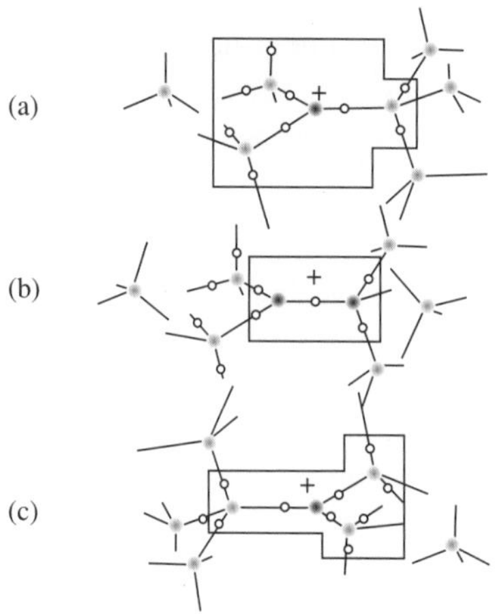

Figure 20.1

D20.5 Passive diffusion of a molecule through a biological membrane via an embedded ion channel protein is expected to be driven by the concentration gradient of the molecule across the membrane; in contrast to the observations of this exercise, the flux is proportional to the gradient and does not reach a maximum that depends on the size of the gradient. This is **Fick's first law of diffusion**. To explain the phenomena of **mediated transport** that exhibits an increased flux at larger gradients until reaching a maximum value at a particular gradient, consider the transport system of Figure 20.2. The transportable molecule locks into a surface hole of a membrane-embedded carrier protein, which undergoes a conformational change to admit the molecule into the hashed protein region in the figure, after which the conformational change squeezes the molecule into the biological cell. The transport may be passive or active. This transport mechanism does provide for a maximum in the flux of the molecule across the membrane. When the gradient is low and the concentration of the transportable molecule outside the cell is low, not all of the carrier protein in the membrane is engaged as a molecule–protein complex. However, as the concentration of the molecule increases there comes a point at which all carrier protein in the membrane of the cell are engaged with transportable molecule. At that point the flux reaches a maximum because a further increase in concentration does not increase the number of molecules being transported.

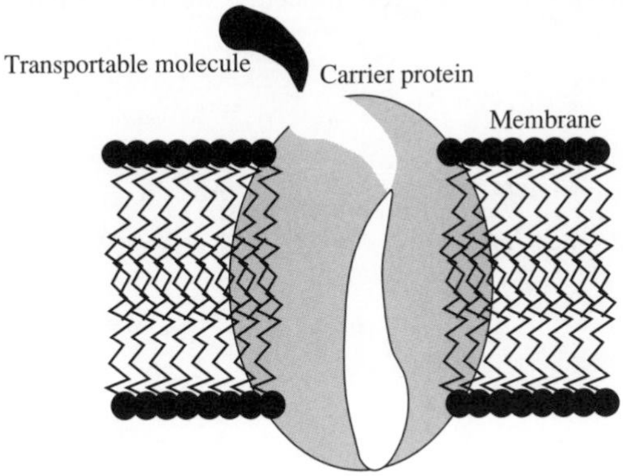

Figure 20.2

Solutions to exercises

E20.1(a) $$\bar{c} = \left(\frac{8RT}{\pi M}\right)^{1/2} \text{[20.7]} \quad \text{and} \quad c = \langle v^2 \rangle^{1/2} = \left(\frac{3RT}{M}\right)^{1/2} \text{[20.3]}$$

The mean translational kinetic energy is: $\langle E_k \rangle_m = \langle \tfrac{1}{2}mv^2 \rangle N_A = \tfrac{1}{2}mN_A\langle v^2 \rangle = \tfrac{1}{2}Mc^2 = \tfrac{3}{2}RT$.

The ratios of species 1 to species 2 at the same temperature are:

$$\frac{\bar{c}_1}{\bar{c}_2} = \left(\frac{M_2}{M_1}\right)^{1/2} \quad \text{and} \quad \frac{\langle E_k \rangle_1}{\langle E_k \rangle_2} = 1$$

(a) $$\frac{\bar{c}_{H_2}}{\bar{c}_{Hg}} = \left(\frac{200.6}{2.016}\right)^{1/2} = \boxed{9.975}$$

(b) The mean translation kinetic energy is independent of molecular mass and depends on temperature alone. Consequently, because the mean translational kinetic energy for a gas is proportional to T, the ratio of mean translational kinetic energies for gases at the same temperature always equals 1.

E20.2(a) (a) On the assumption that the gas is perfect,

$$T = \frac{pV}{nR} = \frac{N_A pV}{NR}$$
$$= \frac{(6.022 \times 10^{23}\ \text{mol}^{-1})(1.00 \times 10^5\ \text{Pa}) \times (1.0 \times 10^{-3}\ \text{m}^3)}{(1.0 \times 10^{23}) \times (8.3145\ \text{J K}^{-1}\ \text{mol}^{-1})} = 72.\bar{4} = \boxed{72\ \text{K}}$$

(b) $c = \langle v^2 \rangle^{1/2} = \left(\dfrac{3RT}{M}\right)^{1/2}$ [20.3]

$$= \left\{\frac{3 \times (8.3145\ \text{J K}^{-1}\,\text{mol}^{-1}) \times (72.\bar{4}\ \text{K})}{2.016 \times 10^{-3}\ \text{kg mol}^{-1}}\right\}^{1/2} = \boxed{94\bar{6}\ \text{m s}^{-1}}$$

(c) The $\boxed{\text{temperature would not be different}}$ if it were O_2 molecules and exerted the same pressure in the same volume, but the root mean square speed of dioxygen molecules would be smaller by a factor of

$$(M_{H_2}/M_{O_2})^{1/2} = (2.016/32.00)^{1/2} = 0.2510$$

E20.3(a) The Maxwell distribution of speeds is $f(v) = 4\pi\left(\dfrac{M}{2\pi RT}\right)^{3/2} v^2 e^{-Mv^2/2RT}$ [20.4]

The factor $M/2RT$ can be evaluated as

$$\frac{M}{2RT} = \frac{28.02 \times 10^{-3}\ \text{kg mol}^{-1}}{2 \times (8.3145\ \text{J K}^{-1}\,\text{mol}^{-1}) \times (500\ \text{K})} = 3.37 \times 10^{-6}\ \text{m}^{-2}\,\text{s}^2$$

Though $f(v)$ varies over the range 290 to 300 m s^{-1}, the variation is small over this small range and we estimate its value to be a constant equal to its value at the center of the range.

$$f(295\ \text{m s}^{-1}) = (4\pi) \times \left(\frac{3.37 \times 10^{-6}\ \text{m}^{-2}\,\text{s}^2}{\pi}\right)^{3/2} \times (295\ \text{m s}^{-1})^2 \times e^{(-3.37\times10^{-6})\times(295)^2} = 9.06 \times 10^{-4}\,\text{m}^{-1}\,\text{s}$$

Therefore, the fraction of molecules in the specified range is

$$\int_{v_{\text{low}}}^{v_{\text{high}}} f\,dv\ [20.5] = f\int_{v_{\text{low}}}^{v_{\text{high}}} dv = f \times \Delta v = (9.06 \times 10^{-4}\,\text{m}^{-1}\,\text{s}) \times (10\ \text{m s}^{-1}) = \boxed{9.06 \times 10^{-3}}$$

corresponding to 0.91%.

COMMENT. This is a rather small percentage and suggests that the approximation of constancy of $f(v)$ over the range is adequate. To test the approximation $f(295\ \text{m s}^{-1})$ and $f(300\ \text{m s}^{-1})$ could be evaluated.

E20.4(a) We derive the desired expression for Δc for a perfect gas by substitution of eqns 20.3 and 20.7 into the definition of Δc and simplifying the result.

$$\Delta c \equiv \{\langle v^2 \rangle - \langle v \rangle^2\}^{1/2} = \{c^2 - \bar{c}^2\}^{1/2}$$

$$= \left\{\left(\frac{3RT}{M}\right) - \left(\frac{8RT}{\pi M}\right)\right\}^{1/2} \quad [20.3, 20.7]$$

$$= \boxed{\left(3 - \frac{8}{\pi}\right)^{1/2}\left(\frac{RT}{M}\right)^{1/2}}$$

Alternatively, the spread of speeds can be written in the attractive form: $\Delta c = \boxed{\left(1 - \dfrac{8}{3\pi}\right)^{1/2} c}$.

E20.5(a) The volume and radius of a spherical container are related by

$$V = \tfrac{4}{3}\pi r^3 \quad \text{or} \quad r = (3V/4\pi)^{1/3}.$$

Thus, $2r = 2 \times \{3 \times (1.00 \times 10^{-3}\,\text{m}^3)/4\pi\}^{1/3} = 0.124\,\text{m}$.

The pressure at which the mean free path λ is comparable to the container diameter $2r$ is

$$\lambda = \frac{kT}{2^{1/2}\sigma p}\,[\text{See note below.}] \quad \text{or} \quad p = \frac{kT}{2^{1/2}\sigma\lambda}$$

$$p = \frac{(1.381 \times 10^{-23}\,\text{J K}^{-1}) \times (298\,\text{K})}{(2^{1/2}) \times (0.36 \times 10^{-18}\,\text{m}^2) \times (0.124\,\text{m})}\,[\text{Table 20.1}] = \boxed{0.0652\,\text{Pa}}\,[1\,\text{J} = 1\,\text{Pa m}^3]$$

Note: The standard, simple kinetic model expression for the mean free path λ estimates that a molecule moving with the mean speed $\bar{c}$ spends a mean time of $1/z$ between collisions, where z is the collision frequency [20.11a°,b°]. (See the IUPAC Compendium of Chemical Terminology in the IUPAC Gold Book; http://goldbook.iupac.org/.) Thus, the model estimates that

$$\lambda = \frac{\bar{c}}{z}\,[\text{20.12 alternative}] = \frac{kT}{2^{1/2}\sigma p}\,[\text{20.13 alternative}]$$

We use this definition throughout our computations.

E20.6(a)

$$\lambda = \frac{kT}{2^{1/2}\sigma p}\,[\text{E20.5(a) note}]$$

$$= \frac{(1.381 \times 10^{-23}\,\text{J K}^{-1}) \times (217\,\text{K})}{(2^{1/2}) \times (0.43 \times 10^{-18}\,\text{m}^2) \times (0.050\,\text{atm}) \times (1.013 \times 10^5\,\text{Pa atm}^{-1})}$$

$$= \boxed{0.97\,\mu\text{m}}$$

E20.7(a) The collision frequency, z, is given by $z = 2^{1/2}\sigma\bar{c}p/kT$ [20.11b and 20.9]. Since the mean speed is dependent on temperature, not pressure, it may be calculated once and used in all parts of this exercise.

$$\bar{c} = \left(\frac{8RT}{\pi M}\right)^{1/2}\,[20.7]$$

$$= \left\{\frac{8 \times (8.3145\,\text{J mol}^{-1}\,\text{K}^{-1}) \times (298\,\text{K})}{\pi \times (39.95 \times 10^{-3}\,\text{kg mol}^{-1})}\right\}^{1/2}$$

$$= \boxed{397\,\text{m s}^{-1}}$$

$$z = \frac{2^{1/2}\sigma\bar{c}p}{kT}\,[\text{20.11b and 20.9}]$$

$$= \frac{2^{1/2} \times (0.36 \times 10^{-18}\,\text{m}^2) \times (397\,\text{m s}^{-1})}{(1.381 \times 10^{-23}\,\text{J K}^{-1}) \times (298\,\text{K})} \times \left(\frac{1.01325 \times 10^5\,\text{Pa}}{1\,\text{atm}}\right) \times p\,[\text{Table 20.1}]$$

$$= (4.98 \times 10^9\,\text{s}^{-1}) \times (p/\text{atm})$$

We need only plug in the pressure in atmospheres to calculate z at that pressure.

(a) $z = \boxed{5.0 \times 10^{10}\,\text{s}^{-1}}$ when $p = 10$ atm

(b) $z = \boxed{5.0 \times 10^{9}\,\text{s}^{-1}}$ when $p = 1$ atm

(c) $z = \boxed{5.0 \times 10^{3}\,\text{s}^{-1}}$ when $p = 10^{-6}$ atm

E20.8(a)

$$\lambda = \frac{kT}{2^{1/2}\sigma p}\ [\text{E20.5(a) note}]$$

$$= \frac{(1.381 \times 10^{-23}\,\text{J K}^{-1}) \times (298.15\,\text{K})}{(2^{1/2}) \times (0.43 \times 10^{-18}\,\text{m}^2) \times (p)}$$

$$= \frac{6.7\overline{7} \times 10^{-3}\,\text{m}}{(p/\text{Pa})} = \frac{6.6\overline{8} \times 10^{-8}\,\text{m}}{p/\text{atm}}$$

(a) When $p = 10$ atm, $\lambda = 6.7 \times 10^{-9}\,\text{m} = \boxed{6.7\,\text{nm}}$.

(b) When $p = 1$ atm, $\lambda = 6.7 \times 10^{-8}\,\text{m} = \boxed{67\,\text{nm}}$.

(c) When $p = 10^{-6}$ atm, $\lambda = 6.7 \times 10^{-2}\,\text{m} = \boxed{6.7\,\text{cm}}$.

E20.9(a) $A = (2.5\,\text{mm}) \times (3.0\,\text{mm}) = 7.5 \times 10^{-6}\,\text{m}^2$

The collision frequency of the Ar gas molecules with surface area A equals $Z_W A$.

$$Z_W A = \frac{pN_A}{(2\pi MRT)^{1/2}} A\,[20.14]$$

$$= \frac{(90\,\text{Pa}) \times (6.022 \times 10^{23}\,\text{mol}^{-1}) \times (7.5 \times 10^{-6}\,\text{m}^2)}{\{2\pi(39.95 \times 10^{-3}\,\text{kg mol}^{-1}) \times (8.3145\,\text{J K}^{-1}\,\text{mol}^{-1}) \times (500\,\text{K})\}^{1/2}}$$

$$= 1.2\overline{6} \times 10^{19}\,\text{s}^{-1}$$

The number of argon molecule collisions within A in time interval t equals $Z_W At$ if p does not change significantly during the period t.

$$Z_W At = (1.2\overline{6} \times 10^{19}\,\text{s}^{-1}) \times (15\,\text{s}) = \boxed{1.9 \times 10^{20}}$$

E20.10(a) The mass loss equals the effusion mass loss multiplied by the time period t:

$m_{\text{loss}} = (\text{rate of effusion}) \times t \times m = (\text{rate of effusion}) \times t \times M/N_A$.

$$m_{\text{loss}} = \left(\frac{pA_0 N_A}{(2\pi MRT)^{1/2}}\right) \times \left(\frac{Mt}{N_A}\right)[20.16]$$

$$= pA_0 t \times \left(\frac{M}{2\pi RT}\right)^{1/2}$$

$$= (0.835\,\text{Pa}) \times \{\pi(1.25 \times 10^{-3}\,\text{m})^2\} \times (2 \times 3600\,\text{s}) \times \left(\frac{0.260\,\text{kg mol}^{-1}}{2\pi(8.3145\,\text{J mol}^{-1}\,\text{K}^{-1}) \times (400\,\text{K})}\right)^{1/2}$$

$$= 1.04 \times 10^{-4}\,\text{kg} = \boxed{104\,\text{mg}}$$

Question. For the same solid shaped in the form of a sphere of radius 0.050 m and suspended in a vacuum, what will be the mass loss in 2.00 h? *Hint*. Make any reasonable approximations.

E20.11(a) The mass loss equals the effusion mass loss multiplied by the time period t:

$$m_{\text{loss}} = (\text{rate of effusion}) \times t \times m = (\text{rate of effusion}) \times t \times M/N_{\text{A}}.$$

$$m_{\text{loss}} = \left(\frac{pA_0N_{\text{A}}}{(2\pi MRT)^{1/2}}\right) \times \left(\frac{Mt}{N_{\text{A}}}\right) [20.16] = pA_0t \times \left(\frac{M}{2\pi RT}\right)^{1/2}$$

Solving for p gives

$$p = \frac{m_{\text{loss}}}{A_0t}\left(\frac{2\pi RT}{M}\right)^{1/2}$$

$$= \frac{285 \times 10^{-6}\,\text{kg}}{\{\pi(0.25 \times 10^{-3}\,\text{m})^2\} \times (400\,\text{s})}\left(\frac{2\pi(8.3145\,\text{J mol}^{-1}\,\text{K}^{-1}) \times (673\,\text{K})}{0.100\,\text{kg mol}^{-1}}\right)^{1/2}$$

$$= \boxed{2.15 \times 10^3\,\text{Pa}}$$

E20.12(a) The pressure of this exercise changes significantly during time period t so it is useful to spend a moment finding an expression for $p(t)$. Mathematically, the rate of effusion is the derivative $-\mathrm{d}N/\mathrm{d}t$. Substitution of the perfect gas law for N, $N = pVN_{\text{A}}/RT$, where V and T are constants, reveals that the rate of effusion can be written as $-(N_{\text{A}}V/RT)\mathrm{d}p/\mathrm{d}t$. This formulation of the rate of effusion, along with eqn 20.16, is used to find $p(t)$.

$$-\left(\frac{N_{\text{A}}V}{RT}\right)\frac{\mathrm{d}p}{\mathrm{d}t} = \frac{pA_0N_{\text{A}}}{(2\pi MRT)^{1/2}} [20.16]$$

$$\frac{\mathrm{d}p}{\mathrm{d}t} = -\frac{pA_0}{V}\left(\frac{RT}{2\pi M}\right)^{1/2}$$

$$\frac{\mathrm{d}p}{p} = -\frac{\mathrm{d}t}{\tau}, \quad \text{where} \quad \tau = \frac{V}{A_0}\left(\frac{2\pi M}{RT}\right)^{1/2}$$

$$\int_{p_0}^{p}\frac{\mathrm{d}p}{p} = -\frac{1}{\tau}\int_0^t \mathrm{d}t, \quad \text{where } p_0 \text{ is the initial pressure}$$

$$\ln\frac{p}{p_0} = -\frac{t}{\tau} \quad \text{or} \quad p(t) = p_0\mathrm{e}^{-t/\tau}$$

The nitrogen gas and carbon dioxide gas data can be used to determine the relaxation time, τ, for each.

$$\tau_{N_2} = \frac{t_{N_2}}{\ln(p_0/p)_{N_2}} = \frac{42\,\text{s}}{\ln(75/50)} = 104\,\text{s}$$

$$\tau_{CO_2} = \frac{t_{CO_2}}{\ln(p_0/p)_{CO_2}} = \frac{52\,\text{s}}{\ln(75/50)} = 128\,\text{s}$$

The above definition of τ shows that it is proportional to $M^{1/2}$. Since the ratio of the relaxation times cancels the constant of proportionality,

$$\left(\frac{M_{CO_2}}{M_{N_2}}\right)^{1/2} = \frac{\tau_{CO_2}}{\tau_{N_2}}$$

$$M_{CO_2} = \left(\frac{\tau_{CO_2}}{\tau_{N_2}}\right)^2 M_{N_2}$$

$$= \left(\frac{128}{104}\right)^2 \times (28.02\ \text{g mol}^{-1}) = \boxed{42.4\ \text{g mol}^{-1}}$$

E20.13(a) In E20.12a it is shown that

$$\ln\frac{p}{p_0} = -\frac{t}{\tau} \quad \text{or} \quad p(t) = p_0 e^{-t/\tau}, \quad \text{where} \quad \tau = \frac{V}{A_0}\left(\frac{2\pi M}{RT}\right)^{1/2}.$$

The relaxation time, τ, of oxygen is calculated with the data.

$$\tau = \left(\frac{3.0\ \text{m}^3}{\pi(0.10 \times 10^{-3}\ \text{m})^2}\right) \times \left\{\frac{2\pi(32.00 \times 10^{-3}\ \text{kg mol}^{-1})}{(8.3145\ \text{J mol}^{-1}\ \text{K}^{-1}) \times (298\ \text{K})}\right\}^{1/2} = 8.6 \times 10^5\ \text{s} = 10.\bar{0}\ \text{days}$$

The time required for the specified pressure decline is calculated with the above eqn.

$$t = \tau \ln(p_0/p) = (10.\bar{0}\ \text{days}) \times \ln(80/70) = \boxed{1.3\ \text{days}}$$

E20.14(a) The flux is

$$J = -\kappa\frac{dT}{dz}[20.20] = -\tfrac{1}{3}\lambda\bar{c}C_{V,m}[\text{A}]\frac{dT}{dz}[20.23] = -\frac{\tfrac{1}{3}\lambda\bar{c}C_{V,m}p}{RT}\frac{dT}{dz}\ [\text{perfect gas, [A]} = p/RT]$$

where the negative sign indicates flow toward lower temperature and

$$\bar{c} = \left(\frac{8RT}{\pi M}\right)^{1/2} [20.7] \quad \text{and} \quad \lambda = \frac{kT}{2^{1/2}\sigma p}[\text{E20.5(a) note}] = \frac{RT}{2^{1/2}N_A\sigma p}.$$

$$J = -\tfrac{1}{3}\left(\frac{RT}{2^{1/2}N_A\sigma p}\right) \times \left(\frac{8RT}{\pi M}\right)^{1/2} \times \left(\frac{C_{V,m}p}{RT}\right) \times \frac{dT}{dz}$$

$$= -\tfrac{2}{3} \times \left(\frac{RT}{\pi M}\right)^{1/2} \times \left(\frac{C_{V,m}}{N_A\sigma}\right) \times \frac{dT}{dz}$$

For argon $M = 39.95\ \text{g mol}^{-1}$, $\sigma = 0.36\ \text{nm}^2$ [Table 20.1], and appendix data tables are used to calculate $C_{V,m}$: $C_{V,m} = C_{p,m} - R = (20.786 - 8.3145)\ \text{J K}^{-1}\ \text{mol}^{-1} = 12.5\ \text{J K}^{-1}\ \text{mol}^{-1}$.

$$J = -\tfrac{2}{3} \times \left(\frac{(8.3145\ \text{J K}^{-1}\ \text{mol}^{-1}) \times (273\ \text{K})}{\pi \times (39.95 \times 10^{-3}\ \text{kg mol}^{-1})}\right)^{1/2} \times \left\{\frac{12.5\ \text{J K}^{-1}\ \text{mol}^{-1}}{(6.022 \times 10^{23}\ \text{mol}^{-1}) \times (0.36 \times 10^{-18}\ \text{m}^2)}\right\}$$

$$\times (2.5\ \text{K m}^{-1})$$

$$= \boxed{-0.013\ \text{J m}^{-2}\ \text{s}^{-1}}$$

E20.15(a) The coefficient of thermal conductivity, κ, is a function of the mean free path, λ, which in turn is a function of the collision cross-section, σ. Hence, reversing the order, σ can be obtained from κ.

$$\kappa = \tfrac{1}{3}\lambda\bar{c}C_{V,\mathrm{m}}[\mathrm{A}]\,[20.23] \quad \text{and} \quad \bar{c} = \left(\frac{8RT}{\pi M}\right)^{1/2}\,[20.7]$$

$$\lambda = \frac{kT}{2^{1/2}\sigma p}\,[\text{E20.5(a) note}] = \frac{1}{2^{1/2}\sigma N_{\mathrm{A}}[\mathrm{A}]}\left[\frac{p}{kT} = N_{\mathrm{A}}[\mathrm{A}]\right].$$

Therefore, $\kappa = \dfrac{\bar{c}C_{V,\mathrm{m}}}{3\times 2^{1/2}\sigma N_{\mathrm{A}}}$ or $\sigma = \dfrac{\bar{c}C_{V,\mathrm{m}}}{3\times 2^{1/2}N_{\mathrm{A}}\kappa}$.

For neon at 273 K, $\bar{c} = \left\{\dfrac{8\times(8.3145\ \mathrm{J\ K^{-1}\ mol^{-1}})\times(273\ \mathrm{K})}{\pi\times(20.18\times 10^{-3}\ \mathrm{kg\ mol^{-1}})}\right\}^{1/2} = 535\ \mathrm{m\ s^{-1}}$.

From Table 20.2, $\kappa = 0.0465\ \mathrm{J\ K^{-1}\ m^{-1}\ s^{-1}}$.

The $C_{V,\mathrm{m}}$ value is calculated with the perfect gas relationship $C_{V,\mathrm{m}} = C_{p,\mathrm{m}} - R$ and the $C_{p,\mathrm{m}}$ value provided in an appendix table ($20.786\ \mathrm{J\ K^{-1}\ mol^{-1}}$). This gives $C_{V,\mathrm{m}} = 12.472\ \mathrm{J\ K^{-1}\ mol^{-1}}$. Alternatively, the $C_{V,\mathrm{m}}$ value is calculated with the equipartition theorem: $C_{V,\mathrm{m}} = \tfrac{3}{2}R = 12.472\ \mathrm{J\ K^{-1}\ mol^{-1}}$.

$$\sigma = \frac{(535\ \mathrm{m\ s^{-1}})\times(12.472\ \mathrm{J\ K^{-1}\ mol^{-1}})}{3\times 2^{1/2}\times(6.022\times 10^{23}\ \mathrm{mol^{-1}})\times(0.0465\ \mathrm{J\ K^{-1}\ m^{-1}\ s^{-1}})}$$

$$= 5.62\times 10^{-20}\ \mathrm{m^2} = \boxed{0.0562\ \mathrm{nm^2}}$$

The experimental value is $0.24\ \mathrm{nm^2}$.

Question. What approximations inherent in the equation used in the solution to this exercise are the likely cause of the factor of 4 difference between the experimental and calculated values of the collision cross-section for neon?

E20.16(a) The thermal energy flux ('heat' flux) is described by: $J(\text{energy}) = -\kappa\dfrac{\mathrm{d}T}{\mathrm{d}z}$ [20.20], where the negative sign indicates flow toward lower temperature. This is the rate of energy transfer per unit area. The total rate of energy transfer across area A is

$$\frac{\mathrm{d}E}{\mathrm{d}t} = AJ(\text{energy}) = -\kappa A\frac{\mathrm{d}T}{\mathrm{d}z}.$$

To calculate the temperature gradient with the given data, we assume that the gradient is in a steady state. Then, recognizing that temperature differences have identical magnitude in Celsius or Kelvin units,

$$\frac{\mathrm{d}T}{\mathrm{d}z} = \frac{\Delta T}{\Delta z} = \frac{\{(-10)-(25)\}\,\mathrm{K}}{5.0\times 10^{-2}\ \mathrm{m}} = -7.0\times 10^{2}\ \mathrm{K\ m^{-1}}.$$

We now assume that the coefficient of thermal conductivity of the gas between the window panes is comparable to that of nitrogen given in Table 20.2: $\kappa \approx 0.0240\ \mathrm{J\ K^{-1}\ m^{-1}\ s^{-1}}$.

Therefore, the rate of outward energy transfer is

$$\frac{\mathrm{d}E}{\mathrm{d}t} \approx -(0.0240\ \mathrm{J\ K^{-1}\ m^{-1}\ s^{-1}}) \times (1.0\ \mathrm{m^2}) \times (-7.0 \times 10^2\ \mathrm{K\ m^{-1}}) \approx 17\ \mathrm{J\ s^{-1}} = \boxed{17\ \mathrm{W}}$$

A 17-W heater is needed to balance this rate of heat loss.

E20.17(a) $\eta = \frac{1}{3}M\lambda\bar{c}p/RT$ [20.24, [A] $= p/RT$ for a perfect gas]

We begin by substituting kinetic theory relationships for λ [E20.5(a) note] and $\bar{c}$ [20.7].

$$\eta = \tfrac{1}{3}M \times \left(\frac{RT}{2^{1/2}\sigma N_{\mathrm{A}}p}\right) \times \left(\frac{8RT}{\pi M}\right)^{1/2} \times \left(\frac{p}{RT}\right) = \left(\frac{M}{3 \times 2^{1/2}\sigma N_{\mathrm{A}}}\right) \times \left(\frac{8RT}{\pi M}\right)^{1/2}$$

We now solve, and compute, the collision cross-section for neon at 273 K.

From Table 20.2, $\eta = 298 \times 10^{-6}\ \mathrm{P} = 298 \times 10^{-7}\ \mathrm{kg\ m^{-1}\ s^{-1}}$.

$$\sigma = \left\{\frac{20.18 \times 10^{-3}\ \mathrm{kg\ mol^{-1}}}{3 \times 2^{1/2} \times (6.022 \times 10^{23}\ \mathrm{mol^{-1}}) \times (298 \times 10^{-7}\ \mathrm{kg\ m^{-1}\ s^{-1}})}\right\}$$

$$\times \left\{\frac{8(8.3145\ \mathrm{J\ K^{-1}\ mol^{-1}}) \times (273\ \mathrm{K})}{\pi(20.18 \times 10^{-3}\ \mathrm{kg\ mol^{-1}})}\right\}^{1/2}$$

$$= 1.42 \times 10^{-19}\ \mathrm{m^2} = \boxed{0.142\ \mathrm{nm^2}}$$

E20.18(a) The flow rate of a compressible gas in a steady-state, Newtonian, laminar flow through a pipe of radius R_{pipe} and length l is described by Poiseuille's formula,

$$\frac{\mathrm{d}V}{\mathrm{d}t} = \frac{(p_1^2 - p_2^2)\pi R_{\mathrm{pipe}}^4}{16 l \eta p_0},$$

where V is the volume flowing, p_1 and p_2 are pressures at each end of the tube, and p_0 is the pressure at which the volume is measured. (This formula is derived below.) Solving for p_1^2 gives

$$p_1^2 = p_2^2 + \left(\frac{16 l \eta p_0}{\pi R_{\mathrm{pipe}}^4}\right) \times \left(\frac{\mathrm{d}V}{\mathrm{d}t}\right)$$

$$= (1.00 \times 10^5)^2\ \mathrm{Pa^2} + \left(\frac{(16) \times (8.50\ \mathrm{m}) \times (176 \times 10^{-7}\ \mathrm{kg\ m^{-1}\ s^{-1}}) \times (1.00 \times 10^5\ \mathrm{Pa})}{\pi \times (5.0 \times 10^{-3}\ \mathrm{m})^4}\right)$$

$$\times \left(\frac{9.5 \times 10^2\ \mathrm{m^3}}{3600\ \mathrm{s}}\right) = 4.22 \times 10^{10}\ \mathrm{Pa^2}$$

Hence, $p_1 = \boxed{205\ \mathrm{kPa}}$ (2.05 bar).

To begin the derivation of the Poiseuille formula consider eqn 20.21 in the form applicable to flow through a cylinder of radius r, where $0 \le r \le R_{\mathrm{pipe}}$: J (x-component of momentum) $= -\eta \times (\mathrm{d}v_x/\mathrm{d}r)$ with x parallel to the tube walls. This momentum flux is the increase in momentum along the radius per time t per area of cylinder surface. As such, the force acting in the x direction is given by

$$(2\pi r l) \times J\,(x\text{ component of momentum}) = -2\pi r l \eta \times (\mathrm{d}v_x/\mathrm{d}r)$$

This force is also given by the pressure difference between the cylinder ends, which is taken to be a constant along the radius with a constant gradient along x so that $(p_2 - p_1)/l = \mathrm{d}p/\mathrm{d}x$, multiplied by the area cross-section of the cylinder: $-\pi r^2(p_2 - p_1)$. Steady, laminar flow requires that these two forces balance. Thus, in this flow model we envision a maximum of flow at $r = 0$ with decreasing flow until it is zero at $r = R$ because of the viscous drag on the surface of each cylinder due to friction with more slowly moving surrounding gas. Equating the two forces and solving for $\mathrm{d}v_x$ gives

$$-2\pi r l\eta \times (\mathrm{d}v_x/\mathrm{d}r) = -(p_2 - p_1)\pi r^2$$

$$\mathrm{d}v_x = \frac{1}{2\eta}\frac{\mathrm{d}p}{\mathrm{d}x} r\,\mathrm{d}r$$

Now, the volume size of a cylinder of length $\mathrm{d}x$ is given by $\mathrm{d}V = \pi r^2\mathrm{d}x$ so $\mathrm{d}x/\mathrm{d}t = (1/\pi r^2)\mathrm{d}V/\mathrm{d}t$. Equating $\mathrm{d}x/\mathrm{d}t$ with $\mathrm{d}v_x$ in our working equation gives

$$\mathrm{d}\left(\frac{\mathrm{d}V}{\mathrm{d}t}\right) = \frac{\pi}{2\eta}\frac{\mathrm{d}p}{\mathrm{d}x} r^3\mathrm{d}r$$

and integration from $r = 0$ to $r = R_{\text{pipe}}$ with no contribution to the gas flow at $r = R_{\text{pipe}}$ gives

$$\frac{\mathrm{d}V}{\mathrm{d}t} = -\frac{\pi R_{\text{pipe}}^4}{8\eta}\frac{\mathrm{d}p}{\mathrm{d}x}$$

To assure conservation of mass in a gaseous volume that expands while moving toward the low-pressure end, we make the substitution $V = nRT/p$ to get

$$\frac{\mathrm{d}n}{\mathrm{d}t} = -\frac{\pi R_{\text{pipe}}^4 p}{8\eta RT}\frac{\mathrm{d}p}{\mathrm{d}x} = -\frac{\pi R_{\text{pipe}}^4}{16\eta RT}\frac{\mathrm{d}p^2}{\mathrm{d}x}, \quad \text{where} \quad \frac{\mathrm{d}p^2}{\mathrm{d}x} = \frac{p_2^2 - p_1^2}{l}$$

because the pressure gradient is independent of x

If we measure the amount of gas at $p = p_0$, we can make the substitution $\mathrm{d}n = (p_0/RT)\mathrm{d}V$ to get a gas flow of

$$\boxed{\frac{\mathrm{d}V}{\mathrm{d}t} = \frac{\pi R_{\text{pipe}}^4}{16\eta p_0 l}(p_1^2 - p_2^2) \quad \text{Poiseuille's formula}}$$

E20.19(a) Taking air to be composed of 20% oxygen and 80% nitrogen, we estimate the average molar mass of air to be: $M = (0.20 \times 32 + 0.80 \times 28)\ \text{g mol}^{-1} = 29\ \text{g mol}^{-1}$.

$$\eta = \tfrac{1}{3}M\lambda\bar{c}[\text{A}]\,[20.24] = \left(\frac{1}{3\sigma N_{\text{A}}}\right) \times \left(\frac{4MRT}{\pi}\right)^{1/2} \quad [\text{E20.5(a) note, 20.7, [A]} = p/RT]$$

$$= \left(\frac{1}{3 \times (0.40 \times 10^{-18}\ \text{m}^2) \times (6.022 \times 10^{23}\ \text{mol}^{-1})}\right)$$

$$\times \left\{\frac{4 \times (29 \times 10^{-3}\ \text{kg mol}^{-1}) \times (8.3145\ \text{J K}^{-1}\ \text{mole}^{-1}) \times T}{\pi}\right\}^{1/2}$$

$$= (7.6\overline{7} \times 10^{-7}\ \text{kg m}^{-1}\,\text{s}^{-1}) \times (T/\text{K})^{1/2}$$

(a) At $T = 273$ K, $\eta = (7.6\overline{7} \times 10^{-7}\ \text{kg m}^{-1}\text{s}^{-1}) \times (273)^{1/2} = 1.2\overline{7} \times 10^{-5}\ \text{kg m}^{-1}\text{s}^{-1} = \boxed{12\overline{7}\ \mu\text{P}}$

(b) At $T = 298$ K, $\eta = (7.6\overline{7} \times 10^{-7}\ \text{kg m}^{-1}\text{s}^{-1}) \times (298)^{1/2} = 1.3\overline{2} \times 10^{-5}\ \text{kg m}^{-1}\text{s}^{-1} = \boxed{13\overline{2}\ \mu\text{P}}$

(c) At $T = 1000$ K, $\eta = (7.6\overline{7} \times 10^{-7}\ \text{kg m}^{-1}\text{s}^{-1}) \times (1000)^{1/2} = 2.4\overline{3} \times 10^{-5}\ \text{kg m}^{-1}\text{s}^{-1} = \boxed{24\overline{3}\ \mu\text{P}}$

E20.20(a) $\kappa = \frac{1}{3}\lambda\bar{c}C_{V,\text{m}}[\text{A}]\,[20.23] = \left(\frac{C_{V,\text{m}}}{3\sigma N_\text{A}}\right) \times \left(\frac{4RT}{\pi M}\right)^{1/2}$ [E20.5(a) note, 20.7, [A] $= p/RT$]

According to the equipartition theorem $C_{V,\text{m}} = \frac{3}{2}R = \frac{3}{2}kN_\text{A}$ for a monatomic gas so

$$\kappa = \left(\frac{k}{2\sigma}\right) \times \left(\frac{4RT}{\pi M}\right)^{1/2}$$

$$= \left(\frac{1.381 \times 10^{-23}\ \text{J K}^{-1}}{2}\right) \times \left(\frac{4 \times (8.3145\ \text{J K}^{-1}\text{mol}^{-1}) \times (300\ \text{K})}{\pi}\right)^{1/2}$$

$$\times \left(\frac{1}{(10^{-18}\ \text{m}^2) \times \sigma/\text{nm}^2}\right) \times \left(\frac{1}{(10^{-3}\ \text{kg mol}^{-1}) \times (M/\text{g mol}^{-1})}\right)^{1/2}$$

$$= (0.0123\ \text{J K}^{-1}\text{m}^{-1}\text{s}^{-1}) \times \left(\frac{1}{\sigma/\text{nm}^2}\right) \times \left(\frac{1}{M/\text{g mol}^{-1}}\right)^{1/2} \quad \text{at} \quad T = 300\ \text{K}$$

$J_\text{energy} = -\kappa\,\text{d}T/\text{d}z\ [20.20] = -\kappa\Delta T/\Delta z$ [steady-state flux]
$= -\kappa \times (310\ \text{K} - 295\ \text{K})/(0.10\ \text{m})$
$= -(150\ \text{K m}^{-1})\kappa$ where the negative indicates an energy flux toward the low temperature

(a) For argon:

$$\kappa = (0.0123\ \text{J K}^{-1}\text{m}^{-1}\text{s}^{-1}) \times \left(\frac{1}{0.36}\right) \times \left(\frac{1}{39.95}\right)^{1/2} = \boxed{5.4\ \text{mJ K}^{-1}\text{m}^{-1}\text{s}^{-1}}$$

The observed value of κ is larger by a factor of 3.

$J_\text{energy} = -(150\ \text{K m}^{-1}) \times (5.4\ \text{mJ K}^{-1}\text{m}^{-1}\text{s}^{-1}) = \boxed{-0.81\ \text{W m}^{-2}}$

Rate of energy flow $= J_\text{energy}A = (-0.81\ \text{W m}^{-2}) \times (0.010\ \text{m}^2) = \boxed{-8.1\ \text{mW}}$

The negative sign indicates flow toward the low temperature.

(b) For helium:

$$\kappa = (0.0123\ \text{J K}^{-1}\text{m}^{-1}\text{s}^{-1}) \times \left(\frac{1}{0.21}\right) \times \left(\frac{1}{4.00}\right)^{1/2} = \boxed{29\ \text{mJ K}^{-1}\text{m}^{-1}\text{s}^{-1}}$$

The observed value of κ is larger by a factor of 5.

$J_\text{energy} = -(150\ \text{K m}^{-1}) \times (29\ \text{mJ K}^{-1}\text{m}^{-1}\text{s}^{-1}) = \boxed{-4.4\ \text{W m}^{-2}}$

Rate of energy flow $= J_\text{energy}A = (-4.4\ \text{W m}^{-2}) \times (0.010\ \text{m}^2) = \boxed{-44\ \text{mW}}$

The negative sign indicates flow toward the low temperature.

COMMENT. The computations of simple kinetic theory give an estimate for the coefficient of thermal conductivity that is low but satisfying in that its simplicity gives a result within an order of magnitude of being correct. To improve the estimate requires removal of the hard-sphere assumption with consideration of more realistic forces of attraction and repulsion. Additionally, the simple utilization of a mean speed has to be replaced with a kinetic theory that recognizes that faster molecules transport more kinetic energy than slower ones.

E20.21(a) $$\kappa = \tfrac{1}{3}\lambda\bar{c}C_{V,\mathrm{m}}[\mathrm{A}]\,[20.23] = \left(\frac{C_{V,\mathrm{m}}}{3\sigma N_\mathrm{A}}\right)\times\left(\frac{4RT}{\pi M}\right)^{1/2}\ [\text{E20.5(a) note, 20.7, }[\mathrm{A}]=p/RT]$$

$$= \left(\frac{12.5\ \mathrm{J\ K^{-1}\ mol^{-1}}}{3\times(0.36\times10^{-18}\ \mathrm{m^2})\times(6.022\times10^{23}\ \mathrm{mol^{-1}})}\right)\times\left(\frac{4\times(8.3145\ \mathrm{J\ K^{-1}\ mol^{-1}})\times(293\ \mathrm{K})}{\pi\times(39.95\times10^{-3}\ \mathrm{kg\ mol^{-1}})}\right)^{1/2}$$

$$= \boxed{5.4\ \mathrm{mJ\ K^{-1}\ m^{-1}\ s^{-1}}}$$

Question. Can the differences between the calculated and experimental values of κ be accounted for by the difference in temperature (293 K here, 273 K in Table 20.2)? If not, what might be responsible for the difference?

E20.22(a) $$D = \tfrac{1}{3}\lambda\bar{c}\,[20.22] = \tfrac{1}{3}\left(\frac{kT}{2^{1/2}\sigma p}\right)\times\left(\frac{8RT}{\pi M}\right)^{1/2}\ [\text{E20.5(a) note, 20.7}]$$

$$= \tfrac{1}{3}\left(\frac{(1.381\times10^{-23}\ \mathrm{J\ K^{-1}})\times(298\ \mathrm{K})}{2^{1/2}\times(0.36\times10^{-18}\ \mathrm{m^2})\times p\times(1\ \mathrm{J\ m^{-3}\ Pa^{-1}})}\right)\times\left(\frac{8\times(8.3145\ \mathrm{J\ K^{-1}\ mol^{-1}})\times(298\ \mathrm{K})}{\pi\times(39.95\times10^{-3}\ \mathrm{kg\ mol^{-1}})}\right)^{1/2}$$

$$= \frac{1.07\ \mathrm{m^2\ s^{-1}}}{p/\mathrm{Pa}}$$

The flux due to diffusion is $J = -D\left(\frac{\mathrm{d}\mathcal{N}}{\mathrm{d}z}\right)$ [20.19], where $\mathcal{N}$ is the number density and z is the direction of the pressure gradient. Dividing both sides by the Avogadro constant converts $\mathcal{N}$ to molar concentration c, while converting the flux to number of moles per unit area per second. Thus, with a negative sign that indicates mass flow from high to low pressure,

$$J = -D\frac{\mathrm{d}c}{\mathrm{d}z} = -D\frac{\mathrm{d}(n/V)}{\mathrm{d}z} = -\frac{D}{RT}\frac{\mathrm{d}p}{\mathrm{d}z}\ [\text{perfect gas law}].$$

For a pressure gradient of $0.10\ \mathrm{atm\ cm^{-1}} = 1.013\times10^{6}\ \mathrm{Pa\ m^{-1}}$,

$$J = \frac{(1.013\times10^{6}\ \mathrm{Pa\ m^{-1}})\times D}{(8.3145\ \mathrm{J\ mol^{-1}\ K^{-1}})\times(298\ \mathrm{K})} = (409\ \mathrm{mol\ m^{-2}\ s^{-1}})\times D/(\mathrm{m^2\ s^{-1}}).$$

(a) $p = 1.00\ \mathrm{Pa}$,

$$D = \frac{1.07\ \mathrm{m^2\ s^{-1}}}{1} = \boxed{1.07\ \mathrm{m^2\ s^{-1}}}\ \text{and}\ J = (409\ \mathrm{mol\ m^{-2}\ s^{-1}})\times(1.07) = \boxed{438\ \mathrm{mol\ m^{-2}\ s^{-1}}}$$

(b) $p = 100\ \mathrm{kPa}$,

$$D = \frac{1.07\ \mathrm{m^2\ s^{-1}}}{1.00\times10^{5}} = \boxed{1.07\times10^{-5}\ \mathrm{m^2\ s^{-1}}}\ \text{and}\ J = (409\ \mathrm{mol\ m^{-2}\ s^{-1}})\times(1.07\times10^{-5})$$

$$= \boxed{4.38\ \mathrm{mmol\ m^{-2}\ s^{-1}}}$$

(c) $p = 10.0$ MPa,

$$D = \frac{1.07\ \text{m}^2\,\text{s}^{-1}}{10.0 \times 10^6} = \boxed{1.07 \times 10^{-7}\ \text{m}^2\,\text{s}^{-1}} \text{ and } J = (409\ \text{mol m}^{-2}\,\text{s}^{-1}) \times (1.07 \times 10^{-7})$$

$$= \boxed{43.8\ \mu\text{mol m}^{-2}\,\text{s}^{-1}}$$

E20.23(a) The molar ionic conductivity is related to mobility by

$$\lambda_\pm = zu_\pm F \text{ [20.36]}$$
$$= 1 \times (7.91 \times 10^{-8}\ \text{m}^2\,\text{s}^{-1}\,\text{V}^{-1}) \times (96485\ \text{C mol}^{-1})$$
$$= \boxed{7.63\ \text{mS m}^2\,\text{mol}^{-1}}$$

E20.24(a) $s = u\mathcal{E}$ [20.34] and $\mathcal{E} = \dfrac{\Delta\phi}{l}$ [20.30]

Therefore,

$$s = u\left(\frac{\Delta\phi}{l}\right) = (7.92 \times 10^{-8}\ \text{m}^2\,\text{s}^{-1}\,\text{V}^{-1}) \times \left(\frac{35.0\ \text{V}}{8.00 \times 10^{-3}\ \text{m}}\right)$$

$$= 3.47 \times 10^{-4}\ \text{m s}^{-1} = \boxed{347\ \mu\text{m s}^{-1}}$$

E20.25(a) The basis for the solution is Kohlrausch's law of independent migration of ions [eqn 20.29]. The limiting molar conductivity of a dissolved salt is the sum of formula weighted limiting molar conductivities of the formula ions at infinite dilution, so

$$\Lambda_\text{m}^\circ = \nu_+\lambda_+ + \nu_-\lambda_- \text{ [20.29]}$$

$$\Lambda_\text{m}^\circ(\text{KCl}) = \lambda(\text{K}^+) + \lambda(\text{Cl}^-) = 14.99\ \text{mS m}^2\,\text{mol}^{-1}$$

$$\Lambda_\text{m}^\circ(\text{KNO}_3) = \lambda(\text{K}^+) + \lambda(\text{NO}_3^-) = 14.50\ \text{mS m}^2\,\text{mol}^{-1}$$

$$\Lambda_\text{m}^\circ(\text{AgNO}_3) = \lambda(\text{Ag}^+) + \lambda(\text{NO}_3^-) = 13.34\ \text{mS m}^2\,\text{mol}^{-1}$$

Hence,

$$\Lambda_\text{m}^\circ(\text{AgCl}) = \Lambda_\text{m}^\circ(\text{AgNO}_3) + \Lambda_\text{m}^\circ(\text{KCl}) - \Lambda_\text{m}^\circ(\text{KNO}_3)$$
$$= (13.34 + 14.99 - 14.50)\ \text{mS m}^2\,\text{mol}^{-1} = \boxed{13.83\ \text{mS m}^2\,\text{mol}^{-1}}$$

E20.26(a) $u_\pm = \dfrac{\lambda_\pm}{zF}$ [20.36]; $z = 1$; $1\ \text{S} = 1\ \Omega^{-1} = 1\ \text{C V}^{-1}\,\text{s}^{-1}$

$$u(\text{Li}^+) = \frac{3.87\ \text{mS m}^2\,\text{mol}^{-1}}{9.6485 \times 10^4\ \text{C mol}^{-1}} = 4.01 \times 10^{-5}\ \text{mS C}^{-1}\,\text{m}^2 = \boxed{4.01 \times 10^{-8}\ \text{m}^2\,\text{V}^{-1}\,\text{s}^{-1}}$$

$$u(\text{Na}^+) = \frac{5.01\ \text{mS m}^2\,\text{mol}^{-1}}{9.6485 \times 10^4\ \text{C mol}^{-1}} = \boxed{5.19 \times 10^{-8}\ \text{m}^2\,\text{V}^{-1}\,\text{s}^{-1}}$$

$$u(\text{K}^+) = \frac{7.35\ \text{mS m}^2\,\text{mol}^{-1}}{9.6485 \times 10^4\ \text{C mol}^{-1}} = \boxed{7.62 \times 10^{-8}\ \text{m}^2\,\text{V}^{-1}\,\text{s}^{-1}}$$

E20.27(a) $D = \frac{uRT}{zF}$ [20.50]; $z = 1$; $1\,\mathrm{C\,V} = 1\,\mathrm{J}$

$$D = \frac{(7.40\times10^{-8}\,\mathrm{m^2\,V^{-1}\,s^{-1}})\times(8.3145\,\mathrm{J\,K^{-1}\,mol^{-1}})\times(298\,\mathrm{K})}{1\times9.6485\times10^{4}\,\mathrm{C\,mol^{-1}}} = \boxed{1.90\times10^{-9}\,\mathrm{m^2\,s^{-1}}}$$

E20.28(a) The linear decay in concentration c has the form $c(x) = c_0 - Ax$, where c_0 is the concentration at $x = 0$ and the constant A is determined by the condition that $c(10\text{ cm}) = \frac{1}{2}c_0$.

$$A = \frac{c_0 - c(x)}{x} = \frac{c_0 - \frac{1}{2}c_0}{10\text{ cm}} = (0.050\text{ cm}^{-1})\times c_0 = (5.0\text{ m}^{-1})\times c_0$$

The thermodynamic force is determined with eqn 20.45.

$$\mathcal{F}(x) = -\frac{RT}{c}\frac{\mathrm{d}c}{\mathrm{d}x}\,[20.45] = -\frac{RT}{c}\frac{\mathrm{d}}{\mathrm{d}x}(c_0 - Ax) = \frac{ART}{c}$$
$$= \frac{(5.0\text{ m}^{-1})\times(8.3145\text{ J K}^{-1}\text{mol}^{-1})\times(298\text{ K})}{1-(0.050\text{ cm}^{-1})\times x}$$
$$= \frac{12.\bar{4}\text{ kN mol}^{-1}}{1-(0.050\text{ cm}^{-1})\times x}$$

$$\mathcal{F}(10\text{ cm}) = \frac{12.\bar{4}\text{ kN mol}^{-1}}{1-(0.050\text{ cm}^{-1})\times(10\text{ cm})} = \boxed{25\text{ kN mol}^{-1}}$$

$\mathcal{F}(20\text{ cm}) = \boxed{\infty}$ because $c = 0$ when $x \geq 20$ cm.

E20.29(a) The Gaussian decay in concentration c has the form $c(x) = c_0 e^{-Ax^2}$, where c_0 is the concentration at $x = 0$ and the constant A is determined by the condition that $c(5\text{ cm}) = \frac{1}{2}c_0$.

$$A = \frac{\ln(c_0/c)}{x^2} = \frac{\ln 2}{(5\text{ cm})^2} = (0.040\text{ cm}^{-2})\times\ln 2 = (4.00\text{ cm}^{-1}\text{ m}^{-1})\times\ln 2$$

The thermodynamic force is determined with eqn 20.45.

$$\mathcal{F}(x) = -\frac{RT}{c}\frac{\mathrm{d}c}{\mathrm{d}x}\,[20.45] = -\frac{RT}{c}\frac{\mathrm{d}}{\mathrm{d}x}c_0 e^{-Ax^2} = 2ARTx$$
$$= 2\times(4.00\text{ m}^{-1})\times\ln 2\times(8.3145\text{ J K}^{-1}\text{mol}^{-1})\times(293\text{ K})\times(x/\text{cm})$$
$$= (13.5\text{ kN mol}^{-1})\times(x/\text{cm})$$

$$\mathcal{F}(5\text{ cm}) = (13.5\text{ kN mol}^{-1})\times(5) = \boxed{67.5\text{ kN mol}^{-1}}$$

E20.30(a) Equation [20.60], $\langle x^2\rangle = 2Dt$, gives the mean square distance travelled in any one dimension in time t. We need the distance travelled from a point in any direction. The distinction here is the distinction between the one-dimensional and three-dimensional diffusion. The mean square three-dimensional distance can be obtained from the one-dimensional mean square distance since motions in the three directions are independent. Since $r^2 = x^2 + y^2 + z^2$ [Pythagorean theorem],

$$\langle r^2\rangle = \langle x^2\rangle + \langle y^2\rangle + \langle z^2\rangle = 3\langle x^2\rangle \text{ [independent motion]}$$
$$= 3\times 2Dt\,[20.60] = 6Dt.$$

Therefore, $t = \dfrac{\langle r^2 \rangle}{6D} = \dfrac{(5.0 \times 10^{-3}\ \text{m})^2}{6 \times (3.17 \times 10^{-9}\ \text{m}^2\ \text{s}^{-1})} = \boxed{1.3 \times 10^3\ \text{s}}$

E20.31(a) $a = \dfrac{kT}{6\pi\eta D}$ [20.52]; $1\ \text{P} = 10^{-1}\ \text{kg m}^{-1}\ \text{s}^{-1}$

$$a = \frac{(1.381 \times 10^{-23}\ \text{J K}^{-1}) \times (298\ \text{K})}{6\pi \times (1.00 \times 10^{-3}\ \text{kg m}^{-1}\ \text{s}^{-1}) \times (5.2 \times 10^{-10}\ \text{m}^2\ \text{s}^{-1})} = 4.2 \times 10^{-10}\ \text{m} = \boxed{0.42\ \text{nm}}$$

E20.32(a) The Einstein–Smoluchowski equation [20.62] relates the diffusion constant to the jump distance λ and time τ required for a jump.

$$D = \frac{\lambda^2}{2\tau}\ [20.62] \quad \text{so} \quad \tau = \frac{\lambda^2}{2D}$$

If the jump distance is about one molecular diameter, or two effective molecular radii, then the jump distance can be obtained by use of the Stokes–Einstein equation [20.52].

$$\lambda = 2a = 2\left(\frac{kT}{6\pi\eta D}\right)[20.52] = \frac{kT}{3\pi\eta D}$$

$$\tau = \frac{1}{2D} \times \left(\frac{kT}{3\pi\eta D}\right)^2$$

$$= \frac{1}{2 \times (2.13 \times 10^{-9}\ \text{m}^2\ \text{s}^{-1})} \times \left\{\frac{(1.381 \times 10^{-23}\ \text{J K}^{-1}) \times (298\ \text{K})}{3\pi \times (0.601 \times 10^{-3}\ \text{kg m}^{-1}\ \text{s}^{-1}) \times (2.13 \times 10^{-9}\ \text{m}^2\ \text{s}^{-1})}\right\}^2 \ \text{[Table 20.4]}$$

$$= 2.73 \times 10^{-11}\ \text{s} = \boxed{27.3\ \text{ps}}$$

COMMENT. In the strictest sense we are dealing with three-dimensional diffusion. However, since we are assuming that only one jump occurs, it is probably an adequate approximation to use an equation derived for one-dimensional diffusion. For three-dimensional diffusion the equation analogous to eqn 20.62 is $\tau = \lambda^2/6D$.

Question. Can you derive the equation? Use an analysis similar to that described in the solution to E20.30a.

E20.33(a) For three-dimensional diffusion we use an equation analogous to eqn 20.60 derived in Exercise 20.30a.

$$\langle r^2 \rangle = 6Dt$$

For iodine in benzene [data from Table 20.7]

$$\langle r^2 \rangle^{1/2} = [6 \times (2.13 \times 10^{-9}\ \text{m}^2\ \text{s}^{-1}) \times (1.0\ \text{s})]^{1/2} = \boxed{113\ \mu\text{m}}$$

For sucrose in water

$$\langle r^2 \rangle^{1/2} = [6 \times (0.522 \times 10^{-9}\ \text{m}^2\ \text{s}^{-1}) \times (1.0\ \text{s})]^{1/2} = \boxed{56\ \mu\text{m}}$$

Solutions to problems

Solutions to numerical problems

P20.1 The time in seconds for a disk to rotate 360° is the inverse of the frequency. The time for it to advance 2° is $\left(\frac{2°}{360°}\right)\times\frac{1}{\nu}$. This is the time required for slots in neighbouring disks to coincide along the atomic beam. For an atom to pass through all neighbouring slots it must have the speed

$$v_x = \text{disk spacing/alignment time} = (1.0\ \text{cm})/\left\{\left(\frac{2°}{360°}\right)\times\frac{1}{\nu}\right\} = 180\ \nu\ \text{cm} = 180\times(\nu/\text{Hz})\ \text{cm s}^{-1}$$

Hence, the distributions of the x component of velocity are

ν/Hz	20	40	80	100	120
v_x/cm s^{-1}	3600	7200	14400	18000	21600
I_{exp} (40 K)	0.846	0.513	0.069	0.015	0.002
I_{exp} (100 K)	0.592	0.485	0.217	0.119	0.057

Theoretically, the velocity distribution in the x direction is

$$f(v_x) = \left(\frac{m}{2\pi kT}\right)^{1/2} e^{-mv_x^2/2kT}\ [20.6 \text{ with } M/R = m/k]$$

Therefore, as $I \propto f$, $I \propto T^{-1/2}e^{-mv_x^2/2kT}$.

$$\text{Since } mv_x^2/2kT = \frac{83.80\times(1.6605\times10^{-27}\ \text{kg})\times\{1.80\times(\nu/\text{Hz})\ \text{m s}^{-1}\}^2}{2\times(1.381\times10^{-23}\ \text{J K}^{-1})\times T} = \frac{(1.63\times10^{-2})\times(\nu/\text{Hz})^2}{T/\text{K}},$$

we can write $I \propto (T/\text{K})^{-1/2}\,e^{-1.63\times10^{-2}\times(\nu/\text{Hz})^2/(T/\text{K})}$ and draw up the following table, obtaining the constant of proportionality by fitting I to the value at $T = 40$ K, $\nu = 80$ Hz.

ν/Hz	20	40	80	100	120
I_{calc} (40 K)	0.80	0.49	(0.069)	0.016	0.003
I_{calc} (100 K)	0.56	0.46	0.209	0.116	0.057

The calculated values are in fair agreement with the experimental data.

P20.3 For discrete, rather than continuous, heights the mean height $\bar{h}$ is (in analogy to Example 20.1):

$$\bar{h} = \langle h\rangle = \sum_i^{\text{all heights}} f_i h_i = \sum_i^{\text{all heights}} \left(\frac{N_i}{N}\right)h_i = \left(\frac{1}{N}\right)\sum_i^{\text{all heights}} N_i h_i$$

where N_i is the number having the height h_i and N is the total number of observations. $N = 53$ in this problem. Likewise, the root mean square height is given by

$$\langle h^2\rangle^{1/2} = \left\{\sum_i^{\text{all heights}} f_i h_i^2\right\}^{1/2} = \left\{\sum_i^{\text{all heights}} \left(\frac{N_i}{N}\right)h_i^2\right\}^{1/2} = \left\{\left(\frac{1}{N}\right)\sum_i^{\text{all heights}} N_i h_i^2\right\}^{1/2}$$

(b) Mean height:

$$\bar{h} = \frac{1}{53}\begin{Bmatrix} 1(1.80) + 2(1.82) + 4(1.84) + 7(1.86) + 10(1.88) \\ + 15(1.90) + 9(1.92) + 4(1.94) + 0(1.96) + 1(1.98) \end{Bmatrix} \text{m} = \boxed{1.8894\ \text{m}}$$

(c) Root mean square height:

$$\langle h^2 \rangle^{1/2} = \left(\frac{1}{53}\right)^{1/2} \begin{Bmatrix} 1(1.80)^2 + 2(1.82)^2 + 4(1.84)^2 + 7(1.86)^2 + 10(1.88)^2 \\ + 15(1.90)^2 + 9(1.92)^2 + 4(1.94)^2 + 0(1.96)^2 + 1(1.98)^2 \end{Bmatrix}^{1/2} \text{m} = \boxed{1.8897\ \text{m}}$$

We have taken the values to be exact and, consequently, have not rounded off so as to show that as always $\langle X^2 \rangle^{1/2} > \langle X \rangle$ for observable X.

P20.5 The number of molecules that escape in unit time is the number per unit time that would have collided with a wall section of area A_0 equal to the area of the small hole. That is,

$$\frac{\mathrm{d}N}{\mathrm{d}t} = -Z_\mathrm{W} A_0 = \frac{-A_0 p}{(2\pi mkT)^{1/2}} \text{ [20.16]}$$

where p is the (constant) vapour pressure of the solid. The change in the number of molecules inside the cell in an interval Δt is therefore $\Delta N = -Z_\mathrm{W} A_0 \Delta t$, and so the mass loss is

$$\Delta w = \Delta N m = -A_0 p \left(\frac{m}{2\pi kT}\right)^{1/2} \Delta t = -A_0 p \left(\frac{M}{2\pi RT}\right)^{1/2} \Delta t$$

Therefore, the vapour pressure of the substance in the cell is

$$p = \left(\frac{-\Delta w}{A \Delta t}\right) \times \left(\frac{2\pi RT}{M}\right)^{1/2}$$

For the vapour pressure of germanium

$$p = \left(\frac{43 \times 10^{-9}\,\text{kg}}{\pi \times (0.50 \times 10^{-3}\,\text{m})^2 \times (7200\,\text{s})}\right) \times \left(\frac{2\pi \times (8.3145\,\text{J K}^{-1}\,\text{mol}^{-1}) \times (1273\,\text{K})}{72.64 \times 10^{-3}\,\text{kg mol}^{-1}}\right)^{1/2}$$

$$= 7.3 \times 10^{-3}\,\text{Pa} = \boxed{7.3\ \text{mPa}}$$

P20.7 $$\kappa \propto \frac{1}{R} \text{ [20.26]}$$

Water makes a contribution 76 mS m^{-1} to an aqueous solution. Therefore,

$$\frac{\kappa(\text{acid soln})}{\kappa(\text{KCl soln})} = \frac{\kappa(\text{acid}) + \kappa(\text{water})}{\kappa(\text{KCl}) + \kappa(\text{water})} = \frac{R(\text{KCl soln})}{R(\text{acid soln})} = \frac{33.21\,\Omega}{300.0\,\Omega}$$

$$\kappa(\text{acid}) = \{\kappa(\text{KCl}) + \kappa(\text{water})\} \times \left(\frac{33.21}{300.0}\right) - \kappa(\text{water})$$

$$= \{1.1639\,\text{S m}^{-1} + 0.076\,\text{S m}^{-1}\} \times \left(\frac{33.21}{300.0}\right) - 0.076\,\text{S m}^{-1}$$

$$= 61.257\,\text{mS m}^{-1}$$

$$\Lambda_m(\text{acid}) = \frac{\kappa}{c}[20.27] = \frac{61.257 \text{ mS m}^{-1}}{1.00 \times 10^{-4} \text{ mol m}^{-3}} = \boxed{0.613 \text{ mS m}^2 \text{ mol}^{-1}}$$

P20.9 The very low solubility of silver chloride means that the molar conductivity equals the limiting value.

$$\begin{aligned}\Lambda_m^\circ &= \nu_+\lambda_+ + \nu_-\lambda_- \ [20.29]\\ &= F\{z_+\nu_+u_+ + z_-\nu_-u_-\}\ [20.36]\\ &= F\{u_+ + u_-\}\ [1\text{:}1 \text{ salt}]\\ &= (96\,485 \text{ C mol}^{-1}) \times (6.24 + 7.91) \times 10^{-8} \text{ m}^2 \text{ s}^{-1} \text{ V}^{-1}\ [\text{Table 20.5}, 1 \text{ S} = 1 \text{ C V}^{-1} \text{s}^{-1}]\\ &= 13.65 \text{ mS m}^2 \text{ mol}^{-1}\end{aligned}$$

$$\begin{aligned}c &= \frac{\kappa}{\Lambda_m}[20.27]\\ &= \frac{0.1887 \text{ mS m}^{-1}}{13.65 \text{ mS m}^2 \text{ mol}^{-1}} = 1.382 \text{ cmol m}^{-3} = \boxed{13.82\ \mu\text{mol dm}^{-3}}\end{aligned}$$

P20.11 We assume that concentrations are low enough to validate application of the law of independent migration of ions so that solution conductance G_i of ion i is proportional to $z_i\nu_iu_ic_i$:

$$G_i \propto \kappa_i\ [20.26] \propto c_i\Lambda_{m,i}^\circ\ [20.27] \propto z_i\nu_iu_ic_i\ [20.38]$$

Thus, since the constant of proportionality cancels when taking a ratio, the proportion of the current that is carried by ion j is

$$\frac{G_j}{\sum_i^{\text{all ions}} G_i} = \frac{z_j\nu_ju_jc_j}{\sum_i^{\text{all ions}} z_i\nu_iu_ic_i}$$

For an aqueous 1.00×10^{-3} mol dm^{-3} hydrochloric acid solution the portion of the current carried by the H^+ ion is:

$$\frac{G_{H^+}}{\sum_i^{\text{all ions}} G_i} = \frac{G_{H^+}}{G_{H^+} + G_{Cl^-}} = \frac{u_{H^+}}{u_{H^+} + u_{Cl^-}} = \frac{36.23}{36.23 + 7.91} = \boxed{0.821}$$

When 1.0 mol dm^{-3} NaCl(aq) is added, the portion of the current carried by the H^+ ion is:

$$\frac{G_{H^+}}{\sum_i^{\text{all ions}} G_i} = \frac{G_{H^+}}{G_{H^+} + G_{Na^+} + G_{Cl^-}} = \frac{u_{H^+}c_{H^+}}{u_{H^+}c_{H^+} + u_{Na^+}c_{Na^+} + u_{Cl^-}c_{Cl^-}}$$

$$= \frac{36.23 \times (0.00100)}{36.23 \times (0.00100) + 5.19 \times (1.00) + 7.91 \times (1.001)} = \boxed{0.00276}$$

P20.13 $$D = \frac{uRT}{zF}[20.50] \quad \text{and} \quad a = \frac{ze}{6\pi\eta u}[20.35]$$

$$D = \frac{(8.3145 \text{ J K}^{-1} \text{mol}^{-1}) \times (298.15 \text{ K}) \times u}{96\,485 \text{ C mol}^{-1}} = 2.569 \times 10^{-2} \text{ V} \times u$$

So $D/(\text{cm}^2\,\text{s}^{-1}) = (2.569\times10^{-2})\times u/(\text{cm}^2\,\text{s}^{-1}\,\text{V}^{-1})$

$$a = \frac{1.602\times10^{-19}\,\text{C}}{(6\pi)\times(0.891\times10^{-3}\,\text{kg m}^{-1}\,\text{s}^{-1})\times u}$$

$$= \frac{9.54\times10^{-18}\,\text{C kg}^{-1}\,\text{m s}}{u}\,[1\,\text{J} = 1\,\text{C V}, 1\,\text{J} = 1\,\text{kg m}^2\,\text{s}^{-2}]$$

$$a/\text{pm} = \frac{954}{u/(10^{-8}\,\text{m}^2\,\text{s}^{-1}\,\text{V}^{-1})}$$

We can now draw up the following table using data from Table 20.5.

	Li^+	Na^+	K^+	Rb^+
$u/(10^{-8}\,\text{m}^2\,\text{s}^{-1}\,\text{V}^{-1})$	4.01	5.19	7.62	7.92
$D/10^{-5}\,\text{cm}^2\,\text{s}^{-1}$	1.03	1.33	1.96	2.04
a/pm	238	184	125	120

The ionic radii themselves (i.e. their crystallographic radii, Table 19.3) are

	Li^+	Na^+	K^+	Rb^+
r_+/pm	59	102	138	149

It would seem that K^+ and Rb^+ have effective hydrodynamic radii that are not larger than their ionic radii. It may be that the exchange rate for the movement of water molecules between the solvent cage of K^+ and Rb^+ and the bulk solvent is so rapid that these ions diffuse with a minimal hydration shell. The effective hydrodynamic and ionic volumes of Li^+ and Na^+ are $\frac{4\pi}{3}a^3$ and $\frac{4\pi}{3}r_+^3$, respectively, and so the volumes occupied by hydrating water molecules are

(a) Li^+: $\Delta V = (\frac{4\pi}{3})\times(238^3 - 59^3)\times10^{-36}\,\text{m}^3 = 5.5\bar{6}\times10^{-29}\,\text{m}^3$

(b) Na^+: $\Delta V = (\frac{4\pi}{3})\times(184^3 - 102^3)\times10^{-36}\,\text{m}^3 = 2.1\bar{6}\times10^{-29}\,\text{m}^3$

The volume occupied by a single H_2O molecule is approximately $\frac{4\pi}{3}\times(150\,\text{pm})^3 = 1.4\times10^{-29}\,\text{m}^3$.

Therefore, Li^+ has about **four** firmly attached H_2O molecules, whereas Na^+ has only **one to two** (according to this analysis).

P20.15 This is essentially one-dimensional diffusion and therefore eqn 20.57 applies:

$$c = \frac{n_0 e^{-x^2/4Dt}}{A(\pi Dt)^{1/2}}\,[20.57]$$

Information for the computation includes:

$$n_0 = \left(\frac{10\,\text{g}}{342\,\text{g mol}^{-1}}\right) = 0.0292\,\text{mol}$$

$$A = \pi R^2 = \pi\times(2.5\,\text{cm})^2 = 19.6\,\text{cm}^2$$

$$D = 0.522\times10^{-9}\,\text{m}^2\,\text{s}^{-1}\,[\text{Table 20.7}]$$

Thus, at $x = 5.0$ cm:

$$\frac{x^2}{4D} = \frac{(5.0 \times 10^{-2}\,\text{m})^2}{4 \times (0.522 \times 10^{-9}\,\text{m}^2\,\text{s}^{-1})} = 1.2 \times 10^6\,\text{s}$$

$$c = \frac{(0.0292\,\text{mol}) \times e^{-1.2\times10^6/(t/\text{s})}}{(19.6 \times 10^{-4}\,\text{m}^2) \times \{\pi \times (0.522 \times 10^{-9}\,\text{m}^2\,\text{s}^{-1}) \times t\}^{1/2} \times (10^3\,\text{dm}^3\,\text{m}^{-3})}$$

$$= \frac{(368\,\text{mol dm}^{-3}) \times e^{-1.2\times10^6/(t/\text{s})}}{(t/\text{s})^{1/2}}$$

(a) For $t = 10$ s, $c = \boxed{0}$.

(b) For $t = 1$ y $= 3.16 \times 10^7$ s, $c = \boxed{0.0630\ \text{mol dm}^{-3}}$.

COMMENT. This problem illustrates the extreme slowness of diffusion through typical macroscopic distances. However, diffusion is rapid enough through distances comparable to the dimensions of a cell. Compare to Problems 20.35 and 20.36.

P20.17 Kohlrausch's law states that the molar conductance of a strong electrolyte varies with the square root of concentration: $\Lambda_m = \Lambda_m^\circ - \mathcal{K}c^{1/2}$ [20.28]. Therefore, a plot of Λ_m against $c^{1/2}$ should be linear with intercept Λ_m°. We draw the following data table and and prepare the plot shown in Figure 20.3.

NaI			KI		
c/(mmol dm^{-3})	$c^{1/2}$	Λ_m/(S cm^2 mol^{-1})	c/(mmol dm^{-3})	$c^{1/2}$	Λ_m/(S cm^2 mol^{-1})
32.02	5.659	50.26	17.68	4.205	42.45
20.28	4.503	51.99	10.88	3.298	45.91
12.06	3.473	54.01	7.19	2.68	47.53
8.64	2.94	55.75	2.67	1.63	51.81
2.85	1.69	57.99	1.28	1.13	54.09
1.24	1.11	58.44	0.83	0.91	55.78
0.83	0.91	58.67	0.19	0.44	57.42

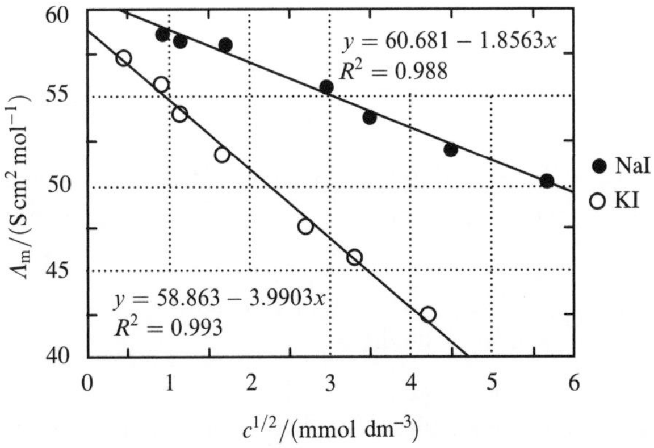

Figure 20.3

Both plots of Figure 20.3 are linear with regression fits that give intercepts such that

$$\Lambda_m^o(\text{NaI}) = \boxed{60.7\ \text{S cm}^2\ \text{mol}^{-1}} \quad \text{and} \quad \Lambda_m^o(\text{KI}) = \boxed{58.9\ \text{S cm}^2\ \text{mol}^{-1}} \text{ in 1,3-dioxolan-2-one.}$$

Since these two electrolytes have a common anion, the difference in conductivity is due to the cations.

$$\lambda(\text{Na}^+) - \lambda(\text{K}^+) = \Lambda_m^o(\text{NaI}) - \Lambda_m^o(\text{KI}) = \boxed{1.8\ \text{S cm}^2\ \text{mol}^{-1}} \text{ in 1,3-dioxolan-2-one.}$$

The quantities in water solution are calculated with eqn 20.29 and the data of Table 20.5:

$$\Lambda_m^o = \lambda_+ + \lambda_- \ [20.29, 1{:}1 \text{ salt}] = (u_+ + u_-)F \ [20.36, z = 1]$$

$$\Lambda_m^o(\text{NaI}) = (96\,485\ \text{C mol}^{-1}) \times (5.19 + 7.96) \times 10^{-8}\ \text{m}^2\ \text{s}^{-1}\ \text{V}^{-1}\ [\text{Table 20.5, } 1\ \text{S} = 1\ \text{C V}^{-1}\ \text{s}^{-1}]$$

$$= 0.0127\ \text{S m}^2\ \text{mol}^{-1} = \boxed{127\ \text{S cm}^2\ \text{mol}^{-1}} \text{ in water}$$

$$\Lambda_m^o(\text{KI}) = (96\,485\ \text{C mol}^{-1}) \times (7.62 + 7.96) \times 10^{-8}\ \text{m}^2\ \text{s}^{-1}\ \text{V}^{-1}$$

$$= 0.0150\ \text{S m}^2\ \text{mol}^{-1} = \boxed{150\ \text{S cm}^2\ \text{mol}^{-1}} \text{ in water}$$

$$\lambda(\text{Na}^+) - \lambda(\text{K}^+) = \Lambda_m^o(\text{NaI}) - \Lambda_m^o(\text{KI}) = \boxed{-23\ \text{S cm}^2\ \text{mol}^{-1}} \text{ in water}$$

The ions are considerably more mobile in water than in 1,3-dioxolan-2-one. Also, the differences between Na^+ and K^+ are minimized, and even inverted, in 1,3-dioxolan-2-one compared to water.

P20.19 The diffusion constant of a C_{60}^- ion in solution is related to the mobility of the ion and to its hydrodynamic radius a in eqn 20.52, the Stokes formula.

$$a = \frac{ze}{6\pi\eta u} \ [20.52, 20.50, \text{and } F = N_A e]$$

$$= \frac{(1) \times (1.602 \times 10^{-19}\ \text{C})}{6\pi(0.93 \times 10^{-3}\ \text{kg m}^{-1}\ \text{s}^{-1}) \times (1.1 \times 10^{-8}\ \text{m}^2\ \text{V}^{-1}\ \text{s}^{-1})} = 8.3 \times 10^{-10}\ \text{m} = \boxed{0.83\ \text{nm}}$$

This is substantially larger than the 0.5 nm van der Waals radius of a Buckminsterfullerene (C_{60}) molecule because the anion attracts a considerable hydration shell through the ion–dipole attraction to water molecules. The Stokes radius reflects the larger effective radius of the combined anion and its hydration shell.

Solutions to theoretical problems

P20.21 We proceed as in *Justification 20.2* except that, instead of taking a product of three one-dimensional distributions in order to get the three-dimensional distribution, we make a product of two one-dimensional distributions.

$$f(v_x, v_y)\text{d}v_x\text{d}v_y = f(v_x)f(v_y)\text{d}v_x\text{d}v_y = \left(\frac{m}{2\pi kT}\right)\text{e}^{-mv^2/2kT}\text{d}v_x\text{d}v_y$$

where $v^2 = v_x^2 + v_y^2$. The probability $f(v)\text{d}v$ that the molecules have a two-dimensional speed, v, in the range v to $v + \text{d}v$ is the sum of the probabilities that it is in any of the area elements $\text{d}v_x\text{d}v_y$ in the circular shell of radius v. The sum of the area elements is the area of the circular shell of radius v and thickness dv, which is $\pi(v+dv)^2 - \pi v^2 = 2\pi v\,dv$. Therefore,

$$\boxed{f(v) = \left(\frac{m}{kT}\right) v e^{-mv^2/2kT}} \left[\frac{M}{R} = \frac{m}{k}\right].$$

The mean speed is determined as

$$\bar{c} = \int_0^\infty v f(v)\,dv = \left(\frac{m}{kT}\right)\int_0^\infty v^2 e^{-mv^2/2kT}\,dv = \left(\frac{m}{kT}\right)\times\left(\frac{\pi^{1/2}}{4}\right)\times\left(\frac{2kT}{m}\right)^{3/2} \text{ [standard integral]}$$

$$= \boxed{\left(\frac{\pi kT}{2m}\right)^{1/2} \text{ or } \left(\frac{\pi RT}{2M}\right)^{1/2}}.$$

P20.23 Rewriting eqn 20.4 with $M/R = m/k$,

$$f(v) = 4\pi\left(\frac{m}{2\pi kT}\right)^{3/2} v^2 e^{-mv^2/2kT}.$$

The proportion of molecules with speeds less than c is

$$P = \int_0^c f(v)\,dv = 4\pi\left(\frac{m}{2\pi kT}\right)^{3/2}\int_0^c v^2 e^{-mv^2/2kT}\,dv$$

Defining $a \equiv m/2kT$,

$$P = 4\pi\left(\frac{a}{\pi}\right)^{3/2}\int_0^c v^2 e^{-av^2}\,dv = -4\pi\left(\frac{a}{\pi}\right)^{3/2}\frac{d}{da}\int_0^c e^{-av^2}\,dv$$

Defining $\chi^2 \equiv av^2$. Then, $dv = a^{-1/2}d\chi$ and

$$P = -4\pi\left(\frac{a}{\pi}\right)^{3/2}\frac{d}{da}\left\{\frac{1}{a^{1/2}}\int_0^{ca^{1/2}} e^{-\chi^2}d\chi\right\}$$

$$= -4\pi\left(\frac{a}{\pi}\right)^{3/2}\left\{-\frac{1}{2}\left(\frac{1}{a}\right)^{3/2}\int_0^{ca^{1/2}} e^{-\chi^2}d\chi + \left(\frac{1}{a}\right)^{1/2}\frac{d}{da}\int_0^{ca^{1/2}} e^{-\chi^2}d\chi\right\}.$$

Then, we use $\int_0^{ca^{1/2}} e^{-\chi^2}d\chi = (\pi^{1/2}/2)\operatorname{erf}(ca^{1/2})$.

$$\frac{d}{da}\int_0^{ca^{1/2}} e^{-\chi^2}d\chi = \left(\frac{dca^{1/2}}{da}\right)\times(e^{-c^2a}) = \frac{1}{2}\left(\frac{c}{a^{1/2}}\right)e^{-c^2a}$$

where we have used $\frac{d}{dz}\int_0^z f(y)dy = f(z)$.

Substituting and cancelling we obtain $P = \operatorname{erf}(ca^{1/2}) - (2ca^{1/2}/\pi^{1/2})e^{-c^2a}$.

Now, $c = (3kT/m)^{1/2}$, so $ca^{1/2} = (3kT/m)^{1/2} \times (m/2kT)^{1/2} = (3/2)^{1/2}$, and

$$P = \operatorname{erf}\left(\sqrt{\frac{3}{2}}\right) - \left(\frac{6}{\pi}\right)^{1/2} e^{-3/2} = 0.92 - 0.31 = \boxed{0.61}$$

Therefore,

(a) $1 - P = \boxed{39\%}$ have a speed greater than the root mean square speed.

(b) $P = \boxed{61\%}$ of the molecules have a speed equal to or less than the root mean square speed.

(c) For the proportions in terms of the mean speed $\bar{c}$, replace c by $\bar{c} = (8kT/\pi m)^{1/2} = (8/3\pi)^{1/2}c$, so $\bar{c}a^{1/2} = 2/\pi^{1/2}$.

Then, $P = \operatorname{erf}(\bar{c}a^{1/2}) - (2\bar{c}a^{1/2}/\pi^{1/2}) \times (e^{-\bar{c}^2 a}) = \operatorname{erf}(2/\pi^{1/2}) - (4/\pi)e^{-4/\pi} = 0.889 - 0.356 = \boxed{0.533}$

That is, $\boxed{53\%}$ of the molecules have a speed equal to or less than the mean, and $\boxed{47\%}$ have a speed greater than the mean.

P20.25 An effusion oven has constant volume, fixed temperature, and effusion hole of area A_0. Gas escapes through the hole, which makes the effusion rate negative.

$$-\frac{dN}{dt} = Z_W A_0 = \frac{pA_0N_A}{(2\pi MRT)^{1/2}} \quad [20.16]$$

For a perfect gas, $pV = nRT = NRT/N_A$, and therefore $N = N_ApV/RT$. Differentiation gives $\frac{dN}{dt} = \frac{N_AV}{RT}\frac{dp}{dt}$.

Substitution into the first equation yields:

$$\frac{N_AV}{RT}\frac{dp}{dt} = -\frac{pA_0N_A}{(2\pi MRT)^{1/2}}$$

$$\frac{dp}{dt} = -\left(\frac{RT}{2\pi M}\right)^{1/2}\frac{A_0}{V}p = -\frac{p}{\tau}, \quad \text{where the time constant is } \tau = \left(\frac{2\pi M}{RT}\right)^{1/2}\frac{V}{A_0}$$

$$\frac{dp}{p} = -\frac{dt}{\tau}$$

$$\int_{p_0}^{p}\frac{dp}{p} = -\frac{1}{\tau}\int_0^t dt$$

$$\ln\left(\frac{p}{p_0}\right) = -\frac{t}{\tau} \quad \text{or} \quad \boxed{p = p_0 e^{-t/\tau}, \text{ where } \tau = \left(\frac{2\pi M}{RT}\right)^{1/2}\frac{V}{A_0}}$$

When $t = t_{1/2}$, $p = \tfrac{1}{2}p_0$. Substitution into the above equation gives

$$\ln\left(\frac{p_0}{2p_0}\right) = -\frac{t_{1/2}}{\tau} \quad \text{or} \quad \boxed{t_{1/2} = \tau \ln(2) = \left(\frac{2\pi M}{RT}\right)^{1/2}\frac{V}{A_0}\ln(2)}$$

The final equation indicates that the half-life for effusive loss is independent of p_0. Furthermore, the half-life increases with both the V/A_0 and $M^{1/2}$ factors. It decreases with the factor $T^{-1/2}$.

P20.27 As discussed in Example 20.5 of the text the probability density that a diffusing molecule has travelled the distance x from the origin in time t is given by the expression

$$P(x) = \frac{1}{(\pi Dt)^{1/2}} e^{-x^2/4Dt}$$

Thus, to find the mean value of any function of x, we needed only weigh the function of x with $P(x)$ and integrate over all possible x.

$$\begin{aligned}\langle x^2\rangle &= \frac{1}{(\pi Dt)^{1/2}} \int_0^\infty x^2 e^{-x^2/4Dt} \\ &= \frac{1}{(\pi Dt)^{1/2}} \times \left\{2\sqrt{\pi}(Dt)^{3/2}\right\} \text{[standard integral]} \\ &= 2Dt\end{aligned}$$

$$\begin{aligned}\langle x^4\rangle &= \frac{1}{(\pi Dt)^{1/2}} \int_0^\infty x^4 e^{-x^2/4Dt} \\ &= \frac{1}{(\pi Dt)^{1/2}} \times \left\{12\sqrt{\pi}(Dt)^{5/2}\right\} \text{[standard integral]} \\ &= 12(Dt)^2\end{aligned}$$

Thus, $\langle x^4\rangle/\langle x^2\rangle^2 = 3$ or $\boxed{\langle x^4\rangle^{1/4}/\langle x^2\rangle^{1/2} = 3^{1/4}}$.

P20.29 $AB \rightleftharpoons A^+ + B^-$

$$K = K_\gamma K_c = K_\gamma \frac{(\alpha c)^2}{(1-\alpha)c} = K_\gamma \frac{\alpha^2 c}{1-\alpha}, \text{ where } \alpha \text{ is the degree of ionization}$$

Thus, $\dfrac{K_\gamma c}{K} = \dfrac{1-\alpha}{\alpha^2}$.

Now, the molar conductivity is related to the degree of ionization by

$$\Lambda_m = (u_+ + u_-)F\alpha = \Lambda_{m,\alpha=1}\alpha, \quad \text{so that} \quad \alpha = \Lambda_m/\Lambda_{m,\alpha=1}$$

Substitution into the equilibrium expression gives

$$K = K_\gamma c\left(\frac{\Lambda_m}{\Lambda_{m,\alpha=1}}\right)^2\left(\frac{1}{1-\dfrac{\Lambda_m}{\Lambda_{m,\alpha=1}}}\right)$$

$$1-\frac{\Lambda_m}{\Lambda_{m,\alpha=1}} = \left(\frac{\Lambda_m}{\Lambda_{m,\alpha=1}}\right)^2 \times \frac{K_\gamma c}{K} = \left(\frac{\Lambda_m}{\Lambda_{m,\alpha=1}}\right)^2\left(\frac{1-\alpha}{\alpha^2}\right)$$

and division by Λ_m gives the desired result.

$$\boxed{\frac{1}{\Lambda_m} = \frac{1}{\Lambda_{m,\alpha=1}} + \frac{(1-\alpha)\Lambda_m}{(\alpha\Lambda_{m,\alpha=1})^2}}$$

We must also examine the meaning of $\Lambda_{m,\alpha=1}$.

$$\Lambda_{m,\alpha=1} = (u_+ + u_-)F = \frac{\kappa}{c_{ions}}[20.27] = \frac{\kappa}{\alpha c},$$ where c_{ions} is the actual ion concentration.

Examination of the above relationship tells us that $\Lambda_{m,\alpha=1}$ is the molar conductivity of a solution for which ionization is complete but the preparation (analytical) concentration is αc. Consequently, by Kohlrausch's law (eqn 20.28) we conclude that

$$\boxed{\Lambda_{m,\alpha=1} = \Lambda_m^\circ - \mathcal{K}(\alpha c)^{1/2}}$$

Solutions to applications: astrophysics and biochemistry

P20.31 The diffusion coefficient for a perfect gas is

$D = \frac{1}{3}\lambda\bar{c}$ [20.22], where $\lambda = (2^{1/2}\sigma\mathcal{N})^{-1}$ [E20.5(a) note] where $\mathcal{N} = p/kT$ is number density.

The mean speed is

$$\bar{c} = \left(\frac{8kT}{\pi m}\right)^{1/2} [20.7] = \left(\frac{8(1.381\times10^{-23}\text{ J K}^{-1})\times(10^4\text{ K})}{\pi(1\,m_u)\times(1.66\times10^{-27}\text{ kg }m_u^{-1})}\right)^{1/2} = 1.46\times10^4\text{ m s}^{-1}$$

Making the approximation that atomic hydrogen collisions have a collision cross-section that is about the same as helium (0.21 nm^2, Table 20.1), we find that the diffusion coefficient is

$$D = \frac{\bar{c}}{3\sigma\mathcal{N}2^{1/2}} = \frac{1.46\times10^4\text{ m s}^{-1}}{3(0.21\times10^{-18}\text{ m}^2)\times(1\times(10^{-2}\text{ m})^{-3})2^{1/2}} = \boxed{1.\bar{6}\times10^{16}\text{ m}^2\text{ s}^{-1}}.$$

The thermal conductivity is

$$\kappa = \frac{\bar{c}C_{V,m}}{3\sigma N_A 2^{1/2}}[20.23] = \frac{(1.46\times10^4\text{ m s}^{-1})\times(20.784-8.3145)\text{ J K}^{-1}\text{ mol}^{-1}}{3(0.21\times10^{-18}\text{ m}^2)\times(6.022\times10^{23}\text{ mol}^{-1})2^{1/2}}$$

$$\kappa = \boxed{0.3\underline{4}\text{ J K}^{-1}\text{ m}^{-1}\text{ s}^{-1}}$$

COMMENT. The validity of these calculations is in doubt because the kinetic theory of gases assumes the Maxwell–Boltzmann distribution, essentially an equilibrium distribution. In such a dilute medium, the times-cales on which particles exchange energy by collision make an assumption of equilibrium unwarranted. It is especially dubious considering that atoms are more likely to interact with photons from stellar radiation than with other atoms.

P20.33 We begin by calculating the nuclei and electron concentrations as well as the nuclei radii.

$$c_H = n_H/V = (\textit{mass fraction})\times(\textit{density})/M_H = 0.36\times(158\text{ g cm}^{-3})/(1.0\text{ g mol}^{-1}) = 57\text{ mol cm}^{-3}$$
$$c_{He} = n_{He}/V = (\textit{mass fraction})\times(\textit{density})/M_{He} = 0.64\times(158\text{ g cm}^{-3})/(4.0\text{ g mol}^{-1}) = 25\text{ mol cm}^{-3}$$
$$c_e = c_H + 2c_{He} = (57+2\times25)\text{ mol cm}^{-3} = 107\text{ mol cm}^{-3}$$
$$c_{total} = c_H + c_{He} + c_e = 189\text{ mol cm}^{-3}$$

The nuclear particle radii are calculated with the formula $r_{nucleus} = 1.4A^{1/3}$ fm where A is the mass number.

$$r_H = (1.4 \times 10^{-15}\ \text{m}) \times 1^{1/3} = 1.4 \times 10^{-15}\ \text{m}$$
$$r_{He} = (1.4 \times 10^{-15}\ \text{m}) \times 4^{1/3} = 2.2 \times 10^{-15}\ \text{m}$$

We will also need the average radius r of a nuclear particle that is participating in a collision because, as shown in Figure 20.4, the excluded volume has radius equal to $2r$.

$$r = 0.36 r_H + 0.64 r_{He}$$
$$= 0.36(1.4\ \text{fm}) + 0.64(2.2\ \text{fm}) = 1.9\ \text{fm} = 1.9 \times 10^{-15}\ \text{m}$$

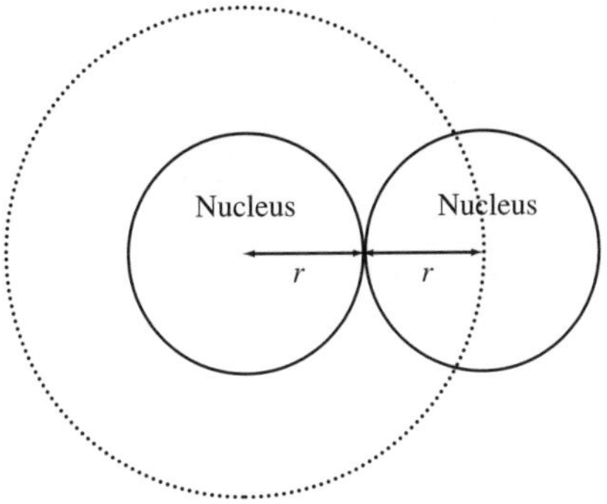

Figure 20.4

(a) The excluded volume of a nuclear collisional pair is estimated to be equal to the volume of the dashed sphere in Figure 20.4. The excluded volume of a single nucleus is $\frac{1}{2}$ of this.

$$\textit{excluded volume per nucleus} = \tfrac{1}{2} \times \tfrac{4}{3}\pi \times (2r)^3$$
$$= \tfrac{1}{2} \times \tfrac{4}{3}\pi \times (2 \times 1.9 \times 10^{-15}\ \text{m})^3 = 1.1 \times 10^{-43}\ \text{m}^3 = 1.1 \times 10^{-37}\ \text{cm}^3$$

The number density $\mathcal{N}$ for the nuclei is

$$\mathcal{N} = (c_H + c_{He}) \times N_A = (82\ \text{mol cm}^{-3}) \times (6.022 \times 10^{23}\ \text{mol}^{-1}) = 4.9 \times 10^{25}\ \text{cm}^{-3}$$

and, therefore, the total excluded volume of all nuclei in 1 cm^3 is

$$\textit{total excluded volume of all nuclei in}\ 1\ \text{cm}^3 = (4.9 \times 10^{25}) \times (1.1 \times 10^{-37}\ \text{cm}^3) = \boxed{5.4 \times 10^{-12}\ \text{cm}^3}$$

This means that the fraction of space that is actually 'occupied' is 5.4×10^{-12}, a value so small that we are justified in saying that the particles occupy a negligibly small volume. This is one of the two conditions that must be satisfied when applying the perfect gas law. The second is that the interaction potential between particles must be a constant that does not depend on particle positions. This condition is satisfied because the nuclei are embedded in a sea of electrons and the electrons move so rapidly that their response to nuclear motion is immediate with the result that potential energy does not change.

(b) The temperature predicted by the perfect gas law is

$$T = \frac{p}{c_{\text{total}}R}$$
$$= \frac{2.5 \times 10^{11}\,\text{atm}}{(189\,\text{mol cm}^{-3}) \times (82.058\,\text{atm cm}^3\,\text{K}^{-1}\,\text{mol}^{-1})} = 1.6 \times 10^7\,\text{K} = \boxed{16\,\text{MK}}$$

(c) To find the temperature predicted by the van der Waals equation of state, we estimate the van der Waals coefficients to be $a = 0$ and

$$b = N_A \times \textit{excluded volume per nucleus}$$
$$= (6.022 \times 10^{23}\,\text{mol}^{-1}) \times (1.1 \times 10^{-37}\,\text{cm}^3) = 6.6 \times 10^{-14}\,\text{cm}^3\,\text{mol}^{-1}$$

Thus, by the van der Waal eos

$$T = \left(\frac{V_m - b}{R}\right) \times \left(p + \frac{a}{V_m^2}\right) = \frac{(V_m - b)p}{R} = \frac{\left(\frac{1}{c_{\text{total}}} - b\right)p}{R}$$

But $1/c_{\text{total}} = (189\,\text{mol cm}^{-3})^{-1} = 0.00529\,\text{cm}^3\,\text{mol}^{-1}$, so we conclude that b is negligibly small compared to V_m. Thus, the van der Waals eos is reduced to the perfect gas law with the prediction that $T = \boxed{16\,\text{MK}}$.

P20.35 Equation [20.60], $\langle x^2 \rangle = 2Dt$, gives the mean square distance travelled in any one dimension in time t. We need the distance travelled from a point in any direction. The distinction here is the distinction between the one-dimensional and three-dimensional diffusion. The mean square three-dimensional distance can be obtained from the one-dimensional mean square distance since motions in the three directions are independent. Since $r^2 = x^2 + y^2 + z^2$ [Pythagorean theorem], $\langle r^2 \rangle = \langle x^2 \rangle + \langle y^2 \rangle + \langle z^2 \rangle = 3\langle x^2 \rangle$ [independent motion] $= 3 \times 2Dt$ [20.60 for $\langle x^2 \rangle$] $= 6Dt$.

Therefore, $t = \dfrac{\langle r^2 \rangle}{6D} = \dfrac{(1.0 \times 10^{-6}\,\text{m})^2}{6(1.0 \times 10^{-11}\,\text{m}^2\,\text{s}^{-1})} = \boxed{1.7 \times 10^{-2}\,\text{s}}$

21 The rates of chemical reactions

Answers to discussion questions

D21.1 The time scales of atomic processes are rapid indeed: according to the following table, a nanosecond is an eternity. Note that the times given here are in some way typical values for times that may vary over two or three orders of magnitude. For example, vibrational wavenumbers can range from about 4400 cm^{-1} (for H_2) to 100 cm^{-1} (for I_2) and even lower, with a corresponding range of associated times. Radiative decay rates of electronic states can vary even more widely: times associated with phosphorescence can be in the millisecond and even second range. A large number of time scales for physical, chemical, and biological processes on the atomic and molecular scale are reported in Figure 2 of A.H. Zewail, Femtochemistry: atomic-scale dynamics of the chemical bond, *Journal of Physical Chemistry A* **104**, 5660 (2000).

Radiative decay of excited electronic states can range from about 10^{-9} to 10^{-4} s—even longer for phosphorescence involving 'forbidden' decay paths. Molecular rotational motion takes place on a scale of 10^{-12} to 10^{-9} s. Molecular vibrations are faster still, about 10^{-14} to 10^{-12} s. Proton transfer reactions occur on a time scale of about 10^{-10} to 10^{-9} s, although protons can hop from molecule to molecule in water even more rapidly (1.5×10^{-12} s, Section 20.7(a)). *Impact on biochemistry I13.1* (Vision) describes several events in vision, including the 200-fs photoisomerization that gets the process started. *Impact on biochemistry I21.1* (Harvesting of light during plant photosynthesis) lists time scales of several energy-transfer and electron-transfer steps in photosynthesis. Initial energy transfer (to a nearby pigment) has a time scale of around 10^{-13} to 5×10^{-12} s, with longer-range transfer (to the reaction centre) taking about 10^{-10} s. Immediate electron transfer is also very fast (about 3 ps), with ultimate transfer (leading to oxidation of water and reduction of plastoquinone) taking from 10^{-10} to 10^{-3} s. The mean time between collisions in liquids is similar to vibrational periods, around 10^{-13} s. One can estimate collision times in liquids very roughly by applying the expression for collisions in gases (Section 20.1(b)) to liquid conditions.

Process	t/ns	Reference
Radiative decay of electronic excited state	1×10^{1}	Section 13.4
Molecular rotational motion	3×10^{-2}	$B\approx 1\ \mathrm{cm^{-1}}$
Molecular vibrational motion	3×10^{-5}	$\tilde{\nu}\approx 1000\ \mathrm{cm^{-1}}$
Proton transfer	0.3	Zewail 2000
Initial chemical reaction of vision*	2×10^{-4}	*Impact* 13.1
Energy transfer in photosynthesis†	1×10^{-3}	Impact 21.1
Electron transfer in photosynthesis	3×10^{-3}	*Impact* 21.1
Collision frequency in liquids	4×10^{-4}	Section 20.1(b)‡

*Photoisomerization of retinal from 11-*cis* to all-*trans*.
†Time from absorption until electron transfer to adjacent pigment.
‡Use formula for gas collision frequency at 300 K, parameters for benzene from data section, and density of liquid benzene.

D21.3 Refer to Table 21.3. We will consider only reactions whose rates depend on the concentration of a single reactant.

In a first-order reaction, the rate of reaction is directly proportional to the concentration of the reactant:

$$v = k_r[\mathrm{A}] \quad \text{and} \quad \ln[\mathrm{A}] = \ln[\mathrm{A}]_0 - k_r t \ [21.12]$$

A plot of the logarithm of reactant concentration against time is a straight line.

In a zero-order reaction, the rate of reaction is constant, independent of the reactant concentration:

$$v = k_r \quad \text{and} \quad [\mathrm{A}] = [\mathrm{A}]_0 - k_r t \quad [\text{based on Table 21.3}]$$

A plot of the reactant concentration itself against time is a straight line.

In a second-order reaction, the rate of reaction is proportional to the square of the reactant concentration:

$$v = k_r[\mathrm{A}]^2 \quad \text{and} \quad \frac{1}{[\mathrm{A}]} = \frac{1}{[\mathrm{A}]_0} + k_r t$$

A plot of the reciprocal of reactant concentration against time is a straight line.

Reaction orders need not be integers, except for elementary reactions. Indeed, reaction orders can change during the course of the reaction. Consider the zero-order reaction mentioned in the text (Section 21.2(c)), the decomposition of phosphine on hot tungsten. As long as enough phosphine is present, the rate of reaction is independent of that concentration; what limits the rate is the availability of catalytic sites on the tungsten. Clearly the integrated rate law for a zero-order reaction cannot be correct at long times, where it would predict negative concentrations. Before that unphysical situation would occur, the concentration of the reactant drops to such an extent that *it* limits the rate of reaction, and the reaction order changes from zero to a non-zero value. The text's treatment of the Lindemann–Hinshelwood mechanism (Section 21.8(a)) also illustrates how a reaction order can change from first to second over the course of a reaction.

D21.5 Yes, a negative activation energy is quite possible for composite reactions. The rate constant of a composite reaction can be a product or ratio of rate constants and equilibrium constants of elementary reactions that contribute to the composite reaction, as illustrated in eqn 21.63. In general, elementary reactions that have a positive activation energy whose rate constants appear in the denominator of a composite rate constant tend to reduce the activation energy of the overall reaction, as illustrated in eqn 21.64. There is no reason why that reduction cannot be to a negative value.

The most common molecular interpretation of the activation energy is as the 'height' of an energy barrier that must be overcome by reactants in order to form products, as discussed in connection with collision theory in Section 22.1(b). Among the limitations of collision theory is that it is only applicable to elementary reactions, those that can possibly occur in one collision or reactive molecular encounter. Thus, it is no surprise that this interpretation fails to apply to composite reactions. The more general interpretation of the activation energy, as a measure of the temperature dependence of the reaction rate (eqn. 21.30), does apply.

Question. Show that the following mechanism leads to an overall negative activation energy for the rate of formation of P if $E_a(2) > E_a(3)$.

$A \rightarrow I \quad k_1$ (slow)
$I \rightarrow B \quad k_2$
$I \rightarrow P \quad k_3$

D21.7

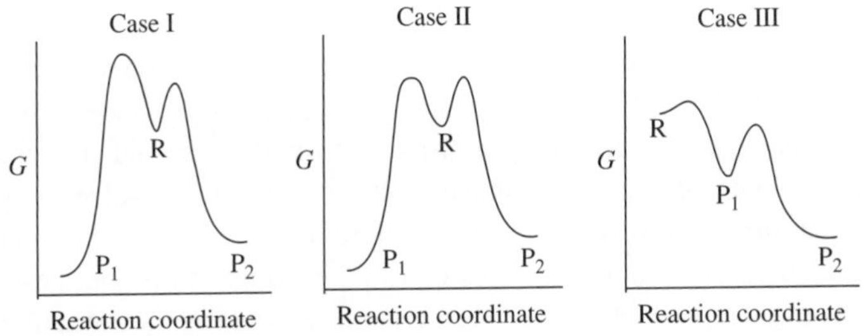

Figures 21.1

Simple diagrams of Gibbs energy against reaction coordinate are useful for distinguishing between kinetic and thermodynamic control of a reaction. For the simple parallel reactions $R \rightarrow P_1$ and $R \rightarrow P_2$, shown in Figure 21.1 as cases I and II, the product P_1 is thermodynamically favoured because the Gibbs energy decreases to a greater extent for its formation. However, the rate at which each product appears does not depend on thermodynamic favorability. Rate constants depend on activation energy. In case I the activation energy for the formation of P_1 is much larger than that for formation of P_2. At low or moderate temperatures the large activation energy may not be readily available and P_1 either cannot form or forms at a slow rate. The much smaller activation energy for P_2 formation is available and, consequently, P_2 is produced even though it is not the thermodynamically favoured product. This is kinetic control. In this case, $[P_2]/[P_1] = k_2/k_1 > 1$ [21.51].

The activation energies for the parallel reactions are equal in Case II and, consequently, the two products appear at identical rates. If the reactions are irreversible, $[P_2]/[P_1] = k_2/k_1 = 1$ at all times. The results are very different for reversible reactions. The activation energy for $P_1 \rightarrow R$ is much larger than that for $P_2 \rightarrow R$ and P_1 accumulates as the more rapid $P_2 \rightarrow R \rightarrow P_1$ occurs. Eventually the ratio $[P_2]/[P_1]$ approaches the equilibrium value

$$\left(\frac{[P_2]}{[P_1]}\right)_{eq} = e^{-(\Delta G_2-\Delta G_1)/RT} < 1$$

This is thermodynamic control.

Case III above represents an interesting consecutive reaction series $R \rightarrow P_1 \rightarrow P_2$. The first step has relatively low activation energy and P_1 rapidly appears. However, the relatively large activation energy for the second step is not available at low and moderate temperatures. By using low or moderate temperatures and short reaction times it is possible to produce more of the thermodynamically less favourable P_1. This is kinetic control. High temperatures and long reaction times will yield the thermodynamically favoured P_2.

The ratio of reaction products is determined by relative reaction rates in kinetic-controlled reactions. Favourable conditions include short reaction times, lower temperatures, and irreversible reactions. Thermodynamic control is favoured by long reaction times, higher temperatures, and reversible reactions. The ratio of products depends on the relative stability of products for thermodynamically controlled reactions.

D21.9 A primary quantum yield is associated with a primary photochemical event in an overall photochemical process. Primary photochemical events are those events that directly involve an excited state after absorbing light energy, events such as fluorescence, phosphorescence, internal conversion, intersystem crossing, or a photochemical reaction that involves the excited state directly. The primary quantum yield from a given primary process, then, is the ratio of the number of the product of that process to the number of photons absorbed (eqn. 21.76a). Secondary processes are those that do not involve an excited state directly, but subsequent events. Examples include a chemical reaction involving a molecule that has received energy transferred from the absorbing molecule or a reaction involving a product of a primary photochemical reaction, which may involve secondary events as well. The overall quantum yield is the ratio of the amount of product (or products) of interest formed, whether by primary or secondary processes, to the amount of photons absorbed. The overall quantum yield of a given product can be determined by measuring the amount of the product(s) formed, the intensity of light employed, and the absorbance of the sample. Primary quantum yields of photochemical processes, such as fluorescence and phosphorescence, can be determined by techniques specific to the process. For example, fluorescence quantum yield can be measured by pulsed laser techniques, in which a short laser pulse is used to excite a molecule and then the fluorescence is detected as the excited molecules decay. In effect, the photons emitted by fluorescence are 'counted' (integrated over time) and compared to the photons absorbed.

Solutions to exercises

E21.1(a) Let the initial amount of ICl be n_{ICl} and the initial amount of H_2 be n_H; the initial amounts of I_2 and HCl are assumed to be zero. Thus, the initial total quantity of gas is $n_{ICl} + n_H$. Let the amount of I_2 formed at any given time be n. In that case, the amount of HCl is $2n$, that of H_2 is $n_H - n$, and the amount of ICl is $n_{ICl} - 2n$. At any given time, then, the total quantity of gas is

$$n_{total} = n_{ICl} - 2n + n_H - n + n + 2n = n_{ICl} + n_H = n_{initial}$$

Thus, there is no change in the amount of gas during the course of the reaction. Since there is no change in volume or temperature either, there is $\boxed{\text{no change in pressure}}$.

COMMENT. Measuring the pressure would *not* be a practical way of monitoring the progress of this reaction.

E21.2(a) $$v = \frac{1}{\nu_J}\frac{d[J]}{dt}\ [21.3b], \quad \text{so} \quad \frac{d[J]}{dt} = \nu_J v$$

Rate of formation of C = $3v = \boxed{8.1\ \text{mol dm}^{-3}\ \text{s}^{-1}}$

Rate of formation of D = $v = \boxed{2.7\ \text{mol dm}^{-3}\ \text{s}^{-1}}$

Rate of consumption of A = $v = \boxed{2.7\ \text{mol dm}^{-3}\ \text{s}^{-1}}$

Rate of consumption of B = $2v = \boxed{5.4\ \text{mol dm}^{-3}\ \text{s}^{-1}}$

E21.3(a) $$v = \frac{1}{\nu_J}\frac{d[J]}{dt}\ [21.3b] = \frac{1}{2}\frac{d[C]}{dt} = \tfrac{1}{2} \times (2.7\ \text{mol dm}^{-3}\ \text{s}^{-1}) = \boxed{1.3\overline{5}\ \text{mol dm}^{-3}\ \text{s}^{-1}}$$

Rate of formation of D = $3v = \boxed{4.0\overline{5}\ \text{mol dm}^{-3}\ \text{s}^{-1}}$

Rate of consumption of A = $2v = \boxed{2.7\ \text{mol dm}^{-3}\ \text{s}^{-1}}$

Rate of consumption of B = $v = \boxed{1.3\overline{5}\ \text{mol dm}^{-3}\ \text{s}^{-1}}$

E21.4(a) The rate is expressed in mol dm^{-3} s^{-1}; therefore

$$\text{mol dm}^{-3}\ \text{s}^{-1} = [k_r] \times (\text{mol dm}^{-3}) \times (\text{mol dm}^{-3}),$$

where $[k_r]$ denotes units of k_r, requires the units to be $\boxed{\text{dm}^3\ \text{mol}^{-1}\ \text{s}^{-1}}$.

(a) Rate of consumption of A = $v = \boxed{k_r[A][B]}$

(b) Rate of formation of C = $3v = \boxed{3k_r[A][B]}$

E21.5(a) Given $\frac{d[C]}{dt} = k_r[A][B][C]$,

the rate of reaction is [21.3b]

$$v = \frac{1}{\nu_J}\frac{d[J]}{dt} = \frac{1}{2}\frac{d[C]}{dt} = \boxed{\frac{1}{2}k_r[A][B][C]}$$

The units of k_r, $[k_r]$, must satisfy

$$\text{mol dm}^{-3}\ \text{s}^{-1} = [k_r] \times (\text{mol dm}^{-3}) \times (\text{mol dm}^{-3}) \times (\text{mol dm}^{-3})$$

Therefore, $[k_r] = \boxed{\text{dm}^6\ \text{mol}^{-2}\ \text{s}^{-1}}$

E21.6(a) (a) For a second-order reaction, denoting the units of k_r by $[k_r]$

$$\text{mol dm}^{-3}\,\text{s}^{-1} = [k_r] \times (\text{mol dm}^{-3})^2; \text{ therefore } \boxed{[k_r] = \text{dm}^3\,\text{mol}^{-1}\,\text{s}^{-1}}$$

For a third-order reaction

$$\text{mol dm}^{-3}\,\text{s}^{-1} = [k_r] \times (\text{mol dm}^{-3})^3; \text{ therefore } \boxed{[k_r] = \text{dm}^6\,\text{mol}^{-2}\,\text{s}^{-1}}$$

(b) For a second-order reaction

$$\text{kPa s}^{-1} = [k_r] \times \text{kPa}^2; \text{ therefore } \boxed{[k_r] = \text{kPa}^{-1}\,\text{s}^{-1}}$$

For a third-order reaction

$$\text{kPa s}^{-1} = [k_r] \times \text{kPa}^3; \text{ therefore } \boxed{[k_r] = \text{kPa}^{-2}\,\text{s}^{-1}}$$

E21.7(a) The rate law is

$$v = k_r[\text{A}]^a \propto p_\text{A}^a = \{p_{\text{A},0}(1-f)\}^a$$

where f is the fraction reacted. That is, concentration and partial pressure are proportional to each other. Thus, we can write

$$\frac{v_1}{v_2} = \frac{p_{\text{A},1}^a}{p_{\text{A},2}^a} = \left(\frac{1-f_1}{1-f_2}\right)^a$$

Taking logarithms

$$\ln\left(\frac{v_1}{v_2}\right) = a\ln\left(\frac{1-f_1}{1-f_2}\right)$$

$$\text{so}\quad a = \frac{\ln\left(\dfrac{v_1}{v_2}\right)}{\ln\left(\dfrac{1-f_1}{1-f_2}\right)} = \frac{\ln\left(\dfrac{1.07}{0.76}\right)}{\ln\left(\dfrac{0.95}{0.80}\right)} = 1.9\overline{9}$$

Hence, the reaction is $\boxed{\text{second order}}$.

COMMENT. Knowledge of the initial pressure is not required for the solution to this exercise. The ratio of pressures was computed using fractions of the initial pressure.

E21.8(a) Table 21.3 gives a general expression for the half-life of a reaction of the type A → P for orders other than 1:

$$t_{1/2} = \frac{2^{n-1}-1}{(n-1)k_r[\text{A}]_0^{n-1}} \propto [\text{A}]_0^{1-n} \propto p_0^{1-n}$$

where the proportionality constants may be functions of the reaction order, the rate constant, or even the temperature, but not of the concentration. Form a ratio of the half-lives at different initial pressures:

$$\frac{t_{1/2}(p_{0,1})}{t_{1/2}(p_{0,2})} = \left(\frac{p_{0,1}}{p_{0,2}}\right)^{1-n} = \left(\frac{p_{0,2}}{p_{0,1}}\right)^{n-1}$$

Hence, $\ln\left(\dfrac{t_{1/2}(p_{0,1})}{t_{1/2}(p_{0,2})}\right) = (n-1)\ln\left(\dfrac{p_{0,2}}{p_{0,1}}\right)$

or $$(n-1) = \frac{\ln\left(\dfrac{410\text{ s}}{880\text{ s}}\right)}{\ln\left(\dfrac{169\text{ Torr}}{363\text{ Torr}}\right)} = 0.999 \approx 1$$

Therefore, $\boxed{n=2}$ in agreement with the result of Exercise 21.7(a).

E21.9(a) $2\,N_2O_5 \rightarrow 4\,NO_2 + O_2 \; v = k_r[N_2O_5]$

Therefore, rate of consumption of $N_2O_5 = 2v = 2k_r[N_2O_5]$

$$\frac{d[N_2O_5]}{dt} = -2k_r[N_2O_5], \quad \text{so} \quad [N_2O_5] = [N_2O_5]_0 e^{-2k_r t}$$

Solve this for t:

$$t = \frac{1}{2k_r}\ln\frac{[N_2O_5]_0}{[N_2O_5]}$$

Therefore, the half-life is:

$$t_{1/2} = \frac{1}{2k_r}\ln 2 = \frac{\ln 2}{(2)\times(3.38\times10^{-5}\,\text{s}^{-1})} = \boxed{1.03\times10^4\,\text{s}}$$

Since the partial pressure of N_2O_5 is proportional to its concentration

$$p(N_2O_5) = p_0(N_2O_5)e^{-2k_r t}$$

(a) $p(N_2O_5) = (500\text{ Torr})\times(e^{-(2\times3.38\times10^{-5}/\text{s})\times(50\text{ s})}) = \boxed{498\text{ Torr}}$

(b) $p(N_2O_5) = (500\text{ Torr})\times(e^{-(2\times3.38\times10^{-5}/\text{s})\times(20\times60\text{ s})}) = \boxed{461\text{ Torr}}$

COMMENT. The half-life formula in Table 21.3 is based on a rate constant for the rate of change of the reactant, that is, based on the assumption that

$$-\frac{d[A]}{dt} = k_r[A]$$

Our expression for the rate of consumption has $2k_r$ instead of k_r, and our expression for $t_{1/2}$ does likewise.

E21.10(a) **Note**: There is a typographical error in the problem statement. It is the concentration of B, not A, that has dropped to 0.020 mol dm^{-3} after 1.0 h.

The integrated rate law is

$$k_r t = \frac{1}{[B]_0 - [A]_0}\ln\left(\frac{[B]/[B]_0}{[A]/[A]_0}\right) \quad [21.19]$$

(a) The stoichiometry of the reaction requires that when

$$\Delta[\mathrm{B}] = (0.020 - 0.050)\ \mathrm{mol\ dm^{-3}} = -0.030\ \mathrm{mol\ dm^{-3}}$$

then $\Delta[\mathrm{A}] = -0.030\ \mathrm{mol\ dm^{-3}}$ as well.

Thus, $[\mathrm{A}] = 0.075\ \mathrm{mol\ dm^{-3}} - 0.030\ \mathrm{mol\ dm^{-3}} = 0.045\ \mathrm{mol\ dm^{-3}}$

when $[\mathrm{B}] = 0.020\ \mathrm{mol\ dm^{-3}}$. Therefore,

$$k_r t = \frac{1}{(0.050 - 0.075)\ \mathrm{mol\ dm^{-3}}} \ln\left(\frac{0.020/0.050}{0.045/0.075}\right)$$

$$k_r \times 1.0\ \mathrm{h} = 16.\bar{2}\ \mathrm{dm^3\ mol^{-1}}$$

$$\text{so,}\quad k_r = \boxed{16.\bar{2}\ \mathrm{dm^3\ mol^{-1}\ h^{-1}}} \times \left(\frac{1\ \mathrm{h}}{3600\ \mathrm{s}}\right) = \boxed{4.5 \times 10^{-3}\ \mathrm{dm^3\ mol^{-1}\ s^{-1}}}$$

(b) The half-life with respect to A is the time required for [A] to fall to $0.0375\ \mathrm{mol\ dm^{-3}}$ (and [B] to $0.0125\ \mathrm{mol\ dm^{-3}}$). We solve eqn. 21.19 for t

$$t_{1/2}(\mathrm{A}) = \left(\frac{1}{(16.\bar{2}\ \mathrm{dm^3\ mol^{-1}\ h^{-1}}) \times (-0.025\ \mathrm{mol\ dm^{-3}})}\right) \times \ln\left(\frac{0.0125/0.050}{0.50}\right)$$

$$= 1.7\bar{1}\ \mathrm{h} = \boxed{6.2 \times 10^3\ \mathrm{s}}$$

Similarly, the half-life with respect to B is the time required for [B] to fall to $0.025\ \mathrm{mol\ dm^{-3}}$ (and [A] to $0.050\ \mathrm{mol\ dm^{-3}}$).

$$t_{1/2}(\mathrm{B}) = \left(\frac{1}{(16.\bar{2}\ \mathrm{dm^3\ mol^{-1}\ h^{-1}}) \times (-0.025\ \mathrm{mol\ dm^{-3}})}\right) \times \ln\left(\frac{0.50}{0.050/0.075}\right)$$

$$= 0.7\bar{1}\ \mathrm{h} = \boxed{2.6 \times 10^3\ \mathrm{s}}$$

COMMENT. This exercise illustrates that there is no unique half-life for reactions other than those of the type $\mathrm{A} \rightarrow \mathrm{P}$.

E21.11(a) The integrated second-order rate law for a reaction of the type $\mathrm{A} + \mathrm{B} \rightarrow$ products is

$$k_r t = \frac{1}{[\mathrm{B}]_0 - [\mathrm{A}]_0} \ln\left(\frac{[\mathrm{B}]/[\mathrm{B}]_0}{[\mathrm{A}]/[\mathrm{A}]_0}\right) [21.19]$$

Introducing $[\mathrm{B}] = [\mathrm{B}]_0 - x$ and $[\mathrm{A}] = [\mathrm{A}]_0 - x$ and rearranging we obtain

$$k_r t = \left(\frac{1}{[\mathrm{B}]_0 - [\mathrm{A}]_0}\right) \ln\left(\frac{[\mathrm{A}]_0([\mathrm{B}]_0 - x)}{([\mathrm{A}]_0 - x)[\mathrm{B}]_0}\right)$$

Solving for x yields, after some rearranging,

$$x = \frac{[\mathrm{A}]_0[\mathrm{B}]_0(\mathrm{e}^{k_r([\mathrm{B}]_0-[\mathrm{A}]_0)t} - 1)}{[\mathrm{B}]_0\mathrm{e}^{([\mathrm{B}]_0-[\mathrm{A}]_0)k_r t} - [\mathrm{A}]_0} = \frac{(0.060) \times (0.110\ \mathrm{mol\ dm^{-3}}) \times (\mathrm{e}^{(0.110-0.060)\times 0.11 \times t/\mathrm{s}} - 1)}{(0.110) \times \mathrm{e}^{(0.110-0.060)\times 0.11 \times t/\mathrm{s}} - 0.060}$$

$$= \frac{(0.060\ \mathrm{mol\ dm^{-3}}) \times (\mathrm{e}^{0.0055t/\mathrm{s}} - 1)}{\mathrm{e}^{0.0055t/\mathrm{s}} - 0.55}$$

(a) After 20 s

$$x=\frac{(0.060\ \text{mol dm}^{-3})\times(e^{0.0055\times 20}-1)}{e^{0.0055\times 20}-0.55}=0.0122\ \text{mol dm}^{-3}$$

which implies that

$$[CH_3COOC_2H_5]=(0.110-0.0122)\ \text{mol dm}^{-3}=\boxed{0.098\ \text{mol dm}^{-3}}$$

(b) After 15 min = 900 s,

$$x=\frac{(0.060\ \text{mol dm}^{-3})\times(e^{0.0055\times 900}-1)}{e^{0.0055\times 900}-0.55}=0.060\ \text{mol dm}^{-3}$$

so $[CH_3COOC_2H_5]=(0.110-0.060)\ \text{mol dm}^{-3}=\boxed{0.050\ \text{mol dm}^{-3}}$

E21.12(a) The rate of consumption of A is

$$-\frac{d[A]}{dt}=2v=2k_r[A]^2\quad[\nu_A=-2]$$

which integrates to $\frac{1}{[A]}-\frac{1}{[A]_0}=2k_rt$ [21.15b with k_r replaced by $2k_r$]

Therefore, $t=\frac{1}{2k_r}\left(\frac{1}{[A]}-\frac{1}{[A]_0}\right)$

$$t=\left(\frac{1}{2\times 4.30\times 10^{-4}\,\text{dm}^3\,\text{mol}^{-1}\,\text{s}^{-1}}\right)\times\left(\frac{1}{0.010\ \text{mol dm}^{-3}}-\frac{1}{0.210\ \text{mol dm}^{-3}}\right)$$

$$=\boxed{1.11\times 10^5\,\text{s}}=\boxed{1.28\ \text{days}}$$

E21.13(a) The reactions whose rate constants are sought are the forward and reverse reactions in the following equilibrium.

$$NH_3(aq)+H_2O(l)\underset{k_r'}{\overset{k_r}{\rightleftharpoons}}NH_4^+(aq)+OH^-(aq)$$

The rate constants are related by

$$K_b=\frac{k_r}{k_r'}=\frac{[NH_4^+][OH^-]}{[NH_3]}=1.78\times 10^{-5}\ \text{mol dm}^{-3}$$

where the concentrations are equilibrium concentrations. (We assign units to K_b, which technically is a pure number, to help us keep track of units in the rate constants. Keeping track of the units makes us realize that k_r is a pseudo-first-order protonation of NH_3 in excess water, for water does not appear in the above expression.) We need one more relationship between the constants, which we can obtain by proceeding as in Example 21.4:

$$\frac{1}{\tau}=k_r+k_r'([NH_4^+]+[OH^-])$$

Substitute into this expression

$$k_r=K_bk_r'\quad\text{and}\quad[NH_4^+]=[OH^-]=(K_b[NH_3])^{1/2}$$

hence, $\dfrac{1}{\tau} = K_b k_r' + 2k_r'(K_b[NH_3])^{1/2} = k_r'\{K_b + 2(K_b[NH_3])^{1/2}\}$

So the reverse rate constant is

$$k_r' = \frac{1}{\tau\{K_b + 2(K_b[NH_3])^{1/2}\}}$$

$$= \frac{1}{7.61\times10^{-9}\text{ s}\{1.78\times10^{-5}\text{ mol dm}^{-3} + 2(1.78\times10^{-5}\times0.15)^{1/2}\text{ mol dm}^{-3}\}}$$

$$= \boxed{4.0\times10^{10}\text{ dm}^{-3}\text{ mol s}^{-1}}$$

and the forward constant is

$$k_r = K_b k_r' = 1.78\times10^{-5}\text{ mol dm}^{-3}\times4.0\times10^{10}\text{ dm}^3\text{ mol}^{-1}\text{ s}^{-1} = \boxed{7.1\times10^5\text{ s}^{-1}}$$

Recall that k_r is the rate constant in the pseudo-first-order rate law

$$-\frac{d[NH_3]}{dt} = k_r[NH_3]$$

Let us call k the rate constant in the bimolecular rate law

$$-\frac{d[NH_3]}{dt} = k[NH_3][H_2O]$$

Setting these two expressions equal to each other yields

$$k = \frac{k_r}{[H_2O]} = \frac{7.1\times10^5\text{ s}^{-1}}{(1000\text{ g dm}^{-3})/(18.02\text{ g mol}^{-1})} = \boxed{1.28\times10^4\text{ dm}^3\text{ mol}^{-1}\text{ s}^{-1}}$$

E21.14(a) Call the rate constant k at temperature T, and the rate constant k' at temperature T':

$$\ln k = \ln A - \frac{E_a}{RT};\quad \ln k' = \ln A - \frac{E_a}{RT'}\ [21.29]$$

Hence, $$E_a = \frac{R\ln\left(\dfrac{k'}{k}\right)}{\left(\dfrac{1}{T}-\dfrac{1}{T'}\right)} = \frac{(8.314\text{ J K}^{-1}\text{ mol}^{-1})\times\ln\left(\dfrac{1.38\times10^{-2}}{2.80\times10^{-3}}\right)}{\dfrac{1}{303\text{ K}}-\dfrac{1}{323\text{ K}}} = \boxed{64.9\text{ kJ mol}^{-1}}$$

For A, we rearrange eqn 21.31:

$$A = k\times e^{E_a/RT} = (2.80\times10^{-3}\text{ mol dm}^{-3}\text{ s}^{-1})\times e^{64.9\times10^3/(8.314\times303)}$$

$$= \boxed{4.32\times10^8\text{ mol dm}^{-3}\text{ s}^{-1}}$$

E21.15(a) Proceed as in Exercise 21.14(a):

$$E_a = \frac{R\ln\left(\dfrac{k_r(T_2)}{k_r(T_1)}\right)}{\left(\dfrac{1}{T_1}-\dfrac{1}{T_2}\right)} = \frac{(8.3145\text{ J K}^{-1}\text{ mol}^{-1})\times\ln 3}{\dfrac{1}{297\text{ K}}-\dfrac{1}{322\text{ K}}} = \boxed{35\text{ kJ mol}^{-1}}$$

E21.16(a) The rate of the overall reaction is

$$v = \frac{d[P]}{dt} = k_2[A][B]$$

however, we cannot have the concentration of an intermediate in the overall rate law.

(i) Assume a pre-equilibrium with

$$K = \frac{[A]^2}{[A_2]}, \quad \text{implying that} \quad [A] = K^{1/2}[A_2]^{1/2}$$

and $v = \boxed{k_2K^{1/2}[A_2]^{1/2}[B]} = k_{\text{eff}}[A_2]^{1/2}[B]$

where $k_{\text{eff}} = k_2K^{1/2}$

(ii) Apply the steady-state approximation:

$$\frac{d[A]}{dt} \approx 0 = 2k_1[A_2] - 2k_1'[A]^2 - k_2[A][B]$$

This is a quadratic equation in [A]

$$[A] = \frac{-b \pm \sqrt{b^2 - 4ac}}{2a} = \frac{k_2[B] \pm \sqrt{k_2^2[B]^2 + 16k_1'k_1[A_2]}}{-4k_1'} = \frac{k_2[B]}{4k_1'}\left(\sqrt{1 + \frac{16k_1'k_1[A_2]}{k_2^2[B]^2}} - 1\right)$$

(In the last step, choose the sign that gives a positive quantity for [A].) Thus, the rate law is

$$v = k_2[A][B] = \boxed{\frac{k_2^2[B]^2}{4k_1'}\left(\sqrt{1 + \frac{16k_1'k_1[A_2]}{k_2^2[B]^2}} - 1\right)}$$

This is a perfectly good rate law, albeit a complicated one. It is not in typical power-law form, but it is a function of reactant concentrations only, with no intermediates. This law simplifies under certain conditions. If $16k_1k_1'[A_2] \gg k_2^2[B]^2$, then

$$\sqrt{1 + \frac{16k_1'k_1[A_2]}{k_2^2[B]^2}} - 1 \approx \frac{4\sqrt{k_1'k_1[A_2]}}{k_2[B]}$$

and $v \approx \dfrac{k_2^2[B]^2}{4k_1'} \times \dfrac{4\sqrt{k_1'k_1[A_2]}}{k_2[B]} = k_2[B]\sqrt{\dfrac{k_1[A_2]}{k_1'}} = \boxed{k_2K^{1/2}[A_2]^{1/2}[B]}$

recovering the pre-equilibrium rate law. If, on the other hand, $16k_1k_1'[A_2] \ll k_2^2[B]^2$, we expand the square root

$$\sqrt{1 + \frac{16k_1'k_1[A_2]}{k_2^2[B]^2}} \approx 1 + \frac{8k_1'k_1[A_2]}{k_2^2[B]^2}$$

and $v \approx \dfrac{k_2^2[B]^2}{4k_1'}\left(1 + \dfrac{8k_1'k_1[A_2]}{k_2^2[B]^2} - 1\right) = \boxed{2k_1[A_2]}$

COMMENT. If the equilibrium is 'fast', the latter condition will not be fulfilled. In fact, this special case amounts to having the first step rate limiting. Note that the full (messy) steady-state approximation is less severe than either the pre-equilibrium or the rate-limiting step approximations, for it includes both as special cases.

E21.17(a) $$\frac{1}{k_r} = \frac{k'_a}{k_a k_b} + \frac{1}{k_a p_A} \text{ [analogous to 21.62]}$$

Therefore, for two different pressures we have

$$\frac{1}{k_r(p_1)} - \frac{1}{k_r(p_2)} = \frac{1}{k_a}\left(\frac{1}{p_1} - \frac{1}{p_2}\right)$$

$$\text{so } k_a = \frac{\frac{1}{p_1} - \frac{1}{p_2}}{\frac{1}{k_r(p_1)} - \frac{1}{k_r(p_2)}} = \frac{\frac{1}{12\text{ Pa}} - \frac{1}{1.30\times10^3\text{ Pa}}}{\frac{1}{2.10\times10^{-5}\text{ s}^{-1}} - \frac{1}{2.50\times10^{-4}\text{ s}^{-1}}} = \boxed{1.9\times10^{-6}\text{ Pa}^{-1}\text{ s}^{-1}}$$

or $\boxed{1.9\text{ MPa}^{-1}\text{ s}^{-1}}$

E21.18(a) Let the steps be

$A + B \rightleftharpoons I$ (fast: k_a, k'_a)

and $I \rightarrow P$ (k_b)

Then, the rate of reaction is

$$v = \frac{d[P]}{dt} = k_b[I]$$

Applying the pre-equilibrium approximation yields

$$\frac{[I]}{[A][B]} = K = \frac{k_a}{k'_a} \quad \text{so} \quad [I] = \frac{k_a[A][B]}{k'_a}$$

$$\text{and} \quad v = \frac{k_a k_b[A][B]}{k'_a} = k_r[A][B] \quad \text{with} \quad k_r = \frac{k_a k_b}{k'_a}$$

Thus, $E_a = E_a(a) = E_a(b) - E'_a(a)$ [21.64] $= (25 + 10 - 38)\text{ kJ mol}^{-1} = \boxed{-3\text{ kJ mol}^{-1}}$.

COMMENT. Activation energies are rarely negative; however, some composite reactions are known to have small, negative activation energies.

E21.19(a) The degree of polymerization is [21.68b]

$$\langle N\rangle = 1 + k_r t[A]_0$$
$$= 1 + (1.39\text{ dm}^3\text{ mol}^{-1}\text{ s}^{-1}) \times 5.00\text{ h} \times 3600\text{ s h}^{-1} \times 1.00\times10^{-2}\text{ mol dm}^{-3}$$
$$= \boxed{251}$$

The fraction condensed is related to the degree of polymerization by

$$\langle N\rangle = \frac{1}{1-p} \quad \text{so} \quad p = \frac{\langle N\rangle - 1}{\langle N\rangle} = \frac{251-1}{251} = \boxed{0.996}$$

E21.20(a) The kinetic chain length varies with concentration as

$$\nu = k_r[\mathrm{M}][\mathrm{I}]^{-1/2} \; [21.74]$$

so the ratio of kinetic chain lengths under different concentrations is

$$\frac{\nu_2}{\nu_1} = \frac{[\mathrm{M}]_2}{[\mathrm{M}]_1} \times \left(\frac{[\mathrm{I}]_1}{[\mathrm{I}]_2}\right)^{1/2} = \frac{1}{4.2} \times \left(\frac{1}{3.6}\right)^{1/2} = \boxed{0.125}$$

E21.21(a) Number of photons absorbed $= \phi^{-1} \times$ number of molecules that react [21.76a]. Therefore,

$$\text{Number absorbed} = \frac{(2.28 \times 10^{-3}\,\text{mol}/2) \times (6.022 \times 10^{23}\,\text{einstein}^{-1})}{2.1 \times 10^{2}\,\text{mol einstein}^{-1}} = \boxed{3.3 \times 10^{18}}$$

E21.22(a) For a source of power output P at wavelength λ, the number of photons (n_λ) generated in a time t is the energy output divided by the energy per mole of photons

$$n_\lambda = \frac{Pt}{h\nu N_\mathrm{A}} = \frac{P\lambda t}{hcN_\mathrm{A}} = \frac{(100\,\mathrm{W}) \times (45\,\mathrm{min}) \times (60\,\mathrm{s\,min^{-1}}) \times (490 \times 10^{-9}\,\mathrm{m})}{(6.626 \times 10^{-34}\,\mathrm{J\,s}) \times (2.998 \times 10^{8}\,\mathrm{m\,s^{-1}}) \times (6.022 \times 10^{23}\,\mathrm{mol^{-1}})}$$

$$= 1.11\,\text{mol} = 1.11\,\text{einstein}.$$

If 40% of the incident photon flux is transmitted, then 60% is absorbed. Therefore,

$$\phi = \frac{\text{molecules decomposed}}{\text{photons absorbed}} = \frac{0.344\,\text{mol}}{0.60 \times 1.11\,\text{einstein}} = \boxed{0.52}$$

E21.23(a) The Stern–Volmer equation (eqn. 21.83) relates the ratio of fluorescence quantum yields in the absence and presence of quenching

$$\frac{\phi_{\mathrm{f},0}}{\phi_\mathrm{f}} = 1 + \tau_0 k_\mathrm{Q}[\mathrm{Q}] = \frac{I_{\mathrm{f},0}}{I_\mathrm{f}}$$

The last equality reflects the fact that fluorescence intensities are proportional to quantum yields. Solve this equation for [Q]:

$$[\mathrm{Q}] = \frac{(I_{\mathrm{f},0}/I_\mathrm{f}) - 1}{\tau_0 k_\mathrm{Q}} = \frac{2-1}{(6.0 \times 10^{-9}\,\mathrm{s}) \times (3.0 \times 10^{8}\,\mathrm{dm^3\,mol^{-1}\,s^{-1}})} = \boxed{0.56\,\mathrm{mol\,dm^{-3}}}$$

E21.24(a) The efficiency of resonance energy transfer is given by [21.85]

$$\eta_\mathrm{T} = 1 - \frac{\phi_\mathrm{f}}{\phi_{\mathrm{f},0}} = 0.10$$

Förster theory relates this quantity to the distance R between donor–acceptor pairs by

$$\eta_\mathrm{T} = \frac{R_0^6}{R_0^6 + R^6} \; [21.86],$$

where R_0 is an empirical parameter listed in Table 21.7. Solving for the distance yields

$$R = R_0\left(\frac{1}{\eta_\mathrm{T}} - 1\right)^{1/6} = (4.9\,\mathrm{nm}) \times \left(\frac{1}{0.10} - 1\right)^{1/6} = \boxed{7.1\,\mathrm{nm}}$$

Solutions to problems

Solutions to numerical problems

P21.1 A simple but practical approach is to make an initial guess at the order by observing whether the half-life of the reaction appears to depend on concentration. If it does not, the reaction is first-order; if it does, refer to Table 21.3 for an expression for the half-life of a reaction of the type A → P for orders other than 1:

$$t_{1/2} = \frac{2^{n-1}-1}{(n-1)k_r[\mathrm{A}]_0^{n-1}} \propto [\mathrm{A}]_0^{1-n}$$

Examination of the data shows that the first half-life is roughly 45 min and the second is about double the first. (Compare the 0–50 min data to the 50–150 min data.) That is, the half-life starting from *half* of the initial concentration is about *twice* the initial half-life, suggesting that the half-life is inversely proportional to initial concentration:

$$t_{1/2} \propto [\mathrm{A}]_0^{-1} = [\mathrm{A}]_0^{1-n} \quad \text{with } n = 2$$

Confirm this suggestion by plotting 1/[A] against time. A second-order reaction will obey

$$\frac{1}{[\mathrm{A}]} = k_r t + \frac{1}{[\mathrm{A}]_0} \quad [21.15b]$$

We draw up the following table (A = NH_4CNO)

t/min	0	20.0	50.0	65.0	150
m(urea)/g	0	7.0	12.1	13.8	17.7
m(A)/g	22.9	15.9	10.8	9.1	5.2
[A]/(mol dm^{-3})	0.381	0.265	0.180	0.152	0.0866
[A]$^{-1}$/(dm^3 mol^{-1})	2.62	3.78	5.56	6.60	11.5

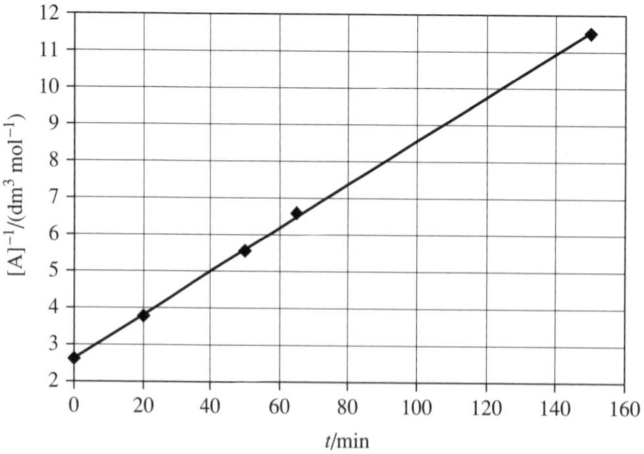

Figure 21.2

The data are plotted in Figure 21.2 and fit closely to a straight line. Hence, the reaction is indeed second order. The rate constant is the slope: $k_r = 0.059\overline{4}\ dm^3\ mol^{-1}\ min^{-1}$. To find [A] at 300 min, use eqn. 21.15(c):

$$[A] = \frac{[A]_0}{1 + k_r t[A]_0} = \frac{0.381\ mol\,dm^{-3}}{1 + (0.059\overline{4}) \times (300) \times (0.381)} = 0.048\overline{9}\ mol\,dm^{-3}$$

The mass of NH_4CNO left after 300 min is

$$m = (0.048\overline{9}\ mol\ dm^{-3}) \times (1.00\ dm^3) \times (60.06\ g\ mol^{-1}) = \boxed{2.94\ g}$$

P21.3 Use the procedure adopted in the solutions to Problems 21.1 and 21.2: is the half-life (or any other similarly defined 'fractional life') constant or does it vary over the course of the reaction? The data are not quite so clear-cut. The half-life appears to be approximately constant at about 10 min: in the interval 0–10 the initial concentration drops by just over half, while in the interval 2–12 the concentration drops by slightly less than half. Another measure would compare the fractional consumption in two equal time intervals. The fractional consumption in the 0–2 interval is about 1/6, while that in the 10–12 interval is about 1/10, suggesting that the fractional consumption is *not* constant over the time the reaction was monitored. We draw up the following table (A = nitrile) in order to examine both first-order and second-order plots (Figure 21.3). The former is a plot of $\ln\left(\frac{[A]}{[A]_0}\right)$ against time (eqn 21.12b); the latter is a plot of 1/[A] against time. Note that the data point given for $t = \infty$ is a typographical error. There is no data available for $t = \infty$, nor is it necessary.

$t/10^3$ s	0	2.00	4.00	6.00	8.00	10.00	12.00
$[A]/(mol\ dm^{-3})$	1.50	1.26	1.07	0.92	0.81	0.72	0.65
$\frac{[A]}{[A]_0}$	1.00	0.840	0.713	0.613	0.540	0.480	0.433
$\ln\left(\frac{[A]}{[A]_0}\right)$	0	−0.174	−0.338	−0.489	−0.616	−0.734	−0.836
$1/[A]$	0.667	0.794	0.935	1.09	1.23	1.39	1.54

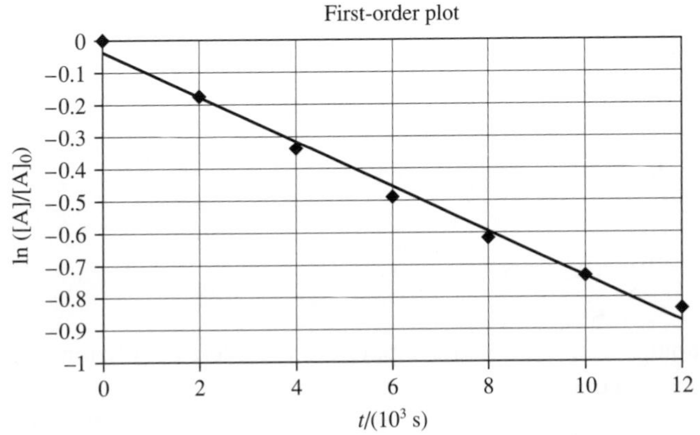

Figure 21.3(a)

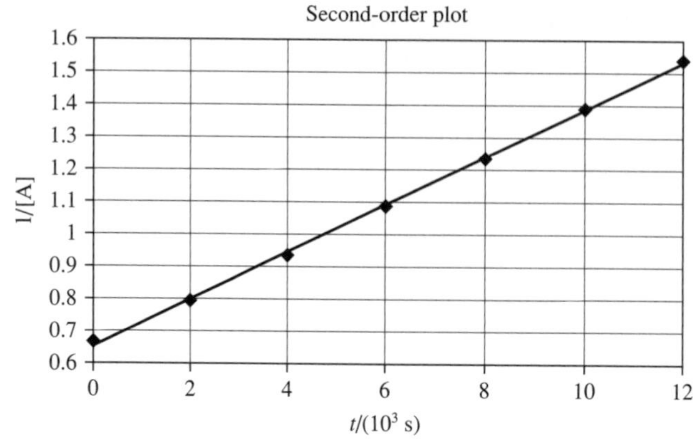

Figure 21.3(b)

The first-order plot is not bad: the correlation coefficient is 0.991. The corresponding first-order rate constant is $k_r = -\text{slope} = \boxed{7.0 \times 10^{-5}\,\text{s}^{-1}}$

The second-order plot looks even better: the correlation coefficient is 0.999. The corresponding second-order rate constant is $k_r = \text{slope} = \boxed{7.3 \times 10^{-5}\,\text{dm}^3\,\text{mol}^{-1}\,\text{s}^{-1}}$.

COMMENT. Based on the given data, the reaction appears to be closer to second order than to first order. (The reaction order need not be an integer.) This conclusion is a tenuous one, however, based on the assumption that the data contain little experimental error. The best course of action for an investigator seeking to establish the reaction order would be further experimentation, following the reaction over a wider range of concentrations.

P21.5 Since both reactions are first order,

$$-\frac{d[A]}{dt} = k_1[A] + k_2[A] = (k_1 + k_2)[A]$$

so $[A] = [A]_0 e^{-(k_1+k_2)t}$ [21.12b with $k_r = k_1 + k_2$]

We are interested in the yield of ketene, CH_2CO; call it K:

$$\frac{d[K]}{dt} = k_2[A] = k_2[A]_0 e^{-(k_1+k_2)t}$$

Integrating yields

$$\int_0^{[K]} d[K] = k_2[A]_0 \int_0^t e^{-(k_1+k_2)t}\,dt$$

$$[K] = \frac{k_2[A]_0}{k_1 + k_2}(1 - e^{-(k_1+k_2)t}) = \frac{k_2}{k_1 + k_2}([A]_0 - [A])$$

The percentage yield is the amount of K produced compared to complete conversion; since the stoichiometry of reaction (2) is one to one, we can write:

$$\%\ \text{yield} = \frac{[K]}{[A]_0} \times 100\% = \frac{k_2}{k_1 + k_2}(1 - e^{-(k_1+k_2)t}) \times 100\%$$

which has its maximum value when the reaction reaches completion

$$\text{max \% yield} = \frac{k_2}{k_1 + k_2} \times 100\% = \frac{4.65\ \text{s}^{-1}}{(3.74 + 4.65)\ \text{s}^{-1}} \times 100\% = \boxed{55.4\%}$$

COMMENT. If we are interested in yield of the desired product (ketene) compared to the products of side reactions (products of reaction 1), it makes sense to define the conversion ratio, the ratio of desired product formed to starting material *reacted*, namely

$$\frac{[\text{K}]}{[\text{A}]_0 - [\text{A}]}$$

which works out in this case to be independent of time

$$\frac{[\text{K}]}{[\text{A}]_0 - [\text{A}]} = \frac{k_2}{k_1 + k_2}$$

If a substance reacts by parallel processes of the same order, then the ratio of the amounts of products will be constant and independent of the extent of the reaction, no matter what the order.

Question. Can you demonstrate the truth of the statement made in the above comment?

P21.7 The stoichiometry of the reaction relates product and reaction concentrations as follows:

$$[\text{A}] = [\text{A}]_0 - 2[\text{B}]$$

When the reaction goes to completion, $[\text{B}] = [\text{A}]_0/2$; hence $[\text{A}]_0 = 0.624\ \text{mol dm}^{-3}$. We can therefore tabulate [A], and examine its half-life. We see that the half-life of A from its initial concentration is approximately 20 min, and that its half-life from the concentration at 20 min is also 20 min. This indicates a first-order reaction. We confirm this conclusion by plotting the data accordingly (in Figure 21.4), using

$$\ln\frac{[\text{A}]_0}{[\text{A}]} = k_\text{A}t \ [21.12\text{b}]$$

which follows from

$$\frac{\text{d}[\text{A}]}{\text{d}t} = -k_\text{A}[\text{A}]$$

t/min	0	10	20	30	40	∞
$[\text{B}]/(\text{mol dm}^{-3})$	0	0.089	0.153	0.200	0.230	0.312
$[\text{A}]/(\text{mol dm}^{-3})$	0.624	0.446	0.318	0.224	0.164	0
$\ln\frac{[\text{A}]}{[\text{A}]_0}$	0	−0.34	−0.67	−1.02	−1.34	

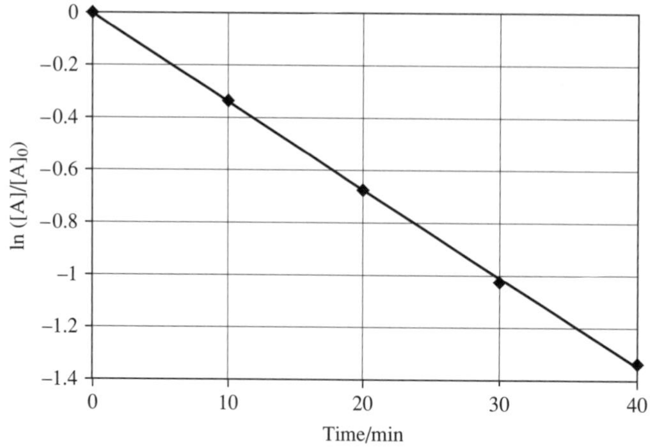

Figure 21.4

The points lie on a straight line, which confirms first-order kinetics. Since the slope of the line is $-3.4\times10^{-2}\,\text{min}^{-1}$, we conclude that $k_A = 3.4\times10^{-2}\,\text{min}^{-1}$. To express the rate law in the form $\upsilon = k_r[A]$ we note that

$$\upsilon = -\frac{1}{2}\frac{d[A]}{dt} = -(\tfrac{1}{2})\times(-k_A[A]) = \tfrac{1}{2}k_A[A]$$

and hence $k_r = \tfrac{1}{2}k_A = \boxed{1.7\times10^{-2}\,\text{min}^{-1}}$

P21.9 If the reaction is first order the concentrations obey

$$\ln\left(\frac{[A]}{[A]_0}\right) = -k_r t \;[21.12b]$$

and, since pressures and concentrations of gases are proportional, the pressures should obey

$$\ln\frac{p_0}{p} = k_r t$$

and $\frac{1}{t}\ln\frac{p_0}{p}$ should be a constant. We test this by drawing up the following table:

p_0/Torr	200	200	400	400	600	600
t/s	100	200	100	200	100	200
p/Torr	186	173	373	347	559	520
$10^4\left(\frac{1}{t/\text{s}}\right)\ln\frac{p_0}{p}$	7.3	7.3	7.0	7.1	7.1	7.2

The values in the last row of the table are virtually constant, and so (in the pressure range spanned by the data) the reaction has first-order kinetics with $k_r = \boxed{7.2\times10^{-4}\,\text{s}^{-1}}$.

P21.11 $2\,\text{HCl} \rightleftharpoons (\text{HCl})_2 \quad K_1 \quad [(\text{HCl})_2] = K_1[\text{HCl}]^2$

$\text{HCl} + \text{CH}_3\text{CH{=}CH}_2 \rightleftharpoons \text{complex} \quad K_2 \quad [\text{complex}] = K_2[\text{HCl}][\text{CH}_3\text{CH{=}CH}_2]$

$(\text{HCl})_2 + \text{complex} \rightarrow \text{CH}_3\text{CHClCH}_3 + 2\,\text{HCl} \quad k_\text{r}$

$$v = \frac{\text{d}[\text{CH}_3\text{CHClCH}_3]}{\text{d}t} = k_\text{r}[(\text{HCl})_2][\text{complex}]$$

Both $(\text{HCl})_2$ and the complex are intermediates, so substitute for them using equilibrium expressions:

$$v = k_\text{r}[(\text{HCl})_2][\text{complex}] = k_\text{r}(K_1[\text{HCl}]^2)(K_2[\text{HCl}][\text{CH}_3\text{CH{=}CH}_2])$$
$$= \boxed{k_\text{r}K_1K_2[\text{HCl}]^3[\text{CH}_3\text{CH{=}CH}_2]},$$

which is third order in HCl and first order in propene. One approach to experimental verification is to look for evidence of proposed intermediates, using infrared spectroscopy to search for $(\text{HCl})_2$, for example.

P21.13 Subtract eqn. 21.29 at T_1 from the same equation at T_2:

$$\ln k_2 - \ln k_1 = \frac{E_\text{a}}{RT_1} - \frac{E_\text{a}}{RT_2}, \quad \text{so} \quad E_\text{a} = R\ln\left(\frac{k_1}{k_2}\right)\left(\frac{1}{T_2} - \frac{1}{T_1}\right)^{-1}$$

T_1/K	300.3	300.3	341.2
T_2/K	341.2	392.2	392.2
$10^{-7}k_1/(\text{dm}^3\,\text{mol}^{-1}\,\text{s}^{-1})$	1.44	1.44	3.03
$10^{-7}k_2/(\text{dm}^3\,\text{mol}^{-1}\,\text{s}^{-1})$	3.03	6.9	6.9
$E_\text{a}/(\text{kJ mol}^{-1})$	15.5	16.7	18.0

The mean is $\boxed{16.7\ \text{kJ mol}^{-1}}$. Compute A from each rate constant, using the mean E_a and

$$A = k_\text{r}\text{e}^{E_\text{a}/RT}$$

T/K	300.3	341.2	392.2
$10^{-7}k_\text{r}/(\text{dm}^3\,\text{mol}^{-1}\,\text{s}^{-1})$	1.44	3.03	6.9
E_a/RT	6.69	5.89	5.12
$10^{-10}A/(\text{dm}^3\,\text{mol}^{-1}\,\text{s}^{-1})$	1.16	1.10	1.16

The mean is $\boxed{1.14 \times 10^{10}\ \text{dm}^3\,\text{mol}^{-1}\,\text{s}^{-1}}$.

P21.15 Using spreadsheet software to evaluate eqn. 21.41, one can draw up a plot like that in Figure 21.5. The curves in this plot represent the concentration of the intermediate [I] as a function of time. They are labelled with the ratio k_a/k_b, where $k_\text{b} = 1\ \text{s}^{-1}$ for all curves and k_a varies. The thickest curve, labelled 10, corresponds to $k_\text{a} = 10\ \text{s}^{-1}$, as specified in part (a) of the problem. As the ratio k_a/k_b gets smaller (or, as the problem puts it, the ratio k_b/k_a gets larger), the concentration profile for I becomes lower, broader, and flatter; that is, [I] becomes more nearly constant over a longer period of time. This is the nature of the $\boxed{\text{steady-state approximation}}$, which becomes more and more valid as consumption of the intermediate becomes fast compared with its formation.

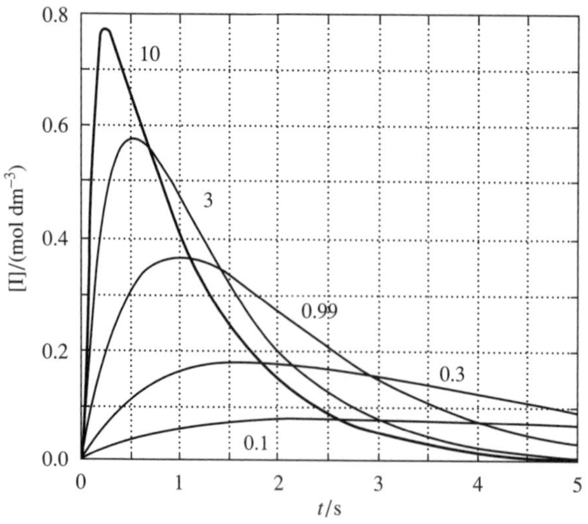

Figure 21.5

P21.17 $\frac{1}{k_r} = \frac{k_a'}{k_a k_b} + \frac{1}{k_a p}$ [analogous to 21.62]

We expect a straight line when $\frac{1}{k_r}$ is plotted against $\frac{1}{p}$. We draw up the following table:

p/Torr	84.1	11.0	2.89	0.569	0.120	0.067
$1/(p/\text{Torr})$	0.012	0.091	0.346	1.76	8.33	14.9
$10^4/(k_r/\text{s}^{-1})$	2.98	2.23	1.54	0.857	0.392	0.303
$10^{-4}/(k_r/\text{s}^{-1})$	0.336	0.448	0.629	1.17	2.55	3.30

These points are plotted in Figure 21.6. There are marked deviations at low pressures, indicating that the Lindemann theory is deficient in that region.

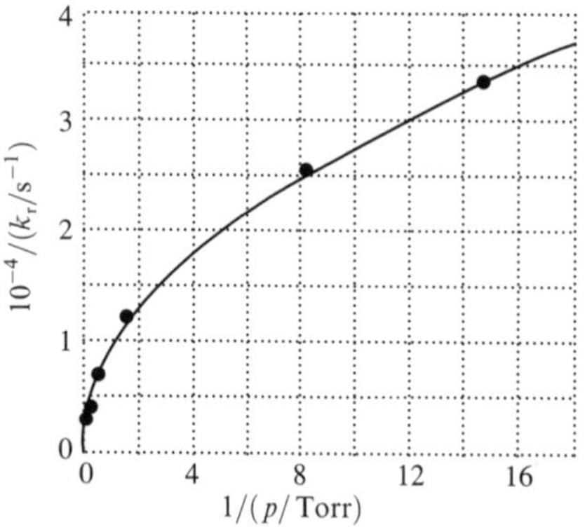

Figure 21.6

P21.19 $M + h\nu_i \rightarrow M^*$, I_{abs} [M = benzophenone]
$M^* + Q \rightarrow M + Q$, k_Q
$M^* \rightarrow M + h\nu_f$, k_f

$$\frac{d[M^*]}{dt} = I_{abs} - k_f[M^*] - k_Q[Q][M^*] \approx 0 \text{ [steady state]}$$

and hence $[M^*] = \dfrac{I_{abs}}{k_f + k_Q[Q]}$

Then, $I_f = k_f[M^*] = \dfrac{k_f I_{abs}}{k_f + k_Q[Q]}$

and so $\boxed{\dfrac{1}{I_f} = \dfrac{1}{I_{abs}} + \dfrac{k_Q[Q]}{k_f I_{abs}}}$

If the exciting light is extinguished, [M*], and hence I_f, decays as $e^{-k_f t}$ in the absence of a quencher. Therefore, we can measure $k_Q/k_f I_{abs}$ from the slope of $1/I_f$ plotted against [Q], and then use k_f to determine k_Q.

We draw up the following table:

10^3[Q]/M	1	5	10
$1/I_f$	2.4	4.0	6.3

The points are plotted in Figure 21.7.

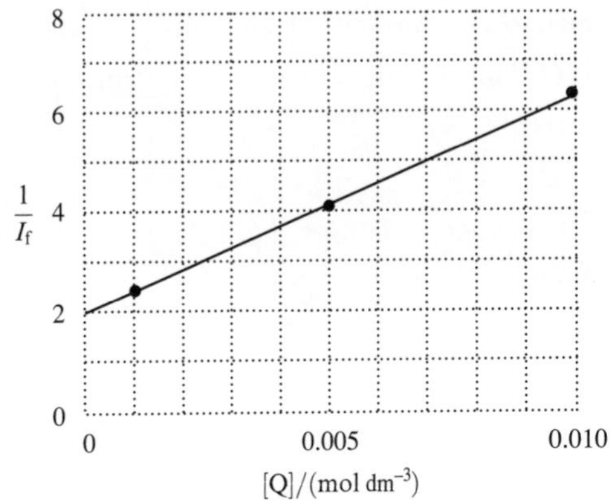

Figure 21.7

The intercept lies at 2.0, and so $I_{abs} = \dfrac{1}{2.0} = 0.50$. The slope is 430, and so

$$\frac{k_Q}{k_f I_{abs}} = 430 \text{ dm}^3 \text{ mol}^{-1}$$

Then, since $I_{abs} = 0.50$ and $k_f = \dfrac{\ln 2}{t_{1/2}}$

$$k_Q = (0.50) \times (430\ \text{dm}^3\ \text{mol}^{-1}) \times \left(\frac{\ln 2}{29 \times 10^{-6}\ \text{s}}\right) = \boxed{5.1 \times 10^6\ \text{dm}^3\ \text{mol}^{-1}\ \text{s}^{-1}}$$

P21.21 $$\eta_T = \frac{R_0^6}{R_0^6 + R^6} \quad \text{or} \quad \frac{1}{\eta_T} = 1 + \left(\frac{R}{R_0}\right)^6 \quad [21.86]$$

A plot of η_T^{-1} vs. R^6 ought to be linear with a slope of $(R_0/\text{nm})^{-6}$ (Figure 21.8).

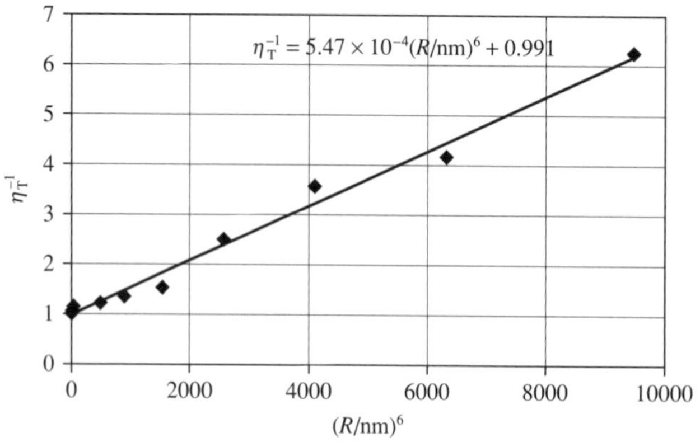

Figure 21.8

The plot appears to be linear with an intercept equal to 1, so we conclude that eqn 21.86 adequately describes the data. The slope of the best-fit line is $(R_0/\text{nm})^{-6}$, so $R_0 = (5.47 \times 10^{-4})^{-1/6}\ \text{nm} = \boxed{3.5\ \text{nm}}$.

Solutions to theoretical problems

P21.23 $\text{A} \rightleftharpoons \text{B}$

$$\frac{d[A]}{dt} = -k_r[A] + k_r'[B] \quad \text{and} \quad \frac{d[B]}{dt} = -k_r'[B] + k_r[A]$$

At all times, $[A] + [B] = [A]_0 + [B]_0$

Therefore, $[B] = [A]_0 + [B]_0 - [A]$

$$\frac{d[A]}{dt} = -k_r[A] + k_r'\{[A]_0 + [B]_0 - [A]\} = -(k_r + k_r')[A] + k_r'([A]_0 + [B]_0)$$

To solve, one must integrate

$$\int \frac{d[A]}{(k_r + k_r')[A] - k_r'([A]_0 + [B]_0)} = -\int dt$$

The solution is $[A] = \boxed{\dfrac{k_r'([A]_0 + [B]_0) + (k_r[A]_0 - k_r'[B]_0)e^{-(k_r+k_r')t}}{k_r + k_r'}}$

The final composition is found by setting $t = \infty$.

$$[\mathrm{A}]_\infty = \boxed{\left(\frac{k_\mathrm{r}'}{k_\mathrm{r}+k_\mathrm{r}'}\right) \times ([\mathrm{A}]_0 + [\mathrm{B}]_0)}$$

and $$[\mathrm{B}]_\infty = [\mathrm{A}]_0 + [\mathrm{B}]_0 - [\mathrm{A}]_\infty = \boxed{\left(\frac{k_\mathrm{r}}{k_\mathrm{r}+k_\mathrm{r}'}\right) \times ([\mathrm{A}]_0 + [\mathrm{B}]_0)}$$

Note that $\boxed{\dfrac{[\mathrm{B}]_\infty}{[\mathrm{A}]_\infty} = \dfrac{k_\mathrm{r}}{k_\mathrm{r}'}}$

P21.25 $$\frac{\mathrm{d}[\mathrm{A}]}{\mathrm{d}t} = -2k_\mathrm{r}[\mathrm{A}]^2[\mathrm{B}], \quad 2\mathrm{A} + \mathrm{B} \rightarrow \mathrm{P}$$

(a) Let x represent [P] at time t, A_0 represent $[\mathrm{A}]_0$, and B_0 represent $[\mathrm{B}]_0$. Then

$$[\mathrm{A}] = A_0 - 2x \quad \text{and} \quad [\mathrm{B}] = B_0 - x = \frac{A_0}{2} - x$$

Therefore, $$\frac{\mathrm{d}[\mathrm{A}]}{\mathrm{d}t} = -2\frac{\mathrm{d}x}{\mathrm{d}t} = -2k_\mathrm{r}(A_0 - 2x)^2 \times (B_0 - x)$$

$$\frac{\mathrm{d}x}{\mathrm{d}t} = k_\mathrm{r}(A_0 - 2x)^2 \times \left(\frac{1}{2}A_0 - x\right) = \frac{1}{2}k_\mathrm{r}(A_0 - 2x)^3$$

$$\frac{1}{2}k_\mathrm{r}t = \int_0^x \frac{\mathrm{d}x}{(A_0 - 2x)^3} = \frac{1}{4} \times \left[\left(\frac{1}{A_0 - 2x}\right)^2 - \left(\frac{1}{A_0}\right)^2\right]$$

Therefore, $$\boxed{k_\mathrm{r}t = \frac{2x(A_0 - x)}{A_0^2(A_0 - 2x)^2}}$$

(b) Now $B_0 = A_0$, so

$$\frac{\mathrm{d}x}{\mathrm{d}t} = k_\mathrm{r}(A_0 - 2x)^2 \times (B_0 - x) = k_\mathrm{r}(A_0 - 2x)^2 \times (A_0 - x)$$

$$k_\mathrm{r}t = \int_0^x \frac{\mathrm{d}x}{(A_0 - 2x)^2 \times (A_0 - x)}$$

We proceed by the method of partial fractions (which is employed in the general case too), and look for the values of α, β, and γ such that

$$\frac{1}{(A_0 - 2x)^2 \times (A_0 - x)} = \frac{\alpha}{(A_0 - 2x)^2} + \frac{\beta}{A_0 - 2x} + \frac{\gamma}{A_0 - x}$$

This requires that

$$\alpha(A_0 - x) + \beta(A_0 - 2x) \times (A_0 - x) + \gamma(A_0 - 2x)^2 = 1$$

Expand and gather terms by powers of x:

$$(A_0\alpha + A_0^2\beta + A_0^2\gamma) - (\alpha + 3\beta A_0 + 4\gamma A_0)x + (2\beta + 4\gamma)x^2 = 1$$

This must be true for all x; therefore

$$A_0\alpha + A_0^2\beta + A_0^2\gamma = 1$$

$$\alpha + 3A_0\beta + 3A_0\gamma = 0$$

and $\quad 2\beta + 4\gamma = 0$

Solving this system yields $\alpha = \dfrac{2}{A_0}$, $\beta = \dfrac{-2}{A_0^2}$, and $\gamma = \dfrac{1}{A_0^2}$

Therefore,

$$k_r t = \int_0^x \left(\frac{(2/A_0)}{(A_0 - 2x)^2} - \frac{(2/A_0^2)}{A_0 - 2x} + \frac{(1/A_0^2)}{A_0 - x} \right) dx$$

$$= \left(\frac{(1/A_0)}{A_0 - 2x} + \frac{1}{A_0^2}\ln(A_0 - 2x) - \frac{1}{A_0^2}\ln(A_0 - x) \right)\Bigg|_0^x$$

$$= \boxed{\left(\frac{2x}{A_0^2(A_0 - 2x)} \right) + \left(\frac{1}{A_0^2} \right) \ln\left(\frac{A_0 - 2x}{A_0 - x} \right)}$$

P21.27 The rate equations are

$$\frac{d[A]}{dt} = -k_a[A] + k_a'[B]$$

$$\frac{d[B]}{dt} = k_a[A] - k_a'[B] - k_b[B] + k_b'[C]$$

$$\frac{d[C]}{dt} = k_b[B] - k_b'[C]$$

These equations are a set of coupled differential equations. Although it is not immediately apparent, they have a closed-form general solution; however, we are looking for the particular circumstances under which the mechanism reduces to the second form given. Since the reaction involves an intermediate, let us explore the result of applying the steady-state approximation to it. Then,

$$\frac{d[B]}{dt} = k_a[A] - k_a'[B] - k_b[B] + k_b'[C] \approx 0$$

and $\quad [B] \approx \dfrac{k_a[A] + k_b'[C]}{k_a' + k_b}$

Therefore, $\quad \dfrac{d[A]}{dt} = -\dfrac{k_a k_b}{k_a' + k_b}[A] + \dfrac{k_a' k_b'}{k_a' + k_b}[C]$

This rate expression may be compared to that given in the text [Section 21.4] for a first-order reaction approaching equilibrium

$$A \underset{k_r'}{\overset{k_r}{\rightleftharpoons}} C$$

Here, $k_r = \dfrac{k_a k_b}{k_a' + k_b}$ and $k_r' = \dfrac{k_a' k_b'}{k_a' + k_b}$

The solutions are $[A] = \left(\dfrac{k_r' + k_r e^{-(k_r'+k_r)t}}{k_r' + k_r}\right) \times [A]_0$ [21.23]

and $[C] = [A]_0 - [A]$

Thus, the conditions under which the first mechanism given reduces to the second are the conditions under which the steady-state approximation holds, namely, when B can be treated as a steady-state intermediate.

P21.29 Let the forward rates be written as

$$r_1 = k_1[A], \quad r_2 = k_2[B], \quad r_3 = k_3[C]$$

and the reverse rates as

$$r_1' = k_1'[B], \quad r_2' = k_2'[C], \quad r_3' = k_3'[D]$$

The net rates are then

$$R_1 = k_1[A] - k_1'[B], \quad R_2 = k_2[B] - k_2'[C], \quad R_3 = k_3[C] - k_3'[D]$$

But $[A] = [A]_0$ and $[D] = 0$, so that the steady-state equations for the net rates of the individual steps are

$$k_1[A]_0 - k_1'[B] = k_2[B] - k_2'[C] = k_3[C]$$

From the second of these equations we find

$$[C] = \frac{k_2[B]}{k_2' + k_3}$$

After inserting this expression for [C] into the first of the steady-state equations we obtain

$$[B] = \frac{k_1[A]_0 + k_2'[C]}{k_1' + k_2} = \frac{k_1[A]_0 + k_2'\left(\dfrac{k_2[B]}{k_2' + k_3}\right)}{k_1' + k_2}$$

which yields, on isolating [B],

$$[B] = [A]_0 \times \frac{k_1}{k_1' + k_2 - \left(\dfrac{k_2 k_2'}{k_2' + k_3}\right)}$$

Thus, at the steady state

$$R_1 = R_2 = R_3 = [\mathrm{A}]_0 k_1 \times \left(1 - \frac{k_1}{k_1' + k_2 - \left(\dfrac{k_2 k_2'}{k_2' + k_3}\right)}\right) = \boxed{\frac{k_1 k_2 k_3 [\mathrm{A}]_0}{k_1' k_2' + k_1' k_3 + k_2 k_3}}$$

COMMENT. At steady state, not only are the net rates of reactions 1, 2, and 3 steady but so are the concentrations [B] and [C]. That is,

$$\frac{\mathrm{d}[\mathrm{B}]}{\mathrm{d}t} = k_1[\mathrm{A}]_0 - (k_1' + k_2)[\mathrm{B}] + k_2'[\mathrm{C}] \approx 0$$

and $$\frac{\mathrm{d}[\mathrm{C}]}{\mathrm{d}t} = k_2[\mathrm{B}] - (k_2' + k_3)[\mathrm{C}] \approx 0$$

In fact, another approach to solving the problem is to solve *these* equations for [B] and [C].

P21.31 The number-average molar mass of the polymer is the average chain length times the molar mass of the monomer

$$\langle M \rangle = \langle N \rangle M_1 = \frac{M_1}{1-p} \; [21.68(\mathrm{a})]$$

The probability P_N that a polymer consists of N monomers is equal to the probability that it has $N-1$ reacted end groups and one unreacted end group. The former probability is p^{N-1}; the latter $1-p$. Therefore, the total probability of finding an N-mer is

$$P_N = p^{N-1}(1-p)$$

We need this probability to get at $\langle M^2 \rangle$, again using number averaging:

$$\langle M^2 \rangle = M_1^2 \langle N^2 \rangle = M_1^2 \sum_N N^2 P_N = M_1^2 (1-p) \sum_N N^2 p^{N-1}$$

$$= M_1^2(1-p)\frac{\mathrm{d}}{\mathrm{d}p} p \frac{\mathrm{d}}{\mathrm{d}p} \sum_N p^N = M_1^2(1-p)\frac{\mathrm{d}}{\mathrm{d}p} p \frac{\mathrm{d}}{\mathrm{d}p}(1-p)^{-1} = \frac{M_1^2(1+p)}{(1-p)^2}$$

Thus, $$\langle M^2 \rangle_N - \langle M \rangle_N^2 = M_1^2 \left(\frac{1+p}{(1-p)^2} - \frac{1}{(1-p)^2}\right) = \frac{pM_1^2}{(1-p)^2}$$

and $$\boxed{(\langle M^2 \rangle_N - \langle M \rangle_N^2)^{1/2} = \frac{p^{1/2} M_1}{1-p}}$$

The time dependence is obtained from

$$p = \frac{k_\mathrm{r} t [\mathrm{A}]_0}{1 + k_\mathrm{r} t [\mathrm{A}]_0} \; [21.67]$$

and $$\langle N \rangle = \frac{1}{1-p} = 1 + k_\mathrm{r} t [\mathrm{A}]_0 \quad [21.68\mathrm{a} \; \& \; 21.68\mathrm{b}]$$

Hence, $$\frac{p^{1/2}}{1-p} = p^{1/2}(1 + k_\mathrm{r} t[\mathrm{A}]_0) = \{k_\mathrm{r} t [\mathrm{A}]_0 (1 + k_\mathrm{r} t [\mathrm{A}]_0)\}^{1/2}$$

and $$(\langle M^2 \rangle_N - \langle M \rangle_N^2)^{1/2} = \boxed{M_1 \{kt[\mathrm{A}]_0(1 + kt[\mathrm{A}]_0)\}^{1/2}}$$

P21.33 In termination by disproportionation, the chain carriers do not combine. The average number of monomers in a polymer molecule equals the average number in a chain carrier when it terminates, namely, the kinetic chain length, ν.

$$\langle N \rangle = \nu = \boxed{k_r[\cdot\mathrm{M}][\mathrm{I}]^{-1/2}}$$

COMMENT. Contrast this result to the reasoning before eqn 21.75, in which the average number is the *sum* of the average numbers of a pair of combining radicals.

P21.35 The rates of the individual steps are

$$\mathrm{A} \rightarrow \mathrm{B} \quad \frac{\mathrm{d[B]}}{\mathrm{d}t} = I_a$$

$$\mathrm{B} \rightarrow \mathrm{A} \quad \frac{\mathrm{d[B]}}{\mathrm{d}t} = -k_r[\mathrm{B}]^2$$

In the photostationary state, $I_a - k_r[\mathrm{B}]^2 = 0$. Hence,

$$[\mathrm{B}] = \boxed{\left(\frac{I_a}{k_r}\right)^{1/2}}$$

This concentration can differ significantly from an equilibrium distribution because changing the illumination may change the rate of the forward reaction without affecting the reverse reaction. Contrast this situation to the corresponding equilibrium expression, in which $[\mathrm{B}]_{\mathrm{eq}}$ depends on a ratio of rate constants for the forward and reverse reactions. In the equilibrium case, the rates of forward and reverse reactions cannot be changed *independently*.

Solutions to applications

P21.37 A simple but practical approach is to make an initial guess at the order by observing whether the half-life of the reaction appears to depend on concentration. If it does not, the reaction is first order; if it does, it may be second order. Examination of the data shows that the half-life is roughly 90 min, but it is not exactly constant. (Compare the 60–150 min data to the 150–240 min data; in both intervals the concentration drops by roughly half. Then examine the 30–120 min interval, where the concentration drops by less than half.) If the reaction is first order, it will obey

$$\ln\left(\frac{c}{c_0}\right) = -k_r t \ [21.12\mathrm{b}]$$

If it is second order, it will obey

$$\frac{1}{c} = k_r t + \frac{1}{c_0} \ [21.15\mathrm{b}]$$

See whether a first-order plot of ln c vs. time or a second-order plot of $1/c$ vs. time has a substantially better fit. We draw up the following table:

t/min	30	60	120	150	240	360	480
c/(ng cm^{-3})	699	622	413	292	152	60	24
(ng cm^{-3})/c	0.00143	0.00161	0.00242	0.00342	0.00658	0.0167	0.0412
ln{c/(ng cm^{-3})}	6.550	6.433	6.023	5.677	5.024	4.094	3.178

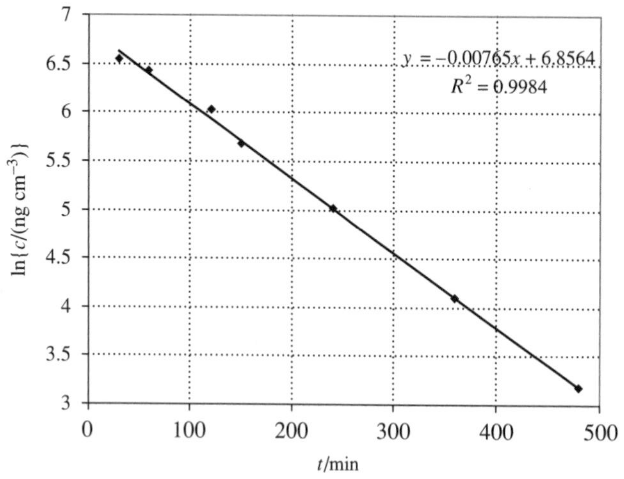

Figure 21.9(a)

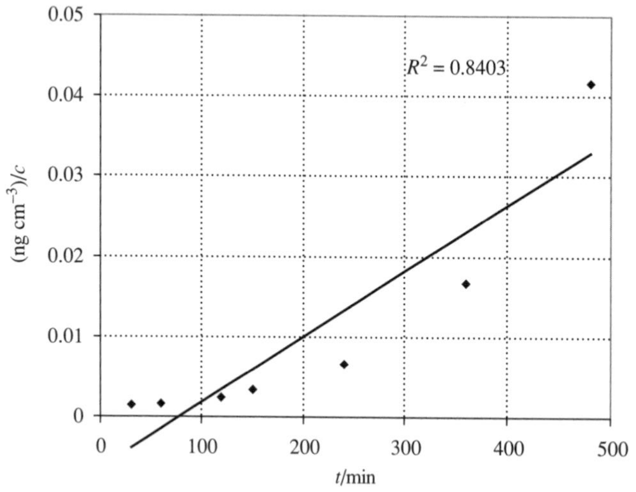

Figure 21.9(b)

The data are plotted in Figures 21.9(a) and (b). The first-order plot fits closely to a straight line with just a hint of curvature near the outset. The second-order plot, conversely, is strongly curved throughout. Hence, the reaction is $\boxed{\text{first order}}$. The rate constant is the negative of the slope of the first-order plot: $k_r = \boxed{0.00765\ \text{min}^{-1}} = \boxed{0.459\ \text{h}^{-1}}$. The half-life is (eqn. 21.13):

$$t_{1/2} = \frac{\ln 2}{k_r} = \frac{\ln 2}{0.459\ \text{h}^{-1}} = \boxed{1.51\ \text{h}} = \boxed{91\ \text{min}}$$

COMMENT. As noted in the problem, the drug concentration is a result of absorption and elimination of the drug, two processes with distinct rates. Elimination is characteristically slower, so the later data points reflect elimination only because absorption is effectively complete by then. The earlier data points, by contrast, reflect both absorption and elimination. It is therefore not surprising that the early points do not adhere so closely to the line so well defined by the later data.

P21.39 The initial rate is

$$v_0 = (3.6\times10^{6}\,\mathrm{dm^9\,mol^{-3}\,s^{-1}})\times(5\times10^{-5}\,\mathrm{mol\,dm^{-3}})^2\times(10^{-5.6}\,\mathrm{mol\,dm^{-3}})^2$$
$$= \boxed{6\times10^{-14}\,\mathrm{mol\,dm^{-3}\,s^{-1}}}$$

The half-life for a second-order reaction is

$$t_{1/2} = \frac{1}{k_{\mathrm{eff}}[\mathrm{HSO_3^-}]_0}\ [21.16]$$

where k_{eff} is the rate constant in the expression

$$-\frac{\mathrm{d}[\mathrm{HSO_3^-}]}{\mathrm{d}t} = 2v = k_{\mathrm{eff}}[\mathrm{HSO_3^-}]^2$$

Comparison to the given rate law and rate constant shows

$$k_{\mathrm{eff}} = 2k_{\mathrm{r}}[\mathrm{H^+}]^2 = 2(3.6\times10^{6}\,\mathrm{dm^9\,mol^{-3}\,s^{-1}})\times(10^{-5.6}\,\mathrm{mol\,dm^{-3}})^2$$
$$= 4.5\times10^{-5}\,\mathrm{dm^3\,mol^{-1}\,s^{-1}}$$

and $$t_{1/2} = \frac{1}{(4.5\times10^{-5}\,\mathrm{dm^3\,mol^{-1}\,s^{-1}})\times(5\times10^{-5}\,\mathrm{mol\,dm^{-3}})} = \boxed{4.\bar{4}\times10^{8}\,\mathrm{s} = 14\,\mathrm{yr}}$$

P21.41 The rate of reaction is the rate at which ozone absorbs photons times the quantum yield. The rate at which ozone absorbs photons is the rate at which photons impinge on the ozone times the fraction of photons absorbed. That fraction is $1 - T$, where T is the transmittance. T is related to the absorbance A by

$$A = -\log T = \varepsilon cL \quad \text{so} \quad 1 - T = 1 - 10^{-\varepsilon cL}$$

and $$1 - T = 1 - 10^{-\{(260\,\mathrm{dm^3\,mol^{-1}\,cm^{-1}})\times(8\times10^{-9}\,\mathrm{mol\,dm^{-3}})\times(10^{5}\,\mathrm{cm})\}} = 0.38$$

If we let F stand for the flux of photons (the rate at which photons impinge on our sample of ozone), then the rate of reaction is

$$v = \phi(1-T)F = (0.94)\times(0.38)\times\frac{(1\times10^{14}\,\mathrm{cm^{-2}\,s^{-1}})\times(1000\,\mathrm{cm^3\,dm^{-3}})}{(6.022\times10^{23}\,\mathrm{mol^{-1}})\times(10^{5}\,\mathrm{cm})}$$
$$= \boxed{5.9\times10^{-13}\,\mathrm{mol\,dm^{-3}\,s^{-1}}}$$

22 Reaction dynamics

Answers to discussion questions

D22.1 The harpoon mechanism accounts for the large steric factor of reactions of the kind $K + Br_2 \rightarrow KBr + Br$ in beams. It is supposed that an electron hops across from K to Br_2 when they are within a certain distance, and then the two resulting ions are drawn together by their mutual Coulombic attraction.

D22.3 The Eyring equation results from the activated complex theory, which is an attempt to account for the rate constants of bimolecular reactions of the form $A + B \rightleftharpoons C^{\ddagger} \rightarrow P$ in terms of the formation of an activated complex. In the formulation of the theory, it is assumed that the activated complex and the reactants are in equilibrium, and the concentration of activated complex is calculated in terms of an equilibrium constant, which in turn is calculated from the partition functions of the reactants and a postulated form of the activated complex. It is further supposed that one normal mode of the activated complex, the one corresponding to displacement along the reaction coordinate, has a very low force constant. Displacement along this mode leads to products, provided that the complex enters a certain configuration of its atoms, known as the transition state. The derivation of the equilibrium constant from the partition functions leads to eqn 22.36 and in turn to eqn 22.37b, the Eyring equation. See Section 22.4 for a more extensive discussion.

The assumption that formation of the transition state always leads to products is highly questionable, so much so that it is not, strictly speaking, part of the theory. (That is, transition-state theory explicitly includes a transmission coefficient to account for the possibility that a transition state may form and the system revert to reactants; however, the transmission coefficient is taken to equal one in the absence of information to the contrary. See eqn 22.32.) Furthermore, the transition state itself is often poorly characterized, therefore its structural parameters and the statistical quantities based on them are highly uncertain.

D22.5 In the kinetic salt effect, the rate of a reaction in solution is changed by modification of the ionic strength of the medium. If the reactant ions have the same charge sign (as in cation/cation or anion/anion reactions), then an increase in ionic strength increases the rate constant. If the reactant ions have opposite signs (as in cation/anion reactions), then an increase in ionic strength decreases the rate constant. In the former case, the effect can be traced to the denser ionic atmosphere that forms round the newly formed and highly charged ion that constitutes the activated complex and the stronger interaction of that ion with the atmosphere. (See the Debye–Huckel theory, *Further Information 5.1.*) In the latter case, the ion corresponding to the activated complex has a lower charge than the reactants and hence it has a more diffuse ionic atmosphere and interacts with it more weakly. In the limit of low ionic strength the rate constant can be expected to follow the relationship

$$\log k_r = \log k_r^{\circ} + 2Az_Az_BI^{1/2} \text{ [22.49]}$$

D22.7 Eqn 22.62 gives an expression that displays many of the influences on the rate constant for electron transfer:

$$k_{et} \propto e^{-\beta r} e^{-\Delta^{\ddagger}G/RT}$$

The rate constant decays exponentially with distance r between donor and acceptor. Exponential decay is consistent with the quantum-mechanical result that the probability of tunnelling (of an electron, in this case) decays exponentially with the thickness of the barrier. The parameter β depends on the medium through which tunnelling must take place, ranging from about 30 nm^{-1} in vacuum to about 9 nm^{-1} for a molecular link. The activation Gibbs energy often has the form

$$\Delta^{\ddagger}G = \frac{(\Delta_r G^{\ominus} + \lambda)^2}{4\lambda} \text{ [22.61]},$$

where λ is the reorganization energy of the redox active species. In systems where the reorganization energy is constant, the dependence of ln k_{et} on $\Delta_r G^{\ominus}$ is an inverted parabola in which the maximum rate occurs when $-\Delta_r G^{\ominus} = \lambda$. See Section 22.9 for a more thorough discussion.

Solutions to exercises

E22.1(a) The collision frequency is

$$z = \sigma \bar{c}_{rel} \mathcal{N} \text{ [20.11a]}$$

$$\text{where } \bar{c}_{rel} = \left(\frac{8kT}{\pi\mu}\right)^{1/2} = 4\left(\frac{kT}{\pi m}\right)^{1/2} \text{ [20.10]},$$

$$\sigma = \pi d^2 \text{ [Section 20.1b]} = 4\pi R^2, \quad \text{and} \quad \mathcal{N} = \frac{p}{kT}$$

$$\text{Therefore, } z = \sigma \bar{c}_{rel} \mathcal{N} = (4\pi R^2) \times \left(\frac{p}{kT}\right) \times 4\left(\frac{kT}{\pi m}\right)^{1/2} = 16pR^2\left(\frac{\pi}{mkT}\right)^{1/2}$$

$$= 16 \times (100 \times 10^3 \text{ Pa}) \times (190 \times 10^{-12} \text{ m})^2$$

$$\times \left(\frac{\pi}{(17.03 \times 1.661 \times 10^{-27} \text{ kg}) \times (1.381 \times 10^{-23} \text{ J K}^{-1}) \times (298 \text{ K})}\right)^{1/2}$$

$$= \boxed{9.49 \times 10^9 \text{ s}^{-1}}$$

The collision density for like molecules is

$$Z = \frac{z\mathcal{N}}{2} \text{ [22.7a]} = \frac{z}{2}\left(\frac{p}{kT}\right) = \frac{9.49 \times 10^9 \text{ s}^{-1}}{2}\left(\frac{100 \times 10^3 \text{ Pa}}{1.381 \times 10^{-23} \text{ J K}^{-1} \times 298 \text{ K}}\right)$$

$$= \boxed{1.15 \times 10^{35} \text{ s}^{-1} \text{ m}^{-3}}$$

For the percentage increase at constant volume, note that $\mathcal{N}$ is constant at constant volume, so the only constant-volume temperature dependence on z (and on Z) is in the speed factor.

$$z \propto T^{1/2} \quad \text{so} \quad \frac{1}{z}\left(\frac{\partial z}{\partial T}\right)_V = \frac{1}{2T} \quad \text{and} \quad \frac{1}{Z}\left(\frac{\partial Z}{\partial T}\right)_V = \frac{1}{2T}$$

Therefore, $$\frac{\delta z}{z} = \frac{\delta Z}{Z} \approx \frac{\delta T}{2T} = \frac{1}{2}\left(\frac{10\text{ K}}{298\text{ K}}\right) = 0.017$$

so both z and Z increase by about $\boxed{1.7\%}$.

E22.2(a) The fraction of collisions having at least E_a along the line of flight may be inferred by dividing out of the collision-theory rate constant (eqn 22.12) those factors that can be identified as belonging to the collision rate: $f = e^{-E_a/RT}$

(a) (i) $$\frac{E_a}{RT} = \frac{10 \times 10^3\text{ J mol}^{-1}}{(8.3145\text{ J K}^{-1}\text{ mol}^{-1}) \times (300\text{ K})} = 4.01 \quad \text{so} \quad f = e^{-4.01} = \boxed{0.018}$$

(ii) $$\frac{E_a}{RT} = \frac{10 \times 10^3\text{ J mol}^{-1}}{(8.3145\text{ J K}^{-1}\text{ mol}^{-1}) \times (1000\text{ K})} = 1.20 \quad \text{so} \quad f = e^{-1.20} = \boxed{0.30}$$

(b) (i) $$\frac{E_a}{RT} = \frac{100 \times 10^3\text{ J mol}^{-1}}{(8.3145\text{ J K}^{-1}\text{ mol}^{-1}) \times (300\text{ K})} = 40.1 \quad \text{so} \quad f = e^{-40.1} = \boxed{3.9 \times 10^{-18}}$$

(ii) $$\frac{E_a}{RT} = \frac{100 \times 10^3\text{ J mol}^{-1}}{(8.3145\text{ J K}^{-1}\text{ mol}^{-1}) \times (1000\text{ K})} = 12.0 \quad \text{so} \quad f = e^{-12.0} = \boxed{6.0 \times 10^{-6}}$$

E22.3(a) A straightforward approach would be to compute $f = e^{-E_a/RT}$ at the new temperature and compare it to that at the old temperature. An approximate approach would be to note that f changes from $f_0 = e^{-E_a/RT}$ to $\exp\left(\dfrac{-E_a}{RT(1+x)}\right)$, where x is the fractional increase in the temperature. If x is small, the exponent changes from $-E_a/RT$ to approximately $-E_a(1-x)/RT$ and f changes from f_0 to

$$f \approx e^{-E_a(1-x)/RT} = e^{-E_a/RT}(e^{-E_a/RT})^{-x} = f_0 f_0^{-x}$$

Thus, the new fraction is the old one times a factor of f_0^{-x}. The increase in f expressed as a percentage is

$$\frac{f - f_0}{f_0} \times 100\% = \frac{f_0 f_0^{-x} - f_0}{f_0} \times 100\% = (f_0^{-x} - 1) \times 100\%$$

(a) (i) $f_0^{-x} = (0.018)^{-10/300} = 1.14$ and the percentage change is $\boxed{14\%}$.

(ii) $f_0^{-x} = (0.30)^{-10/1000} = 1.01$ and the percentage change is $\boxed{1\%}$.

(b) (i) $f_0^{-x} = (3.9 \times 10^{-18})^{-10/300} = 3.8$ and the percentage change is $\boxed{280\%}$ ($\boxed{260\%}$ using the exact approach).

(ii) $f_0^{-x} = (6.0 \times 10^{-6})^{-10/1000} = 1.13$ and the percentage change is $\boxed{13\%}$.

E22.4(a) $k_r = P\sigma\left(\dfrac{8kT}{\pi\mu}\right)^{1/2} N_A e^{-E_a/RT}$ [22.13]

We are not given a steric factor, so assume that $P = 1$.

$$k_r = 0.36 \times (10^{-9}\,\text{m})^2 \times \left(\frac{8 \times (1.381 \times 10^{-23}\,\text{J K}^{-1}) \times (650\,\text{K})}{\pi \times (3.32 \times 10^{-27}\,\text{kg})}\right)^{1/2} \times (6.022 \times 10^{23}\,\text{mol}^{-1})$$

$$\times \exp\left(\frac{-171 \times 10^3\,\text{J}}{(8.3145\,\text{J K}^{-1}\,\text{mol}^{-1}) \times (650\,\text{K})}\right)$$

$$= \boxed{1.03 \times 10^{-5}\,\text{m}^3\,\text{mol}^{-1}\,\text{s}^{-1}}$$

$$= \boxed{1.03 \times 10^{-2}\,\text{dm}^3\,\text{mol}^{-1}\,\text{s}^{-1}}$$

COMMENT. The assumption that $P = 1$ is made because it is convenient in the absence of additional information, not because it is likely to be accurate.

E22.5(a) According to the RRK model,

$$P = \left(1 - \frac{E^*}{E}\right)^{s-1} \quad [22.14a]$$

where E is the available energy, E^* the energy needed to break a bond, and s the number of modes of motion available to accept the energy.

$$P = \left(1 - \frac{350\,\text{kJ mol}^{-1}}{400\,\text{kJ mol}^{-1}}\right)^{5-1} = \boxed{2.4 \times 10^{-4}}$$

E22.6(a) The rate constant for a diffusion-controlled bimolecular reaction is

$k_d = 4\pi R^* D N_A$ [22.18]

where $D = D_A + D_B = 2 \times (5 \times 10^{-9}\,\text{m}^2\,\text{s}^{-1}) = 1.0 \times 10^{-8}\,\text{m}^2\,\text{s}^{-1}$

$k_d = 4\pi \times (0.4 \times 10^{-9}\,\text{m}) \times (1.0 \times 10^{-8}\,\text{m}^2\,\text{s}^{-1}) \times (6.022 \times 10^{23}\,\text{mol}^{-1})$

$k_d = 3 \times 10^7\,\text{m}^3\,\text{mol}^{-1}\,\text{s}^{-1} = \boxed{3 \times 10^{10}\,\text{dm}^3\,\text{mol}^{-1}\,\text{s}^{-1}}$

E22.7(a) The rate constant for a diffusion-controlled bimolecular reaction is

$$k_d = \frac{8RT}{3\eta}\,[22.21] = \frac{8 \times (8.3145\,\text{J K}^{-1}\,\text{mol}^{-1}) \times (298\,\text{K})}{3\eta} = \frac{6.61 \times 10^3\,\text{J mol}^{-1}}{\eta}$$

(a) For water, $\eta = 1.00 \times 10^{-3}\,\text{kg m}^{-1}\,\text{s}^{-1}$

$$k_d = \frac{6.61 \times 10^3\,\text{J mol}^{-1}}{1.00 \times 10^{-3}\,\text{kg m}^{-1}\,\text{s}^{-1}} = 6.61 \times 10^6\,\text{m}^3\,\text{mol}^{-1}\,\text{s}^{-1} = \boxed{6.61 \times 10^9\,\text{dm}^3\,\text{mol}^{-1}\,\text{s}^{-1}}$$

(b) For pentane, $\eta = 2.2 \times 10^{-4}\,\text{kg m}^{-1}\,\text{s}^{-1}$

$$k_d = \frac{6.61 \times 10^3\,\text{J mol}^{-1}}{2.2 \times 10^{-4}\,\text{kg m}^{-1}\,\text{s}^{-1}} = 3.0 \times 10^7\,\text{m}^3\,\text{mol}^{-1}\,\text{s}^{-1} = \boxed{3.0 \times 10^{10}\,\text{dm}^3\,\text{mol}^{-1}\,\text{s}^{-1}}$$

E22.8(a) The rate constant for a diffusion-controlled bimolecular reaction is

$$k_d = \frac{8RT}{3\eta} = \frac{8 \times (8.3145\ \mathrm{J\ K^{-1}\,mol^{-1}}) \times (298\ \mathrm{K})}{3 \times (0.89 \times 10^{-3}\ \mathrm{kg\ m^{-1}\,s^{-1}})}$$

$$= 7.4 \times 10^{6}\ \mathrm{m^3\,mol^{-1}\,s^{-1}} = \boxed{7.4 \times 10^{9}\ \mathrm{dm^3\,mol^{-1}\,s^{-1}}}$$

Since this reaction is elementary bimolecular it is second order; hence

$$t_{1/2} = \frac{1}{2k_d[\mathrm{A}]_0}\ \text{[Table 21.3, with } k_r = 2k_d \text{ because two atoms are consumed]}$$

so $$t_{1/2} = \frac{1}{2 \times (7.4 \times 10^{9}\ \mathrm{dm^3\,mol^{-1}\,s^{-1}}) \times (1.0 \times 10^{-3}\ \mathrm{mol\ dm^{-3}})} = \boxed{6.7 \times 10^{-8}\ \mathrm{s}}$$

E22.9(a) The steric factor, P, is

$$P = \frac{\sigma^*}{\sigma}\ \text{[Section 22.1(c)]}$$

The mean collision cross-section is $\sigma = \pi d^2$ with $d = (d_A + d_B)/2$.

Get the diameters from the collision cross-sections:

$$d_A = (\sigma_A/\pi)^{1/2} \quad \text{and} \quad d_B = (\sigma_B/\pi)^{1/2},$$

so, $$\sigma = \frac{\pi}{4}\left\{\left(\frac{\sigma_A}{\pi}\right)^{1/2} + \left(\frac{\sigma_B}{\pi}\right)^{1/2}\right\}^2 = \frac{(\sigma_A^{1/2} + \sigma_B^{1/2})^2}{4} = \frac{\{(0.95\ \mathrm{nm^2})^{1/2} + (0.65\ \mathrm{nm^2})^{1/2}\}^2}{4} = \boxed{0.79\ \mathrm{nm^2}}$$

Therefore, $P = \dfrac{9.2 \times 10^{-22}\ \mathrm{m^2}}{0.79 \times (10^{-9}\ \mathrm{m})^2} = \boxed{1.16 \times 10^{-3}}$.

E22.10(a) Since the reaction is diffusion controlled, the rate-limiting step is bimolecular and therefore second order; hence

$$\frac{\mathrm{d}[\mathrm{P}]}{\mathrm{d}t} = k_d[\mathrm{A}][\mathrm{B}]$$

where $k_d = 4\pi R^* D N_A$ [22.18] $= 4\pi N_A R^*(D_A + D_B)$

$$= 4\pi N_A \times (R_A + R_B) \times \frac{kT}{6\pi\eta}\left(\frac{1}{R_A} + \frac{1}{R_B}\right)[22.20] = \frac{2RT}{3\eta}(R_A + R_B) \times \left(\frac{1}{R_A} + \frac{1}{R_B}\right),$$

$$k_d = \frac{2 \times (8.3145\ \mathrm{J\ K^{-1}\,mol^{-1}}) \times (313\ \mathrm{K})}{3 \times (2.37 \times 10^{-3}\ \mathrm{kg\ m^{-1}\,s^{-1}})} \times (588 + 1650) \times \left(\frac{1}{588} + \frac{1}{1650}\right)$$

$$= 3.78 \times 10^{6}\ \mathrm{m^3\,mol^{-1}\,s^{-1}} = 3.78 \times 10^{9}\ \mathrm{dm^3\,mol^{-1}\,s^{-1}}.$$

Therefore, the initial rate is

$$\frac{\mathrm{d}[\mathrm{P}]}{\mathrm{d}t} = (3.78 \times 10^{9}\ \mathrm{dm^3\,mol^{-1}\,s^{-1}}) \times (0.150\ \mathrm{mol\ dm^{-3}}) \times (0.330\ \mathrm{mol\ dm^{-3}})$$

$$= \boxed{1.87 \times 10^{8}\ \mathrm{mol\ dm^{-3}\,s^{-1}}}.$$

COMMENT. If the approximation of eqn 22.21 is used, $k_d = 2.93 \times 10^{9}\ \mathrm{dm^3\,mol^{-1}\,s^{-1}}$. In this case the approximation results in a difference of about 20% compared to the expression used above.

E22.11(a) For reactions in solution the relationship between energy and enthalpy of activation is

$$\Delta^{\ddagger}H = E_a - RT \text{ [Brief comment in Section 22.5(a)]}$$

$$k_r = B\mathrm{e}^{\Delta^{\ddagger}S/R}\mathrm{e}^{-\Delta^{\ddagger}H/RT}, \quad B = \left(\frac{kT}{h}\right) \times \left(\frac{RT}{p^{\ominus}}\right) \text{ [22.43]}$$

$$= B\mathrm{e}^{\Delta^{\ddagger}S/R}\mathrm{e}^{-E_a/RT}\,\mathrm{e} = A\mathrm{e}^{-E_a/RT}$$

Therefore, $A = \mathrm{e}\,B\mathrm{e}^{\Delta^{\ddagger}S/R}$, implying that $\Delta^{\ddagger}S = R\left(\ln\frac{A}{B} - 1\right)$

Therefore, since $E_a = 8681\text{ K} \times R$

$$\Delta^{\ddagger}H = E_a - RT = (8681\text{ K} - 303\text{ K})R = (8378\text{ K}) \times (8.3145\text{ J K}^{-1}\text{ mol}^{-1}) = \boxed{+69.66\text{ kJ mol}^{-1}}$$

$$B = \frac{(1.381 \times 10^{-23}\text{ J K}^{-1}) \times (303\text{ K})}{6.626 \times 10^{-34}\text{ J s}} \times \frac{(8.3145\text{ J K}^{-1}\text{ mol}^{-1}) \times (303\text{ K})}{10^5\text{ Pa}}$$

$$= 1.59 \times 10^{11}\text{ m}^3\text{ mol}^{-1}\text{ s}^{-1} = 1.59 \times 10^{14}\text{ dm}^3\text{ mol}^{-1}\text{ s}^{-1}$$

and hence $\Delta^{\ddagger}S = (8.3145\text{ J K}^{-1}\text{ mol}^{-1}) \times \left\{\ln\left(\frac{2.05 \times 10^{13}\text{ dm}^3\text{ mol}^{-1}\text{ s}^{-1}}{1.59 \times 10^{14}\text{ dm}^3\text{ mol}^{-1}\text{ s}^{-1}}\right) - 1\right\}$

$$= \boxed{-25.3\text{ J K}^{-1}\text{ mol}^{-1}}$$

E22.12(a) The enthalpy of activation for a bimolecular solution reaction is [Exercise 22.11(a)]

$$\Delta^{\ddagger}H = E_a - RT = 8.3145\text{ J K}^{-1}\text{ mol}^{-1} \times (9134\text{ K} - 303\text{ K}) = \boxed{+73.4\text{ kJ mol}^{-1}}$$

The entropy of activation is [Exercise 22.11(a)]

$$\Delta^{\ddagger}S = R\left(\ln\frac{A}{B} - 1\right)$$

with $B = \dfrac{kRT^2}{hp^{\ominus}} = 1.59 \times 10^{14}\text{ dm}^3\text{ mol}^{-1}\text{ s}^{-1}$ [Exercise 22.11(a)]

Therefore, $\Delta^{\ddagger}S = 8.3145\text{ J K}^{-1}\text{ mol}^{-1} \times \left\{\ln\left(\frac{7.78 \times 10^{14}}{1.59 \times 10^{14}}\right) - 1\right\} = +4.9\text{ J K}^{-1}\text{ mol}^{-1}$

Hence, $\Delta^{\ddagger}G = \Delta^{\ddagger}H - T\Delta^{\ddagger}S = \{73.4 - (303) \times (4.9 \times 10^{-3})\}\text{kJ mol}^{-1} = \boxed{+71.9\text{ kJ mol}^{-1}}$

E22.13(a) Use eqn 22.44 to relate a bimolecular gas-phase rate constant to activation energy and entropy:

$$k_r = \mathrm{e}^2 B\mathrm{e}^{\Delta^{\ddagger}S/R}\mathrm{e}^{-E_a/RT}$$

where $B = \left(\frac{kT}{h}\right) \times \left(\frac{RT}{p^{\ominus}}\right) \text{[22.43]} = \dfrac{(1.381 \times 10^{-23}\text{ J K}^{-1}) \times (338\text{ K})^2 \times (8.3145\text{ J mol}^{-1}\text{ K}^{-1})}{(6.626 \times 10^{-34}\text{ J s}) \times (10^5\text{ Pa})}$

$$= 1.98 \times 10^{11}\text{ m}^3\text{ mol}^{-1}\text{ s}^{-1}.$$

Solve for the entropy of activation :

$$\Delta^{\ddagger}S = R\left(\ln\frac{k_r}{B} - 2\right) + \frac{E_a}{T}$$

The derivations in Section 22.5(a) are based on a k_r that contains concentration units, whereas the rate constant given here has pressure units. So k_r is not the given rate constant, but

$$k_r = 7.84 \times 10^{-3}\ \text{kPa}^{-1}\ \text{s}^{-1} \times RT$$
$$k_r = 7.84 \times 10^{-3} \times (10^3\ \text{Pa})^{-1}\ \text{s}^{-1} \times 8.3145\ \text{J K}^{-1}\ \text{mol}^{-1} \times 338\ \text{K} = 0.0220\ \text{m}^3\ \text{mol}^{-1}\ \text{s}^{-1}$$

Hence, $$\Delta^{\ddagger}S = 8.3145\ \text{J K}^{-1}\ \text{mol}^{-1} \times \left(\ln\frac{0.0220\ \text{m}^3\ \text{mol}^{-1}\ \text{s}^{-1}}{1.98 \times 10^{11}\ \text{m}^3\ \text{mol}^{-1}\ \text{s}^{-1}} - 2\right) + \frac{58.6 \times 10^3\ \text{J mol}^{-1}}{338\ \text{K}}$$

$$= \boxed{-91\ \text{J K}^{-1}\ \text{mol}^{-1}}.$$

E22.14(a) For a bimolecular gas-phase reaction,

$$\Delta^{\ddagger}S = R\left(\ln\frac{k_r}{B} - 2\right) + \frac{E_a}{T}\ \text{[Exercise 22.13(a)]}$$

$$= R\left(\ln\frac{A}{B} - \frac{E_a}{RT} - 2\right) + \frac{E_a}{T} = R\left(\ln\frac{A}{B} - 2\right)$$

where $B = \dfrac{kRT^2}{hp^{\ominus}}$ [Exercise 22.11(a)]

For two structureless particles, the rate constant is

$$k_r = N_A\sigma^*\left(\frac{8kT}{\pi\mu}\right)^{1/2} e^{-\Delta_r E_0/RT}\ [22.39]$$

The activation energy is [21.30]

$$E_a = RT^2\frac{d\ln k_r}{dT} = RT^2\frac{d}{dT}\left(\ln N_A\sigma^* + \frac{1}{2}\ln\frac{8k}{\pi\mu} + \frac{1}{2}\ln T - \frac{\Delta_r E_0}{RT}\right)$$

$$= RT^2\left(\frac{1}{2T} + \frac{\Delta_r E_0}{RT^2}\right) = \Delta_r E_0 + \frac{RT}{2},$$

so the prefactor is

$$A = k_r e^{E_a/RT} = N_A\sigma^*\left(\frac{8kT}{\pi\mu}\right)^{1/2} e^{-\Delta_r E_0/RT}(e^{\Delta_r E_0/RT}e^{1/2}) = N_A\sigma^*\left(\frac{8kT}{\pi\mu}\right)^{1/2} e^{1/2}$$

Hence,

$$\Delta^{\ddagger}S = R\left\{\ln N_A\sigma^*\left(\frac{8kT}{\pi\mu}\right)^{1/2} + \frac{1}{2} - \ln\frac{kRT^2}{p^{\ominus}h} - 2\right\} = R\left\{\ln\frac{\sigma^* p^{\ominus} h}{(kT)^{3/2}}\left(\frac{8}{\pi\mu}\right)^{1/2} - \frac{3}{2}\right\}$$

For identical particles,

$$\mu = m/2 = (50 \times 1.661 \times 10^{-27}\ \text{kg})/2 = 4.2 \times 10^{-26}\ \text{kg},$$

and hence

$$\Delta^{\ddagger}S = 8.3145\ \mathrm{J\ K^{-1}\ mol^{-1}}$$
$$\times\left\{\ln\frac{0.40\times(10^{-9}\ \mathrm{m})^2\times 10^5\ \mathrm{Pa}\times 6.626\times 10^{-34}\ \mathrm{J\ s}}{(1.381\times 10^{-23}\ \mathrm{J\ K^{-1}}\times 300\ \mathrm{K})^{3/2}}\left(\frac{8}{\pi\times 4.2\times 10^{-26}\ \mathrm{kg}}\right)^{1/2}-\frac{3}{2}\right\}$$
$$= \boxed{-72\ \mathrm{J\ K^{-1}\ mol^{-1}}}$$

E22.15(a) At low pressure, the reaction can be assumed to be bimolecular. (See Section 21.8(a).)

(a) $\Delta^{\ddagger}S = R\left(\ln\frac{A}{B}-2\right)$ [Exercise 22.14(a)]

where $B = \frac{kRT^2}{hp^{\ominus}} = \frac{1.381\times 10^{-23}\ \mathrm{J\ K^{-1}}\times 8.3145\ \mathrm{J\ K^{-1}\ mol^{-1}}\times(298\ \mathrm{K})^2}{6.626\times 10^{-34}\ \mathrm{J\ s}\times 10^5\ \mathrm{Pa}}$
$= 1.54\times 10^{11}\ \mathrm{m^3\ mol^{-1}\ s^{-1}} = 1.54\times 10^{14}\ \mathrm{dm^3\ mol^{-1}\ s^{-1}}$.

Hence, $\Delta^{\ddagger}S = 8.3145\ \mathrm{J\ K^{-1}\ mol^{-1}}\times\left(\ln\frac{4.6\times 10^{12}\ \mathrm{dm^3\ mol^{-1}\ s^{-1}}}{1.54\times 10^{14}\ \mathrm{dm^3\ mol^{-1}\ s^{-1}}}-2\right) = \boxed{-46\ \mathrm{J\ K^{-1}\ mol^{-1}}}$

(b) The enthalpy of activation for a bimolecular gas-phase reaction is

$\Delta^{\ddagger}H = E_a - 2RT$ [Brief comment in Section 22.5(a).]
$\Delta^{\ddagger}H = 10.0\ \mathrm{kJ\ mol^{-1}} - 2\times 8.3145\ \mathrm{J\ mol^{-1}\ K^{-1}}\times 298\ \mathrm{K} = \boxed{+5.0\ \mathrm{kJ\ mol^{-1}}}$

(c) The Gibbs energy of activation at 298 K is

$\Delta^{\ddagger}G = \Delta^{\ddagger}H - T\Delta^{\ddagger}S = 5.0\ \mathrm{kJ\ mol^{-1}} - (298\ \mathrm{K})\times(-46\times 10^{-3}\ \mathrm{kJ\ K^{-1}\ mol^{-1}})$

$\Delta^{\ddagger}G = \boxed{+18.7\ \mathrm{kJ\ mol^{-1}}}$

E22.16(a) Use eqn 22.49 to examine the effect of ionic strength on a rate constant:

$\log k_r = \log k_r^{\circ} + 2Az_Az_BI^{1/2}$

Hence, $\log k_r^{\circ} = \log k_r - 2Az_Az_BI^{1/2} = \log 12.2 - 2\times 0.509\times 1\times(-1)\times(0.0525)^{1/2} = 1.55$

and $k_r^{\circ} = \boxed{36\ \mathrm{dm^6\ mol^{-2}\ min^{-1}}}$

E22.17(a) To solve this exercise requires some information from the previous edition of the textbook. Eqns 22.61 (of the current edition) and 24.81 (of the previous edition) are two expressions that contain the given data, the desired quantity (reorganization energy, λ) and one other unknown quantity ($\Delta^{\ddagger}G$); thus, they constitute a system of two equations in two unknowns. Substituting eqn 22.61

$$\Delta^{\ddagger}G = \frac{(\Delta_rG^{\ominus}+\lambda)^2}{4\lambda}$$

into eqn 24.81 yields:

$$k_{et} = \frac{2H_{AB}^2}{h}\left(\frac{\pi^3}{4\lambda RT}\right)^{1/2}\exp\left(\frac{-\Delta^{\ddagger}G}{RT}\right) = \frac{H_{AB}^2}{h}\left(\frac{\pi^3}{\lambda RT}\right)^{1/2}\exp\left(\frac{-(\Delta_rG^{\ominus}+\lambda)^2}{4\lambda RT}\right)$$

The only unknown in this equation is λ. Putting in the numbers, and recognizing that H_{AB} ought to be an energy

$$H_{AB} \to hcH_{AB} = (6.626 \times 10^{-34}\ \text{J s})(2.998 \times 10^{10}\ \text{cm s}^{-1})(0.03\ \text{cm}^{-1}) = 6 \times 10^{-25}\ \text{J}$$

yields:

$$30.5\ \text{s}^{-1} = \frac{(6 \times 10^{-25}\ \text{J})^2}{6.626 \times 10^{-34}\ \text{J s}}\left(\frac{\pi^3}{\lambda(1.381 \times 10^{-23}\ \text{J K}^{-1})(298\ \text{K})}\right)^{1/2}$$
$$\times \exp\left(\frac{-\{(-0.182\ \text{eV})(1.602 \times 10^{-19}\ \text{J eV}^{-1}) + \lambda\}^2}{4\lambda(1.381 \times 10^{-23}\ \text{J K}^{-1})(298\ \text{K})}\right),$$

where the Boltzmann constant takes the place of the gas constant to put all energies on a molecular rather than a molar scale. One can solve this numerically using the root-finding command of a symbolic mathematics package, or graphically by plotting the right-hand side vs. the (constant) left-hand side and finding the value of λ at which the two lines cross. The reorganization energy turns out to be:

$$\lambda = \boxed{1.\bar{9} \times 10^{-19}\ \text{J}}\ \text{or about}\ \boxed{1.2\ \text{eV}}.$$

E22.18(a) For the same donor and acceptor at different distances, eqn. 22.63 applies:

$$\ln k_{et} = -\beta r + \text{constant}$$

The slope of a plot of $\ln k_{et}$ versus r is $-\beta$. The slope of a line defined by two points is:

$$\text{slope} = \frac{\Delta y}{\Delta x} = \frac{\ln k_{et,2} - \ln k_{et,1}}{r_2 - r_1} = -\beta = \frac{\ln 4.51 \times 10^5 - \ln 2.02 \times 10^5}{(1.23 - 1.11)\ \text{nm}}$$

so $\beta = \boxed{-6.7\ \text{nm}^{-1}}$.

E22.19(a) The conditions are in the limit of large, positive overpotentials, so eqn 22.70 applies:

$$\ln j = \ln j_0 + (1 - \alpha)f\eta$$

where $f = \frac{F}{RT} = \frac{96845\ \text{C mol}^{-1}}{(8.3145\ \text{J K}^{-1}\ \text{mol}^{-1}) \times (298\ \text{K})} = 38.9\ \text{V}^{-1}$

Subtracting this equation from the same relationship between another set of currents and overpotentials, we have

$$\ln \frac{j'}{j} = (1 - \alpha)f(\eta' - \eta)$$

which rearranges to

$$\eta' = \eta + \frac{\ln (j'/j)}{(1-\alpha)f} = (125 \times 10^{-3}\ \text{V}) + \frac{\ln (75/55.0)}{(1 - 0.39) \times (38.9\ \text{V})^{-1}} = \boxed{0.138\ \text{V}}$$

E22.20(a) Take antilogarithms of eqn 22.70; then

$$j_0 = j\,\text{e}^{-(1-\alpha)\eta f} = (55.0\ \text{mA cm}^{-2}) \times \text{e}^{-(1-0.39)\times 0.125\ \text{V}\times 38.9/\text{V}} = \boxed{2.82\ \text{mA cm}^{-2}}.$$

E22.21(a) $O_2(g)$ is produced at the anode in this electrolysis and $H_2(g)$ at the cathode. The net reaction is

$$2\,H_2O(l) \rightarrow 2\,H_2(g) + O_2(g)$$

For a large positive overpotential we use eqn 22.70:

$$\ln j = \ln j_0 + (1-\alpha)f\eta,$$

which implies $\ln\dfrac{j'}{j} = (1-\alpha)f(\eta'-\eta) = (1-0.5)\times(0.60-0.40)\text{ V}\times 38.9\text{ V}^{-1} = 3.\overline{9}$

and $\quad j' = j\mathrm{e}^{3.\overline{9}} = (1.0\text{ mA cm}^{-2})\times(4\overline{9}) = 4\overline{9}\text{ mA cm}^{-2}$

Hence, the anodic current density $\boxed{\text{increases}}$ by roughly a $\boxed{\text{factor of 50}}$ with a corresponding increase in O_2 evolution.

E22.22(a) From Table 22.3, $j_0 = 6.3\times10^{-6}\text{ A cm}^{-2}$ and $\alpha = 0.58$.

(a) According to the Butler–Volmer equation (eqn 22.65)

$$j = j_0(\mathrm{e}^{(1-\alpha)f\eta} - \mathrm{e}^{-\alpha f\eta})$$

so $\quad \dfrac{j}{j_0} = \mathrm{e}^{(1-0.58)\times(38.9/\text{V})\times0.20\text{ V}} - \mathrm{e}^{-0.58\times(38.9/\text{V})\times0.20\text{ V}} = (26.\overline{3} - 0.011) \approx 26$

$$j = (26)\times(6.3\times10^{-6}\text{ A cm}^{-2}) = \boxed{1.7\times10^{-4}\text{ A cm}^{-2}}.$$

(b) Eqn 22.69 (also known as the Tafel equation) corresponds to the neglect of the second exponential above, which is very small for an overpotential of 0.2 V. (Even when it was kept, in part (a), it was negligible.) Hence,

$$j = \boxed{1.7\times10^{-4}\text{ A cm}^{-2}}.$$

The validity of the Tafel equation increases with higher overpotentials, but decreases at lower overpotentials. A plot of j against η becomes linear (non-exponential) as $\eta \rightarrow 0$.

E22.23(a) $$\frac{j}{j_0} = \mathrm{e}^{(1-\alpha)f\eta} - \mathrm{e}^{-\alpha f\eta}\,[22.65] = \mathrm{e}^{(1/2)f\eta} - \mathrm{e}^{-(1/2)f\eta} \quad [\alpha = 0.5]$$

$$= 2\sinh\left(\frac{f\eta}{2}\right) \quad \left[\sinh x = \frac{e^x - e^{-x}}{2}\right]$$

Use $\quad \dfrac{f\eta}{2} = \dfrac{(38.9\text{ V}^{-1})\eta}{2} = 19.5(\eta/\text{V}) = 0.0195(\eta/\text{mV})$

Thus, $\quad j = 2j_0\sinh(\tfrac{1}{2}f\eta) = (1.58\text{ mA cm}^{-2})\times\sinh\left(\dfrac{0.0195\eta}{\text{mV}}\right)$

(a) If $\eta = 10$ mV, then $j = (1.58\text{ mA cm}^{-2})\times\sinh(0.195) = \boxed{0.31\text{ mA cm}^{-2}}$.

(b) If $\eta = 100$ mV, then $j = (1.58\text{ mA cm}^{-2})\times\sinh(1.95) = \boxed{5.44\text{ mA cm}^{-2}}$.

(c) If $\eta = -5.0$ V, then $j = (1.58\text{ mA cm}^{-2})\times\sinh(-97.5) = \boxed{-2\times10^{42}\text{ mA cm}^{-2}}$.

COMMENT. The extraordinarily large magnitude of the answer to (c) strongly suggests that the Butler–Volmer equation is not applicable at such large negative overpotentials!

E22.24(a) The current density of electrons is j_0/e because each one carries a charge of magnitude e. Look up j_0 values in Table 22.3, and recall that $1\ \text{A} = 1\ \text{C s}^{-1}$.

(a) $\text{Pt}\,|\,\text{H}_2\,|\,\text{H}^+ \quad j_0 = 7.9 \times 10^{-3}\ \text{A cm}^{-2}$

$$\frac{j_0}{e} = \frac{0.79 \times 10^{-3}\ \text{A cm}^{-2}}{1.602 \times 10^{-19}\ \text{C}} = \boxed{4.9 \times 10^{15}\ \text{cm}^{-2}\ \text{s}^{-1}}$$

(b) $\text{Pt}\,|\,\text{Fe}^{3+}, \text{Fe}^{2+} \quad j_0 = 2.5 \times 10^{-3}\ \text{mA cm}^{-2}$

$$\frac{j_0}{e} = \frac{2.5 \times 10^{-3}\ \text{A cm}^{-2}}{1.602 \times 10^{-19}\ \text{C}} = \boxed{1.6 \times 10^{16}\ \text{cm}^{-2}\ \text{s}^{-1}}$$

(c) $\text{Pb}\,|\,\text{H}_2\,|\,\text{H}^+ \quad j_0 = 5.0 \times 10^{-12}\ \text{A cm}^{-2}$

$$\frac{j_0}{e} = \frac{5.0 \times 10^{-12}\ \text{A cm}^{-2}}{1.602 \times 10^{-19}\ \text{C}} = \boxed{3.1 \times 10^{7}\ \text{cm}^{-2}\ \text{s}^{-1}}$$

There are approximately $\dfrac{(1.0 \times 10^{-2}\ \text{m})^2}{(280 \times 10^{-12}\ \text{m})^2} = 1.3 \times 10^{15}$ atoms in each square centimeter of surface. The numbers of electrons per atom are therefore $\boxed{3.9\ \text{s}^{-1}}$, $\boxed{12\ \text{s}^{-1}}$, and $\boxed{2.4 \times 10^{-8}\ \text{s}^{-1}}$, respectively. The last corresponds to less than one event per year.

E22.25(a) When the overpotential is small, its relationship to the current density is

$$\eta = \frac{RTj}{Fj_0}\ [22.68] = \frac{j}{fj_0},$$

which implies that the current through surface area S is

$$I = Sj = Sj_0 f\eta$$

An ohmic resistance r obeys $\eta = Ir$, and so we can identify the resistance as

$$r = \frac{\eta}{I} = \frac{1}{Sj_0 f} = \frac{1}{1.0\ \text{cm}^2 \times 38.9\ \text{V}^{-1} \times j_0} = \frac{2.57 \times 10^{-2}\ \Omega}{(j_0/\text{A cm}^{-2})}\ [1\ \text{V} = 1\ \text{A}\ \Omega]$$

(a) $\text{Pt}\,|\,\text{H}_2\,|\,\text{H}^+ \quad j_0 = 7.9 \times 10^{-4}\ \text{A cm}^{-2}$

$$r = \frac{2.57 \times 10^{-2}\ \Omega}{7.9 \times 10^{-4}} = \boxed{33\ \Omega}$$

(b) $\text{Hg}\,|\,\text{H}_2\,|\,\text{H}^+ \quad j_0 = 7.9 \times 10^{-13}\ \text{A cm}^{-2}$

$$r = \frac{2.57 \times 10^{-2}\ \Omega}{7.9 \times 10^{-13}} = \boxed{3.3 \times 10^{10}\ \Omega} = 33\ \text{G}\Omega$$

E22.26(a) We assume $\alpha \approx 0.5$. We look up the standard potential (zero-current, unit activity) of the Zn^{2+}, Zn couple; it is -0.76 V. Zinc will deposit from a solution of unit activity when the potential is below -0.76 V. The hydrogen ion current toward the zinc electrode is then

$$j(\text{H}^+) = j_0 \text{e}^{-\alpha f\eta}\ [22.93] = (50 \times 10^{-12}\ \text{A cm}^{-2}) \times \text{e}^{-0.5\times(38.9/\text{V})\times(-0.76\ \text{V})} = 1.3 \times 10^{-4}\ \text{A cm}^{-2}$$

This is a borderline negligible current. If we consider a current of 1 mA cm^{-2} or greater to be significant (i.e. to produce a significant rate of evolution of H_2), then $\boxed{\text{one can (barely) deposit zinc}}$ under these conditions without a lot of hydrogen evolution.

Solutions to problems

Solutions to numerical problems

P22.1 Comparing the Arrhenius equation (21.31) to eqn. 22.13, with $\sigma^* = P\sigma$ yields

$$A = N_A\sigma^*\left(\frac{8kT}{\pi\mu}\right)^{1/2}$$

where $\mu = m(CH_3)/2 = (15.03\, m_u/2) \times 1.6605 \times 10^{-27}\text{ kg } m_u^{-1} = 1.248 \times 10^{-26}\text{ kg}$

So, $A = \sigma^* \times (6.022 \times 10^{23}\text{ mol}^{-1}) \times \left(\frac{(8) \times (1.381 \times 10^{-23}\text{ J K}^{-1}) \times (298\text{ K})}{(\pi) \times (1.248 \times 10^{-26}\text{ kg})}\right)^{1/2}$

$= (5.52 \times 10^{26}) \times (\sigma^*)\text{ mol}^{-1}\text{ m s}^{-1}$.

(a) $\sigma^* = \frac{2.4 \times 10^{10}\text{ mol}^{-1}\text{ dm}^3\text{ s}^{-1}}{5.52 \times 10^{26}\text{ mol}^{-1}\text{ m s}^{-1}} = \frac{2.4 \times 10^{7}\text{ mol}^{-1}\text{ m}^3\text{ s}^{-1}}{5.52 \times 10^{26}\text{ mol}^{-1}\text{ m s}^{-1}} = \boxed{4.3\overline{5} \times 10^{-20}\text{ m}^2}$

(b) Take $\sigma \approx \pi d^2$ and estimate d as twice the bond length; therefore

$\sigma = (\pi) \times (154 \times 2 \times 10^{-12}\text{ m})^2 = 3.0 \times 10^{-19}\text{ m}^2$

Hence, $P = \frac{\sigma^*}{\sigma} = \frac{4.3\overline{5} \times 10^{-20}}{3.0 \times 10^{-19}} = \boxed{0.15}$.

P22.3 For radical recombination it has been found experimentally that $E_a \approx 0$. (This result is also easy to explain: activation barriers usually correspond to the energy cost of breaking or distorting bonds in reactants, but there are no bonds to be broken in radical recombination.) The maximum rate of recombination is obtained when $P = 1$, and then

$$k_r = P\sigma N_A\left(\frac{8kT}{\pi\mu}\right)^{1/2} e^{-E_a/RT}\ [22.13] = 4\sigma N_A\left(\frac{kT}{\pi m}\right)^{1/2} \quad [E_a = 0, P = 1, \mu = \tfrac{1}{2}m]$$

$\sigma \approx \pi d^2 = \pi \times (308 \times 10^{-12}\text{ m})^2 = 3.0 \times 10^{-19}\text{ m}^2$

Hence, $k_r = (4) \times (3.0 \times 10^{-19}\text{ m}^2) \times (6.022 \times 10^{23}\text{ mol}^{-1})$

$\times \left(\frac{(1.381 \times 10^{-23}\text{ J K}^{-1}) \times (298\text{ K})}{(\pi) \times (15.03 \times 1.6605 \times 10^{-27}\text{ kg})}\right)^{1/2}$

$= 1.6\overline{4} \times 10^{8}\text{ m}^3\text{ mol}^{-1}\text{ s}^{-1} = \boxed{1.6\overline{4} \times 10^{11}\text{ mol}^{-1}\text{ dm}^3\text{ s}^{-1}}$

The rate constant is for the rate law

$v = k_r[CH_3]^2$

Therefore, $-\dfrac{d[CH_3]}{dt} = 2k_r[CH_3]^2$

and the integrated rate law is $\dfrac{1}{[CH_3]} - \dfrac{1}{[CH_3]_0} = 2k_r t$ [21.15b]

For 90% recombination, $[CH_3] = 0.10 \times [CH_3]_0$, which occurs when

$$2k_r t = \frac{10}{[CH_3]_0} - \frac{1}{[CH_3]_0} = \frac{9}{[CH_3]_0} \quad \text{or} \quad t = \frac{9}{2k_r[CH_3]_0}$$

The mole fraction of CH_3 radicals when 10 mol % of ethane is dissociated is

$$\frac{(2)\times(0.10)}{1+0.10} = 0.18$$

The initial partial pressure of CH_3 radicals is thus

$$p_0 = xp_{tot} = (0.18)\times(100\times10^3\ \text{Pa}) = 1.8\times10^4\ \text{Pa}$$

and $[CH_3]_0 = \dfrac{p_0}{RT} = \dfrac{1.8\times10^4\ \text{Pa}}{RT}$

Therefore, $t = \dfrac{9RT}{(2k_r)\times p_0} = \dfrac{(9)\times(8.3145\ \text{J K}^{-1}\ \text{mol}^{-1})\times(298\ \text{K})}{2\times(1.6\bar{4}\times10^8\ \text{m}^3\ \text{mol}^{-1}\ \text{s}^{-1})\times(1.8\times10^4\ \text{Pa})} = 3.7\times10^{-9}\ \text{s} = \boxed{3.7\ \text{ns}}$.

P22.5 $\log k_r = \log k_r^\circ + 2Az_Az_BI^{1/2}$ [22.49]

This expression suggests that we should plot $\log k_r$ against $I^{1/2}$ and determine z_B from the slope, since we know that $|z_A| = 1$. We draw up the following table:

$I/(\text{mol kg}^{-1})$	0.0025	0.0037	0.0045	0.0065	0.0085
$\{I/(\text{mol kg}^{-1})\}^{1/2}$	0.050	0.061	0.067	0.081	0.092
$\log\{k_r/(\text{dm}^3\ \text{mol}^{-1}\ \text{s}^{-1})\}$	0.021	0.049	0.064	0.072	0.100

These points are plotted in Figure 22.1.

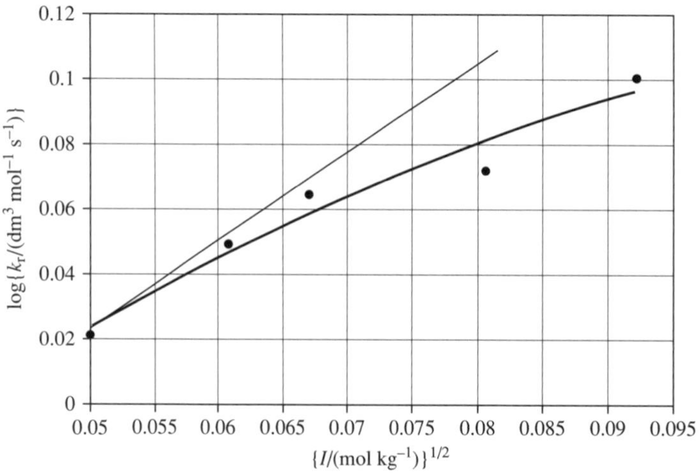

Figure 22.1

The slope of the limiting line in Figure 22.1 is approximately 2.5. Since this slope is equal to $2Az_Az_B \times (\text{mol kg}^{-1})^{1/2} = 1.018z_Az_B$, we have $z_Az_B \approx 2.5$. But $|z_A| = 1$, and so $|z_B| = 2$. Furthermore, z_A and z_B have the same sign because $z_Az_B > 0$. (In fact, the data refer to I^- and $S_2O_8^{2-}$.)

P22.7 Example 22.2 estimates a steric factor within the harpoon mechanism:

$$\frac{\sigma^*}{\sigma} \approx \left(\frac{e^2}{4\pi\varepsilon_0 d(I - E_{ea})}\right)^2$$

Taking $\sigma = \pi d^2$ gives

$$\sigma^* \approx \pi\left(\frac{e^2}{4\pi\varepsilon_0[I(\text{M}) - E_{ea}(\text{X}_2)]}\right)^2$$
$$= \frac{6.5\ \text{nm}^2}{\{(I - E_{ea})/\text{eV}\}^2}$$

Thus, σ^* is predicted to increase as $I - E_{ea}$ decreases. We construct the following table from the data:

σ^*/nm^2	Cl_2	Br_2	I_2
Na	0.45	0.42	0.56
K	0.72	0.68	0.97
Rb	0.77	0.72	1.05
Cs	0.97	0.90	1.34

All values of σ^* in the table are smaller than the experimental ones, but they do show the correct trends down the columns. The variation with E_{ea} across the table is not so good, possibly because the electron affinities used here are poor estimates.

Question. Can you find better values of electron affinities, and do they improve the horizontal trends in the table?

P22.9 The high-overpotential version of the Butler–Volmer equation predicts a linear dependence of j on η:

$$\ln j = \ln j_0 + (1 - \alpha)f\eta \ [22.70]$$

Draw up the following table for regression analysis:

η/mV	50	100	150	200	250
$\ln(j/\text{mA cm}^{-2})$	0.98	2.19	3.40	4.61	5.81

The points are plotted in Figure 22.2.

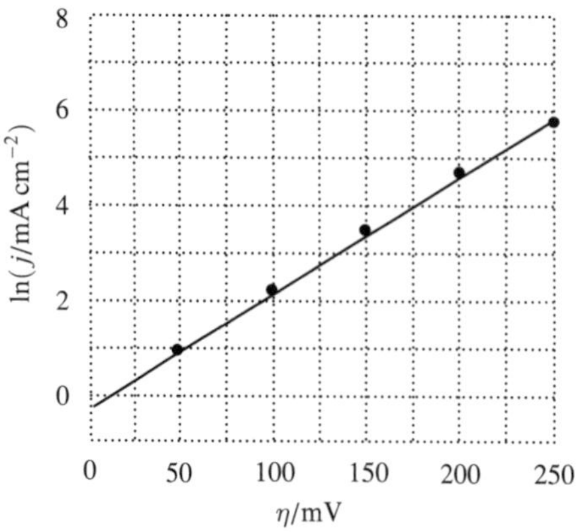

Figure 22.2

The plot is linear over the entire range of data. The *y*-intercept is at –0.25, and so

$$j_0/(\text{mA cm}^{-2}) = e^{-0.25} = \boxed{0.78}.$$

The slope is 0.0243, and so $(1-\alpha)f = 0.0243\ \text{mV}^{-1} = 24.3\ \text{V}^{-1}$. Since $f = 38.9\ \text{V}^{-1}$ at 25°C [Exercise 22.19(a)], it follows that $1-\alpha = 0.62$, and $\alpha = \boxed{0.38}$. If η were large but negative,

$$|j| \approx j_0 e^{-\alpha f\eta}\ [22.71] = (0.78\ \text{mA cm}^{-2})e^{-0.38\eta\times(0.0389/\text{mV})} = (0.78\ \text{mA cm}^{-2})e^{-0.015(\eta/\text{mV})}$$

and we draw up the following table:

η/mV	–50	–100	–150	–200	–250
$\lvert j\rvert/\text{mA cm}^{-2}$	1.65	3.50	7.40	15.7	33.2

P22.11 $Fe^{2+} + 2\,e^- \rightarrow Fe$ $\quad \nu = 2$; $E^{\ominus} = -0.447\ \text{V}$ [*CRC Handbook of Physics and Chemistry*]

(a) The zero-current potential is given by the Nernst equation (6.27):

$$E_0 = E^{\ominus} - \frac{RT}{\nu F}\ln Q$$

$$= E^{\ominus} - \frac{RT}{\nu F}\ln\frac{1}{[Fe^{2+}]} \text{ assuming } \gamma_{Fe^{2+}} = 1$$

$$= -0.447\ \text{V} - \frac{1}{2\times(38.9\ \text{V}^{-1})}\ln\left(\frac{1}{1.70\times10^{-6}}\right) = \boxed{-0.618\ \text{V}} \left[\frac{F}{RT} = 38.9\ \text{V}^{-1}\right]$$

$$\eta = E' - E_0\ [22.66]$$

η values are reported in the table below.

(b) The current density is related to the reaction rate

$$j = \frac{\nu F}{A}\frac{dn_{Fe}}{dt} = \frac{2\times(96845\ \text{C mol}^{-1})}{9.1\ \text{cm}^2}\frac{dn_{Fe}}{dt}$$

The same quantity is given by the Butler–Volmer equation

$$j = j_0(e^{(1-\alpha)f\eta} - e^{-\alpha f\eta})\,[22.65] = j_0 e^{-\alpha f\eta}(e^{f\eta} - 1) = -j_c(e^{f\eta} - 1)\,[22.93]$$

so $j_c = \dfrac{-j}{e^{f\eta} - 1} = \dfrac{-j}{e^{(38.9/V)\eta} - 1}$

j_c values are reported in the following table:

$\dfrac{dn_{Fe}}{dt}/(10^{-12}\ \text{mol s}^{-1})$	$-E'/\text{mV}$	$-\eta/\text{mV}$	$j/\mu\text{A cm}^{-2}$	$j_c/\mu\text{A cm}^{-2}$
1.47	702	84	0.0313	0.0325
2.18	727	109	0.0464	0.0471
3.11	752	134	0.0662	0.0666
7.26	812	194	0.155	0.155

(c) $j_c = j_0 e^{-\alpha f\eta}$ [22.93]

so $\ln j_c = \ln j_0 - \alpha f\eta$

The data are plotted in Figure 22.3. Performing a linear regression analysis of the $\ln j_c$ versus η data, we find

$\ln j_0/\mu\text{A cm}^{-2} = -4.612$ so $j_0 = 9.94 \times 10^{-3}\ \mu\text{A cm}^{-2}$

and $\alpha f = 0.0141\ \text{mV}^{-1}$ so $\alpha = 0.363$

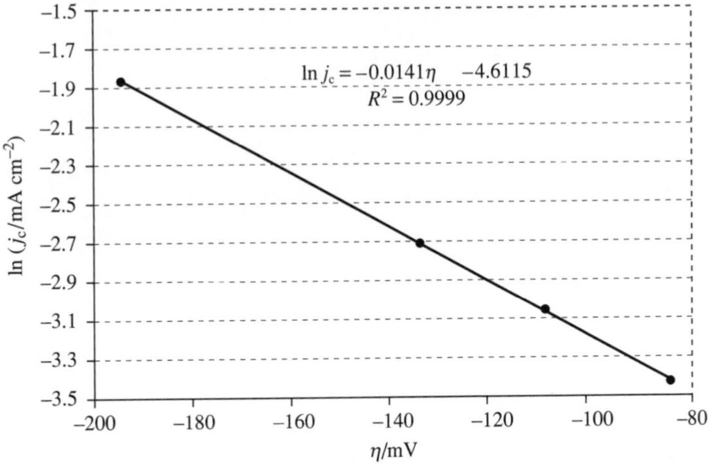

Figure 22.3

$R^2 = 0.9999$, indicating that the plot provides an excellent description of the data.

P22.13 This problem differs somewhat from the simpler one-electron transfers considered in the text. In place of $Ox + e^- \rightarrow Red$ we have here

$In^{3+} + 3\,e^- \rightarrow In$

namely, a three-electron transfer. Therefore, equations that contain the Faraday constant F (or f, which is proportional to F) need to be modified by including the factor z (in this case 3). In place of eqns 22.70 and 22.72, we have

$$\ln j = \ln j_0 + z(1-\alpha)f\eta \quad \text{anode}$$
$$\ln(-j) = \ln j_0 - z\alpha f\eta \quad \text{cathode}$$

We draw up the following table:

$j/(\mathrm{A\ m^{-2}})$	$-E/\mathrm{V}$	η/V	$\ln(j/\mathrm{A\ m^{-2}})$
0	0.388	0	
0.590	0.365	0.023	−0.5276
1.438	0.350	0.038	0.3633
3.507	0.335	0.053	1.255

We do a linear regression of $\ln j$ against η with the following results (see Figure 22.4):

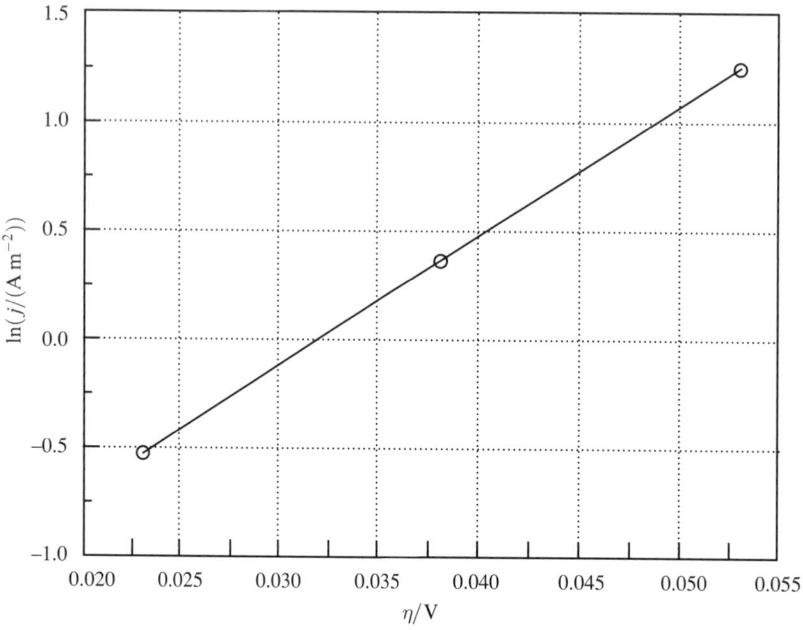

Figure 22.4

$$\text{slope} = z(1-\alpha)f = 59.42\ \mathrm{V^{-1}}, \quad \text{standard deviation} = 0.0154$$
$$y\text{-intercept} = \ln j_0 = -1.894, \quad \text{standard deviation} = 0.0006$$
$$R = 1 \text{ almost exactly}$$

The fit of the three data points to the Tafel equation is almost exact. Solving for α from the slope, we obtain

$$\alpha = 1 - \frac{59.42\ \mathrm{V^{-1}}}{3f} = 1 - \left(\frac{59.42\ \mathrm{V^{-1}}}{3\times(38.9\ \mathrm{V^{-1}})}\right) = \boxed{0.50}$$

which matches the usual value of α exactly.

$$j_0 = e^{-1.894}\ \text{A m}^{-2} = \boxed{0.150\ \text{A m}^{-2}}.$$

The cathodic current density at $E = -0.365$ V is obtained from

$$\begin{aligned}\ln(-j_c) &= \ln j_0 - z\alpha f\eta \\ &= -1.894 - (3 \times 0.50 \times 0.023\ \text{V}) \times (38.9\ \text{V}^{-1}) = -3.2\overline{6}\end{aligned}$$

so, $-j_c = e^{-3.2\overline{6}} = \boxed{0.038\ \text{A m}^{-2}}$

Solutions to theoretical problems

P22.15 We are to show that eqn. 22.26

$$[\text{J}]^* = [\text{J}]e^{-k_r t}$$

is a solution of eqn. 22.25,

$$\frac{\partial[\text{J}]^*}{\partial t} = D\frac{\partial^2[\text{J}]^*}{\partial x^2} - k_r[\text{J}]^*$$

provided that [J] is a solution of

$$\frac{\partial[\text{J}]}{\partial t} = D\frac{\partial^2[\text{J}]}{\partial x^2}$$

Evaluate the derivatives of [J]*:

$$\frac{\partial[\text{J}]^*}{\partial t} = \frac{\partial[\text{J}]}{\partial t}e^{-k_r t} - k_r[\text{J}]e^{-k_r t} \quad \text{and} \quad \frac{\partial^2[\text{J}]^*}{\partial x^2} = \frac{\partial^2[\text{J}]}{\partial x^2}e^{-k_r t}$$

Use the fact that [J] is a solution in the absence of reaction

$$D\frac{\partial^2[\text{J}]^*}{\partial x^2} = D\frac{\partial^2[\text{J}]}{\partial x^2}e^{-k_r t} = \frac{\partial[\text{J}]}{\partial t}e^{-k_r t} = \frac{\partial[\text{J}]^*}{\partial t} + k_r[\text{J}]e^{-k_r t}$$

which gives us back eqn. 22.25, as required.

P22.17 The standard molar partition function is defined in Section 16.8(a) as the molecular partition function at standard pressure. From the checklist of key equations in Chapter 16,

$$\frac{q_m^{\ominus T}}{N_A} = \frac{kT}{p^{\ominus}\Lambda^3} = \frac{kT}{p^{\ominus}} \times \frac{(2\pi mkT)^{3/2}}{h^3} = \frac{(kT)^{5/2}(2\pi m_u)^{3/2}}{p^{\ominus}h^3} \times \left(\frac{m}{m_u}\right)^{3/2}.$$

Substituting the values of the constants yields

$$\begin{aligned}\frac{q_m^{\ominus T}}{N_A} &= \frac{(1.381 \times 10^{-23}\ \text{J K}^{-1})^{5/2}\{2\pi \times (1.6605 \times 10^{-27}\ \text{kg})\}^{3/2}}{(10^5\ \text{Pa}) \times (6.626 \times 10^{-34}\ \text{J s})^3} \times T^{5/2} \times \left(\frac{m}{m_u}\right)^{3/2} \\ &= 0.0260 \times \left(\frac{T}{\text{K}}\right)^{5/2} \times \left(\frac{m}{m_u}\right)^{3/2}\end{aligned}$$

For $T \approx 300$ K, $M = \dfrac{m}{m_u} = 50$ g mol^{-1}, $\dfrac{q_m^{\ominus T}}{N_A} \approx \boxed{1.4 \times 10^7}$

$$q^R(\text{non-linear}) = \frac{1}{\sigma}\left(\frac{kT}{hc}\right)^{3/2}\left(\frac{\pi}{\tilde{A}\tilde{B}\tilde{C}}\right)^{1/2} \; [16.16]$$

Substituting the values of the constants yields

$$q^R(\text{non-linear}) = \left(\frac{1.381 \times 10^{-23}\ \text{J K}^{-1}}{(6.626 \times 10^{-34}\ \text{J s}) \times (2.998 \times 10^{10}\ \text{cm s}^{-1})}\right)^{3/2} \times \frac{\pi^{1/2}}{\sigma} \times \frac{T^{3/2}}{(\tilde{A}\tilde{B}\tilde{C})^{1/2}}$$

$$= \frac{1.027}{\sigma} \times \left(\frac{T}{\text{K}}\right)^{3/2} \times \left(\frac{\text{cm}^{-3}}{\tilde{A}\tilde{B}\tilde{C}}\right)^{1/2}$$

For $T \approx 300$ K, $\tilde{A} \approx \tilde{B} \approx \tilde{C} \approx 2$ cm^{-1}, $\sigma \approx 2$, $q^R(\text{non-linear}) \approx \boxed{900}$

$$q^R(\text{linear}) = \frac{kT}{\sigma hc\tilde{B}} [16.15\text{b}] = \frac{(1.381 \times 10^{-23}\ \text{J K}^{-1})T}{(6.626 \times 10^{-34}\ \text{J s}) \times (2.998 \times 10^{10}\ \text{cm s}^{-1})\sigma\tilde{B}} = \frac{0.695}{\sigma} \times \frac{(T/\text{K})}{(\tilde{B}/\text{cm}^{-1})}$$

For $T \approx 300$ K, $\tilde{B} \approx 1$ cm^{-1}, $\sigma \approx 1$, $q^R(\text{linear}) \approx \boxed{200}$

Energies of most excited vibrational states in small molecules and of most excited electronic states are high enough to make $q^V \approx q^E \approx \boxed{1}$

$$k_r = \frac{\kappa kT}{h}\bar{K}_c^{\ddagger} [22.37\text{b}] = \left(\frac{\kappa kT}{h}\right) \times \left(\frac{RT}{p^{\ominus}}\right) \times \left(\frac{N_A \bar{q}_{C\ddagger}^{\ominus}}{q_A^{\ominus} q_B^{\ominus}}\right) e^{-\Delta_r E_0/RT} [22.36] = A e^{-E_a/RT}$$

If A and B are structureless molecules, then we use estimates from above to evaluate

$$\frac{q_A^{\ominus}}{N_A} = \frac{q_A^{\ominus T}}{N_A} \approx 1.4 \times 10^7 \approx \frac{q_B^{\ominus}}{N_A} = \frac{q_B^{\ominus T}}{N_A}$$

$$\frac{\bar{q}_{C\ddagger}^{\ominus}}{N_A} = \frac{q_{C\ddagger}^{\ominus T} q^R(\text{linear})}{N_A} \approx (2^{3/2}) \times (1.4 \times 10^7) \times (200) = 8 \times 10^9$$

(The factor of $2^{3/2}$ comes from $m_C = m_A + m_B \approx 2m_A$ and $q^T \propto m^{3/2}$.)

$$\frac{RT}{p^{\ominus}} = \frac{(8.3145\ \text{J K}^{-1}\ \text{mol}^{-1}) \times (300\ \text{K})}{10^5\ \text{Pa}} = 2.5 \times 10^{-2}\ \text{m}^3\ \text{mol}^{-1}$$

and $$\frac{\kappa kT}{h} \approx \frac{kT}{h} = \frac{(1.381 \times 10^{-23}\ \text{J K}^{-1}) \times (300\ \text{K})}{6.626 \times 10^{-34}\ \text{J s}} = 6.25 \times 10^{12}\ \text{s}^{-1}$$

Strictly speaking, E_a is not exactly the same as $\Delta_r E_0$, but they are approximately equal—close enough for the purposes of estimating the order of magnitude of the pre-exponential factor. So, once we identify $E_a \approx \Delta_r E_0$, we identify the pre-exponential factor with everything *other than* the exponential term. Therefore, the pre-exponential factor

$$A \approx \frac{(6.25 \times 10^{12}\ \text{s}^{-1}) \times (2.5 \times 10^{-2}\ \text{m}^3\ \text{mol}^{-1}) \times (8 \times 10^9)}{(1.4 \times 10^7)^2}$$

$$\approx 6.3 \times 10^6\ \text{m}^3\ \text{mol}^{-1}\ \text{s}^{-1} = \boxed{6.3 \times 10^9\ \text{dm}^3\ \text{mol}^{-1}\ \text{s}^{-1}}.$$

According to collision theory [22.13]

$$A = P\sigma\left(\frac{8kT}{\pi\mu}\right)^{1/2} N_A$$

Take $\sigma \approx 0.5\ \text{nm}^2 = 5 \times 10^{-19}\ \text{m}^2$ as a typical value for small molecules. Also note that $\mu = m/2$ for identical particles:

$$A = P \times 5 \times 10^{-19}\ \text{m}^2 \times \left(\frac{8 \times (1.381 \times 10^{-23}\ \text{J K}^{-1}) \times (300\ \text{K})}{\pi \times (25m_u) \times (1.661 \times 10^{-27}\ \text{kg}\ m_u^{-1})}\right)^{1/2} \times 6.022 \times 10^{23}\ \text{mol}^{-1}$$

$$= 1.5 \times 10^8\ \text{m}^3\ \text{mol}^{-1}\ \text{s}^{-1} \times P.$$

The values are quite consistent, for they imply $P \approx 0.04$, which is certainly a plausible value.

If A and B are non-linear triatomics, then

$$\frac{q_A^{\ominus}}{N_A} \approx (1.4 \times 10^7) \times (900) = 1.3 \times 10^{10} \approx \frac{q_B^{\ominus}}{N_A}$$

$$\frac{q_{C^\ddagger}^{\ominus}}{N_A} \approx (2^{3/2}) \times (1.4 \times 10^7) \times (900) = 3.6 \times 10^{10}$$

and $$A \approx \frac{(6.25 \times 10^{12}\ \text{s}^{-1}) \times (2.5 \times 10^{-2}\ \text{m}^3\ \text{mol}^{-1}) \times (3.6 \times 10^{10})}{(1.3 \times 10^{10})^2}$$

$$\approx 33\ \text{m}^3\ \text{mol}^{-1}\ \text{s}^{-1} = \boxed{3.3 \times 10^4\ \text{dm}^3\ \text{mol}^{-1}\ \text{s}^{-1}}.$$

Comparison to the expression from collision theory implies $\boxed{P = 2 \times 10^{-7}}$.

P22.19 The diffusion process described is unimolecular, hence first order, and therefore analogous but not identical to the second-order case of Section 22.4. We may write

$$[\text{A}^\ddagger] = K^\ddagger[\text{A}] \quad [\text{analogous to 22.29}]$$

and $$-\frac{\text{d[A]}}{\text{d}t} = k^\ddagger[\text{A}^\ddagger] = \kappa\nu^\ddagger[\text{A}^\ddagger] \approx \nu^\ddagger[\text{A}^\ddagger] = \nu^\ddagger K^\ddagger[\text{A}] = k_r[\text{A}] \quad [22.30\text{–}22.32]$$

Thus, $$k_r \approx \nu^\ddagger K^\ddagger = \nu^\ddagger\left(\frac{kT}{h\nu^\ddagger}\right) \times \left(\frac{\bar{q}^\ddagger}{q}\right)\text{e}^{-\Delta_r E_0/RT}$$

where $\bar{q}^\ddagger$ and q are the (vibrational) partition functions at the top (missing one mode) and foot of the well, respectively. Let the y direction be the direction of diffusion. Hence, for the activated atom the vibrational mode in this direction is lost, and

$\bar{q}^\ddagger = q_x^{\ddagger\text{V}} q_z^{\ddagger\text{V}}$ for the activated atom, and

$q = q_x^\text{V} q_y^\text{V} q_z^\text{V}$ for an atom at the bottom of a well.

For classical vibration, $q^\text{V} \approx \dfrac{kT}{hc\tilde{\nu}}\ [16.21] = \dfrac{kT}{h\nu}$

Hence, $$k_r = \frac{kT}{h}\left(\frac{(kT/h\nu^\ddagger)^2}{(kT/h\nu)^3}\right)\text{e}^{-\Delta_r E_0/RT} = \boxed{\frac{\nu^3}{(\nu^\ddagger)^2}\text{e}^{-\Delta_r E_0/RT}} \approx \frac{\nu^3}{(\nu^\ddagger)^2}\text{e}^{-E_a/RT}$$

(a) If $\nu^{\ddagger} = \nu$, then $k_r \approx \nu e^{-E_a/RT} = 10^{11}\,\mathrm{s^{-1}} \times e^{-60000/(8.3145\times500)} = 5.4\times10^4\,\mathrm{s^{-1}}$

$$D = \frac{\lambda^2}{2\tau}[20.62] \approx \frac{1}{2}\lambda^2 k_r \left[\tau = \frac{1}{k_r}\text{(period for vibration with enough energy)}\right]$$

$$= \tfrac{1}{2}\times(316\times10^{-12}\,\mathrm{m})^2 \times 5.4\times10^4\,\mathrm{s^{-1}} = \boxed{2.7\times10^{-15}\,\mathrm{m^2\,s^{-1}}}$$

(b) If $\nu^{\ddagger} = \nu/2$, then $k_r \approx 4\nu e^{-E_a/RT} = 2.2\times10^5\,\mathrm{s^{-1}}$

$$D = 4\times(2.7\times10^{-15}\,\mathrm{m^2\,s^{-1}}) = \boxed{1.1\times10^{-14}\,\mathrm{m^2\,s^{-1}}}.$$

P22.21 We use the Eyring equation (combining eqns 22.36 and 22.37b) to compute the bimolecular rate constant

$$k_r = \kappa\frac{kT}{h}\left(\frac{RT}{p^{\ominus}}\right)\frac{N_A\bar{q}^{\ominus}_{C\ddagger}}{q^{\ominus}_H q^{\ominus}_{D_2}}\exp\left(\frac{-\Delta_r E_0}{RT}\right) \approx \frac{(RT)^2\bar{q}^{\ominus}_{C\ddagger}}{hp^{\ominus}q^{\ominus}_H q^{\ominus}_{D_2}}\exp\left(\frac{-\Delta_r E_0}{RT}\right)$$

We are to consider a variety of activated complexes, but the reactants (H and D_2) and their partition functions do not change. Consider them first. The partition function of H is solely translational:

$$q^{\ominus}_H = \frac{RT}{p^{\ominus}\Lambda_H^3} \text{ and } \Lambda_H = \left(\frac{h^2}{2\pi kTm_H}\right)^{1/2}, \text{ so } q^{\ominus}_H = \frac{RT(2\pi kTm_H)^{3/2}}{p^{\ominus}h^3}$$

We have neglected the spin degeneracy of H, which will cancel the spin degeneracy of the activated complex.

The partition function of D_2 has a rotational term as well as the translational term.

$$q^{\ominus}_{D_2} = \frac{RT}{p^{\ominus}\Lambda_{D_2}^3}\times\frac{kT}{\sigma hc\tilde{B}_{D_2}} = \frac{RkT^2(2\pi kTm_{D_2})^{3/2}}{2p^{\ominus}h^4c\tilde{B}_{D_2}}$$

We have neglected the vibrational partition function of D_2, which is very close to unity at the temperature in question. The symmetry number σ is 2 for a homonuclear diatomic, and the rotational constant is 30.44 cm^{-1}. Now, the partition function of the activated complex will have a translational piece that is the same regardless of the model:

$$\bar{q}^{\ominus}_{C\ddagger} = q^{T\ominus}_{C\ddagger}\times q^{R}_{C\ddagger}\times\bar{q}^{V}_{C\ddagger}$$

where $$q^{T\ominus}_{C\ddagger} = \frac{RT(2\pi kTm_{HD_2})^{3/2}}{p^{\ominus}h^3}$$

Let us aggregate the model-independent factors into a single term, F where:

$$F = \frac{(RT)^2\bar{q}^{T\ominus}_{C\ddagger}}{hp^{\ominus}q^{\ominus}_H q^{\ominus}_{D_2}}\exp\left(\frac{-\Delta_r E_0}{RT}\right) = \frac{2h^3c\tilde{B}_{D_2}m_{HD_2}^{3/2}}{kT(2\pi m_H m_{D_2}kT)^{3/2}}\exp\left(\frac{-\Delta_r E_0}{RT}\right)$$

$$= h^3c\tilde{B}_{D_2}\left(\frac{5^3}{2m_H^3(4)^3p^3T^3k^5}\right)^{1/2}\exp\left(\frac{-\Delta_r E_0}{RT}\right) = 2.71\times10^4\,\mathrm{dm^3\,mol^{-1}\,s^{-1}}.$$

where we have taken $m_{HD_2} = 5m_H$ and $m_{D_2} = 4m_H$.

Now, $k_r = F \times q^{R}_{C‡} \times \bar{q}^{V}_{C‡}$

The number of vibrational modes in the activated complex is $3 \times 3 - 6 = 3$ for a non-linear complex, one more for a linear complex; however, in either case, one mode is the reaction coordinate, and is removed from the partition function. Therefore, assuming all real vibrations to have the same wave-number $\tilde{\nu}$.

$$\bar{q}^{V}_{C‡} = q^{2}_{\text{mode}} \text{ (non-linear) or } q^{3}_{\text{mode}} \text{ (linear)}$$

$$\text{where} \quad q_{\text{mode}} = \left[1 - \exp\left(\frac{-hc\tilde{\upsilon}}{kT}\right)\right]^{-1} = 1.028$$

if the vibrational wavenumbers are 1000 cm^{-1}. The rotational partition function is

$$q^{R}_{C‡} = \frac{kT}{\sigma hc\tilde{B}} \text{ (linear) or } \frac{1}{\sigma}\left(\frac{kT}{hc}\right)^{3/2}\left(\frac{\pi}{\tilde{A}\tilde{B}\tilde{C}}\right)^{1/2} \text{ (non-linear)}$$

where the rotational constants are related to moments of inertia by

$$\tilde{B} = \frac{\hbar}{4\pi cI}, \text{ where } I = \sum mr^2$$

and r is the distance from an atom to a rotational axis.

(a) The first model for the activated complex is triangular, with two equal sides of

$$s = 1.30 \times 74 \text{ pm} = 96 \text{ pm}$$

and a base of

$$b = 1.20 \times 74 \text{ pm} = 89 \text{ pm}$$

The moment of inertia about the axis of the altitude of the triangle (z-axis) is

$$I_1 = 2m_D(b/2)^2 = m_H b^2, \text{ so } A = \frac{\hbar}{4\pi c m_H b^2} = 21.2 \text{ cm}^{-1}$$

To find the other moments of inertia, we need to find the centre of mass. Clearly it is in the plane of the molecule and on the z-axis; the centre of mass is the position z at which

$$\sum_i m_i(z_i - z) = 0 = 2(2m_H)(0 - z) + m_H(H - z)$$

where H is the height of the triangle,

$$H = [s^2 - (b/2)^2]^{1/2} = 85 \text{ pm}$$

so the centre of mass is $z = H/5$.

The moment of inertia about the axis in the plane of the triangle perpendicular to the altitude is

$$I_2 = 2(2m_H)(H/5)^2 + m_H(4H/5)^2 = (4m_H/5)H^2$$

$$\text{so} \quad \tilde{B} = \frac{\hbar}{4\pi c(4m_H/5)H^2} = 28.3 \text{ cm}^{-1}$$

The distance from the centre of mass to the D atoms is

$$r_D = [(H/5)^2 + (b/2)^2]^{1/2} = 48 \text{ pm}$$

and the moment of inertia about the axis perpendicular to the plane of the triangle is

$$I_3 = 2(2m_H)r_D^2 + m_H(4H/5)^2 = 2(2m_H)[(H/5)^2 + (b/2)^2] + m_H(4H/5)^2$$
$$= (4m_H/5)(s^2 + b^2).$$

so $$\tilde{C} = \frac{\hbar}{4\pi c(4m_H/5)(s^2+b^2)} = 12.2 \text{ cm}^{-1}$$

The rotational partition function is

$$q_{C\ddagger}^{R} = \frac{1}{s}\left(\frac{kT}{hc}\right)^{3/2}\left(\frac{\pi}{\tilde{A}\tilde{B}\tilde{C}}\right)^{1/2} = 47.7$$

(The symmetry number σ is 2 for this model.) The vibrational partition function is

$$\bar{q}_{C\ddagger}^{V} = q_{\text{mode}}^{2} = 1.057$$

So, the rate constant is:

$$k_r = F \times q_{C\ddagger}^{R} \times \bar{q}_{C\ddagger}^{V} = \boxed{1.37 \times 10^{6} \text{ dm}^3 \text{ mol}^{-1} \text{ s}^{-1}}$$

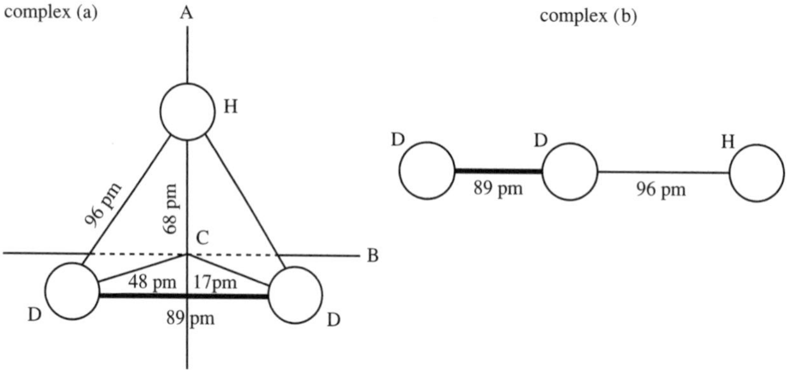

(b) To compute the moment of inertia, we need the centre of mass. Let the terminal D atom be at $x = 0$, the central D atom at $x = b$, and the H atom at $x = b + s$. The centre of mass is the position X at which

$$\sum_i m_i(x_i - X) = 0 = 2m_H(0 - X) + 2m_H(b - X) + m_H(s + b - X)$$
$$5X = 3b + s \text{ so } x = (3b + s)/5$$

The moment of inertia is

$$I = \sum_i m_i(x_i - X)^2 = 2m_H X^2 + 2m_H(b - X)^2 + m_H(s + b - X)^2$$
$$= 3.97 \times 10^{-47} \text{ m kg}^2$$

and $\tilde{B} = \frac{\hbar}{4\pi cI} = 7.06 \text{ cm}^{-1}$

The rotational partition function is

$$q^{R}_{C\ddagger} = \frac{kT}{\sigma hc\tilde{B}} = 39.4$$

(The symmetry number σ is 1 for this model.) The vibrational partition function is

$$\bar{q}^{V}_{C\ddagger} = q^{3}_{\text{mode}} = 1.09$$

So the rate constant is

$$k_r = F \times q^{R}_{C\ddagger} \times \bar{q}^{V}_{C\ddagger} = \boxed{1.16 \times 10^{6} \text{ dm}^{3} \text{ mol}^{-1} \text{ s}^{-1}}$$

(c) Both models are already pretty good, coming within a factor of 3 to 4 of the experimental result, and neither model has much room for improvement. Consider how to try to change either model to reduce the rate constant toward the experimental value. The factor F is model independent. The factor $\bar{q}^{V}_{C\ddagger}$ is nearly at its minimum possible value, 1, so stiffening the vibrational modes will have almost no effect. Only the factor $q^{R}_{C\ddagger}$ is amenable to lowering, and even that not by much. It would be decreased if the rotational constants were increased, which means decreasing the moments of inertia and the bond lengths. Reducing the lengths s and b in the models to the equilibrium bond length of H_2 would only drop k_r to 6.5×10^{5} (model a) or 6.9×10^{5} (model b) $\text{dm}^{3} \text{ mol}^{-1} \text{ s}^{-1}$, even with a stiffening of vibrations. Reducing the HD distance in model (a) to 80% of the H_2 bond length does produce a rate constant of $4.2 \times 10^{5} \text{ dm}^{3} \text{ mol}^{-1} \text{ s}^{-1}$ (assuming stiff vibrations of 2000 cm^{-1}); such a model is not intermediate in structure between reactants and products, however. It appears that the rate constant is rather insensitive to the geometry of the complex.

P22.23 Let η oscillate between η_+ and η_- around a mean value η_0. Then, η_- is large and positive and η_+ still larger:

$$j = j_0 e^{(1-\alpha)\eta f} \,[22.69] = j_0 e^{\eta f/2} \,[\alpha = 1/2]$$

where $f = \frac{F}{RT}$ and η varies as depicted in Figure 22.5(a).

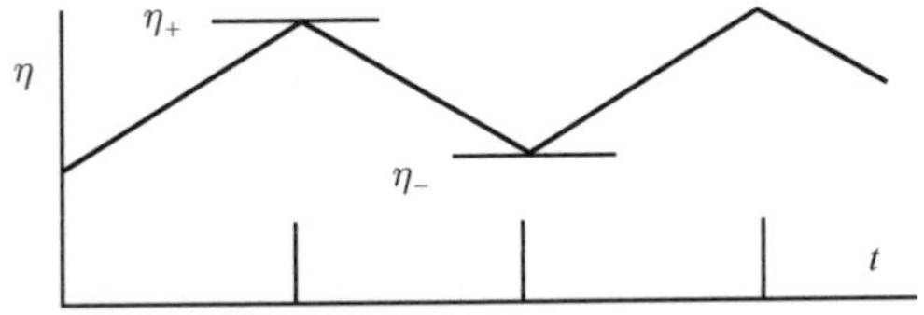

Figure 22.5(a)

Therefore, j is a chain of increasing and decreasing exponential functions. During the increasing phase,

$$j = j_0 e^{(\eta_- + \gamma t)f/2} \propto e^{t/\tau}$$

where $\tau = \dfrac{2RT}{\gamma F}$ and γ is a constant that characterizes the rate at which η rises. During the decreasing phase,

$$j = j_0 e^{(\eta_+ - \gamma t)f/2} \propto e^{-t/\tau}.$$

The variation of j is depicted in Figure 22.5(b).

Figure 22.5(b)

Solutions to applications

P22.25 (a) The rate constant of a diffusion-limited reaction is

$$k_d = \frac{8RT}{3\eta}[22.21] = \frac{8 \times (8.3145\ \mathrm{J\ K^{-1}\ mol^{-1}}) \times (298\ \mathrm{K})}{3 \times (1.06 \times 10^{-3}\ \mathrm{kg\ m^{-1}\ s^{-1}})}$$

$$= 6.23 \times 10^{6}\ \mathrm{m^3\ mol^{-1}\ s^{-1}} = \boxed{6.23 \times 10^{9}\ \mathrm{dm^3\ mol^{-1}\ s^{-1}}}$$

(b) The rate constant is related to the diffusion constants and reaction distance by

$$k_d = 4\pi R^* D N_A\ [22.18]$$

so $$R^* = \frac{k_d}{4\pi D N_A} = \frac{(2.77 \times 10^{9}\ \mathrm{dm^3\ mol^{-1}\ s^{-1}}) \times (10^{-3}\ \mathrm{m^3\ dm^{-3}})}{4\pi \times (1 \times 10^{-9}\ \mathrm{m^2\ s^{-1}}) \times (6.022 \times 10^{23}\ \mathrm{mol^{-1}})} = \boxed{4 \times 10^{-10}\ \mathrm{m}} = 0.4\ \mathrm{nm}$$

P22.27 For a series of reactions with a fixed edge-to-edge distance and reorganization energy, the logarithm of the rate constant is

$$\ln k_{et} = -\frac{\Delta^{\ddagger}G^{\ominus}}{kT} + \text{constant [from 22.62 in molecular units]}$$

Substituting

$$\Delta^{\ddagger}G^{\ominus} = \frac{(\Delta_r G^{\ominus} + \lambda)^2}{4\lambda}[22.61] = \frac{(\Delta_r G^{\ominus})^2 + 2\lambda\Delta_r G^{\ominus} + \lambda^2}{4\lambda}$$

yields

$$\ln k_{et} = -\frac{(\Delta_r G^{\ominus})^2}{4\lambda kT} - \frac{\Delta_r G^{\ominus}}{2kT} + \text{constant},$$

(i.e. quadratic dependence on the reaction free energy).

Draw up the following table:

$\Delta_r G^{\ominus}$/eV	$k_{et}/(10^6\ s^{-1})$	$\ln(k_{et}/s^{-1})$
−0.665	0.657	13.4
−0.705	1.52	14.2
−0.745	1.12	13.9
−0.975	8.99	16.0
−1.015	5.76	15.6
−1.055	10.1	16.1

and plot $\ln k_{et}$ vs. $\Delta_r G^{\ominus}$ (Figure 22.6).

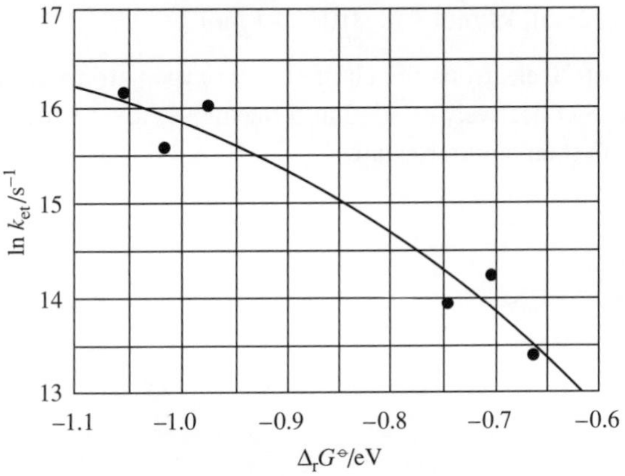

Figure 22.6

The least squares quadratic fit equation is

$$\ln k_{et}/s^{-1} = 3.23 - 21.1(\Delta_r G^{\ominus}/\text{eV}) - 8.48(\Delta_r G^{\ominus}/\text{eV})^2, \quad r^2 = 0.938$$

The coefficient of the quadratic term is

$$-\frac{1}{4\lambda kT} = -\frac{8.48}{\text{eV}^2}$$

$$\text{so} \quad \lambda = \frac{(\text{eV})^2}{4(8.48)kT} = \frac{(1.602\times 10^{-19}\ \text{J eV}^{-1})(\text{eV})^2}{2(8.48)(1.381\times 10^{-23}\ \text{J K}^{-1})(298\ \text{K})} = \boxed{1.15\ \text{eV}}$$

P22.29 $$E^{\ominus} = \frac{-\Delta_r G^{\ominus}}{\nu F}$$

(a) $H_2 + \frac{1}{2}O_2 \rightarrow H_2O$; $\Delta_r G^{\ominus} = -237\ \text{kJ mol}^{-1}$

Since $\nu = 2$,

$$E^{\ominus} = \frac{-(-237\ \text{kJ mol}^{-1})}{2\times(96.48\ \text{kC mol}^{-1})} = \boxed{+1.23\ \text{V}}$$

(b) $CH_4 + 2\,O_2 \rightarrow CO_2 + 2\,H_2O$

$$\Delta_r G^{\ominus} = 2\Delta_f G^{\ominus}(H_2O) + \Delta_f G^{\ominus}(CO_2) - \Delta_f G^{\ominus}(CH_4)$$
$$= \{(2) \times (-237.1) + (-394.4) - (-50.7)\}\ \text{kJ mol}^{-1} = -817.9\ \text{kJ mol}^{-1}$$

The reaction corresponds to the transfer of eight electrons. (The carbon goes from −4 to +4 while each of the four oxygen atoms goes from 0 to −2.) It follows that, for the species in their standard states,

$$E^{\ominus} = \frac{-(-817.9\ \text{kJ mol}^{-1})}{8 \times (96.48\ \text{kC mol}^{-1})} = \boxed{+1.06\ \text{V}}$$

(c) $C_3H_8 + 5\,O_2 \rightarrow 3\,CO_2 + 4\,H_2O$

$$\Delta_r G^{\ominus} = 4\Delta_f G^{\ominus}(H_2O) + 3\Delta_f G^{\ominus}(CO_2) - \Delta_f G^{\ominus}(C_3H_4)$$
$$= \{4 \times (-237.1) + 3 \times (-394.4) - (-23.5)\}\ \text{kJ mol}^{-1} = -2108.1\ \text{kJ mol}^{-1}$$

The reaction corresponds to the transfer of 20 electrons. (Each of the 10 oxygen atoms goes from 0 to −2 while the carbon atoms go from −8 collectively (2 at −3 and one at −2) to +12 collectively (+4 each).) It follows that, for the species in their standard states,

$$E^{\ominus} = \frac{-(-2108.1\ \text{kJ mol}^{-1})}{20 \times (96.48\ \text{kC mol}^{-1})} = \boxed{+1.093\ \text{V}}$$

23 Catalysis

Answers to discussion questions

D23.1 The **Michaelis–Menten mechanism** of enzyme activity models the enzyme with one active site that, weakly and reversibly, binds a substrate in homogeneous solution. It is a three-step mechanism. The first and second steps are the reversible formation of the enzyme–substrate complex (ES). The third step is the decay of the complex into the product. The steady-state approximation is applied to the concentration of the intermediate (ES) and its use simplifies the derivation of the final rate expression. However, the justification for the use of the approximation with this mechanism is suspect, in that both rate constants for the reversible step may not be as large, in comparison to the rate constant for the decay to products, as they need to be for the approximation to be valid. The mechanism clearly indicates that the simplest form of the rate law, $v = v_{max} = k_b[E]_0$, occurs when $[S]_0 \gg K_M$ and the general form of the rate law does seem to match the principal experimental features of enzyme-catalysed reactions. It provides a mechanistic understanding of both the turnover number and catalytic efficiency. The model may be expanded to include multisubstrate reactions and inhibition.

D23.3 As temperature increases we expect the rate of an enzyme-catalysed reaction to increase. However, at a sufficiently high temperature the enzyme **denatures** and a decrease in the reaction rate is observed. Temperature-related denaturation is caused by the action of vigorous vibrational motion, which destroys secondary and tertiary protein structure. Electrostatic, internal hydrogen bonding, and van der Waals interactions that hold the protein in its active, folded shape are broken with the protein unfolding into a **random coil**. The active site and enzymatic activity is lost.

The rate of a particular enzyme-catalysed reaction may also appear to decrease at high temperature in the special case in which an alternative substrate reaction, which has a relatively slow rate at low temperature, has the faster rate increase with increasing temperature. A temperature may be reached at which the alternative reaction predominates.

D23.5 **AFM**, atomic force microscopy, drags a sharp stylus attached to a cantilever across a surface and monitors the deflection of a laser beam from the back of the cantilever. Tiny changes in deflection indicate attraction to or repulsion from atoms on a sample surface. Since no current is involved, both conductive and non-conductive surfaces may be viewed. Surface damage is avoided by using a cantilever that has a very small spring constant. The method does not require a vacuum and it has been applied in liquid environment. Biological polymers may be viewed and nanometre resolutions have been achieved. However, an incorrect probe choice may cause image artifacts and distortions. Thermal drift of adsorbates may result in image distortions during relatively slow surface scans.

FIM, field-ionization microscopy, points a tip, having a point radius of about 50 nm, toward a fluorescent screen in a chamber containing about 1 mTorr to 1 nTorr of either hydrogen or helium. A positive 2–20 kV potential applied to the tip causes the hydrogen or helium gas adsorbate molecules to ionize and accelerate to the fluorescent screen. The image portrays the electrical characteristics of the tip surface and surface diffusion characteristics of the adsorbate are deduced. See the very interesting historical review of the technique in the Nov. 28, 2005 issue of *Chemical and Engineering News* (V. 83, number 48, p. 13–16).

LEED, low-energy electron diffraction, uses electrons with energies in the range 10–200 eV, which ensures diffraction from atoms only on or near the sample surface. Diffraction intensities depend on the vertical location of the atoms. The diffraction pattern is sharp if the surface is well ordered for distances long compared with the wavelength of the incident electrons. Diffuse patterns indicate a poorly ordered surface or the presence of impurities. If the LEED pattern does not correspond to the pattern expected by extrapolation of the bulk to the surface, then either a reconstruction of the surface has occurred or there is order in the arrangement of an adsorbed layer. The interpretation of LEED data can be very complicated.

SAM, scanning Auger electron microscopy, uses a focused 1–5 keV electron beam to probe and map surface composition to a resolution of about 50 nm. The high-energy impact causes the ejection of an electron from a low-lying orbital, and an upper electron falls into it. The energy this releases may result either in the generation of **X-ray fluorescence** or in the ejection of a second electron, the **Auger effect**. The emissions are used to identify chemical constituents at interfaces and surfaces of conducting and semiconducting materials to a depth of 1–5 nm.

SEM, scanning electron microscopy, uses magnetic fields to focus and scan a beam of electrons across a sample surface. Scattered electrons from a small irradiated area are detected and the electrical signal is sent to a video screen. Resolution is typically between 1.5 and 3.0 nm. Non-conductive materials require a thin conductive coating to prevent electrical charging of the sample.

STM, scanning tunnelling microscopy, reveals atomic details of surface and adsorbate structure. Surface chemical reactions can be viewed as they happen. The tip of the STM, which may end in a single atom, can also be used to manipulate adsorbed atoms on a surface, making possible the fabrication of complex and yet very tiny structures, such as nanometre-sized electronic devices. The method is based on the quantum-mechanical tunnelling effect in the presence of a bias voltage between the STM tip and sample surface. A piezoelectric scanner is used to position and move the tip in very close proximity to the surface and the electrical current of tunnelling generates an image of the surface topography with a resolution in the nanometre range. Images of surface electronic states may be generated. A host of very interesting STM images can be viewed at http://www.almaden.ibm.com/vis/stm/gallery.html.

D23.7 Heterogeneous catalysis on a solid surface requires the reacting molecules or fragments to encounter each other by adsorption on the surface. Therefore, the rate of the catalysed reaction is determined by the sticking probabilities of the species on the surface as described by Figure 23.30 of the text.

Solutions to exercises

E23.1(a) The fast, reversible step suggests the pre-equilibrium approximation:

$$K = \frac{[BH^+][A^-]}{[AH][B]} \quad \text{and} \quad [A^-] = \frac{K[AH][B]}{[BH^+]}.$$

Thus, the rate of product formation is

$$\frac{d[P]}{dt} = k_b[AH][A^-] = \boxed{\frac{k_b K[AH]^2[B]}{[BH^+]}}.$$

The application of the steady-state approximation to $[A^-]$ gives a similar, but significantly different, result:

$$\frac{d[A^-]}{dt} = k_a[AH][B] - k_a'[A^-][BH^+] - k_b[A^-][AH] = 0.$$

$$\text{Therefore, } [A^-] = \frac{k_a[AH][B]}{k_a'[BH^+] + k_b[AH]}$$

and the rate of formation of product is

$$\frac{d[P]}{dt} = k_b[AH][A^-] = \frac{k_a k_b[AH]^2[B]}{k_a'[BH^+] + k_b[AH]}.$$

E23.2(a) Since $v = \dfrac{v_{max}}{1 + K_M/[S]_0}$ [23.3a],

$$\begin{aligned} v_{max} &= (1 + K_M/[S]_0)v \\ &= (1 + 0.046/0.105) \times (1.04 \text{ mmol dm}^{-3}\text{ s}^{-1}) \\ &= \boxed{1.50 \text{ mmol dm}^{-3}\text{ s}^{-1}} \end{aligned}$$

E23.3(a)

$$\begin{aligned} k_{cat} &= v_{max}/[E]_0 \text{ [23.4]} \\ &= (0.425 \text{ mmol dm}^{-3}\text{ s}^{-1})/(3.60 \times 10^{-6} \text{ mmol dm}^{-3}) = \boxed{1.18 \times 10^5 \text{ s}^{-1}} \end{aligned}$$

$$\begin{aligned} \eta &= k_{cat}/K_M \text{ [23.5]} \\ &= (1.18 \times 10^5 \text{ s}^{-1})/(0.015 \text{ mol dm}^{-3}) = \boxed{7.9 \times 10^6 \text{ dm}^3\text{ mol}^{-1}\text{ s}^{-1}} \end{aligned}$$

Diffusion limits the **catalytic efficiency**, η, to a maximum of about 10^8–10^9 $\text{dm}^3\text{ mol}^{-1}\text{ s}^{-1}$. Since the catalytic efficiency of this enzyme is significantly smaller than the maximum, the enzyme is not 'catalytically perfect'.

E23.4(a) Eqn 23.7 describes competitive inhibition as the case for which $\alpha = 1 + [I]/K_I$ and $\alpha' = 1$. Thus,

$$v = \frac{v_{max}}{1 + \alpha K_M/[S]_0}.$$

By setting the ratio $v\,([I] = 0)/v\,([I])$ equal to 2 and solving for α, we can subsequently solve for the inhibitor concentration that reduces the catalytic rate by 50%:

$$\frac{v([\mathrm{I}]=0)}{v([\mathrm{I}])}=\frac{1+\alpha K_\mathrm{M}/[\mathrm{S}]_0}{1+K_\mathrm{M}/[\mathrm{S}]_0}=2$$

$$\alpha=\frac{2(1+K_\mathrm{M}/[\mathrm{S}]_0)-1}{K_\mathrm{M}/[\mathrm{S}]_0}$$
$$=\frac{2(1+3.0/0.10)-1}{3.0/0.10}=2.0\overline{3}$$

$$[\mathrm{I}]=(\alpha-1)K_\mathrm{I}$$
$$=1.0\overline{3}\times(2.0\times10^{-5}\ \mathrm{mol\ dm^{-3}})=\boxed{2.0\times10^{-5}\ \mathrm{mol\ dm^{-3}}}$$

E23.5(a) The collision frequency, Z_W, of gas molecules with an ideally smooth surface area is given by eqn 23.9:

$$Z_\mathrm{W}=\frac{p}{(2\pi MkT/N_\mathrm{A})^{1/2}}\ [23.9;\ m=M/N_\mathrm{A}]$$
$$=\frac{p\times\{(\mathrm{kg\ m^{-1}\,s^{-2}})/\mathrm{Pa}\}\times(10^{-4}\ \mathrm{m^2/cm^2})}{\{2\pi\times(1.381\times10^{-23}\ \mathrm{J\ K^{-1}})\times(298.15\ \mathrm{K})\times(\mathrm{kg\ mol^{-1}})/(6.022\times10^{23}\ \mathrm{mol^{-1}})\}^{1/2}\{M/(\mathrm{kg\ mol^{-1}})\}^{1/2}}$$
$$=4.825\times10^{17}\left(\frac{p/\mathrm{Pa}}{\{M/(\mathrm{kg\ mol^{-1}})\}^{1/2}}\right)\mathrm{cm^{-2}\,s^{-1}}\quad \text{at } 25°\mathrm{C}$$

(a) Hydrogen ($M=0.002016\ \mathrm{kg\ mol^{-1}}$)

(i) $p=100\ \mathrm{Pa}$, $Z_\mathrm{W}=\boxed{1.07\times10^{21}\ \mathrm{cm^{-2}\,s^{-1}}}$

(ii) $p=0.10\ \mu\mathrm{Torr}=1.33\times10^{-5}\ \mathrm{Pa}$, $Z_\mathrm{W}=\boxed{1.4\times10^{14}\ \mathrm{cm^{-2}\,s^{-1}}}$

(b) Propane ($M=0.04410\ \mathrm{kg\ mol^{-1}}$)

(i) $p=100\ \mathrm{Pa}$, $Z_\mathrm{W}=\boxed{2.30\times10^{20}\ \mathrm{cm^{-2}\,s^{-1}}}$

(ii) $p=0.10\ \mu\mathrm{Torr}=1.33\times10^{-5}\ \mathrm{Pa}$, $Z_\mathrm{W}=\boxed{3.1\times10^{13}\ \mathrm{cm^{-2}\,s^{-1}}}$

E23.6(a) $A=\pi d^2/4=\pi\,(1.5\ \mathrm{mm})^2/4=1.77\times10^{-6}\ \mathrm{m^2}$

The collision frequency of the Ar gas molecules with surface area A equals $Z_\mathrm{W}A$.

$$Z_\mathrm{W}A=\frac{p}{(2\pi MkT/N_\mathrm{A})^{1/2}}A\ [23.9;\ m=M/N_\mathrm{A}]$$

$$p=(Z_\mathrm{W}A)\times(2\pi MkT/N_\mathrm{A})^{1/2}/A$$
$$=(4.5\times10^{20}\ \mathrm{s^{-1}})\times\{2\pi(39.95\times10^{-3}\ \mathrm{kg\ mol^{-1}})$$
$$\times(1.381\times10^{-23}\ \mathrm{J\ K^{-1}})\times(425\ \mathrm{K})/(6.022\times10^{23}\ \mathrm{mol^{-1}})\}^{1/2}/(1.77\times10^{-6}\ \mathrm{m^2})$$
$$=1.3\times10^4\ \mathrm{Pa}=\boxed{0.13\ \mathrm{bar}}$$

E23.7(a) The further apart the atoms responsible for the pattern, the closer the spots appear in the pattern (see Example 23.3). Doubling the vertical separation between atoms of the unreconstructed face, which has LEED pattern (a), yields a reconstructed surface that gives LEED pattern (b).

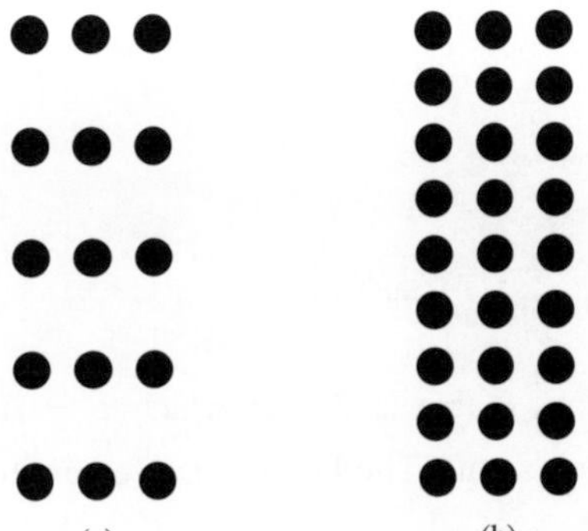
(a) (b)

E23.8(a) Let us assume that the nitrogen molecules are close packed, as shown in Figure 23.1 as spheres, in the monolayer. Then, one molecule occupies the parallelogram area of $2\sqrt{3}\,r^2$, where r is the radius of the adsorbed molecule. Furthermore, let us assume that the collision cross-section of Table 20.1 ($\sigma = 0.43\ \text{nm}^2 = 4\pi r^2$) gives a reasonable estimate of r: $r = (\sigma/4\pi)^{1/2}$. With these assumptions the surface area occupied by one molecule is:

$$A_{\text{molecule}} = 2\sqrt{3}\,(\sigma/4\pi) = \sqrt{3}\,\sigma/2\pi = \sqrt{3}\,(0.43\ \text{nm}^2)/2\pi = 0.12\ \text{nm}^2.$$

In this model the surface area per gram of the catalyst equals $A_{\text{molecule}}N$, where N is the number of adsorbed molecules. N can be calculated with the 0°C data, a temperature that is so high compared to the boiling point of nitrogen that all molecules are likely to be desorbed from the surface as perfect gas.

$$N = \frac{pV}{kT} = \frac{(760\ \text{Torr}) \times (133.3\ \text{Pa/Torr}) \times (3.86 \times 10^{-6}\ \text{m}^3)}{(1.381 \times 10^{-23}\ \text{J K}^{-1}) \times (273.15\ \text{K})} = 1.04 \times 10^{20}$$

$$A_{\text{molecule}}N = (0.12 \times 10^{-18}\ \text{m}^2) \times (1.04 \times 10^{20}) = \boxed{12\ \text{m}^2}$$

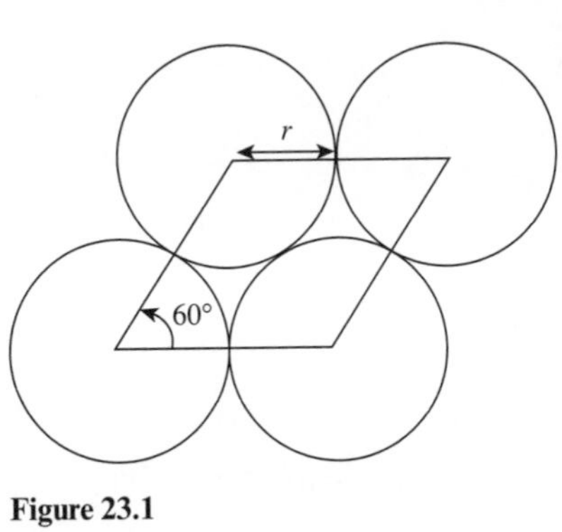

Figure 23.1

E23.9(a)

$$\theta = \frac{V}{V_\infty}\ [23.10] = \frac{V}{V_{\text{mon}}} = \frac{Kp}{1+Kp}\ [23.12]$$

This rearranges to [Example 23.4]

$$\frac{p}{V} = \frac{p}{V_{\text{mon}}} + \frac{1}{KV_{\text{mon}}}.$$

Hence, $\dfrac{p_2}{V_2} - \dfrac{p_1}{V_1} = \dfrac{p_2}{V_{\text{mon}}} - \dfrac{p_1}{V_{\text{mon}}}$.

Solving for V_{mon}:

$$V_{\text{mon}} = \frac{p_2 - p_1}{(p_2/V_2 - p_1/V_1)} = \frac{(760 - 145.4)\ \text{Torr}}{(760/1.443 - 145.4/0.286)\ \text{Torr cm}^{-3}} = \boxed{33.6\ \text{cm}^3}.$$

E23.10(a) The enthalpy of adsorption is typical of $\boxed{\text{chemisorption}}$ (Table 23.2), for which $\tau_0 \approx 10^{-14}$ s [Section 23.24 (b)] because the adsorbate–substrate bond is stiff. The half-life for remaining on the surface is

$$t_{1/2} = \tau_0 e^{E_d/RT}\ [21.24] \approx (10^{-14}\ \text{s}) \times (e^{120\times10^3/(8.3145\times400)})\ [E_d \approx -\Delta_{\text{ad}}H] \approx \boxed{50\ \text{s}}$$

E23.11(a) $$\frac{m_1}{m_2} = \frac{\theta_1}{\theta_2} = \frac{p_1}{p_2} \times \frac{1+Kp_2}{1+Kp_1} \quad [23.10 \text{ and } 23.12]$$

which solves to

$$K = \frac{(m_1p_2/m_2p_1) - 1}{p_2 - (m_1p_2/m_2)} = \frac{(m_1/m_2)\times(p_2/p_1) - 1}{1-(m_1/m_2)} \times \frac{1}{p_2}$$
$$= \frac{(0.44/0.19)\times(3.0/26.0) - 1}{1-(0.44/0.19)} \times \frac{1}{3.0\ \text{kPa}} = 0.19\ \text{kPa}^{-1}$$

Therefore,

$$\theta_1 = \frac{(0.19\ \text{kPa}^{-1})\times(26.0\ \text{kPa})}{(1)+(0.19\ \text{kPa}^{-1})\times(26.0\ \text{kPa})} = \boxed{0.83}\ [23.12] \quad \text{and} \quad \theta_2 = \frac{(0.19)\times(3.0)}{(1)+(0.19)\times(3.0)} = \boxed{0.36}.$$

E23.12(a) $\theta = \dfrac{Kp}{1+Kp}$ [23.12], which implies that $p = \left(\dfrac{\theta}{1-\theta}\right)\dfrac{1}{K}$.

(a) $p = (0.15/0.85)/0.75\ \text{kPa}^{-1} = \boxed{0.24\ \text{kPa}}$

(b) $p = (0.95/0.05)/0.75\ \text{kPa}^{-1} = \boxed{25\ \text{kPa}}$

E23.13(a) $\theta = \dfrac{Kp}{1+Kp}$ [23.12], which implies that $K = \left(\dfrac{\theta}{1-\theta}\right)\times\left(\dfrac{1}{p}\right)$.

Additionally, $\ln\left(\dfrac{K_2}{K_1}\right) = -\dfrac{\Delta_{\text{ad}}H}{R}\left(\dfrac{1}{T_2} - \dfrac{1}{T_1}\right)$ [6.23] $= \dfrac{\Delta_{\text{des}}H}{R}\left(\dfrac{1}{T_2} - \dfrac{1}{T_1}\right)$ $[\Delta_{\text{ad}}H = -\Delta_{\text{des}}H]$.

Since $\theta_2 = \theta_1$, $K_2/K_1 = p_1/p_2$ and

$$\ln\frac{p_1}{p_2} = \frac{\Delta_{\text{des}}H}{R}\left(\frac{1}{T_2} - \frac{1}{T_1}\right) = \left(\frac{10.2\ \text{kJ mol}^{-1}}{8.3145\ \text{J K}^{-1}\text{mol}^{-1}}\right)\times\left(\frac{1}{313\ \text{K}} - \frac{1}{298\ \text{K}}\right) = -0.197$$

which implies that $p_2 = (12\ \text{kPa}) \times (e^{0.197}) = \boxed{15\ \text{kPa}}$.

E23.14(a) $\theta = \dfrac{Kp}{1+Kp}$ [23.12], which implies that $K = \left(\dfrac{\theta}{1-\theta}\right) \times \left(\dfrac{1}{p}\right)$.

Additionally, $\ln\left(\dfrac{K_2}{K_1}\right) = -\dfrac{\Delta_{ad}H}{R}\left(\dfrac{1}{T_2} - \dfrac{1}{T_1}\right)$ [6.23] or $\Delta_{ad}H = -R\ln\left(\dfrac{K_2}{K_1}\right) \times \left(\dfrac{1}{T_2} - \dfrac{1}{T_1}\right)^{-1}$.

Since $\theta_2 = \theta_1$, $K_2/K_1 = p_1/p_2$ and

$$\Delta_{ad}H = -R\ln\left(\frac{p_1}{p_2}\right) \times \left(\frac{1}{T_2} - \frac{1}{T_1}\right)^{-1}$$

$$= -(8.3145\ \text{J K}^{-1}\,\text{mol}^{-1}) \times \ln\left(\frac{490\ \text{kPa}}{3.2 \times 10^3\ \text{kPa}}\right) \times \left(\frac{1}{250\ \text{K}} - \frac{1}{190\ \text{K}}\right)^{-1}$$

$$= \boxed{-12.\bar{4}\ \text{kJ mol}^{-1}}.$$

E23.15(a) The desorption time for a given volume is proportional to the half-life of the absorbed species and, consequently, the ratio of desorption times at two different temperatures is given by:

$$t(2)/t(1) = t_{1/2}(2)/t_{1/2}(1) = e^{E_d/RT_2}/e^{E_d/RT_1}\ [23.24] = e^{E_d(1/T_2 - 1/T_1)/R}.$$

Solving for the activation energy for desorption, E_d, gives:

$$E_d = R\ln\{t(2)/t(1)\}(1/T_2 - 1/T_1)^{-1}$$

$$= (8.3145\ \text{J K}^{-1}\,\text{mol}^{-1}) \times \ln\left(\frac{2.0\ \text{min}}{27\ \text{min}}\right) \times \left(\frac{1}{1978\ \text{K}} - \frac{1}{1856\ \text{K}}\right)^{-1}$$

$$= \boxed{65\bar{1}\ \text{kJ mol}^{-1}}.$$

The desorption time, t, for the same volume at temperature T is given by:

$$t = t(1)e^{E_d(1/T_2 - 1/T_1)/R} = (27\ \text{min})\exp\left\{(65\bar{1} \times 10^3\ \text{J mol}^{-1}) \times \left(\frac{1}{T} - \frac{1}{1856\ \text{K}}\right)/(8.3145\ \text{J K}^{-1}\,\text{mol}^{-1})\right\}$$

$$= (27\ \text{min})\exp\left\{(78.3) \times \left(\frac{1}{T/1000\ \text{K}} - \frac{1}{1.856}\right)\right\}.$$

(a) At 298 K, $t = \boxed{1.6 \times 10^{97}\ \text{min}}$, which is about forever.

(b) At 3000 K, $t = \boxed{2.8 \times 10^{-6}\ \text{min}}$.

E23.16(a) The average time of molecular residence is proportional to the half-life of the absorbed species and, consequently, the ratio of average residence times at two different temperatures is given by:

$$t(2)/t(1) = t_{1/2}(2)/t_{1/2}(1) = e^{E_d/RT_2}/e^{E_d/RT_1}\ [23.24] = e^{E_d(1/T_2 - 1/T_1)/R}.$$

Solving for the activation energy for desorption, E_d, gives:

$$E_d = R\ln\{t(2)/t(1)\}(1/T_2 - 1/T_1)^{-1}$$

$$= (8.3145\ \text{J K}^{-1}\,\text{mol}^{-1}) \times \ln\left(\frac{3.49\ \text{s}}{0.36\ \text{s}}\right) \times \left(\frac{1}{2362\ \text{K}} - \frac{1}{2548\ \text{K}}\right)^{-1}$$

$$= \boxed{61\bar{1}\ \text{kJ mol}^{-1}}.$$

E23.17(a) At 400 K: $t_{1/2} = \tau_0 e^{E_d/RT}$ [23.24] $= (0.10\text{ ps}) \times e^{0.301E_d/\text{kJ mol}^{-1}}$.
At 1000 K: $t_{1/2} = \tau_0 e^{E_d/RT}$ [23.24] $= (0.10\text{ ps}) \times e^{0.120E_d/\text{kJ mol}^{-1}}$.

(a) $E_d = 15\text{ kJ mol}^{-1}$

$t_{1/2}(400\text{ K}) = (0.10\text{ ps}) \times e^{0.301\times15} = \boxed{9.1\text{ ps}}$, $t_{1/2}(1000\text{ K}) = (0.10\text{ ps}) \times e^{0.120\times15} = \boxed{0.60\text{ ps}}$.

(b) $E_d = 150\text{ kJ mol}^{-1}$

$t_{1/2}(400\text{ K}) = (0.10\text{ ps}) \times e^{0.301\times150} = \boxed{4.1\times10^6\text{ s}}$, $t_{1/2}(1000\text{ K}) = (0.10\text{ ps}) \times e^{0.120\times150} = \boxed{6.6\text{ μs}}$.

E23.18(a) $$v = k_r\theta = \frac{k_rKp}{1+Kp}\ [23.26]$$

(a) On gold: $\theta \approx 1$ and $v = k_r\theta \approx$ constant, a $\boxed{\text{zeroth-order}}$ reaction.

(b) On platinum: $\theta \approx Kp$ (as $Kp \ll 1$) so $v = k_rKp$ and the reaction is $\boxed{\text{first order}}$.

Solutions to problems

Solutions to numerical problems

P23.1 We draw up the table below, which includes data rows required for a Lineweaver–Burk plot ($1/v$ against $1/[S]_0$). The linear regression fit is summarized in the Figure 23.2 plot.

[ATP]/(μmol dm^{-3})	0.60	0.80	1.4	2.0	3.0
v/(μmol dm^{-3} s^{-1})	0.81	0.97	1.30	1.47	1.69
1/{[ATP]/(μmol dm^{-3})}	1.67	1.25	0.714	0.500	0.333
1/{v/(μmol dm^{-3} s^{-1})}	1.23	1.03	0.769	0.680	0.592

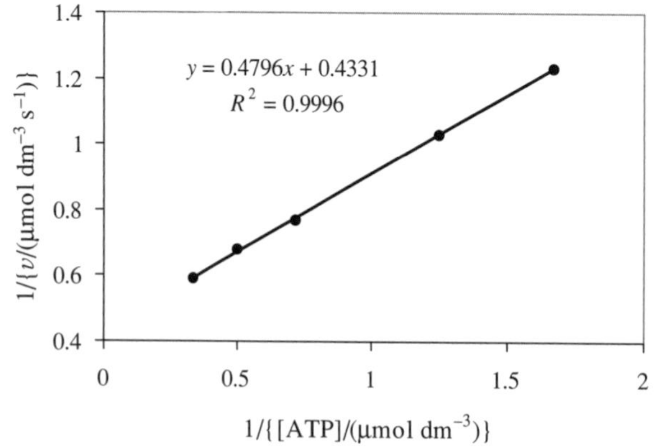

Figure 23.2

$1/v_{max} = intercept$ [23.3b]

$v_{max} = 1/intercept = 1/(0.433\text{ μmol dm}^{-3}\text{ s}^{-1}) = \boxed{2.31\text{ μmol dm}^{-3}\text{ s}^{-1}}$

$k_b = v_{max}/[E]_0$ [23.4] $= (2.31\text{ μmol dm}^{-3}\text{ s}^{-1})/(0.020\text{ μmol dm}^{-3}) = \boxed{115\text{ s}^{-1}}$

$k_{cat} = k_b$ [23.4] $= \boxed{115\text{ s}^{-1}}$

$$K_M = v_{max} \times slope\,[23.3b] = (2.31\ \mu mol\ dm^{-3}\ s^{-1}) \times (0.480\ s) = \boxed{1.11\ \mu mol\ dm^{-3}}$$

$$\eta = k_{cat}/K_M\,[23.5] = (115\ s^{-1})/(1.11\ \mu mol\ dm^{-3}) = \boxed{104\ dm^{-3}\ \mu mol^{-1}\ s^{-1}}$$

P23.3 (a) The dissociation equilibrium may be rearranged to give the following four relationships.

$$[E^-] = K_{E,a}[EH]/[H^+] \qquad [EH_2^+] = [EH][H^+]/K_{E,b}$$

$$[ES^-] = K_{ES,a}[ESH]/[H^+] \quad [ESH_2] = [ESH][H^+]/K_{ES,b}$$

Mass balance provides an equation for [EH]:

$$[E]_0 = [E^-] + [EH] + [EH_2^+] + [ES^-] + [ESH] + [ESH_2]$$

$$= \frac{K_{E,a}[EH]}{[H^+]} + [EH] + \frac{[EH][H^+]}{K_{E,b}} + \frac{K_{ES,a}[ESH]}{[H^+]} + [ESH] + \frac{[ESH][H^+]}{K_{ES,b}}$$

$$[EH] = \frac{[E]_0 - \left\{1 + \dfrac{[H^+]}{K_{ES,b}} + \dfrac{K_{ES,a}}{[H^+]}\right\}[ESH]}{1 + \dfrac{[H^+]}{K_{E,b}} + \dfrac{K_{E,a}}{[H^+]}}$$

$$= \frac{[E]_0 - c_1[ESH]}{c_2} \quad \text{where} \quad c_1 = 1 + \frac{[H^+]}{K_{ES,b}} + \frac{K_{ES,a}}{[H^+]} \text{ and } c_2 = 1 + \frac{[H^+]}{K_{E,b}} + \frac{K_{E,a}}{[H^+]}$$

The steady-state approximation provides an equation for [ESH]:

$$\frac{d[ESH]}{dt} = k_a[EH][S] - k_a'[ESH] - k_b[ESH] = 0$$

$$[ESH] = \frac{k_a}{k_a' + k_b}[EH][S] = K_M^{-1}[EH][S] = K_M^{-1}[S]\left\{\frac{[E]_0 - c_1[ESH]}{c_2}\right\}$$

$$[ESH] = \frac{K_M^{-1}[S][E]_0/c_2}{1 + K_M^{-1}[S]c_1/c_2} = \frac{[E]_0/c_1}{1 + K_M(c_2/c_1)/[S]}$$

The rate law becomes:

$$v = d[P]/dt = k_b[ESH] = \frac{k_b[E]_0/c_1}{1 + K_M(c_2/c_1)/[S]} = \frac{v'_{max}}{1 + K'_M/[S]}$$

where $v'_{max} = k_b[E]_0/\left\{1 + \dfrac{[H^+]}{K_{ES,b}} + \dfrac{K_{ES,a}}{[H^+]}\right\} = v_{max}/\left\{1 + \dfrac{[H^+]}{K_{ES,b}} + \dfrac{K_{ES,a}}{[H^+]}\right\}$

and $K'_M = K_M\left\{1 + \dfrac{[H^+]}{K_{E,b}} + \dfrac{K_{E,a}}{[H^+]}\right\}/\left\{1 + \dfrac{[H^+]}{K_{ES,b}} + \dfrac{K_{ES,a}}{[H^+]}\right\}$.

(b) $v_{max} = 1.0 \times 10^{-6}\ mol\ dm^{-3}\ s^{-1}$; $K_{ES,b} = 1.0 \times 10^{-6}$; $K_{ES,a} = 1.0 \times 10^{-8}$.

Figure 23.3 shows a plot of v'_{max} against pH. The plot indicates a maximum value of v'_{max} at pH = 7.0 for this set of equilibrium and kinetic constants. A formula for the pH of the maximum can be derived by finding the point at which $\dfrac{dv'_{max}}{d[H^+]} = 0$. This gives:

$$[H^+]_{max} = (K_{ES,a}K_{ES,b})^{1/2} = \sqrt{(1.0 \times 10^{-8}\ mol\ dm^{-3})(1.0 \times 10^{-6}\ mol\ dm^{-3})} = 1.0 \times 10^{-7}\ mol\ dm^{-3}$$

which corresponds to $\boxed{pH = 7.0}$.

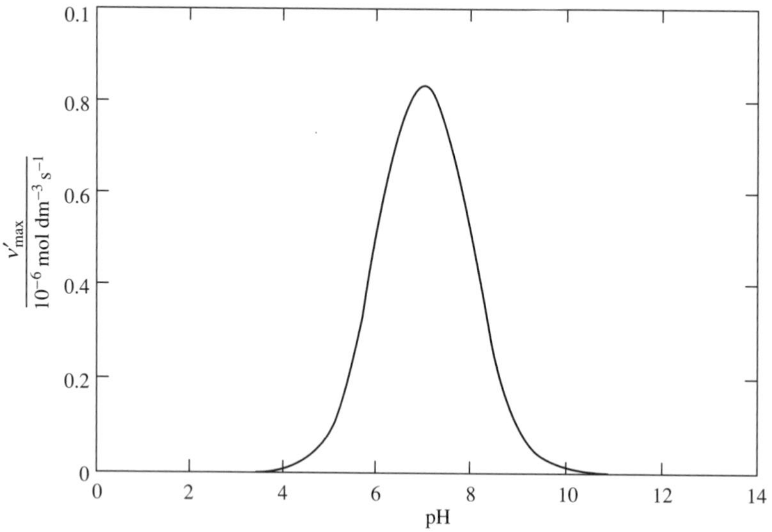

Figure 23.3

(c) $v_{max} = 1.0 \times 10^{-6}\ mol\ dm^{-3}\ s^{-1}$; $K_{ES,b} = 1.0 \times 10^{-4}$; $K_{ES,a} = 1.0 \times 10^{-10}$

Figure 23.4 shows a plot of v'_{max} against pH. The plot once again indicates a maximum value of v'_{max} at pH = 7.0 for this set of equilibrium and kinetic constants. However, the rate is high over a much larger pH range than appeared in part (b). This reflects the behaviour of the term $1 + [H^+]/K_{ES,b} + K_{ES,a}/[H^+]$ in the denominator of the v'_{max} expression. When $K_{ES,b}$ is relatively large, large $[H^+]$ values (low pH) cause growth in the values of v'_{max}. However, when $K_{ES,a}$ is relatively small, very small $[H^+]$ values (high pH) cause a decline in the v'_{max} values.

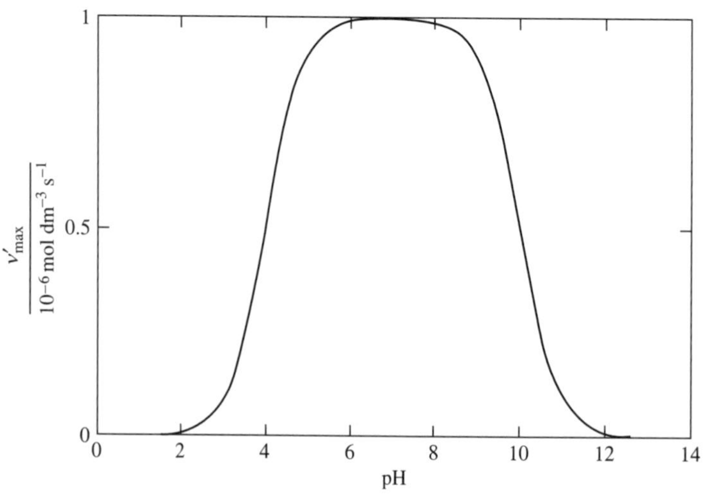

Figure 23.4

P23.5 We explore this problem by examining several different placements of a probe cation on two-dimensional ionic lattices. First, consider the lattice of Figure 23.5(a). The figure shows a dark univalent probe cation in a vacancy within a two-dimensional square ionic lattice of grey univalent cations and white univalent anions. Let $d_0 = 200$ pm be the distance between nearest neighbours and let v_0 be the absolute value of the Coulombic interaction between nearest neighbours.

$$V_0 = \frac{e^2}{4\pi\varepsilon_0 d_0} = \frac{(1.602 \times 10^{-19}\,\text{C})^2}{(1.113 \times 10^{-10}\,\text{J}^{-1}\,\text{C}^2\,\text{m}^{-1}) \times (200 \times 10^{-12}\,\text{m})} = 1.153 \times 10^{-18}\,\text{J}$$

The symmetry of the lattice around the probe cation consists of four regions like that of Figure 23.5(b) so we calculate the total Coulombic interaction of the probe within the lattice quadrant of Figure 23.5(b) and multiply by 4. The calculation is pursued one column at a time and the column interactions are summed.

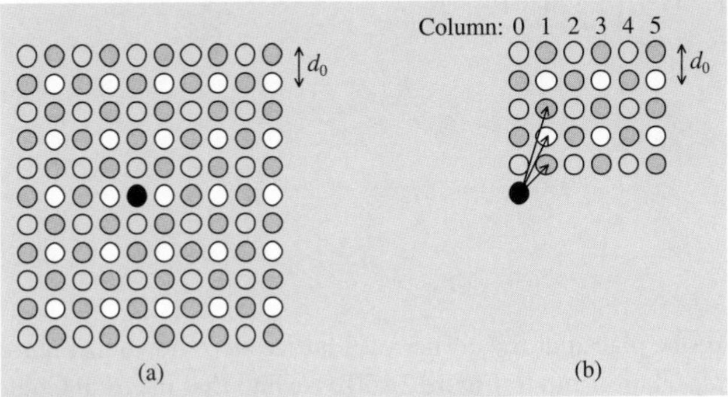

Figure 23.5

Probe-to-column 0 interaction:

$$V_{\text{column 0}} = -V_0 \times \left(1 - \frac{1}{2} + \frac{1}{3} - \frac{1}{4} + \frac{1}{5} \cdots\right) = -V_0 \ln 2 = V_0 \sum_{n=1}^{\infty} \frac{(-1)^{n+0}}{(n^2 + 0^2)^{1/2}} \text{ (useful form)}$$

Probe-to-column 1 interaction using the Pythagorean theorem for the probe–ion distance:

$$V_{\text{column 1}} = V_0 \times \left(\frac{1}{2^{1/2}} - \frac{1}{5^{1/2}} + \frac{1}{10^{1/2}} - \frac{1}{17^{1/2}} \cdots\right) = V_0 \sum_{n=1}^{\infty} \frac{(-1)^{n+1}}{(n^2 + 1^2)^{1/2}}$$

Similarly, the probe-to-column m interaction, using the Pythagorean theorem for the probe–ion distance, is

$$V_{\text{column } m} = V_0 \sum_{n=1}^{\infty} \frac{(-1)^{n+m}}{(n^2 + m^2)^{1/2}} \text{ for } 0 \le m < \infty \text{ (the sum is performed with a calculator or software)}$$

The total interaction for the region shown in Figure 23.5(b) is the sum of the above expression over all columns

$$\begin{aligned} V_{\text{Fig. 23.5(b)}} &= V_0 \sum_{m=0}^{\infty} \sum_{n=1}^{\infty} \frac{(-1)^{n+m}}{(n^2 + m^2)^{1/2}} \\ &= -0.4038 V_0 = (-0.4038) \times (1.153 \times 10^{-18}\,\text{J}) \\ &= -4.656 \times 10^{-19}\,\text{J} \end{aligned}$$

The total Coulombic interaction of the probe cation with the lattice of Figure 23.5(a) is

$$V_{\text{total}} = 4V_{\text{Fig. 23.5(b)}} = 4(-4.656 \times 10^{-19}\text{ J}) = \boxed{-1.862 \times 10^{-18}\text{ J}}$$

where the negative value indicates a net attraction. Following this calculation, the calculations of the next two probe–lattice arrangements is easy.

Now consider the probe–lattice arrangement of Figure 23.6, the probe on terrace arrangement. By examination of the symmetry relationship to Figure 23.5(b) we find that the total Coulombic interaction of the probe cation with the lattice is

$$\begin{aligned} V_{\text{total}} &= 2V_{\text{Fig. 23.5(b)}} - V_{\text{column 0}} \\ &= 2 \times (-0.4038V_0) - (-V_0 \ln 2) \\ &= -0.1145V_0 \\ &= (-0.1145) \times (1.153 \times 10^{-18}\text{ J}) = \boxed{-1.320 \times 10^{-19}\text{ J}} \end{aligned}$$

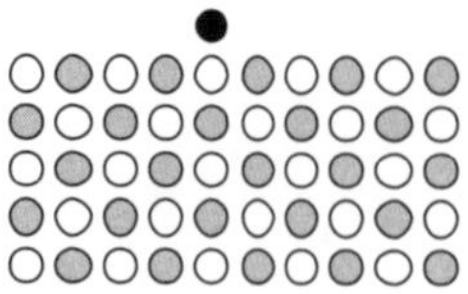

Figure 23.6

Our last computation is for the probe placed at the corner of a lattice step shown in Figure 23.7. Again, by examining the symmetry relationship to Figure 23.5(b) we find that the total Coulombic interaction of the probe cation with the lattice is

$$\begin{aligned} V_{\text{total}} &= 3V_{\text{Fig. 23.5(b)}} - V_{\text{column 0}} \\ &= 3 \times (-0.4038V_0) - (-V_0 \ln 2) \\ &= -0.5183V_0 \\ &= (-0.5183) \times (1.153 \times 10^{-18}\text{ J}) = \boxed{-5.975 \times 10^{-19}\text{ J}} \end{aligned}$$

Once again the negative value indicates a net attraction.

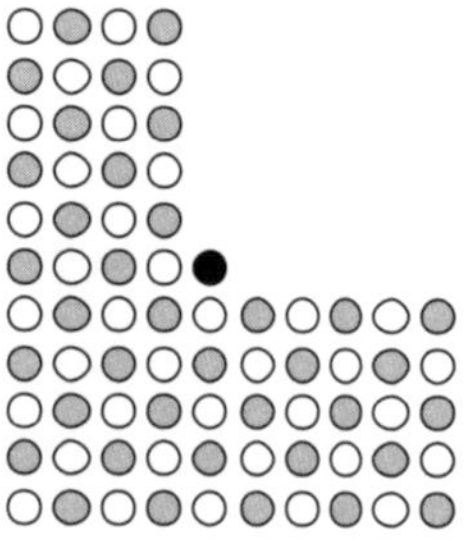

Figure 23.7

P23.7 Refer to Figure 23.8.

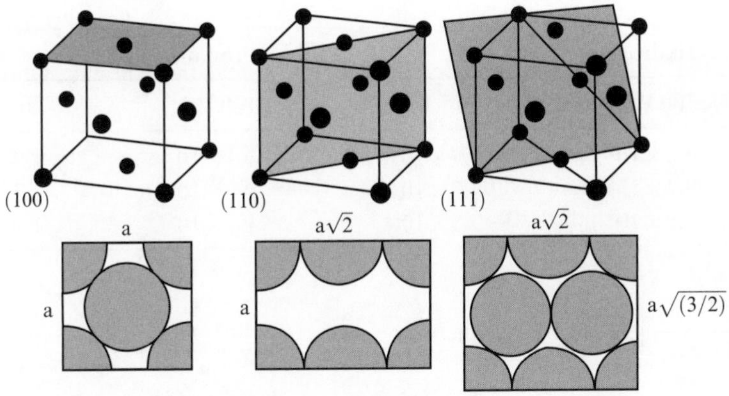

Figure 23.8

The (100) and (110) faces each expose two atoms, and the (111) face exposes four. The areas of the faces of each cell are (a) $(352\ \text{pm})^2 = 1.24 \times 10^{-15}\ \text{cm}^2$, (b) $\sqrt{2} \times (352\ \text{pm})^2 = 1.75 \times 10^{-15}\ \text{cm}^2$, and (c) $\sqrt{3} \times (352\ \text{pm})^2 = 2.15 \times 10^{-15}\ \text{cm}^2$. The numbers of atoms exposed per square centimetre, the surface number density, are therefore

(a) $$\frac{2}{1.24 \times 10^{-15}\ \text{cm}^2} = \boxed{1.61 \times 10^{15}\ \text{cm}^{-2}}$$

(b) $$\frac{2}{1.75 \times 10^{-15}\ \text{cm}^2} = \boxed{1.14 \times 10^{15}\ \text{cm}^{-2}}$$

(c) $$\frac{4}{2.15 \times 10^{-15}\ \text{cm}^2} = \boxed{1.86 \times 10^{15}\ \text{cm}^{-2}}$$

The collision frequency, Z_W, of gas molecules with a surface is given by eqn 23.9:

$$Z_W = \frac{p}{(2\pi MkT/N_A)^{1/2}} \quad [23.9;\ m = M/N_A]$$

$$= \frac{p \times \{(\text{kg m}^{-1}\,\text{s}^{-2})/\text{Pa}\} \times (10^{-4}\ \text{m}^2/\text{cm}^2)}{\{2\pi \times (1.381 \times 10^{-23}\ \text{J K}^{-1}) \times (298.15\ \text{K}) \times (\text{kg mol}^{-1})/(6.022 \times 10^{23}\ \text{mol}^{-1})\}^{1/2}\{M/(\text{kg mol}^{-1})\}^{1/2}}$$

$$= 4.825 \times 10^{17}\left(\frac{p/\text{Pa}}{\{M/(\text{kg mol}^{-1})\}^{1/2}}\right)\text{cm}^{-2}\,\text{s}^{-1} \quad \text{at } 25°\text{C}$$

(a) Hydrogen ($M = 0.002016\ \text{kg mol}^{-1}$)
  (i) $p = 100\ \text{Pa}$, $Z_W = 1.07 \times 10^{21}\ \text{cm}^{-2}\,\text{s}^{-1}$
  (ii) $p = 0.10\ \mu\text{Torr} = 1.33 \times 10^{-5}\ \text{Pa}$, $Z_W = 1.4 \times 10^{14}\ \text{cm}^{-2}\,\text{s}^{-1}$

(b) Propane ($M = 0.04410\ \text{kg mol}^{-1}$)
  (i) $p = 100\ \text{Pa}$, $Z_W = 2.30 \times 10^{20}\ \text{cm}^{-2}\,\text{s}^{-1}$
  (ii) $p = 0.10\ \mu\text{Torr} = 1.33 \times 10^{-5}\ \text{Pa}$, $Z_W = 3.1 \times 10^{13}\ \text{cm}^{-2}\,\text{s}^{-1}$

The frequency of collisions per surface atom, Z, is calculated by dividing Z_W by the surface number densities for the different planes. We can therefore draw up the following table:

	Hydrogen		Propane	
$Z/(\text{atom}^{-1}\ \text{s}^{-1})$	100 Pa	10^{-7} Torr	100 Pa	10^{-7} Torr
(100)	6.6×10^{5}	8.7×10^{-2}	1.4×10^{5}	1.9×10^{-2}
(110)	9.4×10^{5}	1.2×10^{-1}	2.0×10^{5}	2.7×10^{-2}
(111)	5.8×10^{5}	7.5×10^{-2}	1.2×10^{5}	1.7×10^{-2}

P23.9 $$\frac{V}{V_{\text{mon}}}=\frac{cz}{(1-z)\{1-(1-c)z\}}\quad\left[23.16,\ \text{BET isotherm},\ z=\frac{p}{p^*}\right]$$

This rearranges to

$$\frac{z}{(1-z)V}=\frac{1}{cV_{\text{mon}}}+\frac{(c-1)z}{cV_{\text{mon}}}.$$

Therefore, a plot of the left-hand side, $z/(1-z)V$, against z should result in a straight line if the data obeys the BET isotherm. Should it be linear, a linear regression fit of the plot yields values for the intercept and slope, which are related to c and V_{mon} by

$$1/cV_{\text{mon}}=\textit{intercept}\quad\text{and}\quad(c-1)/cV_{\text{mon}}=\textit{slope}$$

Solving for c and V_{mon} yields

$$c=1+\textit{slope}/\textit{intercept}\quad\text{and}\quad V_{\text{mon}}=1/(c\times\textit{intercept})$$

We draw up the following tables:

(a) 0°C, $p^*=429.6$ kPa

p/kPa	14.0	37.6	65.6	79.2	82.7	100.7	106.4
$10^3 z$	32.6	87.5	152.7	184.4	192.4	234.3	247.7
$\dfrac{10^3 z}{(1-z)(V/\text{cm}^3)}$	3.03	7.11	12.1	14.1	15.4	17.7	20.0

(b) 18°C, $p^*=819.7$ kPa

p/kPa	5.3	8.4	14.4	29.2	62.1	74.0	80.1	102.0
$10^3 z$	6.5	10.2	17.6	35.6	75.8	90.3	97.8	124.4
$\dfrac{10^3 z}{(1-z)(V/\text{cm}^3)}$	0.70	1.06	1.74	3.27	6.35	7.58	8.08	10.1

The $z/(1-z)V$ against z points are plotted in Figure 23.9. It is apparent that the plots are linear so we conclude that the data fit the BET isotherm. The linear regression fits are summarized in the figure.

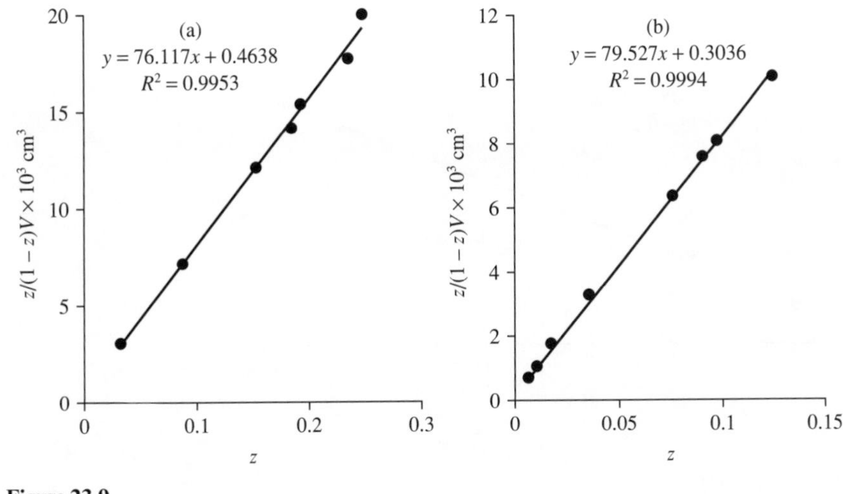

Figure 23.9

(a) $intercept = 0.4638 \times 10^{-3}\ \text{cm}^{-3}$ and $slope = 76.12 \times 10^{-3}\ \text{cm}^{-3}$

$c = 1 + slope/intercept = 1 + 76.12/0.4638 = \boxed{165}$

$V_{\text{mon}} = 1/(c \times intercept) = 1/(165 \times 0.4638 \times 10^{-3}\ \text{cm}^{-3})$

$= \boxed{13.1\ \text{cm}^3}$

(b) $intercept = 0.3036 \times 10^{-3}\ \text{cm}^{-3}$ and $slope = 79.53 \times 10^{-3}\ \text{cm}^{-3}$

$c = 1 + slope/intercept = 1 + 79.53/0.3036 = \boxed{263}$

$V_{\text{mon}} = 1/(c \times intercept) = 1/(263 \times 0.3036 \times 10^{-3}\ \text{cm}^{-3})$

$= \boxed{12.5\ \text{cm}^3}$

P23.11 $\theta = c_1 p^{1/c_2}$ [Freundlich isotherm, 23.20]

We adapt the Freundlich gas isotherm to a liquid by noting that $w_a \propto \theta$ and replacing p by $[A]/c^{\ominus}$, the concentration of the acid divided by the standard concentration $c^{\ominus} = 1$ mol dm^3. Then, $w_a = c_1([A]/c^{\ominus})^{1/c_2}$ (with c_1, c_2 modified constants), and hence $\ln(w_a/\text{g}) = \ln(c_1/\text{g}) + \dfrac{1}{c_2} \times \ln([A]/c^{\ominus})$. A plot of $\ln(w_a/\text{g})$ against $\ln([A]/c^{\ominus})$ is predicted to be linear in a Freundlich isotherm with intercept $\ln(c_1/\text{g})$ and slope $1/c_2$. We draw up the following table and prepared the desired plot, shown in Figure 23.10.

$[A]/(\text{mol dm}^{-3})$	0.05	0.10	0.50	1.0	1.5
w_a/g	0.04	0.06	0.12	0.16	0.19
$\ln([A]/\text{mol dm}^{-3})$	−3.00	−2.30	−0.693	−0.00	0.405
$\ln(w_a/\text{g})$	−3.22	−2.81	−2.12	−1.83	−1.66

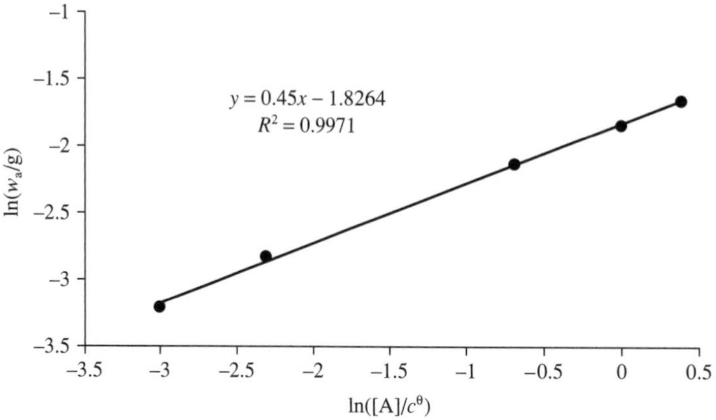

Figure 23.10

Since the plot is linear, the linear regression fit shown in Figure 23.10 is appropriate and the Freundlich coefficients are

$$c_1 = e^{intercept}\,g = e^{-1.83} = \boxed{0.16\ g}$$

$$c_2 = 1/slope = 1/0.45 = \boxed{2.2}$$

P23.13 Taking the natural logarithm of the isotherm $c_{ads} = Kc_{sol}^{1/n}$ gives

$$\ln c_{ads} = \ln K + (\ln c_{sol})/n$$

so a plot of $\ln c_{ads}$ versus $\ln c_{sol}$ would have a slope of $1/n$ and a y-intercept of $\ln K$. The transformed data and plot are shown in Figure 23.11.

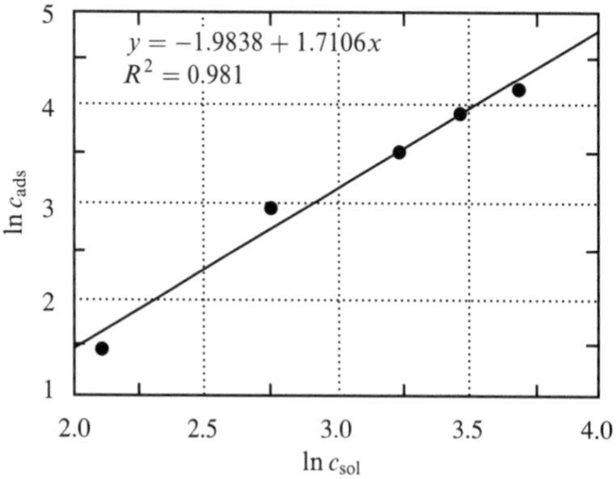

Figure 23.11

$c_{sol}/(mg\ g^{-1})$	8.26	15.65	25.43	31.74	40.00
$c_{ads}/(mg\ g^{-1})$	4.4	19.2	35.2	52.0	67.2
$\ln c_{sol}$	2.11	2.75	3.24	3.46	3.69
$\ln c_{ads}$	1.48	2.95	3.56	3.95	4.21

$$K = e^{-1.9838}\ \text{mg g}^{-1} = \boxed{0.138\ \text{mg g}^{-1}} \quad \text{and} \quad n = 1/1.71 = \boxed{0.58}$$

In order to express this information in terms of fractional coverage, the amount of adsorbate corresponding to monolayer coverage must be known. This saturation point, however, has no special significance in the Freundlich isotherm (i.e. it does not correspond to any limiting case).

P23.15 The Langmuir isotherm is

$$\theta = \frac{Kp}{1+Kp}\ [23.12] = \frac{n}{n_\infty}, \quad \text{so} \quad n(1+Kp) = n_\infty Kp \quad \text{and} \quad \frac{p}{n} = \frac{p}{n_\infty} + \frac{1}{Kn_\infty}$$

so a plot of p/n against p should be a straight line with slope $1/n_\infty$ and y-intercept $1/Kn_\infty$. The transformed data and plot (Figure 23.12) follow.

p/kPa	31.00	38.22	53.03	76.38	101.97	130.47	165.06	182.41	205.75	219.91
n/(mol kg^{-1})	1.00	1.17	1.54	2.04	2.49	2.90	3.22	3.30	3.35	3.36
$\frac{p/n}{\text{kPa mol}^{-1}\text{ kg}}$	31.00	32.67	34.44	37.44	40.95	44.99	51.26	55.28	61.42	65.45

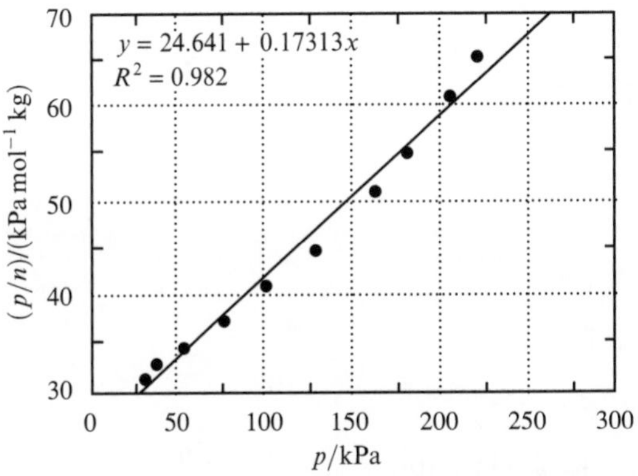

Figure 23.12

The plot seems to show some systematic variation from a straight line but a linear regression fit, shown in Figure 23.12 as a box insert, gives a satisfactory description of the trend so we conclude that

$$n_\infty = \frac{1}{slope} = \frac{1}{0.173\ \text{mol}^{-1}\ \text{kg}} = \boxed{5.78\ \text{mol kg}^{-1}}$$

$$K = \frac{1}{intercept \times n_\infty} = \frac{1}{(24.64\ \text{kPa mol}^{-1}\ \text{kg}) \times (5.78\ \text{mol kg}^{-1})} = \boxed{7.02\ \text{MPa}^{-1}}$$

Solutions to theoretical problems

P23.17 (a) $A + P \rightarrow P + P$ autocatalytic step, $v = k_r[A][P]$

Let $[A] = [A]_0 - x$ and $[P] = [P]_0 + x$.

We substitute these definitions into the rate expression, simplify, and integrate.

$$v = -\frac{d[A]}{dt} = k_r[A][P] - \frac{d([A]_0 - x)}{dt}$$
$$= k_r([A]_0 - x)([P]_0 + x)$$

$$\frac{dx}{([A]_0 - x)([P]_0 + x)} = k_r\,dt$$

$$\frac{1}{[A]_0 + [P]_0}\left(\frac{1}{[A]_0 - x} + \frac{1}{[P]_0 + x}\right)dx = k_r\,dt$$

$$\frac{1}{[A]_0 + [P]_0}\int_0^x\left(\frac{1}{[A]_0 - x} + \frac{1}{[P]_0 + x}\right)dx = k_r\int_0^t dt$$

$$\frac{1}{[A]_0 + [P]_0}\left\{\ln\left(\frac{[A]_0}{[A]_0 - x}\right) + \ln\left(\frac{[P]_0 + x}{[P]_0}\right)\right\} = k_r t$$

$$\ln\left\{\left(\frac{[A]_0}{[P]_0}\right)\left(\frac{[P]_0 + x}{[A]_0 - x}\right)\right\} = k_r([A]_0 + [P]_0)t$$

$$\ln\left\{\left(\frac{[A]_0}{[P]_0}\right)\left(\frac{[P]}{[A]_0 + [P]_0 - [P]}\right)\right\} = k_r([A]_0 + [P]_0)t$$

$$\ln\left\{\left(\frac{1}{b}\right)\left(\frac{[P]}{[A]_0 + [P]_0 - [P]}\right)\right\} = at, \quad \text{where} \quad a = k_r([A]_0 + [P]_0) \quad \text{and} \quad b = \frac{[P]_0}{[A]_0}$$

$$\frac{[P]}{[A]_0 + [P]_0 - [P]} = be^{at}$$

$$[P] = ([A]_0 + [P]_0)be^{at} - be^{at}[P]$$

$$(1 + be^{at})[P] = [P]_0\left(1 + \frac{[A]_0}{[P]_0}\right)be^{at} = [P]_0\left(1 + \frac{1}{b}\right)be^{at} = [P]_0(b + 1)e^{at}$$

$$\boxed{\frac{[P]}{[P]_0} = (b + 1)\frac{e^{at}}{1 + be^{at}}}.$$

(b) See Figure 23.13a.

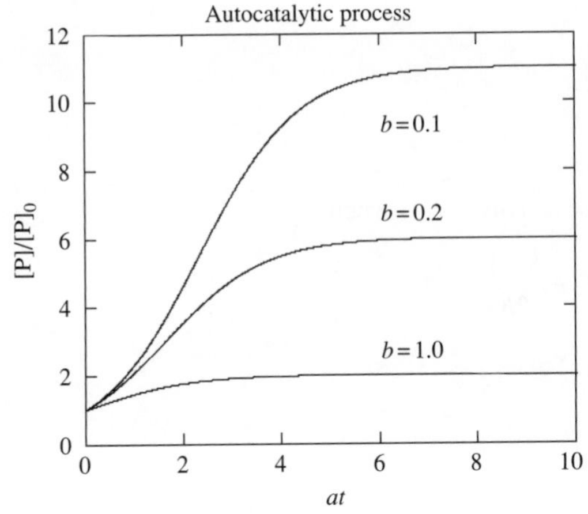

Figure 23.13(a)

The growth to [P] reaches a maximum at very long times. As $t \to \infty$, the exponential term in the denominator of $[P]/[P]_0 = (b + 1)e^{at}/(1 + be^{at})$ becomes so large that the denominator becomes be^{at}. Thus, $([P]/[P]_0)_{max} = (b + 1)e^{at}/(be^{at}) = (b + 1)/b$, where $b = [P]_0/[A]_0$ and this maximum occurs as $t \to \infty$.

The autocatalytic curve $[P]/[P]_0 = (b + 1)e^{at}/(1 + be^{at})$ has a shape that is very similar to that of the first-order process $[P]/[A]_0 = 1 - e^{-k_r t}$. However, $[P]_{max} = [A]_0$ at $t \to \infty$ for the first-order process, whereas $[P]_{max} = (1 + 1/b)[P]_0$ for the autocatalytic mechanism. In a series of experiments at fixed $[A]_0$ and assorted $[P]_0$, only the autocatalytic mechanism will show variation in $[P]_{max}$. Another difference is that the autocatalytic curve is initially concave up, which gives an overall sigmoidal curve, whereas the first-order curve is concave down. See Figure 23.13(b).

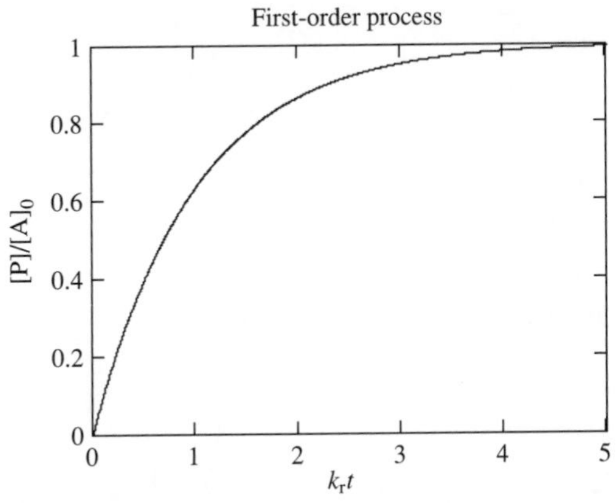

Figure 23.13(b)

(c) Let $[\mathrm{P}]_{v_{\max}}$ be the concentration of P at which the reaction rate is a maximum and let $t_{\max}$ be the corresponding time.

$$v = k_r[\mathrm{A}][\mathrm{P}] = k_r([\mathrm{A}]_0 - x)([\mathrm{P}]_0 + x)$$
$$= k_r\{[\mathrm{A}]_0[\mathrm{P}]_0 + ([\mathrm{A}]_0 - [\mathrm{P}]_0)x - x^2\}$$

$$\frac{\mathrm{d}v}{\mathrm{d}t} = k_r([\mathrm{A}]_0 - [\mathrm{P}]_0 - 2x)$$

The reaction rate is a maximum when $\mathrm{d}v/\mathrm{d}t = 0$. This occurs when

$$x = [\mathrm{P}]_{v_{\max}} - [\mathrm{P}]_0 = \frac{[\mathrm{A}]_0 - [\mathrm{P}]_0}{2} \quad \text{or} \quad \frac{[\mathrm{P}]_{v_{\max}}}{[\mathrm{P}]_0} = \frac{b+1}{2b}$$

Substitution into the final equation of part (a) gives:

$$\frac{[\mathrm{P}]_{v_{\max}}}{[\mathrm{P}]_0} = \frac{b+1}{2b} = (b+1)\frac{\mathrm{e}^{at_{\max}}}{1 + b\mathrm{e}^{at_{\max}}}$$

Solving for $t_{\max}$:

$$1 + b\mathrm{e}^{at_{\max}} = 2b\mathrm{e}^{at_{\max}}$$
$$\mathrm{e}^{at_{\max}} = b^{-1}$$
$$at_{\max} = \ln(b^{-1}) = -\ln(b)$$
$$t_{\max} = \boxed{-\frac{1}{a}\ln(b)}$$

(d) $$\frac{\mathrm{d}[\mathrm{P}]}{\mathrm{d}t} = k_r[\mathrm{A}]^2[\mathrm{P}]$$

$$[\mathrm{A}] = A_0 - x, \quad [\mathrm{P}] = P_0 + x, \quad \frac{\mathrm{d}[\mathrm{P}]}{\mathrm{d}t} = \frac{\mathrm{d}x}{\mathrm{d}t} = k_r(A_0 - x)^2(P_0 + x)$$

$$\int_0^x \frac{\mathrm{d}x}{(A_0 - x)^2(P_0 + x)} = k_r t$$

Solve the integral by partial fractions:

$$\frac{1}{(A_0 - x)^2(P_0 + x)} = \frac{\alpha}{(A_0 - x)^2} + \frac{\beta}{A_0 - x} + \frac{\gamma}{P_0 + x}$$
$$= \frac{\alpha(P_0 + x) + \beta(A_0 - x)(P_0 + x) + \gamma(A_0 - x)^2}{(A_0 - x)^2(P_0 + x)}$$

$$\left.\begin{aligned} P_0\alpha + A_0P_0\beta + A_0^2\gamma &= 1 \\ \alpha + (A_0 - P_0)\beta - 2A_0\gamma &= 0 \\ -\beta + \gamma &= 0 \end{aligned}\right\}$$

This set of simultaneous equations solves to

$$\alpha = \frac{1}{A_0 + P_0}, \quad \beta = \gamma = \frac{\alpha}{A_0 + P_0}.$$

Therefore,

$$k_r t = \left(\frac{1}{A_0 + P_0}\right) \int_0^x \left[\left(\frac{1}{A_0 - x}\right)^2 + \left(\frac{1}{A_0 + P_0}\right)\left(\frac{1}{A_0 - x} + \frac{1}{P_0 - x}\right)\right] dx$$

$$= \left(\frac{1}{A_0 + P_0}\right)\left\{\left(\frac{1}{A_0 - x}\right) - \left(\frac{1}{A_0}\right) + \left(\frac{1}{A_0 + P_0}\right)\left[\ln\left(\frac{A_0}{A_0 - x}\right) + \ln\left(\frac{P_0 + x}{P_0}\right)\right]\right\}$$

$$= \left(\frac{1}{A_0 + P_0}\right)\left[\left(\frac{x}{A_0(A_0 - x)}\right) + \left(\frac{1}{A_0 + P_0}\right)\ln\left(\frac{A_0(P_0 + x)}{(A_0 - x)P_0}\right)\right].$$

Therefore, with $y = \dfrac{x}{A_0}$ and $p = \dfrac{P_0}{A_0}$,

$$A_0(A_0 + P_0)kt = \boxed{\left(\frac{y}{1 - y}\right) + \left(\frac{1}{1 - p}\right)\ln\left(\frac{p + y}{p(1 - y)}\right)}$$

The maximum rate occurs at

$$\frac{dv_P}{dt} = 0, \quad v_P = k_r[A]^2[P]$$

and hence at the solution of

$$2k_r\left(\frac{d[A]}{dt}\right)[A][P] + k_r[A]^2\frac{d[P]}{dt} = 0$$

$$-k_r[A][P]v_P + k_r[A]^2 v_P = 0 \quad [\text{as } v_A = -v_P]$$

$$k_r[A]([A] - 2[P])v_P = 0$$

This is, the rate is a maximum when $[A] = 2[P]$, which occurs at

$$A_0 - x = 2P_0 + 2x, \quad \text{or} \quad x = \tfrac{1}{3}(A_0 - 2P_0); \quad y = \tfrac{1}{3}(1 - 2p)$$

Substituting this condition into the integrated rate law gives

$$A_0(A_0 + P_0)k_r t_{max} = \left(\frac{1}{1 + p}\right)\left(\frac{1}{2}(1 - 2p) + \ln\frac{1}{2p}\right)$$

or $\boxed{(A_0 + P_0)^2 k t_{max} = \tfrac{1}{2} - p - \ln 2p}$.

(e) $\dfrac{d[P]}{dt} = k_r[A][P]^2$

$$\frac{dx}{dt} = k_r(A_0 - x)(P_0 + x)^2 \quad [x = P - P_0]$$

$$kt = \int_0^x \frac{dx}{(A_0 - x)(P_0 + x)^2}$$

Integrate by partial fractions (as in part (d)):

$$k_r t = \left(\frac{1}{A_0+P_0}\right)\int_0^x \left\{\left(\frac{1}{P_0+x}\right)^2 + \left(\frac{1}{A_0+P_0}\right)\left[\frac{1}{P_0+x}+\frac{1}{A_0-x}\right]\right\} dx$$

$$= \left(\frac{1}{A_0+P_0}\right)\left\{\left(\frac{1}{P_0}-\frac{1}{P_0+x}\right)+\left(\frac{1}{A_0+P_0}\right)\left[\ln\left(\frac{P_0+x}{P_0}\right)+\ln\left(\frac{A_0}{A_0-x}\right)\right]\right\}$$

$$= \left(\frac{1}{A_0+P_0}\right)\left[\left(\frac{x}{P_0(P_0+x)}\right)+\left(\frac{1}{A_0+P_0}\right)\ln\left(\frac{(P_0+x)A_0}{P_0(A_0-x)}\right)\right]$$

Therefore, with $y = \dfrac{x}{[\mathrm{A}]_0}$ and $p = \dfrac{P_0}{A_0}$,

$$A_0(A_0+P_0)k_r t = \boxed{\left(\frac{y}{p(p+y)}\right)+\left(\frac{1}{1+p}\right)\ln\left(\frac{p+y}{p(1-y)}\right)}$$

The rate is maximum when

$$\frac{dv_P}{dt} = 2k_r[\mathrm{A}][\mathrm{P}]\left(\frac{d[\mathrm{P}]}{dt}\right)+k_r\left(\frac{d[\mathrm{A}]}{dt}\right)[\mathrm{P}]^2$$

$$= 2k_r[\mathrm{A}][\mathrm{P}]v_P - k_r[\mathrm{P}]^2 v_P = k_r[\mathrm{P}](2[\mathrm{A}]-[\mathrm{P}])v_P = 0$$

That is, at $[\mathrm{A}] = \frac{1}{2}[\mathrm{P}]$.

On substitution of this condition into the integrated rate law, we find

$$A_0(A_0+P_0)k_r t_{max} = \left(\frac{2-p}{2p(1+p)}\right)+\left(\frac{1}{1+p}\right)\ln\frac{2}{p}$$

$$\text{or } (A_0+P_0)^2 k_r t_{max} = \boxed{\frac{2-p}{2p}+\ln\frac{2}{p}}$$

P23.19 Michaelis–Menten mechanism: $\mathrm{E+S \rightleftharpoons ES}$ k_a, k_a'

$\mathrm{ES \rightarrow P + E}$ k_b

Assuming a rapid pre-equilibrium of E, S, and ES for the reversible step implies that

$$K = \frac{k_a}{k_a'} = \frac{[\mathrm{ES}]}{[\mathrm{E}][\mathrm{S}]} \quad \text{and} \quad [\mathrm{ES}] = K[\mathrm{E}][\mathrm{S}]$$

but the law of mass balance demands that $[\mathrm{E}] = [\mathrm{E}]_0 - [\mathrm{ES}]$ so $[\mathrm{ES}] = K([\mathrm{E}]_0 - [\mathrm{ES}])[\mathrm{S}]$ and, solving for [ES], we find that

$$[\mathrm{ES}] = \frac{[\mathrm{E}]_0}{1+\dfrac{1}{K[\mathrm{S}]_0}}$$

where the free substrate concentration has been replaced by $[S]_0$ because the substrate is typically in large excess relative to the enzyme. Now substitute the latter expression into the Michaelis–Menten expression for the rate of product P formation:

$$v = k_b[ES] = \frac{k_b[E]_0}{1+\dfrac{1}{K[S]_0}}, \quad \text{where} \quad v_{max} = k_b[E]_0$$

$$\boxed{v = \frac{v_{max}}{1+\dfrac{1}{K[S]_0}} \quad \text{rate law based on rapid pre-equilibrium approximation}}$$

With $K_M = (k_a' + k_b)/k_a$ the steady-state approximation (eqn 23.1) gives:

$$v = \frac{v_{max}}{1+\dfrac{K_M}{[S]_0}} \quad \text{rate law based on steady-state approximation}$$

Inspection reveals that the two approximations are identical when $K_M = 1/K$, which implies that

$$(k_a' + k_b)/k_a = k_a'/k_a \quad \text{or} \quad \boxed{k_a' \gg k_b}$$

P23.21 (a) We add to the Michaelis–Menten mechanism the inhibition by the substrate

$$SES \rightleftharpoons ES + S \quad K_I = [ES][S]/[SES]$$

where the inhibited enzyme, SES, forms when S binds to ES and thereby prevents the formation of product. This inhibition might possibly occur when S is at a very high concentration. Enzyme mass balance is written in terms of [ES], K_I, K_M (= [E][S]/[ES]), and [S]. (For practical purposes the free substrate concentration is replaced by $[S]_0$ because the substrate is typically in large excess relative to the enzyme.)

$$[E]_0 = [E] + [ES] + [SES]$$

$$= \frac{K_M[ES]}{[S]} + [ES] + \frac{[ES][S]}{K_I} = \left(1+\frac{K_M}{[S]}+\frac{[S]}{K_I}\right)[ES]$$

Thus,

$$[ES] = \frac{[E]_0}{\left(1+\dfrac{K_M}{[S]}+\dfrac{[S]}{K_I}\right)}$$

and the expression for the rate of product formation becomes

$$v = k_b[ES] = \frac{v_{max}}{1+\dfrac{K_M}{[S]_0}+\dfrac{[S]_0}{K_I}} \quad \text{where} \quad v_{max} = k_b[E]_0$$

The denominator term $[S]_0/K_I$ reflects a reduced reaction rate caused by inhibition as the concentration of S becomes very large.

(b) To examine the effect that substrate inhibition has on the double reciprocal **Lineweaver–Burk plot** of $1/v$ against $1/[S]_0$, take the inverse of the above rate expression and compare it to the uninhibited expression [23.3b]:

$$\frac{1}{v}=\frac{1}{v_{max}}+\left(\frac{K_M}{v_{max}}\right)\frac{1}{[S]_0} \quad [21.4b]$$

The inverse of the inhibited rate law is

$$\frac{1}{v}=\frac{1}{v_{max}}+\left(\frac{K_M}{v_{max}}\right)\frac{1}{[S]_0}+\left(\frac{[S]_0^2}{v_{max}K_I}\right)\frac{1}{[S]_0}$$

$$=\frac{1}{v_{max}}+\left(\frac{K_M}{v_{max}}+\frac{[S]_0^2}{v_{max}K_I}\right)\frac{1}{[S]_0}.$$

The uninhibited and inhibited line shapes are sketched in Figure 23.14.

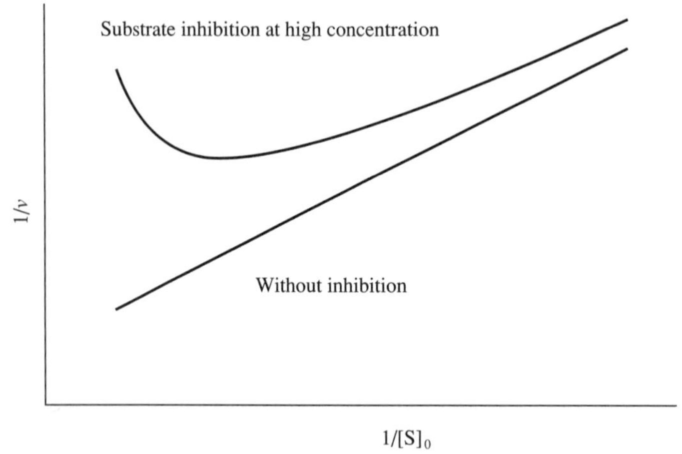

Figure 23.14

Comparing the two expressions, we see that the two curves match at high values of $1/[S]_0$. However, as the concentration of $[S]_0$ increases ($1/[S]_0$ decreases) the $1/v$ curve with inhibition curves upward because the reaction rate is decreasing.

Solutions to applications: chemical engineering and environmental science

P23.23 For the Langmuir and BET isotherm (using $p^* = 200$ kPa) tests we draw up the following table [eqn 23.12, Example 23.4, eqn 23.16, and Example 23.6]:

p/kPa	13.3	26.7	40.0	53.3	66.7	80.0
$p/V/(\text{kPa cm}^{-3})$	0.743	0.809	0.851	0.877	0.886	0.876
$10^3 z$	67	134	200	267	334	400
$\dfrac{10^3 z}{(1-z)(V/\text{cm}^3)}$	3.98	4.67	5.32	5.98	6.65	7.30

p/v is plotted against p in Figure 23.15(a), and $z/(1-z)v$ is plotted against z in Figure 23.15(b).

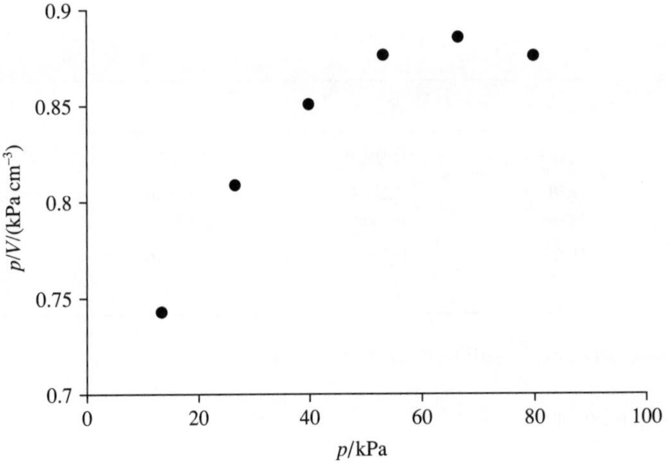

Figure 23.15(a)

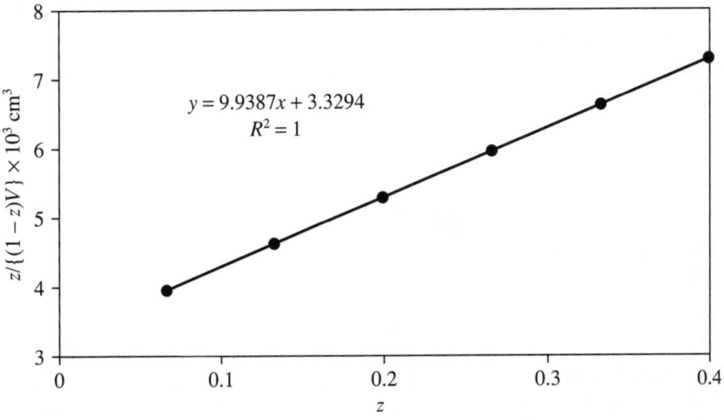

Figure 23.15(b)

We see that the plot of p/V against p is non-linear so we reject the proposition that it is described by a Langmuir isotherm. The plot of $z/(1-z)V$ against z appears to be linear so we accept the proposition that $\boxed{\text{it is described by the BET isotherm}}$ and the linear regression fit is summarized in Figure 23.15(b).

The BET isotherm has an intercept of $3.33 \times 10^{-3}\ \text{cm}^{-3}$ and a slope of $9.94 \times 10^{-3}\ \text{cm}^{-3}$. Since $1/cv_{\text{mon}}$ equals the intercept of a BET isotherm and $(c-1)/cV_{\text{mon}}$ equals the slope, we find that

$$c = 1 + slope/intercept = 1 + 9.94/3.33 = \boxed{3.96}$$

and

$$v_{\text{mon}} = 1/(c \times intercept) = 1/(3.96 \times 3.33 \times 10^{-3}\ \text{cm}^{-3}) = \boxed{75.8\ \text{cm}^3}$$

P23.25 (a) $$\frac{1}{q_{\mathrm{VOC,RH=0}}}=\frac{1+bc_{\mathrm{VOC}}}{abc_{\mathrm{VOC}}}=\frac{1}{abc_{\mathrm{VOC}}}+\frac{1}{a}$$

Parameters of regression fit:

θ/°C	$1/a$	$1/ab$	R	a	b/ppm^{-1}
33.6	9.07	709.8	0.9836	0.110	0.0128
41.5	10.14	890.4	0.9746	0.0986	0.0114
57.4	11.14	1599	0.9943	0.0898	0.00697
76.4	13.58	2063	0.9981	0.0736	0.00658
99	16.82	4012	0.9916	0.0595	0.00419

The linear regression fit is generally good at all temperatures with

$\boxed{R\text{ values in the range 0.975 to 0.991}}$

(b) $$\ln a=\ln k_{\mathrm{a}}-\frac{\Delta_{\mathrm{ad}}H}{R}\frac{1}{T}\quad\text{and}\quad \ln b=\ln k_{\mathrm{b}}-\frac{\Delta_{\mathrm{b}}H}{R}\frac{1}{T}$$

Linear regression analysis of $\ln a$ versus $1/T$ gives the intercept $\ln k_{\mathrm{a}}$ and slope $-\Delta_{\mathrm{ad}}H/R$, while a similar statement can be made for a $\ln b$ versus $1/T$ plot. The temperature must be in Kelvin.

For $\ln a$ versus $1/T$:

$\ln k_{\mathrm{a}}=-5.605$, standard deviation $=0.197$

$-\Delta_{\mathrm{ad}}H/R=1043.2\ \mathrm{K}$, standard deviation $=65.4\ \mathrm{K}$

$R=0.9942$ [good fit]

$$k_{\mathrm{a}}=\mathrm{e}^{-5.605}=\boxed{3.68\times10^{-3}}$$

$$\Delta_{\mathrm{ad}}H=-(8.31451\ \mathrm{J\ K^{-1}mol^{-1}})\times(1043.2\ \mathrm{K})$$
$$=\boxed{-8.67\ \mathrm{kJ\ mol^{-1}}}$$

For $\ln b$ versus $1/T$:

$\ln(k_b/(\mathrm{ppm}^{-1}))=-10.550$, standard deviation $=0.713$

$-\Delta_{\mathrm{b}}H/R=1895.4\ \mathrm{K}$, standard deviation $=236.8\ \mathrm{K}$

$R=0.9774$ [good fit]

$$k_{\mathrm{b}}=\mathrm{e}^{-10.550}\ \mathrm{ppm}^{-1}=\boxed{2.62\times10^{-5}\ \mathrm{ppm}^{-1}}$$

$$\Delta_{\mathrm{b}}H=-(8.31451\ \mathrm{J\ K^{-1}\ mol^{-1}})\times(1895.4\ \mathrm{K})$$

$$\boxed{\Delta_{\mathrm{b}}H=-15.7\ \mathrm{kJ\ mol^{-1}}}$$

(c) k_{a} may be interpreted to be the maximum adsorption capacity at an adsorption enthalpy of zero, while k_{b} is the maximum affinity in the case for which the adsorbant–surface bonding enthalpy is zero.

P23.27 We write the isotherms in the following forms where q is milligrams of solvent sorbed per gram of ground rubber (g_R) and K, K_F, K_L, and M are empirical constants.

Linear isotherm: $q = K \times (c_{eq}/\text{mg dm}^{-3})$

Freundlich isotherm: $q = K_F \times (c_{eq}/\text{mg dm}^{-3})^{1/n}$ or $\ln(q/\text{mg g}_R^{-1}) = \ln(K_F/\text{mg g}_R^{-1}) + \frac{1}{n}\ln(c_{eq}/\text{mg dm}^{-3})$

Langmuir isotherm: $q = K_L M c_{eq}/(1 + M c_{eq})$ or $\dfrac{1}{q} = \dfrac{1}{K_L} + \left(\dfrac{1}{K_L M}\right) \times \dfrac{1}{c_{eq}}$

(a) K unit: mg g_R^{-1} [g_R = mass (grams) of rubber]

K_F unit: mg g_R^{-1}

K_L unit: mg g_R^{-1}

M unit: mg^{-1} dm^3

(b) Determination of best description of data.

Data analysis with the linear sorption isotherm. Since $K = q/c_{eq}$, we calculate q/c_{eq} and calculate the average and standard deviation.

$K = 0.126$ mg g_R^{-1} [standard deviation = 0.041 mg g_R^{-1}]

The standard deviation is a large percentage of K, which may indicate some combination of random and/or systematic variation. A test of a relationship is often facilitated by checking the appearance of a plot. To further test the linear hypothesis between q and c, we prepare the plot, shown in Figure 23.16, with the constraint that the intercept equal zero. The plot appears to have a systematic non-linear component and only 88% of the variation is explained by the linear regression. However, with the small number of data points random error could give this appearance so we tentatively reject the linear hypothesis.

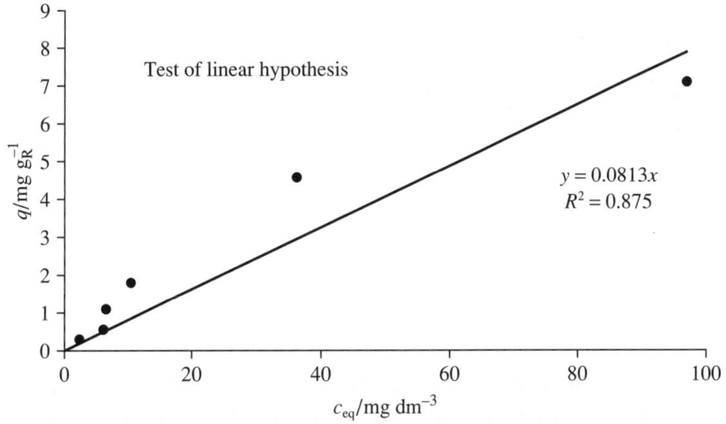

Figure 23.16

Data analysis with the Freundlich isotherm. To test the Freundlich hypothesis between q and c, we prepare the plot of $\ln q$ against $\ln c_{eq}$, shown in Figure 23.17. The appearance of the plot appears to be linear with considerable random scatter but there is no definitive visual indicator of systematic non-linearity. Ninety-four per cent of the variation is explained by the linear regression fit.

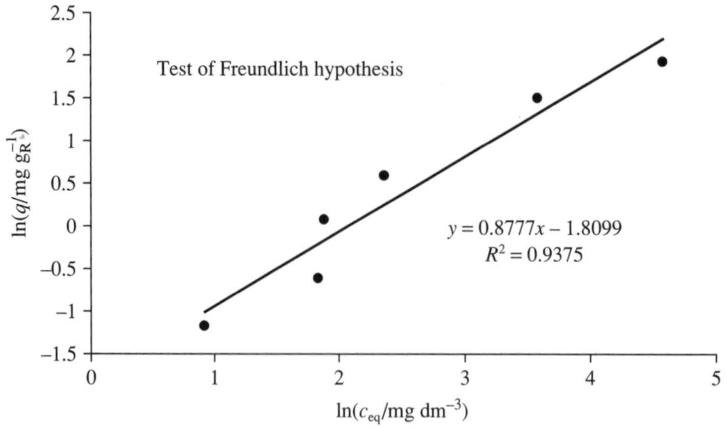

Figure 23.17

Data analysis with the Langmuir isotherm. To test the Langmuir hypothesis between q and c, we prepare the plot of $1/q$ against $1/c_{eq}$, shown in Figure 23.18. The appearance of the plot appears to be linear with considerable random scatter but there is no definitive visual indicator of systematic non-linearity. Ninety-four per cent of the variation is explained by the linear regression fit. However, the negative, but very small, intercept of the regression fit implies either that $K_L < 0$ or that K_L is so small that it has been swamped by the random scatter. We further recognize that a Langmuir isotherm must have a positive K_L because it is an equilibrium constant. The data, such as it is, does not firmly support the Langmuir isotherm hypothesis.

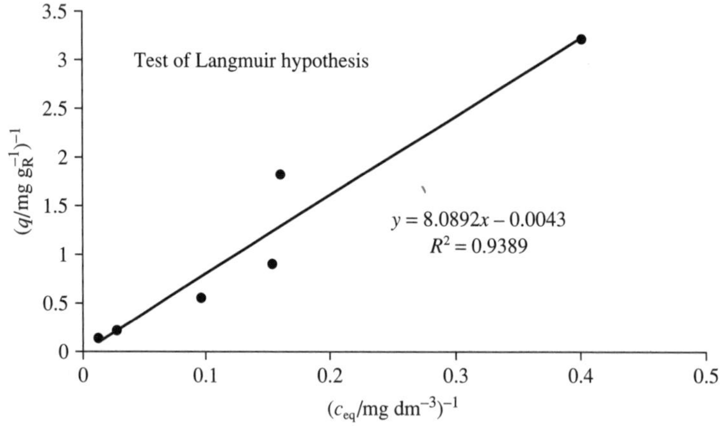

Figure 23.18

Summary: Only the Freundlich isotherm gives an adequate description of the data. The empirical constants for this isotherm are calculated with the linear regression fit shown as an insert in the Freundlich plot.

Freundlich constants:

$$K_F = e^{intercept}\ \text{mg}\ g_R^{-1} = e^{-1.81}\ \text{mg}\ g_R^{-1} = \boxed{0.164\ \text{mg}\ g_R^{-1}}$$

$$n = 1/slope = 1/0.878 = \boxed{1.14}$$

(c) The ratio of rubber-to-charcoal Freundlich sorption isotherms is $\dfrac{q_{\text{rubber}}}{q_{\text{charcoal}}} = \dfrac{0.164c_{\text{eq}}^{0.878}}{c_{\text{eq}}^{1.6}} = \boxed{0.164c_{\text{eq}}^{-0.72}}$

The sorption efficiency of ground rubber is much less than that of activated charcoal and drops significantly with increasing concentration. The only advantage of the ground rubber is its exceedingly low cost relative to activated charcoal, which might convert to a lower cost per gram of contaminant adsorbed.